NOTATION

$\sum_{i=1}^{n} a_i$	$a_1 + a_2 + \ldots + a_n$
$\prod_{i=1}^{n} a_i$	$a_1 a_2 \ldots a_n$
$\mathbb{R}, \mathbb{R}^n$	real numbers, n-space
$\mathbb{C}$	complex numbers
$\begin{bmatrix} a_{11} & a_{12} & \cdots & a_{1n} \\ a_{21} & a_{22} & \cdots & a_{2n} \\ \vdots & \vdots & \ddots & \vdots \\ a_{m1} & a_{m2} & \cdots & a_{mn} \end{bmatrix}$	$m \times n$ matrix
$[a_{ij}], 1 \le i, \le m, 1 \le j \le n$	compact notation for an $m \times n$ matrix
$A^{m \times n}$	A has dimension $m \times n$
0	zero vector, matrix,or scalar, depending on the context
$\det(A), |A|$	determinant of square matrix
A^{-1}	inverse of square matrix
A^T	transpose
$\begin{bmatrix} a_1 \\ \vdots \\ a_n \end{bmatrix}, \begin{bmatrix} a_1 & \ldots & a_n \end{bmatrix}^T$	Column vector
$\begin{bmatrix} a_1 & \cdots & a_n \end{bmatrix}$	Row vector
$\mathrm{diag}(a_1, a_2, \ldots, a_n)$	Diagonal matrix
$\mathrm{rank}(A)$	rank of matrix
$\mathrm{null}(A)$	null-space of matrix
$M_{ij}(A)$	minor for row i, column j
$|x|, |z|$	absolute value of real number x and modulus of complex number z
$\bar{z}$	complex conjugate
λ	eigenvalue
$\langle x, y \rangle = x^T y = y^T x$	inner product of x and y
$x \times y$	cross product of x and y in $\mathbb{R}^3$
$\overrightarrow{AB}$	vector from point A to point B
$\|\cdot\|$	any vector or matrix norm
$\|A\|_2$	Euclidean (2-) norm of vector or matrix
$\|A\|_1$	1-norm of vector or matrix
$\|A\|_\infty$	Infinity norm of vector or matrix
$\langle f, g \rangle = \int_a^b f(x) g(x)\, dx$	L^2 inner product
$|A|$	$|A| = [|a_{ij}|] 1 \le i \le m, 1 \le j \le n.$
$\mathrm{proj}_u(v)$	orthogonal projection of vector v onto vector u
$\sigma_{max}/\sigma_{min}$	Maximum (minimum) singular values of a matrix
$\approx$ or $\cong$	approximately equal to
$x \oplus y, x \ominus y, x \otimes y, x \oslash y$	floating point addition, subtraction, multiplication, division
$\mathrm{fl}(x)$	floating point approximation to x
$\in$	contained in
$\kappa(A)$	condition number of matrix
δx	small perturbation in x
$\triangle A, \delta A$	small perturbation in matrix
$\implies$	follows from
$A^{\ddagger}$	pseudoinverse $\left((A^T A)^{-1} A^T \right)$
$x_i^{(k)}$	i^{th} component of vector x_k

$\tilde{\Sigma}$	$m \times n$ diagonal matrix of the form 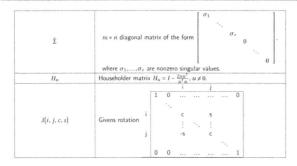 where $\sigma_1, \ldots, \sigma_r$ are nonzero singular values.	
H_u	Householder matrix $H_u = I - \frac{2uu^T}{u^Tu}$, $u \neq 0$.	
$J(i, j, c, s)$	Givens rotation	$\begin{array}{cccccccc} & & & i & & j & & \\ 1 & 0 & \ldots & \ldots & \ldots & \ldots & \ldots & 0 \\ & \ddots & & & & & & \\ i & & & c & & s & & \\ & & & \vdots & \ddots & \vdots & & \\ j & & & -s & & c & & \\ & & & & & & \ddots & \\ 0 & 0 & \ldots & \ldots & \ldots & \ldots & \ldots & 1 \end{array}$

Matrix Decompositions

Square or Rectangular Matrices

Diagonalization

If A has a full set of eigenvalues $\lambda_1, \lambda_2, \ldots, \lambda_n$ and corresponding eigenvectors $v_1, v_2, \ldots, v_n$, then $V^{-1}AV = D$, where $V = \begin{bmatrix} v_1 & v_2 & \ldots & v_{n-1} & v_n \end{bmatrix}$ and $D = \text{diag}(\lambda_1, \lambda_2, \ldots, \lambda_n)$. If A is symmetric, V is orthogonal.

QR

$A^{m \times n} = Q^{m \times m} R^{m \times n}$

Reduced QR

$A = Q^{m \times n} R^{n \times n}$, $m > n$

SVD

$A^{m \times n} = U^{m \times m} \tilde{\Sigma}^{m \times n} \left(V^{n \times n}\right)^T = U \text{diag}(\sigma_1, \sigma_2, \ldots, \sigma_r, 0, \ldots, 0) V^T$

Reduced SVD

$A^{m \times n} = U^{m \times n} \tilde{\Sigma}^{n \times n} \left(V^{n \times n}\right)^T$, $m > n$

Square Matrices

LU

$PA = LU$

Cholesky

$A = R^T R$

Schur's Triangularization

$A = PTP^T$

Arnoldi

$A^{n \times n} Q_m(:, 1 : m) = Q_{m+1}^{n \times (m+1)} H_m^{(m+1) \times m}$

$A^{n \times n} Q^{n \times m} = Q^{n \times m} H^{m \times m} + f_m^{n \times 1} e_m^T$

Lanczos

$A^{n \times n} Q_m(:, 1 : m) = Q_{m+1}^{n \times (m+1)} T_m^{(m+1) \times m}$

$A^{n \times n} Q^{n \times m} = Q^{n \times m} T^{m \times m} + f_m^{n \times 1} e_m^T$

Lower Case Greek Alphabet

name	character	name	character	name	character
alpha	α	iota	ι	rho	ρ
beta	β	kappa	κ	sigma	σ
gamma	γ	lambda	λ	tau	τ
delta	δ	mu	μ	upsilon	υ
epsilon	ϵ	nu	ν	phi	ϕ
zeta	ζ	xi	ξ	chi	χ
eta	η	omicron	o	psi	ψ
theta	θ	pi	π	omega	ω

Frequently Used Upper Case Greek Letters

name	character
gamma	Γ
delta	Δ
theta	Θ
lambda	Λ
pi	Π
sigma	Σ
omega	Ω

Numerical Linear Algebra with Applications

Numerical Linear Algebra with Applications
Using MATLAB

By

William Ford
Department of Computer Science
University of the Pacific

AMSTERDAM • BOSTON • HEIDELBERG • LONDON
NEW YORK • OXFORD • PARIS • SAN DIEGO
SAN FRANCISCO • SINGAPORE • SYDNEY • TOKYO
Academic Press is an imprint of Elsevier

Academic Press is an imprint of Elsevier
32 Jamestown Road, London NW1 7BY, UK
525 B Street, Suite 1800, San Diego, CA 92101-4495, USA
225 Wyman Street, Waltham, MA 02451, USA
The Boulevard, Langford Lane, Kidlington, Oxford OX5 1GB, UK

First edition 2015

Library of Congress Cataloging-in-Publication Data
A catalog record for this book is available from the Library of Congress

British Library Cataloguing in Publication Data
A catalogue record for this book is available from the British Library

For information on all Academic Press publications
visit our web site at store.elsevier.com

ISBN: 978-0-12-394435-1

Dedication

**Dedicated to my friend the late Paul Burdick
for all the wonderful conversations we had
and to
Dr. Ravi Jain, a great Dean and friend**

Contents

List of Figures   xiii
List of Algorithms   xvii
Preface   xix

1. Matrices   1

1.1. Matrix Arithmetic   1
  1.1.1. Matrix Product   2
  1.1.2. The Trace   5
  1.1.3. MATLAB Examples   6
1.2. Linear Transformations   7
  1.2.1. Rotations   7
1.3. Powers of Matrices   11
1.4. Nonsingular Matrices   13
1.5. The Matrix Transpose and Symmetric
     Matrices   16
1.6. Chapter Summary   18
1.7. Problems   19
  1.7.1. MATLAB Problems   22

2. Linear Equations   25

2.1. Introduction to Linear Equations   25
2.2. Solving Square Linear Systems   27
2.3. Gaussian Elimination   28
  2.3.1. Upper-Triangular Form   29
2.4. Systematic Solution of Linear Systems   31
2.5. Computing the Inverse   34
2.6. Homogeneous Systems   36
2.7. Application: A Truss   37
2.8. Application: Electrical Circuit   39
2.9. Chapter Summary   40
2.10. Problems   42
  2.10.1. MATLAB Problems   43

3. Subspaces   47

3.1. Introduction   47
3.2. Subspaces of $\mathbb{R}^n$ 47
3.3. Linear Independence   49
3.4. Basis of a Subspace   50
3.5. The Rank of a Matrix   51
3.6. Chapter Summary   55
3.7. Problems   56
  3.7.1. MATLAB Problems   57

4. Determinants   59

4.1. Developing the Determinant of a 2 × 2
     and a 3 × 3 Matrix   59
4.2. Expansion by Minors   60
4.3. Computing a Determinant Using Row
     Operations   64
4.4. Application: Encryption   71
4.5. Chapter Summary   73
4.6. Problems   74
  4.6.1. MATLAB Problems   76

5. Eigenvalues and Eigenvectors   79

5.1. Definitions and Examples   79
5.2. Selected Properties of Eigenvalues
     and Eigenvectors   83
5.3. Diagonalization   84
  5.3.1. Powers of Matrices   88
5.4. Applications   89
  5.4.1. Electric Circuit   89
  5.4.2. Irreducible Matrices   91
  5.4.3. Ranking of Teams Using
         Eigenvectors   94
5.5. Computing Eigenvalues and Eigenvectors
     using MATLAB   95
5.6. Chapter Summary   96
5.7. Problems   97
  5.7.1. MATLAB Problems   99

6. Orthogonal Vectors
   and Matrices   103

6.1. Introduction   103
6.2. The Inner Product   104
6.3. Orthogonal Matrices   107
6.4. Symmetric Matrices and Orthogonality   109
6.5. The L^2 Inner Product   110
6.6. The Cauchy-Schwarz Inequality   111
6.7. Signal Comparison   112
6.8. Chapter Summary   113
6.9. Problems   114
  6.9.1. MATLAB Problems   116

7. **Vector and Matrix Norms**   119

    7.1. Vector Norms   119
       7.1.1. Properties of the 2-Norm   121
       7.1.2. Spherical Coordinates   123
    7.2. Matrix Norms   126
       7.2.1. The Frobenius Matrix Norm   127
       7.2.2. Induced Matrix Norms   127
    7.3. Submultiplicative Matrix Norms   131
    7.4. Computing the Matrix 2-Norm   132
    7.5. Properties of the Matrix 2-Norm   136
    7.6. Chapter Summary   138
    7.7. Problems   140
       7.7.1. MATLAB Problems   142

8. **Floating Point Arithmetic**   145

    8.1. Integer Representation   145
    8.2. Floating-Point Representation   147
       8.2.1. Mapping from Real Numbers
           to Floating-Point Numbers   148
    8.3. Floating-Point Arithmetic   150
       8.3.1. Relative Error   150
       8.3.2. Rounding Error Bounds   151
    8.4. Minimizing Errors   155
       8.4.1. Avoid Adding a Huge Number to a
           Small Number   155
       8.4.2. Avoid Subtracting Numbers That Are
           Close   155
    8.5. Chapter Summary   156
    8.6. Problems   158
       8.6.1. MATLAB Problems   160

9. **Algorithms**   163

    9.1. Pseudocode Examples   163
       9.1.1. Inner Product of Two Vectors   164
       9.1.2. Computing the Frobenius
           Norm   164
       9.1.3. Matrix Multiplication   164
       9.1.4. Block Matrices   165
    9.2. Algorithm Efficiency   166
       9.2.1. Smaller Flop Count Is Not
           Always Better   168
       9.2.2. Measuring Truncation Error   168
    9.3. The Solution to Upper and Lower
       Triangular Systems   168
       9.3.1. Efficiency Analysis   170
    9.4. The Thomas Algorithm   171
       9.4.1. Efficiency Analysis   173
    9.5. Chapter Summary   174
    9.6. Problems   175
       9.6.1. MATLAB Problems   177

10. **Conditioning of Problems
and Stability of Algorithms**   181

    10.1. Why Do We Need Numerical
        Linear Algebra?   181
    10.2. Computation Error   183
       10.2.1. Forward Error   183
       10.2.2. Backward Error   184
    10.3. Algorithm Stability   185
       10.3.1. Examples of Unstable
           Algorithms   186
    10.4. Conditioning of a Problem   187
    10.5. Perturbation Analysis for Solving a
       Linear System   190
    10.6. Properties of the Matrix Condition
       Number   193
    10.7. MATLAB Computation of a Matrix
       Condition Number   195
    10.8. Estimating the Condition Number   195
    10.9. Introduction to Perturbation Analysis
       of Eigenvalue Problems   196
    10.10. Chapter Summary   197
    10.11. Problems   199
       10.11.1. MATLAB Problems   200

11. **Gaussian Elimination and the *LU*
Decomposition**   205

    11.1. *LU* Decomposition   205
    11.2. Using *LU* to Solve Equations   206
    11.3. Elementary Row Matrices   208
    11.4. Derivation of the *LU* Decomposition   210
       11.4.1. Colon Notation   214
       11.4.2. The *LU* Decomposition
           Algorithm   216
       11.4.3. *LU* Decomposition Flop
           Count   217
    11.5. Gaussian Elimination with Partial
       Pivoting   218
       11.5.1. Derivation of $PA = LU$ 219
       11.5.2. Algorithm for Gaussian
           Elimination with Partial
           Pivoting   223
    11.6. Using the *LU* Decomposition to Solve
       $Ax_i = b_i, 1 \le i \le k$ 225
    11.7. Finding A^{-1} 226
    11.8. Stability and Efficiency of Gaussian
       Elimination   227
    11.9. Iterative Refinement   228
    11.10. Chapter Summary   230
    11.11. Problems   232
       11.11.1. MATLAB Problems   236

12. Linear System Applications   241

12.1. Fourier Series   241
    12.1.1. The Square Wave   243
12.2. Finite Difference Approximations   244
    12.2.1. Steady-State Heat and Diffusion   245
12.3. Least-Squares Polynomial Fitting   247
    12.3.1. Normal Equations   249
12.4. Cubic Spline Interpolation   252
12.5. Chapter Summary   256
12.6. Problems   257
    12.6.1. MATLAB Problems   260

13. Important Special Systems   263

13.1. Tridiagonal Systems   263
13.2. Symmetric Positive Definite Matrices   267
    13.2.1. Applications   269
13.3. The Cholesky Decomposition   269
    13.3.1. Computing the Cholesky
          Decomposition   270
    13.3.2. Efficiency   272
    13.3.3. Solving $Ax = b$ If A Is Positive
          Definite   272
    13.3.4. Stability   273
13.4. Chapter Summary   273
13.5. Problems   274
    13.5.1. MATLAB Problems   277

14. Gram-Schmidt Orthonormalization   281

14.1. The Gram-Schmidt Process   281
14.2. Numerical Stability of the
    Gram-Schmidt Process   284
14.3. The QR Decomposition   287
    14.3.1. Efficiency   289
    14.3.2. Stability   290
14.4. Applications of the QR Decomposition   290
    14.4.1. Computing the Determinant   291
    14.4.2. Finding an Orthonormal Basis for
          the Range of a Matrix   291
14.5. Chapter Summary   292
14.6. Problems   292
    14.6.1. MATLAB Problems   293

15. The Singular Value Decomposition   299

15.1. The SVD Theorem   299
15.2. Using the SVD to Determine
    Properties of a Matrix   302
    15.2.1. The Four Fundamental
          Subspaces of a Matrix   304
15.3. SVD and Matrix Norms   306
15.4. Geometric Interpretation of the SVD   307
15.5. Computing the SVD Using MATLAB   308

15.6. Computing A^{-1} 309
15.7. Image Compression Using the SVD   310
    15.7.1. Image Compression Using
          MATLAB   311
    15.7.2. Additional Uses   313
15.8. Final Comments   314
15.9. Chapter Summary   314
15.10. Problems   316
    15.10.1. MATLAB Problems   317

16. Least-Squares Problems   321

16.1. Existence and Uniqueness of
    Least-Squares Solutions   322
    16.1.1. Existence and Uniqueness
          Theorem   322
    16.1.2. Normal Equations and
          Least-Squares Solutions   324
    16.1.3. The Pseudoinverse, $m \geq n$ 324
    16.1.4. The Pseudoinverse, $m < n$ 325
16.2. Solving Overdetermined Least-Squares
    Problems   325
    16.2.1. Using the Normal Equations   326
    16.2.2. Using the QR Decomposition   327
    16.2.3. Using the SVD   329
    16.2.4. Remark on Curve Fitting   332
16.3. Conditioning of Least-Squares
    Problems   332
    16.3.1. Sensitivity when using the
          Normal Equations   333
16.4. Rank-Deficient Least-Squares Problems   333
    16.4.1. Efficiency   338
16.5. Underdetermined Linear Systems   338
    16.5.1. Efficiency   341
16.6. Chapter Summary   341
16.7. Problems   342
    16.7.1. MATLAB Problems   343

**17. Implementing the QR
Decomposition   351**

17.1. Review of the QR Decomposition
    Using Gram-Schmidt   351
17.2. Givens Rotations   352
    17.2.1. Zeroing a Particular Entry in a
          Vector   353
17.3. Creating a Sequence of Zeros in a
    Vector Using Givens Rotations   355
17.4. Product of a Givens Matrix with a
    General Matrix   356
17.5. Zeroing-Out Column Entries in a
    Matrix Using Givens Rotations   357
17.6. Accurate Computation of the Givens
    Parameters   358

17.7. The Givens Algorithm for the *QR*
Decomposition 359
17.7.1. The Reduced *QR*
Decomposition 361
17.7.2. Efficiency 362
17.8. Householder Reflections 362
17.8.1. Matrix Column Zeroing Using
Householder Reflections 365
17.8.2. Implicit Computation with
Householder Reflections 367
17.9. Computing the *QR* Decomposition
Using Householder Reflections 368
17.9.1. Efficiency and Stability 372
17.10. Chapter Summary 373
17.11. Problems 373
17.11.1. MATLAB Problems 376

18. The Algebraic Eigenvalue Problem 379

18.1. Applications of the Eigenvalue
Problem 379
18.1.1. Vibrations and Resonance 380
18.1.2. The Leslie Model in
Population Ecology 383
18.1.3. Buckling of a Column 386
18.2. Computation of Selected Eigenvalues
and Eigenvectors 388
18.2.1. Additional Property of a
Diagonalizable Matrix 389
18.2.2. The Power Method for
Computing the Dominant
Eigenvalue 390
18.2.3. Computing the Smallest
Eigenvalue and Corresponding
Eigenvector 393
18.3. The Basic *QR* Iteration 394
18.4. Transformation to Upper Hessenberg
Form 395
18.4.1. Efficiency and Stability 400
18.5. The Unshifted Hessenberg *QR*
Iteration 400
18.5.1. Efficiency 403
18.6. The Shifted Hessenberg *QR* Iteration 403
18.6.1. A Single Shift 404
18.7. Schur's Triangularization 405
18.8. The Francis Algorithm 409
18.8.1. Francis Iteration of
Degree One 409
18.8.2. Francis Iteration of Degree Two 413
18.9. Computing Eigenvectors 420
18.9.1. Hessenberg Inverse Iteration 421
18.10. Computing Both Eigenvalues
and Their Corresponding
Eigenvectors 423

18.11. Sensitivity of Eigenvalues to
Perturbations 424
18.11.1. Sensitivity of Eigenvectors 427
18.12. Chapter Summary 428
18.13. Problems 430
18.13.1. MATLAB Problems 432

19. The Symmetric Eigenvalue Problem 439

19.1. The Spectral Theorem and Properties
of a Symmetric Matrix 439
19.1.1. Properties of a Symmetric Matrix 440
19.2. The Jacobi Method 440
19.2.1. Computing Eigenvectors Using
the Jacobi Iteration 444
19.2.2. The Cyclic-by-Row Jacobi
Algorithm 444
19.3. The Symmetric *QR* Iteration Method 446
19.3.1. Tridiagonal Reduction of a
Symmetric Matrix 449
19.3.2. Orthogonal Transformation to a
Diagonal Matrix 451
19.4. The Symmetric Francis Algorithm 452
19.4.1. Theoretical Overview and
Efficiency 453
19.5. The Bisection Method 453
19.5.1. Efficiency 457
19.5.2. Matrix *A* Is Not Unreduced 457
19.6. The Divide-and-Conquer Method 458
19.6.1. Using dconquer 461
19.7. Chapter Summary 461
19.8. Problems 463
19.8.1. MATLAB Problems 465

20. Basic Iterative Methods 469

20.1. Jacobi Method 469
20.2. The Gauss-Seidel Iterative Method 470
20.3. The SOR Iteration 471
20.4. Convergence of the Basic Iterative
Methods 473
20.4.1. Matrix Form of the Jacobi
Iteration 473
20.4.2. Matrix Form of the Gauss-Seidel
Iteration 473
20.4.3. Matrix Form for SOR 474
20.4.4. Conditions Guaranteeing
Convergence 474
20.4.5. The Spectral Radius and Rate of
Convergence 476
20.4.6. Convergence of the Jacobi and
Gauss-Seidel Methods for
Diagonally Dominant Matrices 477
20.4.7. Choosing ω for SOR 478
20.5. Application: Poisson's Equation 478

20.6. Chapter Summary 481
20.7. Problems 483
    20.7.1. MATLAB Problems 486

21. Krylov Subspace Methods 491

21.1. Large, Sparse Matrices 491
    21.1.1. Storage of Sparse Matrices 492
21.2. The CG Method 493
    21.2.1. The Method of Steepest Descent 493
    21.2.2. From Steepest Descent to CG 497
    21.2.3. Convergence 501
21.3. Preconditioning 501
21.4. Preconditioning for CG 503
    21.4.1. Incomplete Cholesky Decomposition 503
    21.4.2. SSOR Preconditioner 506
21.5. Krylov Subspaces 508
21.6. The Arnoldi Method 509
    21.6.1. An Alternative Formulation of the Arnoldi Decomposition 511
21.7. GMRES 512
    21.7.1. Preconditioned GMRES 514
21.8. The Symmetric Lanczos Method 516
    21.8.1. Loss of Orthogonality with the Lanczos Process 516
21.9. The MINRES Method 519
21.10. Comparison of Iterative Methods 520
21.11. Poisson's Equation Revisited 521
21.12. The Biharmonic Equation 523
21.13. Chapter Summary 524
21.14. Problems 526
    21.14.1. MATLAB Problems 528

22. Large Sparse Eigenvalue Problems 533

22.1. The Power Method 533
22.2. Eigenvalue Computation Using the Arnoldi Process 534
    22.2.1. Estimating Eigenvalues Without Restart or Deflation 535
    22.2.2. Estimating Eigenvalues Using Restart 536
    22.2.3. A Restart Method Using Deflation 537
    22.2.4. Restart Strategies 539
22.3. The Implicitly Restarted Arnoldi Method 540
    22.3.1. Convergence of the Arnoldi Iteration 544

22.4. Eigenvalue Computation Using the Lanczos Process 544
    22.4.1. Mathematically Provable Properties 546
22.5. Chapter Summary 547
22.6. Problems 548
    22.6.1. MATLAB Problems 548

23. Computing the Singular Value Decomposition 551

23.1. Development of the One-Sided Jacobi Method for Computing the Reduced SVD 551
    23.1.1. Stability of Singular Value Computation 554
23.2. The One-Sided Jacobi Algorithm 555
    23.2.1. Faster and More Accurate Jacobi Algorithm 557
23.3. Transforming a Matrix to Upper-Bidiagonal Form 558
23.4. Demmel and Kahan Zero-Shift QR Downward Sweep Algorithm 559
23.5. Chapter Summary 565
23.6. Problems 565
    23.6.1. MATLAB Problems 566

A. Complex Numbers 569

A.1. Constructing the Complex Numbers 569
A.2. Calculating with Complex Numbers 570
A.3. Geometric Representation of $\mathbb{C}$ 571
A.4. Complex Conjugate 571
A.5. Complex Numbers in MATLAB 573
A.6. Euler's Formula 575
A.7. Problems 575
    A.7.1. MATLAB Problems 576

B. Mathematical Induction 579

B.1. Problems 581

C. Chebyshev Polynomials 583

C.1. Definition 583
C.2. Properties 584
C.3. Problems 584
    C.3.1. MATLAB Problems 585

Glossary 587
Bibliography 595
Index 597

List of Figures

Fig. 0.1	NLALIB hierarchy.	xxv
Fig. 1.1	Matrix multiplication.	3
Fig. 1.2	Rotating the xy-plane.	7
Fig. 1.3	Rotated line	8
Fig. 1.4	Rotate three-dimensional coordinate system.	9
Fig. 1.5	Translate a point in two dimensions.	9
Fig. 1.6	Rotate a line about a point.	10
Fig. 1.7	Rotation about an arbitrary point.	11
Fig. 1.8	Undirected graph.	12
Fig. 2.1	Polynomial passing through four points.	26
Fig. 2.2	Inconsistent equations.	31
Fig. 2.3	Truss.	38
Fig. 2.4	Electrical circuit.	39
Fig. 2.5	Truss problem.	45
Fig. 2.6	Circuit problem.	45
Fig. 3.1	Subspace spanned by two vectors.	48
Fig. 4.1	Geometrical interpretation of the determinant.	75
Fig. 5.1	Direction of eigenvectors.	80
Fig. 5.2	Circuit with an inductor.	89
Fig. 5.3	Currents in the RL circuit.	92
Fig. 5.4	Digraph of an irreducible matrix.	93
Fig. 5.5	Hanowa matrix.	101
Fig. 6.1	Distance between points.	104
Fig. 6.2	Equality, addition, and subtraction of vectors.	104
Fig. 6.3	Scalar multiplication of vectors.	104
Fig. 6.4	Vector length.	106
Fig. 6.5	Geometric interpretation of the inner product.	106
Fig. 6.6	Law of cosines.	106
Fig. 6.7	Triangle inequality.	112
Fig. 6.8	Signal comparison.	112
Fig. 6.9	Projection of one vector onto another.	115
Fig. 7.1	Effect of an orthogonal transformation on a vector.	122
Fig. 7.2	Spherical coordinates.	123
Fig. 7.3	Orthonormal basis for spherical coordinates.	124
Fig. 7.4	Point in spherical coordinate basis and Cartesian coordinates.	125
Fig. 7.5	Function specified in spherical coordinates.	126
Fig. 7.6	Effect of a matrix on vectors.	128
Fig. 7.7	Unit spheres in three norms.	129
Fig. 7.8	Image of the unit circle.	133
Fig. 8.1	Floating-point number system.	149
Fig. 8.2	Map of IEEE double-precision floating-point.	150
Fig. 9.1	Matrix multiplication.	165
Fig. 10.1	Forward and backward errors.	184
Fig. 10.2	The Wilkinson polynomial.	187
Fig. 10.3	Ill-conditioned Cauchy problem.	188
Fig. 10.4	Conditioning of a problem.	189
Fig. 11.1	LU decomposition of a matrix.	206
Fig. 11.2	$k \times k$ submatrix.	215
Fig. 11.3	Gaussian elimination flop count.	217
Fig. 12.1	Square wave with period 2π.	244
Fig. 12.2	Fourier series converging to a square wave.	244
Fig. 12.3	The heat equation: a thin rod insulated on its sides.	245

Fig. 12.4	Numerical solution of the heat equation: subdivisions of the x and t axes.	245
Fig. 12.5	Numerical solution of the heat equation:locally related points in the grid.	246
Fig. 12.6	Grid for the numerical solution of the heat equation.	246
Fig. 12.7	Graph of the solution for the heat equation problem.	247
Fig. 12.8	Linear least-squares approximation.	250
Fig. 12.9	Quadratic least-squares approximation.	251
Fig. 12.10	Estimating absolute zero.	252
Fig. 12.11	Linear interpolation.	253
Fig. 12.12	Cubic splines.	253
Fig. 12.13	Cubic spline approximation.	256
Fig. 12.14	Sawtooth wave with period 2π.	258
Fig. 13.1	Conductance matrix.	270
Fig. 14.1	Vector orthogonal projection.	282
Fig. 14.2	Removing the orthogonal projection.	282
Fig. 14.3	Result of the first three steps of Gram-Schmidt.	283
Fig. 15.1	The four fundamental subspaces of a matrix.	305
Fig. 15.2	SVD rotation and distortion.	308
Fig. 15.3	(a) Lena (512×512) and (b) lena using 35 modes.	312
Fig. 15.4	Lena using 125 modes.	312
Fig. 15.5	Singular value graph of lena.	313
Fig. 15.6	SVD image capture.	314
Fig. 16.1	Geometric interpretation of the least-squares solution.	322
Fig. 16.2	An overdetermined system.	322
Fig. 16.3	Least-squares estimate for the power function.	329
Fig. 16.4	The reduced SVD for a full rank matrix.	330
Fig. 16.5	Velocity of an enzymatic reaction.	332
Fig. 16.6	Underdetermined system.	339
Fig. 17.1	Givens matrix.	353
Fig. 17.2	Givens rotation.	354
Fig. 17.3	Householder reflection.	363
Fig. 17.4	Linear combination associated with Householder reflection.	363
Fig. 17.5	Householder reflection to a multiple of e_1.	366
Fig. 17.6	Transforming an $m \times n$ matrix to upper triangular form using householder reflections.	369
Fig. 17.7	Householder reflections and submatrices.	369
Fig. 17.8	Householder reflection for a submatrix.	369
Fig. 18.1	Tacoma Narrows Bridge collapse.	380
Fig. 18.2	Mass-spring system.	380
Fig. 18.3	Solution to a system of ordinary differential equations.	382
Fig. 18.4	Populations using the Leslie matrix.	387
Fig. 18.5	Column buckling.	387
Fig. 18.6	Deflection curves for critical loads P_1, P_2, and P_3.	389
Fig. 18.7	Reduced Hessenberg matrix.	401
Fig. 18.8	Inductive step in Schur's triangularization.	407
Fig. 18.9	Schur's triangularization.	407
Fig. 18.10	Eigenvalues of a 2×2 matrix as shifts.	413
Fig. 18.11	Springs problem.	430
Fig. 19.1	Bisection.	454
Fig. 19.2	Interlacing.	454
Fig. 19.3	Bisection method: λ_k located to the left.	456
Fig. 19.4	Bisection method: λ_k located to the right.	457
Fig. 19.5	Bisection and multiple eigenvalues.	458
Fig. 19.6	Secular equation.	460
Fig. 20.1	SOR spectral radius.	479
Fig. 20.2	Region in the plane.	479
Fig. 20.3	Five-point stencil.	480
Fig. 20.4	Poisson's equation. (a) Approximate solution and (b) analytical solution.	481
Fig. 20.5	One-dimensional Poisson equation grid.	484
Fig. 20.6	One-dimensional red-black GS.	486
Fig. 21.1	Examples of sparse matrices. (a) Positive definite: structural problem, (b) symmetric indefinite: quantum chemistry problem, and (c) nonsymmetric: computational fluid dynamics problem.	492
Fig. 21.2	Steepest descent. (a) Quadratic function in steepest descent and (b) gradient and contour lines.	495
Fig. 21.3	Steepest descent. (a) Deepest descent zigzag and (b) gradient contour lines.	495
Fig. 21.4	2-Norm and A-norm convergence.	498
Fig. 21.5	CG vs. steepest descent. (a) Density plot for symmetric positive definite sparse matrix CGDES and (b) residuals of CG and steepest descent.	502

Fig. 21.6	Cholesky decomposition of a sparse symmetric positive definite matrix.	504
Fig. 21.7	CG vs. PRECG.	506
Fig. 21.8	Arnoldi projection from $\mathbb{R}^n$ into $\mathbb{R}^m$, $m \ll n$.	509
Fig. 21.9	Arnoldi decomposition form 1.	511
Fig. 21.10	Arnoldi decomposition form 2.	512
Fig. 21.11	Large nonsymmetric matrix.	515
Fig. 21.12	Lanczos decomposition.	516
Fig. 21.13	Lanczos process with and without reorthogonalization. (a) Lanczos without reorthogonalization and (b) Lanczos with reorthogonalization.	518
Fig. 21.14	Large sparse symmetric matrices.	520
Fig. 21.15	Iterative method decision tree.	521
Fig. 21.16	Poisson's equation grid for $n = 4$.	521
Fig. 21.17	Estimating the normal derivative.	523
Fig. 21.18	36×36 biharmonic matrix density plot.	524
Fig. 21.19	The biharmonic equation. (a) Biharmonic equation numerical solution and (b) biharmonic equation true solution.	525
Fig. 21.20	(a) Electrostatic potential fields induced by approximately 15 randomly placed point charges (b) contour plot of randomly placed point charges.	532
Fig. 22.1	Nonsymmetric sparse matrix used in a chemical engineering model	534
Fig. 22.2	Eigenvalues and Ritz values of a random sparse matrix.	536
Fig. 23.1	Demmel and Kahan zero-shift QR downward sweep.	562
Fig. A.1	Complex addition and subtraction.	572
Fig. A.2	Complex conjugate.	572
Fig. A.3	Riemann zeta function.	577
Fig. C.1	The first five Chebyshev polynomials.	584

List of Algorithms

Algorithm 9.1	Inner Product of Two Vectors	164
Algorithm 9.2	Frobenius Norm	164
Algorithm 9.3	Product of Two Matrices	165
Algorithm 9.4	Solving an Upper Triangular System	169
Algorithm 9.5	Solving a Lower Triangular System	169
Algorithm 9.6	The Thomas Algorithm	173
Algorithm 11.1	*LU* Decomposition Without a Zero on the Diagonal	216
Algorithm 11.2	Gaussian Elimination with Partial Pivoting	223
Algorithm 11.3	Solve $Ax = b$ for Multiple Right-Hand Sides	226
Algorithm 11.4	Iterative Improvement	229
Algorithm 12.1	Cubic Spline Approximation	255
Algorithm 13.1	Computing the *LU* Decomposition of a Tridiagonal Matrix	265
Algorithm 13.2	Solve a Factored Tridiagonal System	266
Algorithm 13.3	The Cholesky Decomposition	271
Algorithm 14.1	Classical Gram-Schmidt	285
Algorithm 14.2	Modified Gram-Schmidt	286
Algorithm 14.3	Modified Gram-Schmidt *QR* Decomposition	288
Algorithm 16.1	Least-Squares Solution Using the Normal Equations	327
Algorithm 16.2	Solving the Least-Squares Problem Using the QR Decomposition	328
Algorithm 16.3	Solving the Least-Squares Problem Using the SVD	330
Algorithm 16.4	Minimum Norm Solution to the Least-Squares Problem	336
Algorithm 16.5	Solution of Full-Rank Underdetermined System Using QR Decomposition	340
Algorithm 17.1	Product of a Givens Matrix J with a General Matrix A	356
Algorithm 17.2	Computing the Givens Parameters	359
Algorithm 17.3	Givens *QR* Decomposition	360
Algorithm 17.4	Zero Out Entries in the First Column of a Matrix using a Householder Reflection	368
Algorithm 17.5	Computation of *QR* Decomposition Using Householder Reflections	371
Algorithm 18.1	The Power Method	391
Algorithm 18.2	Transformation to Upper Hessenberg Form	400
Algorithm 18.3	Unshifted Hessenberg *QR* Iteration	402
Algorithm 18.4	Single Shift Using the Francis Iteration of Degree One	413
Algorithm 18.5	Implicit Double-Shift *QR*	419
Algorithm 18.6	Inverse Iteration to Find Eigenvector of an Upper Hessenberg Matrix	422
Algorithm 18.7	Compute the Condition Number of the Eigenvalues of a Matrix	426
Algorithm 19.1	Jacobi Method for Computing All Eigenvalues of a Real Symmetric Matrix	445
Algorithm 19.2	Orthogonal Reduction of a Symmetric Matrix to Tridiagonal Form	450
Algorithm 20.1	SOR Iteration	472
Algorithm 21.1	Steepest Descent	496
Algorithm 21.2	Conjugate Gradient	500
Algorithm 21.3	Preconditioned Conjugate Gradient	505
Algorithm 21.4	Arnoldi Process	511
Algorithm 21.5	GMRES	513
Algorithm 21.6	Incomplete *LU* Decomposition	514
Algorithm 21.7	Lanczos Method	517
Algorithm 21.8	MINRES	519
Algorithm 22.1	The Implicitly Restarted Arnoldi Process	543
Algorithm 23.1	One-Sided Jacobi Algorithm	555
Algorithm 23.2	Reduction of a Matrix to Upper-bidiagonal Form	559
Algorithm 23.3	Demmel and Kahan Zero-Shift *QR* Downward Sweep.	563

Preface

This book is intended for an advanced undergraduate or a first-year graduate course in numerical linear algebra, a very important topic for engineers and scientists. Many of the numerical methods used to solve engineering and science problems have linear algebra as an important component. Examples include spline interpolation, estimation using least squares, and the solution of ordinary and partial differential equations. It has been said that, next to calculus, linear algebra is the most important component in engineering problem solving. In computer science, linear algebra is a critical as well. The Google matrix is an example, as is computer graphics where matrices are used for rotation, projection, rescaling, and translation. Applications to engineering and science are provided throughout the book.

Two important problems in a customary applied linear algebra course are the solution of general linear algebraic systems $Ax = b$, where A is an $m \times n$ matrix, and the computation of eigenvalues and their associated eigenvectors. If the system is square ($m = n$) and nonsingular, the student is taught how to find a solution to $Ax = b$ using Cramer's Rule, Gaussian elimination and multiplication by the inverse. In many areas of application, such as statistics and signal processing, A is square and singular or $m \neq n$. In these situations, the transformation to reduced row echelon form produces no solution or infinitely many, and this is just fine in a theoretical sense, but is not helpful for obtaining a useful solution. Eigenvalues are often discussed late in the course, and the student learns to compute eigenvalues by finding the roots of the characteristic polynomial, never done in practice.

A study of numerical linear algebra is different from a study of linear algebra. The problem is that many of the theoretical linear algebra methods are not practical for use with a computer. To be used on a computer, an algorithm, a method for solving a problem step by step in a finite amount of time, must be developed that deals with the advantages and problems of using a computer. Any such algorithm must be efficient and not use too much computer memory. For instance, Cramer's Rule is not practical for matrices of size 4×4 or greater, since it performs far to many operations. Since a digital computer performs arithmetic in binary with a fixed number of digits, errors occur when entering data and performing computations. For instance, 1/3 cannot be represented exactly in binary, and its binary representation must be approximated. In addition, computation using the operations of addition, subtraction, multiplication, and division rarely can be done exactly, resulting in errors. An algorithm must behave properly in the presence of these inevitable errors; in other words, small errors during computation should produce small errors in the output. For example, the use of Gaussian elimination to solve an $n \times n$ linear system should use a method known as partial pivoting to control errors. When the matrix A is $m \times n$, $m \neq n$, a solution must be obtained in the sense of least-squares, and the efficient and implementation of least-squares presents challenges. The eigenvalue problem is of primary importance in engineering and science. In practice, eigenvalues are not found by finding the roots of a polynomial, since polynomial root finding is very prone to error. Algorithms have been developed for accurate solution of the eigenvalue problem on a computer.

In the book, algorithms are stated using pseudocode, and MATLAB is the vehicle used for algorithm implementation. MATLAB does a superb job of dealing with numeric computation and is used in most engineering programs. Accompanying the text is a library of MATLAB functions and programs, named NLALIB, that implements most of the algorithms discussed in the book. Many examples in the book include computations using MATLAB, as do many exercises. In some cases, a problem will require the student to write a function or program using the MATLAB programming language. If the student is not familiar with MATLAB or needs a refresher, the MathWorks Web site www.mathworks.com provides access to tutorials. There are also many free online tutorials.

If the reader does not have access to MATLAB, it is possible to use GNU Octave, a system primarily intended for numerical computations. The Octave language is quite similar to MATLAB so that most programs are easily portable.

This book is novel, in that there is no assumption the student has had a course in linear algebra. Engineering students who have completed the usual mathematics sequence, including ordinary differential equations, are well prepared. The prerequisites for a computer science student should include at least two semesters of calculus and a course in discrete mathematics. Chapters 1-6 supply an introduction to the the basics of linear algebra. A thorough knowledge of these chapters prepares the student very well for the remainder of the book. If the student has had a course in applied or theoretical linear algebra, these chapters can be used for a quick review.

Throughout the book, proofs are provided for most of the major results. In proofs, the author has made an effort to be clear, to the point of including more detail than normally provided in similar books.It is left to the instructor to determine how much emphasis should be given to the proofs.

The exercises include routine pencil and paper computations. Exercises of this type force the student to better understand the workings of an algorithm. There are some exercises involving proofs. Hints are provided if a proof will be challenging for most students. In the problems for each chapter, there are exercises to be done using MATLAB.

TOPICS

Chapters 1-6 provide coverage of applied linear algebra sufficient for reading the remainder of the book.

Chapter 1: Matrices

The chapter introduces matrix arithmetic and the very important topic of linear transformations. Rotation matrices provide an interesting and useful example of linear transformations. After discussing matrix powers, the concept of the matrix inverse and transpose concludes the chapter.

Chapter 2: Linear Equations

This chapter introduces Gaussian elimination for the solution of linear systems $Ax = b$ and for the computation of the matrix inverse. The chapter also introduces the relationship between the matrix inverse and the solution to a linear homogeneous equation. Two applications involving a truss and an electrical circuit conclude the chapter.

Chapter 3: Subspaces

This chapter is, by its very nature, somewhat abstract. It introduces the concepts of subspaces, linear independence, basis, matrix rank, range, and null space. Although the chapter may challenge some readers, the concepts are essential for understanding many topics in the book, and it should be covered thoroughly.

Chapter 4: Determinants

Although the determinant is rarely computed in practice, it is often used in proofs of important results. The chapter introduces the determinant and its computation using expansion by minors and by row elimination. The chapter ends with an interesting application of the determinant to text encryption.

Chapter 5: Eigenvalues and Eigenvectors

This is a very important chapter, and its results are used throughout the book. After defining the eigenvalue and an associated eigenvector, the chapter develops some of their most important properties, including their use in matrix diagonalization. The chapter concludes with an application to the solution of systems of ordinary differential equations and the problem of ranking items using eigenvectors.

Chapter 6: Orthogonal Vectors and Matrices

This chapter introduces the inner product and its association with orthogonal matrices. Orthogonal matrices play an extremely important role in matrix factorization. The L^2 inner product of functions is briefly introduced to emphasize the general concept of an inner product.

Chapter 7: Vector and Matrix Norms

The study of numerical linear algebra begins with this chapter. The analysis of methods in numerical linear algebra relies heavily on the concept of vector and matrix norms. This chapter develops the 2-norm, the 1-norm, and the infinity norm for vectors. A development of matrix norms follows, the most important being matrix norms associated with a vector norm, called subordinate norms. The infinity and 1-norms are easy to compute, but the connection between their computation and the mathematical definition of the a matrix norm is somewhat complex. A MATLAB program motivates the process for the computation of the infinity norm, and the chapter contains a complete proof verifying the algorithm for computing the infinity norm. The 2-norm is the most useful matrix norm and by far the most difficult to compute. After motivating the computation process with a MATLAB program, the chapter provides a proof that the 2-norm is the square root of the largest singular value of the matrix and develops properties of the matrix 2-norm.

Chapter 8: Floating Point Arithmetic

The chapter presents the representation of integer and floating point data in a computer, discusses the concepts of overflow and underflow, and explains why roundoff errors occur that cannot be avoided. There is a careful discussion concerning the concepts of absolute and relative error measurement and why relative error is normally used. The chapter presents a mathematical analysis of floating point errors for addition and states results for other operations. The chapter concludes with a discussion of situations where a careful choice of algorithm can minimize errors. This chapter is critical for understanding the remaining chapters. The only content that can be reasonably omitted is the mathematical discussion of floating point errors.

Chapter 9: Algorithms

The algorithms in the book are presented using pseudocode, and the pseudocode is quite complete. It is intended that in most cases the conversion between pseudocode and MATLAB should not be difficult. The chapter introduces the concept of algorithm efficiency by computing the the number of floating point operations, called the flop count, or representing it using big-O notation. The presentation of algorithms for matrix multiplication, the solution to upper and lower triangular systems, and the Thomas algorithm for the solution of a tridiagonal system are the primary examples. Included is a brief discussion of block matrices and basic block matrix operations.

Chapter 10: Conditioning of Problems and the Stability of Algorithms

The chapter introduces the concept of stability and the conditioning. An algorithm is unstable if small changes in the data can cause large changes in the result of the computation. An algorithm may be stable, but the data supplied to the algorithm can be ill-conditioned. For instance, some matrices are very sensitive to errors during Gaussian elimination. After discussing examples and introducing some elementary perturbation analysis using backward and forward error, the chapter develops the condition number of a matrix and its properties. The condition number of a matrix plays an important role as we develop algorithms in the remainder of the book. This material is at the heart of numerical linear algebra and should be covered at least intuitively. There are a number of problems involving numerical experiments, and some of these should be done in order to appreciate the issues involved.

Chapter 11: Gaussian Elimination and the LU Factorization

This chapter introduces the LU decomposition of a square matrix. The LU decomposition uses Gaussian elimination, but is not a satisfactory algorithm without using partial pivoting to minimize errors. The LU decomposition properly computed can be used to solve systems of the form $Ax_i = b_i$, $1 \leq i \leq k$. The somewhat expensive Gaussian elimination algorithm need be used only once. After its computation, many solutions $\{x_i\}$ are quickly found using forward and back substitution.

Chapter 12: Linear Systems Applications

Four applications that involve linear systems comprise this chapter. A discussion of Fourier series introduces the concept of an infinite dimensional vector space and provides an application for the L^2 inner product introduced in Chapter 6. A second application involves finite difference approximations for the heat equation. Finite difference techniques are important when approximating the solution to boundary value problems for ordinary and partial differential equations. Chapter 16 discusses least-squares problems. As a tune-up for this chapter, the third application develops approximation by polynomial least-squares. The last application is a discussion of cubic spline interpolation. Using this process, a series of cubic polynomials are fitted between each pair of data points over an interval $a \leq x \leq b$, with the requirement that the curve obtained be twice differentiable. These cubic splines can then be used to very accurately estimate the data at other points in the interval. The computation of cubic splines involves the solution of a tridiagonal system of equations, and the Thomas algorithm presented in Chapter 9 works very well.

Chapter 13: Important Special Systems

Numerical linear algebra is all about computing solutions to problems accurately and efficiently. As a result, algorithms must be developed that take advantage of a special structure or properties of a matrix. This chapter discusses the factorization of a tridiagonal matrix and the Cholesky factorization of a symmetric positive definite matrix. In both cases, the matrix factorization leads to more efficient means of solving a linear system having a coefficient matrix of one of these types.

Chapter 14: Gram-Schmidt Orthonormalization

The Gram-Schmidt algorithm for computing an orthonormal basis is time-honored and important. It becomes critical in the development of algorithms such as the singular value and Arnoldi decompositions. The chapter carefully develops the QR decomposition using Gram-Schmidt. Although the decomposition is not normally done this way, it serves to demonstrate that this extremely important tool exists. As a result, the MATLAB algorithm qr can be used with some understanding until efficient methods for the QR decomposition are explained.

Chapter 15: The Singular Value Decomposition

The singular value decomposition (SVD) is perhaps the most important result in numerical linear algebra. Its uses are many, including providing a method for estimating matrix rank and the solution of least-squares problems. This chapter proves the SVD theorem and provides applications. Perhaps the most interesting application is the use of the SVD in image compression. Practical algorithms for the computation of the SVD are complex, and are left to Chapter 23.

Chapter 16: Least Squares Problems

Approximation using least-squares has important applications in statistics and many other areas. For instance, data collected by sensor networks is often analyzed using least-squares in order to approximate events taking place. Least-squares problems arise when the data requires the solution to an $m \times n$ system $Ax = b$, where $m \neq n$. Normally, there is no solution $\bar{x}$ such that $A\bar{x} = b$, or there are infinitely many solutions, so we seek a solution that minimizes the Euclidean norm of $Ax - b$. Least-squares provides an excellent application for the QR factorization and the SVD.

Chapter 17: Implementing the QR Factorization

The QR factorization using the Gram-Schmidt process was developed in Chapter 14. This chapter presents two other approaches to the factorization, the use of Givens rotations and Householder reflections. In each case, the algorithm is more stable than Gram-Schmidt. Also, we will have occasion to use Givens rotations and Householder reflections for other purposes, such as the computation of eigenvalues. If a detailed presentation is not required, these ideas have a nice geometrical interpretation.

Chapter 18: The Algebraic Eigenvalue Problem

The applications of the eigenvalue problem are vast. The chapter begins by presenting three applications, a problem in vibration and resonance, the Leslie model in population biology, and the buckling of a column. The accurate computation of eigenvalues and their associated eigenvectors is difficult. The power and inverse power methods are developed for computing the largest and smallest eigenvalues of a matrix. These methods are important but have limited use. The chapter discusses the QR iteration for the computation of all the eigenvalues and their associated eigenvectors of a real matrix whose eigenvalues are distinct. The development is detailed and includes the use of the shifted Hessenberg QR iteration. The chapter also develops the computation of eigenvectors using the Hessenberg inverse iteration. The method used in most professional implementations is the implicit QR iteration, also known as the Francis iteration. The chapter develops the algorithm for the computation of both the real and complex eigenvalues of a real matrix.

Chapter 19: The Symmetric Eigenvalue Problem

If a matrix is symmetric, an algorithm can exploit its symmetry and compute eigenvalues faster and more accurately. Fortunately, many very important problems in engineering and science involve symmetric matrices. The chapter develops five methods for the computation of eigenvalues and their associated eigenvectors, the Jacobi method, the symmetric QR iteration method, the Francis algorithm, the bisection method, and the divide and conquer method.

Chapter 20: Basic Iterative Methods

Iterative methods are used for the solution of large, sparse, systems, since ordinary Gaussian elimination operations will destroy the sparse structure of the matrix. This chapter presents the classical Jacobi, Gauss-Seidel, and SOR methods, along with discussion of convergence. The chapter concludes with the application of iterative methods to the solution of the two-dimensional Poisson equation.

Chapter 21: Krylov Subspace Methods

This is a capstone chapter, and should be covered, at least in part, in any numerical linear algebra course. The conjugate gradient method (CG) for the solution of large, sparse symmetric positive definite systems is presented. This method is one of the jewels of numerical linear algebra and has revolutionized the solution of many very large problems. The presentation motivates the algorithm and provides mathematical details that explain why it works. The conjugate gradient method is a Krylov subspace method, although the book does not develop it using this approach. However, the next algorithm presented is the general minimum residual method (GMRES) for the iterative solution of large, sparse, general matrices, and it is approached as a Krylov subspace method. The Krylov subspace-based minimum residual (MINRES) method for the solution of large, sparse, symmetric, non-positive definite matrices is the last method presented. If a matrix is ill-conditioned, CG, GMRES, and MINRES do not perform well. The solution is to precondition the system before applying an iterative method. The chapter presents preconditioning techniques for CG and GMRES. After presenting a chart detailing approaches to large, sparse problems, the chapter concludes with another approach to the Poisson equation and a discussion of the biharmonic equation that is one of the most important equations in applied mechanics.

Chapter 22: Large Sparse Eigenvalue Problems

The chapter discusses the use of the Arnoldi and Lanczos processes to find a few eigenvalues of large, sparse matrices. Two approaches are discussed, explicit and implicit restarting. The mathematics behind the performance of these methods is beyond the scope of the text, but the algorithms are presented and MATLAB implementations provided. Various exercises test the methods and clearly demonstrate the challenge of this problem.

Chapter 23: Computing the Singular Value Decomposition

The chapter develops two methods for computing the SVD, the one-sided Jacobi method, and the Demmel and Kahan zero-shift QR downward sweep algorithm. Developing the two methods requires a knowledge of many results from earlier chapters.

Appendices A, B, and C

Appendix A provides a discussion of complex numbers so that a reader unfamiliar with the topic will be able to acquire the knowledge necessary when the book uses basic results from the theory of complex numbers. Appendix B presents a brief discussion of mathematical induction, and Appendix C presents an overview of Chebyshev polynomials. Although these polynomials are not used within any proof in the book, they are referenced in theorems whose proofs are provided by other sources.

INTENDED AUDIENCE

Numerical linear algebra is often a final chapter in a standard linear algebra text, and yet is of paramount importance for engineers and scientists. The book covers many of the most important topics in numerical linear algebra, but is not intended to be encyclopedic. However, there are many references to material not covered in the book. Also, it is the author's hope that the material is more accessible as a first course than existing books, and that the first six chapters provide material sufficient for the book to be used without a previous course in applied linear algebra. The book is also is very useful for self-study and can serve as a reference for engineers and scientists. It can also serve as an entry point to more advanced books, such as James Demmel's book [1] or the exhaustive presentation of the topic by Golub and Van Loan [2].

WAYS TO USE THE BOOK

The instructor will need to decide how much theory should be covered; namely, how much emphasis will be placed on understanding the proofs and doing problems involving proofs. If the students are not experienced with proofs, one approach is to explain methods and theorems as intuitively as possible, supporting the discussion with numerical examples in class, and having the students do numerous numerical exercises both in class and in assignments. For instance, using Jacobi rotations to compute the eigenvalues of a real symmetric matrix is easily explained using simple diagrams and running a MATLAB program included with the software distribution graphically demonstrates how the method performs a reduction to a diagonal matrix. This approach works well with engineering students who have little or no experience with theorems and proofs. They will learn how to solve problems, large and small, using the appropriate methods.

If the audience consists of students who are mathematics majors or who have significant mathematical training, then some proofs should be covered and assignments should include proofs. Some of these exercises include hints to get the student started. The author believes that for a student to stare at the hypothesis and conclusion only to give up in frustration makes no sense, when a simple hint will kick start the process.

Of course, the amount of material that the instructor can cover depends on the background of the students. Mathematics majors will likely have taken a theoretical or applied linear algebra course. After optionally reviewing the material in Chapters 1-6 the study of numerical linear algebra can begin. The following is a list of suggestions for various chapters that outlines material that can be omitted, covered lightly, or must be covered.

- In Chapter 7, proofs that justify methods for computing matrix norms can be omitted, but MATLAB programs that motivate the methods should be discussed.
- Chapter 8 is essential to an understanding of numerical linear algebra. It presents storage formats for integers and floating point numbers and shows why the finite precision arithmetic used by a computer leads to roundoff error. Some examples are provided that show how rearranging the order of computation can help to reduce error.
- Chapter 10 that discusses the stability and conditioning of algorithms should be covered at least intuitely. There are numerous examples and problems in the book that illustrate the problems that can occur with floating point arithmetic.
- In Chapter 11, the LU decomposition must be presented, and the student should use it to solve a number of problems. If desired, the use of elementary row matrices to prove why the LU decomposition works can be omitted. It is very important the student understand that multiple systems can be solved with only one LU decomposition. The efficiency of many algorithms depends on it.
- The instructor can choose among the applications in Chapter 12, rather than covering the entire chapter.
- In Chapter 13, factoring tridiagonal matrices can be safely omitted, but positive definite matrices and the Cholesky decomposition must be covered.
- The Gram-Schmidt orthogonalization method and its use in forming the QR decomposition is important and not particularly difficult, so it should be covered.
- Except for the proof of the SVD theorem, all of Chapter 15 should be presented. The use of the SVD for image compression excites students and is just plain fun.
- In Chapter 16, rank-deficient and underdetermined least-squares can be omitted, since the majority of applications involve full rank overdetermined systems.
- It is recommended that Chapter 17 concerning the computation of the QR decomposition using Givens rotations and Householder reflections be covered. These tools are needed later in the book when discussing the eigenvalue problem. Both of these methods can be explained intuitively, supported by MATLAB programs from NLALIB, so the instructor can omit many of the details if desired.
- Chapter 18 discusses the general algebraic eigenvalue problem, and should be covered in part. Certainly it is important to discuss the power and inverse power methods and the QR iteration with and without shifts and deflation. The Francis,

or implicit QR iteration, is used in practice with both single and double implicit shifts. The details are complex, but an overview can be presented, followed by numerical experiments.

- The Spectral Theorem is used throughout the book, and its proof in Chapter 19 can be omitted with no harm. The Jacobi method for computing the eigenvalues and eigenvectors of a symmetric matrix can be covered thoroughly or intuitively. There are a number of programming and mathematical issues involved, but the idea is quite simple, and an intuitive explanation will suffice. Certainly the symmetric QR iteration method should be covered. If the Francis algorithm was covered in Chapter 18, it makes sense to present the single shift Francis algorithm. The bisection method is interesting and not difficult, so covering it is a good option. The chapter concludes with the complex divide-and-conquer method, and it is optional. NLALIB contains a C implementation of the algorithm using the MATLAB MEX interface, and it might be interesting demonstrate the algorithm's performance on a large, dense, symmetric matrix.
- The author feels that some coverage of iterative methods is very important since many engineering and science students will deal with projects that involve large, sparse matrices. The classical material on the Jacobi, Gauss-Seidel, and SOR iterations in Chapter 20 can be covered quickly by not presenting convergence theorems.
- The conjugate gradient method (CG) in Chapter 21 should be introduced and the student should gain experience using it and the preconditioned CG to solve large systems. The approach to its development is through the method of steepest descent. That algorithm is simple and can be supported by geometrical arguments. CG is an improvement of steepest descent, and the mathematical details can be skipped if desired. The application of Krylov subspace methods to develop the Arnoldi and Lanczos decompositions is somewhat technical, but the results are very important. At a minimum, the student should work some exercises that involve using NLALIB to execute some decompositions. It is then easy to see how these decompositions lead to the GMRES and MINRES methods. The software distribution contains a number of large, sparse matrices used in actual applications. These are used for examples and exercises in the book.
- Chapter 22 is very interesting both from a practical and theoretical standpoint. However, the material is challenging and can be left to more advanced courses. A possibility is using the chapter as an introduction to such books as Refs. [3–6].
- The SVD is used from Chapter 15 on, so the student is very familiar with its applications. Chapter 23 contains two methods for computing the SVD, and this material can be left to a subsequent course.

MATLAB LIBRARY

NLALIB is an essential supplement to the book. Figure 0.1 shows the structure of the library, in which almost all major algorithms are implemented as functions. As is customary, directory names are abbreviations; for instance, the subdirectory geneigs contains demonstration software for methods to compute the eigenvalues of a general, non-sparse, matrix. The book provides many examples of matrices from actual applications or matrices designed for testing purposes, and these matrices are included in NLALIB in MATLAB matrix format.

SUPPLEMENTS

At http://textbooks.elsevier.com/web/Manuals.aspx?isbn=9780123944351, the instructor will find supplements that include the solution to every problem in the book, laboratory exercises that can be used after lectures for more interactive

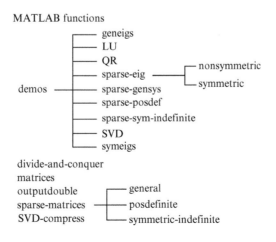

FIGURE 0.1 NLALIB hierarchy.

learning, and a complete set of PowerPoint slides. For students, Elsevier provides the Web site http://booksite.elsevier.com/978012394435 that provides students with review questions and solutions. The author also provides the Web site http://ford-book.info that provides a summary of the book and links to associated Web sites.

ACKNOWLEDGMENTS

Although I have co-authored a number of books in computer science, this is my first mathematics text. I am indebted to Patrcia Osborn, the acquisitions editor at Elsevier who had faith in the book and ushered it through the acceptance phase. I would also like to thank the reviewers of the book, Jakub Kurzak, Ph.D., Research Director, Innovative Computing Laboratory, University of Tennessee, Rajan Bhatt, Ph.D., research engineer at the University of Iowa Center for Computer Aided Design, and Zhongshan Li, Ph.D., Professor and Graduate Director of Mathematics, Georgia State University. Their comments were helpful and encouraging. I also appreciate the efforts of the editorial project managers, Jill Cetel, Jessica Vaughn, and Paula Callaghan, who were invaluable in helping me through what is a difficult process. I also wish to thank Anusha Sambamoorthy, Project Manager, Book Production, Chennai, Elsevier, for her help with getting the book into print.

Dr Keith Matthews, Honorary Research Consultant, University of Queensland, gave me permission to use his online textbook, *Elementary Linear Algebra*, as a starting point for the material in the first six chapters. While I have made many changes to suit the needs of the book, his generosity saved me much time.

I thank the University of the Pacific, School of Engineering and Computer Science, Stockton, California, for providing resources and a sabbatical leave during the development of the book. I also appreciate comments made by engineering graduate students as I used the manuscript in an advanced computation course. Lastly, I am indebted to William Topp, Ph.D., with whom I have coauthored a number of books. He provided encouragement and consolation during the lengthy process of developing the text.

Chapter 1

Matrices

You should be familiar with

- Two- and three-dimensional geometry
- Elementary functions

Linear algebra is a branch of mathematics that is used by engineers and applied scientists to design and analyze complex systems. Civil engineers use linear algebra to design and analyze load-bearing structures such as bridges. Mechanical engineers use linear algebra to design and analyze suspension systems, and electrical engineers use it to design and analyze electrical circuits. Electrical, biomedical, and aerospace engineers use linear algebra to enhance X-rays, tomographs, and images from space. This introduction is intended to serve as a basis for the study of numerical linear algebra, the study of procedures used on a computer to perform linear algebra computations, most notably matrix operations. As you will see, there is a big difference between theoretical linear algebra and applying linear algebra on a computer and obtaining reliable results. It is assumed only that the reader has completed one or more calculus courses and has had some exposure to vectors and matrices, although the text provides a review of the basic concepts. It will be helpful but not necessary if the reader has taken a course in discrete mathematics that provided some exposure to mathematical proofs.

Section 1.1 discusses matrix operations, including matrix multiplication and that matrix multiplication obeys many of the familiar laws of arithmetic apart from the commutative law. While matrix multiplication is most often performed on a computer, it is necessary to understand its definition, fundamental properties, and applications. For instance, a linear system of equations is elegantly expressed in matrix form. This section also introduces the matrix trace operator and the very useful fact that trace $(AB) =$ trace (BA) for square matrices A and B. This section concludes with a presentation of basic MATLAB operations for executing these fundamental matrix operations.

A linear transformation is an absolutely critical concept in linear algebra, and Section 1.2 presents the concept and shows how a linear transformation performs a rotation of a figure in the xy-plane or in three-dimensional space. This application of linear transformations is fundamental to computer graphics.

Section 1.3 discusses powers of matrices and shows the connection between matrix powers and the number of possible paths between two vertices of a graph. This section also presents the interesting Fibonacci matrix.

Section 1.4 introduces the matrix inverse and a number of its properties. It is shown that a linear system has a unique solution when its coefficient matrix has an inverse.

Section 1.5 discusses the matrix transpose and this motivates the definition of a symmetric matrix. As we will see in later chapters, symmetric matrices have many applications in engineering and science.

1.1 MATRIX ARITHMETIC

A matrix is a rectangular array of numbers with m rows and n columns. The symbol $\mathbb{R}^{m \times n}$ denotes the collection of all $m \times n$ matrices whose entries are real numbers. Matrices will usually be denoted by capital letters, and the notation $A = [a_{ij}]$ specifies that the matrix is composed of entries a_{ij} located in the ith row and jth column of A.

A vector is a matrix with either one row or one column; for instance,

$$x = \begin{bmatrix} 1 \\ -2 \\ 6 \end{bmatrix}$$

is a column vector, and

$$y = \begin{bmatrix} 6 & -1 & 3 \end{bmatrix}$$

is a row vector. The elements of a vector require only one subscript. For the vector x, $x_2 = -4$.

Numerical Linear Algebra with Applications. http://dx.doi.org/10.1016/B978-0-12-394435-1.00001-6

Example 1.1.   The formula $a_{ij} = 1/(i+j)$ for $1 \leq i \leq 3, 1 \leq j \leq 4$ defines a 3×4 matrix $A = [a_{ij}]$, namely,

$$A = \begin{bmatrix} \frac{1}{2} & \frac{1}{3} & \frac{1}{4} & \frac{1}{5} \\ \frac{1}{3} & \frac{1}{4} & \frac{1}{5} & \frac{1}{6} \\ \frac{1}{4} & \frac{1}{5} & \frac{1}{6} & \frac{1}{7} \end{bmatrix}.$$

The first column of A is the column vector $\begin{bmatrix} \frac{1}{2} \\ \frac{1}{3} \\ \frac{1}{4} \end{bmatrix}$.   ■

Definition 1.1 (Equality of matrices).   Matrices A and B are said to be equal if they have the same size and their corresponding elements are equal; i.e., A and B have dimension $m \times n$, and $A = [a_{ij}], B = [b_{ij}]$, with $a_{ij} = b_{ij}$ for $1 \leq i \leq m, 1 \leq j \leq n$.

Definition 1.2 (Addition of matrices).   Let $A = [a_{ij}]$ and $B = [b_{ij}]$ be of the same size. Then $A + B$ is the matrix obtained by adding corresponding elements of A and B; that is,

$$A + B = [a_{ij}] + [b_{ij}] = [a_{ij} + b_{ij}].$$

Definition 1.3 (Scalar multiple of a matrix).   Let $A = [a_{ij}]$ and t be a number (*scalar*). Then tA is the matrix obtained by multiplying all elements of A by t; that is,

$$tA = t[a_{ij}] = [ta_{ij}].$$

Definition 1.4 (Negative of a matrix).   Let $A = [a_{ij}]$. Then $-A$ is the matrix obtained by replacing the elements of A by their negatives; that is,

$$-A = -[a_{ij}] = [-a_{ij}].$$

Definition 1.5 (Subtraction of matrices).   Matrix subtraction is defined for two matrices $A = [a_{ij}]$ and $B = [b_{ij}]$ of the same size, in the usual way; that is,

$$A - B = [a_{ij}] - [b_{ij}] = [a_{ij} - b_{ij}].$$

Definition 1.6 (The zero matrix).   Each $m \times n$ matrix, all of whose elements are zero, is called the *zero* matrix (of size $m \times n$) and is denoted by the symbol 0.

The matrix operations of addition, scalar multiplication, negation and subtraction satisfy the usual laws of arithmetic. (In what follows, s and t are arbitrary scalars and A, B, C are matrices of the same size.)

1. $(A + B) + C = A + (B + C)$;
2. $A + B = B + A$;
3. $0 + A = A$;
4. $A + (-A) = 0$;
5. $(s + t)A = sA + tA$, $(s - t)A = sA - tA$;
6. $t(A + B) = tA + tB$, $t(A - B) = tA - tB$;
7. $s(tA) = (st)A$;
8. $1A = A$, $0A = 0$, $(-1)A = -A$;
9. $tA = 0 \Rightarrow t = 0$ or $A = 0$.

Other similar properties will be used when needed.

1.1.1   Matrix Product

Definition 1.7 (Matrix product).   Let $A = [a_{ij}]$ be a matrix of size $m \times p$ and $B = [b_{jk}]$ be a matrix of size $p \times n$ (i.e., the number of columns of A equals the number of rows of B). Then AB is the $m \times n$ matrix $C = [c_{ik}]$ whose (i,j)th element is defined by the formula

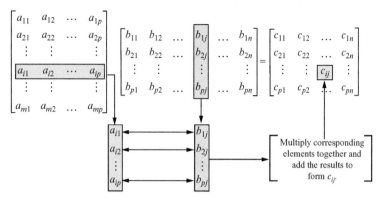

FIGURE 1.1 Matrix multiplication.

$$c_{ij} = \sum_{k=1}^{p} a_{ik}b_{kj} = a_{i1}b_{1j} + \cdots + a_{ip}b_{pj}.$$

A way to look at this is that c_{ij} is the sum of the products of corresponding elements from row i of A and column j of B. For hand computation, fix on row 1 of A. Form the sum of products of corresponding elements from row 1 of A and column 1 of B, then the sum of products of corresponding elements from row 1 of A and column 2 of B, and so forth, until forming the sum of the products of corresponding elements of row 1 of A and column n of B. This computes the first row of the product matrix C. Now use row 2 of A in the same fashion to compute the second row of C. Continue until you have all m rows of C (Figure 1.1).

Example 1.2.

$$\begin{bmatrix} 1 & 2 \\ 3 & 4 \end{bmatrix}\begin{bmatrix} 5 & 6 \\ 7 & 8 \end{bmatrix} = \begin{bmatrix} 1 \times 5 + 2 \times 7 & 1 \times 6 + 2 \times 8 \\ 3 \times 5 + 4 \times 7 & 3 \times 6 + 4 \times 8 \end{bmatrix} = \begin{bmatrix} 19 & 22 \\ 43 & 50 \end{bmatrix},$$

$$\begin{bmatrix} 5 & 6 \\ 7 & 8 \end{bmatrix}\begin{bmatrix} 1 & 2 \\ 3 & 4 \end{bmatrix} = \begin{bmatrix} 23 & 34 \\ 31 & 46 \end{bmatrix} \neq \begin{bmatrix} 1 & 2 \\ 3 & 4 \end{bmatrix}\begin{bmatrix} 5 & 6 \\ 7 & 8 \end{bmatrix},$$

$$\begin{bmatrix} 1 \\ 2 \end{bmatrix}\begin{bmatrix} 3 & 4 \end{bmatrix} = \begin{bmatrix} 3 & 4 \\ 6 & 8 \end{bmatrix},$$

$$\begin{bmatrix} 3 & 4 \end{bmatrix}\begin{bmatrix} 1 \\ 2 \end{bmatrix} = \begin{bmatrix} 11 \end{bmatrix},$$

$$\begin{bmatrix} 1 & -1 \\ 1 & -1 \end{bmatrix}\begin{bmatrix} 1 & -1 \\ 1 & -1 \end{bmatrix} = \begin{bmatrix} 0 & 0 \\ 0 & 0 \end{bmatrix}.$$

∎

Remark 1.1.   Matrix multiplication is a computationally expensive operation. On a computer, multiplication is a much more time-consuming operation than addition. Consider computing the product of an $m \times k$ matrix A and a $k \times n$ matrix B. The computation of $(AB)_{ij}$ requires calculating k products. This must be done n times to form each row of AB, so the computation of a row of AB requires kn multiplications. There are m rows in AB, so the total number of multiplications is $m(kn) = mkn$. If A and B are both $n \times n$ matrices, n^3 multiplications must be performed. For example, if the matrices have dimension 10×10, the computation of their product requires 1000 multiplications. To multiply two 100×100 matrices involves computing 1,000,000 products. A matrix most of whose entries are zero is called *sparse*. There are faster ways to multiply *sparse* matrices, and we will deal with these matrices in Chapters 21 and 22.

Theorem 1.1.   *Matrix multiplication obeys many of the familiar laws of arithmetic apart from the commutative law.*

1. $(AB)C = A(BC)$ *if* A, B, C *are* $m \times p, p \times k, k \times n$, *respectively;*
2. $t(AB) = (tA)B = A(tB)$, $A(-B) = (-A)B = -(AB)$;
3. $(A+B)C = AC + BC$ *if* A *and* B *are* $m \times n$ *and* C *is* $n \times p$;
4. $D(A+B) = DA + DB$ *if* A *and* B *are* $m \times n$ *and* D *is* $p \times m$.

We prove the associative law only:

Proof. Assume that A is an $m \times p$ matrix, B is a $p \times k$ matrix, and C is a $k \times n$ matrix. Observe that $(AB)C$ and $A(BC)$ are both of size $m \times n$.

Let $A = [a_{iq}], B = [b_{ql}], C = [c_{lj}]$. Then

$$((AB)C)_{ij} = \sum_{q=1}^{k}(AB)_{iq}c_{qj} = \sum_{q=1}^{k}\left(\sum_{l=1}^{p}a_{il}b_{lq}\right)c_{qj}$$

$$= \sum_{q=1}^{k}\sum_{l=1}^{p}a_{il}b_{lq}c_{qj}.$$

Similarly,

$$(A(BC))_{ij} = \sum_{l=1}^{p}\sum_{q=1}^{k}a_{il}b_{lq}c_{qj}.$$

However, the double summations are equal. Sums of the form

$$\sum_{q=1}^{k}\sum_{l=1}^{p}d_{lq} \quad \text{and} \quad \sum_{l=1}^{p}\sum_{q=1}^{k}d_{lq}$$

represent the sum of the kp elements of the rectangular array $[d_{lq}]$, by rows and by columns, respectively. Consequently, $((AB)C)_{ij} = (A(BC))_{ij}$ for $1 \leq i \leq m, 1 \leq j \leq n$. Hence, $(AB)C = A(BC)$. $\qquad\square$

One of the primary uses of matrix multiplication is formulating a system of equations as a matrix problem. The system of m linear equations in n unknowns

$$a_{11}x_1 + a_{12}x_2 + \cdots + a_{1n}x_n = b_1$$
$$a_{21}x_1 + a_{22}x_2 + \cdots + a_{2n}x_n = b_2$$
$$\vdots$$
$$a_{m1}x_1 + a_{m2}x_2 + \cdots + a_{mn}x_n = b_m$$

is equivalent to a single-matrix equation

$$\begin{bmatrix} a_{11} & a_{12} & \cdots & a_{1n} \\ a_{21} & a_{22} & \cdots & a_{2n} \\ \vdots & \vdots & & \vdots \\ a_{m1} & a_{m2} & \cdots & a_{mn} \end{bmatrix} \begin{bmatrix} x_1 \\ x_2 \\ \vdots \\ x_n \end{bmatrix} = \begin{bmatrix} b_1 \\ b_2 \\ \vdots \\ b_m \end{bmatrix},$$

that is, $Ax = b$, where $A = [a_{ij}]$ is the *coefficient matrix* of the system, $x = \begin{bmatrix} x_1 \\ x_2 \\ \vdots \\ x_n \end{bmatrix}$ is the *vector of unknowns* and

$b = \begin{bmatrix} b_1 \\ b_2 \\ \vdots \\ b_m \end{bmatrix}$ is the *vector of constants.*

Another useful matrix equation equivalent to the above system of linear equations is

$$x_1\begin{bmatrix} a_{11} \\ a_{21} \\ \vdots \\ a_{m1} \end{bmatrix} + x_2\begin{bmatrix} a_{12} \\ a_{22} \\ \vdots \\ a_{m2} \end{bmatrix} + \cdots + x_n\begin{bmatrix} a_{1n} \\ a_{2n} \\ \vdots \\ a_{mn} \end{bmatrix} = \begin{bmatrix} b_1 \\ b_2 \\ \vdots \\ b_m \end{bmatrix}.$$

We will begin a study of $n \times n$ linear systems in Chapter 2 and continue the study throughout the book. In Chapter 16, most of the systems we deal with will have dimension $m \times n$, where $m \neq n$.

Example 1.3. The system

$$\begin{aligned} x + y + z &= 1, \\ x - y + z &= 0, \\ 3x + 5y - z &= 2 \end{aligned}$$

is equivalent to the matrix equation

$$\begin{bmatrix} 1 & 1 & 1 \\ 1 & -1 & 1 \\ 3 & 5 & -1 \end{bmatrix} \begin{bmatrix} x \\ y \\ z \end{bmatrix} = \begin{bmatrix} 1 \\ 0 \\ 2 \end{bmatrix}$$

and to the equation

$$x \begin{bmatrix} 1 \\ 1 \\ 3 \end{bmatrix} + y \begin{bmatrix} 1 \\ -1 \\ 5 \end{bmatrix} + z \begin{bmatrix} 1 \\ 1 \\ -1 \end{bmatrix} = \begin{bmatrix} 1 \\ 0 \\ 2 \end{bmatrix}.$$

The solution to the system is $\begin{bmatrix} 0.0000 \\ 0.5000 \\ 0.5000 \end{bmatrix}$:

$$\begin{aligned} (1)\,0.0000 + (1)\,0.5000 + (1)\,0.5000 &= 1, \\ (1)\,0.0000 - (1)\,0.5000 + (1)\,0.5000 &= 0, \\ 3\,(0.0000) + 5\,(0.5000) - (1)\,0.5000 &= 2. \end{aligned}$$ ∎

1.1.2 The Trace

The trace is a matrix operation that is frequently used in matrix formulas, and it is very simple to compute.

Definition 1.8. If A is an $n \times n$ matrix, the trace of A, written trace (A), is the sum of the diagonal elements; that is,

$$\text{trace}\,(A) = a_{11} + a_{22} + \cdots + a_{nn} = \sum_{i=1}^{n} a_{ii}.$$

Example 1.4. If $A = \begin{bmatrix} 5 & 8 & 12 & -1 \\ 7 & 4 & -8 & 7 \\ 0 & 3 & -6 & 5 \\ -1 & -9 & 4 & 3 \end{bmatrix}$, then trace $(A) = 5 + 4 + (-6) + 3 = 6$. ∎

There are a number of relationships satisfied by trace. For instance, trace $(A + B)$ = trace (A) + trace (B) (Problem 1.22(a)), and trace (cA) = c trace (A), where c is a scalar (Problem 1.22(b)). A more complex relationship is the trace of product of two matrices.

Theorem 1.2. *If A is an $n \times n$ matrix and B is an $n \times n$ matrix, then* trace (AB) = trace (BA).

Proof. By the definition of matrix multiplication,

$$\text{trace}\,(AB) = \sum_{i=1}^{n} (AB)_{ii} = \sum_{i=1}^{n} \left(\sum_{k=1}^{n} a_{ik}b_{ki} \right) = \sum_{k=1}^{n} \left(\sum_{i=1}^{n} b_{ik}a_{ki} \right) = \text{trace}\,(BA). \qquad \square$$

1.1.3   MATLAB Examples

There are numerous examples throughout this book that involve the use of MATLAB. It is fundamental to our use of MATLAB that you are familiar with "*vectorization.*" When performing vector or matrix operations using the operators "*", "/", and "^", it may be necessary to use the dot operator ("."). As stated in the preface, the reader is expected to be familiar with MATLAB, but it is not necessary to be an expert.

Example 1.5.   *Matrix Operations*
The operators $+$, $-$, $*$ work as expected in MATLAB, and the command `trace` computes the trace of a matrix.

```
>> A = [1 5 1;2 -1 6;1 0 3]

A =
      1     5     1
      2    -1     6
      1     0     3
>> B = [2 3 0;3 -1 7;4 8 9]

B =
      2     3     0
      3    -1     7
      4     8     9
>> 5*A -10*B + 3*A*B

ans =
     48    13   137
     55   170   101
      7     1     6
>> trace(A + B)

ans =
     13
>> 7*trace(A + B)

ans =
     91
>> trace(A*B)

ans =
    103
>> trace(B*A)

ans =
    103
>> A.*B

ans =
      2    15     0
      6     1    42
      4     0    27
```

■

1.2   LINEAR TRANSFORMATIONS

Throughout this book we will assume that matrices have elements that are real numbers. The real numbers include the integers $(\dots, -2, -1, 0, 1, 2, \dots)$, which are a subset of the *rational numbers* (p/q), where p and q are positive integers, $q \neq 0$. The remaining numbers are called *irrational numbers*; for instance, π and e are irrational numbers. We will use the symbol $\mathbb{R}$ to denote the collection of real numbers. An *n-dimensional column vector* is an $n \times 1$ matrix. The collection of all n-dimensional column vectors is denoted by $\mathbb{R}^n$.

Every matrix is associated with a type of function called a *linear transformation*.

Definition 1.9 (Linear transformation).   We can associate an $m \times n$ matrix A with the function $T_A : \mathbb{R}^n \to \mathbb{R}^m$, defined by $T_A(x) = Ax$ for all $x \in \mathbb{R}^n$. More explicitly, using components, the above function takes the form

$$
\begin{aligned}
y_1 &= a_{11}x_1 + a_{12}x_2 + \cdots + a_{1n}x_n \\
y_2 &= a_{21}x_1 + a_{22}x_2 + \cdots + a_{2n}x_n \\
&\vdots \\
y_m &= a_{m1}x_1 + a_{m2}x_2 + \cdots + a_{mn}x_n,
\end{aligned}
$$

where $y_1, y_2, \dots, y_m$ are the components of the column vector $T_A(x)$, in other words $y = Ax$.

A linear transformation has the property that

$$
T_A(sx + ty) = sT_A(x) + tT_A(y)
$$

for all $s, t \in \mathbb{R}$ and all n-dimensional column vectors x, y. This is true because

$$
T_A(sx + ty) = A(sx + ty) = s(Ax) + t(Ay) = sT_A(x) + tT_A(y).
$$

1.2.1   Rotations

One well-known example of a linear transformation arises from rotating the (x, y)-plane in two-dimensional Euclidean space counterclockwise about the origin $(0, 0)$ through θ radians. A point (x, y) will be transformed into the point $(\overline{x}, \overline{y})$. By referring to Figure 1.2, the coordinates of the rotated point can be found using a little trigonometry.

$$
\begin{aligned}
\overline{x} &= d\cos(\theta + \alpha) = d\cos(\theta)\cos(\alpha) - d\sin(\theta)\sin(\alpha) = x\cos(\theta) - y\sin(\theta), \\
\overline{y} &= d\sin(\theta + \alpha) = d\sin(\theta)\cos(\alpha) + d\cos(\theta)\sin(\alpha) = x\sin(\theta) + y\cos(\theta).
\end{aligned}
$$

The equations in matrix form are

$$
R = \begin{bmatrix} \overline{x} \\ \overline{y} \end{bmatrix} = \begin{bmatrix} \cos\theta & -\sin\theta \\ \sin\theta & \cos\theta \end{bmatrix} \begin{bmatrix} x \\ y \end{bmatrix}.
$$

Example 1.6.   Rotate the line $y = 5x + 1$ an angle of $30°$ counterclockwise. Graph the original and the rotated line.

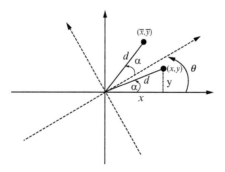

FIGURE 1.2 Rotating the *xy*-plane.

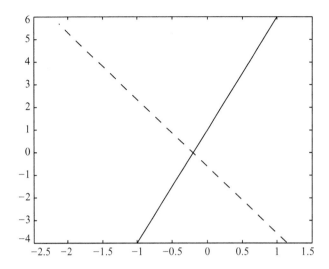

FIGURE 1.3 Rotated line

Since $30°$ is $\pi/6$ radians, the rotation matrix is

$$\begin{bmatrix} \cos\frac{\pi}{6} & -\sin\frac{\pi}{6} \\ \sin\frac{\pi}{6} & \cos\frac{\pi}{6} \end{bmatrix} = \begin{bmatrix} 0.866 & -0.5 \\ 0.5 & 0.866 \end{bmatrix}.$$

Now compute the rotation.

$$\begin{bmatrix} \bar{x} \\ \bar{y} \end{bmatrix} = \begin{bmatrix} 0.866 & -0.5 \\ 0.5 & 0.866 \end{bmatrix} \begin{bmatrix} x \\ 5x+1 \end{bmatrix} = \begin{bmatrix} -1.634x - 0.5 \\ 4.83x + 0.866 \end{bmatrix}.$$

Choose two points on the line $y = 5x + 1$, say $(0, 1)$ and $(1, 6)$, apply the transformation to these points, and determine two points on the line.

$$\begin{bmatrix} \bar{x_1} \\ \bar{y_1} \end{bmatrix} = \begin{bmatrix} -0.5 \\ 0.866 \end{bmatrix},$$
$$\begin{bmatrix} \bar{x_2} \\ \bar{y_2} \end{bmatrix} = \begin{bmatrix} -2.134 \\ 5.696 \end{bmatrix}.$$

Figure 1.3 is the graph of the original and the rotated line.   ∎

In three-dimensional Euclidean space, the equations

$$\bar{x} = x\cos\theta - y\sin\theta, \quad \bar{y} = x\sin\theta + y\cos\theta, \quad \bar{z} = z,$$
$$\bar{x} = x, \quad \bar{y} = y\cos\phi - z\sin\phi, \quad \bar{z} = y\sin\phi + z\cos\phi,$$
$$\bar{x} = x\cos\psi - z\sin\psi, \quad \bar{y} = y, \quad \bar{z} = x\sin\psi + z\cos\psi$$

correspond to rotations about the positive z-, x-, and y-axes, counterclockwise through θ, ϕ, ψ radians, respectively.

The product of two matrices is related to the product of the corresponding linear transformations:

If A is $m \times k$ and B is $k \times n$, the linear transformation $T_A T_B$ first performs transformation T_B, and then T_A. For instance, we might rotate about the x-axis, followed by a rotation about the z-axis. This transformation is in fact equal to the linear transformation T_{AB}, since

$$T_A T_B(x) = A(Bx) = (AB)x = T_{AB}(x).$$

The following example is useful for producing rotations in three-dimensional animated design (see Ref. [7, pp. 97-112]).

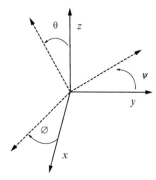

FIGURE 1.4 Rotate three-dimensional coordinate system.

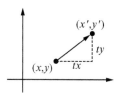

FIGURE 1.5 Translate a point in two dimensions.

Example 1.7. The linear transformation resulting from successively rotating three-dimensional space about the positive z, x, and y-axes, counterclockwise through θ, ϕ, ψ radians, respectively (Figure 1.4), is equal to T_{ABC}, where

$$
C = \begin{bmatrix} \cos\theta & -\sin\theta & 0 \\ \sin\theta & \cos\theta & 0 \\ 0 & 0 & 1 \end{bmatrix}, \quad B = \begin{bmatrix} 1 & 0 & 0 \\ 0 & \cos\phi & -\sin\phi \\ 0 & \sin\phi & \cos\phi \end{bmatrix}, \quad A = \begin{bmatrix} \cos\psi & 0 & -\sin\psi \\ 0 & 1 & 0 \\ \sin\psi & 0 & \cos\psi \end{bmatrix}.
$$

The matrix ABC is somewhat complex:

$$
\begin{aligned}
A(BC) &= \begin{bmatrix} \cos\psi & 0 & -\sin\psi \\ 0 & 1 & 0 \\ \sin\psi & 0 & \cos\psi \end{bmatrix} \begin{bmatrix} \cos\theta & -\sin\theta & 0 \\ \cos\phi\sin\theta & \cos\phi\cos\theta & -\sin\phi \\ \sin\phi\sin\theta & \sin\phi\cos\theta & \cos\phi \end{bmatrix} \\
&= \begin{bmatrix} \cos\psi\cos\theta - \sin\psi\sin\phi\sin\theta & -\cos\psi\sin\theta - \sin\psi\sin\phi\sin\theta & -\sin\psi\cos\phi \\ \cos\phi\sin\theta & \cos\phi\cos\theta & -\sin\phi \\ \sin\psi\cos\theta + \cos\psi\sin\phi\sin\theta & -\sin\psi\sin\theta + \cos\psi\sin\phi\cos\theta & \cos\psi\cos\phi \end{bmatrix}.
\end{aligned}
$$
∎

Now consider a new problem. Reposition a point (x, y) along a straight line a distance of (tx, ty), where t is a scalar. The new location of the point is $(x + tx, y + ty)$ (Figure 1.5).

To determine a linear transformation, we use a 3×3 matrix

$$
T = \begin{bmatrix} 1 & 0 & tx \\ 0 & 1 & ty \\ 0 & 0 & 1 \end{bmatrix}
$$

and multiply T by the column vector $\begin{bmatrix} x \\ y \\ 1 \end{bmatrix}$. Then

$$
\begin{bmatrix} 1 & 0 & tx \\ 0 & 1 & ty \\ 0 & 0 & 1 \end{bmatrix} \begin{bmatrix} x \\ y \\ 1 \end{bmatrix} = \begin{bmatrix} x + tx \\ y + ty \\ 1 \end{bmatrix}.
$$

In order to combine translation and rotation using matrix multiplication, we need to create a 3×3 matrix that performs a two-dimensional rotation. Define

$$R = \begin{bmatrix} \cos\theta & -\sin\theta & 0 \\ \sin\theta & \cos\theta & 0 \\ 0 & 0 & 1 \end{bmatrix}.$$

Now,

$$\begin{bmatrix} \cos\theta & -\sin\theta & 0 \\ \sin\theta & \cos\theta & 0 \\ 0 & 0 & 1 \end{bmatrix} \begin{bmatrix} x \\ y \\ 1 \end{bmatrix} = \begin{bmatrix} x\cos\theta - y\sin\theta \\ x\sin\theta + y\cos\theta \\ 1 \end{bmatrix}.$$

In each case, we can ignore the z component of 1.

Example 1.8. Now we can perform an interesting and practical matrix calculation. Take the line $y = 5x + 1$ and rotate it $30°$ counterclockwise about the point $(2, 11)$.

To solve this problem, first translate the point $(2, 11)$ to $(0, 0)$, rotate $30°$ counterclockwise, and then translate the point from the origin back to $(2, 11)$ (Figure 1.6).

Here are the matrices involved.

$$T_1 = \begin{bmatrix} 1 & 0 & -2 \\ 0 & 1 & -11 \\ 0 & 0 & 1 \end{bmatrix}.$$ Translate $(2, 11)$ to the origin using $tx = -2$ and $ty = -11$.

$$R = \begin{bmatrix} \cos\frac{\pi}{6} & -\sin\frac{\pi}{6} & 0 \\ \sin\frac{\pi}{6} & \cos\frac{\pi}{6} & 0 \\ 0 & 0 & 1 \end{bmatrix}.$$ Rotate $30°$

$$T_2 = \begin{bmatrix} 1 & 0 & 2 \\ 0 & 1 & 11 \\ 0 & 0 & 1 \end{bmatrix}.$$ Translate back to $(2, 11)$ using $tx = 2$ and $ty = 11$.

Compute the product $F = T_2 R T_1$.

$$F = \begin{bmatrix} 0.8660 & -0.5000 & 5.7679 \\ 0.5000 & 0.8660 & 0.4737 \\ 0 & 0 & 1.0000 \end{bmatrix}.$$

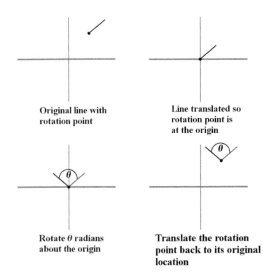

Original line with rotation point

Line translated so rotation point is at the origin

Rotate θ radians about the origin

Translate the rotation point back to its original location

FIGURE 1.6 Rotate a line about a point.

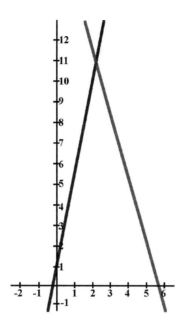

FIGURE 1.7 Rotation about an arbitrary point.

By computing the two points $F \begin{bmatrix} 0 \\ 1 \\ 1 \end{bmatrix}$ and $F \begin{bmatrix} 2 \\ 11 \\ 1 \end{bmatrix}$ on the rotated line and using the two point formula for the equation of a line, we obtain the equation $y = -2.9561x + 16.9121$.

Figure 1.7 is a plot of both the original and the rotated line.   ■

1.3   POWERS OF MATRICES

Definition 1.10 (The identity matrix).   The $n \times n$ matrix $I = [\delta_{ij}]$, defined by $\delta_{ij} = 1$ if $i = j$, $\delta_{ij} = 0$ if $i \neq j$, is called the $n \times n$ *identity* matrix of order n. In other words, the columns of the identity matrix of order n are the vectors

$$e_1 = \begin{bmatrix} 1 \\ 0 \\ \vdots \\ 0 \\ 0 \end{bmatrix}, e_2 = \begin{bmatrix} 0 \\ 1 \\ \vdots \\ 0 \\ 0 \end{bmatrix}, \ldots, e_n = \begin{bmatrix} 0 \\ 0 \\ \vdots \\ 0 \\ 1 \end{bmatrix}.$$

For example, $I = \begin{bmatrix} 1 & 0 \\ 0 & 1 \end{bmatrix}$ and $I = \begin{bmatrix} 1 & 0 & 0 \\ 0 & 1 & 0 \\ 0 & 0 & 1 \end{bmatrix}$. The identity matrix plays a critical role in linear algebra. When any $n \times n$ matrix A is multiplied by the identity matrix, either on the left or the right, the result is A. Thus, the identity matrix acts like 1 in the real number system. For example,

$$\begin{bmatrix} 2 & 6 & 1 \\ 7 & 2 & 9 \\ -1 & 5 & -4 \end{bmatrix} \begin{bmatrix} 1 & 0 & 0 \\ 0 & 1 & 0 \\ 0 & 0 & 1 \end{bmatrix} = \begin{bmatrix} 2(1) + 6(0) + 1(0) & 2(0) + 6(1) + 1(0) & 2(0) + 6(0) + 1(1) \\ 7 & 2 & 9 \\ -1 & 5 & -4 \end{bmatrix} = \begin{bmatrix} 2 & 6 & 1 \\ 7 & 2 & 9 \\ -1 & 5 & -4 \end{bmatrix}.$$

Definition 1.11 (kth power of a matrix).   If A is an $n \times n$ matrix, we define A^k as follows: $A^0 = I$ and $A^k = \underbrace{A \times A \times A \cdots A \times A}_{A \text{ occurs } k \text{ times}}$ for $k \geq 1$.

For example, $A^4 = A \times A \times A \times A$. Compute from left to right as follows:

$$A^2 = A \times A, \quad A^3 = (A)^2 \times A, \quad A^4 = (A)^3 \times A.$$

Example 1.9. The MATLAB exponentiation operator ^ applies to matrices.

```
>> A = [1 1;1 0]
A =
1 1
1 0
>> A^8
ans =
34 21
21 13
```

A is known as the *Fibonacci matrix*, since it generates elements from the famous Fibonacci sequence

$$0, 1, 1, 2, 3, 5, 8, 13, 21, 34, \ldots$$

■

Example 1.10. Let $A = \begin{bmatrix} 7 & 4 \\ -9 & -5 \end{bmatrix}$. Let's investigate powers of A and see if we can find a formula for A^n.

$$A^2 = \begin{bmatrix} 7 & 4 \\ -9 & -5 \end{bmatrix}\begin{bmatrix} 7 & 4 \\ -9 & -5 \end{bmatrix} = \begin{bmatrix} 13 & 8 \\ -18 & -11 \end{bmatrix},$$

$$A^3 = \begin{bmatrix} 13 & 8 \\ -18 & -11 \end{bmatrix}\begin{bmatrix} 7 & 4 \\ -9 & -5 \end{bmatrix} = \begin{bmatrix} 19 & 12 \\ -27 & -17 \end{bmatrix},$$

$$A^4 = \begin{bmatrix} 19 & 12 \\ -27 & -17 \end{bmatrix}\begin{bmatrix} 7 & 4 \\ -9 & -5 \end{bmatrix} = \begin{bmatrix} 25 & 16 \\ -36 & -23 \end{bmatrix},$$

$$A^5 = \begin{bmatrix} 25 & 16 \\ -36 & -23 \end{bmatrix}\begin{bmatrix} 7 & 4 \\ -9 & -5 \end{bmatrix} = \begin{bmatrix} 31 & 20 \\ -45 & -29 \end{bmatrix}.$$

The elements in positions $(1, 2)$ and $(2, 1)$ follow a pattern. The element in position $(1, 2)$ is always $4n$, and the element at position $(2, 1)$ is always $-9n$. The element at $(1, 1)$ is $6n + 1$, so we only need the pattern for the entry at $(2, 2)$. It is always one (1) more than $-6n$, so it has the value $1 - 6n$. Here is the formula for A^n.

$$A^n = \begin{bmatrix} 1 + 6n & 4n \\ -9n & 1 - 6n \end{bmatrix} \quad \text{if } n \geq 1.$$

This is not a mathematical proof, just an example of pattern recognition. The result can be formally proved using mathematical induction (see Appendix B). ■

Our final example of matrix powers is a result from *graph theory*. A graph is a set of *vertices* and connections between them called *edges*. You have seen many graphs; for example, a map of the interstate highway system is a graph, as is the airline route map at the back of those boring magazines you find on airline flights. Consider the simple graph in Figure 1.8. A *path* from one vertex v to another vertex w is a sequence of edges that connect v and w. For instance, here are three paths from A to F: A-B-F, A-B-D-F, and A-B-C-E-B-F. The length of a path between v and w is the number of edges that must

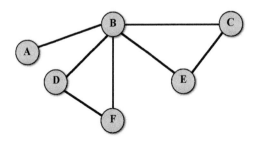

FIGURE 1.8 Undirected graph.

be crossed in moving from one to the other. For instance, in our three paths, the first has length 2, the second has length 3, and the third has length 5.

If a graph has n vertices, the *adjacency matrix* of the graph is an $n \times n$ matrix that specifies the location of edges. The concept is best illustrated by displaying the adjacency matrix for our six vertex graph, rather than giving a mathematical definition.

$$\text{Adj} = \begin{array}{c} \\ A \\ B \\ C \\ D \\ E \\ F \end{array} \begin{array}{c} \begin{array}{cccccc} A & B & C & D & E & F \end{array} \\ \left[\begin{array}{cccccc} 0 & 1 & 0 & 0 & 0 & 0 \\ 1 & 0 & 1 & 1 & 1 & 1 \\ 0 & 1 & 0 & 0 & 1 & 0 \\ 0 & 1 & 0 & 0 & 0 & 1 \\ 0 & 1 & 1 & 0 & 0 & 0 \\ 0 & 1 & 0 & 1 & 0 & 0 \end{array} \right] \end{array}$$

A one (1) occurs in row A, column B, so there is an edge connecting A and B. Similarly, a one is in row E, column C, so there is an edge connecting E and C. There is no edge between A and D, so row A, column D contains zero (0).

There is a connection between the adjacency matrix of a graph and the number of possible paths between two vertices. Clearly, Adj^1 specifies all the paths of length 1 from one vertex to another (an edge).

If Adj *is the adjacency matrix for a graph, then* Adj^k *defines the number of possible paths of length k between any two vertices.* We will not attempt to prove this, but will use our graph as an example.

$$\text{Adj}^2 = \begin{bmatrix} 1 & 0 & 1 & 1 & 1 & 1 \\ 0 & 5 & 1 & 1 & 1 & 1 \\ 1 & 1 & 2 & 1 & 1 & 1 \\ 1 & 1 & 1 & 2 & 1 & 1 \\ 1 & 1 & 1 & 1 & 2 & 1 \\ 1 & 1 & 1 & 1 & 1 & 2 \end{bmatrix},$$

$$\text{Adj}^3 = \begin{bmatrix} 0 & 5 & 1 & 1 & 1 & 1 \\ 5 & 4 & 6 & 6 & 6 & 6 \\ 1 & 6 & 2 & 2 & 3 & 2 \\ 1 & 6 & 2 & 2 & 2 & 3 \\ 1 & 6 & 3 & 2 & 2 & 2 \\ 1 & 6 & 2 & 3 & 2 & 2 \end{bmatrix}.$$

By looking at Adj^2, we see that there is one path of length 2 between C and E, C-B-E, and two paths of length 2 connecting E to E (E-C-E, E-B-E). There are five (5) paths of length 3 between B and A (B-A-B-A, B-D-B-A, B-C-B-A, B-E-B-A, B-F-B-A). Note that if we reverse each path of length three from B to A, we have a path that starts at A and ends at B. Look carefully at Adj, Adj^2, and Adj^3 and notice that the entry at position (i, j) is always the same as the entry at (j, i). Such a matrix is termed *symmetric*. If you exchange rows and columns, the matrix remains the same. There are many applications of symmetric matrices in science and engineering.

1.4 NONSINGULAR MATRICES

Definition 1.12 (Nonsingular matrix).   An $n \times n$ matrix A is called *nonsingular* or *invertible* if there exists an $n \times n$ matrix B such that

$$AB = BA = I.$$

If A does not have an inverse, A is called *singular*.

A matrix B such that $AB = BA = I$ is called an *inverse* of A. There can only be one inverse, as Theorem 1.3 shows.

Theorem 1.3.   *A matrix A can have only one inverse.*

Proof. Assume that $AB = I$, $BA = I$, and $CA = AC = I$. Then,
$C(AB) = (CA)B$, and $CI = IB$, so $C = B$. ☐

When determining if B is the inverse of A, it is only necessary to verify $AB = I$ or $BA = I$. This is important because an procedure that computes the inverse of A need only to verify the product in one direction or the other.

Theorem 1.4.   *If B is a matrix such that $BA = I$, then $AB = I$. Similarly, if $AB = I$, then $BA = I$.*

Proof. Assume that $BA = I$. Then $ABAB = AIB = AB$, and $ABAB - AB = 0$. Factor out AB, and $AB(AB - I) = 0$. Either $AB = 0$ or $AB - I = 0$. If $AB = 0$, then $ABA = 0(A) = 0$. But, $BA = I$, and so it follows that $A = 0$. The product of any matrix with the zero matrix is the zero matrix, so $BA = I$ is not possible. Thus, $AB - I = 0$, or $AB = I$. The fact that $AB = I$ implies $BA = I$ is handled in the same fashion.   □

If we denote the inverse by A^{-1}, then

$$\left(A^{-1}\right) A = I,$$
$$A\left(A^{-1}\right) = I,$$

and it follows that

$$(A^{-1})^{-1} = A.$$

This says the inverse of A^{-1} is A itself.

The inverse has a number of other properties that play a role in developing results in linear algebra. For instance,

$$\left(B^{-1}A^{-1}\right)(AB) = B^{-1}\left(A^{-1}A\right)B = B^{-1}IB = B^{-1}B = I,$$

and so

$$(AB)^{-1} = B^{-1}A^{-1}.$$

By Theorem 1.4, we do not need to verify that $(AB)\left(B^{-1}A^{-1}\right) = I$.

Remark 1.2.   The above result generalizes to a product of m nonsingular matrices: If $A_1, \ldots, A_m$ are nonsingular $n \times n$ matrices, then the product $A_1 \ldots A_m$ is also nonsingular. Moreover,

$$(A_1 \ldots A_m)^{-1} = A_m^{-1} \ldots A_1^{-1},$$

so the inverse of a product equals the product of the inverses *in the reverse order.*

Example 1.11.   If A and B are $n \times n$ matrices satisfying $A^2 = B^2 = (AB)^2 = I$, show that $AB = BA$. This says that A, B, and AB are each their own inverse, or $(AB)^{-1} = AB$, $B^{-1} = B$, and $A^{-1} = A$. Now,

$$AB = (AB)^{-1} = B^{-1}A^{-1} = BA,$$

and so $AB = BA$.   ■

Normally, is it not true that $AB = BA$ for $n \times n$ matrices A and B. For instance,

$$\begin{bmatrix} 1 & 2 \\ -1 & 5 \end{bmatrix} \begin{bmatrix} 6 & 1 \\ -7 & 4 \end{bmatrix} = \begin{bmatrix} -8 & 9 \\ -41 & 19 \end{bmatrix},$$

and

$$\begin{bmatrix} 6 & 1 \\ -7 & 4 \end{bmatrix} \begin{bmatrix} 1 & 2 \\ -1 & 5 \end{bmatrix} = \begin{bmatrix} 5 & 17 \\ -11 & 6 \end{bmatrix}.$$

Remark 1.3.   We will show how to compute the inverse in Chapter 2; however, it is computationally expensive. The inverse is primarily a tool for developing other results.

A matrix having an inverse guarantees that a linear system has a unique solution.

Theorem 1.5. *If the coefficient matrix A of a system of n equations in n unknowns has an inverse, then the system $Ax = b$ has the unique solution $x = A^{-1}b$.*

Proof. **1.** (Uniqueness) Assume $Ax = b$. Then

$$
\begin{aligned}
A^{-1}(Ax) &= A^{-1}b, \\
\left(A^{-1}A\right)x &= A^{-1}b, \\
Ix &= A^{-1}b, \\
x &= A^{-1}b.
\end{aligned}
$$

2. (Existence) Assume $x = A^{-1}b$. Then

$$
Ax = A\left(A^{-1}b\right) = \left(AA^{-1}\right)b = Ib = b. \qquad \square
$$

A linear system $Ax = 0$ is said to be *homogeneous*. If A is nonsingular, then $x = A^{-1}0 = 0$, so the system has only 0 as its solution. It is said to have only the *trivial solution*.

Example 1.12. Consider the homogeneous system

$$
\begin{bmatrix} 1 & 3 \\ 1 & 2 \end{bmatrix} x = 0.
$$

A simple calculation verifies that $\begin{bmatrix} 1 & 3 \\ 1 & 2 \end{bmatrix}^{-1} = \begin{bmatrix} -2 & 3 \\ 1 & -1 \end{bmatrix}$, and so

$$
\begin{bmatrix} -2 & 3 \\ 1 & -1 \end{bmatrix}\begin{bmatrix} 1 & 3 \\ 1 & 2 \end{bmatrix} x = \begin{bmatrix} -2 & 3 \\ 1 & -1 \end{bmatrix}\begin{bmatrix} 0 \\ 0 \end{bmatrix},
$$
$$
Ix = x = 0. \qquad \blacksquare
$$

There are some cases where there is an explicit formula for the inverse matrix. In particular, we can demonstrate a formula for the inverse of a 2×2 matrix subject to a condition.

Let $A = \begin{bmatrix} a & b \\ c & d \end{bmatrix}$ with $ad - bc \neq 0$ and let $B = (1/(ad - bc))\begin{bmatrix} d & -b \\ -c & a \end{bmatrix}$. Perform a direct calculation of AB.

$$
\begin{aligned}
\begin{bmatrix} a & b \\ c & d \end{bmatrix}\left(\frac{1}{ad - bc}\right)\begin{bmatrix} d & -b \\ -c & a \end{bmatrix} &= \left(\frac{1}{ad - bc}\right)\begin{bmatrix} a & b \\ c & d \end{bmatrix}\begin{bmatrix} d & -b \\ -c & a \end{bmatrix} \\
&= \left(\frac{1}{ad - bc}\right)\begin{bmatrix} ad - bc & -ab + ab \\ cd - cd & ad - bc \end{bmatrix} = \begin{bmatrix} 1 & 0 \\ 0 & 1 \end{bmatrix} = I
\end{aligned}
$$

Remark 1.4. The expression $ad - bc$ is called the *determinant* of A and is denoted by $\det(A)$. Later we will see that A has an inverse if and only if $\det A \neq 0$.

The MATLAB function that computes the inverse of an $n \times n$ matrix is `inv`. If A is an $n \times n$ matrix, then

```
>> B = inv(A);
```

computes A^{-1}.

Example 1.13. The following MATLAB statements demonstrate the use of `inv` and verify that $(AB)^{-1} = B^{-1}A^{-1}$ for two particular matrices.

```
>> format rational;
>> A = [1 3 -1; 4 1 6; 0 2 3]
A =
1 3 -1
4 1 6
0 2 3
```

```
>> A_inv = inv(A)
A_inv =
9/53  11/53 -19/53
12/53 -3/53 10/53
-8/53 2/53  11/53

>> B = [1 4 0; 3 5 1; 2 -7 8]
B =
1  4 0
3  5 1
2 -7 8

>> B_inv = inv(B)
B_inv =
-47/41 32/41 -4/41
22/41 -8/41  1/41
31/41 -15/41 7/41

>> inv(A*B)
ans =
-7/2173 -621/2173 1169/2173
94/2173 268/2173 -487/2173
43/2173 400/2173 -662/2173

>> B_inv * A_inv
ans =
-7/2173 -621/2173 1169/2173
94/2173 268/2173 -487/2173
43/2173 400/2173 -662/2173
```

■

1.5 THE MATRIX TRANSPOSE AND SYMMETRIC MATRICES

There is another property of a matrix that we will use extensively, the matrix transpose.

Definition 1.13 (The transpose of a matrix). Let A be an $m \times n$ matrix. Then A^T, the *transpose* of A, is the matrix obtained by interchanging the rows and columns of A. In other words if $A = [a_{ij}]$, then $\left(A^T\right)_{ij} = a_{ji}$. Consequently A^T is an $n \times m$ matrix.

For instance, if

$$
A = \begin{bmatrix} 1 & 9 & 0 \\ 3 & 7 & 15 \\ 4 & 8 & 1 \\ -7 & 12 & 3 \end{bmatrix},
$$

then

$$
A^T = \begin{bmatrix} 1 & 3 & 4 & -7 \\ 9 & 7 & 8 & 12 \\ 0 & 15 & 1 & 3 \end{bmatrix}.
$$

Theorem 1.6. *The transpose operation has the following properties:*

1. $\left(A^T\right)^T = A$;
2. $(A \pm B)^T = A^T \pm B^T$ *if A and B are $m \times n$;*
3. $(sA)^T = sA^T$ *if s is a scalar;*
4. $(AB)^T = B^T A^T$ *if A is $m \times k$ and B is $k \times n$;*

5. *A is nonsingular, then* A^T *is also nonsingular and* $\left(A^T\right)^{-1} = \left(A^{-1}\right)^T$.

6. $x^T x = x_1^2 + \cdots + x_n^2$ *if* $x = [x_1, \ldots, x_n]^T$ *is a column vector.*

Proof. We will verify 5 and 6 and leave the remaining properties to the exercises.

Property 5: $A^T \left(A^{-1}\right)^T = \left(A^{-1}A\right)^T$ by property 4. Therefore, $A^T \left(A^{-1}\right)^T = I^T = I$.

Property 6: $x^T x = \begin{bmatrix} x_1 & \cdots & x_n \end{bmatrix} \begin{bmatrix} x_1 \\ \vdots \\ x_n \end{bmatrix} = x_1^2 + \cdots + x_n^2$. $\qquad\square$

There is a frequently occurring class of matrices defined in terms of the transpose operation.

Definition 1.14 (Symmetric matrix). A matrix A is *symmetric* if $A^T = A$. In other words A is square ($n \times n$) and $a_{ji} = a_{ij}$ for all $1 \le i \le n, 1 \le j \le n$. Another way of looking this is that when the rows and columns are interchanged, the resulting matrix is A. For instance,

$$A = \begin{bmatrix} a & b \\ b & c \end{bmatrix}$$

is a general 2×2 symmetric matrix.

Example 1.14. $A = \begin{bmatrix} 1 & 8 & 12 & 3 \\ 8 & 5 & 1 & 10 \\ 12 & 1 & 6 & 9 \\ 3 & 10 & 9 & 2 \end{bmatrix}$ is a symmetric matrix. Notice row 1 has the same entries as column 1, row 2 has the same entries as column 2, and so forth. $\qquad\blacksquare$

The following proposition proves a property of $A^T A$ that is critical for the computation of what are termed singular values in Chapter 7.

Proposition 1.1. *If A is an* $m \times n$ *matrix, then* $A^T A$ *is a symmetric matrix.*

Proof. $A^T A$ is an $n \times n$ matrix, since A^T has dimension $n \times m$, and A has dimension $m \times n$. $A^T A$ is symmetric, since

$$\left(A^T A\right)^T = A^T \left(A^T\right)^T = A^T A. \qquad\square$$

Example 1.15. In MATLAB, the transpose operator is $'$.

```
>> A = [1 8 -1; 3 -9 15; -1 5 3]
A =
      1      8     -1
      3     -9     15
     -1      5      3
>> A_TA = A'*A
A_TA =
     11    -24     41
    -24    170   -128
     41   -128    235
>> A_TA - A_TA'
ans =
      0      0      0
      0      0      0
      0      0      0
```
$\qquad\blacksquare$

1.6   CHAPTER SUMMARY

Matrix Arithmetic

This chapter defines a matrix, introduces matrix notation, and presents matrix operations, including matrix multiplication. To multiply matrices A and B, the number of columns of A must equal the number of rows of B. It should be emphasized that multiplication of large matrices, most of whose elements are nonzero, is a time-consuming computation. When matrices in applications are very large, they normally consist primarily of zero entries, and are termed sparse. Although matrix multiplication obeys most of the laws of arithmetic, it is not commutative; that is, if A and B are $n \times n$ matrices, AB is rarely equal to BA.

A vector is a matrix having one row or one column. In this book, we will primarily use column vectors such as

$$\begin{bmatrix} -1 \\ 3 \\ 7 \end{bmatrix}.$$

The trace of an $n \times n$ matrix is the sum of its diagonal elements a_{ii}, $1 \le i \le n$, or trace $(A) = \sum_{i=1}^{n} a_{ii}$. The trace occurs in many matrix formulas, and we will encounter it in later chapters. It is important to note that even though $AB \ne BA$ in general, in fact trace $(AB) =$ trace (BA).

A primary topic in this book is the solution of linear systems of equations, and we write them using matrix notation; for instance, the system

$$\begin{aligned} x_1 - x_2 + 5x_3 &= 1, \\ -2x_1 + 4x_2 + x_3 &= 0, \\ 7x_1 - 2x_2 - 6x_3 &= 8 \end{aligned}$$

using matrix notation is

$$\begin{bmatrix} 1 & -1 & 5 \\ -2 & 4 & 1 \\ 7 & -2 & -6 \end{bmatrix} \begin{bmatrix} x_1 \\ x_2 \\ x_3 \end{bmatrix} = \begin{bmatrix} 1 \\ 0 \\ 8 \end{bmatrix}.$$

Linear Transformations

If A is an $m \times n$ matrix, a linear transformation is a mapping from $\mathbb{R}^n$ to $\mathbb{R}^m$ defined by $y = Ax$. We will deal with linear transformations throughout the remainder of this book. An excellent example is a two-dimensional linear transformation of the form

$$y = \begin{bmatrix} \cos \theta & -\sin \theta \\ \sin \theta & \cos \theta \end{bmatrix} \begin{bmatrix} x_1 \\ x_2 \end{bmatrix}$$

that rotates the vector $\begin{bmatrix} x_1 \\ x_2 \end{bmatrix}$ counter clockwise through angle θ. Such linear transformations perform rotation, displacement, and scaling of objects in computer graphics.

Powers of Matrices

There are numerous applications of matrix powers, A^k, $k \ge 0$. Given an undirected graph, powers of the adjacency matrix provide a count of the number of paths between any two vertices. We will see in Chapters 21 and 22 that a sequence of the form $Ax_0, A^2x_0, \dots, A^{k-1}x_0$ forms the basis for series of important methods that solve linear systems and compute eigenvalues of large, sparse matrices. We will discuss solving linear systems in Chapter 2 and eigenvalues in Chapter 5.

Nonsingular Matrices

The inverse of a matrix is of great theoretical importance in linear algebra. An $n \times n$ matrix A has inverse A^{-1} if

$$A^{-1}A = AA^{-1} = I,$$

where I is the identity matrix

$$\begin{bmatrix} 1 & 0 & \cdots & \cdots & 0 \\ 0 & 1 & 0 & \cdots & 0 \\ 0 & 0 & \ddots & \cdots & 0 \\ \vdots & \vdots & \ddots & 1 & \vdots \\ 0 & 0 & \cdots & 0 & 1 \end{bmatrix}.$$

When it exists, the inverse is unique, and the matrix is termed nonsingular. Not all matrices have an inverse, and such matrices are said to be singular. A linear system $Ax = b$ has a unique solution $x = A^{-1}b$ when A is nonsingular. A very important result is that a homogeneous system of the form $Ax = 0$ has only the solution $x = 0$ if A is nonsingular.

The Matrix Transpose and Symmetric Matrices

The transpose of an $m \times n$ matrix A, named A^T, is the $n \times m$ matrix obtained by exchanging the rows and columns of A. Theorem 1.6 lists properties of the transpose.

An important class of matrices are the symmetric matrices. A square matrix A is symmetric if $A^T = A$, and this means that $a_{ij} = a_{ji}$, $1 \le i, j \le n$. Many problems in engineering and science involve symmetric matrices, and entire sections of this book deal with them. As you will see, when a problem involves a symmetric matrix, this normally leads to a faster and more accurate solution.

It is of the utmost importance that you remember the relationship $(AB)^T = B^T A^T$, as we will use it again and again throughout this book. Here is an interesting fact we will use beginning in Chapter 7. If A is any $m \times n$ matrix, then $A^T A$ is symmetric, since

$$\left(A^T A\right)^T = A^T \left(A^T\right)^T = A^T A.$$

1.7 PROBLEMS

1.1 For

$$A = \begin{bmatrix} 1 & 8 & -1 \\ 0 & 6 & -7 \\ 2 & 4 & 12 \end{bmatrix}, \quad B = \begin{bmatrix} 6 & -1 & 25 \\ 14 & -6 & 0 \\ -9 & 15 & 25 \end{bmatrix}$$

compute the following:
a. $A - B$
b. $8A$
c. $5A + 7B$

1.2 Using the matrices A and B from Problem 1.1, compute AB. Do not use a computer program. Do it with pencil and paper.

1.3 Let A, B, C, D be matrices defined by

$$A = \begin{bmatrix} 3 & 0 \\ -1 & 2 \\ 1 & 1 \end{bmatrix}, \quad B = \begin{bmatrix} 1 & 5 & 2 \\ -1 & 1 & 0 \\ -4 & 1 & 3 \end{bmatrix},$$

$$C = \begin{bmatrix} -3 & -1 \\ 2 & 1 \\ 4 & 3 \end{bmatrix}, \quad D = \begin{bmatrix} 4 & -1 \\ 2 & 0 \end{bmatrix}.$$

Which of the following matrices are defined? Compute those matrices which are defined.

$$A + B, A + C, AB, BA, CD, DC, D^2.$$

1.4 Let $A = \begin{bmatrix} -1 & 0 & 1 \\ 0 & 1 & 1 \end{bmatrix}$. Show that if B is a 3×2 matrix such that $AB = I$, then

$$B = \begin{bmatrix} a & b \\ -a-1 & 1-b \\ a+1 & b \end{bmatrix}$$

for suitable numbers a and b. Use the associative law to show that $(BA)^2 B = B$.

1.5 If $A = \begin{bmatrix} a & b \\ c & d \end{bmatrix}$, show that $A^2 - (a+d)A + (ad - bc)I_2 = 0$.

1.6 A square matrix $D = [d_{ij}]$ is called *diagonal* if $d_{ij} = 0$ for $i \neq j$; that is the *off-diagonal* elements are zero. Show that premultiplication of a matrix A by a diagonal matrix D results in matrix DA whose rows are the rows of A multiplied by the respective diagonal elements of D.

1.7 Write the following linear algebraic system in matrix form.

$$5x_1 + 6x_2 - x_3 + 2x_4 = 1,$$
$$-x_1 + 2x_2 + x_3 - 9x_4 = 8,$$
$$2x_1 - x_3 = -3,$$
$$3x_2 + 28x_3 - 2x_4 = 0.$$

1.8 Write this matrix equation as a system of equations.

$$\begin{bmatrix} 1 & 0 & 9 \\ -8 & 3 & 45 \\ 12 & -6 & 55 \end{bmatrix} x = \begin{bmatrix} 1 \\ 0 \\ 1 \end{bmatrix}.$$

1.9 Define the linear transformation $T_A : \mathbb{R}^5 \to \mathbb{R}^5$ using the matrix

$$A = \begin{bmatrix} 1 & 7 & 0 & 0 & 0 \\ 4 & 5 & 8 & 0 & 0 \\ 0 & 6 & 1 & 1 & 0 \\ 0 & 0 & 7 & 3 & -9 \\ 0 & 0 & 0 & 1 & 2 \end{bmatrix}.$$

A is termed a *tridiagonal matrix*, since the only non-zero entries are along the main diagonal, the diagonal below, and the diagonal above.

a. Compute $A \begin{bmatrix} 0 \\ 1 \\ -1 \\ 3 \\ 2 \end{bmatrix}$.

b. Compute $A \begin{bmatrix} 6 \\ 0 \\ 1 \\ 3 \\ 0 \end{bmatrix}$.

c. Compute the general product

$$\begin{bmatrix} a_{11} & a_{12} & 0 & 0 \\ a_{21} & a_{22} & a_{23} & 0 \\ 0 & a_{32} & a_{33} & a_{34} \\ 0 & 0 & a_{43} & a_{44} \end{bmatrix} \begin{bmatrix} x_1 \\ x_2 \\ x_3 \\ x_4 \end{bmatrix}.$$

d. Propose a formula for the product $y = Ax$ of an $n \times n$ tridiagonal matrix A and an $n \times 1$ column vector x. Use the result of part (c) to help you formulate your answer.

1.10 Rotate the line $y = -x + 3$ 30° counterclockwise about the origin, and graph the two lines on the same set of axes.

1.11 Rotate the line $y = -x + 3$ 60° counterclockwise about the point $(4, -1)$, and graph the two lines on the same set of axes.

1.12 Let $A = \begin{bmatrix} 1 & 4 \\ -3 & 1 \end{bmatrix}$. Show that A is nonsingular by verifying that

$$A^{-1} = \begin{bmatrix} \frac{1}{13} & -\frac{4}{13} \\ \frac{3}{13} & \frac{1}{13} \end{bmatrix}.$$

1.13 Let $A = \begin{bmatrix} 1 & 3 \\ 2 & 0 \end{bmatrix}$

  a. Find the inverse of the matrix
  b. Use A^{-1} to solve the system

$$x_1 + 3x_2 = 6,$$
$$2x_1 - 9x_2 = 1.$$

1.14 If $A = \begin{bmatrix} 1 & 4 \\ -3 & 1 \end{bmatrix}$.

  a. Verify that $A^2 - 2A + 13I = 0$
  b. Show that $A^{-1} = -\frac{1}{13}(A - 2I)$.

1.15 Let $A = \begin{bmatrix} 1 & 1 & -1 \\ 0 & 0 & 1 \\ 2 & 1 & 2 \end{bmatrix}$. Verify that $A^3 = 3A^2 - 3A + I$.

1.16 Let A be an $n \times n$ matrix.
  a. If $A^2 = 0$, prove that A is singular. Start by assuming A^{-1} exists. Compute $A^{-1}(A)^2$ and deduce that A must be singular.
  b. If $A^2 = A$ and $A \neq I$, prove that A is singular. Use the same logic as in part (a).

1.17 If $X = \begin{bmatrix} 1 & 2 \\ 3 & 4 \\ 5 & 6 \end{bmatrix}$ and $Y = \begin{bmatrix} -1 \\ 3 \\ 4 \end{bmatrix}$, find XX^T, X^TX, YY^T, Y^TY.

1.18 For matrices A and B, show that $(AB)^T = B^TA^T$.

$$A = \begin{bmatrix} 1 & 4 & -1 \\ 0 & 7 & 1 \\ 1 & 7 & 2 \end{bmatrix}, \quad B = \begin{bmatrix} 1 & 2 & 6 \\ 1 & -7 & 3 \\ 0 & 1 & 2 \end{bmatrix}.$$

1.19 If A is a symmetric $n \times n$ matrix and B is $n \times m$, prove that B^TAB is a symmetric $m \times m$ matrix.

1.20 Show that $A = \begin{bmatrix} 1 & 0 & 0 \\ 0 & 0 & 1 \\ 0 & 1 & 0 \end{bmatrix}$ is its own inverse.

1.21 It is not usually the case for $n \times n$ matrices A and B that $AB = BA$. For instance,

$$A = \begin{bmatrix} 1 & 2 \\ 0 & 3 \end{bmatrix}, \quad B = \begin{bmatrix} 7 & 3 \\ 1 & 8 \end{bmatrix}, \quad AB = \begin{bmatrix} 9 & 19 \\ 3 & 24 \end{bmatrix}, \quad BA = \begin{bmatrix} 7 & 23 \\ 1 & 26 \end{bmatrix}.$$

Let A and B be $n \times n$ diagonal matrices:

$$A = \begin{bmatrix} a_{11} & 0 & \cdots & 0 \\ 0 & a_{22} & \cdots & 0 \\ \vdots & \vdots & \ddots & \vdots \\ 0 & 0 & \cdots & a_{nn} \end{bmatrix}, \quad B = \begin{bmatrix} b_{11} & 0 & \cdots & 0 \\ 0 & b_{22} & \cdots & 0 \\ \vdots & \vdots & \ddots & \vdots \\ 0 & 0 & \cdots & b_{nn} \end{bmatrix}.$$

Show that $AB = BA$.

1.22 Prove the following formulas are satisfied by the trace matrix operator.

    a. trace $(A + B)$ = trace (A) + trace (B)

    b. trace (cA) = c trace (A), where c is a scalar.

1.23 For an arbitrary $n \times 1$ column vector and an $n \times n$ matrix A, show that $x^T A x$ is a real number. This is called a *quadratic form*. For $x = [1\ 3\ 9]^T$, compute $x^T A x$ for the matrix

$$A = \begin{bmatrix} 1 & -8 & 3 \\ 4 & 0 & -1 \\ 3 & 5 & 7 \end{bmatrix}.$$

1.24 Prove the following properties of the matrix transpose operator.

    a. $\left(A^T\right)^T = A$.

    b. $(A \pm B)^T = A^T \pm B^T$ if A and B are $m \times n$.

    c. $(sA)^T = sA^T$ if s is a scalar.

1.25 Prove that $(AB)^T = B^T A^T$ if A is $m \times n$ and B is $n \times p$. Hint: Use the definition of matrix multiplication and the fact that taking the transpose means element a_{ij} of A is the element at row j, column i of A^T.

1.7.1 MATLAB Problems

1.26 For this exercise, use the MATLAB command `format rational` so the computations are done using rational arithmetic. Find the inverse of matrices A and B.

$$A = \begin{bmatrix} 1 & 4 & 1 \\ 1 & 3 & 2 \\ -1 & 2 & 7 \end{bmatrix}, \quad B = \begin{bmatrix} 1 & 0 & 1 \\ 2 & 5 & 12 \\ -9 & 1 & 1 \end{bmatrix}$$

Verify that $(AB)^{-1} = B^{-1}A^{-1}$.

1.27 Use MATLAB to find the inverse of the matrix $A = \begin{bmatrix} 1 & 3 & -1 & -9 \\ 0 & 3 & 0 & 1 \\ 12 & 8 & -11 & 0 \\ 2 & 1 & 5 & 3 \end{bmatrix}.$

1.28 The $n \times n$ Hilbert matrices are defined by $H(i,j) = 1/(i+j-1), 1 \le i,j \le n$. For instance, here is the 5×5 Hilbert matrix.

$$H = \begin{bmatrix} 1 & \frac{1}{2} & \frac{1}{3} & \frac{1}{4} & \frac{1}{5} \\ \frac{1}{2} & \frac{1}{3} & \frac{1}{4} & \frac{1}{5} & \frac{1}{6} \\ \frac{1}{3} & \frac{1}{4} & \frac{1}{5} & \frac{1}{6} & \frac{1}{7} \\ \frac{1}{4} & \frac{1}{5} & \frac{1}{6} & \frac{1}{7} & \frac{1}{8} \\ \frac{1}{5} & \frac{1}{6} & \frac{1}{7} & \frac{1}{8} & \frac{1}{9} \end{bmatrix}.$$

Systems of the form $Hx = b$, where H is a Hilbert matrix are notoriously difficult to solve because they are *ill-conditioned*. This means that a solution can change widely with only a small change in the elements of b or H. Chapter 10 discusses ill-conditioned matrices. The MATLAB command `hilb` builds an $n \times n$ Hilbert matrix. For instance to find the 6×6 Hilbert matrix, execute

```
>> H = hilb(6);
```

    a. The command

```
format shortg
```

    causes output of the best of fixed or floating point format with 5 digits. Using this format, compute the inverse of the 6×6 Hilbert matrix. What makes you suspicious that it is ill-conditioned?

    b. The exact inverse of any Hilbert matrix consists entirely of integer entries. Using the Symbolic Toolbox will provide an exact answer. If your MATLAB distribution has this software, use the help system to determine how to use the commands `syms` and `sym`. Determine the exact value of H^{-1}.

1.29 **a.** Write a MATLAB `function t = tr(A)` that computes the trace of matrix A. Test to make sure A is a square matrix.

   b. Use `tr` to compute the trace for the matrix of Problem 1.9 and the Hilbert matrix of order 15 (Problem 1.28). Verify your result by using the MATLAB command `trace`.

1.30 This problem uses the result of Problem 1.9(d).

   a. Write a MATLAB `function y = triprod(A,x)` that forms the product of an $n \times n$ tridiagonal matrix A with an $n \times 1$ column vector x.

   b. Test the function using the matrix and vectors specified in Problem 1.9, parts (a), and (b).

Chapter 2

Linear Equations

You should be familiar with

- Matrix arithmetic
- Linear transformations
- The matrix transpose and symmetric matrices

The solution of linear systems of equations is of primary importance in linear algebra. The problem of solving a linear system arises in almost all areas of engineering and science, including the structure of materials, statics and dynamics, the design and analysis of circuits, quantum physics, and computer graphics. The solution to linear systems also hides under the surface in many methods. For instance, a standard tool for data fitting is cubic splines. The fit is found by finding the solution to a system of linear equations.

This chapter introduces the basics of solving linear equations using Gaussian elimination. Since most applications deal with systems that have the same number of equations as unknowns (*square systems*), we will restrict our discussion to these systems. The approach we take is naive, in that it will ignore the numerical problems involved with performing Gaussian elimination on a computer. As we will see in Chapter 8, errors occur due to the fact that a digital computer performs arithmetic operations with a fixed number of digits. The issue of solving systems accurately is discussed in Chapter 11.

2.1 INTRODUCTION TO LINEAR EQUATIONS

A *linear equation* in n unknowns $x_1, x_2, \ldots, x_n$ is an equation of the form

$$a_1 x_1 + a_2 x_2 + \cdots + a_n x_n = b,$$

where $a_1, a_2, \ldots, a_n, b$ are given real numbers. For example, with x and y instead of x_1 and x_2, the linear equation $2x + 3y = 6$ describes the line passing through the points $(3, 0)$ and $(0, 2)$.

Similarly, with x, y, and z instead of x_1, x_2, and x_3, the linear equation $2x + 3y + 4z = 12$ describes the plane passing through the points $(6, 0, 0)$, $(0, 4, 0)$, $(0, 0, 3)$.

A *system* of n linear equations in n unknowns $x_1, x_2, \ldots, x_n$ is a family of equations

$$
\begin{aligned}
a_{11}x_1 + a_{12}x_2 + \cdots + a_{1n}x_n &= b_1 \\
a_{21}x_1 + a_{22}x_2 + \cdots + a_{2n}x_n &= b_2 \\
&\vdots \\
a_{n1}x_1 + a_{n2}x_2 + \cdots + a_{nn}x_n &= b_n.
\end{aligned}
$$

We wish to determine if such a system has a solution, that is to find out if there exist numbers $x_1, x_2, \ldots, x_n$ that satisfy each of the equations simultaneously. We say that the system is *consistent* if it has a solution. Otherwise, the system is called *inconsistent*.

Geometrically, solving a system of linear equations in two (or three) unknowns is equivalent to determining whether or not a family of lines (or planes) has a common point of intersection.

Numerical Linear Algebra with Applications. http://dx.doi.org/10.1016/B978-0-12-394435-1.00002-8

Example 2.1. Solve the system

$$2x + 3y = 6,$$
$$x - y = 2.$$

Multiply the second equation by 3 and add the result to the first equation.

$$5x + 0y = 12 \Longrightarrow x = 12/5.$$

Now substitute $x = 12/5$ into the second equation to obtain

$$y = x - 2 = \frac{12}{5} - 2 = \frac{2}{5}.$$

The solution to the system is $x = 12/5, y = 2/5$. ■

In Example 2.1, we solved the problem by dealing directly with the equations in symbolic form. This approach would be tedious and virtually unworkable for a large number of equations. We will develop a means of solving systems by using the matrix form of the equation, as discussed in Section 1.1.1.

Example 2.2. Find a polynomial of the form $y = a_0 + a_1x + a_2x^2 + a_3x^3$ which passes through the points $(-3, -2), (-1, 2), (1, 5), (2, 1)$ (Figure 2.1).

When x has the values $-3, -1, 1, 2$, then y takes corresponding values $-2, 2, 5, 1$ and we get four equations in the unknowns a_0, a_1, a_2, a_3:

$$a_0 - 3a_1 + 9a_2 - 27a_3 = -2,$$
$$a_0 - a_1 + a_2 - a_3 = 2,$$
$$a_0 + a_1 + a_2 + a_3 = 5,$$
$$a_0 + 2a_1 + 4a_2 + 8a_3 = 1.$$

We will learn how to solve such a system using matrix techniques. Essentially, the process is the same as in Example 2.1. We eliminate unknowns from equations until we find one value. Using it, we are able to determine the other unknown values. The process is called *Gaussian elimination*. For our problem, the unique solution is

$$a_0 = 93/20, \quad a_1 = 221/120,$$
$$a_2 = -23/20, \quad a_3 = -41/120,$$

and the required polynomial is

$$y = \frac{93}{20} + \frac{221}{120}x - \frac{23}{20}x^2 - \frac{41}{120}x^3.$$ ■

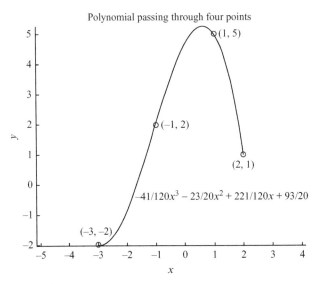

FIGURE 2.1 Polynomial passing through four points.

2.2 SOLVING SQUARE LINEAR SYSTEMS

Note that the system 2.1 can be written concisely as

$$\sum_{j=1}^{n} a_{ij}x_j = b_i, \quad i = 1, 2, \ldots, n. \tag{2.1}$$

We will deal with the system in matrix form. The matrix

$$\begin{bmatrix} a_{11} & a_{12} & \cdots & a_{1n} \\ a_{21} & a_{22} & \cdots & a_{2n} \\ \vdots & \vdots & \ddots & \vdots \\ a_{n1} & a_{n2} & \cdots & a_{nn} \end{bmatrix}$$

is called the *coefficient matrix* of the system, while the matrix

$$\left[\begin{array}{cccc|c} a_{11} & a_{12} & \cdots & a_{1n} & b_1 \\ a_{21} & a_{22} & \cdots & a_{2n} & b_2 \\ \vdots & \vdots & \ddots & \vdots & \vdots \\ a_{n1} & a_{n2} & \cdots & a_{nn} & b_n \end{array} \right]$$

is called the *augmented matrix* of the system.

We show how to solve any square system of linear equations, using the Gaussian elimination process. Consider the following 3 × 3 system of equations:

$$2x_1 + 3x_2 - x_3 = 8,$$
$$-x_1 + 6x_2 + 5x_3 = 1,$$
$$9x_1 - 5x_2 = 4.$$

In matrix form, the system is written as $Ax = b$, where

$$A = \begin{bmatrix} 2 & 3 & -1 \\ -1 & 6 & 5 \\ 9 & -5 & 0 \end{bmatrix},$$

$$b = \begin{bmatrix} 8 \\ 1 \\ 4 \end{bmatrix}.$$

The augmented matrix for the system is

$$\left[\begin{array}{ccc|c} 2 & 3 & -1 & 8 \\ -1 & 6 & 5 & 1 \\ 9 & -5 & 0 & 4 \end{array} \right].$$

The matrix is really just a compact way of writing the system of equations. We now need to define some terms. The following operations are the ones used on the augmented matrix during Gaussian elimination and will not change the solution to the system. Note that performing these operations on the matrix is equivalent to performing the same operations directly on the equations.

Definition 2.1 (Elementary row operations). Three types of *elementary row operations* can be performed on matrices:

1. Interchanging two rows:

   $R_i \leftrightarrow R_j$ interchanges rows i and j.

2. Multiplying a row by a nonzero scalar:

   $R_i \rightarrow tR_i$ multiplies row i by the nonzero scalar t.

3. Adding a multiple of one row to another row:

   $R_j \rightarrow R_j + tR_i$ adds t times row i to row j.

Definition 2.2 (Row equivalence). Matrix A is *row-equivalent* to matrix B if B is obtained from A by a sequence of elementary row operations.

Example 2.3. Consider the 3×3 system

$$
\begin{aligned}
x + 2y &= 1 \\
2x + y + z &= 3 \ . \\
x - y + 2z &= 2
\end{aligned}
$$

In matrix form, the system is

$$
\begin{bmatrix} 1 & 2 & 0 \\ 2 & 1 & 1 \\ 1 & -1 & 2 \end{bmatrix} \begin{bmatrix} x \\ y \\ z \end{bmatrix} = \begin{bmatrix} 1 \\ 3 \\ 2 \end{bmatrix} .
$$

The augmented matrix is

$$
\left[\begin{array}{ccc|c} 1 & 2 & 0 & 1 \\ 2 & 1 & 1 & 3 \\ 1 & -1 & 2 & 2 \end{array} \right] .
$$

Now begin row operations. In each case, we show the result if the operations are performed directly on the system rather than using matrix operations. You will see that performing these operations on the matrix is equivalent to performing the same operations directly on the equations. Working from left to right,

$$
A = \left[\begin{array}{ccc|c} 1 & 2 & 0 & 1 \\ 2 & 1 & 1 & 3 \\ 1 & -1 & 2 & 2 \end{array} \right] \xrightarrow{R_2 = R_2 - 2R_1} \left[\begin{array}{ccc|c} 1 & 2 & 0 & 1 \\ 0 & -3 & 1 & 1 \\ 1 & -1 & 2 & 2 \end{array} \right]
$$

$$
\begin{aligned}
x \ +2y \qquad &= 1 \qquad\qquad x \ +2y \qquad &= 1 \\
2x \ +y \ +z &= 3 \quad\to\quad \qquad -3y \ +z &= 1 \\
x \ -y \ +2z &= 2 \qquad\qquad x \ -y \ +2z &= 2
\end{aligned}
$$

$$
\xrightarrow{R_2 \leftrightarrow R_3} \left[\begin{array}{ccc|c} 1 & 2 & 0 & 1 \\ 1 & -1 & 2 & 2 \\ 0 & -3 & 1 & 1 \end{array} \right] \xrightarrow{R_1 = 2R_1} \left[\begin{array}{ccc|c} 2 & 4 & 0 & 2 \\ 1 & -1 & 2 & 2 \\ 0 & -3 & 1 & 1 \end{array} \right] = B
$$

$$
\begin{aligned}
x \ +2y \qquad &= 1 \qquad\qquad 2x \ +4y \qquad &= 2 \\
x \ -y \ +2z &= 2 \quad\to\quad x \ -y \ +2z &= 2 \\
-3y \ +z &= 1 \qquad\qquad -3y \ +z &= 1
\end{aligned}
$$

Thus, A is row-equivalent to B. Clearly, B is also row-equivalent to A, by performing the inverse row-operations $R_1 \to \frac{1}{2}R_1, R_2 \leftrightarrow R_3, R_2 \to R_2 + 2R_1$ on B. ■

It is not difficult to prove that if A and B are row-equivalent augmented matrices of two systems of linear equations, then the two systems have the same solution sets—a solution of the one system is a solution of the other. For example, the systems whose augmented matrices are A and B in the above example are, respectively,

$$
\begin{cases} x + 2y = 1 \\ 2x + y + z = 3 \\ x - y + 2z = 2 \end{cases} \quad \text{and} \quad \begin{cases} 2x + 4y = 2 \\ x - y + 2z = 2 \\ -3y + z = 1 \end{cases}
$$

and these systems have precisely the same solutions.

2.3 GAUSSIAN ELIMINATION

The augmented matrix for a system of n equations and n unknowns is

$$
\left[\begin{array}{cccc|c} a_{11} & a_{12} & \ldots & a_{1n} & b_1 \\ a_{21} & a_{22} & \ldots & a_{2n} & b_2 \\ \vdots & \vdots & \ddots & \vdots & \vdots \\ a_{n1} & a_{n2} & \ldots & a_{nn} & b_n \end{array} \right]
$$

Gaussian elimination performs row operations on the augmented matrix until the portion corresponding to the coefficient matrix is reduced to upper-triangular form.

Definition 2.3. An $n \times n$ matrix A whose entries are of the form

$$U_{ij} = \begin{cases} a_{ij}, & i \leq j \\ 0, & i > j \end{cases}$$

is called an *upper triangular* matrix.

$$U = \begin{bmatrix} a_{11} & a_{12} & \ldots & a_{1n} \\ 0 & a_{22} & \ldots & a_{2n} \\ \vdots & \vdots & \ddots & \vdots \\ 0 & 0 & \ldots & a_{nn} \end{bmatrix}$$

Starting with the matrix $A = \begin{bmatrix} a_{11} & a_{12} & \ldots & a_{1n} & \vline & b_1 \\ a_{21} & a_{22} & \ldots & a_{2n} & \vline & b_2 \\ \vdots & \vdots & \ddots & \vdots & \vline & \vdots \\ a_{n1} & a_{n2} & \ldots & a_{nn} & \vline & b_n \end{bmatrix}$, row elimination produces a matrix in upper triangular form

$$\begin{bmatrix} c_{11} & c_{12} & \ldots & c_{1n} & \vline & b_1' \\ 0 & c_{22} & \ldots & c_{2n} & \vline & b_2' \\ \vdots & \vdots & \ddots & \vdots & \vline & \vdots \\ 0 & 0 & \ldots & c_{nn} & \vline & b_n' \end{bmatrix},$$

which is easy to solve.

2.3.1 Upper-Triangular Form

In upper-triangular form, a simple procedure known as *back substitution* determines the solution. Since the linear algebraic systems corresponding to the original and final augmented matrix have the same solution, the solution to the upper-triangular system

$$\begin{bmatrix} c_{11} & c_{12} & & \ldots & & c_{1n} & \vline & b_1' \\ 0 & c_{22} & & \ldots & & c_{2n} & \vline & b_2' \\ \vdots & \vdots & & \ddots & & \vdots & \vline & \vdots \\ 0 & \ldots & c_{n-1,n-1} & c_{n-1,n} & \vline & b_{n-1}' \\ 0 & 0 & & \ldots & & c_n & \vline & b_n' \end{bmatrix}$$

begins with

$$x_n = \frac{b_n'}{c_{nn}}$$

followed by

$$x_{n-1} = \frac{b_{n-1}' - c_{n-1,n} x_n}{c_{n-1,n-1}}.$$

In general,

$$x_i = \frac{b_i' - \sum_{j=i+1}^{n} c_{ij} x_j}{c_{ii}}, \quad i = n-1, n-2, \ldots, 1.$$

We now formally describe the *Gaussian elimination* procedure. Start with matrix A and produce matrix B in upper-triangular form which is row-equivalent to A. If A is the augmented matrix of a system of linear equations, then applying back substitution to B determines the solution to the system. It is also possible that there is no solution to the system, and the row-reduction process will make this evident.

Begin at element a_{11}. If $a_{11} = 0$, exchange rows so $a_{11} \neq 0$. Now make all the elements below a_{11} zero by subtracting a multiple of row 1 from row i, $2 \leq i \leq n$. The multiplier used for row i is

$$\frac{a_{i1}}{a_{11}}.$$

The matrix is now in this form

$$\begin{bmatrix} a_{11} & a_{12} & \cdots & a_{1n} & b_1 \\ 0 & a'_{22} & \cdots & a'_{2n} & b'_2 \\ \vdots & \vdots & \ddots & \vdots & \vdots \\ 0 & a'_{n2} & \cdots & a'_{nn} & b'_n \end{bmatrix}.$$

Now perform the same process of elimination by using a'_{22} and multipliers a'_{i2}/a'_{22}, making a row exchange if necessary, so that all the elements below a'_{22} are 0,

$$\begin{bmatrix} a_{11} & a_{12} & \cdots & \cdots & a_{1n} & b_1 \\ 0 & a'_{22} & \cdots & \cdots & a'_{2n} & b'_2 \\ \vdots & 0 & a''_{33} & \vdots & a''_{3n} & b_3 \\ 0 & 0 & a''_{43} & \ddots & a''_{4n} & b_4 \\ \vdots & \vdots & \vdots & \ddots & \vdots & \vdots \\ 0 & 0 & a''_{n3} & \cdots & a''_{nn} & b'_n \end{bmatrix}$$

Repeat this process until the matrix is in upper-triangular form, and then execute back substitution to compute the solution.

Example 2.4. Solve the system

$$x_1 + x_2 - x_3 = 1,$$
$$2x_1 + x_2 + x_3 = 0,$$
$$-x_1 - 2x_2 + 3x_3 = 2.$$

Row reduce the augmented matrix to upper-triangular form.

$$\begin{bmatrix} 1 & 1 & -1 & 1 \\ 2 & 1 & 1 & 0 \\ -1 & -2 & 3 & 2 \end{bmatrix} \xrightarrow{R_2 = R_2 - 2R_1} \begin{bmatrix} 1 & 1 & -1 & 1 \\ 0 & -1 & 3 & -2 \\ -1 & -2 & 3 & 2 \end{bmatrix}$$

$$\begin{bmatrix} 1 & 1 & -1 & 1 \\ 0 & -1 & 3 & -2 \\ -1 & -2 & 3 & 2 \end{bmatrix} \xrightarrow{R_3 = R_3 - (-1)R_1} \begin{bmatrix} 1 & 1 & -1 & 1 \\ 0 & -1 & 3 & -2 \\ 0 & -1 & 2 & 3 \end{bmatrix}$$

$$\begin{bmatrix} 1 & 1 & -1 & 1 \\ 0 & -1 & 3 & -2 \\ 0 & -1 & 2 & 3 \end{bmatrix} \xrightarrow{R_3 = R_3 - (1)R_2} \begin{bmatrix} 1 & 1 & -1 & 1 \\ 0 & -1 & 3 & -2 \\ 0 & 0 & -1 & 5 \end{bmatrix}$$

Execute back substitution.

$$(-1)x_3 = 5, \quad x_3 = -5,$$
$$-x_2 + 3(-5) = -2, \quad x_2 = -13,$$
$$x_1 + (1)(-13) - (-5) = 1, \quad x_1 = 9.$$

Final solution: $x_1 = 9, x_2 = -13, x_3 = -5$

When computing a solution "by hand," it is a good idea to verify that the solution is correct.

$$9 + (-13) - (-5) = 1,$$
$$2(9) + (-13) + (-5) = 0,$$
$$-(9) - 2(-13) + 3(-5) = 2.$$

■

2.4 SYSTEMATIC SOLUTION OF LINEAR SYSTEMS

Suppose a system of n linear equations in n unknowns $x_1, \ldots, x_n$ has augmented matrix A and that A is row-equivalent to a matrix B in upper-triangular form. Then A and B have dimension $n \times (n+1)$.

Case 1: if we perform elementary row operations on the augmented matrix of the system and get a matrix with one of its rows equal to $[0\,0\,0 \ldots 0\,b]$, where $b \neq 0$, or a row of the form $[0\,0\,0 \ldots 0]$, then the system is said to be *inconsistent*. In this situation, there may be no solution of infinitely many solutions. In Figure 2.2, lines in the plane illustrate these two situations.

Case 2: There is a unique solution if Case 1 does not occur.

Example 2.5. Solve the system

$$
\begin{aligned}
x_1 + 2x_2 &= 1, \\
2x_1 + x_2 + x_3 &= 0, \\
-x_1 + 6x_2 + 3x_3 &= 1.
\end{aligned}
$$

The augmented matrix of the system is

$$
A = \begin{bmatrix} 1 & 2 & 0 & 1 \\ 2 & 1 & 1 & 0 \\ -1 & 6 & 3 & 1 \end{bmatrix},
$$

which is row equivalent to the upper-triangular matrix

$$
B = \begin{bmatrix} 1 & 2 & 0 & 1 \\ 0 & -3 & 1 & -2 \\ 0 & 0 & \frac{17}{3} & -\frac{10}{3} \end{bmatrix}.
$$

Back substitution gives the solution

$$
x_1 = \frac{1}{17}, \quad x_2 = \frac{8}{17}, \quad x_3 = -\frac{10}{17}.
$$

∎

Solve the system

$$
\begin{aligned}
2x_1 + 2x_2 - 2x_3 &= 5, \\
7x_1 + 7x_2 + x_3 &= 10, \\
5x_1 + 5x_2 - x_3 &= 5.
\end{aligned}
$$

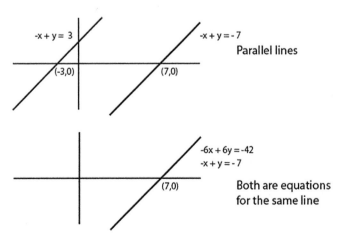

FIGURE 2.2 Inconsistent equations.

The augmented matrix is

$$A = \begin{bmatrix} 2 & 2 & -2 & | & 5 \\ 7 & 7 & 1 & | & 10 \\ 5 & 5 & -1 & | & 5 \end{bmatrix},$$

which is row equivalent to

$$B = \begin{bmatrix} 2 & 2 & -2 & | & 5 \\ 0 & 0 & 8 & | & -15/2 \\ 0 & 0 & 0 & | & -15/4 \end{bmatrix}.$$

The system is inconsistent, since the last row is $[0\,0\,0 - 15/4]$, which implies that

$$(0)\,x_1 + (0)\,x_2 + (0)\,x_3 = -15/4.$$

The system has no solution. ∎

Example 2.6. Solve the system

$$\begin{aligned} x_1 + 2x_2 + x_3 &= -1, \\ 7x_1 + 4x_2 + 4x_3 &= 5, \\ 6x_1 + 2x_2 + 3x_3 &= 6. \end{aligned} \qquad (2.2)$$

The augmented matrix is

$$\begin{bmatrix} 1 & 2 & 1 & | & -1 \\ 7 & 4 & 4 & | & 5 \\ 6 & 2 & 3 & | & 6 \end{bmatrix}.$$

Now,

$$\begin{bmatrix} 1 & 2 & 1 & | & -1 \\ 7 & 4 & 4 & | & 5 \\ 6 & 2 & 3 & | & 6 \end{bmatrix} \xrightarrow{R_2 = R_2 - (7)\,R_1} \begin{bmatrix} 1 & 2 & 1 & | & -1 \\ 0 & -10 & -3 & | & 12 \\ 6 & 2 & 3 & | & 6 \end{bmatrix}$$

$$\begin{bmatrix} 1 & 2 & 1 & | & -1 \\ 0 & -10 & -3 & | & 12 \\ 6 & 2 & 3 & | & 6 \end{bmatrix} \xrightarrow{R_3 = R_3 - (6)\,R_1} \begin{bmatrix} 1 & 2 & 1 & | & -1 \\ 0 & -10 & -3 & | & 12 \\ 0 & -10 & -3 & | & 12 \end{bmatrix}$$

$$\begin{bmatrix} 1 & 2 & 1 & | & -1 \\ 0 & -10 & -3 & | & 12 \\ 0 & -10 & -3 & | & 12 \end{bmatrix} \xrightarrow{R_3 = R_3 - (1)R_2} \begin{bmatrix} 1 & 2 & 1 & | & -1 \\ 0 & -10 & -3 & | & 12 \\ 0 & 0 & 0 & | & 0 \end{bmatrix}$$

Note that any values of x_1, x_2, and x_3 will satisfy

$$(0)\,x_1 + (0)\,x_2 + (0)\,x_3 = 0.$$

Let's continue backward:

$$-10x_2 - 3x_3 = 12,$$

so

$$x_2 = \frac{-3x_3 - 12}{10}.$$

Substitute this relationship into the first equation of 2.2, and after a little algebra obtain

$$x_1 = \frac{-2x_3 + 7}{5}.$$

We have determined x_1 and x_2 in terms of x_3. The complete solution is

$$x_1 = \frac{-2x_3 + 7}{5}, \quad x_2 = \frac{-3x_3 - 12}{10},$$

with x_3 arbitrary, so there are infinitely many solutions.   ■

Example 2.7.   For which rational numbers a and b does the following system have (i) no solution, (ii) a unique solution, (iii) infinitely many solutions?

$$x_1 - 2x_2 + 3x_3 = 4,$$
$$2x_1 - 3x_2 + ax_3 = 5,$$
$$3x_1 - 4x_2 + 5x_3 = b.$$

The augmented matrix of the system is

$$\begin{bmatrix} 1 & -2 & 3 & 4 \\ 2 & -3 & a & 5 \\ 3 & -4 & 5 & b \end{bmatrix}$$

$$\begin{bmatrix} 1 & -2 & 3 & 4 \\ 2 & -3 & a & 5 \\ 3 & -4 & 5 & b \end{bmatrix} \xrightarrow[R_3 = R_3-(3)R_1]{R_2 = R_2-(2)R_1} \begin{bmatrix} 1 & -2 & 3 & 4 \\ 0 & 1 & a-6 & -3 \\ 0 & 2 & -4 & b-12 \end{bmatrix}$$

$$\xrightarrow{R_3 = R_3 - 2R_2} \begin{bmatrix} 1 & -2 & 3 & 4 \\ 0 & 1 & a-6 & -3 \\ 0 & 0 & -2a+8 & b-6 \end{bmatrix} = B.$$

Case 1. $a \neq 4$. Then $-2a+8 \neq 0$ and back substitution gives a unique solution with

$$x_3 = \frac{b-6}{8-2a}.$$

Case 2. $a = 4$. Then

$$B = \begin{bmatrix} 1 & -2 & 3 & 4 \\ 0 & 1 & -2 & -3 \\ 0 & 0 & 0 & b-6 \end{bmatrix}.$$

If $b \neq 6$ we get no solution, whereas if $b = 6$ then

$$x_2 - 2x_3 = -3, \quad x_2 = -3 + 2x_3,$$
$$x_1 - 2(-3 + 2x_3) + 3x_3 = 4, \quad x_1 = -2 + x_3.$$

The complete solution is

$x_1 = -2 + x_3, x_2 = -3 + 2x_3$, with x_3 arbitrary.   ■

The MATLAB operator for solving the linear algebraic system $Ax = b$ is "\". Use the syntax A\b to obtain the solution x. If MATLAB detects that A may be singular, you will get an error message.

```
>> A = [1 -2 3;2 -3 4;3 -4 5];
>> b = [4 5 7]';
>> A\b
Warning: Matrix is close to singular or badly scaled.
         Results may be inaccurate. RCOND = 2.312965e-018.

ans =
  1.0e+015 *

   -4.5036
   -9.0072
   -4.5036
>> B = [1 5 -1;3 5 2;1 5 3]
```

```
B =
     1      5     -1
     3      5      2
     1      5      3
>> B\b
ans =
   -0.6250
    1.0750
    0.7500
```

2.5 COMPUTING THE INVERSE

It is often useful to represent a matrix as a sequence of column vectors. If $v_1, v_2, \ldots, v_n$ are $n \times 1$ column vectors, the matrix $A = [v_1 \, v_2 \, \ldots \, v_n]$ is the $n \times n$ matrix with first column v_1, second column $v_2, \ldots$, and last column v_n.

Given the three column vectors $v_1 = \begin{bmatrix} 1 \\ -1 \\ 4 \end{bmatrix}, v_2 = \begin{bmatrix} 6 \\ 0 \\ 8 \end{bmatrix}, v_3 = \begin{bmatrix} 5 \\ 3 \\ 7 \end{bmatrix}$, then $A = [v_1 \, v_2 \, v_3] = \begin{bmatrix} 1 & 6 & 5 \\ -1 & 0 & 3 \\ 4 & 8 & 7 \end{bmatrix}$.

This notation can also be used to represent a matrix product:

$$\begin{bmatrix} a_{11} & \cdots & a_{1i} & \cdots & a_{1n} \\ a_{21} & \cdots & \vdots & \cdots & a_{2n} \\ a_{31} & \cdots & a_{ii} & \cdots & a_{3n} \\ \vdots & \cdots & \vdots & \ddots & \vdots \\ a_{n1} & \cdots & a_{ni} & \cdots & a_{nn} \end{bmatrix} \begin{bmatrix} b_{11} & \cdots & b_{1i} & \cdots & b_{1n} \\ b_{21} & \cdots & \vdots & \cdots & b_{2n} \\ b_{31} & \cdots & b_{ii} & \cdots & b_{3n} \\ \vdots & \cdots & \vdots & \ddots & \vdots \\ b_{n1} & \cdots & b_{ni} & \cdots & b_{nn} \end{bmatrix} = \begin{bmatrix} A \begin{bmatrix} b_{11} \\ b_{21} \\ b_{31} \\ \vdots \\ b_{n1} \end{bmatrix} \cdots A \begin{bmatrix} b_{1i} \\ \vdots \\ b_{ii} \\ \vdots \\ b_{ni} \end{bmatrix} \cdots A \begin{bmatrix} b_{1n} \\ b_{2n} \\ b_{3n} \\ \vdots \\ b_{nn} \end{bmatrix} \end{bmatrix}. \tag{2.3}$$

An example with a 2×2 matrix is easily generalized to the $n \times n$ case.

$$AB = \begin{bmatrix} a_{11} & a_{12} \\ a_{21} & a_{22} \end{bmatrix} \begin{bmatrix} b_{11} & b_{12} \\ b_{21} & b_{22} \end{bmatrix} = \begin{bmatrix} a_{11}b_{11} + a_{12}b_{21} & a_{11}b_{12} + a_{12}b_{22} \\ a_{21}b_{11} + a_{22}b_{21} & a_{21}b_{12} + a_{22}b_{22} \end{bmatrix} = \begin{bmatrix} A \begin{bmatrix} b_{11} \\ b_{21} \end{bmatrix} A \begin{bmatrix} b_{12} \\ b_{22} \end{bmatrix} \end{bmatrix}.$$

Equation 2.3 can be used to develop a method for the computation of A^{-1}. Solve n linear equations.

$$Ax_i = \begin{bmatrix} 0 \\ 0 \\ \vdots \\ 1 \\ \vdots \\ 0 \\ 0 \end{bmatrix}, \quad 1 \le i \le n$$

for column vectors x_i. In other words, find the solutions of

$$Ax_1 = \begin{bmatrix} 1 \\ 0 \\ 0 \\ \vdots \\ 0 \end{bmatrix}, Ax_2 = \begin{bmatrix} 0 \\ 1 \\ 0 \\ \vdots \\ 0 \end{bmatrix}, Ax_3 = \begin{bmatrix} 0 \\ 0 \\ 1 \\ \vdots \\ 0 \end{bmatrix}, \ldots, Ax_n = \begin{bmatrix} 0 \\ 0 \\ 0 \\ \vdots \\ 1 \end{bmatrix}.$$

Now form the $n \times n$ matrix B whose first column is $x_1 = \begin{bmatrix} x_{11} \\ x_{21} \\ x_{31} \\ \vdots \\ x_{n1} \end{bmatrix}$, whose second column is $x_2 = \begin{bmatrix} x_{12} \\ x_{22} \\ x_{32} \\ \vdots \\ x_{n2} \end{bmatrix}$, ..., and whose

last column is $x_n = \begin{bmatrix} x_{1n} \\ x_{2n} \\ x_{3n} \\ \vdots \\ x_{nn} \end{bmatrix}$, and we have

$$AB = A \begin{bmatrix} x_{11} & \cdots & x_{1i} & \cdots & x_{1n} \\ x_{21} & \cdots & \vdots & \cdots & x_{2n} \\ x_{31} & \cdots & x_{ii} & \cdots & x_{3n} \\ \vdots & \cdots & \vdots & \ddots & \vdots \\ x_{n1} & \cdots & x_{ni} & \cdots & x_{nn} \end{bmatrix} = [Ax_1\, Ax_2\, \ldots\, Ax_i\, \ldots\, Ax_n] = \left[\begin{bmatrix} 1 \\ 0 \\ 0 \\ \vdots \\ 0 \end{bmatrix} \begin{bmatrix} 0 \\ 1 \\ 0 \\ \vdots \\ 0 \end{bmatrix} \cdots \begin{bmatrix} 0 \\ 0 \\ 0 \\ \vdots \\ 1 \end{bmatrix} \right] = I.$$

This calculation is most conveniently done by forming the augmented matrix

$$\left[\begin{array}{ccccc|cccc} a_{11} & a_{12} & \cdots & a_{1,n-1} & a_{1n} & 1 & 0 & \cdots & 0 \\ a_{21} & a_{22} & \cdots & a_{2,n-1} & a_{2n} & 0 & 1 & \cdots & 0 \\ \vdots & \vdots & \ddots & \vdots & \vdots & \vdots & \vdots & \ddots & \vdots \\ a_{n-1,1} & a_{n-1,2} & \cdots & a_{n-1,n-1} & a_{n-1,n} & 0 & 0 & \cdots & \vdots \\ a_{n1} & a_{n2} & \cdots & a_{n,n-1} & a_{nn} & 0 & 0 & \cdots & 0 \\ & & & & & 0 & 0 & \cdots & 1 \end{array} \right]$$

and row-reducing the coefficient matrix to upper-triangular form, all the while performing the row operations on the augmented portion of the matrix. Then perform back substitution n times to find $x_1, x_2, \ldots, x_n$.

Example 2.8. Find the inverse of the matrix

$$A = \begin{bmatrix} 1 & 0 & 2 \\ 1 & 3 & 0 \\ 2 & 1 & 5 \end{bmatrix}.$$

$$\left[\begin{array}{ccc|ccc} 1 & 0 & 2 & 1 & 0 & 0 \\ 1 & 3 & 0 & 0 & 1 & 0 \\ 2 & 1 & 5 & 0 & 0 & 1 \end{array} \right] \xrightarrow[\substack{R_2 = R_2 - (1)R_1 \\ R_3 = R_3 - (2)R_1}]{} = \left[\begin{array}{ccc|ccc} 1 & 0 & 2 & 1 & 0 & 0 \\ 0 & 3 & -2 & -1 & 1 & 0 \\ 0 & 1 & 1 & -2 & 0 & 1 \end{array} \right]$$

$$\xrightarrow[R_3 = R_3 - \left(\frac{1}{3}\right)R_2]{} \left[\begin{array}{ccc|ccc} 1 & 0 & 2 & 1 & 0 & 0 \\ 0 & 3 & -2 & -1 & 1 & 0 \\ 0 & 0 & \frac{5}{3} & -\frac{5}{3} & -\frac{1}{3} & 1 \end{array} \right]$$

Now do back substitution.

$$\frac{5}{3}x_{31} = -\frac{5}{3}, \quad x_{31} = -1,$$

$$3x_{21} - 2x_{31} = -1, \quad x_{21} = \frac{2x_{31} - 1}{3} = -1,$$

$$x_{11} + 0\,(x_{21}) + 2x_{31} = 1, \quad x_{11} = 1 - 2x_{31} = 3.$$

The first column of A^{-1} is $\begin{bmatrix} 3 \\ -1 \\ -1 \end{bmatrix}$. Perform back substitution two more times to obtain

$$A^{-1} = \begin{bmatrix} 3 & \frac{2}{5} & -\frac{6}{5} \\ -1 & \frac{1}{5} & \frac{2}{5} \\ -1 & -\frac{1}{5} & \frac{3}{5} \end{bmatrix}.$$

■

Since not every matrix has an inverse, this process may fail. The matrix is singular if during back substitution you obtain a row of zeros in the coefficient matrix.

Example 2.9. $A = \begin{bmatrix} 1 & 2 & 3 \\ 1 & 0 & 1 \\ 3 & 4 & 7 \end{bmatrix}$

$$\begin{bmatrix} 1 & 2 & 3 & | & 1 & 0 & 0 \\ 1 & 0 & 1 & | & 0 & 1 & 0 \\ 3 & 4 & 7 & | & 0 & 0 & 1 \end{bmatrix} \xrightarrow[R_3 = R_3 - 3R_1]{R_2 = R_2 - R_1} \begin{bmatrix} 1 & 2 & 3 & | & 1 & 0 & 0 \\ 0 & -2 & -2 & | & -1 & 1 & 0 \\ 0 & -2 & -2 & | & -3 & 0 & 1 \end{bmatrix}$$

$$\xrightarrow{R_3 = R_3 - R_2} \begin{bmatrix} 1 & 2 & 3 & | & 1 & 0 & 0 \\ 0 & -2 & -2 & | & -1 & 1 & 0 \\ 0 & 0 & 0 & | & -2 & -1 & 1 \end{bmatrix}$$

There is a row of zeros. A is singular. ■

2.6 HOMOGENEOUS SYSTEMS

An $n \times n$ system of homogeneous linear equations

$$a_{11}x_1 + a_{12}x_2 + \cdots + a_{1n}x_n = 0$$
$$a_{21}x_1 + a_{22}x_2 + \cdots + a_{2n}x_n = 0$$
$$\vdots$$
$$a_{n1}x_1 + a_{n2}x_2 + \cdots + a_{nn}x_n = 0$$

is always consistent since $x_1 = 0, \ldots, x_n = 0$ is a solution. This solution is called the *trivial solution*, and any other solution is called a *nontrivial solution*. For example, consider the homogeneous system

$$x_1 - x_2 = 0,$$
$$x_1 + x_2 = 0.$$

Using the augmented matrix, we have

$$\begin{bmatrix} 1 & -1 & | & 0 \\ 1 & 1 & | & 0 \end{bmatrix} \xrightarrow{R_2 = R_2 - R_1} \begin{bmatrix} 1 & -1 & | & 0 \\ 0 & 2 & | & 0 \end{bmatrix},$$

so $x_1 = x_2 = 0$, and the system has only the trivial solution. Notice that is really not necessary to attach the column of zeros.

Example 2.10. Solve the homogeneous system

$$x_1 + 2x_2 + x_3 = 0,$$
$$5x_1 + 2x_2 + 7x_3 = 0,$$
$$2x_1 + 3x_3 = 0.$$

$$\begin{bmatrix} 1 & 2 & 1 \\ 5 & 2 & 7 \\ 2 & 0 & 3 \end{bmatrix} \xrightarrow[\begin{subarray}{l} R_2 = R_2 - 5R_1 \\ R_3 = R_3 - 2R_1 \end{subarray}]{} \begin{bmatrix} 1 & 2 & 1 \\ 0 & -8 & 2 \\ 0 & -4 & 1 \end{bmatrix} \xrightarrow[]{R_3 = R_3 - \left(\frac{1}{2}\right)R_2} \begin{bmatrix} 1 & 2 & 1 \\ 0 & -8 & 2 \\ 0 & 0 & 0 \end{bmatrix},$$

so the system has the solution $x_1 = -\frac{3}{2}x_3, x_2 = x_3/4$, with x_3 arbitrary. Choosing $x_3 = 1$ gives rise to the nontrivial solution

$$x_1 = -\frac{3}{2}, \quad x_2 = \frac{1}{4}, \quad x_3 = 1. \qquad \blacksquare$$

Recall that in Chapter 1, we showed that if A is nonsingular, then the homogeneous system has only the trivial solution. We are now in a position to show that the reverse is also true.

Theorem 2.1. *If the homogeneous system $Ax = 0$ has only the trivial solution, then A is nonsingular; that is A^{-1} exists.*

Proof. During row-reduction of the augmented matrix used to compute A^{-1}, there cannot be a row of zeros, or $Ax = 0$ would have an infinite number of solutions. As a result, back substitution will produce the inverse, and A is nonsingular. $\square$

2.7 APPLICATION: A TRUSS

A truss is a structure normally containing triangular units constructed of straight members with ends connected at joints referred to as pins. Trusses are the primary structural component of many bridges. External forces and reactions to those forces are considered to act only at the pins and result in internal forces in the members, which are either tensile or compressive. Civil engineers design trusses and must determine the forces at the pins of a truss so it will remain static under a load. Figure 2.3 depicts a truss under a load of 1500 units. It is allowed to move slightly horizontally at pin 1 and is static at pin 5. The figure names the pins, member forces and reaction forces. There are seven member forces labeled D, E, F, G, H, I, and J. A positive value for a member force means that it is a tensile force so the member force is directed away from the pins at its ends. The three reaction forces are labeled as A, B, and C. These are due to a roller and a pinned support, and a positive value for a reaction force means that it acts in the direction shown, while a negative value means the assumed sense is wrong and should be the opposite.

The values of all these unknown (internal and reaction) forces can be found by solving a system of equations. The truss is in equilibrium, so each pin of the truss contributes two equations to the system. One equation expresses the fact that the x components of the forces on that joint add to zero, and the other equation expresses the fact that the y components do also. This truss contains 5 joints, labeled 1, 2, 3, 4, 5, and thus yields 10 equations, so the resulting system of equations is 10×10. There is a unique solution (meaning the truss is stable) as long as the coefficient matrix is nonsingular. If the truss consists only of triangles, then this is guaranteed by the laws of statics.

At pin 5, each force is either in the horizontal or vertical direction. At the remaining pins, there are both horizontal and vertical components of force, so the necessary angles must be known. If we apply the equilibrium rules, the equations for determining the truss forces are

Equation	A	B	C	D	E	F	G	H	I	J	RHS
$1 - x$	0	0	0	$-\cos(48.4)$	0	-1	0	0	0	0	0
$1 - y$	-1	0	0	$\sin(48.4)$	0	0	0	0	0	0	0
$2 - x$	0	0	0	$-\cos(48.4)$	$\cos(60.9)$	0	1	0	0	0	0
$2 - y$	0	0	0	$-\sin(48.4)$	$-\sin(60.9)$	0	0	0	0	0	0
$3 - x$	0	0	0	0	$-\cos(60.9)$	-1	0	$\cos(45.0)$	0	1	0
$3 - y$	0	0	0	0	$-\sin(60.9)$	0	0	$-\sin(45.0)$	0	0	1500
$4 - x$	0	0	0	0	0	0	-1	$-\cos(45.0)$	0	0	0
$4 - y$	0	0	0	0	0	0	0	$-\sin(45.0)$	-1	0	0
$5 - x$	0	-1	0	0	0	0	0	0	0	-1	0
$5 - y$	0	0	-1	0	0	0	0	0	1	0	0

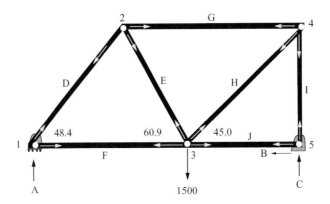

FIGURE 2.3 Truss.

These equations correspond to the matrix formulation

$$
\begin{bmatrix}
0 & 0 & 0 & -0.6639 & 0 & -1 & 0 & 0 & 0 & 0 \\
-1 & 0 & 0 & 0.7478 & 0 & 0 & 0 & 0 & 0 & 0 \\
0 & 0 & 0 & -0.6639 & 0.4863 & 0 & 1 & 0 & 0 & 0 \\
0 & 0 & 0 & -0.7478 & -0.8738 & 0 & 0 & 0 & 0 & 0 \\
0 & 0 & 0 & 0 & -0.4863 & -1 & 0 & 0.7071 & 0 & 1 \\
0 & 0 & 0 & 0 & -0.8738 & 0 & 0 & -0.7071 & 0 & 0 \\
0 & 0 & 0 & 0 & 0 & 0 & -1 & -0.7071 & 0 & 0 \\
0 & 0 & 0 & 0 & 0 & 0 & 0 & -0.7071 & -1 & 0 \\
0 & -1 & 0 & 0 & 0 & 0 & 0 & 0 & 0 & -1 \\
0 & 0 & -1 & 0 & 0 & 0 & 0 & 0 & 1 & 0
\end{bmatrix}
\begin{bmatrix}
A \\ B \\ C \\ D \\ E \\ F \\ G \\ H \\ I \\ J
\end{bmatrix}
=
\begin{bmatrix}
0 \\ 0 \\ 0 \\ 0 \\ 0 \\ 1500 \\ 0 \\ 0 \\ 0 \\ 0
\end{bmatrix}.
$$

Note that the coefficient matrix contains mostly zeros, so it is a sparse matrix.

The software distribution contains two files, TRUSS.mat and TRUSS.txt. TRUSS.mat is the truss coefficient matrix in internal MATLAB format, and TRUSS.txt is a text file representation of the matrix. After reading in the matrix using either file and applying the MATLAB "\" operator, the solution is

$$
\begin{bmatrix}
A \\ B \\ C \\ D \\ E \\ F \\ G \\ H \\ I \\ J
\end{bmatrix}
=
\begin{bmatrix}
613.66 \\ 0.0 \\ 886.34 \\ 820.62 \\ -702.29 \\ -544.81 \\ 886.34 \\ -1253.5 \\ 886.34 \\ 0.0
\end{bmatrix}.
$$

```
>> load Truss
>> format shortg % output looks better
>> rhs = zeros(10,1);
>> rhs(6) = 1500;
>> Truss\rhs

ans =
        613.66
             0
        886.34
        820.62
       -702.29
```

```
-544.81
 886.34
-1253.5
 886.34
      0
```

2.8   APPLICATION: ELECTRICAL CIRCUIT

Figure 2.4 is a diagram of a DC electric circuit containing batteries and resistors. The voltage across a resistor in a circuit is determined using Ohm's Law, $V = RI$, where R is the resistance and I is the current. We would like to determine the currents i_i, i_2, i_3. The currents can be determined by using *Kirchhoff's rules*, which state that

1. At any junction point in a circuit where the current can divide, the sum of the currents into the junction must equal the sum of the currents out of the junction.
2. When any closed loop in the circuit is traversed, the sum of the changes in voltage must equal zero.

Rule 1 is called the *Kirchhoff's Current Law*, and is a consequence of conservation of charge. Rule 2 is termed the *Kirchhoff's Voltage Law*, and is a consequence of the conservation of energy. In the circuit of Figure 2.4, there are two junction points A and B and two loops. What we must do is apply Kirchhoff's rules to obtain a system of equations that will allow us to find the current values. Before we can begin, the direction of the current in each branch must be assigned. Once this is done, appropriate signs must be given to each resistor and voltage. Label the side of a resistor on which the current enters as positive $(+)$ and the side on which the current exits as negative $(-)$. It can be difficult to determine in which direction the current actually flows. If the direction is not correct, the current will be negative in that branch. Note the choices made for the circuit of Figure 2.4. It is now time to apply Kirchhoff's rules as follows:

1. Apply the first rule to all but one junction point. Each time you use the first rule, a current not already used must be included; otherwise, you will have redundant equations.
2. Apply the second rule to enough loops so the currents in the loop equations and those in the junction equations equal the number of unknown currents.

We will now determine the equations for the circuit in Figure 2.4.

Apply rule 1 at junction A: $i_1 + i_3 = i_2$.

In the two applications of the loop rule, note that the signs in each equation are determined by whether the current moves from $+$ to $-$ or $-$ to $+$.

Apply rule 2 to loop 1: $-V_1 + R_1 i_1 + R_2 i_2 + V_2 + R_4 i_1 = 0$, and $(R_1 + R_4) i_1 + R_2 i_2 = V_1 - V_2$
Apply rule 2 to loop 2: $R_2 i_2 + V_2 - V_3 + R_3 i_3 = 0$, and $R_2 i_2 + R_3 i_3 = V_3 - V_2$

We now have three equations and three unknowns:

$$\begin{aligned} i_1 - i_2 + i_3 &= 0, \\ (R_1 + R_4) i_1 + R_2 i_2 &= V_1 - V_2, \\ R_2 i_2 + R_3 i_3 &= V_3 - V_2. \end{aligned} \tag{2.4}$$

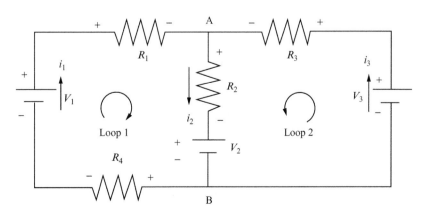

FIGURE 2.4 Electrical circuit.

Choose the following values for the batteries and the resistors:

Components	V_1	V_2	V_3	R_1	R_2	R_3	R_4
	2 V	3 V	5 V	1 Ω	2 Ω	5 Ω	3 Ω

Equation 2.4 then becomes

$$i_1 - i_2 + i_3 = 0,$$
$$4i_1 + 2i_2 = -1,$$
$$2i_2 + 5i_3 = 2.$$

In matrix form, the system is $\begin{bmatrix} 1 & -1 & 1 \\ 4 & 2 & 0 \\ 0 & 2 & 5 \end{bmatrix} \begin{bmatrix} i_1 \\ i_2 \\ i_3 \end{bmatrix} = \begin{bmatrix} 0 \\ -1 \\ 2 \end{bmatrix}$, and the currents are

$$i_1 = -0.2895, \quad i_2 = 0.0789, \quad i_3 = 0.3684.$$

Note that i_1 flows in the direction opposite to what is shown in Figure 2.4.

2.9 CHAPTER SUMMARY

Introduction to Linear Equations

A system *of n* linear equations in *n* unknowns $x_1, x_2, \ldots, x_n$ is a family of equations

$$a_{11}x_1 + a_{12}x_2 + \cdots + a_{1n}x_n = b_1$$
$$a_{21}x_1 + a_{22}x_2 + \cdots + a_{2n}x_n = b_2$$
$$\vdots$$
$$a_{n1}x_1 + a_{n2}x_2 + \cdots + a_{nn}x_n = b_n$$

Determine if there exist numbers $x_1, x_2, \ldots, x_n$ which satisfy each of the equations simultaneously. To determine a solution using Gaussian elimination, multiply an equation by a constant and subtract from another equation in order to eliminate an unknown. Do this in an organized fashion until obtaining an equation containing only one unknown. Compute it and then execute a process of substitution to determine the remaining unknowns. It is tedious to do this by writing down equations, and we want an approach that we can implement on a computer. The solution is to write the system in matrix form $Ax = b$, where A is the matrix of coefficients and to perform matrix operations, something a computer can do very well.

Solving Square Linear Equations

Given a matrix A, there are three fundamental operations we can perform:

- Multiply a row by constant.
- Exchange two rows.
- Multiply a row by a constant and subtract it from another row.

These are termed elementary row operations, and a matrix produced by one or more of these operations is said to be row equivalent to the original matrix. The approach we take to solving linear systems is to attach the right-hand sides as an additional column to form an $n \times (n + 1)$ matrix. This is called the augmented matrix. Performing elementary row operations with this matrix is equivalent to performing the same operations directly on the equations. If we transform the augmented matrix

$$\left[\begin{array}{ccccc|c} a_{11} & a_{22} & \cdots & a_{1,n-1} & a_{1n} & b_1 \\ a_{21} & a_{22} & \cdots & a_{2,n-1} & a_{2n} & b_2 \\ \vdots & \vdots & \ddots & \vdots & \vdots & \vdots \\ & & & \ddots & & b_{n-1} \\ \vdots & \vdots & \vdots & & \vdots & b_n \\ a_{n1} & a_{n2} & \cdots & a_{n,n-1} & a_{nn} & \end{array}\right]$$

to a matrix

$$\left[\begin{array}{ccccc|c} \bar{a}_{11} & \bar{a}_{12} & \cdots & \bar{a}_{1,n-1} & \bar{a}_{1n} & \bar{b}_1 \\ \bar{a}_{21} & \bar{a}_{22} & \cdots & \bar{a}_{2,n-1} & \bar{a}_{2n} & \bar{b}_2 \\ \vdots & \vdots & \ddots & \vdots & \vdots & \vdots \\ \vdots & \vdots & \vdots & \ddots & \vdots & \vdots \\ \bar{a}_{n1} & \bar{a}_{n2} & \cdots & \bar{a}_{n,n-1} & \bar{a}_{nn} & \bar{b}_n \end{array}\right],$$

the two systems have the same solutions.

Gaussian Elimination

To execute Gaussian elimination, create the augmented matrix and perform row operations that reduce the coefficient matrix to upper-triangular form. The solution to the upper-triangular system is the same as the solution to the original linear system. Solve the upper-triangular system by back substitution, as long as the element at position (n, n) is not zero. The unknown x_n is immediately available using the last row of the augmented matrix. Using x_n in the equation represented by row $n - 1$, we find x_{n-1}, and so forth, until determining x_1. If position (n, n) is zero, then the entire last row of the coefficient matrix is zero, and there is either no solution or infinitely many solutions.

Systematic Solution of Linear Equations

One method of solving a linear system $Ax = b$ is reduction to upper triangular form. If $a_{11} = 0$, exchange rows so it is nonzero. Multiply row 1 by a_{21}/a_{11} and subtract from row 2. That zeros-out the element in row 2, column 1. Now multiply row 1 by a_{31}/a_{11} and subtract from row 3, zeroing out the element in row 3, column 1. Continue until zeroing-out a_{1n}. The first column now has the form

$$\left[\begin{array}{c} a_{11} \\ 0 \\ \vdots \\ 0 \end{array}\right].$$

Move down one row and over one column to position $(2, 2)$. Make sure the element there, $\bar{a}_{22}$, is nonzero; if not, swap rows. Using $\bar{a}_{22}$, zero-out the elements at positions $(3, 2)$, $(4, 2)$, ..., $(n, 2)$. Move to row 3, column 3 and repeat the process. After processing $n - 1$ columns the augmented matrix is in upper-triangular form

$$\left[\begin{array}{cccc|c} c_{11} & c_{12} & \cdots & c_{1n} & b_1' \\ 0 & c_{22} & \cdots & c_{2n} & b_2' \\ \vdots & \vdots & \ddots & \vdots & \vdots \\ 0 & 0 & \cdots & c_{nn} & b_n' \end{array}\right].$$

Perform back substitution to compute the unique solution, if it exists.

If the last row of the reduced coefficient matrix is all zeros, and the corresponding element of the augmented column is nonzero, the system has no solution; otherwise, there are infinitely many solutions.

Computing the Inverse

To compute the inverse, attach the n columns of the identity matrix to form an augmented matrix. By performing elementary row operations on the entire augmented matrix, reduce the coefficient matrix portion to upper-triangular form. Perform back substitution once for every attached column that was produced from the identity matrix. The solution obtained from the

original right-hand side $\begin{bmatrix} 1 \\ 0 \\ \vdots \\ 0 \end{bmatrix}$ is the first column of the inverse. Continue in the same fashion to obtain columns $2 - n$ of

the inverse.

Homogeneous Systems

To solve a system of the form $Ax = 0$, there is no reason to form the augmented matrix, since all components will remain zero during row elimination. After reduction to upper-triangular form, if the element in position (n, n) is nonzero, the system has the unique solution $x = 0$; otherwise, there is an infinite number of solutions, and the matrix A is singular.

Applications

A truss presents a problem in statics. Forces must balance so the truss remains intact under a given load or loads. At the pins, the force-in must equal the force-out. If there are k pins, a system with $2k$ equations and $2k$ unknowns results, where the right-hand side is formed from the loads. A solution of the system gives the forces at the pins.

If an electrical circuit has one or more batteries and consists entirely of resistors, the currents in the circuit are found by solving a linear system determined by using the relationship $V = RI$ for every resistor and applying Kirchhoff's rules.

2.10 PROBLEMS

2.1 Find the unique solution to each system of linear equations.
  a. $2x + y = 3$
  $x - y = 1$
  b. $x_1 + 2x_2 + x_3 = 1$
  $x_2 - x_3 = 0$
  $x_1 + 2x_2 + 2x_3 = 1$

2.2 Solve the following systems of linear equations by reducing the augmented matrix to upper-triangular form:
  a. $x_1 + x_2 + x_3 = 2$
  $2x_1 + 3x_2 - x_3 = 8$
  $x_1 - x_2 - x_3 = -8$
  b. $2x_2 + 3x_3 - 4x_4 = 1$
  $2x_3 + 3x_4 = 4$
  $2x_1 + 2x_2 - 5x_3 + 2x_4 = 4$
  $2x_1 - 6x_3 + 9x_4 = 7$

2.3 Show that the following system is consistent if $c = 2a - 3b$ and solve the system in this case.

$$2x - y + 3z = a$$
$$3x + y - 5z = b$$
$$-5x - 5y + 21z = c$$

2.4 Solve the homogeneous system

$$-3x_1 + x_2 + x_3 + x_4 = 0$$
$$x_1 - 3x_2 + x_3 + x_4 = 0$$
$$x_1 + x_2 - 3x_3 + x_4 = 0$$
$$x_1 + x_2 + x_3 - 3x_4 = 0.$$

2.5 For which rational numbers λ does the homogeneous system

$$x + (\lambda - 3)y = 0$$
$$(\lambda - 3)x + y = 0$$

have a nontrivial solution?

2.6 Let $A = \begin{bmatrix} a & b \\ c & d \end{bmatrix}$. Show that A is row-equivalent to $\begin{bmatrix} 1 & 0 \\ 0 & 1 \end{bmatrix}$ if $ad - bc \neq 0$, but is row-equivalent to a matrix whose second row is zero, if $ad - bc = 0$.

2.7 For which rational numbers a does the following system have (i) no solutions, (ii) exactly one solution, (iii) infinitely many solutions?

$$x_1 + 2x_2 - 3x_3 = 4$$
$$3x_1 - x_2 + 5x_3 = 2$$
$$4x_1 + x_2 + (a^2 - 14)x_3 = a+2$$

2.8 Find the rational number k for which the matrix $A = \begin{bmatrix} 1 & 2 & k \\ 3 & -1 & 1 \\ 5 & 3 & -5 \end{bmatrix}$ is singular.

2.9 Show that the matrix $A = \begin{bmatrix} 1 & a & b \\ -a & 1 & c \\ -b & -c & 1 \end{bmatrix}$ is nonsingular by demonstrating that A is row-equivalent to I.

2.10 Find the inverse, if it exists, for each matrix.

a. $\begin{bmatrix} 1 & 3 \\ 0 & 2 \end{bmatrix}$

b. $\begin{bmatrix} 1 & 3 \\ 2 & 6 \end{bmatrix}$

c. $\begin{bmatrix} 1 & -1 & 2 \\ 0 & 1 & 3 \\ 0 & 4 & 2 \end{bmatrix}$

d. $\begin{bmatrix} 1 & 2 & 0 & 0 \\ 1 & 3 & -1 & 0 \\ 0 & -1 & 1 & 3 \\ 0 & 0 & 2 & 3 \end{bmatrix}$

e. $\begin{bmatrix} 1 & 2 & 0 & 0 \\ 1 & 1 & -1 & 0 \\ 0 & -1 & 1 & 3 \\ 0 & 0 & 2 & 3 \end{bmatrix}$

(e) and (d) differ only in row 2, column 2. Note the huge difference!

f. What type of matrices are those in (d) and (e)?

2.10.1   MATLAB Problems

2.11 Use MATLAB to solve the linear algebraic system

$$\begin{bmatrix} 1 & 3 & 8 & 0 \\ -1 & -12 & 3 & 1 \\ 15 & 3 & 5 & 6 \\ 55 & 2 & 35 & 5 \end{bmatrix} \begin{bmatrix} x_1 \\ x_2 \\ x_3 \\ x_4 \end{bmatrix} = \begin{bmatrix} 1 \\ 0 \\ 2 \\ 3 \end{bmatrix}.$$

2.12 Solve the system $Hx = [1\ 1\ \ldots\ 1]^T$, where H is the 20×20 Hilbert matrix discussed in Problem 1.28 of Chapter 1. Now solve the system $Hx = [0.99\ 0.99\ \ldots\ 0.99]^T$. Discuss the results.

In algebra, the polynomial $x^2 - 5x + 6$ can be factored as $(x - 3)(x - 2)$. Under the right conditions, a matrix can also be factored. Exercises 2.13 and 2.14 are designed to introduce you to *matrix factorization*, a topic of great importance in numerical linear algebra.

2.13 A *bidiagonal matrix* is a matrix with nonzero entries along the main diagonal and either the diagonal above or the diagonal below. The matrix $B1$ is an *upper bidiagonal matrix* and $B2$ is a *lower bidiagonal matrix*.

$$B1 = \begin{bmatrix} 5 & 1 & 0 & 0 \\ 0 & 5 & 2 & 0 \\ 0 & 0 & 5 & 3 \\ 0 & 0 & 0 & 5 \end{bmatrix}, \quad B2 = \begin{bmatrix} 5 & 0 & 0 & 0 \\ -1 & 5 & 0 & 0 \\ 0 & -2 & 5 & 0 \\ 0 & 0 & -3 & 5 \end{bmatrix}.$$

A *tridiagonal matrix* has only nonzero entries along the main diagonal and the diagonals above and below. T is a tridiagonal matrix.

$$T = \begin{bmatrix} 5 & -1 & 0 & 0 & 0 \\ 1 & 5 & -1 & 0 & 0 \\ 0 & 1 & 5 & -1 & 0 \\ 0 & 0 & 1 & 5 & -1 \\ 0 & 0 & 0 & 1 & 5 \end{bmatrix}.$$

We will show in Chapter 13 that a nonsingular tridiagonal matrix can be factored into the product of a lower bidiagonal matrix and an upper bidiagonal matrix. The lower bidiagonal matrix has ones on its diagonal.

a. Using pencil and paper, verify that

$$\begin{bmatrix} 1 & 2 & 0 \\ 3 & -1 & 1 \\ 0 & 5 & 2 \end{bmatrix} = \begin{bmatrix} 1 & 0 & 0 \\ 3 & 1 & 0 \\ 0 & -5/7 & 1 \end{bmatrix} \begin{bmatrix} 1 & 2 & 0 \\ 0 & -7 & 1 \\ 0 & 0 & 19/7 \end{bmatrix}.$$

b. Using the MATLAB command `diag`, build the tridiagonal matrix T as follows:

```
>> a = ones(4,1);
>> b = 5*ones(5,1);
>> c = -ones(4,1);
>> T = diag(a,-1) + diag(b) + diag(c,1);
```

The book software distribution supplies the function `trifact` that factors a tridiagonal matrix. Enter the following MATLAB statements and then verify that `T = LU`.

```
>> [L U] = trifact(T);
```

You will learn how to efficiently program `trifact` in Chapter 13.

2.14 If A is an $n \times n$ matrix, consider the product $x^T A x$. If vector x is an $n \times 1$ column vector, then x^T is a $1 \times n$ row vector. The product is of dimension $(1 \times n)(n \times n)(n \times 1) = 1 \times 1$, or a scalar. A symmetric matrix with the property that $x^T A x > 0$ for all $x \neq 0$ is said to be *positive definite*. Positive definite matrices play a role in many fields of engineering and science. We will study these matrices in Chapter 13 and subsequent chapters.

a. Show that the matrix $A = \begin{bmatrix} 2 & 1 \\ 1 & 2 \end{bmatrix}$ is positive definite by showing that

$$[x_1 \ x_2] \, A \begin{bmatrix} x_1 \\ x_2 \end{bmatrix} > 0 \quad \text{for all} \quad \begin{bmatrix} x_1 \\ x_2 \end{bmatrix} \neq \begin{bmatrix} 0 \\ 0 \end{bmatrix}.$$

b. A positive definite matrix can be uniquely factored into the product $R^T R$, where R is an upper-triangular matrix.

If $R = \begin{bmatrix} \sqrt{2} & \frac{1}{\sqrt{2}} \\ 0 & \sqrt{\frac{3}{2}} \end{bmatrix}$ Show that

$$\begin{bmatrix} 2 & 1 \\ 1 & 2 \end{bmatrix} = \begin{bmatrix} \sqrt{2} & 0 \\ \frac{1}{\sqrt{2}} & \sqrt{\frac{3}{2}} \end{bmatrix} \begin{bmatrix} \sqrt{2} & \frac{1}{\sqrt{2}} \\ 0 & \sqrt{\frac{3}{2}} \end{bmatrix}.$$

c. The MATLAB command `gallery` produces many different kinds of matrices to use for testing purposes. Enter the command

```
>> A = gallery('moler',5)
```

It generates a 5×5 positive-definite matrix. The MATLAB command `chol(A)` computes the matrix R. Use it to find the factorization $R^T R$ of A.

2.15 Given the truss problem in Figure 2.5, find the forces necessary to hold the truss in equilibrium.

2.16 **a.** Set up a linear system to determine the currents in Figure 2.6.

  b. Solve for the currents using the following values:

Components	V_1	V_2	R_1	R_2	R_3
	5 V	2 V	2 Ω	4 Ω	1 Ω

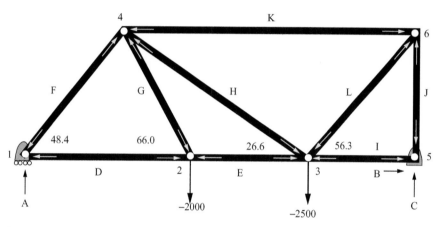

FIGURE 2.5 Truss problem.

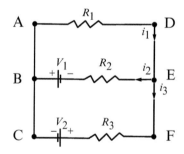

FIGURE 2.6 Circuit problem.

Chapter 3

Subspaces

You should be familiar with

- Vector operations
- Row reducing a matrix
- Solving linear systems using row elimination

3.1 INTRODUCTION

Throughout this chapter, we will be studying $\mathbb{R}^n$, the set of n-dimensional column vectors with real-valued components. We continue our study of matrices by considering a class of subsets of $\mathbb{R}^n$ called *subspaces*. These arise naturally, for example, when we solve a system of n linear homogeneous equations in n unknowns. The *row space, column space*, and *null space* of the coefficient matrix play a role in many applications. We also study the concept of linear independence of a set of vectors, which gives rise to the concept of subspace dimension.

Remark 3.1. We will use the mathematical symbol $\in$ that means "contained in." For instance, $u \in \mathbb{R}^2$ means that u is a vector in the plane, so $u = \begin{bmatrix} u_1 \\ u_2 \end{bmatrix}$, where u_1, u_2 are real numbers.

3.2 SUBSPACES OF $\mathbb{R}^n$

Definition 3.1. A subset S of $\mathbb{R}^n$ is called a subspace of $\mathbb{R}^n$ if

1. The zero vector belongs to S (i.e., $0 \in S$);
2. If $u \in S$ and $v \in S$, then $u + v \in S$ (S is said to be closed under vector addition);
3. If $u \in S$ and $t \in \mathbb{R}$, then $tu \in S$ (S is said to be closed under scalar multiplication).

$\mathbb{R}^n$ is a subspace of itself, and we call $\mathbb{R}^n$ a *vector space*. The complete definition of a vector space is very general, and we will not provide it here. In Chapter 12, we will introduce Fourier series as an example of a vector space whose elements are functions.

Example 3.1. Let A be an $n \times n$ matrix. Then the set of vectors $x \in \mathbb{R}^n$ satisfying $Ax = 0$ is a subspace of $\mathbb{R}^n$ called the *null space* of A and is denoted by $N(A)$.
Verify each property of a subspace.

1. $A \times 0 = 0$, so $0 \in N(A)$.
2. If $x, y \in N(A)$, then $Ax = 0$ and $Ay = 0$, so $A(x + y) = Ax + Ay = 0 + 0 = 0$ and $x + y \in N(A)$.
3. If $x \in N(A)$ and $t \in \mathbb{R}$, then $A(tx) = t(Ax) = t0 = 0$, so $tx \in N(A)$. ∎

Example 3.2. If $A = \begin{bmatrix} 1 & 0 \\ 0 & 1 \end{bmatrix}$, the only solution to $Ax = 0$ is $x = 0$, so $N(A) = \{0\}$, the set consisting of just the zero vector.
The matrix $A = \begin{bmatrix} 1 & 2 \\ 2 & 4 \end{bmatrix}$ row reduces to $\begin{bmatrix} 1 & 2 \\ 0 & 0 \end{bmatrix}$, so $x_1 + 2x_2 = 0$, and $x_1 = -2x_2$. $N(A)$ is the set of all scalar multiples of $\begin{bmatrix} -2 \\ 1 \end{bmatrix}$, a line in the plane. ∎

Numerical Linear Algebra with Applications. http://dx.doi.org/10.1016/B978-0-12-394435-1.00003-X

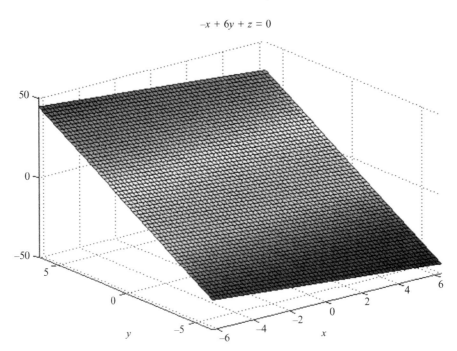

FIGURE 3.1 Subspace spanned by two vectors.

In the vector space $\mathbb{R}^3$, take the vectors $\begin{bmatrix} 1 & 1 & 5 \end{bmatrix}^T$ and $\begin{bmatrix} 2 & -1 & -8 \end{bmatrix}^T$ and form all possible linear combinations $c_1 \begin{bmatrix} 1 & 1 & 5 \end{bmatrix}^T + c_2 \begin{bmatrix} 2 & -1 & -8 \end{bmatrix}^T$. This set of vectors is a plane in $\mathbb{R}^3$ (Figure 3.1). ∎

Definition 3.2. If $x_1, \ldots, x_m$ is a set of vectors in $\mathbb{R}^n$, then an expression of the form $c_1 x_1 + \cdots + c_m x_m$ is said to be a *linear combination* of $x_1, \ldots, x_m$.

Theorem 3.1. *Let $x_1, \ldots, x_m \in \mathbb{R}^n$. Then the set consisting of all linear combinations*

$$c_1 x_1 + \cdots + c_m x_m,$$

where $c_1, \ldots, c_m \in \mathbb{R}$ is a subspace of $\mathbb{R}^n$. This subspace is called the subspace spanned by $x_1, \ldots, x_m$ and is denoted by

$$\text{span} \{x_1, \ldots, x_m\}.$$

Proof. To show that the set of all linear combinations of $x_1, x_2, \ldots, x_m$ is a subspace, we must verify properties 1, 2, and 3 of Definition 3.1.

Property 1: $0 = 0x_1 + \cdots + 0x_m$, so $0 \in \text{span} \{x_1, \ldots, x_m\}$.

Property 2: If $x, y \in \text{span} \{x_1, \ldots, x_m\}$, then $x = c_1 x_1 + \cdots + c_m x_m$ and $y = d_1 x_1 + \cdots + d_m x_m$, so

$$\begin{aligned} x + y &= (c_1 x_1 + \cdots + c_m x) + (d_1 x_1 + \cdots + d_m x_m) \\ &= (c_1 + d_1) x_1 + \cdots + (c_m + d_m) x_n \end{aligned}$$

and $x + y \in \text{span} \{x_1, \ldots, x_m\}$.

Property 3: If $x \in \text{span} \{x_1, \ldots, x_m\}$ and $t \in \mathbb{R}^n$, then $x = c_1 x_1 + \cdots + c_m x_m$, $tx = t(c_1 x_1 + \cdots + c_m x_m) = (tc_1) x_1 + \cdots + (tc_m) x_m \in \text{span} \{x_1, \ldots, x_m\}$. □

Definition 3.3. If A is an $n \times n$ matrix, the subspace spanned by the columns of A is a subspace of $\mathbb{R}^n$, called the *column space* of A. Also, the subspace spanned by the rows of A is a subspace of $\mathbb{R}^n$ called the *row space* of A.

Example 3.3. The $n \times n$ identity matrix has columns

$$e_1 = \begin{bmatrix} 1 & 0 & \ldots & 0 \end{bmatrix}^T, e_2 = \begin{bmatrix} 0 & 1 & \ldots & 0 \end{bmatrix}^T, \ldots, e_{n-1} = \begin{bmatrix} 0 & \ldots & 1 & 0 \end{bmatrix}^T, \begin{bmatrix} 0 & \ldots & 0 & 1 \end{bmatrix}^T.$$

Since $\begin{bmatrix} x_1 & x_2 & \dots & x_{n-1} & x_n \end{bmatrix}^{\mathrm{T}} = x_1 e_1 + x_2 e_2 + \cdots + x_{n-1} e_{n-1} + x_n e_n$, the column space of I is $\mathbb{R}^n$. The e_i are called the *standard basis vectors*. Any vector in $\mathbb{R}^n$ can be written as a linear combination of the standard basis vectors. In a similar fashion, the rows of I span $\mathbb{R}^n$. ∎

Example 3.4. The equation $2x - 3y + 5z = 0$ defines a relationship between the components of a vector $\begin{bmatrix} x & y & z \end{bmatrix}^{\mathrm{T}}$. Find the subspace S of $\mathbb{R}^3$ spanned by all such vectors. If $[x, y, z]^{\mathrm{T}} \in S$, then $x = \frac{3}{2}y - \frac{5}{2}z$, so

$$\begin{bmatrix} x \\ y \\ z \end{bmatrix} = \begin{bmatrix} \frac{3}{2}y - \frac{5}{2}z \\ y \\ z \end{bmatrix} = y \begin{bmatrix} \frac{3}{2} \\ 1 \\ 0 \end{bmatrix} + z \begin{bmatrix} -\frac{5}{2} \\ 0 \\ 1 \end{bmatrix}.$$

Thus, S is the subspace spanned by $\begin{bmatrix} \frac{3}{2} \\ 1 \\ 0 \end{bmatrix}$ and $\begin{bmatrix} -\frac{5}{2} \\ 0 \\ 1 \end{bmatrix}$. This subspace is not $\mathbb{R}^3$. Consider the vector $\begin{bmatrix} 1 \\ 2 \\ 3 \end{bmatrix}$ and determine

if it can be written as a linear combination of $\begin{bmatrix} \frac{3}{2} \\ 1 \\ 0 \end{bmatrix}$ *and* $\begin{bmatrix} -\frac{5}{2} \\ 0 \\ 1 \end{bmatrix}$. There must be scalars c_1 and c_2 such that

$$c_1 \begin{bmatrix} \frac{3}{2} \\ 1 \\ 0 \end{bmatrix} + c_2 \begin{bmatrix} -\frac{5}{2} \\ 0 \\ 1 \end{bmatrix} = \begin{bmatrix} 1 \\ 2 \\ 3 \end{bmatrix}.$$

This requires that

$$\frac{3}{2}c_1 - \frac{5}{2}c_2 = 1.$$

We must have $c_1 = 2$ and $c_2 = 3$, but

$$\frac{3}{2}(2) - \frac{5}{2}(3) = -4\frac{1}{2} \neq 1.$$

The two vectors do not span $\mathbb{R}^3$. In general, it takes n vectors to span $\mathbb{R}^n$. ∎

3.3 LINEAR INDEPENDENCE

Definition 3.4. The concept of linear independence of a set of vectors in $\mathbb{R}^n$ is extremely important in linear algebra and its applications.

Vectors $x_1, \dots, x_m$ in $\mathbb{R}^n$ are said to be *linearly dependent* if there exist scalars $c_1, \dots, c_m$, *not all zero*, such that

$$c_1 x_1 + \cdots + c_m x_m = 0. \tag{3.1}$$

Suppose $c_i \neq 0$. Then, $x_i = -(c_1 x_1 + c_2 x_2 + \cdots + c_{i-1} x_{i-1} + c_{i+1} x_{i+1} + \cdots + c_m x_m)/c_i$. The vector x_i can be written as a linear combination of the remaining vectors; in other words, it is dependent on them. The vectors $x_1, \dots, x_m$ are called *linearly independent* if they are not linearly dependent. To test for linear independence, Equation 3.1 is a linear homogeneous equation with unknowns $\begin{bmatrix} c_1 & c_2 & \dots & c_{m-1} & c_m \end{bmatrix}^{\mathrm{T}}$. The vectors are linearly independent if the system has only the trivial solution $c_1 = 0, \dots, c_m = 0$. Conversely, if $x_1, x_2, \dots, x_m$ are linearly independent, then the homogeneous system has only the trivial solution.

Example 3.5. Are the following three vectors in $\mathbb{R}^3$ linearly independent or dependent?

$$x_1 = \begin{bmatrix} 1 \\ 2 \\ 3 \end{bmatrix}, \quad x_2 = \begin{bmatrix} -1 \\ 1 \\ 2 \end{bmatrix}, \quad x_3 = \begin{bmatrix} -1 \\ 7 \\ 12 \end{bmatrix}.$$

Form

$$c_1 \begin{bmatrix} 1 \\ 2 \\ 3 \end{bmatrix} + c_2 \begin{bmatrix} -1 \\ 1 \\ 2 \end{bmatrix} + c_3 \begin{bmatrix} -1 \\ 7 \\ 12 \end{bmatrix} = \begin{bmatrix} 0 \\ 0 \\ 0 \end{bmatrix}.$$

This corresponds to the homogeneous system.

$$\begin{bmatrix} 1 & -1 & -1 \\ 2 & 1 & 7 \\ 3 & 2 & 12 \end{bmatrix} \begin{bmatrix} c_1 \\ c_2 \\ c_3 \end{bmatrix} = 0$$

$$\begin{bmatrix} 1 & -1 & -1 \\ 2 & 1 & 7 \\ 3 & 2 & 12 \end{bmatrix} \xrightarrow[R3 = R3 - 3R1]{R2 = R2 - 2R1} \begin{bmatrix} 1 & -1 & -1 \\ 0 & 3 & 9 \\ 0 & 5 & 15 \end{bmatrix} \xrightarrow{R3 = R3 - (5/3)R2} \begin{bmatrix} 1 & -1 & -1 \\ 0 & 3 & 9 \\ 0 & 0 & 0 \end{bmatrix}.$$

The row of zeros in the row-reduced matrix indicates that there are infinitely many solutions to the homogeneous system, so x_1, x_2, x_3 are linearly dependent. ∎

Example 3.6. Are the vectors $u = \begin{bmatrix} 1 \\ 2 \\ -1 \end{bmatrix}$, $v = \begin{bmatrix} 1 \\ -1 \\ 3 \end{bmatrix}$, $w = \begin{bmatrix} 1 \\ 2 \\ 3 \end{bmatrix}$ linearly independent? Let $c_1 u + c_2 v + c_3 w = 0$.
This corresponds to the linear homogeneous system

$$\begin{bmatrix} 1 & 1 & 1 \\ 2 & -1 & 2 \\ -1 & 3 & 3 \end{bmatrix} \begin{bmatrix} c_1 \\ c_2 \\ c_3 \end{bmatrix} = 0.$$

$$\begin{bmatrix} 1 & 1 & 1 \\ 2 & -1 & 2 \\ -1 & 3 & 3 \end{bmatrix} \xrightarrow[R3 = R3 - (-1)R1]{R2 = R2 - 2R1} \begin{bmatrix} 1 & 1 & 1 \\ 0 & -3 & 0 \\ 0 & 4 & 4 \end{bmatrix} \xrightarrow{R3 = R3 - (-4/3)R2} \begin{bmatrix} 1 & 1 & 1 \\ 0 & -3 & 0 \\ 0 & 0 & 4 \end{bmatrix}$$

The final system in the row-elimination process has the unique solution $c_1 = c_2 = c_3 = 0$, so u, v, and w are linearly independent. ∎

3.4 BASIS OF A SUBSPACE

We now come to the fundamental concept of a *basis* for a subspace.

Definition 3.5. Vectors $x_1, \ldots, x_m$ belonging to a subspace S are said to form a *basis* for S if

1. $x_1, \ldots, x_m$ span S.
2. $x_1, \ldots, x_m$ are linearly independent.

Example 3.7. The standard basis vectors $e_1, \ldots, e_n$ form a basis for $\mathbb{R}^n$. This is the reason for the term "standard basis."
If $x = \begin{bmatrix} x_1 \\ \vdots \\ x_n \end{bmatrix}$, then $x = x_1 e_1 + x_2 e_2 + \cdots + x_n e_n$, so e_1, e_2, ..., e_n span $\mathbb{R}^n$. They are linearly independent, since if

$$c_1 e_1 + c_2 e_2 + \cdots + c_n e_n = \begin{bmatrix} c_1 \\ c_2 \\ \vdots \\ c_n \end{bmatrix} = 0, \text{ then } c_1 = c_2 = \cdots = c_n = 0.$$ ∎

A subspace normally has more than one basis; for instance, let u, v, and w be the linearly independent vectors of Example 3.6. To show that the vectors are a basis for $\mathbb{R}^3$, it is necessary to show that the vectors span $\mathbb{R}^3$. Let x be any vector in $\mathbb{R}^3$. There must be a linear combination of u, v, and w that equals x; in other words, there must be scalars c_1, c_2, c_3, such that $c_1 u + c_2 v + c_3 w = x$. This is a system of linear equations

$$\begin{bmatrix} 1 & 1 & 1 \\ 2 & -1 & 2 \\ -1 & 3 & 3 \end{bmatrix} \begin{bmatrix} c_1 \\ c_2 \\ c_3 \end{bmatrix} = \begin{bmatrix} x_1 \\ x_2 \\ x_3 \end{bmatrix}.$$

Form the augmented matrix $\begin{bmatrix} 1 & 1 & 1 & x_1 \\ 2 & -1 & 2 & x_2 \\ -1 & 3 & 3 & x_3 \end{bmatrix}$. The row-reduction operations performed in Example 3.6 show that there

is a unique solution for $\begin{bmatrix} c_1 \\ c_2 \\ c_3 \end{bmatrix}$, and so u, v, and w form a basis for $\mathbb{R}^3$.

There are some important properties of a basis that are stated in Theorem 3.2.

Theorem 3.2.   *If S is a subspace of $\mathbb{R}^n$, then*

1. *Each vector in a basis for S must be nonzero.*
2. *If u is a vector in S, there is one and only one way to write u as a linear combination of basis vectors for S.*
3. *A subspace span $\{x_1, \ldots, x_m\}$, where at least one of $x_1, \ldots, x_m$ is nonzero, has a basis $v_1, \ldots, v_p$, where $p \leq m$.*

Proof. To prove 1, assume that $x_1, \ldots, x_m$ is a basis for S, and that $x_1 = 0$. Then we have the nontrivial linear combination $1(x_1) + 0(x_2) + \cdots + 0(x_m) = 0$, and $x_1, \ldots, x_m$ are linearly dependent.

Let $x_1, \ldots, x_m$ be a basis for S. Assume that $u = c_1 x_1 + c_2 x_2 + \cdots + c_m x_m$ and that $u = d_1 x_1 + d_2 x_2 + \cdots + d_m x_m$. Subtract the two equations to obtain

$$(c_1 - d_1) x_1 + (c_2 - d_2) x_2 + \cdots + (c_m - d_m) x_m = 0.$$

Since $x_1, \ldots, x_m$ are linearly independent, $(c_1 - d_1) = 0$, $(c_2 - d_2) = 0$, $\ldots$, $(c_m - d_m) = 0$, and $c_i = d_i$, $1 \leq i \leq m$. We have proved 2.

Statement 3 tells us that the span of any set of vectors has a basis as long as all the vectors are not zero. Scan $x_1, \ldots, x_m$ from left to right and let x_{j_1} be the first nonzero vector. If $j_1 = m$ or all the vectors x_k, $j_1 + 1 \leq k \leq m$ are multiples of x_{j_1}, then $p = 1$. Otherwise, let x_{j_2} be the next nonzero vector following x_{j_1} such that x_{j_2} is not a multiple of x_{j_1}. If $j_2 = m$ or if all the vectors x_k, $j_2 + 1 \leq k \leq m$ are linear combinations of x_{j_1} and x_{j_2}, then $p = 2$. This argument will eventually terminate with a set of vectors $x_{j_1}, x_{j_2}, \ldots, x_{j_p}$ that must be linearly independent. Any vectors among $x_1, \ldots, x_m$ that were combinations of other vectors have been eliminated, so $x_{j_1}, x_{j_2}, \ldots, x_{j_p}$ spans S, and is a basis for S. $\square$

Example 3.8.   Let x and y be linearly independent vectors in $\mathbb{R}^n$. Consider the subspace span $\{0, 2x, x, -y, x + y\}$. Apply the technique used in proving part 3 of Theorem 3.1. Skip over 0 and record $2x$ as a nonzero vector. Move right and discard x because it is a multiple of $2x$. The next vector, $-y$, is not a multiple of x because x and y are linearly independent. The final vector $x + y$ is a linear combination of $2x$ and $-y$. The subspace $\{0, 2x, x, -y, x + y\}$ has a basis $2x$, $-y$.   ■

By using arguments similar to those in Theorem 3.2, the following claims can be proved. The proofs will not be provided, but the interested reader can consult Strang [8, pp. 175-175] for very readable arguments.

Claim 3.1.   If S is a subspace of $\mathbb{R}^n$, then

1. Any two bases for a subspace S must contain the same number of elements. This number is called the *dimension* of S and is written dim S.
2. If a subspace has dimension m, then any set of m linearly independent vectors is a basis for S.

3.5   THE RANK OF A MATRIX

In this section, we will determine how to find a basis for the row space of a matrix and discuss the relationship between the row space and the column space of a matrix.

Definition 3.6.   The number of elements in a basis for the row space is called the *rank* of the matrix.

The rank is defined for any $m \times n$ matrix, but we will deal only with square matrices for now. The process we develop to find the rank of a matrix will involve row reductions, but we will go beyond just getting to upper-triangular form and will also "zero out" as many elements in the upper triangle as we can. The process is illustrated with examples and is based upon the following lemma and a theorem that follows from it.

Lemma 3.1. *Subspaces* span $\{x_1, x_2, \ldots, x_r\}$ *and* span $\{y_1, y_2, \ldots, y_s\}$ *are equal if each of $x_1, x_2, \ldots, x_r$ is a linear combination of $y_1, y_2, \ldots, y_s$ and each of $y_1, y_2, \ldots, y_s$ is a linear combination of $x_1, x_2, \ldots, x_r$.*

Proof. Let $x = c_1 x_1 + \ldots + c_r$. Since each of $x_1, x_2, \ldots, x_r$ is a linear combination of $y_1, y_2, \ldots, y_s$, it follows that x is a linear combination of $y_1, y_2, \ldots, y_s$. Similarly, if $y = d_1 y_1 + \cdots + d_s y_s$, then y is a linear combination of $x_1, x_2, \ldots, x_r$. This shows the two subspaces are equal. $\square$

Theorem 3.3. *The row space of a matrix is the same as the row space of any matrix derived from it using row reduction.*

Proof. Suppose that matrix B is obtained from matrix A by a sequence of elementary row operations. Then each row of B is a linear combination of the rows of A. But A can be obtained from B by a sequence of elementary row operations, so each row of A is a linear combination of the rows of B. By Lemma 3.1, the two row spaces are equal. $\square$

Example 3.9. Let $A = \begin{bmatrix} 1 & 2 \\ 2 & 4 \end{bmatrix}$. The upper-triangular form for A is $B = \begin{bmatrix} 1 & 2 \\ 0 & 0 \end{bmatrix}$, and we cannot eliminate any more elements. The row space of A and B is the same and consists of all multiples of the vector $\begin{bmatrix} 1 & 2 \end{bmatrix}$. Hence, $\begin{bmatrix} 1 & 2 \end{bmatrix}$ is a basis for the row space of A and the rank of A is 1.

Find the row space and rank of $A = \begin{bmatrix} 1 & 2 \\ 3 & 4 \end{bmatrix}$. We begin row reduction with

$$\begin{bmatrix} 1 & 2 \\ 3 & 4 \end{bmatrix} \xrightarrow{R2 = R2 - 3R1} \begin{bmatrix} 1 & 2 \\ 0 & -2 \end{bmatrix}.$$

Continue on and use the -2 in row 2, column 2 to eliminate the element above it in row 1, column 2. Also, remember we can multiply any row by a constant during row reduction.

$$\begin{bmatrix} 1 & 2 \\ 0 & -2 \end{bmatrix} \xrightarrow{R1 = R1 - (-1)R2} \begin{bmatrix} 1 & 0 \\ 0 & -2 \end{bmatrix} \xrightarrow{R2 = R2/(-2)} \begin{bmatrix} 1 & 0 \\ 0 & 1 \end{bmatrix}$$

The row space consists of all linear combinations of the vectors $\begin{bmatrix} 1 & 0 \end{bmatrix}$ and $\begin{bmatrix} 0 & 1 \end{bmatrix}$, and so the row space is $\mathbb{R}^2$, and the rank of A is 2. ∎

Example 3.10. Consider $A = \begin{bmatrix} 1 & 0 & 2 \\ 0 & 1 & 3 \\ 1 & 4 & 5 \end{bmatrix}$. Perform row reductions to determine a basis for the row space and the rank of A.

$$\begin{bmatrix} 1 & 0 & 2 \\ 0 & 1 & 3 \\ 1 & 4 & 5 \end{bmatrix} \xrightarrow{R3 = R3 - (1)R1} \begin{bmatrix} 1 & 0 & 2 \\ 0 & 1 & 3 \\ 0 & 4 & 3 \end{bmatrix} \xrightarrow{R3 = R3 - 4R2} \begin{bmatrix} 1 & 0 & 2 \\ 0 & 1 & 3 \\ 0 & 0 & -9 \end{bmatrix}$$

This is upper-triangular form, but continue eliminating as many elements as we can. To make things easier, divide row 3 by -9.

$$\begin{bmatrix} 1 & 0 & 2 \\ 0 & 1 & 3 \\ 0 & 0 & -9 \end{bmatrix} \xrightarrow{R3 = R3/(-9)} \begin{bmatrix} 1 & 0 & 2 \\ 0 & 1 & 3 \\ 0 & 0 & 1 \end{bmatrix} \xrightarrow[R1 = R1 - 2R3]{R2 = R2 - 3R3} \begin{bmatrix} 1 & 0 & 0 \\ 0 & 1 & 0 \\ 0 & 0 & 1 \end{bmatrix}$$

The row space of A is $\mathbb{R}^3$, and the rank of A is 3. Note that this matrix has no null space; in other words, the null space of A is the empty set. ∎

Example 3.11. Let $A = \begin{bmatrix} 2 & 5 & -4 & 1 \\ 3 & 8 & -9 & 2 \\ 1 & 1 & 7 & -1 \\ 1 & 2 & 1 & 0 \end{bmatrix}$. Perform row reductions to determine a basis for the row space and the rank

of A.

$$\begin{bmatrix} 2 & 5 & -4 & 1 \\ 3 & 8 & -9 & 2 \\ 1 & 1 & 7 & -1 \\ 1 & 2 & 1 & 0 \end{bmatrix} \xrightarrow[\substack{R2 = R2 - (3/2)\,R1 \\ R3 = R3 - (1/2)\,R1 \\ R4 = R4 - (1/2)\,R1}]{} \begin{bmatrix} 2 & 5 & -4 & 1 \\ 0 & \frac{1}{2} & -3 & \frac{1}{2} \\ 0 & -\frac{3}{2} & 9 & -\frac{3}{2} \\ 0 & -\frac{1}{2} & 3 & -\frac{1}{2} \end{bmatrix} \xrightarrow[\substack{R3 = R3 - (-3)\,R2 \\ R4 = R4 - (-1)\,R2}]{} \begin{bmatrix} 2 & 5 & -4 & 1 \\ 0 & \frac{1}{2} & -3 & \frac{1}{2} \\ 0 & 0 & 0 & 0 \\ 0 & 0 & 0 & 0 \end{bmatrix}$$

Now eliminate the 5 in row 1, column 2.

$$\begin{bmatrix} 2 & 5 & -4 & 1 \\ 0 & \frac{1}{2} & -3 & \frac{1}{2} \\ 0 & 0 & 0 & 0 \\ 0 & 0 & 0 & 0 \end{bmatrix} \xrightarrow[R1 = R1 - (10)\,R2]{} \begin{bmatrix} 2 & 0 & 26 & -4 \\ 0 & \frac{1}{2} & -3 & \frac{1}{2} \\ 0 & 0 & 0 & 0 \\ 0 & 0 & 0 & 0 \end{bmatrix}.$$

We cannot go any further without removing the zero we just produced; however, let's multiply row 1 by $\frac{1}{2}$ and row 2 by 2 to make the leading element of each nonzero row 1.

$$\begin{bmatrix} 2 & 0 & 26 & -4 \\ 0 & \frac{1}{2} & -3 & \frac{1}{2} \\ 0 & 0 & 0 & 0 \\ 0 & 0 & 0 & 0 \end{bmatrix} \xrightarrow[\substack{R1 = (1/2)\,R1 \\ R2 = (2)\,R2}]{} \begin{bmatrix} 1 & 0 & 13 & -2 \\ 0 & 1 & -6 & 1 \\ 0 & 0 & 0 & 0 \\ 0 & 0 & 0 & 0 \end{bmatrix}.$$

The rank of A is 2, and $\begin{bmatrix} 1 & 0 & 13 & -2 \end{bmatrix}, \begin{bmatrix} 0 & 1 & -6 & 1 \end{bmatrix}$ are a basis for the row space. ∎

This process of finding the rank of a matrix seems somewhat disorganized. In reality, the rank of a matrix is found by computing the singular value decomposition (SVD). We will discuss the SVD in Chapter 15, and will show how to compute it accurately and efficiently in Chapter 23.

Example 3.12.   We have computed the row space for some matrices, and now we will find the null space and nullity of a matrix. Let A be the matrix of Example 3.11. The null space of A is the subspace of vectors x such that $Ax = 0$. Using the computations in Exercise 3.11, we have

$$A = \begin{bmatrix} 2 & 5 & -4 & 1 \\ 3 & 8 & -9 & 2 \\ 1 & 1 & 7 & -1 \\ 1 & 2 & 1 & 0 \end{bmatrix} \implies \begin{bmatrix} 1 & 0 & 13 & -2 \\ 0 & 1 & -6 & 1 \\ 0 & 0 & 0 & 0 \\ 0 & 0 & 0 & 0 \end{bmatrix}.$$

This says that

$$x_1 + 13x_3 - 2x_4 = 0,$$
$$x_2 - 6x_3 + x_4 = 0.$$

From these equations,
$x_1 = -13x_3 + 2x_4,\ x_2 = 6x_3 - x_4$. The variables x_4 and x_3 are arbitrary, so a basis for the null space of A is

$$z_1 = \begin{bmatrix} -13 \\ 6 \\ 1 \\ 0 \end{bmatrix}, \quad z_2 = \begin{bmatrix} 2 \\ -1 \\ 0 \\ 1 \end{bmatrix}.$$

Note the presence of two zero rows and a nullity of two. ∎

Find the rank of a matrix using the MATLAB command `rank(A)`. The command `Z = null(A)` is used to find a basis for the null space of a matrix as a set of column vectors that form the matrix Z. If you only want the nullity of the matrix, obtain the number of columns of the matrix returned by `null` using `size(null(A),2)`.

Example 3.13. Note that in the MATLAB output, $(A*Z(:,1))$' should be a row vector of zeros, but very small nonzero vector components are the result. In Chapter 8, we will discuss why this occurs.

```
>> A = [2 5 -4 1;3 8 -9 2;1 1 7 -1;1 2 1 0];
>> rank(A)

ans =
     2
>> Z = null(A);
>> (A*Z(:,1))'

ans =
  1.0e-014 *

  -0.0860   -0.1277   -0.0583   -0.0472

>> size(null(A),2)

ans =
     2
```

■

Notice that in Example 3.13, the rank of the matrix is 2 and the nullity is 2, so rank + nullity = 4, the matrix size. This is an example of the relation between the dimension of the row space of a matrix and its null space.

Theorem 3.4. *If A is an n×n matrix, the rank of A plus the nullity of A equals n.*

We will prove this result in Chapter 15 when we present the singular value decomposition of a matrix.

Remark 3.2. An $n \times n$ matrix A is said to have *full rank* if the rank of A is n. It follows from Theorem 3.4 that if a matrix has full rank, then there is no null space.

Example 3.14. Let A be the matrix in Examples 3.12 and 3.13. As stated in Definition 3.1, the column space of a matrix A is the subspace spanned by the columns of A. We will find a basis for the column space of A, the subspace spanned by $\begin{bmatrix} 2 \\ 3 \\ 1 \\ 1 \end{bmatrix}, \begin{bmatrix} 5 \\ 8 \\ 1 \\ 2 \end{bmatrix}, \begin{bmatrix} -4 \\ -9 \\ 7 \\ 1 \end{bmatrix}, \begin{bmatrix} 1 \\ 2 \\ -1 \\ 0 \end{bmatrix}$. Form A^{T}, the matrix whose rows are the columns of A and perform row operations on A^{T}.

$$\begin{bmatrix} 2 & 3 & 1 & 1 \\ 5 & 8 & 1 & 2 \\ -4 & -9 & 7 & 1 \\ 1 & 2 & -1 & 0 \end{bmatrix} \xrightarrow[\substack{R2 = R2 - (5/2)R1 \\ R3 = R3 - (-2)R1 \\ R4 = R4 - (1/2)R1}]{} \begin{bmatrix} 2 & 3 & 1 & 1 \\ 0 & 1/2 & -3/2 & -1/2 \\ 0 & -3 & 9 & 3 \\ 0 & 1/2 & -3/2 & -1/2 \end{bmatrix}$$

$$\begin{bmatrix} 2 & 3 & 1 & 1 \\ 0 & 1/2 & -3/2 & -1/2 \\ 0 & -3 & 9 & 3 \\ 0 & 1/2 & -3/2 & -1/2 \end{bmatrix} \xrightarrow[\substack{R3 = R3 - (-6)R2 \\ R4 = R4 - (1)R2}]{} \begin{bmatrix} 2 & 3 & 1 & 1 \\ 0 & 1/2 & -3/2 & -1/2 \\ 0 & 0 & 0 & 0 \\ 0 & 0 & 0 & 0 \end{bmatrix}$$

$$\begin{bmatrix} 2 & 3 & 1 & 1 \\ 0 & 1/2 & -3/2 & -1/2 \\ 0 & 0 & 0 & 0 \\ 0 & 0 & 0 & 0 \end{bmatrix} \xrightarrow[R1 = R1 - (6)R2]{} \begin{bmatrix} 2 & 0 & 10 & 4 \\ 0 & 1/2 & -3/2 & -1/2 \\ 0 & 0 & 0 & 0 \\ 0 & 0 & 0 & 0 \end{bmatrix}$$

This is as far as we can go, so a basis for the column space of A is

$$\begin{bmatrix} 2 \\ 0 \\ 10 \\ 4 \end{bmatrix}, \begin{bmatrix} 0 \\ 1/2 \\ -3/2 \\ -1/2 \end{bmatrix}.$$

■

Notice that the dimension of the column space of A in Example 3.14 is the same as the dimension of the row space of A. This is true for any $m \times n$ matrix, and we will prove this in Chapter 15.

3.6 CHAPTER SUMMARY

Subspaces of $\mathbb{R}^n$

$\mathbb{R}^n$ is a vector space, and subsets of $\mathbb{R}^n$, including $\mathbb{R}^n$ itself, are called subspaces. A subspace satisfies three properties:

- The zero vector is in the subspace.
- If x is a vector in the subspace, then cx is in the subspace, where c is a number.
- If x and y are in the subspace, then so is $x + y$.

If $x_1, x_2, \ldots, x_k$ are in $\mathbb{R}^n, k \leq n$, then the set of all linear combinations $c_1 x_1 + c_2 x_2 + \cdots + c_k x_k$ is a subspace, and is said to be the subspace, S, of $\mathbb{R}^n$ spanned by $x_1, \ldots, x_k$, and we write

$$S = \text{span}\{x_1, \ldots, x_k\}.$$

As an example, let $x_1 = \begin{bmatrix} -1 \\ 0 \\ 5 \end{bmatrix}$ and $x_2 = \begin{bmatrix} 3 \\ 1 \\ -8 \end{bmatrix}$. The subspace spanned by these vectors is a plane in three-space.

Important examples of subspaces include the column space, row space, and the null space of a matrix.

Linear Independence

A collection of vectors is linearly independent if no vector can be written as a linear combination of the others. For instance, if $u = \begin{bmatrix} 1 \\ 0 \\ -1 \end{bmatrix}, v = \begin{bmatrix} 2 \\ 7 \\ 3 \end{bmatrix}, w = \begin{bmatrix} 1 \\ 0 \\ 0 \end{bmatrix}$, then these vectors are linearly independent. But, if we let $v = \begin{bmatrix} 4 \\ 0 \\ -1 \end{bmatrix}$, then $v = u + 3w$, and the vectors are linearly dependent.

In general, a set of vectors $v_1, \ldots, v_k$ is linearly independent if

$$c_1 v_1 + c_2 v_2 + \cdots + c_k v_k = 0$$

only when $c_1 = c_2 = \cdots = c_k = 0$. This means that the homogeneous system

$$\begin{bmatrix} v_1 & v_2 & \ldots & v_{k-1} & v_k \end{bmatrix} \begin{bmatrix} c_1 \\ c_2 \\ \vdots \\ c_{k-1} \\ c_k \end{bmatrix} = 0$$

has only the zero solution.

Basis of a Subspace

The basis for a subspace is a set of vectors that span the subspace and are linearly independent. For instance, the standard basis for $\mathbb{R}^n$ is

$$e_1 = \begin{bmatrix} 1 & \ldots & 0 & 0 \end{bmatrix}^{\text{T}}, e_2 = \begin{bmatrix} 0 & 1 & \ldots & 0 \end{bmatrix}^{\text{T}}, \ldots, e_n = \begin{bmatrix} 0 & \ldots & 0 & 1 \end{bmatrix}^{\text{T}}.$$

A subspace can have more than one basis. For instance, the vectors u, v, w in Section "Linear Independence" are a basis for $\mathbb{R}^3$, as is e_1, e_2, e_3. However, all bases for a subspace have the same number of vectors, and this number is the dimension of the subspace.

Matrix Rank

The rank of a matrix is the dimension of the subspace spanned by its rows. As we will prove in Chapter 15, the dimension of the column space is equal to the rank. This has important consequences; for instance, if A is an $m \times n$ matrix and $m \geq n$, then rank $(A) \leq n$, but if $m < n$, then rank $(A) \leq m$. It follows that if a matrix is not square, either its columns or its rows must be linearly dependent.

For small square matrices, perform row elimination in order to obtain an upper-triangular matrix. If a row of zeros occurs, the rank of the matrix is less than n, and it is singular. As we will see in Chapters 7, 15, and 23, finding the rank of an arbitrary matrix is somewhat complex and relies on the computation of what are termed its singular values.

For any $m \times n$ matrix, rank $(A) +$ nullity $(A) = n$. Thus, if A is $n \times n$, then for A to be nonsingular, nullity (A) must be zero.

3.7 PROBLEMS

3.1 Which of the following subsets of $\mathbb{R}^2$ are subspaces?

a. $\begin{bmatrix} x \\ y \end{bmatrix}$ satisfying $x = 2y$;

b. $\begin{bmatrix} x \\ y \end{bmatrix}$ satisfying $x = 2y$ and $2x = y$;

c. $\begin{bmatrix} x \\ y \end{bmatrix}$ satisfying $x = 2y+1$;

d. $\begin{bmatrix} x \\ y \end{bmatrix}$ satisfying $xy = 0$;

e. $\begin{bmatrix} x \\ y \end{bmatrix}$ satisfying $x \geq 0$ and $y \geq 0$.

3.2 Determine if $x_1 = \begin{bmatrix} 1 \\ 0 \\ 1 \\ 2 \end{bmatrix}$, $x_2 = \begin{bmatrix} 0 \\ 1 \\ 1 \\ 2 \end{bmatrix}$, $x_3 = \begin{bmatrix} 1 \\ 1 \\ 1 \\ 3 \end{bmatrix}$, and $x_4 = \begin{bmatrix} 0 \\ 1 \\ 4 \\ 5 \end{bmatrix}$ are linearly independent in $\mathbb{R}^4$.

3.3 For which real numbers λ are the following vectors linearly independent in $\mathbb{R}^3$?

$$x_1 = \begin{bmatrix} \lambda \\ -1 \\ -1 \end{bmatrix}, \quad x_2 = \begin{bmatrix} -1 \\ \lambda \\ -1 \end{bmatrix}, \quad x_3 = \begin{bmatrix} -1 \\ -1 \\ \lambda \end{bmatrix}.$$

3.4 Find a basis for the row space of the following matrix. What is the rank of A?

$$A = \begin{bmatrix} 1 & 1 & 2 & 0 \\ 2 & 2 & 5 & 0 \\ 0 & 0 & 0 & 1 \\ 8 & 11 & 19 & 0 \end{bmatrix}.$$

3.5 Find a basis for the row space of the following matrix. What is the rank of A? Find a basis for the column space of A.

$$A = \begin{bmatrix} 1 & 0 & 1 & 0 \\ 0 & 1 & 0 & 1 \\ 1 & 1 & 1 & 1 \\ 0 & 0 & 1 & 1 \end{bmatrix}.$$

3.6 Determine the rank of A, and find a basis for the column space of A.

$$A = \begin{bmatrix} 1 & 0 & 1 & -1 \\ 0 & 1 & 2 & -4 \\ 1 & 0 & 1 & -1 \\ 1 & 1 & 3 & -5 \end{bmatrix}.$$

3.7 Find a basis for the subspace S of $\mathbb{R}^3$ defined by the equation

$$x + 2y + 3z = 0.$$

Verify that $y_1 = [-1, -1, 1]^T \in S$, and find a basis for S that includes y_1.

3.8 Find the null space of the matrix

$$A = \begin{bmatrix} 1 & 7 & 2 \\ -1 & 23 & 8 \\ 3 & 6 & 1 \end{bmatrix}.$$

3.9 If A is a 4 × 4 nonsingular matrix, what can you say about the columns of A?

3.10 This problem deals with the very important concept of *matrix range*.
   a. The *range* of an $n \times n$ matrix A is set of all vectors Ax as x varies through all vectors in $\mathbb{R}^n$. Describe the range in terms of the rows or columns of A. How would you find a basis for the range of A?
   b. Find a basis for the range of the matrix $A = \begin{bmatrix} 1 & 6 & 2 \\ -1 & 3 & 0 \\ -2 & 15 & 2 \end{bmatrix}$.

3.11 The *left nullspace* of a matrix A is the set of all vectors x such that $x^T A = 0$.
   a. Find the left nullspace of the matrix A in Problem 3.10.
   b. Find the nullspace of A^T.
   c. Prove that if A is an $n \times n$ matrix, the left nullspace of A is equal to the nullspace of A^T.

3.12 If $u_1, u_2, \ldots, u_n$ is a basis for $\mathbb{R}^n$ and A is an invertible $n \times n$ matrix, show that $Au_1, Au_2, \ldots, Au_n$ is also a basis for $\mathbb{R}^n$.

3.13 If v_1, v_2, v_3 are linearly independent, for what values of c are the vectors $v_2 - v_1$, $cv_3 - v_2$ and $v_1 - v_3$ linearly independent?

3.14 Prove that if A is an $n \times n$ matrix, rank $(A) = $ rank $\left(A^T \right)$.

3.15 Given $n \times n$ matrices A and B, prove that rank $(AB) \leq$ min {rank (A), rank (B)}.
   Hint: Does $ABx = A(Bx)$ give you a portion of the required result? The result of Problem 3.14 says that rank $(AB) = $ rank $\left([AB]^T \right)$

3.16 **a.** If a matrix A has dimension $m \times n$, with $m > n$, what is the maximum rank of A?
   b. If a matrix A has dimension $m \times n$ with $m < n$, what is the maximum rank of A?
   c. Assuming that the MATLAB function rand returns a matrix of random numbers, if the computation

$$\text{rank (rand } (k, 5))$$

is executed for $k = 1, 2, \ldots, 8$, what output do you expect?

3.7.1 MATLAB Problems

3.17 An $n \times n$ matrix is *rank deficient* if its rank is less than n. The 8 × 8 rosser matrix is often used for testing, and we will have occasion to use it later in this book. Assign A to be the rosser matrix using the MATLAB command A = rosser.
   a. Is A rank deficient?
   b. Using your answer to (a), is A invertible? Try to find the inverse using the MATLAB command inv(A).
   c. Is A symmetric?

3.18 Let $D = \begin{bmatrix} 1 & 3 & 7 & 1 & 1 & 1 & 5 & 0 \\ 2 & 2 & 6 & 0 & 1 & 7 & 3 & 1 \\ 1 & 1 & 3 & 1 & 0 & 2 & 1 & 0 \end{bmatrix}$.

The MATLAB functions rank and null apply to any matrix. Obtain the rank of a matrix D using rank(D). The function null has two variations, and for this problem use null(A,'r'). The parameter "r" specifies that MATLAB is to use row reduction to determine a basis for the null space. Use MATLAB to find the following:

a. the nullity of D.

b. a basis for the null space of D

c. the rank of D

3.19 MATLAB has a suite of matrices used for testing software and is accessed using the gallery command. The first argument is a string naming the matrix.

a. Assign A to be the clement matrix using the command A = gallery('clement', 5). What is the rank of A? Is A rank deficient?

b. Is A singular?

c. Execute the commands B = A; B(1,1) = 1.0e-14; B(2,2) = 1.0e-14; B(3,3) = 1.0e-14. What is the rank of B?

d. Using the MATLAB command inv, find the inverse of B.

e. Is it reasonable to say that B is nearly rank deficient?

3.20 Let $A = \begin{bmatrix} 1 & 2 & 3 & 4 \\ 5 & 6 & 7 & 8 \\ 9 & 10 & 11 & 12 \\ 13 & 14 & 15 & 16 \end{bmatrix}$.

a. What is the rank of A?

b. Find the rank of $U = \begin{bmatrix} 1 & 1 & 0 & 0 \\ 0 & 1 & 1 & 0 \\ 0 & 0 & 1 & 1 \\ 0 & 0 & 0 & 1 \end{bmatrix}$.

c. What is the rank of UA and AU? Does this support the result of Problem 3.15?

3.21 Let $A = \begin{bmatrix} 1 & 2 & 3 & 4 \\ 5 & 6 & 7 & 8 \\ 9 & 10 & 11 & 12 \\ 13 & 14 & 15 & 16 \end{bmatrix}$, $B = \begin{bmatrix} -1 & 1 & 2 \\ 4 & 5 & 6 \\ 7 & 8 & 9 \\ 10 & 11 & 12 \end{bmatrix}$.

a. Compute the rank of A and the rank of B.

b. Compute the rank of the square matrices $A^T A$ and $B^T B$.

c. Is there a general relationship you can postulate by considering the specific results in parts (a) and (b)?

Chapter 4

Determinants

You should be familiar with

- Matrix row elimination
- Linear system of equations

The determinant is defined for any $n \times n$ matrix and produces a scalar value. You have probably dealt with determinants before, possibly while using Cramer's rule. The determinant has many theoretical uses in linear algebra. Among these is the definition of eigenvalues and eigenvectors, as we will see in Chapter 5. In Section 4.1, we will develop a formula for the inverse of a matrix that involves a determinant. A matrix is invertible if its determinant is not zero. In vector calculus, the Jacobian matrix is the matrix of all first-order partial derivatives of a multivariate function. The determinant of the Jacobian matrix, called the Jacobian, is used in multivariable calculus.

Although the determinant has theoretical uses, because of the complexity of its calculation, a determinant is seldom used in practice unless the matrix size is small. In this chapter, we will begin with the definition of the determinant and then discuss its evaluation using expansion by minors and Gaussian elimination. This chapter ends with an interesting example where determinants play a role in file encryption.

4.1 DEVELOPING THE DETERMINANT OF A 2 × 2 AND A 3 × 3 MATRIX

The determinant of an $n \times n$ matrix is the sum of all possible products of n elements formed by choosing one element from each row in the order 1, 2, ..., n in different columns along with the proper sign. The sign is found by writing down the sequence of column indices in each product and counting the number of interchanges necessary to put the column indices in the order 1, 2, ..., n. If the number of interchanges is even, the sign is $+$; otherwise, the sign is $-$.

Example 4.1. Find the determinant of the general 2 × 2 matrix $A = \begin{bmatrix} a_{11} & a_{12} \\ a_{21} & a_{22} \end{bmatrix}$. First list the products without the sign. Note that in the first product we chose a_{11} from row 1 and a_{22} from row 2. In the second product, we chose a_{12} from row 1 and a_{21} from row 2.

$a_{11}a_{22} \quad a_{12}a_{21}$

In the first product, the sequence of column indices is 1, 2, so its sign is $+$. In the second product, the column indices are in the order 2, 1, so its sign is $-$. Thus, the determinant of A, $\det(A)$, is

$$\det(A) = a_{11}a_{22} - a_{12}a_{21}.$$ ■

Example 4.2 finds the formula for the determinant of a 3 × 3 matrix. After reading through the example, you will see why we will not give the formula for the determinant of a 4 × 4 matrix.

Example 4.2. Let $A = \begin{bmatrix} a_{11} & a_{12} & a_{13} \\ a_{21} & a_{22} & a_{23} \\ a_{31} & a_{32} & a_{33} \end{bmatrix}$. Here is the sum of products, each with its sign.

$$a_{11}a_{22}a_{33} - a_{11}a_{23}a_{32} - a_{12}a_{21}a_{33} + a_{12}a_{23}a_{31} + a_{13}a_{21}a_{32} - a_{13}a_{22}a_{31} =$$
$$a_{11}(a_{22}a_{33} - a_{23}a_{32}) - a_{12}(a_{21}a_{33} - a_{23}a_{31}) + a_{13}(a_{21}a_{32} - a_{22}a_{31})$$ ■

Note that the number of products to be evaluated in a 2 × 2 matrix is $2 = 2(1) = 2!$, and in a 3 × 3 matrix is $6 = 3(2)(1) = 3!$. In general, to evaluate the determinant of an $n \times n$ matrix involves choosing one of n elements in row 1, then one of $(n-1)$ elements in row 2, ..., and 1 element in row n, for a total of $n(n-1)(n-2)...(2)(1) = n!$ products.

Numerical Linear Algebra with Applications. http://dx.doi.org/10.1016/B978-0-12-394435-1.00004-1

For instance, to evaluate the determinant of a 15×15 matrix requires the computation of $15! = 1,307,674,368,000$ products. This is an unbelievably huge number of products, and it would be futile to try to use the definition of a determinant even with a supercomputer.

Remark 4.1.   Another and sometimes more convenient notation for the determinant of a square matrix A is $|A|$.

Using the definition is "messy," but from Examples 4.1 and 4.2, we can introduce the concept of evaluating a determinant using expansion by minors. Looking at the result of Example 4.2, each factor in parentheses is the determinant of a 2×2 matrix by the result of Example 4.1. Thus, the determinant of a 3×3 matrix $A = \begin{bmatrix} a_{11} & a_{12} & a_{13} \\ a_{21} & a_{22} & a_{23} \\ a_{31} & a_{32} & a_{33} \end{bmatrix}$ has the value

$$a_{11} \begin{vmatrix} a_{22} & a_{23} \\ a_{32} & a_{33} \end{vmatrix} - a_{12} \begin{vmatrix} a_{21} & a_{23} \\ a_{31} & a_{33} \end{vmatrix} + a_{13} \begin{vmatrix} a_{21} & a_{22} \\ a_{31} & a_{32} \end{vmatrix}. \tag{4.1}$$

Looking at Equation 4.1, note that the multipliers of the determinants move through row 1 in the order column 1, 2, and 3 and alternate in sign. Each term contains the determinant of the 2×2 matrix obtained by crossing out the row and column of the multiplier. This process can be generalized to what is termed *expansion by minors*.

4.2   EXPANSION BY MINORS

In Section 4.1, we found a formula for the determinant of a 2×2 and a 3×3 matrix and indicated that those results can be generalized to compute the determinant of an $n \times n$ matrix. This process, stated in Theorem 4.1 without proof, is said to be *recursive* because it involves computing determinants of smaller matrices until the problem reduces to evaluating determinants of 2×2 matrices.

Theorem 4.1.   *Let $M_{ij}(A)$ (or simply M_{ij} if there is no ambiguity) denote the determinant of the $(n-1) \times (n-1)$ submatrix of A formed by deleting the ith row and jth column of A. Assume that the determinant function has been defined for matrices of size $(n-1) \times (n-1)$. Then the determinant of the $n \times n$ matrix A is defined by what we call the first-row Laplace expansion:*

$$\begin{aligned} |A| &= a_{11}M_{11}(A) - a_{12}M_{12}(A) + \cdots + (-1)^{1+n}M_{1n}(A) \\ &= \sum_{j=1}^{n} (-1)^{1+j} a_{1j}M_{1j}. \end{aligned}$$

The values M_{ij} are termed minors, and the evaluation process in Theorem 4.1 is an example of expansion by minors.

Example 4.3.   Compute the determinant of the 3×3 matrix $\begin{bmatrix} 1 & 2 & -1 \\ 0 & 4 & 1 \\ 3 & 5 & -9 \end{bmatrix}$.

$$\begin{vmatrix} 1 & 2 & -1 \\ 0 & 4 & 1 \\ 3 & 5 & -9 \end{vmatrix} = (1) \begin{vmatrix} 4 & 1 \\ 5 & -9 \end{vmatrix} - (2) \begin{vmatrix} 0 & 1 \\ 3 & -9 \end{vmatrix} + (-1) \begin{vmatrix} 0 & 4 \\ 3 & 5 \end{vmatrix}$$

$$= (1)(-41) - (2)(-3) + (-1)(-12) = -23. \qquad \blacksquare$$

Example 4.4.   A matrix and its transpose have equal determinants; that is $|A^T| = |A|$ (Problem 4.19).

$$\begin{vmatrix} 1 & 0 & 2 \\ 1 & 2 & 5 \\ 3 & -1 & 1 \end{vmatrix} = (1) \begin{vmatrix} 2 & 5 \\ -1 & 1 \end{vmatrix} - (0) \begin{vmatrix} 1 & 5 \\ 3 & 1 \end{vmatrix} + (2) \begin{vmatrix} 1 & 2 \\ 3 & -1 \end{vmatrix} = 7 - 14 = -7,$$

$$\begin{vmatrix} 1 & 1 & 3 \\ 0 & 2 & -1 \\ 2 & 5 & 1 \end{vmatrix} = (1) \begin{vmatrix} 2 & -1 \\ 5 & 1 \end{vmatrix} - (1) \begin{vmatrix} 0 & -1 \\ 2 & 1 \end{vmatrix} + (3) \begin{vmatrix} 0 & 2 \\ 2 & 5 \end{vmatrix} = 7 - 2 - 12 = -7. \qquad \blacksquare$$

Sometimes the calculation of the determinant by minors (generally a tedious process) is simple.

Theorem 4.2. *If a row of a matrix is zero, then the value of the determinant is 0.*

Proof. Assume the matrix has the form $\begin{bmatrix} a_{11} & a_{12} & a_{13} & \ldots & a_{1,n-1} & a_{1n} \\ a_{21} & a_{22} & a_{23} & \ldots & a_{2,n-1} & a_{2n} \\ \vdots & \vdots & \ddots & \ldots & \ldots & \vdots \\ 0 & 0 & 0 & \ldots & \ldots & 0 \\ a_{n1} & a_{n2} & a_{n3} & \ldots & \ldots & a_{nn} \end{bmatrix}$. When we expand by minors across the first row, each $(n-1) \times (n-1)$ matrix has a row of zeros. This continues as we proceed with ever smaller matrices. Finally, we will arrive at a set of 2×2 matrices each with a row of zeros, and such a matrix has a determinant of 0. $\square$

Example 4.5. Let $A = \begin{bmatrix} 1 & -1 & 6 \\ 0 & 0 & 0 \\ -8 & 9 & 10 \end{bmatrix}$.

$$\begin{vmatrix} 1 & -1 & 6 \\ 0 & 0 & 0 \\ -8 & 9 & 10 \end{vmatrix} = (1) \begin{vmatrix} 0 & 0 \\ 9 & 10 \end{vmatrix} - (-1) \begin{vmatrix} 0 & 0 \\ -8 & 10 \end{vmatrix} + (6) \begin{vmatrix} 0 & 0 \\ -8 & 9 \end{vmatrix} = 0.$$ $\blacksquare$

Another determinant it is easy to evaluate is that of a *lower triangular matrix*.

$$\begin{vmatrix} a_{11} & 0 & \ldots & 0 \\ a_{21} & a_{22} & \ldots & 0 \\ \vdots & \vdots & & \ddots \\ a_{n1} & a_{n2} & \ldots & a_{nn} \end{vmatrix}$$

Its determinant is the product of the diagonal elements, $a_{11}a_{22}\ldots a_{nn}$. We can see this by observing the sequence of expansion by minors.

$$\begin{vmatrix} a_{11} & 0 & \ldots & 0 \\ a_{21} & a_{22} & \ldots & 0 \\ \vdots & \vdots & & \ddots \\ a_{n1} & a_{n2} & \ldots & a_{nn} \end{vmatrix} = a_{11} \begin{vmatrix} a_{22} & 0 & \ldots & 0 \\ a_{32} & a_{33} & \ldots & 0 \\ \vdots & \vdots & & \ddots \\ a_{n2} & a_{n3} & \ldots & a_{nn} \end{vmatrix} = a_{11}a_{22} \begin{vmatrix} a_{33} & 0 & \ldots & 0 \\ a_{43} & a_{44} & \ldots & 0 \\ \vdots & \vdots & & \ddots \\ a_{n3} & a_{n4} & \ldots & a_{nn} \end{vmatrix} = \cdots$$

We continue this process until evaluating a 2×2 matrix. At this point, we are done and have computed the value $a_{11}a_{22}\ldots a_{nn}$.

We will leave it to the exercises, but the same result applies to the determinant of an *upper triangular matrix*.

$$|A| = a_{11}a_{22}\ldots a_{nn}.$$

A special case that plays a role in applications is when A is a *diagonal matrix*. If

$$A = \text{diag}(a_{11},\ldots,a_{nn}) = \begin{bmatrix} a_{11} & & & & \\ & a_{22} & & & \\ & & \ddots & & \\ & & & a_{n-1,n-1} & \\ & & & & a_{nn} \end{bmatrix},$$

then $|A| = a_{11}\ldots a_{nn}$. In particular, for a scalar matrix product tI, we have $\det(tI) = t^n$.

Remark 4.2. A very useful fact concerning determinants is that $\det(AB) = (\det A)(\det B)$. We will not provide a proof but will use this result numerous times.

Example 4.6. Let $A = \begin{bmatrix} 1 & 4 & 5 \\ 0 & -9 & 12 \\ 0 & 0 & 2 \end{bmatrix}$ and $B = \begin{bmatrix} 8 & 0 & 0 \\ 100 & 2 & 0 \\ 2 & 6 & -1 \end{bmatrix}$. Note that A is upper triangular and B is lower triangular, so

$$|A|\,|B| = [(1)(-9)(2)][(8)(2)(-1)] = 288.$$

The matrix AB is

$$AB = \begin{bmatrix} 418 & 38 & -5 \\ -876 & 54 & -12 \\ 4 & 12 & -2 \end{bmatrix}.$$

This determinant would be a chore to evaluate using expansion by minors, so we will use the MATLAB function det.

```
>> det(C)
ans =
   288.0000
```
∎

The evaluation of an $n \times n$ matrix was presented in terms of the first-row expansion. Actually, we can expand the determinant along any row or column, and we call this *expansion by minors*.

$$\det A = \sum_{j=1}^{n} (-1)^{i+j} a_{ij} M_{ij}(A)$$

is the ith row expansion and

$$\det A = \sum_{i=1}^{n} (-1)^{i+j} a_{ij} M_{ij}(A)$$

is the jth column expansion.

Remark 4.3.   The expression $(-1)^{i+j}$ obeys the chessboard pattern of signs:

$$\begin{bmatrix} + & - & + & \cdots \\ - & + & - & \cdots \\ + & - & + & \cdots \\ \vdots & & & \end{bmatrix}.$$

Example 4.7.   Evaluate $\begin{vmatrix} 1 & 1 & -1 \\ 2 & 0 & 3 \\ 8 & -7 & 1 \end{vmatrix}$ using expansion by minors across row 2.

$$\begin{vmatrix} 1 & 1 & -1 \\ 2 & 0 & 3 \\ 8 & -7 & 1 \end{vmatrix} = -(2) \begin{vmatrix} 1 & -1 \\ -7 & 1 \end{vmatrix} + (0) \begin{vmatrix} 1 & -1 \\ 8 & 1 \end{vmatrix} - (3) \begin{vmatrix} 1 & 1 \\ 8 & -7 \end{vmatrix} = 12 + 0 + 45 = 57.$$

Using the Laplace expansion, we get

$$\begin{vmatrix} 1 & 1 & -1 \\ 2 & 0 & 3 \\ 8 & -7 & 1 \end{vmatrix} = (1) \begin{vmatrix} 0 & 3 \\ -7 & 1 \end{vmatrix} - (1) \begin{vmatrix} 2 & 3 \\ 8 & 1 \end{vmatrix} + (-1) \begin{vmatrix} 2 & 0 \\ 8 & -7 \end{vmatrix} = 21 + 22 + 14 = 57.$$
∎

As we have said, the determinant is primarily a theoretical tool, and one matrix built by using the determinant is useful in that regard.

Definition 4.1 (Cofactor).   The (i,j) cofactor of A, denoted by $C_{ij}(A)$ (or C_{ij} if there is no ambiguity), is defined by

$$C_{ij}(A) = (-1)^{i+j} M_{ij}(A).$$

Remark 4.4.   Notice that $C_{ij}(A)$, like $M_{ij}(A)$, does not depend on a_{ij}.

Definition 4.2 (Adjoint). If $A = [a_{ij}]$ is an $n \times n$ matrix, the *adjoint* of A, denoted by adj A, is the transpose of the matrix of cofactors. Hence,

$$\text{adj } A = \begin{bmatrix} C_{11} & C_{21} & \cdots & C_{n1} \\ C_{12} & C_{22} & \cdots & C_{n2} \\ \vdots & & & \vdots \\ C_{1n} & C_{2n} & \cdots & C_{nn} \end{bmatrix}.$$

Example 4.8. Let $A = \begin{bmatrix} 1 & 2 & 3 \\ 4 & 5 & 6 \\ 8 & 8 & 9 \end{bmatrix}$.

$$\text{adj } (A) = \begin{bmatrix} C_{11} & C_{21} & C_{31} \\ C_{12} & C_{22} & C_{32} \\ C_{13} & C_{23} & C_{33} \end{bmatrix}$$

$$= \begin{bmatrix} \begin{vmatrix} 5 & 6 \\ 8 & 9 \end{vmatrix} & -\begin{vmatrix} 2 & 3 \\ 8 & 9 \end{vmatrix} & \begin{vmatrix} 2 & 3 \\ 5 & 6 \end{vmatrix} \\[2mm] -\begin{vmatrix} 4 & 6 \\ 8 & 9 \end{vmatrix} & \begin{vmatrix} 1 & 3 \\ 8 & 9 \end{vmatrix} & -\begin{vmatrix} 1 & 3 \\ 4 & 6 \end{vmatrix} \\[2mm] \begin{vmatrix} 4 & 5 \\ 8 & 8 \end{vmatrix} & -\begin{vmatrix} 1 & 2 \\ 8 & 8 \end{vmatrix} & \begin{vmatrix} 1 & 2 \\ 4 & 5 \end{vmatrix} \end{bmatrix}$$

$$= \begin{bmatrix} -3 & 6 & -3 \\ 12 & -15 & 6 \\ -8 & 8 & -3 \end{bmatrix}.$$

In addition, compute the product of A and its adjoint:

$$A \, (\text{adj } (A)) = \begin{bmatrix} 1 & 2 & 3 \\ 4 & 5 & 6 \\ 8 & 8 & 9 \end{bmatrix} \begin{bmatrix} -3 & 6 & -3 \\ 12 & -15 & 6 \\ -8 & 8 & -3 \end{bmatrix} = \begin{bmatrix} -3 & 0 & 0 \\ 0 & -3 & 0 \\ 0 & 0 & -3 \end{bmatrix} = -3I.$$

Also compute the determinant of A.

$$|A| = \begin{vmatrix} 5 & 6 \\ 8 & 9 \end{vmatrix} - 2\begin{vmatrix} 4 & 6 \\ 8 & 9 \end{vmatrix} + 3\begin{vmatrix} 4 & 5 \\ 8 & 8 \end{vmatrix} = -3 + 24 - 24 = -3. \qquad \blacksquare$$

Example 4.8 illustrates a very interesting property of the adjoint. The product of a matrix A and its adjoint is a diagonal matrix whose diagonal entries are det (A). We will not prove this result, but will use it to develop a formula for computing the inverse of a matrix.

Theorem 4.3. *The inverse of a matrix is related to the adjoint by the relation*

$$A^{-1} = \frac{1}{\det A} \text{adj } (A). \tag{4.2}$$

Proof. Since

$$A \times \text{adj } (A) = \begin{bmatrix} \det A & 0 & & & 0 \\ 0 & \det A & & & \\ & & \ddots & & \\ & & & \det A & \\ 0 & & & & \det A \end{bmatrix},$$

we have $A \times adj\,(A) = (\det A)\,I$. Thus, $A((1/\det A)\text{adj}(A)) = I$, and $A^{-1} = (1/\det A)\text{adj}(A)$ $\qquad \square$

Example 4.9. We computed the determinant and the adjoint of the matrix $A = \begin{bmatrix} 1 & 2 & 3 \\ 4 & 5 & 6 \\ 8 & 8 & 9 \end{bmatrix}$ in Example 4.8.

Equation 4.2 computes the inverse of A.

$$A^{-1} = -\frac{1}{3} \begin{bmatrix} -3 & 6 & -3 \\ 12 & -15 & 6 \\ -8 & 8 & -3 \end{bmatrix}.$$ ∎

4.3 COMPUTING A DETERMINANT USING ROW OPERATIONS

A determinant has properties that allow its computation without resorting to expansion by minors. Example 4.10 demonstrates the properties.

Example 4.10.

1. A determinant is a linear function of each row separately.

If two rows are added, with all other rows remaining the same, the determinants are added.

$$\begin{vmatrix} 2 & 3 & 4 \\ -1 & -2 & -3 \\ -4 & -3 & -4 \end{vmatrix} + \begin{vmatrix} 5 & 6 & 7 \\ -1 & -2 & -3 \\ -4 & -3 & -4 \end{vmatrix} = 2 + 8 = 10,$$

$$\begin{vmatrix} 2+5 & 3+6 & 4+7 \\ -1 & -2 & -3 \\ -4 & -3 & -4 \end{vmatrix} = \begin{vmatrix} 7 & 9 & 11 \\ -1 & -2 & -3 \\ -4 & -3 & -4 \end{vmatrix} = 10.$$

If a row of A is multiplied by a scalar t, then the determinant of the modified matrix is $t \det A$.

$$\begin{vmatrix} 1 & 4 & 0 \\ (7)2 & (7)5 & (7)1 \\ 1 & 0 & 0 \end{vmatrix} = (1) \begin{vmatrix} 4 & 0 \\ 35 & 7 \end{vmatrix} = 28 = (7) \begin{vmatrix} 1 & 4 & 0 \\ 2 & 5 & 1 \\ 1 & 0 & 0 \end{vmatrix} = (7)(4)$$

2. When two rows of a matrix are equal, the determinant is zero.

$$\begin{vmatrix} 1 & 0 & 1 \\ 2 & 1 & 8 \\ 1 & 0 & 1 \end{vmatrix} = (1) \begin{vmatrix} 1 & 8 \\ 0 & 1 \end{vmatrix} - (2) \begin{vmatrix} 0 & 1 \\ 0 & 1 \end{vmatrix} + (1) \begin{vmatrix} 0 & 1 \\ 1 & 8 \end{vmatrix} = 1 - 0 + (1)(-1) = 0$$

3. If two rows of a matrix are exchanged, the determinant changes sign.

$$\begin{vmatrix} 1 & 0 & 0 \\ 2 & 5 & 1 \\ 1 & 4 & 0 \end{vmatrix} = (1) \begin{vmatrix} 5 & 1 \\ 4 & 0 \end{vmatrix} = -4$$

$$\begin{vmatrix} 1 & 4 & 0 \\ 2 & 5 & 1 \\ 1 & 0 & 0 \end{vmatrix} = (1) \begin{vmatrix} 5 & 1 \\ 0 & 0 \end{vmatrix} - (4) \begin{vmatrix} 2 & 1 \\ 1 & 0 \end{vmatrix} = 4$$

4. If a multiple of a row is subtracted from another row, the value of the determinant remains unchanged.

$$\begin{vmatrix} 1 & 4 & 0 \\ 2 & 5 & 1 \\ 1 & 0 & 0 \end{vmatrix} \xrightarrow{R2 = R2 - 8R1} \begin{vmatrix} 1 & 4 & 0 \\ -6 & -27 & 1 \\ 1 & 0 & 0 \end{vmatrix} = (1) \begin{vmatrix} 4 & 0 \\ -27 & 1 \end{vmatrix} = 4$$ ∎

Theorem 4.5 formally states the properties demonstrated in Example 4.10.

Theorem 4.4.

1. *A determinant is a linear function of each row separately.*
2. *If two rows of a matrix are equal, the determinant is zero.*

3. *If two rows of a matrix are interchanged, the determinant changes sign.*

4. *If a multiple of a row is subtracted from another row, the value of the determinant is unchanged.*

Proof. **1.** Assume matrices A and B as follows:

$$A = \begin{bmatrix} a_{11} & a_{12} & \ldots & a_{1,n-1} & a_{1n} \\ \vdots & \ddots & \ldots & \ldots & \vdots \\ a_{i1} & a_{i2} & \ldots & a_{i,n-1} & a_{in} \\ \vdots & \vdots & \vdots & \vdots & \vdots \\ a_{n1} & a_{n2} & \ldots & a_{n,n-1} & a_{nn} \end{bmatrix}, \quad B = \begin{bmatrix} a_{11} & a_{12} & \ldots & a_{1,n-1} & a_{1n} \\ \vdots & \ddots & \ldots & \ldots & \vdots \\ a'_{i1} & a'_{i2} & \ldots & a'_{i,n-1} & a'_{in} \\ \vdots & \vdots & \vdots & \vdots & \vdots \\ a_{n1} & a_{n2} & \ldots & a_{n,n-1} & a_{nn} \end{bmatrix}.$$

Then

$$\begin{vmatrix} a_{11} & a_{12} & \ldots & a_{1,n-1} & a_{1n} \\ \vdots & \ddots & \ldots & \ldots & \vdots \\ a_{i1}+a'_{i1} & a_{i2}+a'_{i2} & \ldots & a_{i,n-1}+a'_{i,n-1} & a_{in}+a'_{in} \\ \vdots & \vdots & \vdots & \vdots & \vdots \\ a_{n1} & a_{n2} & \ldots & a_{n,n-1} & a_{nn} \end{vmatrix} =$$

$$\sum_{k=1}^{n} (-1)^{i+k} \left(a_{ik} + a'_{ik} \right) M_{ik} = \sum_{k=1}^{n} (-1)^{i+k} a_{ik} M_{ik} + \sum_{k=1}^{n} (-1)^{i+k} a'_{ik} M_{ik} =$$

$$|A| + |B|$$

The proof that $\det (tA) = t \det (A)$ is left to Problem 4.13.

2. See Problem 4.15.

3. See Problem 4.14.

4. Subtract a multiple, t, of row j from row i.

$$\begin{bmatrix} a_{11} & a_{12} & \ldots & a_{1,n-1} & a_{1n} \\ \vdots & \ddots & \ldots & \ldots & \vdots \\ a_{i1}-ta_{j1} & a_{i2}-ta_{j2} & \ldots & a_{i,n-1}-ta_{j,n-1} & a_{in}-ta_{jn} \\ \vdots & \vdots & \vdots & \vdots & \vdots \\ a_{n1} & a_{n2} & \ldots & a_{n,n-1} & a_{nn} \end{bmatrix}$$

$$= \begin{bmatrix} a_{11} & a_{12} & \ldots & a_{1,n-1} & a_{1n} \\ \vdots & \ddots & \ldots & \ldots & \vdots \\ a_{i1} & a_{i2} & \ldots & a_{i,n-1} & a_{in} \\ \vdots & \vdots & \vdots & \vdots & \vdots \\ a_{n1} & a_{n2} & \ldots & a_{n,n-1} & a_{nn} \end{bmatrix} + \begin{bmatrix} a_{11} & a_{12} & \ldots & a_{1,n-1} & a_{1n} \\ \vdots & \ddots & \ldots & \ldots & \vdots \\ -ta_{j1} & -ta_{j2} & \ldots & -ta_{j,n-1} & -ta_{jn} \\ \vdots & \vdots & \vdots & \vdots & \vdots \\ a_{n1} & a_{n2} & \ldots & a_{n,n-1} & a_{nn} \end{bmatrix} = \text{(property 1)}$$

$$A - t \begin{bmatrix} a_{11} & a_{12} & \ldots & a_{1,n-1} & a_{1n} \\ \vdots & \ddots & \ldots & \ldots & \vdots \\ a_{j1} & a_{j2} & \ldots & a_{j,n-1} & a_{jn} \\ \vdots & \vdots & \vdots & \vdots & \vdots \\ a_{n1} & a_{n2} & \ldots & a_{n,n-1} & a_{nn} \end{bmatrix} = A - t \times 0 = A \text{ (properties 1 and 2)}$$

Property 2 applies because row j appears twice in $\begin{bmatrix} a_{11} & a_{12} & \ldots & a_{1,n-1} & a_{1n} \\ \vdots & \ddots & \ldots & \ldots & \vdots \\ a_{j1} & a_{j2} & \ldots & a_{j,n-1} & a_{jn} \\ \vdots & \vdots & \vdots & \vdots & \vdots \\ a_{n1} & a_{n2} & \ldots & a_{n,n-1} & a_{nn} \end{bmatrix}.$ $\square$

The properties in Theorem 4.4 provide a means for calculating a determinant without using expansion by minors. Reduce the matrix to upper triangular form, recording any sign changes caused by row interchanges, together with any factors taken out of a row. Then compute the product of the diagonal elements. Here are some examples.

Example 4.11. Evaluate the determinant

$$\begin{vmatrix} 1 & 2 & 3 \\ 4 & 5 & 6 \\ 8 & 8 & 9 \end{vmatrix}.$$

Using row operations $R_2 \to R_2 - 4R_1$ and $R_3 \to R_3 - 8R_1$ gives

$$\begin{vmatrix} 1 & 2 & 3 \\ 4 & 5 & 6 \\ 8 & 8 & 9 \end{vmatrix} = \begin{vmatrix} 1 & 2 & 3 \\ 0 & -3 & -6 \\ 0 & -8 & -15 \end{vmatrix}.$$

Now perform the row operation $R_3 \to R_3 - \frac{8}{3}R_2$

$$\begin{vmatrix} 1 & 2 & 3 \\ 4 & 5 & 6 \\ 8 & 8 & 9 \end{vmatrix} = \begin{vmatrix} 1 & 2 & 3 \\ 0 & -3 & -6 \\ 0 & 0 & 1 \end{vmatrix} = (1)(-3)(1) = -3.$$ ∎

Example 4.12. Evaluate the determinant $\begin{vmatrix} 1 & 1 & 2 & 1 \\ 3 & 1 & 4 & 5 \\ 7 & 6 & 1 & 2 \\ 1 & 1 & 3 & 4 \end{vmatrix}.$

Begin by using row operations to zero-out the elements in column 1, rows 2-4.

$$\begin{vmatrix} 1 & 1 & 2 & 1 \\ 3 & 1 & 4 & 5 \\ 7 & 6 & 1 & 2 \\ 1 & 1 & 3 & 4 \end{vmatrix} = \begin{vmatrix} 1 & 1 & 2 & 1 \\ 0 & -2 & -2 & 2 \\ 0 & -1 & -13 & -5 \\ 0 & 0 & 1 & 3 \end{vmatrix} \quad \text{(factor} - 2 \text{ from row 2)}$$

$$= -2 \begin{vmatrix} 1 & 1 & 2 & 1 \\ 0 & 1 & 1 & -1 \\ 0 & -1 & -13 & -5 \\ 0 & 0 & 1 & 3 \end{vmatrix} \quad \text{(add row 2 to row 3)}$$

$$= -2 \begin{vmatrix} 1 & 1 & 2 & 1 \\ 0 & 1 & 1 & -1 \\ 0 & 0 & -12 & -6 \\ 0 & 0 & 1 & 3 \end{vmatrix} \quad \text{(swap rows 3 and 4 to put 1 on the main diagonal)}$$

$$= 2 \begin{vmatrix} 1 & 1 & 2 & 1 \\ 0 & 1 & 1 & -1 \\ 0 & 0 & 1 & 3 \\ 0 & 0 & -12 & -6 \end{vmatrix} (R_4 \to R_4 + 12R_3)$$

$$= 2 \begin{vmatrix} 1 & 1 & 2 & 1 \\ 0 & 1 & 1 & -1 \\ 0 & 0 & 1 & 3 \\ 0 & 0 & 0 & 30 \end{vmatrix} \quad \text{(multiply the diagonal elements)} = 60.$$ ∎

In Chapter 12, we will introduce the Vandermonde matrix $V = \begin{bmatrix} 1 & x_1 & x_1^2 & \cdots & x_1^{n-1} \\ 1 & x_2 & x_2^2 & \cdots & x_2^{n-1} \\ 1 & x_3 & x_3^2 & \cdots & x_3^{n-1} \\ \vdots & \vdots & \vdots & \ddots & \vdots \\ 1 & x_n & x_n^2 & \cdots & x_n^{n-1} \end{bmatrix}$ in connection with finding a

polynomial that approximates data in the least-squares sense. The determinant of the Vandermonde matrix has a simple formula, and Example 4.13 demonstrates this for a 3×3 matrix.

Example 4.13. Find the determinant of the Vandermonde matrix $V = \begin{bmatrix} 1 & x_1 & x_1^2 \\ 1 & x_2 & x_2^2 \\ 1 & x_3 & x_3^2 \end{bmatrix}$.

$$\begin{bmatrix} 1 & x_1 & x_1^2 \\ 1 & x_2 & x_2^2 \\ 1 & x_3 & x_3^2 \end{bmatrix} \xrightarrow{\substack{R2 = R2 - (1)R1 \\ R3 = R3 - (1)R1}} \begin{bmatrix} 1 & x_1 & x_1^2 \\ 0 & x_2 - x_1 & x_2^2 - x_1^2 \\ 0 & x_3 - x_1 & x_3^2 - x_1^2 \end{bmatrix} = \begin{bmatrix} 1 & x_1 & x_1^2 \\ 0 & x_2 - x_1 & (x_2 - x_1)(x_2 + x_1) \\ 0 & x_3 - x_1 & (x_3 - x_1)(x_3 + x_1) \end{bmatrix}$$

Noting that only one entry in column 1 is nonzero, expand by minors down the first column.

$$\det \begin{bmatrix} 1 & x_1 & x_1^2 \\ 0 & x_2 - x_1 & (x_2 - x_1)(x_2 + x_1) \\ 0 & x_3 - x_1 & (x_3 - x_1)(x_3 + x_1) \end{bmatrix} = \det \begin{bmatrix} x_2 - x_1 & (x_2 - x_1)(x_2 + x_1) \\ x_3 - x_1 & (x_3 - x_1)(x_3 + x_1) \end{bmatrix} =$$

$$(x_2 - x_1)(x_3 - x_1)(x_3 + x_1) - (x_3 - x_1)(x_2 - x_1)(x_2 + x_1) =$$

$$(x_2 - x_1)(x_3 - x_1)[x_3 + x_1 - (x_2 + x_1)] = (x_2 - x_1)(x_3 - x_1)(x_3 - x_2) \qquad \blacksquare$$

During row reduction from A to B, multiplying a row by a scalar and subtracting from another row does not change the determinant, $\det B = \det A$; however, a row may be multiplied by a scalar or two rows exchanged. In either of these cases, it follows that $\det B = c \det A$, where $c \neq 0$. Hence, $\det B \neq 0$ if and only if $\det A \neq 0$ and $\det B = 0$ if and only if $\det A = 0$. This logic leads to a useful theoretical result.

Theorem 4.5.
1. *A is nonsingular if and only if $\det A \neq 0$;*
2. *A is singular if and only if $\det A = 0$;*
3. *The homogeneous system $Ax = 0$ has a nontrivial solution if and only if* $\det A = 0$.

Proof.
1. Perform a sequence of elementary row operations reducing A to B, where we intend for B to be the identity matrix. B cannot have a zero row, for otherwise the homogeneous equation $Ax = 0$ will have a nonzero solution. Thus, the reduction must be successful, so $B = I$, and it follows that $\det A = c \det I = c \neq 0$, so $\det A \neq 0$. Since $A^{-1} = (1/|A|)\text{adj}(A)$, A^{-1} exists, and A is nonsingular.
2. The previous statement logically implies that "$\det A = 0$ if and only if A is singular."
3. If $Ax = 0$ has a nontrivial solution, then reduction to upper triangular form must produce a zero row, so $\det A = 0$. If $Ax = 0$ has a unique solution, then after reduction to upper triangular form there can be no zeros on the diagonal, or there will be infinitely many solutions. Thus $\det A \neq 0$. $\qquad \square$

Example 4.14.   This example uses MATLAB statements to illustrate Theorem 4.5. Note that in addition to having a zero determinant, the nullity of B is nonzero, also indicating that B is singular.

```
>> A = [1 6 25;16 32 19;56 53 5];
>> det(A)

ans =
       -18543

>> A\[1 2 5]'

ans =
    0.0910
```

```
    -0.0053
     0.0376

>> B = [1 3 2;5 14 7;2 5 1];
>> det(B)

ans =
 -3.3307e-015

>> size(null(B),2) % compute the nullity of B

ans =
     1

>> B\[1 2 5]'
Warning: Matrix is close to singular or badly scaled.
         Results may be inaccurate. RCOND = 4.587698e-018.

ans =
  1.0e+016 *

     1.2610
    -0.5404
     0.1801
```

Note that in the MATLAB results, det B is very small but not zero as it would be if we did the calculation by hand. The MATLAB command det uses the row elimination method to compute the determinant. As you will see in Chapter 8, a computer in general does not perform exact arithmetic, and small errors are made during the elimination process. ∎

Example 4.15. Find numbers a for which the following homogeneous system has a nontrivial solution and solve the system for these values of a:

$$
\begin{aligned}
x - 2y + 3z &= 0, \\
ax + 3y + 2z &= 0, \\
6x + y + az &= 0.
\end{aligned}
$$

The determinant of the coefficient matrix is

$$
\Delta = \begin{vmatrix} 1 & -2 & 3 \\ a & 3 & 2 \\ 6 & 1 & a \end{vmatrix} = \begin{vmatrix} 1 & -2 & 3 \\ 0 & 3+2a & 2-3a \\ 0 & 13 & a-18 \end{vmatrix}
$$

$$
= \begin{vmatrix} 3+2a & 2-3a \\ 13 & a-18 \end{vmatrix}
$$

after expanding by minors using the first column. The value of the determinant is then

$$
(3+2a)(a-18) - 13(2-3a) = 2a^2 + 6a - 80 = 2(a+8)(a-5).
$$

So $\Delta = 0 \Leftrightarrow a = -8$ or $a = 5$ and these values of a are the only values for which the given homogeneous system has a nontrivial solution.

If $a = -8$, reduction of the coefficient matrix to upper triangular form gives

$$
\begin{bmatrix} 1 & 0 & -1 \\ 0 & 1 & -2 \\ 0 & 0 & 0 \end{bmatrix},
$$

and so the complete solution is $x = z, y = 2z$, with z arbitrary, and the null space is spanned by $\begin{bmatrix} 1 \\ 2 \\ 1 \end{bmatrix}$. The case of $a = 5$ is left to the exercises. ∎

If the Symbolic Toolbox is available in your MATLAB distribution, there are many possibilities for performing symbolic manipulation with linear algebra. The following is a partial quotation from the help system for the toolbox.

Symbolic objects are a special MATLAB® data type introduced by the Symbolic Math Toolbox™ software. They allow you to perform mathematical operations in the MATLAB workspace analytically, without calculating numeric values. You can use symbolic objects to perform a wide variety of analytical computations: ...

Example 4.16 shows how to solve the problem in Example 4.15 using the Symbolic Math Toolbox. This example is entirely optional. We will introduce a small number of additional examples using the toolbox at various points in the book.

Example 4.16. The command `syms` constructs symbolic objects, in this case a and A. The command `det` operating on a symbolic matrix does symbolic calculations. Notice that it computed the same determinant that was found in Example 4.15. The command `solve` finds the solutions to det(A) = 0. After assigning a = -8, the first of the two values, to locations (2,1) and (3,3), the command `null(A)` finds the null space of the matrix A. The result is that the null space is spanned by $\begin{bmatrix} 1 \\ 2 \\ 1 \end{bmatrix}$, precisely the result of Example 4.15. The command `colspace(A)` finds a basis for the column space of A. There are two vectors, so the rank of A is 2.

```
>> syms a A
>> A = [1 -2 3;a 3 2;6 1 a]

A =
[ 1, -2, 3]
[ a,  3, 2]
[ 6,  1, a]

>> D = det(A)

D =
2*a^2 + 6*a - 80

>> vals = solve(D)

vals =
 -8
  5

>> A(2,1) = vals(1);
>> A(3,3) = vals(1);
>> null(A)

ans =
 1
 2
 1

>> colspace(A)

ans =
[  1,  0]
[  0,  1]
[ -2, -1]
```                                                                                        ■

To finish this section, we present an old (1750) method of solving a system of *n* equations in *n* unknowns called *Cramer's rule*. It is useful for solving 2 × 2 and 3 × 3 systems, but otherwise is too computationally expensive to use for larger systems. It does have theoretical uses in areas of mathematics such as differential equations.

Theorem 4.6. *The system of n linear equations in n unknowns $x_1, \ldots, x_n$*

$$a_{11}x_1 + a_{12}x_2 + \cdots + a_{1n}x_n = b_1$$
$$a_{21}x_1 + a_{22}x_2 + \cdots + a_{2n}x_n = b_2$$

$$\vdots$$

$$a_{n1}x_1 + a_{n2}x_2 + \cdots + a_{nn}x_n = b_n$$

has a unique solution if $\Delta = \det \begin{bmatrix} a_{11} & a_{12} & \cdots & a_{1,n-1} & a_{1n} \\ a_{21} & a_{22} & \cdots & a_{2,n-1} & a_{2n} \\ \vdots & \vdots & \ddots & \vdots & \vdots \\ a_{n-1,1} & a_{n-1,2} & \ddots & a_{n-1,n-1} & a_{n-1,n} \\ a_{n1} & a_{n2} & \cdots & a_{n,n-1} & a_{nn} \end{bmatrix} \neq 0$, *namely,*

$$x_1 = \frac{\Delta_1}{\Delta}, \ x_2 = \frac{\Delta_2}{\Delta}, \ldots, x_n = \frac{\Delta_n}{\Delta},$$

where Δ_i *is the determinant of the matrix formed by replacing the ith column of the coefficient matrix A by the entries* $b_1, b_2, \ldots, b_n$.

Proof. Suppose the coefficient determinant $\Delta \neq 0$. Then A^{-1} exists and is given by $A^{-1} = (1/\Delta)\mathrm{adj}A$, and the system has the unique solution

$$\begin{bmatrix} x_1 \\ x_2 \\ \vdots \\ x_n \end{bmatrix} = A^{-1} \begin{bmatrix} b_1 \\ b_2 \\ \vdots \\ b_n \end{bmatrix} = \frac{1}{\Delta} \begin{bmatrix} C_{11} & C_{21} & \cdots & C_{n1} \\ C_{12} & C_{22} & \cdots & C_{n2} \\ \vdots & & & \vdots \\ C_{1n} & C_{2n} & \cdots & C_{nn} \end{bmatrix} \begin{bmatrix} b_1 \\ b_2 \\ \vdots \\ b_n \end{bmatrix} \tag{4.3}$$

$$= \frac{1}{\Delta} \begin{bmatrix} b_1 C_{11} + b_2 C_{21} + \cdots + b_n C_{n1} \\ b_1 C_{12} + b_2 C_{22} + \cdots + b_n C_{n2} \\ \vdots \\ b_1 C_{1n} + b_2 C_{2n} + \cdots + b_n C_{nn} \end{bmatrix}. \tag{4.4}$$

Consider

$$\Delta_i = \begin{vmatrix} \cdots & a_{1,i-1} & b_1 & a_{1,i+1} & \cdots \\ \cdots & a_{2,i-1} & b_2 & a_{2,i+1} & \cdots \\ \vdots & \vdots & \vdots & \vdots & \vdots \\ \vdots & \vdots & \vdots & \vdots & \vdots \\ \cdots & a_{n,i-1} & b_n & a_{n,i+1} & \cdots \end{vmatrix} = b_1 C_{1i} + b_2 C_{2,i} + \cdots + b_n C_{ni}. \tag{4.5}$$

The value of Δ_i in Equation 4.5 is the entry in row i in Equation 4.4. Hence,

$$\begin{bmatrix} x_1 \\ x_2 \\ \vdots \\ x_n \end{bmatrix} = \frac{1}{\Delta} \begin{bmatrix} \Delta_1 \\ \Delta_2 \\ \vdots \\ \Delta_n \end{bmatrix} = \begin{bmatrix} \Delta_1/\Delta \\ \Delta_2/\Delta \\ \vdots \\ \Delta_n/\Delta \end{bmatrix}. \qquad \square$$

Example 4.17. Use Cramer's rule to find the solution to the system $\begin{bmatrix} 1 & -1 & 5 \\ 8 & 3 & 12 \\ -1 & -9 & 2 \end{bmatrix} \begin{bmatrix} x_1 \\ x_2 \\ x_3 \end{bmatrix} = \begin{bmatrix} 5 \\ 3 \\ 1 \end{bmatrix}$. We will perform the calculations using the MATLAB function det and compare the result with the solution using the operator "\."

```
>> A = [1 -1 5;8 3 12;-1 -9 2];
>> b = [5 1 3]';
>> delta = det(A);
>> x1 = det([b A(:,2) A(:,3)])/delta

x1 =
   -2.1970

>> x2 = det([A(:,1) b A(:,3)])/delta

x2 =
    0.2414

>> x3 = det([A(:,1) A(:,2) b])/delta

x3 =
    1.4877

>> A\b

ans =
   -2.1970
    0.2414
    1.4877
```

■

4.4 APPLICATION: ENCRYPTION

We begin this section with an interesting and useful result about matrices having integer entries. We are able to prove this from our work with adjoints and determinants.

Theorem 4.7. *If an $n \times n$ matrix A has all integer entries and $\det(A) = \pm 1$, then A^{-1} exists and has all integer entries.*

Proof. We have

$$A^{-1} = \frac{1}{\det A} \text{adj} A,$$

where adj$A$ is the adjoint of A. Recall that the adjoint of A is the transpose of the matrix of cofactors of A

$$\text{adj}(A) = \begin{bmatrix} C_{11} & C_{21} & \dots & C_{n1} \\ C_{12} & C_{22} & \dots & C_{n2} \\ \vdots & & & \vdots \\ C_{1n} & C_{2n} & \dots & C_{nn} \end{bmatrix}.$$

Each cofactor $C_{ij} = (-1)^{i+j} M_{ij}$, where M_{ij} is the minor for the entry in row i, column j. In this situation, each minor is the determinant of a matrix of integer entries, so every cofactor is an integer. Since $\det A = \pm 1$, it follows that A^{-1} has only integer entries. □

There are many ways to encrypt a message, and the use of encryption has become particularly significant in recent years (e.g., due to the explosion of financial transactions on Internet). One way to encrypt or code a message uses invertible matrices. Consider an $n \times n$ invertible matrix A. Convert the message into an $n \times m$ matrix B so that AB can be computed. Send the message generated by AB. The receiving end will need to know A^{-1} in order to decrypt the message sent using

$$A^{-1}(AB) = B.$$

Keep in mind that whenever an undesired intruder finds A, we must be able to change it. So we should have a mechanical way of generating simple matrices A which are invertible and have simple inverse matrices. Note that, in general, the inverse of a matrix involves fractions which are not easy to send in an electronic form. The optimal situation is to have both A and its inverse have integer entries. In fact, we can use Theorem 4.7 to generate such a class of matrices. One practical way is to start with an upper triangular matrix with entries of ± 1 on the diagonal and integer entries. Since the determinant of an upper triangular matrix is the product of its diagonal elements, such a matrix will have determinant ± 1. Now use elementary row operations to alter the matrix. Do not multiply rows with nonintegers while doing elementary row operations. Recall that adding a multiple of one row to another does not change the value of a determinant, and swapping two rows just changes the sign of the determinant. Here is an example.

Example 4.18.   Consider the matrix

$$\begin{bmatrix} 1 & 2 & 9 \\ 0 & -1 & 3 \\ 0 & 0 & 1 \end{bmatrix}.$$

Add the first row to both the second and third rows to obtain

$$\begin{bmatrix} 1 & 2 & 9 \\ 1 & 1 & 12 \\ 1 & 2 & 10 \end{bmatrix}.$$

Now add the second row to the third

$$\begin{bmatrix} 1 & 2 & 9 \\ 1 & 1 & 12 \\ 2 & 3 & 22 \end{bmatrix}.$$

Finally, add rows two and three together, multiply the sum by -2 and add to the first row. We obtain the matrix

$$A = \begin{bmatrix} -5 & -6 & -59 \\ 1 & 1 & 12 \\ 2 & 3 & 22 \end{bmatrix}.$$

Verify that $\det A = -1$. This must be the case since the original upper triangular matrix has determinant -1 and we only added multiples of one row to another. The inverse of A is

$$A^{-1} = \begin{bmatrix} 14 & 45 & 13 \\ -2 & -8 & -1 \\ -1 & -3 & -1 \end{bmatrix}.$$

Now consider the message
   TODAY IS A GOOD DAY
To every letter we will associate a number. An easy way to do that is to associate 0 to a space, 1 to "A," 2 to "B," etc. Another way is to associate 0 to a blank or space, 1 to "A," -1 to "B," 2 to "C," -2 to "D," etc. Let us use the second choice. We encode our message as follows:

T	O	D	A	Y		I	S		A		G	O	O	D		D	A	Y
-10	8	-2	1	13	0	5	10	0	1	0	4	8	8	-2	0	-2	1	13

Now we rearrange these numbers into a matrix B. For example, sequence the numbers by columns, adding zeros in the last column if necessary. The matrix B must have three rows for the product AB to be defined.

$$B = \begin{bmatrix} -10 & 1 & 5 & 1 & 8 & 0 & 13 \\ 8 & 13 & 10 & 0 & 8 & -2 & 0 \\ -2 & 0 & 0 & 4 & -2 & 1 & 0 \end{bmatrix}$$

Now perform the product AB. We get

$$\begin{bmatrix} -5 & -6 & -59 \\ 1 & 1 & 12 \\ 2 & 3 & 22 \end{bmatrix} \begin{bmatrix} -10 & 1 & 5 & 1 & 8 & 0 & 13 \\ 8 & 13 & 10 & 0 & 8 & -2 & 0 \\ -2 & 0 & 0 & 4 & -2 & 1 & 0 \end{bmatrix} = \begin{bmatrix} 120 & -83 & -85 & -241 & 30 & -47 & -65 \\ -26 & 14 & 15 & 49 & -8 & 10 & 13 \\ -40 & 41 & 40 & 90 & -4 & 16 & 26 \end{bmatrix}$$

The encrypted message to be sent is

120 -26 -40 -83 14 41 -85 15 40 -241 49 90 30 -8 -4 -47 10 16 -65 13 26 ∎

4.5   CHAPTER SUMMARY

Development of the Determinant Concept

The determinant of an $n \times n$ matrix is a concept used primarily for theoretical purposes and is the basis for the definition of eigenvalues, the subject of Chapters 5, 18, 19, 22, and 23. The original definition of a determinant is a sum of permutations with an attached sign. This definition is rarely used to evaluate a determinant. A determinant can be evaluated using a process known as expansion by minors.

Expansion by Minors

Expansion by minors is a simple way to evaluate the determinant of a 2×2 or a 3×3 matrix. For larger values of n, the method is not practical, but we will see it is very useful in proving important results.

The minor, $M_{ij}(A)$, is the determinant of the $(n-1) \times (n-1)$ submatrix of A formed by deleting the ith row and jth column of A. Expansion by minors is a recursive process. The determinant of an $n \times n$ matrix is a linear combination of the minors obtained by expansion down any row or any column. By continuing this process, the problem reduces to the evaluation of 2×2 matrices, where

$$\det \begin{bmatrix} a_{11} & a_{12} \\ a_{21} & a_{22} \end{bmatrix} = a_{11}a_{22} - a_{12}a_{21}.$$

Important properties of determinants include

- $\det A^T = \det A$
- $\det(AB) = \det(A)\det(B)$
- If a row or column of A is zero, $\det A = 0$.
- The determinant of an upper or lower triangular matrix is the product of its diagonal elements.
- A cofactor $C_{ij}(A) = (-1)^{i+j}M_{ij}(A)$. The adjoint is the transpose of the matrix of cofactors, and it follows that

$$A^{-1} = \frac{\text{adj}(A)}{\det A}.$$

Like determinants in general, this result is useful for theoretical purposes. We make use of it in Section 4.4.

Computing a Determinant Using Row Operations

The following facts about determinants allow the computation using elementary row operations.

- If two rows are added, with all other rows remaining the same, the determinants are added, and $\det(tA) = t\det(A)$ where t is a constant.
- If two rows of a matrix are equal, the determinant is zero.

- If two rows of a matrix are interchanged, the determinant changes sign.
- If a multiple of a row is subtracted from another row, the value of the determinant is unchanged.

Apply these rules and reduce the matrix to upper triangular form. The determinant is the product of the diagonal elements. This is how MATLAB computes det(A). As we will see in Chapter 8, errors inherent in floating point arithmetic may produce an answer that is close to, but not equal to the true result.

Using row operations on a determinant, we can show that

- A is nonsingular if and only if $\det A \neq 0$;
- A is singular if and only if $\det A = 0$;
- The system $Ax = 0$ has a nontrivial solution if and only if $\det A = 0$.

Although the chapter developed Cramer's rule, it should be used for theoretical use only.

Application of Determinants to Encryption

Let A be an $n \times n$ matrix. Using the result $A^{-1} = \text{adj}\,(A)/\det A$, the inverse of a matrix with integer entries has integer entries. Form an upper triangular matrix with integer entries, all of whose diagonal entries are ± 1. Subtract integer multiples of one row from another and swap rows to "jumble up" the matrix, keeping the determinant to be ± 1. Find the inverse. Encode the message as a sequence of integers stored in an $n \times p$ matrix B, and transmit AB. Capture the encoded message by forming $A^{-1}(AB) = B$.

4.6 PROBLEMS

4.1 Compute $\begin{vmatrix} 1 & -1 & 2 \\ 2 & 3 & 1 \\ 5 & 1 & -1 \end{vmatrix}$ using the definition of a determinant as the sum over all permutations of the set $\{1, 2, 3\}$.

4.2 Using pencil and paper, evaluate the following determinants:

(a) $\begin{vmatrix} 1 & -1 & 5 \\ 7 & 1 & 0 \\ 3 & 3 & 2 \end{vmatrix}$ (b) $\begin{vmatrix} -1 & 4 & 5 \\ 9 & 1 & -2 \\ -4 & 1 & 3 \end{vmatrix}$

4.3 In the matrix A, an x means "any value." Show that the determinant of A is 0 regardless of the x values.

$$A = \begin{bmatrix} x & x & x & x & x \\ 0 & 0 & 0 & x & x \\ 0 & 0 & 0 & x & x \\ 0 & 0 & 0 & x & x \\ 0 & 0 & 0 & x & x \end{bmatrix}.$$

4.4 Compute the inverse of the matrix $A = \begin{bmatrix} 1 & 0 & -2 \\ 3 & 1 & 4 \\ 5 & 2 & -3 \end{bmatrix}$ by first computing the adjoint matrix.

4.5 Let $P_i = (x_i, y_i), i = 1, 2, 3$. If x_1, x_2, x_3 are distinct, show that there is precisely one curve of the form $y = ax^2 + bx + c$ passing through P_1, P_2, and P_3.

4.6 By considering the determinant of the coefficient matrix, find a relationship between a and b for which the following system has exactly one solution:

$$\begin{aligned} x - 2y + bz &= 3, \\ ax + 2z &= 2, \\ 5x + 2y &= 1. \end{aligned}$$

4.7 Use Cramer's rule to solve the system

$$\begin{aligned} -2x + 3y - z &= 1 \\ x + 2y - z &= 4 \\ -2x - y + z &= -3 \end{aligned}$$

4.8 Show that

$$\begin{vmatrix} 1 & 1 & 1 & 1 \\ r & 1 & 1 & 1 \\ r & r & 1 & 1 \\ r & r & r & 1 \end{vmatrix} = (1-r)^3.$$

4.9 Find a 3×3 matrix consisting entirely of values 1 and -1 that has the largest possible determinant.

4.10 Complete Example 4.15 for $a = 5$.

4.11 Complete Example 4.16 for $a = 5$.

4.12 Prove that the determinant of an upper triangular matrix is the product of its diagonal elements.

4.13 Prove that if a row of a square matrix A is multiplied by a scalar t, then the determinant of the modified matrix is $t \det (A)$.

4.14 In this problem, you will prove that if two rows of a matrix are interchanged, the determinant changes sign. Represent the matrix as a column of rows:

By using this representation, explain the validity of each step in the proof.

a. Add row j to row i.

b. Subtract row i from row j.

c. Add row j to row i.

d. Multiply row j by -1, and the proof is complete.

4.15 Prove that if two rows of a matrix are equal, the determinant is zero. Hint: Exchanging the two equal rows causes the sign to change.

4.16 Using the fact that $|AB| = |A|\,|B|$ show that if A is invertible, then $|A^{-1}| = 1/|A|$.

4.17 Show that if $A^{\mathrm{T}} = -A$ and n is odd, then $|A| = 0$. Hint: $|-A| = (-1)^n\,|A|$.

4.18 Show that if $A^2 + I = 0$, then must n be even. Hint: $|A^2| = |A|^2$.

4.19 Prove that $\det A = \det A^{\mathrm{T}}$ using mathematical induction.

a. Show the statement is true for $n = 2$. This is the base case.

b. Assume the statement is true for any n, and show it is true for $n+1$. Hint: Expansion by minors requires evaluating the determinants of $n \times n$ matrices.

4.20 If A is a real matrix, there is a geometrical interpretation of $\det A$ (Figure 4.1). Let $P = (x_1, y_1)$ and $Q = (x_2, y_2)$ be points in the plane, forming a triangle with the origin $O = (0, 0)$.

1. Let (r_1, θ_1), (r_1, θ_2) be the polar coordinate representation of (x_1, y_1), (x_2, y_2), and $\alpha = \theta_2 - \theta_1$. Show that apart from sign, $\frac{1}{2}\begin{vmatrix} x_1 & y_1 \\ x_2 & y_2 \end{vmatrix}$ is the area of the triangle OPQ.

2. Find the area of a parallelogram as a determinant.

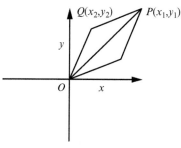

FIGURE 4.1 Geometrical interpretation of the determinant.

4.6.1 MATLAB Problems

4.21 Each matrix is singular. Verify this using the determinant, and apply the MATLAB command `null` to find the nullity of each. What is the rank of A and B?

$$A = \begin{bmatrix} 1 & 19 & -122 \\ 3 & 57 & -366 \\ -1 & -19 & 122 \end{bmatrix}, \quad B = \begin{bmatrix} 1 & 0.25 & -9.25 \\ 3 & 0.75 & -27.75 \\ -17 & 4.25 & 216.75 \end{bmatrix}$$

4.22 Evaluate the following determinants:

$$(a) \begin{vmatrix} 246 & 427 & 327 \\ 1014 & 543 & 443 \\ -342 & 721 & 621 \end{vmatrix} \quad (b) \begin{vmatrix} 1 & 2 & 3 & 4 \\ -2 & 1 & -4 & 3 \\ 3 & -4 & -1 & 2 \\ 4 & 3 & -2 & -1 \end{vmatrix}.$$

4.23 Using the Symbolic Toolbox, express the determinant of the matrix

$$B = \begin{bmatrix} 1 & 1 & 2 & 1 \\ 1 & 2 & 3 & 4 \\ 2 & 4 & 7 & 2t+6 \\ 2 & 2 & 6-t & t \end{bmatrix}$$

as polynomial in t, and determine the values of t for which B^{-1} exists.

4.24 Use the Symbolic Math Toolbox to find the column space of each matrix in Problem 4.21.

4.25 Use MATLAB to compute the determinant of

$$A = \begin{bmatrix} 20 & -34 & 8 & 12 & 3 \\ -99 & 17 & 23 & 67 & 10 \\ 1 & 0 & 3 & 9 & 18 \\ 3 & 5 & 0 & 9 & 11 \\ 7 & 1 & 53 & 5 & 55 \end{bmatrix}.$$

4.26 Using MATLAB, execute

```
>> A = rosser
```

This creates a matrix we will use numerous times throughout the book for testing purposes.

a. Using `rank`, verify that the matrix is singular.

b. Compute its determinant using `det`. Is the output correct?

c. Compute the determinant of A using the Symbolic Toolbox as follows:

```
>> syms A;
>> A = sym(rosser);
>> det(A)
```

d. Given the fact that MATLAB uses row operations to compute the determinant in part (b), suggest a reason for the large difference between the results of (b) and (c).

4.27 The MATLAB command `rand(n)` builds an $n \times n$ matrix containing pseudo random values in the range $0 < x < 1$ drawn from the standard uniform distribution. This means that it is equally probable you will obtain a number from any collection of subintervals of $0 < x < 1$ of the same length; for instance, the chance of obtaining a number from the interval $0.25 \leq x \leq 0.35$ has the same probability as obtaining a number from the interval $0.60 \leq x \leq 0.70$. Find the determinant of random matrices of order 5, 10, 25, 50, 100, 250, 400, and 500. Does a pattern develop? If you got `Inf` as output, what does it mean?

4.28 This interesting problem was described on a MathWorks Web page. The famous Fibonacci numbers are generated from the sequence

$$f_0 = 0, \quad f_1 = 1, \quad f_n = f_{n-1} + f_{n-2}, \quad n \geq 2.$$

Most students have had some exposure to complex numbers. If not, consult Appendix A. In the complex number system, the number $i = \sqrt{-1}$, so $i^2 = -1$, $i^3 = -i$, $i^4 = 1$, and so forth. MATLAB deals naturally with complex numbers. Create a tridiagonal matrix, with ones on the main diagonal, and i on the first sub and super diagonals with the anonymous function

```
fibmat = @(n) eye(n) + diag(repmat(sqrt(-1),n-1,1),1) + diag(repmat(sqrt(-1),n-1,1),-1);
```

The command spy(A) produces a figure placing "*" in locations that are nonzero and leaving the remainder of the figure blank. Verify that the matrix has a tridiagonal pattern by executing the following commands.

```
>> spy(fibmat(5));
>> figure(2);
>> spy(fibmat(10));
>> figure(3);
>> spy(fibmat(25));
```

Compute the determinants of the sequence of matrices fibmat(1), fibmat(2), ..., fibmat(10) by executing the loop

```
for n = 1:10
    det(fibmat(n))
end
```

Comment on the results.

4.29 In Chapter 11, we will show that if no row interchanges are performed, an $n \times n$ matrix can be factored into a product $A = LU$, where L is a lower triangular matrix with ones on its diagonal and U is the upper triangular matrix obtained from Gaussian elimination. For instance, if

$$A = \begin{bmatrix} 1 & 4 & 3 \\ 2 & 9 & 12 \\ -1 & -9 & 3 \end{bmatrix},$$

then

$$A = \begin{bmatrix} 1 & 0 & 0 \\ 2 & 1 & 0 \\ -1 & -5 & 1 \end{bmatrix} \begin{bmatrix} 1 & 4 & 3 \\ 0 & 1 & 6 \\ 0 & 0 & 36 \end{bmatrix}.$$

a. Develop a function

```
function d = ludet(L,U)
```

that takes the factors L and U of a matrix A and computes det A. Check your function by computing the determinant of the matrix A using ludet and MATLAB's det.

b. The function lugauss in the software distribution computes the LU decomposition of a matrix without using row exchanges. It must be the case that during row elimination, a zero never appears on the diagonal. Create three matrices, of dimension 3×3, 4×4, and 5×5, and use ludet to compute the determinant of each matrix. Verify your results using det.

4.30 The $n \times n$ Hilbert matrices are defined by $H(i,j) = 1/(i+j-1), 1 \le i,j \le n$. These matrices are famous because they are ill-conditioned. A small change in a matrix entry or the right-hand side vector can cause the system $Hx = b$ to have a very different solution. We will study ill-conditioning in Chapter 10. For now, we will look at the determinants of Hilbert matrices and their inverses. The MATLAB command hilb(n) builds the $n \times n$ Hilbert matrix.

a. Compute the determinant of the Hilbert matrices of order 5, 10, 15, and 25. What appears to happen as the order increases?

b. It can be shown that the inverse of a Hilbert matrix consists entirely of integers. Compute the inverse of the Hilbert matrix of order 5.

c. Compute the determinant of the inverse for each of the Hilbert matrices of order 5, 10, 15, and 25. Relate your results to those of part (a).

4.31 Encode the message "I LOVE LINEAR ALGEBRA" using the technique described in Section 4.4. Verify that the coded message decodes correctly.

Chapter 5

Eigenvalues and Eigenvectors

You should be familiar with

- Solution to homogeneous systems of equations
- Polynomials and their roots
- Nonsingular and singular matrices
- Diagonal matrices

Let A be an $n \times n$ matrix. For a large number of problems in engineering and science, it is necessary to find vector v such that Av is a multiple, λ, of v; in other words, $Av = \lambda v$. Av is parallel to v, and λ either stretches or shrinks v. The value λ is an eigenvalue and v is an eigenvector associated with λ. Computing eigenvalues and eigenvectors is one of the most important problems in numerical linear algebra. Eigenvalues are critical in such fields as structural mechanics, nuclear physics, biology, the solution of differential equations, computer science, and so on. Eigenvalues play a critical role in the study of vibrations, where they represent the natural frequencies of a system. When vibrating structures begin to have larger and larger amplitudes of vibration, they can have serious problems. Some examples include the wobbling of the Millennium Bridge over the River Thames in London and the collapse of the Tacoma Narrows Bridge in the state of Washington. These are examples of a phenomenon known as *resonance*. A mathematical analysis of a general model for vibrating structures shows that when the system is excited by a harmonic force that depends on time, the system approaches a resonance state when the forces approach or reach a particular eigenvalue. We will discuss specific applications of eigenvalues at various places in the remainder of this book.

Note that some applications involve matrices with complex entries and vectors spaces of complex numbers. We do not deal with matrices of this type in the book; however, many of the techniques we discuss can be adapted for use with complex vectors and matrices. The reader can consult books such as in Refs. [1, 2, 9] for details.

5.1 DEFINITIONS AND EXAMPLES

If A is an $n \times n$ matrix, in order to find eigenvalue λ and an associated eigenvector v, it must be the case that $Av = \lambda v$, and this is equivalent to the homogeneous system

$$(A - \lambda I)\, v = 0. \tag{5.1}$$

We know from Theorem 4.5 that

$$\det (A - \lambda I) = 0 \tag{5.2}$$

in order that the system 5.1 have a nonzero solution. The determinant of $A - \lambda I$ is a polynomial of degree n, so Equation 5.2 is a problem of finding roots of the polynomial

$$p\,(\lambda) = \det (A - \lambda I)\,.$$

Remark 5.1. A polynomial $p\,(\lambda) = a_n \lambda^n + a_{n-1} \lambda^{n-1} + \cdots + a_2 \lambda^2 + a_1 \lambda + a_0$ of degree n has exactly n roots, and any complex roots occur in conjugate pairs. (If you are unfamiliar with complex numbers, see Appendix A.) The polynomial may have one or more roots of multiplicity two or more. This means that the polynomial has a factor $(x - r)^k$, $2 \le k \le n$. The root r is counted k times.

Definition 5.1 officially defines the eigenvalue problem and introduces some terms.

Definition 5.1. If A is an $n \times n$ matrix, the polynomial $p\,(\lambda) = \det (A - \lambda I)$ is called the *characteristic polynomial* of A, and the equation $p\,(\lambda) = 0$ is termed the *characteristic equation*. If λ is a root of p, it is termed an *eigenvalue* of A, and if v

Numerical Linear Algebra with Applications. http://dx.doi.org/10.1016/B978-0-12-394435-1.00005-3

is a nonzero column vector satisfying $Av = \lambda v$, it is an *eigenvector* of A. We say that v is an eigenvector corresponding to the eigenvalue λ.

When multiplied by a matrix A, an arbitrary vector v normally changes direction and length. Eigenvectors are special. The product Av may expand, shrink, or leave the length of v unchanged, but Av will always point in the same direction as v or in the opposite direction.

Example 5.1. Let $A = \begin{bmatrix} -0.4707 & 0.7481 \\ 1.7481 & 1.4707 \end{bmatrix}$, $w = \begin{bmatrix} 1 & -1 \end{bmatrix}^T$, $v = \begin{bmatrix} 0.2898 & 0.9571 \end{bmatrix}^T$.

$$Aw = \begin{bmatrix} -1.2188 & 0.2774 \end{bmatrix}^T,$$
$$Av = \begin{bmatrix} 0.5796 & 1.9142 \end{bmatrix}^T.$$

Note that $Av = \begin{bmatrix} 0.5796 & 1.9142 \end{bmatrix}^T = 2.0v$, and v is an eigenvector of A corresponding to eigenvalue 2.0. Figure 5.1 is a graph of w, Aw, and v, Av. Note how Aw has a different direction than w, but Av points in the same direction as v. ■

The following steps show how to use Definition 5.1 to find the eigenvalues and eigenvectors of an $n \times n$ matrix. It must be noted that in the case of a root, λ, with multiplicity of two or more, there may be only one eigenvector associated with λ. In practice, the accurate computation of eigenvalues and associated eigenvectors is a complex task and is not generally done this way for reasons we will explain in later chapters.

Given an $n \times n$ matrix A:

1. Find the polynomial $p(\lambda) = \det(A - \lambda I)$.
2. Compute the n roots of $p(\lambda) = 0$. These are the eigenvalues λ_1, λ_2, ...,λ_n of A.
3. For each distinct λ_i, find an eigenvector x_i such that

$$Ax_i = \lambda_i x$$

by solving

$$(A - \lambda I)v = 0$$

If λ_i is a multiple root, there may be only one associated eigenvector. If not, compute the distinct eigenvectors.

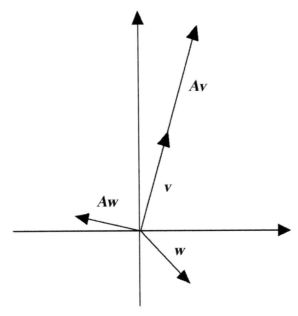

FIGURE 5.1 Direction of eigenvectors.

Example 5.2. Find the eigenvalues of $A = \begin{bmatrix} 2 & 1 \\ 1 & 2 \end{bmatrix}$ and all the eigenvectors. The characteristic polynomial is $\det\left(\begin{bmatrix} 2-\lambda & 1 \\ 1 & 2-\lambda \end{bmatrix}\right) = (2-\lambda)^2 - 1 = \lambda^2 - 4\lambda + 3$, so the eigenvalues are the roots of the characteristic equation $\lambda^2 - 4\lambda + 3 = (\lambda - 1)(\lambda - 3) = 0$. Hence, $\lambda = 1$ and $\lambda = 3$ are the eigenvalues of A. To find the eigenvector corresponding to $\lambda = 1$, solve the homogeneous system $\begin{bmatrix} 2-(1) & 1 \\ 1 & 2-(1) \end{bmatrix}\begin{bmatrix} x \\ y \end{bmatrix} = \begin{bmatrix} 1 & 1 \\ 1 & 1 \end{bmatrix}\begin{bmatrix} x \\ y \end{bmatrix} = \begin{bmatrix} 0 \\ 0 \end{bmatrix}$, which corresponds to the equations

$$x + y = 0,$$
$$x + y = 0.$$

These two equations simply say that the sum of x and y must be 0, so $x = -y$. Consider y to be a parameter that varies through all real numbers not equal to 0. Consequently, the eigenvectors corresponding to $\lambda = 1$ are the vectors of the form $\begin{bmatrix} x \\ y \end{bmatrix} = \begin{bmatrix} -y \\ y \end{bmatrix} = y\begin{bmatrix} -1 \\ 1 \end{bmatrix}$, with $y \neq 0$. Choose $y = 1$ to obtain a specific eigenvector $\begin{bmatrix} -1 \\ 1 \end{bmatrix}$.

Taking $\lambda = 3$ gives the two equations

$$-x + y = 0,$$
$$x - y = 0.$$

These equations require that $x = y$. Again, considering y to be a parameter, $\begin{bmatrix} x \\ y \end{bmatrix} = \begin{bmatrix} y \\ y \end{bmatrix} = y\begin{bmatrix} 1 \\ 1 \end{bmatrix}$. Let $y = 1$ to obtain the eigenvector $\begin{bmatrix} 1 \\ 1 \end{bmatrix}$. ■

Summary: The final result is

λ	Eigenvectors
1	$\begin{bmatrix} -1 \\ 1 \end{bmatrix}$
3	$\begin{bmatrix} 1 \\ 1 \end{bmatrix}$

■

Although more work is involved, the same procedure can be performed to compute the eigenvalues and corresponding eigenvectors of a 3×3 matrix.

Example 5.3. Let $A = \begin{bmatrix} 4 & 8 & 3 \\ 0 & -1 & 0 \\ 0 & -2 & 2 \end{bmatrix}$. Determine the characteristic polynomial:

$$\det(A - \lambda I) = \det\begin{bmatrix} 4-\lambda & 8 & 3 \\ 0 & -1-\lambda & 0 \\ 0 & -2 & 2-\lambda \end{bmatrix}$$

$$= (4-\lambda)\det\left(\begin{bmatrix} -1-\lambda & 0 \\ -2 & 2-\lambda \end{bmatrix}\right) = (4-\lambda)(-1-\lambda)(2-\lambda).$$

The roots of the characteristic polynomial are $\lambda_1 = 4$, $\lambda_2 = -1$, $\lambda_3 = 2$. We will find three eigenvectors $x_1 = \begin{bmatrix} x_{11} \\ x_{21} \\ x_{31} \end{bmatrix}$, $x_2 = \begin{bmatrix} x_{12} \\ x_{22} \\ x_{32} \end{bmatrix}$, $x_3 = \begin{bmatrix} x_{13} \\ x_{23} \\ x_{33} \end{bmatrix}$ by finding nonzero solutions to

$$\begin{bmatrix} 4-\lambda & 8 & 3 \\ 0 & -1-\lambda & 0 \\ 0 & -2 & 2-\lambda \end{bmatrix} x = 0. \tag{5.3}$$

for each value of λ.

$\lambda_1 = 4$ in Equation 5.3:

Solve the homogeneous system $\begin{bmatrix} 0 & 8 & 3 \\ 0 & -5 & 0 \\ 0 & -2 & -2 \end{bmatrix} \begin{bmatrix} x_{11} \\ x_{21} \\ x_{31} \end{bmatrix} = 0$. Using Gaussian elimination, we have

$\begin{bmatrix} 0 & 8 & 3 \\ 0 & -5 & 0 \\ 0 & -2 & -2 \end{bmatrix} \xrightarrow{R3 = R3 - 2/5\,R2} = \begin{bmatrix} 0 & 8 & 3 \\ 0 & -5 & 0 \\ 0 & 0 & -2 \end{bmatrix}$. Thus, $x_{31} = x_{21} = 0$. The first row specifies that $(0)x_1 + 8(0) +$

$3(0) = 0$. The component x_{11} is not constrained. Any value of x_{11} will work. Choose $x_{11} = 1$ to obtain the eigenvector
$x_1 = \begin{bmatrix} 1 \\ 0 \\ 0 \end{bmatrix}$.

$\lambda_2 = -1$ in Equation 5.3:

The homogeneous system we need to solve is $\begin{bmatrix} 5 & 8 & 3 \\ 0 & 0 & 0 \\ 0 & -2 & 3 \end{bmatrix} \begin{bmatrix} x_{12} \\ x_{22} \\ x_{32} \end{bmatrix} = 0$. Exchange rows 2 and 3 to obtain the system

$\begin{bmatrix} 5 & 8 & 3 \\ 0 & -2 & 3 \\ 0 & 0 & 0 \end{bmatrix} \begin{bmatrix} x_{12} \\ x_{22} \\ x_{32} \end{bmatrix} = 0$. The second row of the system specifies that $-2x_{22} + 3x_{32} = 0$, so $x_{22} = 3/2x_{32}$. The first row

requires that $5x_{12} + 8(3/2x_{32}) + 3x_{32} = 0$, and $x_{12} = -3x_{32}$. This gives a general eigenvector of $x_{32} \begin{bmatrix} -3 \\ 3/2 \\ 1 \end{bmatrix}$. If we choose

$x_{32} = 1$, the eigenvector is $x_2 = \begin{bmatrix} -3 \\ 3/2 \\ 1 \end{bmatrix}$.

$\lambda_3 = 2$ in Equation 5.3:

Solve the homogeneous system $\begin{bmatrix} 2 & 8 & 3 \\ 0 & -3 & 0 \\ 0 & -2 & 0 \end{bmatrix} \begin{bmatrix} x_{13} \\ x_{23} \\ x_{33} \end{bmatrix} = 0$. Gaussian elimination gives $\begin{bmatrix} 2 & 8 & 3 \\ 0 & -3 & 0 \\ 0 & -2 & 0 \end{bmatrix}$

$\xrightarrow{R3 = R3 - (-2/3)\,R2} = \begin{bmatrix} 2 & 8 & 3 \\ 0 & -3 & 0 \\ 0 & 0 & 0 \end{bmatrix}$. The second row requires that $x_{23} = 0$. Row 1 specifies that $2x_{13} + 8(0) + 3x_{33} = 0$,

and $x_{13} = -3/2x_{33}$. This gives the general eigenvector $x_3 = x_{33} \begin{bmatrix} -3/2 \\ 0 \\ 1 \end{bmatrix}$. By choosing $x_{33} = 1$, the eigenvector is

$x_3 = \begin{bmatrix} -3/2 \\ 0 \\ 1 \end{bmatrix}$. ∎

Example 5.4. Let $A = \begin{bmatrix} 6 & 12 & 19 \\ -9 & -20 & -33 \\ 4 & 9 & 15 \end{bmatrix}$. The characteristic polynomial of A is $(\lambda + 1)(\lambda - 1)^2$, so $\lambda = 1$ is a multiple
root. To find an eigenvector(s) associated with $\lambda = 1$, we need to solve

$$\begin{bmatrix} 5 & 12 & 19 \\ -9 & -21 & -33 \\ 4 & 9 & 14 \end{bmatrix} \begin{bmatrix} x_1 \\ x_2 \\ x_3 \end{bmatrix} = \begin{bmatrix} 0 \\ 0 \\ 0 \end{bmatrix}.$$

After Gaussian elimination, we obtain the upper-triangular matrix $\begin{bmatrix} 5 & 12 & 19 \\ 0 & 3/5 & 6/5 \\ 0 & 0 & 0 \end{bmatrix}$, and a solution to the upper-triangular

system is $x_3 \begin{bmatrix} 1 \\ -2 \\ 1 \end{bmatrix}$. There is only one linearly independent eigenvector associated with $\lambda = 1$. ∎

There are cases where an eigenvalue of multiplicity k does produce k linearly independent eigenvectors.

Example 5.5. If $A = \begin{bmatrix} 1 & 1 & 1 \\ 1 & 1 & 1 \\ 1 & 1 & 1 \end{bmatrix}$, the characteristic equation is $\lambda^2 (\lambda - 3)$, and $\lambda = 0$ is an eigenvalue of multiplicity 2. After performing Gaussian elimination, the homogeneous equation is

$$\begin{bmatrix} 1 & 1 & 1 \\ 0 & 0 & 0 \\ 0 & 0 & 0 \end{bmatrix} \begin{bmatrix} x_1 \\ x_2 \\ x_3 \end{bmatrix} = \begin{bmatrix} 0 \\ 0 \\ 0 \end{bmatrix},$$

and its solution is $x_1 = -x_2 - x_3$, where x_2 and x_3 are arbitrary. Thus, any solution of the homogeneous system is of the form

$$x = c_1 \begin{bmatrix} -1 \\ 1 \\ 0 \end{bmatrix} + c_2 \begin{bmatrix} -1 \\ 0 \\ 1 \end{bmatrix}.$$

$\begin{bmatrix} -1 \\ 1 \\ 0 \end{bmatrix}$ and $\begin{bmatrix} -1 \\ 0 \\ 1 \end{bmatrix}$ are linearly independent eigenvectors. ∎

Remark 5.2. Note that the matrix in Example 5.5 is symmetric. Whenever an $n \times n$ real matrix is symmetric, it has n linearly independent eigenvectors, even if its characteristic equation has roots of multiplicity 2 or more. This will be proved in Chapter 19.

5.2 SELECTED PROPERTIES OF EIGENVALUES AND EIGENVECTORS

There are some properties of eigenvalues and eigenvectors you should know, and developing them will support your understanding of the eigenvalue problem. First, there is a relation between the eigenvalues of a matrix and whether the matrix is invertible.

Proposition 5.1. *An $n \times n$ matrix A is singular if and only if it has a 0 eigenvalue.*

Proof. If A is singular, by Theorem 4.3, $Ax = 0$ has a solution $x \neq 0$. Thus, $Ax = (0) x = 0$, and $\lambda = 0$ is an eigenvalue.

If A has a eigenvalue $\lambda = 0$, then there exists a vector $x \neq 0$ such that $Ax = \lambda x = 0$, and the homogeneous system $Ax = 0$ has a nontrivial solution. If A is nonsingular, then $x = A^{-1}0 = 0$ is the unique solution. Thus, A is singular. □

Note that if v is an eigenvector of A corresponding to eigenvalue λ and α is a constant, then $A (\alpha v) = \alpha A v = \alpha (\lambda v) = \lambda (\alpha v)$, and αv is an eigenvector of A. This causes us to suspect that the set of eigenvectors corresponding to λ is a subspace.

Lemma 5.1. *Together with 0, the eigenvectors corresponding to λ form a subspace called an eigenspace.*

Proof. To show that a set of vectors form a subspace S, we must show that 0 is in S, that αv is in S for any constant α and any v in S, and that if v_1, v_2 are in S, then so is $v_1 + v_2$. Let S be the set containing the zero vector and all eigenvectors of A corresponding to eigenvalue λ. By hypothesis, 0 is in S. We already showed that if v is an eigenvector corresponding to λ, then so is αv for any constant α. If v_1 and v_2 are eigenvectors of A, then $A (v_1 + v_2) = Av_1 + Av_2 = \lambda (v_1 + v_2)$, and $v_1 + v_2$ is in S, so S is a subspace. □

The next result will support your understanding of the relationship between eigenvalues and the roots of the characteristic equation.

Proposition 5.2. *If A is an $n \times n$ matrix, then $\det A = \Pi_{i=1}^{n} \lambda_i$.*

Proof. First assume that all the eigenvalues $\lambda_1, \lambda_2, \ldots, \lambda_n$ of A are distinct; in other words $\lambda_k \neq \lambda_j, k \neq j$. The characteristic polynomial $p (\lambda)$ is of degree n, and is the determinant of

$$
\begin{bmatrix}
a_{11} - \lambda & a_{12} & \cdots & & \cdots & a_{1n} \\
a_{21} & a_{22} - \lambda & \cdots & & \cdots & a_{2n} \\
\vdots & \vdots & \ddots & & \cdots & \vdots \\
\vdots & \vdots & & \cdots & a_{n-1,n-1} - \lambda & a_{n-1,n} \\
a_{n1} & a_{n2} & \cdots & & a_{n,n-1} & a_{nn} - \lambda
\end{bmatrix}.
$$

Expansion by minors shows us that the leading term of $p(\lambda)$ is $(-1)^n \lambda^n$. By Remark 5.1,

$$
\det(A - \lambda I) = p(\lambda) = (-1)^n (\lambda - \lambda_1)(\lambda - \lambda_2) \ldots (\lambda - \lambda_n).
$$

Let $\lambda = 0$, and we have $\det A = (-1^n)(-1)^n \lambda_1 \lambda_2 \ldots \lambda_n = \Pi_{i=1}^n \lambda_i$.

Now assume that one or more eigenvalues are repeated. In this case, $p(\lambda)$ has one or more factors of the form $(\lambda - \lambda_i)^k$, where $k \geq 2$. Think of such a factor as

$$
(\lambda - \lambda_{i1})(\lambda - \lambda_{i2}) \ldots (\lambda - \lambda_{ik}),
$$

where $\lambda_{i1} = \lambda_{i2} = \cdots = \lambda_{ik}$. The same argument we just gave shows that $\det A = \Pi_{i=1}^n \lambda_i$. $\square$

5.3 DIAGONALIZATION

In this section, we will show that, under the right conditions, we can use the eigenvectors of a matrix to transform it into a diagonal matrix of eigenvalues. The process is termed diagonalization, and is an important concept in matrix algebra; in fact, it is critical to developing results such as the singular value decomposition (Chapter 15), and computing eigenvalues of symmetric matrices (Chapter 19).

Definition 5.2. Matrix B is *similar* to matrix A if there exists a nonsingular matrix X such that

$$
B = X^{-1}AX.
$$

Example 5.6. Let $A = \begin{bmatrix} 1 & -1 & 2 \\ -2 & 1 & 1 \\ -1 & 3 & 1 \end{bmatrix}$ and $X = \begin{bmatrix} 1 & 2 & 1 \\ 2 & 3 & 3 \\ 4 & 7 & 6 \end{bmatrix}$. Now, $X^{-1} = \begin{bmatrix} 3 & 5 & -3 \\ 0 & -2 & 1 \\ -2 & -1 & 1 \end{bmatrix}$, so

$$
B = X^{-1}AX = \begin{bmatrix} 14 & 27 & 23 \\ 1 & 2 & 0 \\ -9 & -18 & -13 \end{bmatrix}
$$

is similar to A. ∎

Definition 5.3. The $n \times n$ matrix A is *diagonalizable* if it is similar to a diagonal matrix. We also say that A can be *diagonalized*.

Example 5.7. Let $A = \begin{bmatrix} -3 & 6 & -2 \\ -12 & 7 & 0 \\ -24 & 16 & -1 \end{bmatrix}$, and X be the matrix in Example 5.6. Then,

$$
X^{-1}AX = \begin{bmatrix} 1 & 0 & 0 \\ 0 & -1 & 0 \\ 0 & 0 & 3 \end{bmatrix},
$$

and A is diagonalizable. ∎

Remark 5.3. Determining whether a matrix can be diagonalized and performing the diagonalization requires that we prove some results.

The fact that $\det(AB) = \det(A)\det(B)$ can be used to show the relation between the determinant of A and that of its inverse.

Lemma 5.2.   *If A is invertible, then* $\det(A)\det(A^{-1}) = 1$.

Proof. Since A is invertible, $AA^{-1} = I$. The determinant of a product is the product of the determinants, so

$$\det(AA^{-1}) = \det(A)\det(A^{-1}) = \det(I) = 1.$$

We can use Lemma 5.2 to prove the following useful result.   □

Theorem 5.1.   *Two similar matrices have the same eigenvalues.*

Proof. Assume A and B are similar matrices. Then, $B = X^{-1}AX$ for some nonsingular matrix X. It follows that

$$\det(B - \lambda I) = \det\left(X^{-1}AX - \lambda X^{-1}X\right) = \det\left(X^{-1}[A - \lambda I]X\right)$$
$$= \det\left(X^{-1}\right)\det(A - \lambda I)\det(X) = \det\left(X^{-1}\right)\det(X)\det(A - \lambda I) = \det(A - \lambda I).$$

A and B have the same characteristic polynomial and thus the same eigenvalues.   □

If A is diagonalizable, then there exist matrices X and D such that $D = X^{-1}AX$. A is similar to D and thus has the same eigenvalues as D. Now, D has the form

$$D = \begin{bmatrix} d_{11} & & & & \\ & d_{22} & & & \\ & & \ddots & & \\ & & & d_{n-1,n-1} & \\ & & & & d_{nn} \end{bmatrix},$$

so the eigenvalues of D are the roots of

$$\det\left(\begin{bmatrix} d_{11} - \lambda & & & & \\ & d_{22} - \lambda & & & \\ & & \ddots & & \\ & & & d_{n-1,n-1} - \lambda & \\ & & & & d_{nn} - \lambda \end{bmatrix}\right) = (d_{11} - \lambda)(d_{22} - \lambda)\ldots(d_{nn} - \lambda).$$

The eigenvalues of D are $\{d_{11}, d_{22}, \ldots, d_{nn}\}$.

The next result is useful in its own right, and we will have occasion to apply it a number of times in this book.

Theorem 5.2.   *Eigenvectors $v_1, v_2, \ldots, v_i$ that correspond to distinct eigenvalues are linearly independent.*

Overview:

The proof is algebraic. If $c_1v_1 + c_2v_2 + \cdots + c_iv_i = 0$ and we show that $c_k = 0$, $1 \leq k \leq i$, then $v_1, v_2, \ldots, v_i$ are linearly independent. Pairs of equations are created and subtracted, and in the process v_i is eliminated. Continue the process and eliminate v_{i-1} and so forth until arriving at an equation $Kc_1v_1 = 0$, $K \neq 0$. Since $v_1 \neq 0$, $c_1 = 0$. The same process can be used to show that $c_2 = 0, \ldots, c_i = 0$.

Proof. Suppose that

$$c_1v_1 + c_2v_2 + \cdots + c_iv_i = 0. \tag{5.4}$$

Multiply by A, noting that $Av_i = \lambda_iv_i$, to obtain

$$c_1\lambda_1v_1 + c_2\lambda_2v_2 + \cdots + c_i\lambda_iv_i = 0. \tag{5.5}$$

Multiply Equation 5.4 by λ_i and subtract from Equation 5.5 to obtain Equation 5.6 that does not involve v_i.

$$c_1(\lambda_1 - \lambda_i)v_1 + c_2(\lambda_2 - \lambda_i)v_2 + c_3(\lambda_3 - \lambda_i)v_3$$
$$+ \cdots + c_{i-1}(\lambda_{i-1} - \lambda_i)v_{i-1} = 0. \tag{5.6}$$

Multiply Equation 5.6 by A to obtain

$$c_1 (\lambda_1 - \lambda_i) \lambda_1 v_1 + c_2 (\lambda_2 - \lambda_i) \lambda_2 v_2 + c_3 (\lambda_3 - \lambda_i) \lambda_3 v_3$$
$$+ \cdots + c_{i-1} (\lambda_{i-1} - \lambda_i) \lambda_{i-1} v_{i-1} = 0. \tag{5.7}$$

Multiply Equation 5.6 by λ_{i-1} and subtract from Equation 5.7 to get an equation not involving v_{i-1}.

$$c_1 (\lambda_1 - \lambda_i)(\lambda_1 - \lambda_{i-1}) v_1 + c_2 (\lambda_2 - \lambda_i)(\lambda_2 - \lambda_{i-1}) v_2$$
$$+ \cdots + c_{i-2} (\lambda_{i-2} - \lambda_i)(\lambda_{i-2} - \lambda_{i-1}) v_{i-2} = 0.$$

If we continue by eliminating v_{i-2}, v_{i-3}, and so forth until eliminating v_2, we are left with

$$(\lambda_1 - \lambda_2)(\lambda_1 - \lambda_3) \ldots (\lambda_1 - \lambda_i) c_1 v_1 = 0.$$

Now $(\lambda_1 - \lambda_2) \neq 0$, $(\lambda_1 - \lambda_3) \neq 0$, ..., $(\lambda_1 - \lambda_i) \neq 0$ by hypothesis, and so $c_1 = 0$. In a similar fashion, we can show that $c_2 = c_3 = \cdots = c_i = 0$, and thus v_1, v_2, ..., v_i are linearly independent. $\qquad\square$

If A has n linearly independent eigenvectors, we are now in a position to develop a method for diagonalizing A.

Theorem 5.3. *Suppose the $n \times n$ matrix A has n linearly independent eigenvectors v_1, v_2, ..., v_n. Place the eigenvectors as columns of the eigenvector matrix $X = [v_1, v_2, \ldots v_n]$. Then*

$$X^{-1}AX = D = \begin{bmatrix} \lambda_1 & & & 0 \\ & \lambda_2 & & \\ & & \ddots & \\ 0 & & & \lambda_n \end{bmatrix},$$

and A can be diagonalized.

Proof.

$$AX = A \begin{bmatrix} v_1 & v_2 & \ldots & v_n \end{bmatrix} = \begin{bmatrix} Av_1 & Av_2 & \ldots & Av_n \end{bmatrix}$$
$$= \begin{bmatrix} \lambda_1 v_1 & \lambda_2 v_2 & \ldots & \lambda_n v_n \end{bmatrix}$$
$$= \begin{bmatrix} v_1 & v_2 & \ldots & v_n \end{bmatrix} \begin{bmatrix} \lambda_1 & & & 0 \\ & \lambda_2 & & \\ & & \ddots & \\ 0 & & & \lambda_n \end{bmatrix} = XD. \tag{5.8}$$

Since v_1, v_2, ..., v_n are linearly independent, X is invertible. From Equation 5.8, we have

$$D = X^{-1}AX. \qquad\square$$

To diagonalize matrix A we need to know that it has n linearly independent eigenvalues. Having distinct eigenvalues does the trick.

Theorem 5.4. *If an $n \times n$ matrix A has distinct eigenvalues, it can be diagonalized.*

Proof. Let v_1, v_2, ..., v_n be eigenvectors of A corresponding to eigenvalues λ_1, λ_2, ..., λ_n, respectively. By Theorem 5.2, v_1, v_2, ..., v_n are linearly independent and, by Theorem 5.3, A can be diagonalized. $\qquad\square$

Remark 5.4. If the matrix does not have n linearly independent eigenvectors, it cannot be diagonalized.

Example 5.8. Let $A = \begin{bmatrix} 4 & 8 & 3 \\ 0 & -1 & 0 \\ 0 & -2 & 2 \end{bmatrix}$ be the matrix of Example 5.3. We found that the eigenvalues are $\lambda_1 = 4$, $\lambda_2 = -1$,

and $\lambda_3 = 2$. By Theorem 5.4, A can be diagonalized. In Example 5.3, we found eigenvectors $v_1 = \begin{bmatrix} 1 \\ 0 \\ 0 \end{bmatrix}$, $v_2 = \begin{bmatrix} -3 \\ 3/2 \\ 1 \end{bmatrix}$, and $v_3 = \begin{bmatrix} -3/2 \\ 0 \\ 1 \end{bmatrix}$ corresponding to λ_1, λ_2, and λ_3, respectively. To diagonalize A, form the eigenvector matrix

$$X = \begin{bmatrix} v_1 & v_2 & v_3 \end{bmatrix} = \begin{bmatrix} 1 & -3 & -3/2 \\ 0 & 3/2 & 0 \\ 0 & 1 & 1 \end{bmatrix}.$$

It is straightforward to verify that $X^{-1}AX = \begin{bmatrix} 4 & 0 & 0 \\ 0 & -1 & 0 \\ 0 & 0 & 2 \end{bmatrix}$. ∎

Remark 5.2 states that a real symmetric matrix A always has n linearly independent eigenvectors and so can be diagonalized. If the characteristic equation of a nonsymmetric matrix A has a factor $(\lambda - \lambda_i)^k$, $k \geq 2$ there must be k linearly independent eigenvectors associated with eigenvalue λ_i for A to be diagonalizable.

Example 5.9. Let $A = \begin{bmatrix} 6 & 12 & 19 \\ -9 & -20 & -33 \\ 4 & 9 & 15 \end{bmatrix}$ be the matrix of Example 5.4. The characteristic polynomial of A is $(\lambda + 1)(\lambda - 1)^2$, so $\lambda = 1$ is a multiple eigenvalue with multiplicity $k = 2$. We found that there is only one linearly independent eigenvector associated with $\lambda = 1$, so A cannot be diagonalized. ∎

In summary, the procedure for diagonalizing a matrix A can be done in series of steps:

---

1. Form the characteristic polynomial $p(\lambda) = \det(A - \lambda I)$ of A.
2. Find the roots of p. If there are complex roots, the matrix cannot be diagonalized in $\mathbb{R}^{n \times n}$.
3. For each eigenvalue λ_i of multiplicity k_i, find k_i linearly independent eigenvectors. If this is not possible, A cannot be diagonalized.
4. Form the matrix $X = \begin{bmatrix} v_1 & v_2 & \dots & v_{n-1} & v_n \end{bmatrix}$ whose columns are eigenvectors of A corresponding to eigenvalues λ_1, λ_2, ... λ_{n-1}, λ_n. Then, $D = X^{-1}AX$, where D is the diagonal matrix with $\lambda_1, \lambda_2, \dots \lambda_{n-1}, \lambda_n$ on its diagonal.

Example 5.10. Let $A = \begin{bmatrix} -5 & 2 & 0 \\ 0 & 1 & 0 \\ 2 & -1 & 1 \end{bmatrix}$. Its eigenvalues are $\lambda_1 = 1$, $\lambda_2 = 1$, $\lambda_3 = -5$. For A to be diagonalizable, there must be two linearly independent eigenvectors corresponding to $\lambda = 1$. We must solve the homogeneous system

$$\begin{bmatrix} -6 & 2 & 0 \\ 0 & 0 & 0 \\ 2 & -1 & 0 \end{bmatrix} x = 0.$$

Its solution space is all multiples of $\begin{bmatrix} 0 \\ 0 \\ 1 \end{bmatrix}$. This subspace has dimension 1, so A is not diagonalizable. ∎

Example 5.11. Let $A = \begin{bmatrix} 1 & 0 & 0 & 0 \\ 4 & 3 & 0 & 0 \\ -2 & 2 & 0 & 0 \\ 5 & -1 & 0 & 0 \end{bmatrix}$. By performing expansion by minors down column 4, we see that the determinant of A is 0. Thus, A is not invertible and has an eigenvalue of 0. In fact, the eigenvalues of A are $\lambda_1 = \lambda_2 = 0$, $\lambda_3 = 3$, $\lambda_4 = 1$, so the multiplicity of the 0 eigenvalue is 2. The homogeneous system

$$\begin{bmatrix} 1-0 & 0 & 0 & 0 \\ 4 & 3-0 & 0 & 0 \\ -2 & 2 & 0-0 & 0 \\ 5 & -1 & 0 & 0-0 \end{bmatrix} x = 0$$

can be row-reduced to the problem

$$\begin{bmatrix} 1 & 0 & 0 & 0 \\ 0 & -1 & 0 & 0 \\ 0 & 0 & 0 & 0 \\ 0 & 0 & 0 & 0 \end{bmatrix} x = 0,$$

which has the two solutions $x_1 = \begin{bmatrix} 0 & 0 & 1 & 0 \end{bmatrix}^{\mathrm{T}}$, $x_2 = \begin{bmatrix} 0 & 0 & 0 & 1 \end{bmatrix}^{\mathrm{T}}$. Now, $\begin{bmatrix} 0 & -3 & -2 & 1 \end{bmatrix}^{\mathrm{T}}$ is an eigenvector corresponding to $\lambda_3 = 3$, and $\begin{bmatrix} -1/6 & 1/3 & 1 & -7/6 \end{bmatrix}^{\mathrm{T}}$ corresponds to $\lambda_4 = 1$. A can be diagonalized by

$$X = \begin{bmatrix} 0 & 0 & 0 & -1/6 \\ 0 & 0 & -3 & 1/3 \\ 1 & 0 & -2 & 1 \\ 0 & 1 & 1 & -7/6 \end{bmatrix}. \qquad \blacksquare$$

Remark 5.5. The procedure we have presented for diagonalizing a matrix is based upon finding the roots of the characteristic equation. As we will see in the beginning of Chapter 10, the procedure is never done this way in practice because the roots of a polynomial are generally difficult to compute accurately.

5.3.1 Powers of Matrices

If a matrix A can be diagonalized, computing A^n is greatly simplified. Since $D = X^{-1}AX$, $A = XDX^{-1}$, and

$$A^2 = \left(XDX^{-1}\right)\left(XDX^{-1}\right) = (XD)\,I\left(DX^{-1}\right) = XD^2X^{-1}.$$

Continuing, we have

$$A^3 = A^2A = \left(XD^2X^{-1}\right)\left(XDX^{-1}\right) = XD^3X^{-1},$$

and in general by mathematical induction (Appendix B)

$$A^n = XD^nX^{-1}.$$

Example 5.12. The matrix $F = \begin{bmatrix} 1 & 1 \\ 1 & 0 \end{bmatrix}$ is called the Fibonacci matrix because its powers can be used to compute the Fibonacci numbers

$$f_0 = 0, \quad f_1 = 1,$$
$$f_n = f_{n-1} + f_{n-2}, \quad n \geq 2.$$

The first few numbers in the sequence are 0, 1, 2, 3, 5, 8, 13, 21, 34, 55, 89. F is symmetric, so it can be diagonalized. The eigenvalues of F are (verify)

$$\lambda_1 = \frac{1 + \sqrt{5}}{2}, \quad \lambda_2 = \frac{1 - \sqrt{5}}{2},$$

and the corresponding eigenvectors are (verify)

$$v_1 = \begin{bmatrix} \frac{1+\sqrt{5}}{2} \\ 1 \end{bmatrix}, \quad v_2 = \begin{bmatrix} \frac{1-\sqrt{5}}{2} \\ 1 \end{bmatrix}.$$

The eigenvalue $(1 + \sqrt{5})/2 = 1.61803\ldots$ is called the *Golden ratio* and was known to the ancient Greeks. Some artists and architects believe the Golden ratio makes the most pleasing and beautiful shape. Using the eigenvalues and eigenvectors of F, we have

$$F = \begin{bmatrix} \frac{1+\sqrt{5}}{2} & \frac{1-\sqrt{5}}{2} \\ 1 & 1 \end{bmatrix} \begin{bmatrix} \frac{1+\sqrt{5}}{2} & 0 \\ 0 & \frac{1-\sqrt{5}}{2} \end{bmatrix} \begin{bmatrix} \frac{1+\sqrt{5}}{2} & \frac{1-\sqrt{5}}{2\sqrt{5}} \\ 1 & 1 \end{bmatrix}^{-1}$$

$$= \begin{bmatrix} \frac{1+\sqrt{5}}{2} & \frac{1-\sqrt{5}}{2} \\ 1 & 1 \end{bmatrix} \begin{bmatrix} \frac{1+\sqrt{5}}{2} & 0 \\ 0 & \frac{1-\sqrt{5}}{2} \end{bmatrix} \begin{bmatrix} \frac{1}{\sqrt{5}} & -\frac{1-\sqrt{5}}{2\sqrt{5}} \\ -\frac{1}{\sqrt{5}} & \frac{1+\sqrt{5}}{2\sqrt{5}} \end{bmatrix}.$$

Now $F^n = \begin{bmatrix} \frac{1+\sqrt{5}}{2} & \frac{1-\sqrt{5}}{2} \\ 1 & 1 \end{bmatrix} \begin{bmatrix} \left(\frac{1+\sqrt{5}}{2}\right)^n & 0 \\ 0 & \left(\frac{1-\sqrt{5}}{2}\right)^n \end{bmatrix} \begin{bmatrix} \frac{1}{\sqrt{5}} & -\frac{1-\sqrt{5}}{2\sqrt{5}} \\ -\frac{1}{\sqrt{5}} & \frac{1+\sqrt{5}}{2\sqrt{5}} \end{bmatrix}$. Using MATLAB to compute F^{50} gives

$$\begin{bmatrix} 20365011074 & 12586269025 \\ 12586269025 & 7778742049 \end{bmatrix}.$$

It can be shown that

$$F^n = \begin{bmatrix} f_{n+1} & f_n \\ f_n & f_{n-1} \end{bmatrix},$$

and

$$f_n = \frac{1}{\sqrt{5}}\left[\left(\frac{1+\sqrt{5}}{2}\right)^{n+1} - \left(\frac{1-\sqrt{5}}{2}\right)^{n+1}\right].$$

The latter formula is quite remarkable since each term in the formula involves $\sqrt{5}$. From F^{50}, we deduce that $f_{51} = 20365011074$, $f_{50} = 12586269025$, and $f_{49} = 7778742049$. ∎

5.4 APPLICATIONS

In the introduction of this chapter, we noted that eigenvalues and eigenvectors have very significant applications to engineering and science. In this section, two applications are outlined. Later in this book, when we have more understanding of eigenvalues and eigenvectors, other applications will be presented in more detail.

5.4.1 Electric Circuit

In Section 2.8, we solved for the currents in a circuit that involved three batteries and four resistors (Figure 2.4). The resistors obey Ohm's Law, $V = RI$, where V is the voltage, R is the resistance, and I is the current. To determine the currents, we had to solve a system of linear algebraic equations. We will now add two inductors to the circuit (Figure 5.2). The relationship between the voltage $v(t)$ across an inductor with inductance L and the current $x(t)$ passing through it is described by the relation $v(t) = L dx/dt$. In other words, the voltage across an inductor is proportional to the rate of change of the current. As a result, the problem of determining the currents becomes a system of differential equations

$$x_1 - x_3 + x_2 = 0,$$

$$(R_1 + R_4)x_1 + R_2 x_3 + L_2 \frac{dx_1}{dt} = V_1 - V_2,$$

$$R_2 x_3 + R_3 x_2 + L_1 \frac{dx_2}{dt} = V_3 - V_2.$$

Choose the following values for the batteries, the resistors, and the inductors.

Components	V_1	V_2	V_3	R_1	R_2	R_3	R_4	L_1	L_2
	2 V	3 V	5 V	1 Ω	2 Ω	5 Ω	3 Ω	1 H	1 H

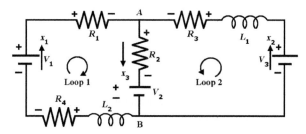

FIGURE 5.2 Circuit with an inductor.

The equations for current flow then become

$$x_1 - x_3 + x_2 = 0,$$

$$4x_1 + 2x_2 + \frac{dx_1}{dt} = -1,$$

$$2x_2 + 5x_3 + \frac{dx_2}{dt} = 2.$$

Solving for x_3 in terms of x_1 and x_2 results in the following system of differential equations:

$$\frac{dx_1}{dt} = -4x_1 - 2x_2 - 1,$$

$$\frac{dx_2}{dt} = -5x_1 - 7x_2 + 2,$$

which after conversion to matrix form is

$$\frac{dx}{dt} = Ax + b, \qquad (5.9)$$

where $A = \begin{bmatrix} -4 & -2 \\ -5 & -7 \end{bmatrix}$ and $b = \begin{bmatrix} -1 \\ 2 \end{bmatrix}$.

To find the general solution to a 2×2 system of first-order differential equations with constant coefficients, first find a general solution to the homogeneous system

$$\frac{dx_h}{dt} = Ax_h, \qquad (5.10)$$

and then determine a particular solution, $x_p(t)$ to Equation 5.9. The function $x(t) = x_h(t) + x_p(t)$ is a solution to Equation 5.9, since

$$\frac{dx}{dt} = \frac{dx_h}{dt} + \frac{dx_p}{dt} = Ax_h(t) + \left(Ax_p(t) + b\right) = A\left(x_h(t) + x_p(t)\right) + b = Ax + b.$$

For a proof that $x(t)$ is a general solution to Equation 5.9, see Ref. [10] or any book on elementary differential equations.

To determine a general solution to Equation 5.9, let $x_h(t) = vf(t)$, where v is a vector and f varies with time. Substituting $x_h(t)$ into Equation 5.10 results in

$$vf'(t) = f(t) Av,$$

so

$$Av = \left(\frac{f'(t)}{f(t)}\right) v.$$

For a fixed t, this is an eigenvalue problem, where $\lambda = f'(t)/f(t)$. If the eigenvalues of A are distinct, there are two eigenvalues λ_1 and λ_2 corresponding to linearly independent eigenvectors v_1 and v_2. For $i = 1, 2$, let

$$\lambda_i = \frac{f'(t)}{f(t)},$$

so

$$f'(t) = \lambda_i f(t),$$

for which a solution is

$$f(t) = c_i e^{\lambda_i t}.$$

Thus, the general solution to the homogeneous equation (5.10) is

$$x_h(t) = c_1 v_1 e^{\lambda_1 t} + c_2 v_2 e^{\lambda_2 t}.$$

It remains to determine a particular solution $x_p(t)$. The right-hand side of Equation 5.9 contains the constant vector b, so we will try a solution of the form $x_p(t) = w$, where w is a constant vector. Substituting this into Equation 5.9 gives $0 = Aw + b$ and, assuming A is nonsingular, w is the unique solution to

$$Aw = -b.$$

We now have the general solution

$$x(t) = c_1 v_1 e^{\lambda_1 t} + c_2 v_2 e^{\lambda_2 t} + w,$$

and are able to solve our problem

$$\frac{dx}{dt} = \begin{bmatrix} -4 & -2 \\ -5 & -7 \end{bmatrix} \begin{bmatrix} x_1 \\ x_2 \end{bmatrix} + \begin{bmatrix} -1 \\ 2 \end{bmatrix}.$$

The eigenvalues of A are $\lambda_1 = -2$ and $\lambda_2 = -9$, with corresponding eigenvectors $v_1 = \begin{bmatrix} -1 \\ 1 \end{bmatrix}$ and $v_2 = \begin{bmatrix} \frac{2}{5} \\ 1 \end{bmatrix}$, so

$$x_h(t) = c_1 v_1 e^{-2t} + c_2 v_2 e^{-9t}.$$

To find $x_p(t)$, solve the system

$$Aw = \begin{bmatrix} 1 \\ -2 \end{bmatrix}.$$

The unique solution is $w = \begin{bmatrix} -0.61111 \\ 0.72222 \end{bmatrix}$, and so the general solution is

$$x(t) = c_1 e^{-2t} \begin{bmatrix} -1 \\ 1 \end{bmatrix} + c_2 e^{-9t} \begin{bmatrix} \frac{2}{5} \\ 1 \end{bmatrix} + \begin{bmatrix} -0.61111 \\ 0.72222 \end{bmatrix}.$$

Assume that at $t = 0$, $x_1(0) = x_2(0) = 0$, so

$$c_1 \begin{bmatrix} -1 \\ 1 \end{bmatrix} + c_2 \begin{bmatrix} \frac{2}{5} \\ 1 \end{bmatrix} + \begin{bmatrix} -0.61111 \\ 0.72222 \end{bmatrix} = 0,$$

that results in the system

$$\begin{bmatrix} -1 & \frac{2}{5} \\ 1 & 1 \end{bmatrix} \begin{bmatrix} c_1 \\ c_2 \end{bmatrix} = \begin{bmatrix} 0.61111 \\ -0.72222 \end{bmatrix},$$

whose solution is

$$c_1 = -0.64286, \quad c_2 = -0.079365.$$

The solution to Equation 5.9 is

$$x(t) = -0.64286 \, e^{-2t} \begin{bmatrix} -1 \\ 1 \end{bmatrix} - 0.079365 \, e^{-9t} \begin{bmatrix} \frac{2}{5} \\ 1 \end{bmatrix} + \begin{bmatrix} -0.61111 \\ 0.72222 \end{bmatrix},$$

and the solution to the circuit problem is

$$x_1(t) = 0.64286 \, e^{-2t} - 0.031746 \, e^{-9t} - 0.61111,$$
$$x_2(t) = -0.64286 \, e^{-2t} - 0.079365 \, e^{-9t} + 0.72222,$$
$$x_3(t) = -0.11111 e^{-9t} + 0.11111,$$

whose graph is shown in Figure 5.3.

Note that the solution to the homogeneous equation (the transient solution) dies out quickly (Figure 5.3), leaving the particular solution (the steady state).

5.4.2 Irreducible Matrices

Our aim in Section 5.4.2 is to show how the eigenvalue problem can be used to create a ranking method. In our case, we will develop a simple method for ranking sports teams, but more advanced methods of ranking using eigenvalues are in use. For instance, sophisticated methods are used to rank NFL teams. Many of these ranking methods require that the matrix be irreducible.

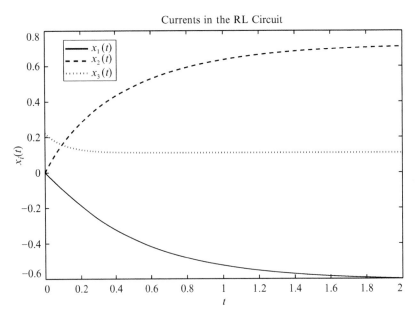

FIGURE 5.3 Currents in the *RL* circuit.

Definition 5.4. An $n \times n$ matrix A is reducible if its indices 1 to n can be divided into two disjoint nonempty sets $S = \{i_1, i_2, \ldots, i_\alpha\}$, $T = \{j_1, j_2, \ldots, j_\beta\}$, $\alpha + \beta = n$, such that

$$a_{i_p, j_q} = 0,$$

for $1 \le p \le \alpha$ and $1 \le q \le \beta$.

Example 5.13. Let $A = \begin{bmatrix} 0 & 0 & 1 & 0 & 0 \\ 1 & 0 & 0 & 2 & 3 \\ 8 & 0 & 0 & 0 & 9 \\ 5 & 0 & 0 & 0 & 0 \\ 0 & 0 & 0 & 12 & 0 \end{bmatrix}$. Let $S = \{1, 3, 4, 5\}$ and $T = \{2\}$. Now, $a_{12} = a_{32} = a_{42} = a_{52} = 0$, so A is reducible.   ∎

For a matrix A to be *irreducible*, it must not be possible to perform such a partitioning. We will discuss two ways other using the definition to show that a matrix is irreducible, one involving simple graph theory, and the other an algebraic approach. For the graph approach, let the set $V = \{1, 2, \ldots, n\}$ and create vertices labeled "1," "2," ..., "n." Connect vertex i to vertex j by a directed arc when $a_{ij} \ne 0$. Such a structure is called a *digraph*. For instance, consider the matrix T

$$T = \begin{bmatrix} 0 & 1 & 0 & 1 & 0 \\ 1 & 0 & 0 & 0 & 1 \\ 0 & 1 & 0 & 1 & 1 \\ 0 & 0 & 1 & 0 & 1 \\ 0 & 0 & 1 & 0 & 0 \end{bmatrix}. \tag{5.11}$$

It is convenient to label the rows and columns:

$$T = \begin{array}{c} \\ 1 \\ 2 \\ 3 \\ 4 \\ 5 \end{array} \begin{array}{ccccc} 1 & 2 & 3 & 4 & 5 \\ \hline \begin{bmatrix} 0 & 1 & 0 & 1 & 0 \\ 1 & 0 & 0 & 0 & 1 \\ 0 & 1 & 0 & 1 & 1 \\ 0 & 0 & 1 & 0 & 1 \\ 0 & 0 & 1 & 0 & 0 \end{bmatrix} \end{array}$$

Figure 5.4 is the digraph for T.

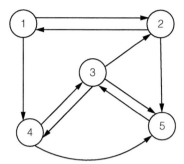

FIGURE 5.4 Digraph of an irreducible matrix.

A matrix is irreducible if beginning at any vertex, arcs can be followed to any other vertex, so that the partitioning in Definition 5.4 is not possible. This type of structure is called a *strongly connected digraph*. Our matrix T is irreducible.

The matrices we deal with for ranking purposes will consist entirely of nonnegative elements (all entries ≥ 0). Such a matrix A is said to be *nonnegative*, and we write $A \geq 0$. If all the elements of A are positive, then $A > 0$. An algebraic approach to verify that a nonnegative matrix is irreducible is specified by the following theorem. For a proof, see Ref. [11].

Theorem 5.5. *A is a nonnegative irreducible $n \times n$ matrix if and only if*

$$(I + A)^{n-1} > 0.$$

Example 5.14. Perform this computation for the matrix T (5.11) and for the matrix A of Example 5.13.

$$(I + T)^4 = \begin{bmatrix} 9 & 14 & 17 & 14 & 22 \\ 9 & 13 & 14 & 12 & 19 \\ 9 & 18 & 26 & 18 & 32 \\ 5 & 13 & 22 & 14 & 26 \\ 4 & 9 & 14 & 9 & 17 \end{bmatrix}, \quad (I + A)^4 = \begin{bmatrix} 9 & 0 & 8 & 4 & 7 \\ 19 & 1 & 12 & 11 & 9 \\ 12 & 0 & 9 & 7 & 8 \\ 8 & 0 & 7 & 2 & 4 \\ 7 & 0 & 4 & 4 & 2 \end{bmatrix}$$

T is irreducible and A is not. ∎

Now we need to see how eigenvalues/eigenvectors are connected with irreducible matrices. The key is the *Perron-Frobenius theorem* for irreducible matrices (for a proof, see Ref. [12]). In the theorem, an eigenvector of matrix A is said to be *simple* if its corresponding eigenvalue is not a multiple root of the characteristic equation for A.

Theorem 5.6. *If the $n \times n$ matrix A has nonnegative entries, then there exists an eigenvector r with nonnegative entries, corresponding to a positive eigenvalue λ. Furthermore, if the matrix A is irreducible, the eigenvector r has strictly positive entries, is unique and simple, and the corresponding eigenvalue is the largest eigenvalue of A in absolute value.*

You know how to compute the distance between two vectors. If v is a vector in $\mathbb{R}^n$, then the length of v is written as length$(v) = \sqrt{v_1^2 + v_2^2 + \cdots + v_n^2}$. The eigenvector r in Theorem 5.6 is computed by the formula

$$r = \lim_{n \to \infty} \frac{A^n r_0}{\text{length}(A^n r_0)},$$

for any nonnegative vector r_0. For the purposes of computing the ranking vector, the book software distribution contains a MATLAB function, `perronfro`, that takes the matrix as an argument and returns an approximation to r and the corresponding eigenvalue λ. This process of computing an eigenvector is called the power method and will be discussed in Chapter 18.

Example 5.15. The matrix T (5.11) is irreducible, and $T \geq 0$, so the Perron-Frobenius theorem applies. The following MATLAB statements find the unique eigenvector and the corresponding largest eigenvalue in magnitude.

```
>> [r lambda] = perronfro(T)

r =
    0.4366
    0.3829
    0.5945
    0.4643
    0.3064

lambda =
    1.9404
```

∎

5.4.3 Ranking of Teams Using Eigenvectors

Have you ever wondered how the Google search engine orders the results of a search? It uses a very large matrix and applies the PageRank process, which involves computing eigenvectors. We will not attempt to explain the process (see Refs. [13, 14]) but rather will present a much simpler procedure that is related to the *PageRank process*.

This discussion derives from Ref. [12], and the paper presents other ranking schemes. The problem is to rank things in order of importance based on some measure of the influence that they have over each other. Suppose that a set of n football teams represented by variables x_i, $1 \leq i \leq n$, are to be ranked. We assume that each team played every other team, and that elements $\{r_{ij}\}$ are weights used in ranking, where i refers to team i, j to team j and $r_{ii} = 0$. The ranking of team i is proportional to the sum of the rankings of the remaining teams weighted by r_{ij}, so

$$x_i = k \sum_{j=1}^{n} r_{ij} x_j, \quad 1 \leq i \leq n, \tag{5.12}$$

where k is the constant of proportionality. We can write Equations 5.12 in the matrix form $kRx = x$, where $R = [r_{ij}]$. This is an eigenvalue/eigenvector problem!

$$Rx = \frac{1}{k}x. \tag{5.13}$$

Theorem 5.6 applies to our eigenvalue problem 5.13 if the matrix R is irreducible. We have the problem of defining the r_{ij} so this is the case. There are many ways to do this, the simplest of which is to let $r_{ij} = 1$ if team i defeats team j or $r_{ij} = 0$ if team i loses to team j. The problem with this assignment is that the losing team gets no credit at all if the score is close, and the winning team gets no extra benefit if it scores many more points than the losing team. Also, this assignment will result in a row of 0s if a team loses all of its games, and such a matrix is not irreducible (convince yourself of this). A better approach is to base the value of r_{ij} on the score of the game. Let S_{ij} be the number of points scored by team i when it played team j, and define $r_{ij} = S_{ij}/(S_{ij} + S_{ji})$. This is an improvement but has the problem that if a game ends in a score like 6-0, the losing team gets no credit at all even though the score was close. We will settle on the following definition of r_{ij}:

$$r_{ij} = \begin{cases} \dfrac{S_{ij} + 1}{S_{ij} + S_{ji} + 2}, & i \neq j, \\ 0, & i = j. \end{cases} \tag{5.14}$$

The losing team gets some credit, there cannot be a zero row, and so R will be irreducible. For an example, assume that eight teams played each other and Table 5.1 contains the scores. For instance, when teams 1 and 2 played, team 1 scored 14 points and team 2 scored 7 points.

Applying Equation 5.14 to the data in Table 5.1 gives the matrix

$$R = \begin{bmatrix} 0.0000 & 0.6522 & 0.3333 & 0.5806 & 0.4717 & 0.2000 & 0.4800 & 0.4286 \\ 0.3478 & 0.0000 & 0.3191 & 0.7442 & 0.6133 & 0.1071 & 0.5556 & 0.6304 \\ 0.6667 & 0.6809 & 0.0000 & 0.5513 & 0.1818 & 0.3000 & 0.2222 & 0.5303 \\ 0.4194 & 0.2558 & 0.4487 & 0.0000 & 0.4286 & 0.8000 & 0.4815 & 0.5000 \\ 0.5283 & 0.3867 & 0.8182 & 0.5714 & 0.0000 & 0.2632 & 0.3333 & 0.6000 \\ 0.8000 & 0.8929 & 0.7000 & 0.2000 & 0.7368 & 0.0000 & 0.4375 & 0.8000 \\ 0.5200 & 0.4444 & 0.7778 & 0.5185 & 0.6667 & 0.5625 & 0.0000 & 0.5814 \\ 0.5714 & 0.3696 & 0.4697 & 0.5000 & 0.4000 & 0.2000 & 0.4186 & 0.0000 \end{bmatrix}.$$

The book software distribution contains a function `rankmatrix` that takes the matrix S of scores and returns the ranking matrix R obtained by applying Equation 5.14. Then apply the function `perronfro` to R to obtain the ranking vector.

TABLE 5.1 Ranking Teams

Team	1	2	3	4	5	6	7	8
1	0	14	3	17	24	0	35	2
2	7	0	14	31	45	2	29	28
3	7	31	0	42	7	17	7	34
4	12	10	34	0	20	31	12	14
5	27	28	35	27	0	14	15	20
6	3	24	41	7	41	0	13	35
7	38	23	27	13	31	17	0	49
8	3	16	30	14	13	8	35	0

Example 5.16.   Assuming the scores in Table 5.1, the following command sequence finds the ranking vector. The largest component of the ranking vector is the top-rated team, the second largest the second rated team, and so forth.

```
>> R = rankmatrix(S);
>> [r lambda] = perronfro(R)

r =
    0.3198
    0.3330
    0.3134
    0.3506
    0.3448
    0.4404
    0.4050
    0.2981

lambda =

    3.9342
```

By looking at the vector r, we see that the teams are ranked from first to last as follows:

$$6 \quad 7 \quad 4 \quad 5 \quad 2 \quad 1 \quad 3 \quad 8$$

∎

5.5   COMPUTING EIGENVALUES AND EIGENVECTORS USING MATLAB

The computation of eigenvalues and eigenvectors in MATLAB is done by the function `eig(A)`. To obtain the eigenvalues and associated eigenvectors, call it using the format

```
>> [V,D] = eig(A);
```

D is a diagonal matrix of eigenvalues and V is a matrix whose columns are the corresponding eigenvectors; for instance, if the eigenvalue/eigenvector pairs are

$$\lambda_1 = -12.2014, \quad v_1 = \begin{bmatrix} 0.0278 \\ 0.4670 \\ -0.8838 \end{bmatrix},$$

$$\lambda_2 = 1.3430, \quad \begin{bmatrix} 0.9925 \\ 0.0357 \\ 0.1167 \end{bmatrix},$$

$$\lambda_3 = 5.8584, \quad \begin{bmatrix} 0.5911 \\ 0.7333 \\ 0.3359 \end{bmatrix},$$

then

$$
V = \begin{bmatrix} 0.0278 & 0.9925 & 0.5911 \\ 0.4670 & 0.0357 & 0.7333 \\ -0.8838 & 0.1167 & 0.3359 \end{bmatrix}
$$

and

$$
D = \begin{bmatrix} -12.2014 & 0 & 0 \\ 0 & 1.3430 & 0 \\ 0 & 0 & 5.8584 \end{bmatrix}.
$$

If you want only the eigenvalues, call eig as follows:

```
>> E = eig(A);
```

Example 5.17. Compute the eigenvalues and eigenvectors for the matrix $B = \begin{bmatrix} 1 & 6 & 3 \\ -1 & 4 & 9 \\ 12 & 35 & 1 \end{bmatrix}$.   ∎

```
>> [V E] = eig(B)
V =
    0.0118    0.9119    0.2500
    0.4211   -0.3220    0.4278
   -0.9069    0.2545    0.8686

E =
  -15.4092        0        0
        0  -0.2812        0
        0        0  21.6905
>> eig(B)

ans =
  -15.4092
   -0.2812
   21.6905
```
                                                                                            ∎

5.6 CHAPTER SUMMARY

Defining Eigenvalues and Their Associated Eigenvectors

λ is an eigenvalue of $n \times n$ matrix A, and $v \neq 0$ is an eigenvector if $Av = \lambda v$; in other words, Av is parallel to v and either shrinks or contracts it. The relationship $Av = \lambda v$ is equivalent to $(A - \lambda I)v = 0$, and in order for there to be a nontrivial solution, we must have

$$
\det(A - \lambda I) = 0.
$$

This is called the characteristic equation, and the polynomial

$$
p(\lambda) = \det(A - \lambda I)
$$

is the characteristic polynomial. The eigenvalues are the roots of the characteristic polynomial, and an eigenvector associated with an eigenvalue λ is a solution to the homogeneous system

$$
(A - \lambda I)v = 0.
$$

The process of finding the eigenvalues and associated eigenvectors would seem to be

Locate the roots $\lambda_1, \lambda_2, \ldots, \lambda_n$ of p and find a nonzero solution to $(A - \lambda_i I)v_i = 0$ for each λ_i.

There is a serious problem with this approach. If p has degree five or more, the eigenvalues must be approximated using numerical techniques, since there is no analytical formula for roots of such polynomials. We will see in Chapter 10 that polynomial root finding can be difficult. A small change in a polynomial coefficient can cause large changes in its roots.

Selected Properties of Eigenvalues and Eigenvectors

A matrix with a 0 eigenvalue is singular, and every singular matrix has a 0 eigenvalue. If we can find the eigenvalues of A accurately, then $\det A = \prod_{i=1}^{n} \lambda_i$. If we happen to need the determinant, this result can be useful.

Matrix Diagonalization

Square matrices A and B are similar if there exists an invertible matrix X such that $B = X^{-1}AX$, and similar matrices have the same eigenvalues. The eigenvalues of A are the diagonal elements of B, and we are said to have diagonalized A. As we will see in later chapters, diagonalization is a primary tool for developing many results.

To diagonalize a matrix requires that we find n linearly independent eigenvectors. If the matrix has n distinct eigenvalues, then it has a basis of n eigenvectors. Form X by making its columns the eigenvectors, keeping the eigenvalues in the same order in the diagonal matrix. If a matrix is symmetric, it has n linearly independent eigenvectors, even in the presence of eigenvalues of multiplicity two or more. Furthermore, the matrix X is orthogonal. If a matrix does not have n linearly independent eigenvectors, it cannot be diagonalized.

If a matrix A is diagonalizable, then it is simple to compute powers of A, since

$$A^k = XD^kX^{-1} = X \begin{bmatrix} \sigma_1^k & & & \\ & \sigma_2^k & & \\ & & \ddots & \\ & & & \sigma_n^k \end{bmatrix} X^{-1}.$$

Applications

The applications of eigenvalues are vast, including such areas as the solution of differential equations, structural mechanics, and the study of vibrations, where they represent the natural frequencies of a system.

In electrical engineering, when a circuit contains resistors, inductors, and batteries, there results a system of first-order differential equations of the form $dx/dt = Ax + b$, and the eigenvalues of A are required for the solution.

A very interesting application of eigenvalues and eigenvectors is in the theory of ranking. The text provides a simple example of ranking teams in a tournament.

Using MATLAB to Compute Eigenvalues and Eigenvectors

The computation of eigenvalues or both eigenvalues and eigenvectors using MATLAB is straightforward. To compute just the eigenvalues, use the format

```
>> E = eig(A);
```

and to find the eigenvectors and a diagonal matrix of eigenvalues, use

```
>> [V,D] = eig(A);
```

If A has distinct eigenvalues, then $V^{-1}AV = D$. If A has n linearly independent eigenvectors, this is also true. If A is symmetric, then things are even nicer, since $P^{T}AP = D$, where P is orthogonal.

5.7 PROBLEMS

5.1 Find the eigenvalues and associated eigenvectors for the matrix $\begin{bmatrix} 1 & 3 \\ 0 & 9 \end{bmatrix}$.

5.2 Find the eigenvalues and associated eigenvectors for the matrix $\begin{bmatrix} 1 & 2 & 1 \\ 6 & -1 & 0 \\ -1 & -2 & -1 \end{bmatrix}$.

5.3 Let $A = \begin{bmatrix} 1 & 2 & 1 \\ 0 & 2 & 1 \\ 3 & 0 & 2 \end{bmatrix}$. Verify that $\left(A^{-1}\right)^{T} = \left(A^{T}\right)^{-1}$.

5.4 Let $A = \begin{bmatrix} 1 & 4 \\ 2 & 3 \end{bmatrix}$, $B = \begin{bmatrix} -1 & 7 \\ 1 & -4 \end{bmatrix}$. Verify that AB and BA have the same eigenvalues.

5.5 Show that the following properties hold for similar matrices A and B.

    a. A is similar to A.

    b. If B is similar to A, then A is similar to B.

    c. If A is similar to B and B is similar to C, then A is similar to C; in other words, similarity is transitive.

5.6 Let $A = \begin{bmatrix} 1/2 & 1/2 & 0 \\ 1/4 & 1/4 & 1/2 \\ 1/4 & 1/4 & 1/2 \end{bmatrix}$.

    a. Verify that $\det(A - \lambda I)$, the characteristic polynomial of A, is given by $(\lambda - 1)\lambda(\lambda - \frac{1}{4})$.

    b. Diagonalize A.

5.7 Assume A can be diagonalized. Under what conditions will

$$\lim_{k \longrightarrow \infty} A^k = 0?$$

5.8 Solve the first-order system of differential equations with initial conditions:

$$\frac{dx_1}{dt} = -3x_1 + x_2,$$

$$\frac{dx_2}{dt} = 2x_1 - 4x_2 + 1,$$

$$x_1(0) = 1, \quad x_2(0) = 0.$$

5.9 Draw the digraph for each matrix and determine which matrices are irreducible.

    a. $\begin{bmatrix} 0 & 1 & 1 & 0 & 0 \\ 1 & 0 & 0 & 0 & 0 \\ 0 & 0 & 0 & 1 & 0 \\ 0 & 0 & 0 & 0 & 1 \\ 0 & 1 & 0 & 0 & 0 \end{bmatrix}$

    b. $\begin{bmatrix} 0 & 0 & 0 & 1 & 0 \\ 1 & 0 & 1 & 1 & 1 \\ 0 & 0 & 0 & 1 & 0 \\ 1 & 0 & 1 & 0 & 1 \\ 0 & 0 & 0 & 1 & 0 \end{bmatrix}$

5.10

    a. Show that $A = \begin{bmatrix} a & 1 \\ 0 & a \end{bmatrix}$ has only one eigenvalue $\lambda = a$ of multiplicity two and that all eigenvectors are multiples of $e_1 = \begin{bmatrix} 1 \\ 0 \end{bmatrix}$.

    b. Consider a general version of part (a), $A^{n \times n} = \begin{bmatrix} a & 1 & & & 0 \\ & a & 1 & & \\ & & a & \ddots & \\ & & & \ddots & 1 \\ 0 & & & & a \end{bmatrix}$. Show that A has one eigenvalue of multiplicity n and that all eigenvectors are multiples of $e_1 = \begin{bmatrix} 1 \\ 0 \\ \vdots \\ 0 \end{bmatrix}$.

5.11 Using matrix of Problem 5.2, verify that the eigenvalues of A^T and A are equal. Can you say the same about the eigenvectors?

5.12
    a. If A and B are $n \times n$ upper-triangular matrices, are the eigenvalues of $A + B$ the sum of the eigenvalues and A and B?

    b. Is part (a) true for two arbitrary $n \times n$ matrices A and B?

5.13 If A is an $n \times n$ nonsingular matrix, prove that $(A^{-1})^{\mathrm{T}} = (A^{\mathrm{T}})^{-1}$.

5.14 If A and B are $n \times n$ matrices, prove that AB and BA have the same eigenvalues. Hint: You must show that every eigenvalue of AB is an eigenvalue of BA, and every eigenvalue of BA is an eigenvalue of AB. Suppose λ is an eigenvalue of AB. Then $ABx = \lambda x$, so $(BA) Bx = \lambda (Bx)$.

5.15 Show that the trace of a 3×3 matrix is equal to the sum of its eigenvalues using the following steps:

    a. Find the characteristic polynomial $p (\lambda)$ and show that the coefficient of λ^2 is the trace of A.

    b. Explain why $p (\lambda) = (-1) (\lambda - \lambda_1) (\lambda - \lambda_2) (\lambda - \lambda_3)$, where the λ_i are the eigenvalues of A.

    c. Show that the coefficient of λ^2 is $\lambda_1 + \lambda_2 + \lambda_3$, and argue that this completes the proof.

5.16 Assume A is a real $n \times n$ matrix with a complex eigenvalue λ, and v is an associated eigenvector. If $\bar{v}$ is the complex conjugate of v, show that $\bar{v}$ is an eigenvector of A associated with eigenvalue $\bar{\lambda}$.

5.17 Prove that A and A^{T} have the same eigenvalues. Hint: By Problem 4.19, $\det A = \det A^{\mathrm{T}}$. Apply this result to the characteristic equation of A.

5.7.1 MATLAB Problems

5.18 The MATLAB function `eigshow` is a graphical demonstration of eigenvalues and eigenvectors. When invoked by `eigshow(A)`, where A is a 2×2 matrix, a graphical dialog appears. Do not press the button labeled **eig/(svd)**. You will see two vectors, a unit vector x, and the vector Ax. We will show in Chapter 15 that as the tip of x traces out the unit circle $x_1^2 + x_2^2 = 1$ the tip of the vector Ax traces out an ellipse whose center is the center of the circle. Move x with the mouse until x and Ax are parallel, if you can. If you are successful, Ax is a multiple of x, so $Ax = kx$. Since x has length 1, length $(Ax) = |k|$ and k is an eigenvalue of A corresponding to the eigenvector x, and its magnitude is $|k|$. Run `eigshow` for each of the following three matrices and estimate the eigenvalues, if you can make Ax parallel to x. There are three possibilities: there are two distinct eigenvalues, a double eigenvalue, and two complex conjugate eigenvalues.

$$A = \begin{bmatrix} 1 & -1 \\ -1 & 2 \end{bmatrix}, \quad B = \begin{bmatrix} 5 & 1 \\ -1 & 3 \end{bmatrix}, \quad C = \begin{bmatrix} 1 & 4 \\ -1 & 3 \end{bmatrix}$$

5.19 Diagonalize $A = \begin{bmatrix} 26 & 48 & 8 \\ 35 & 28 & 13 \\ 45 & 7 & 43 \end{bmatrix}$.

5.20 Use MATLAB to find the eigenvalues and eigenvectors of the matrix

$$A = \begin{bmatrix} 1 & 6 & 0 & -1 & 5 \\ 5 & -9 & 22 & 2 & 1 \\ 0 & 1 & 3 & 5 & 7 \\ 9 & 0 & -4 & -7 & -1 \\ 3 & 5 & 2 & 15 & 35 \end{bmatrix}.$$

Note that two of its eigenvalues and its corresponding eigenvectors are complex.

5.21 In MATLAB, `W = wilkinson(n)` returns one of Wilkinson's $n \times n$ eigenvalue test matrices. The matrix is symmetric and tridiagonal, with pairs of nearly, but not exactly, equal eigenvalues. The most frequently used case is `wilkinson(21)`. Its two largest eigenvalues are both about 10.746; they agree to 14, but not to 15, decimal places. Find the eigenvalues of the matrices `wilkinson(11)` and `wilkinson(21)`.

5.22

Team	1	2	3	4	5	6
1	0	17	25	25	10	30
2	38	0	24	48	21	29
3	20	31	0	14	24	17
4	36	3	25	0	24	45
5	24	30	13	14	0	0
6	28	24	20	10	23	0

a. Rank the teams whose scores are given in the table using Equation 5.14.

b. Use the simple scheme $a_{ij} = 1$ if team i defeats team j and $a_{ij} = 0$ if team i loses to team j. Compare the results of part (a).

c. Use the formula $a_{ij} = S_{ij}/(S_{ij} + S_{ji})$, and compare the results with those of parts (a) and (b).

5.23 Let $A = \begin{bmatrix} 1 & 2 & 3 \\ -3 & -7 & -4 \\ -1 & -3 & 1 \end{bmatrix}$.

a. Perform the computation

```
>> EA = eig(A);
>> EAINV = eig(inv(A));
```

b. Does part (a) motivate a general result concerning the eigenvalues of A and A^{-1}? If your answer is yes, prove it.

5.24 The Cayley-Hamilton theorem is an interesting result in theoretical linear algebra. It says that any $n \times n$ matrix satisfies its own characteristic equation. For instance, if the characteristic polynomial for a matrix A is $\lambda^3 + 3\lambda^2 - \lambda + 1$, then $A^3 + 3A^2 - A + I = 0$. Verify the Cayley-Hamilton theorem for each matrix. *Note*: The MATLAB function `poly(A)` returns a vector containing the coefficients of A's characteristic polynomial from highest to lowest power of λ.

a. $A = \begin{bmatrix} 1 & -1 \\ 3 & 5 \end{bmatrix}$

b. $A = \begin{bmatrix} 1 & 0 & 5 \\ 2 & 1 & -6 \\ 0 & 2 & 3 \end{bmatrix}$

c. $A = \begin{bmatrix} 1 & -5 & 2 & 55 & 12 \\ 0 & 4 & 13 & 6 & -8 \\ 0 & 0 & 18 & 1 & -56 \\ 0 & 0 & 0 & -7 & 88 \\ 0 & 0 & 0 & 0 & 5 \end{bmatrix}$

5.25 The MATLAB statement

```
>> F = gallery('frank',n,1);
```

returns an $n \times n$ matrix. Create the matrix F = `gallery('frank',15,1)` and perform the following computations:

a. The determinant of any Frank matrix is 1. Verify this for F.

b. If n is odd, 1 is an eigenvalue. Verify this for F.

c. Some of the eigenvalues of F are sensitive to changes in the entries of F. Perturb the entries of F by executing the statement

```
>> F = F + 1.0e-8*ones(15,15);
```

and compute the eigenvalues. Comment on the change in eigenvalues between the original F and the perturbed F.

5.26 If a matrix has eigenvalues of multiplicity greater than 1, generally those eigenvalues are more sensitive to small changes in the matrix. This means that small changes in the matrix might cause significant changes in its eigenvalues.

a. Build the matrix $A = \begin{bmatrix} 1 & 0 & 0 \\ 1 & 1 & 0 \\ 5 & -3 & 1 \end{bmatrix}$ and execute

```
>> [V D] = eig(A);
```

b. How many linearly independent eigenvectors does A have?

c. Build the matrix $B = \begin{bmatrix} 0.9999 & 0 & 0 \\ 0.9999 & 0.9998 & 0 \\ 4.9999 & -3.0001 & 1.0001 \end{bmatrix}$ and show it has three distinct eigenvalues.

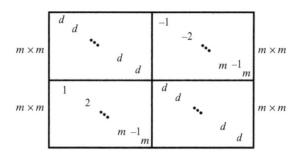

FIGURE 5.5 Hanowa matrix.

d. Let $\delta = \begin{bmatrix} 0 & 1.0 \times 10^{-6} & 1.0 \times 10^{-6} \\ 0 & 0 & 1.0 \times 10^{-6} \\ 0 & 0 & 0 \end{bmatrix}$ and compute the eigenvalues of $B + \delta$. Comment on the results and propose a relationship between A and $B + \delta B$ that might account for what you see.

5.27 A *block structured matrix* is built by putting together submatrices, where each submatrix is a block. An example is a Hanowa matrix. If m is an integer and d is a real number, then a $(2m) \times (2m)$ Hanowa matrix has block structure (Figure 5.5). We will discuss block matrices in Section 9.1.4.

a. Using the MATLAB functions `eye` and `diag`, construct a 6×6 Hanowa matrix H with $d = 3$.

b. Find the eigenvalues of H.

c. Using $d = 3$, build Hanowa matrices of dimensions 10×10 and 20×20 and compute their eigenvalues.

d. From your results in parts (b) and (c), propose a formula for the eigenvalues of an $n \times n$ Hanowa matrix.

Chapter 6

Orthogonal Vectors and Matrices

You should be familiar with

- Distance between points in two- and three-space
- Geometric interpretation of vector addition and subtraction
- Simple geometry and trigonometry
- Rotation matrices
- Real symmetric matrices
- Computation of $\int_a^b f(t) g(t) \, dt$ (for Section 6.5)

6.1 INTRODUCTION

We will have occasion in the book to use two- and three-dimensional vectors as examples. If we are discussing a property or operation that applies to all vectors, we can use vectors in $\mathbb{R}^2$ and $\mathbb{R}^3$ as illustrations, since we can visualize the results, whereas that is not possible for a vector in $\mathbb{R}^n$, $n \geq 4$.

In three-dimensional space, *points* are defined as ordered triples of real numbers and the *distance* between points $P_1 = (x_1, y_1, z_1)$ and $P_2 = (x_2, y_2, z_2)$ is defined by the formula (Figure 6.1)

$$d = \sqrt{(x_2 - x_1)^2 + (y_2 - y_1)^2 + (z_2 - z_1)^2}.$$

Directed line segments $\overrightarrow{P_1P_2}$ (Figure 6.1) are introduced as three-dimensional column vectors: If $P_1 = (x_1, y_1, z_1)$ and $P_2 = (x_2, y_2, z_2)$, then

$$\overrightarrow{P_1P_2} = \begin{bmatrix} x_2 - x_1 \\ y_2 - y_1 \\ z_2 - z_1 \end{bmatrix}.$$

If P is a point, we let $P = \overrightarrow{OP}$ and call P the *position vector* of P, where O is the origin.

There are geometrical interpretations of equality, addition, subtraction, and scalar multiplication of vectors (Figure 6.2).

1. Equality of vectors: Suppose A, B, C, D are distinct points such that no three are collinear. Then $\overrightarrow{AB} = \overrightarrow{CD}$ if and only if $\overrightarrow{AB} \parallel \overrightarrow{CD}$ and $\overrightarrow{AC} \parallel \overrightarrow{BD}$.
2. Addition of vectors obeys the *parallelogram law*: Let A, B, C be non-collinear. Then

$$\overrightarrow{AB} + \overrightarrow{AC} = \overrightarrow{AD},$$

   where D is the point such that $\overrightarrow{AB} \parallel \overrightarrow{CD}$ and $\overrightarrow{AC} \parallel \overrightarrow{BD}$
3. The difference of two vectors $\overrightarrow{AB} - \overrightarrow{AC}$ is a vector whose start is the tip of $\overrightarrow{AC}$ and whose tip coincides with the tip of $\overrightarrow{AB}$.
4. Scalar multiplication of vectors (Figure 6.3): Let $\overrightarrow{AP} = t\overrightarrow{AB}$, where A and B are distinct points. Then P is on the line AB, and
   a. $P = A$ if $t = 0$, $P = B$ if $t = 1$;
   b. P is between A and B if $0 < t < 1$;
   c. B is between A and P if $t > 1$;
   d. A is between P and B if $t < 0$.

Numerical Linear Algebra with Applications. http://dx.doi.org/10.1016/B978-0-12-394435-1.00006-5

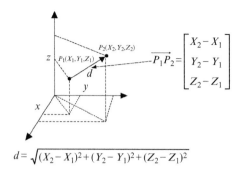

$$d = \sqrt{(X_2 - X_1)^2 + (Y_2 - Y_1)^2 + (Z_2 - Z_1)^2}$$

FIGURE 6.1 Distance between points.

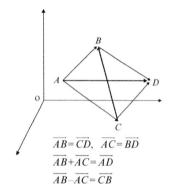

$$\overrightarrow{AB} = \overrightarrow{CD}, \quad \overrightarrow{AC} = \overrightarrow{BD}$$
$$\overrightarrow{AB} + \overrightarrow{AC} = \overrightarrow{AD}$$
$$\overrightarrow{AB} - \overrightarrow{AC} = \overrightarrow{CB}$$

FIGURE 6.2 Equality, addition, and subtraction of vectors.

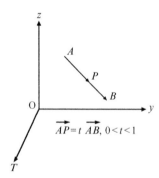

$$\overrightarrow{AP} = t\,\overrightarrow{AB}, \ 0 < t < 1$$

FIGURE 6.3 Scalar multiplication of vectors.

6.2 THE INNER PRODUCT

Along with matrix multiplication, the inner product is an important operator in linear algebra. It defines vector length, orthonormal bases, the L^2 matrix norm, projections, and Householder reflections. We will study these and many more constructs that use the inner product.

Definition 6.1. Given two vectors $x = \begin{bmatrix} x_1 \\ \vdots \\ x_n \end{bmatrix}$ and $\begin{bmatrix} y_1 \\ \vdots \\ y_n \end{bmatrix}$ in $\mathbb{R}^n$, we define the *inner product* of x and y, written $\langle x, y \rangle$ to be the real number

$$\langle x, y \rangle = x_1 y_1 + x_2 y_2 + \cdots + x_n y_n = \sum_{i=1}^{n} x_i y_i.$$

Note that $x^T y = \begin{bmatrix} x_1 & \cdots & x_n \end{bmatrix} \begin{bmatrix} y_1 \\ \vdots \\ y_n \end{bmatrix} = \langle x, y \rangle$, so we can compute the inner product as the matrix product, $x^T y$. Since $\sum_{i=1}^{n} x_i y_i = \sum_{i=1}^{n} y_i x_i$, $y^T x$ is another way to compute the inner product.

Remark 6.1.   In many books, the notation $x \cdot y$ to refers to the inner product, and it is called the *dot product*. We will seldom use this notation in the book.

Example 6.1.   In $\mathbb{R}^3$, if $x = \begin{bmatrix} a_1 \\ b_1 \\ c_1 \end{bmatrix}$ and $y = \begin{bmatrix} a_2 \\ b_2 \\ c_2 \end{bmatrix}$, then $\langle x, y \rangle = a_1 a_2 + b_1 b_2 + c_1 c_2$. For instance, if $u = \begin{bmatrix} 4 \\ 6 \\ 4 \end{bmatrix}$, $v = \begin{bmatrix} -9 \\ 7 \\ -1 \end{bmatrix}$, then $\langle u, v \rangle = 4\,(-9) + 6\,(7) + 4\,(-1) = 2$ ∎

There are properties of the inner product that we will use throughout the book.

Theorem 6.1.   *The inner product has the following properties:*

1. $\langle x, y + z \rangle = \langle x, y \rangle + \langle x, z \rangle$
2. $\langle cx, y \rangle = \langle x, cy \rangle = c\,\langle x, y \rangle$, *where c is a scalar*
3. $\langle x, y \rangle = \langle y, x \rangle$
4. $\langle x, 0 \rangle = 0$
5. $\langle x, x \rangle = \sum_{i=1}^{n} x_i^2$
6. *If* $\langle x, x \rangle = 0$, *then* $x = 0$

Proof. We will prove properties 5 and 6, leaving the remaining properties to the exercises.

Property 5: $\langle x, x \rangle = \begin{bmatrix} x_1 & \cdots & x_n \end{bmatrix}^T \begin{bmatrix} x_1 \\ \vdots \\ x_n \end{bmatrix} = x_1^2 + x_2^2 + \cdots + x_{n-1}^2 + x_n^2 = \sum_{i=1}^{n} x_i^2$

Property 6: If $\langle x, x \rangle = 0$, then from Property 5 $x_1^2 + x_2^2 + \cdots + x_{n-1}^2 + x_n^2 = 0$. The only way this can occur is if $x_i = 0$, $1 \le i \le n$.   □

There are many occasions where we will need to deal with the length of a vector, so we need a compact notation for vector length. A vector u in the plane has length $\sqrt{x^2 + y^2}$, and a vector u in three-dimensional space has length $\sqrt{x^2 + y^2 + z^2}$ (Figure 6.4). Using Property 5 of Theorem 6.1, we see that in either two or three dimensions length $(u) = \sqrt{\langle u, u \rangle}$. The notation $\|u\|_2$ will be used to specify the length of vector u. This notation will be fully developed in Chapter 7 when we discuss vector norms.

There is a nice geometric interpretation for the inner product of vectors in $\mathbb{R}^2$ and $\mathbb{R}^3$. For simplicity, we will consider vectors in $\mathbb{R}^2$, but the same reasoning applies to three-dimensional vectors. Suppose that θ is the angle between vectors u and v such that $0 \le \theta \le \pi$ as shown in Figure 6.5. It follows that $\langle u, v \rangle = \|u\|_2 \|v\|_2 \cos \theta$. In Figure 6.6, the three vectors form the triangle AOB. Note that the length of each side is the length of the vector forming that side. The law of cosines tells us that

$$\|u - v\|_2^2 = \|u\|_2^2 + \|v\|_2^2 - 2 \|u\|_2 \|v\|_2 \cos \theta. \tag{6.1}$$

Using the properties of inner products, we can write the left-hand side of Equation 6.1 as

$$\|u - v\|_2^2 = \langle u - v, u - v \rangle = \langle u, u \rangle - 2 \langle u, v \rangle + \langle v, v \rangle = \|u\|_2^2 - 2 \langle u, v \rangle + \|v\|_2^2.$$

Equating the rewritten left-hand side of Equation 6.1 with the right-hand side gives

$$\|u\|_2^2 - 2 \langle u, v \rangle + \|v\|_2^2 = \|u\|^2 + \|v\|_2^2 - 2 \|u\|_2 \|v\|_2 \cos \theta. \tag{6.2}$$

After cancelation of terms in Equation 6.2, we have the result

$$\langle u, v \rangle = \|u\|_2 \|v\|_2 \cos\theta. \tag{6.3}$$

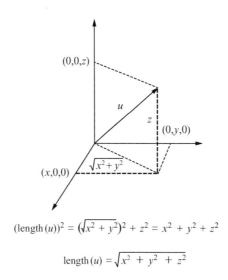

$$(\text{length}\,(u))^2 = \left(\sqrt{x^2 + y^2}\right)^2 + z^2 = x^2 + y^2 + z^2$$

$$\text{length}\,(u) = \sqrt{x^2 + y^2 + z^2}$$

FIGURE 6.4 Vector length.

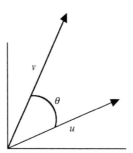

FIGURE 6.5 Geometric interpretation of the inner product.

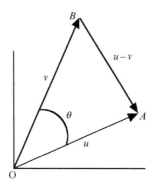

FIGURE 6.6 Law of cosines.

This formula is usually used to determine the angle between two vectors, not to compute the inner product.

Example 6.2.   Determine the angle between $u = \begin{bmatrix} 3 \\ -4 \\ -1 \end{bmatrix}$ and $v = \begin{bmatrix} 0 \\ 5 \\ 2 \end{bmatrix}$.

$\langle u, v \rangle = -22$, $\|u\|_2 = \sqrt{26}$, $\|v\|_2 = \sqrt{29}$. The angle between the two vectors is given by

$$\cos \theta = \frac{\langle u, v \rangle}{\|u\|_2 \|v\|_2} = \frac{-22}{\sqrt{26}\sqrt{29}} = -0.8011927,$$

so

$\theta = \cos^{-1}(-0.8011927) = 2.5 \, \text{rad} = 143.24°.$ ∎

Another application of the inner product is to determine whether two vectors are perpendicular or parallel. Vectors u and v are perpendicular, when the angle θ between them is $\pi/2$. Assume u and v are nonzero. The cosine of $\pi/2$ is 0, so by Equation 6.3, $\langle u, v \rangle = 0$, and u and v are perpendicular. Vectors u and v are *parallel* when the angle between them is either 0 radians (pointing in the same direction) or π radians (pointing in opposite directions). Since $\cos(0) = 1$ and $\cos(\pi) = -1$, it follows from Equation 6.3 that either

$$\langle u, v \rangle = \|u\|_2 \|v\|_2 \ (\theta = 0) \quad \text{or} \quad \langle u, v \rangle = -\|u\|_2 \|v\|_2 \ (\theta = \pi)$$

implies that u and v are parallel.

Example 6.3.   Determine if the following vectors are parallel, perpendicular, or neither.

a. $u = \begin{bmatrix} 6 \\ -2 \\ -1 \end{bmatrix}, v = \begin{bmatrix} 2 \\ 5 \\ 2 \end{bmatrix}$

$\langle u, v \rangle = 6(2) - 2(5) - 1(2) = 0$

u and v are perpendicular.

b. $u = \begin{bmatrix} 2 \\ -1 \end{bmatrix}, v = \begin{bmatrix} -1/2 \\ 1/4 \end{bmatrix}$.

$\langle u, v \rangle = 2(-1/2) + (-1)(1/4) = -\frac{5}{4}$

Compute their lengths, and test to see if they are parallel.

$\|u\|_2 = \sqrt{5}$ and $\|v\|_2 = \sqrt{5/16} = \frac{\sqrt{5}}{4}$. Now, $\langle u, v \rangle = -\frac{5}{4} = -\sqrt{5}\left(\frac{\sqrt{5}}{4}\right) = -\|u\|_2 \|v\|_2$

The two vectors are parallel.   ∎

6.3   ORTHOGONAL MATRICES

Vectors u and v are called *orthogonal* if $\langle u, v \rangle = 0$. We briefly mentioned orthogonal matrices in Chapter 5, and will now provide a formal definition. Many tools in numerical linear algebra involve orthogonal matrices, such as the *QR* decomposition (introduced in Chapter 14) and the singular value decomposition (SVD) (introduced in Chapter 15). Over the course of this book, we will see that orthogonal matrices are the most beautiful of all matrices, and that they have an intimate relation with orthogonal vectors.

Definition 6.2.   An $n \times n$ matrix P is orthogonal if $P^{\mathrm{T}} = P^{-1}$.

The simplest example of an orthogonal matrix is the 2×2 rotation matrix introduced in Chapter 1.

Example 6.4.   A rotation matrix $P = \begin{bmatrix} \cos \theta & -\sin \theta \\ \sin \theta & \cos \theta \end{bmatrix}$ is orthogonal, since

$$P^{\mathrm{T}}P = \begin{bmatrix} \cos \theta & \sin \theta \\ -\sin \theta & \cos \theta \end{bmatrix} \begin{bmatrix} \cos \theta & -\sin \theta \\ \sin \theta & \cos \theta \end{bmatrix} = \begin{bmatrix} \cos^2 \theta + \sin^2 \theta & 0 \\ 0 & \cos^2 \theta + \sin^2 \theta \end{bmatrix} = \begin{bmatrix} 1 & 0 \\ 0 & 1 \end{bmatrix}.$$
    ∎

Orthogonal matrices of many sizes occur in applications, from 2×2 to 1000×1000, and larger.

Example 6.5. Let $P = \begin{bmatrix} 0.00 & -0.80 & -0.60 \\ 0.80 & -0.36 & 0.48 \\ 0.60 & 0.48 & -0.64 \end{bmatrix}$. To verify that P is an orthogonal matrix, form $P^{\mathrm{T}}P$.

$$\begin{bmatrix} 0.00 & 0.80 & 0.60 \\ -0.80 & -0.36 & 0.48 \\ -0.60 & 0.48 & -0.64 \end{bmatrix} \begin{bmatrix} 0.00 & -0.80 & -0.60 \\ 0.80 & -0.36 & 0.48 \\ 0.60 & 0.48 & -0.64 \end{bmatrix} = \begin{bmatrix} 1 & 0 & 0 \\ 0 & 1 & 0 \\ 0 & 0 & 1 \end{bmatrix}.$$

Now, take a look at the columns of P.

$$\left\| \begin{array}{c} 0.00 \\ 0.80 \\ 0.60 \end{array} \right\|_2 = \left\| \begin{array}{c} -0.80 \\ -0.36 \\ 0.48 \end{array} \right\|_2 = \left\| \begin{array}{c} -0.60 \\ 0.48 \\ -0.64 \end{array} \right\|_2 = 1,$$

so each column has length 1 (a *unit vector*). Take the inner product of columns 1 and 2.

$$\left\langle \begin{bmatrix} 0.00 \\ 0.80 \\ 0.60 \end{bmatrix}, \begin{bmatrix} -0.80 \\ -0.36 \\ 0.48 \end{bmatrix} \right\rangle = 0.00\,(-0.80) + 0.80\,(-0.36) + 0.60\,(0.48) = 0.$$

Verify that the two remaining inner products are also zero. In summary, for this matrix, the columns of P are orthogonal, and each column has length 1. A set of orthogonal vectors, each with unit length, are said to be *orthonormal*. It is not a coincidence that the columns are orthonormal. ∎

Theorem 6.2. *Let P be an $n \times n$ real matrix. Then P is an orthogonal matrix if and only if the columns of P are orthogonal and have unit length.*

Proof. Let $P = \begin{bmatrix} a_{11} & \cdots & a_{1i} & \cdots & a_{1n} \\ a_{21} & \cdots & \vdots & \cdots & a_{2n} \\ \vdots & & a_{ii} & & \vdots \\ \vdots & & \vdots & & \vdots \\ a_{n1} & & a_{ni} & & a_{nn} \end{bmatrix}$, $P^{\mathrm{T}}P = I$. View P as $P = \begin{bmatrix} v_1 & \cdots & v_i & \cdots & v_n \end{bmatrix}$, where $v_i = \begin{bmatrix} a_{1i} \\ a_{2i} \\ \vdots \\ a_{n-1,i} \\ a_{ni} \end{bmatrix}$,

$1 \leq i \leq n$ are the columns of P. Then, $P^{\mathrm{T}} = \begin{bmatrix} v_1^{\mathrm{T}} \\ v_2^{\mathrm{T}} \\ \vdots \\ v_{n-1}^{\mathrm{T}} \\ v_n^{\mathrm{T}} \end{bmatrix}$, where v_i^{T}, $1 \leq i \leq n$ are the rows of P^{T}.

Thus,

$$P^{\mathrm{T}}P = \begin{bmatrix} v_1^{\mathrm{T}} \\ v_2^{\mathrm{T}} \\ \vdots \\ v_{n-1}^{\mathrm{T}} \\ v_n^{\mathrm{T}} \end{bmatrix} \begin{bmatrix} v_1 & \cdots & v_i & \cdots & v_n \end{bmatrix} = \begin{bmatrix} v_1^{\mathrm{T}}v_1 & \cdots & v_1^{\mathrm{T}}v_i & \cdots & v_1^{\mathrm{T}}v_n \\ v_2^{\mathrm{T}}v_1 & \ddots & \vdots & \cdots & v_2^{\mathrm{T}}v_n \\ \vdots & \vdots & v_i^{\mathrm{T}}v_i & \vdots & \vdots \\ \vdots & \vdots & \vdots & \ddots & \vdots \\ v_n^{\mathrm{T}}v_1 & \cdots & v_n^{\mathrm{T}}v_i & \cdots & v_n^{\mathrm{T}}v_n \end{bmatrix} \tag{6.4}$$

If $P^{\mathrm{T}}P = I = \begin{bmatrix} 1 & & & & 0 \\ & 1 & & & \\ & & \ddots & & \\ & & & 1 & \\ 0 & & & & 1 \end{bmatrix}$, then Equation 6.4 implies that $\langle v_i, v_j \rangle = \begin{cases} 1, & i = j \\ 0, & i \neq j \end{cases}$, so the columns of P are

orthogonal. Since $\langle v_i, v_i \rangle = 1$, $\|v_i\|_2^2 = 1$, and the columns of P have unit length.

If the columns of P are orthogonal and of unit length, Equation 6.4 implies that $P^TP = I$, and P is an orthogonal matrix. $\qquad\square$

Orthogonal matrices have other interesting properties. Among them is the fact the their determinant is always ± 1.

Theorem 6.3. *If P is orthogonal,* $\det P = \pm 1$.

Proof. Recall that the determinant of a product is the product of the determinants, and $\det P^T = \det P$. Then,

$$\det(I) = \det\left(P^TP\right) = \left(\det P^T\right)(\det P) = (\det P)(\deg P) = (\det P)^2,$$

so $(\det P)^2 = \det I = 1$, and $\det P = \pm 1$. $\qquad\square$

Remark 6.2.   If the determinant of an orthogonal matrix is 1, we say it is a *proper orthogonal matrix.*

6.4   SYMMETRIC MATRICES AND ORTHOGONALITY

In this and later chapters, we will discover many interesting and useful facts about symmetric matrices; in particular, many computations can be done faster and more accurately for a symmetric matrix. A good example is the computation of the eigenvalues of a symmetric matrix. We will begin right here with Theorem 6.4 that tells us the relationship between any two distinct eigenvalues and the corresponding eigenvectors of a real symmetric matrix.

Theorem 6.4.   *If A is a real symmetric matrix, then any two eigenvectors corresponding to distinct eigenvalues are orthogonal.*

Proof. Let λ_1 and λ_2 be distinct eigenvalues with associated eigenvectors v_1 and v_2. Then, $Av_1 = \lambda_1 v_1$ and $Av_2 = \lambda_2 v_2$. Take the inner product of the first equation by v_2 and the inner product of the second equation by v_1:

$$v_2^T(Av_1) = \lambda_1\langle v_2, v_1\rangle, \quad (Av_2)^T v_1 = \lambda_2\langle v_2, v_1\rangle. \tag{6.5}$$

In Equation 6.5, $(Av_2)^T v_1 = v_2^T A^T v_1$, so Equation 6.5 becomes

$$v_2^T(Av_1) = \lambda_1\langle v_2, v_1\rangle, \quad v_2^T A^T v_1 = \lambda_2\langle v_2, v_1\rangle. \tag{6.6}$$

Since $A^T = A$, in Equation 6.6, we have

$$v_2^T(Av_1) = \lambda_1\langle v_2, v_1\rangle, \quad v_2^T(Av_1) = \lambda_2\langle v_2, v_1\rangle,$$

and

$$\lambda_1\langle v_2, v_1\rangle = \lambda_2\langle v_2, v_1\rangle. \tag{6.7}$$

Equation 6.7 gives

$$(\lambda_1 - \lambda_2)\langle v_2, v_1\rangle = 0.$$

Since $\lambda_1 \neq \lambda_2$, $\langle v_2, v_1\rangle = 0$, and v_1, v_2 are orthogonal. $\qquad\square$

Example 6.6.   Let A be the symmetric matrix $A = \begin{bmatrix} 3 & 1 & -1 \\ 1 & 3 & -1 \\ -1 & -1 & 5 \end{bmatrix}$. The eigenvalues of A are $\lambda_1 = 2, \lambda_2 = 3, \lambda_3 = 6$, and eigenvectors corresponding to the eigenvalues are

$$\begin{bmatrix} 1 \\ -1 \\ 0 \end{bmatrix}, \quad \begin{bmatrix} 1 \\ 1 \\ 1 \end{bmatrix}, \quad \begin{bmatrix} -1 \\ -1 \\ 2 \end{bmatrix},$$

respectively. The three eigenvectors are mutually orthogonal, and you also should note that the eigenvectors are linearly independent, so they are a basis for $\mathbb{R}^3$. As a result, the matrix $X = \begin{bmatrix} 1 & 1 & 1 \\ -1 & 1 & -1 \\ 0 & 1 & 2 \end{bmatrix}$ is invertible. If we form the product $X^{-1}AX$, the result is

$$X^{-1}AX = D,$$

where D is the diagonal matrix $D = \begin{bmatrix} 2 & 0 & 0 \\ 0 & 3 & 0 \\ 0 & 0 & 6 \end{bmatrix}$ with the eigenvalues of A on the diagonal. In other words, A is diagonalizable. Let's go one step further and build a matrix, P, whose columns are those of X converted to a unit vector. Do this by dividing each column vector by its length, and obtain $P = \begin{bmatrix} 0.7071 & 0.5774 & -0.4082 \\ -0.7071 & 0.5774 & -0.4082 \\ 0.0000 & 0.5774 & 0.8165 \end{bmatrix}$. By Theorem 6.2, P is an orthogonal matrix. Now compute $P^{T}AP$ and you will again get D. Thus, A is diagonalizable using an orthogonal matrix. ∎

Real symmetric matrices have wonderful properties. We can get a hint of this by taking a look at the nonsymmetric matrix $A = \begin{bmatrix} 1 & 2 & 4 \\ 0 & 1 & -1 \\ 0 & 0 & 2 \end{bmatrix}$. Since $\det \begin{bmatrix} 1-\lambda & 2 & 4 \\ 0 & 1-\lambda & -1 \\ 0 & 0 & 2-\lambda \end{bmatrix} = (1-\lambda)(1-\lambda)(2-\lambda)$, the eigenvalues of A are $\lambda_1 = 1$, $\lambda_2 = 1$, $\lambda_3 = 2$. A computation shows that the eigenvectors corresponding to the eigenvalues are

$$v_1 = \begin{bmatrix} 1 \\ 0 \\ 0 \end{bmatrix}, \quad v_2 = \begin{bmatrix} -1 \\ 0 \\ 0 \end{bmatrix}, \quad v_3 = \begin{bmatrix} 2 \\ -1 \\ 1 \end{bmatrix}.$$

The eigenvectors span a subspace of dimension two. This is caused by the duplicate eigenvalue 1, and this matrix cannot be diagonalized.

Example 6.7. Let $A = \begin{bmatrix} 1 & 1 & 1 & 1 & 1 \\ 1 & 1 & 1 & 1 & 1 \\ 1 & 1 & 1 & 1 & 1 \\ 1 & 1 & 1 & 1 & 1 \\ 1 & 1 & 1 & 1 & 1 \end{bmatrix}$. A is symmetric and has characteristic polynomial $p(\lambda) = \lambda^4(\lambda - 5)$, so A has four eigenvalues of 0. Despite this, there are five linearly independent eigenvectors. Use the MATLAB command `[V D] = eig(A)` and note there are four values of 0 on the diagonal of D. Verify the following:

- V is an orthogonal matrix.
- The rank of V is 5, so the columns of V are linearly independent and form a basis for $\mathbb{R}^5$.
- $V^{T}AV = D$

Despite the fact that A has four equal eigenvalues, it can be diagonalized. ∎

6.5 THE L^2 INNER PRODUCT

We have presented the inner product for the vector space $\mathbb{R}^n$, and showed that it satisfies the properties in Theorem 6.1. The general concept of an inner product extends beyond Euclidean space to any vector space for which an inner product can be defined. In particular, there are many applications for vector spaces whose elements are functions, and such vector spaces normally have infinite dimension. Chapter 12 presents Fourier series to illustrate this concept. Fourier series is one of the most useful topics in engineering and science. The applications of Fourier series include heat conduction, signal processing, analysis of sound waves, seismic imaging, and solving differential equations. The inner product used with Fourier series and many other vector spaces of functions is the L^2 inner product.

Definition 6.3. If functions $f(t)$ and $g(t)$ are continuous on the interval $a \le t \le b$, the L^2 *inner product* is

$$\langle f, g \rangle_{L^2} = \int_a^b f(t)\, g(t)\, \mathrm{d}t.$$

It is not difficult to show that $\langle \cdot, \cdot \rangle_{L^2}$ satisfies the requirements for an inner product. For instance,

$$\langle cf, g \rangle_{L^2} = \int_a^b (cf(t)) \, g(t) \, dt = \int_a^b f(t) \, (cg(t)) \, dt = \langle f, cg \rangle_{L^2} = c \int_a^b f(t) \, g(t) \, dt = c \, \langle f, g \rangle_{L^2},$$

so

$$\langle cf, g \rangle_{L^2} = \langle f, cg \rangle_{L^2} = c \, \langle f, g \rangle_{L^2}.$$

Proving the remaining properties is left to the exercises.

The length of a vector u is $\sqrt{\langle u, u \rangle}$. We can also define the length or size of a function over the interval $a \le t \le b$ by

$$\|f\|_{L_2} = \sqrt{\langle f, f \rangle_{L^2}} = \sqrt{\int_a^b f^2(t) \, dt}.$$

A function f is normalized if $\langle f, f \rangle_{L^2} = 1$, and two functions f and g are orthogonal if $\langle f, g \rangle_{L^2} = 0$. A Fourier series consists of an infinite sequence of normalized trigonometric functions that are mutually orthogonal with respect to the L^2 norm.

Example 6.8. Given the functions $f(t) = (1/\sqrt{\pi}) \sin(5t)$ and $g(t) = (1/\sqrt{\pi}) \cos(3t)$, compute $\langle f, g \rangle_{L^2}$ and $\|f\|_{L_2}$.

a. $\langle f, g \rangle_{L^2} = (1/\pi) \int_0^{2\pi} \sin(5t) \cos(3t) \, dt = -(1/4\pi) \left(\cos 2t + \frac{1}{4} \cos 8t \right) \Big|_0^{2\pi} = 0.$ (f and g are orthogonal.)

b. $\|f\|_{L_2} = \sqrt{(1/\pi) \int_0^{2\pi} \sin^2 5t \, dt} = \sqrt{(1/\pi) \left[(t/2) - \frac{1}{20} \sin 10t \right] \Big|_0^{2\pi}} = 1$ ∎

Similarly, $\|g\|_2 = 1$, so f and g are orthogonal and have unit length using the L^2 inner product.

6.6 THE CAUCHY-SCHWARZ INEQUALITY

The *Cauchy-Schwarz inequality* is one of the most widely used inequalities in mathematics, and will have occasion to use it in proofs. We can motivate the result by assuming that vectors u and v are in $\mathbb{R}^2$ or $\mathbb{R}^3$. In either case, $\langle u, v \rangle = \|u\|_2 \|v\|_2 \cos \theta$. If $\theta = 0$ or $\theta = \pi$, $|\langle u, v \rangle| = \|u\|_2 \|v\|_2$. This occurs when u and v are parallel, or when $v = cu$ for some scalar multiple c. For $0 < \theta < \pi$, $|\cos \theta| < 1$, so $|\langle u, v \rangle| \le ||u||_2 ||v||_2$.

Theorem 6.5 (Cauchy-Schwarz inequality). *For any n-dimensional vectors u and v,*

$$|\langle u, v \rangle| \le ||u||_2 ||v||_2,$$

and equality occurs if and only if $v = cu$.

For a proof, see (15, p. 316).

Remark 6.3. The Cauchy-Schwarz inequality applies to any vector space that has an inner product; for instance, it applies to a vector space that uses the L^2-norm.

Recall in high school geometry you were told that the sum of the lengths of two sides of a triangle is greater than the third side. This is an instance of the *triangle inequality* that follows by using the Cauchy-Schwarz inequality:

$$\|u + v\|_2^2 = \|u\|_2^2 + 2 \langle u, v \rangle + \|v\|_2^2 \le \|u\|_2^2 + 2 \|u\|_2 \|v\|_2 + \|v\|_2^2 = (\|u\|_2 + \|v\|_2)^2,$$

and

$$||u + v||_2 \le ||u||_2 + ||v||_2.$$

The triangle inequality holds for any number of dimensions, but is easily visualized in $\mathbb{R}^3$. In Figure 6.7, note the progression from a normal triangle to the three sides collapsing into a line, corresponding to $v = cu$. In this case, $\|x + y\|_2 = \|x\|_2 + \|y\|_2$.

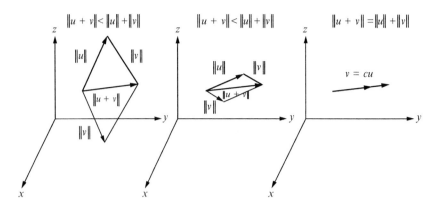

FIGURE 6.7 Triangle inequality.

6.7 SIGNAL COMPARISON

There is a particularly interesting implication of the Cauchy-Schwartz inequality [89].We ask the question "When is an expression of the form $|\langle u/\|u\|_2, v/\|v\|_2\rangle|$ a maximum?" Note that both $u/\|u\|_2$ and $v/\|v\|_2$ are unit vectors. By the Cauchy-Schwartz inequality, we know that $|\langle u/\|u\|_2, v/\|v\|_2\rangle| \leq \|u/\|u\|_2\| \|v/\|v\|_2\| = 1$ and that $|\langle u/\|u\|_2, v/\|v\|_2\rangle| = 1$ if and only if $v/\|v\|_2 = cu/\|u\|_2$ for some scalar c. Hence, $|\langle u/\|u\|_2, v/\|v\|_2\rangle|$ attains a maximum when $v/\|v\|_2 = cu/\|u\|_2$ for some c. Now suppose we collect numerous samples of scalars of the form $|\langle u/\|u\|_2, v/\|v\|_2\rangle|$. The largest values will occur when $v = cu$. This result is very useful in developing *matched filter detector techniques*. When dealing with signals, we replace vectors by functions. Use the L^2 inner product to compare functions, and the Cauchy-Schwarz inequality applies to the L^2 inner product. We want to find the member signal in a set S of signals that most closely matches a target signal v. Define $f(u,v) = |\langle u/\|u\|_2, v/\|v\|_2\rangle|$. To find the best matching signal we need to evaluate

$$u_{\max} = \max_{u \in S} f(u,v).$$

The value $u_{\max}$ that produces the maximum value of $f(u,v)$ is not necessarily unique, so there may be more than one matching signal in S. It is possible that among the current members of S, the signal, u_{max}, giving the maximum value of $f(u,v)$ may be small and a poor match for the target signal v. A solution is to set a threshold and return no matching signals if $f(u,v)$ is below the threshold. There also may be a signal u that produces an very high value of $f(u,v)$, well above the actual match desired. This corresponds to a local maximum, and there are techniques to filter out local maxima.

Example 6.9.   Here is the target signal and a set of three candidate signals (Figure 6.8). An application of the technique we have outlined will determine that among the candidate signals (c) is the best match for $f(t)$.   ■

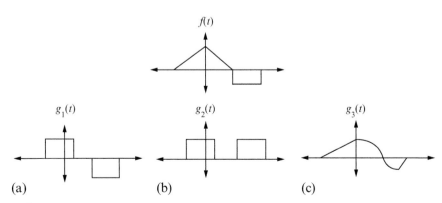

FIGURE 6.8 Signal comparison.

6.8 CHAPTER SUMMARY

The Inner Product

The inner product of two vectors is often called the dot product, although we will seldom use the term in the text. If x and y are vectors in $\mathbb{R}^n$, the inner product of x and y is $\langle x, y \rangle = \sum_{i=1}^{n} x_i y_i$, alternatively written $\langle x, y \rangle = x^T y$. Since the inner product is commutative, we can also write $\langle x, y \rangle = y^T x$. This expression for the inner product will be useful in many places throughout the book.

Among the most important properties of the inner product is that $\langle x, x \rangle = \sum_{i=1}^{n} x_i^2$, and so the length of a vector can be expressed as $\text{length}(x) = \sqrt{\langle x, x \rangle} = \|x\|_2$. Also, if $\langle x, x \rangle = 0$, $x = 0$.

In two and three dimensions, the inner product has the geometric interpretation

$$\langle x, y \rangle = \|x\|_2 \|y\|_2 \cos \theta,$$

where θ is the angle between x and y.

Orthogonal Matrices

Orthogonal matrices are the most beautiful of all matrices. A matrix P is orthogonal if $P^T P = I$, or the inverse of P is its transpose. Alternatively, a matrix is orthogonal if and only if its columns are orthonormal, meaning they are orthogonal and of unit length. An interesting property of an orthogonal matrix P is that $\det P = \pm 1$. As an example, rotation matrices are orthogonal.

Orthogonal matrices are involved in some of the most important decompositions in numerical linear algebra, the QR decomposition (Chapter 14), and the SVD (Chapter 15). The fact that orthogonal matrices are involved makes them invaluable tools for many applications.

Symmetric Matrices and Orthogonality

Symmetric matrices can always be diagonalized with an orthogonal matrix; in other words, there is an orthogonal matrix of eigenvectors such that $P^T A P = D$, where D is a diagonal matrix of eigenvalues. This allows us to develop method for computing their eigenvalues more rapidly than we can find eigenvalues for nonsymmetric matrices. We begin the development of this diagonalization result by showing that any eigenvectors of a symmetric matrix corresponding to distinct eigenvalues are orthogonal.

The L^2 Inner Product

The inner product can be extended to functions by defining

$$\langle f, g \rangle_{L^2} = \int_a^b f(t) g(t) \, dt,$$

so that

$$\|f\|_{L_2}^2 = \int_a^b f^2(t) \, dt.$$

There are important sequences of functions that are orthogonal under the L^2 inner product. Chapter 12 looks at Fourier series, where the functions are trigonometric.

The Cauchy-Schwarz Inequality

The inequality,

$$|\langle x, y \rangle| \le \|x\|_2 \|y\|_2$$

applies to any vector space with an inner product, and is called the Cauchy-Schwarz inequality. Among other things, it can be used to prove the triangle inequality

$$\|x + y\|_2 \le \|x\|_2 + \|y\|_2.$$

Although we will use the Cauchy-Schwarz inequality in later chapters as a theoretical tool, it has applications in matched filter detector techniques. Given function $f(t)$, it can be used to determine the best match to $f(t)$ among a set of candidate signals.

6.9 PROBLEMS

6.1 Compute the distance between the specified points.

    a. $\begin{bmatrix} 1 & -6 & 7 \end{bmatrix}^T, \begin{bmatrix} 3 & 2 & 1 \end{bmatrix}^T$

    b. $\begin{bmatrix} -1 & 4 & -9 & 12 & 15 \end{bmatrix}^T, \begin{bmatrix} 2 & -8 & 0 & -7 & 3 \end{bmatrix}^T$

6.2 Draw the vectors in $\mathbb{R}^2$.

    a. $\begin{bmatrix} 1 \\ -9 \end{bmatrix} + 4 \begin{bmatrix} -14 \\ 2 \end{bmatrix}$

    b. $\begin{bmatrix} 5 \\ 1 \end{bmatrix} - 3 \begin{bmatrix} -1 \\ 2 \end{bmatrix}$

6.3 Find the inner product of each vector pair.

    a. $\begin{bmatrix} 1 & -5 & 2 \end{bmatrix}^T, \begin{bmatrix} -10 & 1 & -8 \end{bmatrix}^T$

    b. $\begin{bmatrix} 17 & 0 & -4 & 12 & 3 \end{bmatrix}^T, \begin{bmatrix} 1 & -1 & 5 & 9 & 2 \end{bmatrix}$

6.4 Determine if each pair of vectors is orthogonal.

    a. $\begin{bmatrix} 1 & -1 & 2 \end{bmatrix}^T, \begin{bmatrix} 1 & -1 & -1 \end{bmatrix}^T$

    b. $\begin{bmatrix} 1 & -2 & 5 & 7 \end{bmatrix}^T, \begin{bmatrix} -1 & 2 & 1 & 1 \end{bmatrix}^T$

6.5 Normalize each vector in Problem 6.4.

6.6 If u and v are unit vectors, compute the following:

    a. $\langle u + v, u - v \rangle$

    b. $\langle u + v, u + v \rangle$

6.7 Find a vector parallel to the vector $\begin{bmatrix} -1 \\ 2 \\ 5 \\ 7 \end{bmatrix}$ and a vector orthogonal to it.

6.8 Find the angle between the vectors $\begin{bmatrix} -1 & 2 & 5 \end{bmatrix}^T$ and $\begin{bmatrix} 1 & -8 & 2 \end{bmatrix}^T$.

6.9 What is the length of the 10-dimensional vector $u = \begin{bmatrix} 1 & -1 & 1 & -1 & 2 & 1 & 1 & 1 & 1 & 1 \end{bmatrix}$? Find a vector orthogonal to u and normalize each vector.

6.10 Verify the Cauchy-Schwarz and triangle inequalities for the vectors $x = \begin{bmatrix} 1 \\ -1 \\ 2 \end{bmatrix}, y = \begin{bmatrix} 2 \\ 3 \\ -4 \end{bmatrix}$.

6.11 Explain the result of applying the Cauchy-Schwarz inequality to the vectors $u = \begin{bmatrix} 0.88 \\ -1.55 \\ 2.68 \end{bmatrix}$, $v = \begin{bmatrix} -7.92 \\ 13.95 \\ -24.12 \end{bmatrix}$.

6.12 Prove the *parallelogram law* $2 \|u\|_2^2 + 2 \|v\|_2^2 = \|u + v\|_2^2 + \|u - v\|_2^2$. Explain where its name comes from by considering vectors u, v in the plane.

6.13 Determine if all possible pairings of the following vectors are parallel, orthogonal or neither.

$$u = \begin{bmatrix} -1 \\ 4 \\ 2 \end{bmatrix}, v = \begin{bmatrix} 1/2 \\ 1/16 \\ 1/8 \end{bmatrix}, w = \begin{bmatrix} -64 \\ -8 \\ 16 \end{bmatrix}$$

6.14 Show that the triangle formed by the points $(-3, 5, 6)$, $(-2, 7, 9)$, and $(2, 1, 7)$ is a 30-60-90 triangle.

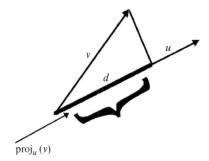

FIGURE 6.9 Projection of one vector onto another.

6.15 One of the primary applications of the inner product is the projection of one vector onto another. Looking at the Figure 6.9, develop a formula for the vector that is a projection of vector v onto vector u.

6.16 Prove parts 1, 2, 3, and 4 of Theorem 6.1.

6.17 Prove that if x and y are vectors in $\mathbb{R}^n$, then $\langle Ax, y \rangle = \langle x, A^{\mathrm{T}}y \rangle$.

Problems 6.18–6.24 deal with the *cross product*, which is defined for three dimensions as follows:

Definition 6.4. Let $i = \begin{bmatrix} 1 \\ 0 \\ 0 \end{bmatrix}$, $j = \begin{bmatrix} 0 \\ 1 \\ 0 \end{bmatrix}$, $k = \begin{bmatrix} 0 \\ 0 \\ 1 \end{bmatrix}$ be the standard basis for $\mathbb{R}^3$. The cross product of $u = \begin{bmatrix} u_1 \\ u_2 \\ u_3 \end{bmatrix}$,

and $v = \begin{bmatrix} v_1 \\ v_2 \\ v_3 \end{bmatrix}$, written $u \times v$ is the vector

$$u \times v = \begin{vmatrix} i & j & k \\ u_1 & u_2 & u_3 \\ v_1 & v_2 & v_3 \end{vmatrix}. \tag{6.8}$$

Equation 6.8 is not a normal determinant. Treat vectors i, j, and k as scalars for the computation and then consider them vectors.

6.18 Show that

$$u \times v = (u_2 v_3 - u_3 v_2)\, i + (u_3 v_1 - u_1 v_3)\, j + (u_1 v_2 - u_2 v_1)\, k.$$

6.19 What is the relationship between $u \times v$ and $v \times u$?

6.20 Show that $u \times v$ is perpendicular to both u and v.

6.21 Show that $u \times u = 0$.

6.22 For each pair of vectors, compute the cross product.

a. $\begin{bmatrix} 1 \\ -1 \\ 3 \end{bmatrix}$, $\begin{bmatrix} -7 \\ 12 \\ 1 \end{bmatrix}$

b. $\begin{bmatrix} a \\ -a \\ b \end{bmatrix}$, $\begin{bmatrix} -b \\ a \\ a \end{bmatrix}$, where a and b are scalars.

6.23 The equation of a plane is determined by three non-collinear points A, B and C. Take the two vectors $\vec{AB}$ and $\vec{AC}$ in the plane and use the cross product to find a vector n perpendicular to the plane. Pick an arbitrary point $P : (x, y, z)$ in the plane and require that $\langle n, \vec{PA} \rangle = 0$. This generates what is called the *normal equation of a plane*.
a. Draw a figure that illustrates this process.
b. Find the equation of the plane containing the points $A : (-1, 2, 3)$, $B\,(5, 1, 2)$, and $C : (-7, 1, 3)$.

6.24 Using the process of computing the normal equation to a plane in Problem 6.23, find an equation for a plane involving a determinant.

6.25 Show that if P is an orthogonal matrix, and x and y are vectors in $\mathbb{R}^n$, then $\langle Px, Py \rangle = \langle x, y \rangle$.

6.26 Show that if A and B are orthogonal matrices, then AB and BA are orthogonal matrices.

6.27 The matrix A has orthogonal columns. Convert it to an orthogonal matrix by normalizing the columns.

$$A = \begin{bmatrix} -1 & 1 & -3 \\ -1 & 1 & 3 \\ 2 & 1 & 0 \end{bmatrix}$$

6.28 Find one eigenvalue and its corresponding eigenvector of the symmetric matrix A by hand. Compute the other two using MATLAB. Using MATLAB, show that any two eigenvectors corresponding distinct eigenvalues are orthogonal, and diagonalize A with an orthogonal matrix.

$$A = \begin{bmatrix} 0 & 1 & -1 \\ 1 & 1 & 1 \\ -1 & 1 & 0 \end{bmatrix}$$

6.29 Find the eigenvalues of matrix A. Can you diagonalize it? Explain.

$$A = \begin{bmatrix} -1 & 1 & 1 \\ -1 & 1 & 1 \\ 1 & -1 & -1 \end{bmatrix}$$

6.30 Find the eigenvalues of A. The matrix A has two equal eigenvalues, but it still has three linearly independent eigenvectors. Diagonalize A with an orthogonal matrix using MATLAB.

$$A = \begin{bmatrix} 0 & 1 & -1 \\ 1 & 0 & 1 \\ -1 & 1 & 0 \end{bmatrix}$$

6.31 Find the L^2 inner product of $f(t) = \sin(\pi t)$, $g(t) = \cos(3\pi t)$, $0 \leq t \leq 2\pi$. Also compute $\langle f, f \rangle^2$.

6.32 Over the interval $0 \leq t \leq 2\pi$, show that $\langle \cos it/\sqrt{\pi}, \cos jt/\sqrt{\pi} \rangle_{L^2} = 0$ for $i, j \geq 1$, $i \neq j$ and that $\langle \cos it/\sqrt{\pi}, \cos it/\sqrt{\pi} \rangle_{L^2} = 1$, $i \geq 1$.

6.33 A *permutation matrix* is a matrix obtained by swapping one or more rows of the identity matrix. Prove that a permutation matrix is orthogonal.

6.34 The vector *outer product*, $u \otimes v$, takes an $m \times 1$ column vector, u, an $n \times 1$ column vector, v, and returns an $m \times n$ matrix obtained by multiplying each element of u by each element of v. In particular, $(u \times v)_{ij} = u_i v_j$. The product $u \otimes v$ is also called the *tensor product*. The outer product computes the inertial tensor in rigid body dynamics, performs transform operations in digital signal processing and digital image processing, and has applications in statistics.
  a. Show that $u \otimes v = u^T v$.
  b. Show that if $A = u \otimes v$, then $Av = u \|v\|_2^2$

6.9.1 MATLAB Problems

6.35 a. Develop `function d = veclength(v)` that computes the length of a vector v.
  b. Use your function to find the length of each vector.

  i. $$\begin{bmatrix} -1 \\ 2 \\ 4 \\ 12 \\ -3 \end{bmatrix}$$

  ii. $$\begin{bmatrix} -1 \\ 35 \\ 52 \\ 6 \end{bmatrix}$$

6.36 The SVD discussed in Chapter 15 says that if A is any $n \times n$ matrix, there exist orthogonal matrices U and V and a diagonal matrix Σ such that $A = U\Sigma V^T$.
  a. Show that $A^T A = V\Sigma^2 V^T$
  b. Prove that the eigenvalues of $A^T A$ are the squares of the elements on the diagonal of Σ.

c. Let $A = \begin{bmatrix} 1 & -1 & 5 & 0 & 3 \\ 5 & -1 & 3 & 6 & 1 \\ 8 & -9 & 2 & 7 & 4 \\ 8 & 4 & -3 & 5 & 1 \\ -1 & -4 & 3 & 0 & 2 \end{bmatrix}$. Execute the MATLAB command `[U S V] = svd(A)` and verify that

$A = USV^T$ by computing `veclength(A-U*S*V')` if you did Problem 6.35; otherwise use `norm(A-U*S*V')`.

d. Find the eigenvalues of A^TA and verify the result of part (b). The elements on the diagonal of S are sorted in descending order. Sort the eigenvalues, E, of A^TA using `sort(E, 'descend')` before computing $\|E - \text{diag}(S)\|_2$.

6.37 A floating point number is a number that contains a fractional part, such as 0.3, 234.56819, and 1.56×10^{-8}. Because a computer generally cannot perform floating point calculations exactly, errors, called round-off errors, are introduced during computation. Chapter 8 discusses round-off errors and their effect on the accuracy of computer calculations. For example, if the exact value of calculation is 0.0, the computed result may be 3.0×10^{-16}. Using MATLAB, compute the inner product of u and v. Find the inner product using exact arithmetic, and comment on the results.

$$u = \begin{bmatrix} 3.2 & -1.5 & 6.3 & -2.5 \end{bmatrix}^T, \quad v = \begin{bmatrix} 4.3 & 0 & 1.8 & 10.04 \end{bmatrix}^T.$$

6.38 Show that each matrix is orthogonal in two different ways, using the definition and by directly showing that the columns have unit length and are orthogonal.

a. $P1 = \begin{bmatrix} -0.40825 & 0.43644 & 0.80178 \\ -0.8165 & 0.21822 & -0.53452 \\ -0.40825 & -0.87287 & 0.26726 \end{bmatrix}$

b. $P2 = \begin{bmatrix} -0.51450 & 0.48507 & 0.70711 \\ -0.68599 & -0.72761 & 0.0000 \\ 0.51450 & -0.48507 & 0.70711 \end{bmatrix}$

6.39 Are any of the two matrices orthogonal?

a. $P1 = \begin{bmatrix} -0.58835 & 0.70206 & 0.40119 \\ -0.78446 & -0.37524 & -0.49377 \\ -0.19612 & -0.60523 & 0.77152 \end{bmatrix}$

b. $P2 = \begin{bmatrix} -0.47624 & -0.4264 & 0.30151 \\ 0.087932 & 0.86603 & -0.40825 \\ -0.87491 & -0.26112 & 0.86164 \end{bmatrix}$

6.40 In Chapter 14, we will begin the study of the QR decomposition. A special case of this decomposition states that for any $n \times n$ matrix A, there exists an $n \times n$ orthogonal matrix Q and an $n \times n$ upper triangular matrix R such that

$$A = QR.$$

The MATLAB command

```
[Q R] = qr(A)
```

computes the factors Q and R. Find the QR decomposition of the matrix in Problem 6.36(c). Verify that Q is orthogonal.

6.41 For a matrix whose elements are complex numbers, there is a definition analogous to the transpose of a real matrix. A^*, called the *conjugate transpose*, is the matrix obtained by taking the complex conjugate of the entries of A and exchanging rows and columns. A matrix is said to be *Hermitian* if $A^* = A$.

a. Find the conjugate transpose of the matrix.

$$\begin{bmatrix} 1-i & 3+i & 7 & 8-3i \\ 6+7i & 4-i & i & 1+i \\ 2-3i & 6+i & 3 & 9+i \\ -1-i & 10+i & 7 & 12+2i \end{bmatrix}$$

b. Using MATLAB, verify your result of part (a).

      c. Investigate whether the MATLAB function `eig` applies to a complex matrix by using it with the matrix in part (a).

      d. Prove that if A is Hermitian, its diagonal entries are real numbers.

6.42   **a.** Write a function `c = mycross(u,v)` that computes the cross product of vectors v and w.

      b. Execute the function with two vectors. In each case, compare the result with the MATLAB function `cross`.

6.43   The inner product, or tensor product, is defined in Problem 6.34.

      a. Write a function `t = tensor(u,v)` that computes the inner product of $m \times 1$ vector u and $n \times 1$ vector v.

      b. Test the function for two pairs of vectors, one pair giving a 5×5 matrix and another giving a 6×4 matrix.

Chapter 7

Vector and Matrix Norms

You should be familiar with

- Basic two- and three-dimensional geometry
- Vector and matrix properties
- The inner product
- Partial derivatives (for Section 7.1.2)
- Eigenvalues
- Symmetric matrices
- Orthogonal matrices

Chapters 1–6 provide sufficient background for the remainder of this book. This chapter actually begins our study of numerical linear algebra that, as we have stated, is very different from theoretical linear algebra. This is because we use a computer and are concerned with numerical accuracy and execution time. Since we are dealing with a numerical subject, it is natural to assume there must be a means of measuring the length of a vector and the "size" of a matrix. In each case, we define what is termed a *norm*. Vector norms have applications in many areas, including signal processing, quantum information theory, measuring deflections, and determining convergence of sequences of vectors. We studied the solution of square linear algebraic systems in Chapter 2. In some cases, the coefficient matrix is sensitive to changes in data; for instance, if there are small changes to the vector b in the system $Ax = b$ due to experimental error, the solution may differ widely, leading to incorrect results. In such a case, the matrix is said to be *ill-conditioned*. The matrix norm plays a critical role in determining if a matrix is ill-conditioned. In addition, there are many applications of matrix norms to specific disciplines such as structural analysis and input-output response in electrical engineering problems.

We begin with a definition of a vector norm and develop some examples of vector norms. These norms are important in their own right, and we will see that some frequently used matrix norms are derived from a vector norm.

7.1 VECTOR NORMS

A vector norm gives us a way of measuring vector length. You are already familiar with the most-used vector norm, the formula for the length of a vector u in $\mathbb{R}^n$. In Chapter 6, we used the notation $\|u\|_2$ for the length of a vector, where

$$\|u\|_2 = \sqrt{u_1^2 + u_2^2 + \cdots + u_n^2}. \tag{7.1}$$

Figure 6.4 graphically shows why this function computes the length of a three-dimensional vector. Let's examine some properties of this length function.

- Since $u_i^2 \geq 0$, $1 \leq i \leq n$, $\|u_2\| = 0$ if and only if $u = 0$.
- If α is a scalar,

$$\|\alpha u\|_2 = \sqrt{(\alpha u_1)^2 + (\alpha u_2)^2 + \cdots + (\alpha u_n)^2}$$
$$= \sqrt{\alpha^2 \left[u_1^2 + u_2^2 + \cdots + u_n^2 \right]} = |\alpha| \sqrt{u_1^2 + u_2^2 + \cdots + u_n^2}$$

- In Chapter 6, we developed the triangle inequality for vectors x, y in $\mathbb{R}^n$, which states that $\|x + y\|_2 \leq \|x\|_2 + \|y\|_2$.

Any function that takes a vector argument, computes a real number, and satisfies these three conditions is called a vector norm, so $\|u\|_2 = \sqrt{u_1^2 + u_2^2 + \cdots + u_n^2}$ is our first vector norm.

Numerical Linear Algebra with Applications. http://dx.doi.org/10.1016/B978-0-12-394435-1.00007-7

Definition 7.1. A function $\|\cdot\|: \mathbb{R}^n \to \mathbb{R}$ is a *norm* provided:

1. $\|x\| \geq 0$ for all $x \in \mathbb{R}^n$; $\|x\| = 0$ if and only if $x = 0$ (*positivity*);
2. $\|\alpha x\| = |\alpha| \, \|x\|$ for all $\alpha \in \mathbb{R}$ (*scaling*);
3. $\|x + y\| \leq \| x \| + \|y\|$ for all $x, y \in \mathbb{R}^n$ (*triangle inequality*).

In this book, the only vector norms we will use are the *p-norms* defined by

$$\|x\|_p = \left(\sum_{i=1}^{n} |x_i|^p \right)^{1/p},$$

for $p = 1, 2, \ldots$ The values $p = 1, 2$, and ∞ are the most commonly used norms. Equation 7.1 corresponds to $p = 2$ and we will refer to it as the *Euclidean norm* or the *2-norm*, and indicate this using the notation

$$\|u\|_2 = \sqrt{u_1^2 + u_2^2 + \cdots + u_n^2} = \sqrt{u^T u}.$$

For $p = 1$ and ∞, the norms are

$$\|x\|_1 = \sum_{i=1}^{n} |x_i|$$

$$\|x\|_\infty = \max_{i=1,\ldots,n} |x_i|$$

Problem 7.9 justifies the formula for $\|x\|_\infty$.

Example 7.1. Let $u = \begin{bmatrix} -1 \\ -9 \\ 2 \end{bmatrix}$.

$$\|u\|_1 = |-1| + |-9| + |2| = 12,$$

$$\|u\|_2 = \sqrt{(-1)^2 + (-9)^2 + (2)^2} = \sqrt{86} = 9.2736,$$

$$\|u\|_\infty = \max\{|-1|, |-9|, |2|\} = 9,$$

$$\|u\|_5 = \left((|-1|)^5 + (|-9|)^5 + (|2|)^5 \right)^{1/5} = 9.0010. \qquad \blacksquare$$

We have already shown that $\|.\|_2$ is a norm, but we should not take for granted that the 1- and ∞-norms satisfy the three requirements for a norm. Theorem 7.1 shows that $\|\cdot\|_\infty$ satisfies the required properties. Showing that $\|.\|_1$ is a norm is left to the exercises.

Theorem 7.1. $\|.\|_\infty$ *is a norm.*

Proof. *Positivity*: Clearly, $\|x\|_\infty \geq 0$ for all $x \neq 0$. If

$$\|x\|_\infty = \max_{1 \leq i \leq} |x_i| = 0,$$

then $x = 0$. If $x = 0$, then all its components are 0.

Scaling:

$$\|\alpha x\|_\infty = \max_{1 \leq i \leq n} |\alpha x_i| = |\alpha| \max_{1 \leq i \leq n} |x_i| = |\alpha| \, \|x\|_\infty.$$

Triangle inequality:

$$\|x + y\|_\infty = \max_{1 \leq i \leq n} |x_i + y_i| \leq \max_{1 \leq i \leq n} (|x_i| + |y_i|) = \max_{1 \leq i \leq n} |x_i| + \max_{1 \leq i \leq n} |y_i| = \|x\|_\infty + \|y\|_\infty. \qquad \square$$

Which norm to use can depend on the application. Also, all three norms are equivalent, which means that each norm is bounded below and above by a multiple of one of the other norms.

Lemma 7.1.

$$\|x\|_\infty \le \|x\|_2 \le \sqrt{n}\,\|x\|_\infty \,,$$
$$\|x\|_\infty \le \|x\|_1 \le n\,\|x\|_\infty \,,$$
$$\|x\|_2 \le \|x\|_1 \le \sqrt{n}\,\|x\|_2 \,.$$

Proof. We will prove the first of the three inequalities. Proofs of the remaining ones are left to the exercises.

Assume that the maximum absolute value of the vector components occurs at index i, so $|x_i| = \|x\|_\infty$. Now, $\|x\|_\infty = \sqrt{x_i^2} \le \sqrt{x_1^2 + x_2^2 + \cdots + x_i^2 + \cdots + x_n^2} = \|x\|_2$.

$$\|x\|_2 = \sqrt{x_1^2 + x_2^2 + \cdots + x_i^2 + \cdots + x_n^2} \le \sqrt{n\,(x_i)^2} = \sqrt{n}\,|x_i| = \sqrt{n}\,\|x\|_\infty \qquad \square$$

Example 7.2. Let $x = \begin{bmatrix} 1 \\ 4 \\ -9 \end{bmatrix}$. Then,

$\|x\|_\infty = 9$, $\|x\|_1 = 14$, $\|x\|_2 = \sqrt{86}$.

Now test each inequality in Lemma 7.1.

$9 \le \sqrt{86} \le 9\sqrt{3},\ \sqrt{86} \le 14 \le \sqrt{258},\ 9 \le 14 \le 27$

The 2-norm is more computationally expensive than the ∞- or the 1-norm. If an application requires the computation of a norm many times, it could be advantageous to use the ∞- or the 1-norm. ∎

Example 7.3. The MATLAB `norm` command will compute norms of a vector. For instance, if $v = \begin{bmatrix} 1 \\ -7 \\ 2 \end{bmatrix}$, the following

MATLAB statements compute the ∞-norm, the 1-norm, and the 2-norm of v. ∎

```
>> norm(v, 1)
ans =
10.0000
>> norm(v, 'inf')
ans =
7
>> norm(v, 2)
ans =
7.3485
```
∎

7.1.1 Properties of the 2-Norm

The 2-norm is the norm most frequently used in applications, and there are good reasons why this is true. There are many relationships satisfied by the 2-norm, and one of the most frequently used is the *Cauchy-Schwarz inequality*:

$$|\langle x, y \rangle| = \left| x^\mathrm{T} y \right| \le \|x\|_2 \, \|y\|_2 \,,$$

with equality holding when x and y are collinear (Theorem 6.5). Another relationship involving the Euclidean norm is the *Pythagorean Theorem* for orthogonal x and y,

$$\|x + y\|_2^2 = \|x\|_2^2 + \|y\|_2^2 \,.$$

The vector 2-norm enjoys yet another property: it is *orthogonally invariant*. This means that for any $n \times n$ orthogonal matrix P

$$\|Px\|_2 = \|x\|_2$$

for all x in $\mathbb{R}^n$, since

$$\|Px\|_2^2 = (Px)^\mathrm{T} Px = x^\mathrm{T} P^\mathrm{T} Px = x^\mathrm{T} Ix = x^\mathrm{T} x = \|x\|_2^2 \,.$$

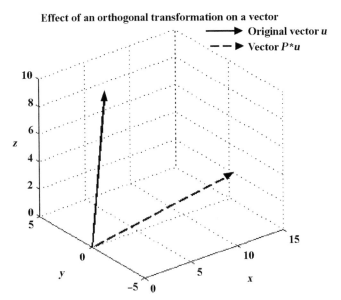

FIGURE 7.1 Effect of an orthogonal transformation on a vector.

In other words, multiplication of a vector, x, by an orthogonal matrix will likely rotate x, but the resulting vector Px has the same length. This is one of the reasons that orthogonal matrices are useful in computer graphics. Figure 7.1 shows the result of applying the orthogonal transformation $P = \begin{bmatrix} 0.4082 & 0.5774 & 0.7071 \\ 0.4082 & 0.5774 & -0.7071 \\ 0.8165 & -0.5774 & 0 \end{bmatrix}$ to $u = \begin{bmatrix} 8 \\ 5 \\ 12 \end{bmatrix}$

producing the vector $v = Pu = \begin{bmatrix} 14.6380 \\ -2.3325 \\ 3.6452 \end{bmatrix}$. Note that $\|u\|_2 = \sqrt{8^2 + 5^2 + 12^2} = 15.2643$, and $\|Pu\|_2 = \sqrt{(14.6380)^2 + (-2.3325)^2 + (3.6452)^2} = 15.2643$.

Theorem 7.2 gives us another important property. A set of orthogonal vectors is a basis for the subspace spanned by those vectors.

Theorem 7.2. *If the nonzero vectors $u_1, u_2, \ldots, u_k$ in $\mathbb{R}^n$ are orthogonal, they form a basis for a k-dimensional subspace of $\mathbb{R}^n$.*

Proof. Let

$$c_1 u_1 + c_2 u_2 + \cdots + c_{i-1} u_{i-1} + c_i u_i + c_{i+1} u_{i+1} + \cdots + c_k u_k = 0.$$

Choose any i, $1 \leq i \leq k$. Then,

$$c_1 \langle u_1, u_i \rangle + c_2 \langle u_2, u_i \rangle + \ldots + c_{i-1} \langle u_{i-1}, u_i \rangle + c_i \langle u_i, u_i \rangle + c_{i+1} \langle u_{i+1}, u_i \rangle + \ldots + c_k \langle u_k, u_i \rangle$$
$$= c_1 (0) + c_2 (0) + \cdots + c_{i-1} (0) + c_i (1) + c_{i+1} (0) + \cdots + c_k (0) = c_i = 0.$$

Since $c_i = 0$, $1 \leq i \leq k$, the u_i are linearly independent, and thus are a basis. $\qquad\square$

If the vectors in a basis $u_1, u_2, \ldots, u_n$ are mutually orthogonal and each vector has unit length $\left(\langle u_i, u_j \rangle = \begin{cases} 0 & i \neq j \\ 1 & i = j \end{cases} \right)$, we say the basis is *orthonormal*. We proved in Theorem 6.2 that if P is an $n \times n$ real matrix, then P is an orthogonal matrix if and only if the columns of P are orthogonal and have unit length. It follows from Theorem 7.2 that the columns of an $n \times n$ orthogonal matrix P are an orthonormal basis for $\mathbb{R}^n$. This fact has a natural physical interpretation. The vector

$$Ux = \begin{bmatrix} u_{11} & u_{12} & \dots & u_{1n} \\ u_{21} & u_{22} & \dots & u_{2n} \\ \vdots & \vdots & \ddots & \vdots \\ u_{n1} & u_{n2} & \dots & u_{nn} \end{bmatrix} \begin{bmatrix} x_1 \\ x_2 \\ \vdots \\ x_n \end{bmatrix} = \begin{bmatrix} u_1 & u_2 & \dots & u_n \end{bmatrix} \begin{bmatrix} x_1 \\ x_2 \\ \vdots \\ x_n \end{bmatrix} = \sum_{i=1}^{n} x_i u_i$$

is a representation of the vector x in the coordinate system whose axes are given by $u_1, u_2, \dots, u_n$. The statement $\|Ux\|_2 = \|x\|_2$ simply means "the length of x does not change when we convert from the standard orthonormal basis $e_1 = [\, 1 \;\; 0 \;\; \dots \;\; 0 \;\; 0 \,]^{\mathrm{T}}, e_2 = [\, 0 \;\; 1 \;\; \dots \;\; 0 \;\; 0 \,]^{\mathrm{T}}, \dots, e_n = [\, 0 \;\; 0 \;\; \dots \;\; 0 \;\; 1 \,]^{\mathrm{T}}$ to the new orthonormal basis $u_1, u_2, \dots, u_n$."

7.1.2 Spherical Coordinates

A very good example of this change of basis is the spherical coordinate system used in geography, astronomy, three-dimensional computer games, vibration problems, and many other areas. The representation for a point in space is given by three coordinates (r, θ, ϕ). Fix a point O in space, called the origin, and construct the usual standard basis $i = \begin{bmatrix} 1 \\ 0 \\ 0 \end{bmatrix}, j = \begin{bmatrix} 0 \\ 1 \\ 0 \end{bmatrix}, k = \begin{bmatrix} 0 \\ 0 \\ 1 \end{bmatrix}$ centered at O. The r coordinate of a point P is the length of the line segment from O to P, θ is the angle between the direction of vector k and P, and ϕ is the angle between the i direction and the projection of $\overrightarrow{OP}$ onto the ij plane. In order for coordinates to be unique we require, $r \geq 0, 0 \leq \theta \leq \pi$, and $0 \leq \phi \leq 2\pi$ (Figure 7.2). The name spherical coordinates comes from the fact that the equation of a sphere in this coordinate system is simply $r = a$, where a is the radius of the sphere. An application of trigonometry shows that rectangular coordinates are obtained from spherical coordinates (r, θ, ϕ) as follows:

$$x = r \sin \theta \cos \phi,$$
$$y = r \sin \theta \sin \phi,$$
$$z = r \cos \theta.$$

The position vector for a point, P, in space is $\overrightarrow{P} = x\mathbf{i} + y\mathbf{j} + z\mathbf{k}$. Now write the vector with x, y, and z replaced by their equivalents in spherical coordinates.

$$\overrightarrow{P} = r \sin \theta \cos \phi \mathbf{i} + r \sin \theta \sin \phi \mathbf{j} + r \cos \theta \mathbf{k}.$$

Our aim is to develop a basis in spherical coordinates. Such a basis must have a unit vector $\mathbf{e_r}$ in the direction of r, $\mathbf{e_\theta}$ in the direction of θ, and $\mathbf{e_\phi}$ in the direction of ϕ (Figure 7.3) such that the position vector

$$\overrightarrow{P} = r\mathbf{e_r} + \theta \mathbf{e_\theta} + \phi \mathbf{e_\phi}.$$

The vectors $\mathbf{e_r}, \mathbf{e_\theta}, \mathbf{e_\phi}$ change direction as the point P moves. If θ and ϕ are fixed and we increase r, $\mathbf{e_r}$ is a unit vector in the direction of change in r. This means we take the partial derivative.

$$\frac{\partial \overrightarrow{P}}{\partial r} = \frac{\partial \, (r \sin \theta \cos \phi \mathbf{i} + r \sin \theta \sin \phi \mathbf{j} + r \cos \theta \mathbf{k})}{\partial r}$$

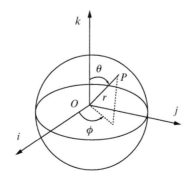

FIGURE 7.2 Spherical coordinates.

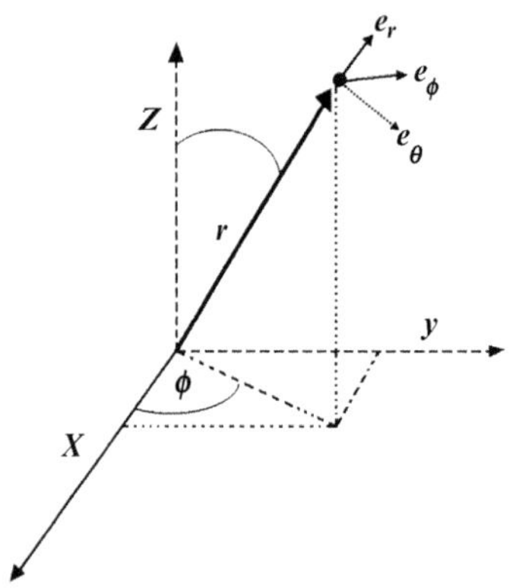

FIGURE 7.3 Orthonormal basis for spherical coordinates.

$$= \sin\theta\cos\phi\mathbf{i} + \sin\theta\sin\phi\mathbf{j} + \cos\theta\mathbf{k}$$

$$= \begin{bmatrix} \sin\theta\cos\phi \\ \sin\theta\sin\phi \\ \cos\theta \end{bmatrix},$$

and

$$\mathbf{e_r} = \frac{\frac{\partial\overrightarrow{\mathbf{P}}}{\partial\mathbf{r}}}{\left\|\frac{\partial\overrightarrow{P}}{\partial\mathbf{r}}\right\|}.$$

Similarly,

$$\mathbf{e}_\theta = \frac{\frac{\partial\overrightarrow{\mathbf{P}}}{\partial\theta}}{\left\|\frac{\partial\overrightarrow{\mathbf{P}}}{\partial\theta}\right\|}, \quad \mathbf{e}_\phi = \frac{\frac{\partial\overrightarrow{\mathbf{P}}}{\partial\phi}}{\left\|\frac{\partial\overrightarrow{\mathbf{P}}}{\partial\phi}\right\|}.$$

After performing the differentiation and division, the result is

$$e_r = \begin{bmatrix} \sin\theta\cos\phi \\ \sin\theta\sin\phi \\ \cos\theta \end{bmatrix}, \quad e_\theta = \begin{bmatrix} \cos\theta\cos\phi \\ \cos\theta\sin\phi \\ -\sin\theta \end{bmatrix}, \quad e_\phi = \begin{bmatrix} -\sin\phi \\ \cos\phi \\ 0 \end{bmatrix}.$$

Now,

$$\overrightarrow{P} = \begin{bmatrix} x \\ y \\ z \end{bmatrix} = r\mathbf{e_r} + \theta\mathbf{e}_\theta + \phi\mathbf{e}_\phi = r\begin{bmatrix} \sin\theta\cos\phi \\ \sin\theta\sin\phi \\ \cos\theta \end{bmatrix} + \theta\begin{bmatrix} \cos\theta\cos\phi \\ \cos\theta\sin\phi \\ -\sin\theta \end{bmatrix} + \phi\begin{bmatrix} -\sin\phi \\ \cos\phi \\ 0 \end{bmatrix},$$

and

$$x = (\sin\theta\cos\phi)\,r + (\cos\theta\cos\phi)\,\theta - (\sin\phi)\,\phi,$$
$$y = (\sin\theta\sin\phi)\,r + (\cos\theta\sin\phi)\,\theta + (\cos\phi)\,\phi,$$
$$z = (\cos\theta)\,r - (\sin\theta)\,\theta.$$

In matrix form

$$\begin{bmatrix} x \\ y \\ z \end{bmatrix} = \begin{bmatrix} \sin\theta\cos\phi & \cos\theta\cos\phi & -\sin\phi \\ \sin\theta\sin\phi & \cos\theta\sin\phi & \cos\phi \\ \cos\theta & -\sin\theta & 0 \end{bmatrix}\begin{bmatrix} r \\ \theta \\ \phi \end{bmatrix} = P\begin{bmatrix} r \\ \theta \\ \phi \end{bmatrix}. \tag{7.2}$$

The matrix P is orthogonal, as we can see by applying simple trigonometry, so $P^{-1} = P^{\mathrm{T}}$ and

$$\begin{bmatrix} r \\ \theta \\ \phi \end{bmatrix} = \begin{bmatrix} \sin\theta\cos\phi & \sin\theta\sin\phi & \cos\theta \\ \cos\theta\cos\phi & \cos\theta\sin\phi & -\sin\theta \\ -\sin\phi & \cos\phi & 0 \end{bmatrix}\begin{bmatrix} x \\ y \\ z \end{bmatrix},$$

Furthermore, $\{e_r, e_\theta, e_\phi\}$ is an orthonormal basis. It must be noted that the basis is a local basis, since the basis vectors change. Applications include the analysis of vibrating membranes, rotational motion, and the Schrodinger equation for the hydrogen atom. In Equation 7.2, the *xyz*-coordinate system is fixed, but the $r\theta\phi$-coordinate system moves. If we choose $r = 1, \theta = \pi/4, \phi = \pi/4$, the orthogonal matrix in Equation 7.2 is

$$P = \begin{bmatrix} 0.5000 & 0.5000 & -0.7071 \\ 0.5000 & 0.5000 & 0.7071 \\ 0.7071 & -0.7071 & 0 \end{bmatrix},$$

and

$$P \begin{bmatrix} 1 \\ \frac{\pi}{4} \\ \frac{\pi}{4} \end{bmatrix} = \begin{bmatrix} 0.33734 \\ 1.4481 \\ 0.15175 \end{bmatrix}.$$

This says that in the coordinate system defined by the spherical basis with $r = 1, \theta = \pi/4, \phi = \pi/4$, the vector $\begin{bmatrix} 1 \\ \frac{\pi}{4} \\ \frac{\pi}{4} \end{bmatrix}$

corresponds to the vector $\begin{bmatrix} 0.33734 \\ 1.4481 \\ 0.15175 \end{bmatrix}$ in the Cartesian coordinate system (Figure 7.4).

```
P =

          0.5           0.5      -0.70711
          0.5           0.5       0.70711
      0.70711      -0.70711             0
>> P*[1 theta phi]'

ans =

      0.33734
       1.4481
      0.15175
```

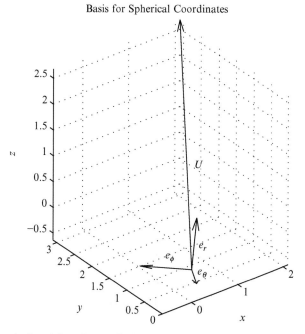

FIGURE 7.4 Point in spherical coordinate basis and Cartesian coordinates.

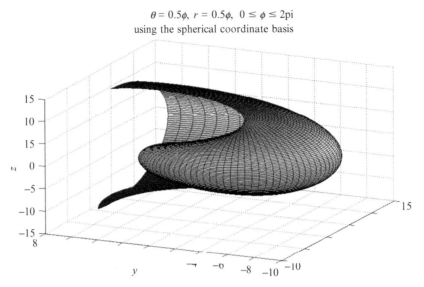

FIGURE 7.5 Function specified in spherical coordinates.

Example 7.4. Apply the change of basis from spherical to Cartesian coordinates to graph the surface formed by the equations $\theta = \frac{1}{2}\phi$, $r = 2\phi$, $0 \le \phi \le 2\pi$. Unfortunately, there is no fixed standard for spherical coordinates. We have defined spherical coordinates as commonly used in physics. MATLAB switches the roles of θ and ϕ; furthermore, ϕ is the angle between the projection of $\overrightarrow{OP}$ onto the xy-plane and OP. As a result, $-(\pi/2) \le \phi \le \pi/2$. The book software distribution contains a MATLAB function

```
[x y z] = sph2rect(r, theta, phi)
```

that uses our definition of spherical coordinates. It takes `r`, `theta`, `phi` in the local basis and returns Cartesian coordinates suitable for graphing a function using the MATLAB `surf` or `mesh` functions. The resulting graph is shown in Figure 7.5.

```
>> phi = linspace(0,2*pi);
>> theta = 0.5*phi;
>> [theta phi] = meshgrid(theta, phi);
>> r = 2*phi;
>> [x y z] = sph2rect(r, theta, phi);
>> surf(x,y,z);
```

■

7.2 MATRIX NORMS

We used vector norms to measure the length of a vector, and we will develop matrix norms to measure the size of a matrix. The size of a matrix is used in determining whether the solution, x, of a linear system $Ax = b$ can be trusted, and determining the convergence rate of a vector sequence, among other things. We define a matrix norm in the same way we defined a vector norm.

Definition 7.2. A function $\|\cdot\| : \mathbb{R}^{m \times n} \to \mathbb{R}$ is a *matrix norm* provided:

1. $\|A\| \ge 0$ for all $A \in \mathbb{R}^{m \times n}$; $\|A\| = 0$ if and only if $A = 0$ (*positivity*);
2. $\|\alpha A\| = |\alpha| \|A\|$ for all $\alpha \in \mathbb{R}$ (*scaling*);
3. $\|A + B\| \le \|A\| + \|B\|$ for all $A, B \in \mathbb{R}^{m \times n}$ (*triangle inequality*)

7.2.1 The Frobenius Matrix Norm

One of the oldest and simplest matrix norms is the *Frobenius norm*, sometimes called the *Hilbert-Schmidt norm*. It is defined as the square root of the sum of the squares of all the matrix entries, or

$$\|A\|_F = \left(\sum_{i=1}^{m}\sum_{j=1}^{n} a_{ij}^2\right)^{1/2}.$$

Clearly, it measure the "size" of matrix A. A matrix with small (large) entries will have a small (large) Frobenius norm, but we need to prove it is actually a matrix norm.

Theorem 7.3. $\|\cdot\|_F$ *is a matrix norm.*

Proof. Positivity: Clearly, $\|A\|_F \geq 0$, and $\|A\|_F = 0$ if and only if $A = 0$.

Scaling: $\|\alpha A\| = \left(\sum_{i=1}^{m}\sum_{j=1}^{n}(\alpha a_{ij})^2\right)^{1/2} = \left(\alpha^2 \sum_{i=1}^{m}\sum_{j=1}^{n} a_{ij}^2\right)^{1/2} = |\alpha|\left(\sum_{i=1}^{m}\sum_{j=1}^{n} a_{ij}^2\right)^{1/2} = |\alpha|\,\|A\|_F.$

Triangle inequality: Consider the $n \times n$ matrix A to be a vector in $\mathbb{R}^{n^2}$ by forming the column vector $v_A = \begin{bmatrix} a_{11} & \cdots & a_{m1} & a_{12} & \cdots & a_{m2} & \cdots & a_{1n} & \cdots & a_{mn} \end{bmatrix}^T$. Similarly, form the vector v_B from matrix B. Then,

$$\|A + B\|_F = \left(\sum_{i=1}^{n}\sum_{j=1}^{n}(a_{ij} + b_{ij})^2\right)^{1/2} = \|v_A + v_B\|_2 \leq \|v_A\| + \|v_B\| = \|A\|_F + \|B\|_F$$

by applying the triangle inequality to the vectors v_A and v_B. $\qquad\square$

Example 7.5. If $A = \begin{bmatrix} -1 & 2 & 5 \\ -1 & 2 & 7 \\ 23 & 4 & 12 \end{bmatrix}, B = \begin{bmatrix} 1 & 1 & 0 \\ 2 & -6 & 3 \\ 1 & 1 & 2 \end{bmatrix}$, then

$\|A\|_F = \sqrt{(-1)^2 + 2^2 + 5^2 + \cdots + 12^2} = 27.8029$, $\|B\|_F = 7.5498$, and

$\|A + B\|_F = \left\|\begin{bmatrix} 0 & 3 & 5 \\ 1 & -4 & 10 \\ 24 & 5 & 14 \end{bmatrix}\right\|_F = 30.7896$. Note that $30.7896 < 27.8029 + 7.5498 = 35.3527$ as expected by the triangle inequality. $\qquad\blacksquare$

7.2.2 Induced Matrix Norms

The most frequently used class of norms are the *induced matrix norms*, that are defined in terms of a vector norm.

Definition 7.3. Assume $\|\cdot\|$ is a vector norm, A is an $m \times n$ matrix, and x an $n \times 1$ vector. Then the matrix norm of A induced by $\|\cdot\|$ is

$$\|A\| = \max_{x \neq 0} \frac{\|Ax\|}{\|x\|}. \tag{7.3}$$

The definition measures the norm of a matrix by finding the largest size of Ax relative to x. If $\|Ax\|$ becomes large for a particular range of vectors, x, that do not have large norms, then $\|A\|$ will be large. An induced matrix norm measures the maximum amount the matrix product Ax can stretch (or shrink) a vector relative to the vector's original length. Figure 7.6 illustrates the effect of a matrix on two vectors in $\mathbb{R}^2$.

Remark 7.1. We use the notation $\|A\|_p$ to denote that the norm of A is derived from the p-norm in Equation 7.3.
   An induced matrix norm is often called a subordinate matrix norm.

We have commented that orthogonal matrices are beautiful things, and when it comes to their norms, they do not disappoint.

Lemma 7.2. *If P is an orthogonal matrix, then $\|P\|_2 = 1$.*

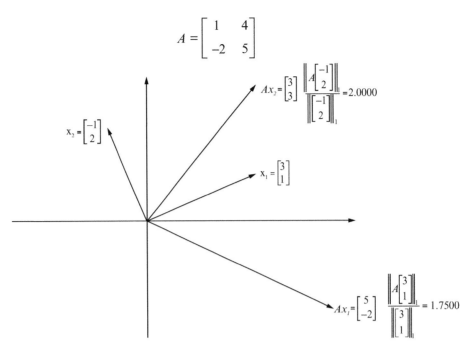

FIGURE 7.6 Effect of a matrix on vectors.

Proof. An orthogonal matrix maintains the 2-norm of x when forming Px, so $\|Px\|_2 = \|x\|_2$. It follows that

$$\|P\|_2 = \max_{x\neq 0} \frac{\|Ax\|_2}{\|x\|_2} = \frac{\|x\|_2}{\|x\|_2} = 1. \qquad \square$$

Not all matrix norms are induced. The *Frobenius norm* is not induced by any vector norm (Problem 7.10).

Another, perhaps easier, way to understand the concept of an induced matrix norm is to use the scaling property of a vector norm as follows:

$$\|A\| = \max_{x\neq 0} \frac{\|Ax\|}{\|x\|} = \max_{x\neq 0} \left\| \left(\frac{1}{\|x\|} \right) Ax \right\| = \max_{x\neq 0} \left\| A\left(\frac{x}{\|x\|} \right) \right\| = \max_{\|x\|=1} \|Ax\| . \qquad (7.4)$$

Equation 7.4 says that to compute induced norm $\|A\|$, find the maximum value of $\|Ax\|$, where x ranges over the unit sphere $\|x\| = 1$. It is helpful to view the *unit* sphere *of a norm*, which is possible for $\mathbb{R}^2$ and $\mathbb{R}^3$. For vectors in $\mathbb{R}^2$, the unit spheres for the ∞-, 1-, and 2-norm have equations

$$-1 \leq x \leq 1, |y| = 1, \quad -1 \leq y \leq 1, |x| = 1,$$
$$\|(x, y)\|_1 = |x| + |y| = 1,$$
$$\|(x, y)\|_2 = \sqrt{x^2 + y^2} = 1.$$

Figure 7.7 shows a graph of all three unit spheres.

You may not have seen a definitions like Equations 7.3 and 7.4. For instance, if $A = \begin{bmatrix} 1 & -8 \\ -1 & 3 \end{bmatrix}$, then

$$\|A\|_\infty = \max_{\|x\|_\infty=1} [\max (|x_1 - 8x_2|, |-x_1 + 3x_2|)] .$$

Computing this value seems complex. Let's run a numerical experiment so we can make an educated guess for the value of the induced infinity matrix norm. We need x to vary over the unit sphere, which is the set of points (x, y) in the plane such that $\max \{|x|, |y|\} = 1$. This is a square (Figure 7.7). Example 7.6 estimates $\|Ax\|_\infty$ using the function `approxinfnorm(A)`. The function generates 2500 random vectors on each side of the square and finds the maximum value of $\|Ax\|_\infty$ among the 10,000 values. Following the estimation of $\|A\|_\infty$, the example applies the function to the matrix $B = \begin{bmatrix} 0.25 & -0.75 \\ -0.75 & 0.30 \end{bmatrix}$.

Example 7.6.   Experimentally estimate $\|\cdot\|_\infty$ for two matrices.

```
function max = approxinfnorm(A)
%APPROXINFNORM Generate 10,000 random values on the unit circle for the
%infinity norm in the plane. Return the maximum value of norm(A*x, 'inf').
%input  : Matrix A
%output : real value max
    max = 0.0;
    for i = 1:10000
        r = 1 - 2*rand;
        if i <= 2500
            x = [1.0 r]';
        elseif i <= 5000
            x = [r, 1.0]';
        elseif i <= 7500
            x = [-1.0 r]';
        else
            x = [r -1.0]';
        end

        bvalue = norm(A*x,'inf');
        if bvalue > max
            max = bvalue;
        end
    end
>> approxinfnorm(A)

ans =
    9.0000

>> approxinfnorm(A)

ans =
    8.9998

>> approxinfnorm(B)

ans =
    1.0500

>> approxinfnorm(B)

ans =
    1.0497
```

■

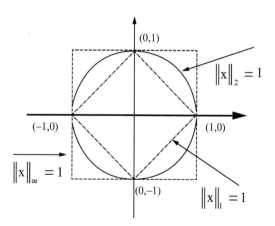

FIGURE 7.7 Unit spheres in three norms.

Notice that in Example 7.6, the experimental results using matrix A indicate an actual value of 9.0, which is the sum of the absolute value of the entries in row 1. For matrix B, the experiment indicates that the norm is the sum of the absolute values of the entries in row 2. In fact, here is the way to compute $\|A\|_\infty$ for any $m \times n$ matrix. Find the sum of the absolute values of the elements in each row of A and take $\|A\|_\infty$ be the maximum of these sums; in other words,

$$\|A\|_\infty = \max_{1 \le k \le n} \sum_{j=1}^{n} |a_{kj}|.$$

How do we get from the definition of the infinity-induced matrix norm

$$\max_{\|x\|=1} \|Ax\|_\infty$$

to this simple expression? Theorem 7.4 answers that question. The proof is somewhat technical, and is here for the interested reader.

Theorem 7.4. *If A is an $m \times n$ matrix,*

$$\|A\|_\infty = \max_{1 \le i \le m} \sum_{j=1}^{n} |a_{ij}|.$$

Proof. We can assume $A \ne 0$, since the result is certainly true for $A = 0$.

$$Ax = \begin{bmatrix} a_{11}x_1 + \cdots + a_{1n}x_n \\ a_{21}x_1 + \cdots + a_{2n}x_n \\ \vdots \\ a_{m-1,1}x_1 + \cdots + a_{n-1,n}x_n \\ a_{m1}x_1 + \cdots + a_{mn}x_n \end{bmatrix},$$

and

$$\|Ax\|_\infty = \max_{1 \le i \le m} \left| \sum_{j=1}^{n} a_{ij}x_j \right|.$$

Using the fact that $|x_i| \le \|x\|_\infty$, $1 \le i \le n$ for vector x, it follows that

$$\|Ax\|_\infty \le \max_{1 \le i \le m} \sum_{j=1}^{n} |a_{ij}|\,|x_j| \le \|x\|_\infty \max_{1 \le i \le m} \sum_{j=1}^{n} |a_{ij}|. \tag{7.5}$$

By the definition of an induced matrix norm and Equation 7.5,

$$\|A\|_\infty = \max_{\|x\|_\infty=1} \|Ax\|_\infty \le \max_{1 \le i \le m} \sum_{j=1}^{n} |a_{ij}|. \tag{7.6}$$

Let $i_{\max}$ be the row index that gives the maximum sum in Equation 7.6 so that

$$\max_{1 \le i \le m} \sum_{j=1}^{n} |a_{ij}| \le \sum_{j=1}^{n} |a_{i_{\max}j}|. \tag{7.7}$$

Let vector $x^{\max} \in \mathbb{R}^n$ be defined by $x_j^{\max} = 0$ if $a_{i_{\max}j} = 0$ and $x_j^{\max} = a_{i_{\max}j}/|a_{i_{\max}j}|$ if $a_{i_{\max}j} \ne 0$. Since $A \ne 0$, $x^{\max} \ne 0$, and $\|x^{\max}\|_\infty = 1$. Furthermore, using Equation 7.7 we have

$$\|A\|_\infty = \max_{\|x\|_\infty=1} \|Ax\|_\infty \ge \left| \sum_{j=1}^{n} a_{i_{\max}j}x_j^{\max} \right| = \sum_{j=1}^{n} \frac{(a_{i_{\max}j})^2}{|a_{i_{\max}j}|} = \sum_{j=1}^{n} |a_{i_{\max}j}| \ge \max_{1 \le i \le m} \sum_{j=1}^{n} |a_{ij}|. \tag{7.8}$$

Now, Equation 7.6 says

$$\|A\|_\infty \le \max_{1 \le i \le m} \sum_{j=1}^{n} |a_{ij}|,$$

and Equation 7.8 says

$$\|A\|_\infty \geq \max_{1 \leq i \leq m} \sum_{j=1}^{n} |a_{ij}|,$$

and so

$$\|A\|_\infty = \max_{1 \leq i \leq m} \sum_{j=1}^{n} |a_{ij}|.$$

The same type of argument (Problem 7.5) shows that the 1-norm is the maximum absolute column sum, or

$$\|A\|_1 = \max_{1 \leq k \leq n} \sum_{i=1}^{m} |a_{ik}|. \qquad \Box$$

Example 7.7. If $A = \begin{bmatrix} -2 & 1 & -8 & 1 \\ 0 & -4 & -21 & 18 \\ -33 & 16 & -6 & 20 \\ 14 & -20 & -18 & 5 \\ 8 & -1 & 12 & 16 \end{bmatrix}$, then

$$\|A\|_\infty = 33 + 16 + 6 + 20 = 75, \quad \|A\|_1 = 8 + 21 + 6 + 18 + 12 = 65. \qquad \blacksquare$$

MATLAB computes matrix norms using the same command, `norm`, that it uses for a vector. For a matrix, in addition to the ∞-, 1-, and 2-norms, the Frobenius norm is available. We apply those norms to the matrix of Example 7.7.

```
>> norm(A,'inf')
ans =
    75
>> norm(A,1)
ans =
    65
>> norm(A,'fro')
ans =
   63.5767
```

Remark 7.2. From the definition of an induced norm, $(\|Ax\|/\|x\|) \leq \|A\|$, and so

$$\|Ax\| \leq \|A\| \, \|x\| \qquad (7.9)$$

We will have occasion to use Equation 7.9 numerous times throughout the remainder of the book.

7.3 SUBMULTIPLICATIVE MATRIX NORMS

Example 7.8. Let $A = \begin{bmatrix} -1 & 2 & 5 \\ -1 & 2 & 7 \\ 23 & 4 & 12 \end{bmatrix}$ and $B = \begin{bmatrix} 1 & 1 & 0 \\ 2 & -6 & 3 \\ 1 & 1 & 2 \end{bmatrix}$. The product $AB = \begin{bmatrix} 8 & -8 & 16 \\ 10 & -6 & 20 \\ 43 & 11 & 36 \end{bmatrix}$, and its Frobenius

norm is $\|AB\|_F = 64.6993$. The product of the Frobenius norms is $\|A\|_F \|B\|_F = 209.9071$, and so

$$\|AB\|_F \leq \|A\|_F \|B\|_F. \qquad \blacksquare$$

The inequality $\|AB\|_F \leq \|A\|_F \|B\|_F$ is not a coincidence. Our first matrix norm, the Frobenius norm is sub-multiplicative.

Definition 7.4. If the matrix norm $\|.\|$ satisfies $\|AB\| \leq \|A\| \|B\|$ for all matrices A and B in $\mathbb{R}^{n \times n}$, it said to be *sub-multiplicative*.

Theorem 7.5.   *The Frobenius matrix norm is sub-multiplicative.*

Proof. Let $C = AB$. Then, $c_{ij} = \sum_{k=1}^{n} a_{ik}b_{kj}$, and

$$\|AB\|_F^2 = \sum_{i=1}^{n}\sum_{j=1}^{n} c_{ij}^2 = \sum_{i=1}^{n}\sum_{j=1}^{n}\left(\sum_{k=1}^{n} a_{ik}b_{kj}\right)^2$$

$$= \sum_{i=1}^{n}\sum_{j=1}^{n}\left(\begin{bmatrix} a_{i1} \\ a_{i2} \\ \vdots \\ a_{i,n-1} \\ a_{in} \end{bmatrix}^{\mathrm{T}}\begin{bmatrix} b_{1j} \\ b_{2j} \\ \vdots \\ b_{n-1,j} \\ b_{nj} \end{bmatrix}\right)^2 \leq \sum_{i=1}^{n}\sum_{j=1}^{n}\left(\left\|\begin{bmatrix} a_{i1} \\ a_{i2} \\ \vdots \\ a_{i,n-1} \\ a_{in} \end{bmatrix}\right\|_2 \left\|\begin{bmatrix} b_{1j} \\ b_{2j} \\ \vdots \\ b_{n-1,j} \\ b_{nj} \end{bmatrix}\right\|_2\right)^2 \qquad (7.10)$$

$$= \sum_{i=1}^{n}\sum_{j=1}^{n}\left[\left(\sum_{k=1}^{n} a_{ik}^2\right)\left(\sum_{k=1}^{n} b_{kj}^2\right)\right] = \|A\|_F^2 \|B\|_F^2. \qquad (7.11)$$

We used the Cauchy-Schwarz inequality in Equation 7.10. Verifying Equation 7.11 is left to the exercises. $\square$

The induced matrix norms are sub-multiplicative:

$$\|AB\| = \max_{x\neq 0}\frac{\|ABx\|}{\|x\|} = \max_{x\neq 0}\frac{\|A(Bx)\|}{\|Bx\|}\frac{\|Bx\|}{\|x\|} \leq \left(\max_{x\neq 0}\frac{\|Ax\|}{\|x\|}\right)\left(\max_{x\neq 0}\frac{\|Bx\|}{\|x\|}\right) = \|A\|\,\|B\|$$

There exist norms that satisfy the three basic matrix norm axioms, but are not submultiplicative; for instance,

$$\|A\| = \max_{1\leq i,j\leq n}|a_{ij}|$$

satisfies the positivity, scaling, and triangle inequality properties, but is not sub-multiplicative (see Exercise 7.6).

7.4 COMPUTING THE MATRIX 2-NORM

The most useful norm for many applications is the induced matrix 2-norm (often called the *spectral norm*):

$$\|A\|_2 = \max_{x\neq 0}\frac{\|Ax\|_2}{\|x\|_2}$$

or

$$\|A\|_2 = \max_{\|x\|_2=1}\|Ax\|.$$

It would seem reasonable that this norm would be more difficult to find, since

$$\|x\|_2 = \sqrt{x_1^2 + x_2^2 + \cdots + x_n^2}$$

is a more complex calculation than $\|x\|_\infty$ or $\|x\|_1$. In fact, it is a nonlinear optimization problem with constraints. In $\mathbb{R}^2$, $\|A\|_2$ is the maximum value of $\|Ax\|_2$ for x on the unit circle. As you can see in Figure 7.8, the image of the matrix $A = \begin{bmatrix} 1 & -8 \\ -1 & 3 \end{bmatrix}$ as x varies over the unit circle is an ellipse. The problem is to find the largest value of $\|Ax\|_2$ on this ellipse.

Let's examine Figure 7.8 in more detail. A semi-major axis of the ellipse is the longest line from the center to a point on the ellipse, and the length of the semi-major axis for our ellipse is 8.6409. Create A with MATLAB and use the `norm` command to compute its 2-norm.

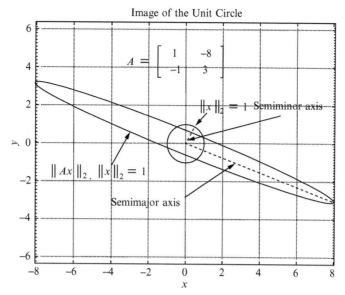

FIGURE 7.8 Image of the unit circle.

```
>> A = [1 -8;-1 3];
>> norm(A)

ans =
    8.6409
```

This is not a coincidence. We will show in Chapter 15 that the norm of a matrix is the length of a semi-major axis of an ellipsoid formed from the image of the unit sphere in k-dimensional space, $k \leq m$.

If A is an $m \times n$ matrix, A^TA is of size $n \times n$ and since $(A^TA)^T = A^TA$, it is also symmetric. The method for computing $\|A\|_2$ exploits properties of A^TA. The following is a summary of the process, followed by an example.

- The eigenvalues of a symmetric matrix are real.
- A^TA is symmetric, so it has real eigenvalues. Furthermore, it can be shown that the eigenvalues of A^TA are nonnegative (≥ 0).
- The square roots of the eigenvalues of A^TA are termed *singular values* of A. The norm of an $m \times n$ matrix, A, is the largest singular value.

Example 7.9. Find the 2-norm of $A = \begin{bmatrix} 1 & 13 & 5 & -9 \\ 12 & 55 & 5 & -6 \\ 18 & 90 & 1 & -1 \\ 3 & 0 & 2 & 3 \end{bmatrix}$ using the MATLAB commands `eig` and `norm`.

```
>> E = eig(A'*A)

E =
   1.0e+004 *

   0.0000
   0.0021
   0.0131
   1.1802

>> sqrt(max(E))

ans =

  108.6373

>> norm(A,2)

ans =
```

```
    108.6373
>> norm(A)  % default without second argument is the 2-norm
ans =
    108.6373
```
∎

Remark 7.3. The fact that

$$\|A\|_2 = \max_{\|x\|_2=1} \|Ax\|_2 = \max\left(\sqrt{s_i}\right),$$

where s_i are the eigenvalues of $A^T A$, is really a theoretical result that will lead to methods for efficiently computing $\|A\|_2$. Computing the norm like we did in Example 7.9 is too slow and prone to errors. Chapter 8 provides justification for this statement.

It is useful to note that

$$\langle Bx, y \rangle = (Bx)^T y = x^T B^T y = \langle x, B^T y \rangle.$$

This says that you can move matrix B from one side of an inner product to the other by replacing B by B^T. Now, if B is symmetric $B = B^T$, and we have

$$\langle Bx, y \rangle = \langle x, By \rangle. \tag{7.12}$$

The remainder of this section mathematically derives the computation of $\|A\|_2$ from the eigenvalues of $A^T A$, and can be skipped if the reader does not need to see the details.

The proof of Lemma 7.3 uses the concept of the conjugate of a complex number and the conjugate transpose of a complex matrix (Definition A.3).

Lemma 7.3. *The eigenvalues of a symmetric matrix are real, and the corresponding eigenvectors can always be assumed to be real.*

Proof. Suppose λ is an eigenvalue of the symmetric matrix A, and u is a corresponding eigenvector. We know that the eigenvalues of an $n \times n$ matrix with real coefficients can be complex and, if so, occur in complex conjugate pairs $a+ib$ and $a-ib$. Since λ might be complex, the vector u may also be a complex vector. Because u is an eigenvector with eigenvalue λ,

$$Au = \lambda u$$

Now take the conjugate transpose of both sides of the latter equation and we have

$$u^* A = \overline{\lambda} u^* \tag{7.13}$$

Multiply 7.13 by u on the right, and

$$\begin{aligned}
\left(u^* A\right) u &= \overline{\lambda} u^* u \\
u^* (Au) &= \overline{\lambda} u^* u \\
u^* (\lambda u) &= \overline{\lambda} u^* u \\
\lambda u^* u &= \overline{\lambda} u^* u
\end{aligned} \tag{7.14}$$

From 7.14, there results

$$\left(\lambda - \overline{\lambda}\right) u^* u = 0.$$

Now, $u^* u > 0$, since u is an eigenvector and cannot be 0. It follows that

$$\lambda - \overline{\lambda} = 0,$$

and $\lambda = \overline{\lambda}$ means that λ is real.

This finishes the first portion of the proof. We now need to show that for any eigenvalue λ, there is a corresponding real eigenvector. Assume that u is an eigenvector of A, so

$$Au = \lambda u.$$

If we take the complex conjugate of both sides, we obtain

$$A\overline{u} = \lambda\overline{u}.$$

By adding the two equations, we have

$$A(u + \overline{u}) = \lambda(u + \overline{u}).$$

Thus, $u + \overline{u}$ is an eigenvector of A. If $u = x + iy$, then $u + \overline{u} = (x + iy) + (x - iy) = 2x$, which is real. $\square$

Remark 7.4. If A is a symmetric square matrix, all its eigenvalues are real, and A has real eigenvectors; however, this does not mean that A has no complex eigenvectors. For instance, consider the symmetric matrix

$$A = \begin{bmatrix} 1 & -8 \\ -8 & 1 \end{bmatrix}.$$

A has eigenvalues $\lambda = -7$ and $\lambda = 9$. For $\lambda = 9$, $u = \begin{bmatrix} -0.7071 \\ 0.7071 \end{bmatrix}$ is an eigenvector, but so is $(1 - 9i)\begin{bmatrix} -0.7071 \\ 0.7071 \end{bmatrix}$. Lemma 7.3 says we can always find a real eigenvector for each real eigenvalue. If A is not symmetric, it may have complex eigenvalue λ, in which case a corresponding eigenvector will be complex.

We are getting closer to deriving the formula for $\|A\|_2$. Because A^TA is an $n \times n$ symmetric matrix, Lemma 7.3 says it has real eigenvalues. The following lemma shows that the eigenvalues of A^TA are in fact always greater than or equal to 0.

Lemma 7.4. *If A is an $m \times n$ real matrix, then the eigenvalues of the $n \times n$ matrix A^TA are nonnegative.*

Proof. The eigenvalues of A^TA are real from Lemma 7.3. Let λ be an eigenvalue of A^TA and $u \neq 0$ be a corresponding eigenvector, so that

$$\left(A^TA\right)u = \lambda u.$$

Take the inner product of this equality with u to obtain

$$\left\langle \left(A^TA\right)u,\, u \right\rangle = \lambda \|u\|_2^2$$

from which we arrive at

$$\lambda = \frac{\left\langle \left(A^TA\right)u,\, u \right\rangle}{\|u\|_2^2}.$$

Note that $\left\langle \left(A^TA\right)u,\, u \right\rangle = \left(A^T(Au)\right)^T u = (Au)^T Au = \langle Au,\, Au \rangle$, and so

$$\lambda = \frac{\|Au\|_2^2}{\|u\|_2^2} \geq 0. \qquad \square$$

We are almost in a position to compute the 2-norm of a matrix, but first we need to define the singular values of A.

Definition 7.5. The *singular values* $\{\sigma_i\}$ of an $m \times n$ matrix A are the square roots of the eigenvalues of A^TA.

Remark 7.5. We can always compute $\sqrt{\lambda}$, where λ is an eigenvalue of A^TA because Lemmas 7.3 and 7.4 guarantee λ is real and nonnegative.

Example 7.10. Let A be the 4×2 matrix $\begin{bmatrix} 2 & 5 \\ 1 & 4 \\ -1 & 6 \\ 7 & 8 \end{bmatrix}$. $A^TA = \begin{bmatrix} 55 & 64 \\ 64 & 141 \end{bmatrix}$, and the eigenvalues of AA^T are $\lambda_1 = 175.1$, and $\lambda_2 = 20.896$, so the singular values are $\sigma_1 = 13.233$, $\sigma_2 = 4.571$. $\blacksquare$

Before proving how to compute $\|A\|_2$ in Theorem 7.7, we state a result concerning symmetric matrices called the *spectral theorem* that we will prove in Chapter 19. It says that any real symmetric matrix A can be diagonalized with a real orthogonal matrix.

Theorem 7.6 (Spectral theorem).   If A is a real symmetric matrix, there exists an orthogonal matrix P such that

$$D = P^\mathrm{T}AP,$$

where D is a diagonal matrix containing the eigenvalues of A, and the columns of P are an orthonormal set of eigenvalues that form a basis for $\mathbb{R}^n$.

We are now in a position to prove how to compute $\|A\|_2$.

Theorem 7.7.   *If A is an $m \times n$ matrix, $\|A\|_2$ is the square root of the largest eigenvalue of $A^\mathrm{T}A$.*

Proof. The symmetric matrix $A^\mathrm{T}A$ is diagonalizable $(D = P^\mathrm{T}A^\mathrm{T}AP)$, and its eigenvalues λ_i are nonnegative real numbers. Let $y = P^\mathrm{T}x$, and

$$\|A\|_2^2 = \max_{x \neq 0} \frac{\|Ax\|_2^2}{\|x\|_2^2} = \max_{x \neq 0} \frac{(Ax)^\mathrm{T}Ax}{\|x\|_2^2}$$

$$= \max_{x \neq 0} \frac{x^\mathrm{T}A^\mathrm{T}Ax}{\|x\|_2^2} = \max_{x \neq 0} \frac{x^\mathrm{T}PDP^\mathrm{T}x}{\|x\|_2^2}$$

$$= \max_{x \neq 0} \frac{\left(P^\mathrm{T}x\right)^\mathrm{T} D \left(P^\mathrm{T}x\right)}{\left\|P^\mathrm{T}x\right\|_2^2} = \max_{y \neq 0} \frac{y^\mathrm{T}Dy}{\|y\|_2^2}$$

$$= \max_{y \neq 0} \frac{\sum_{i=1}^n \lambda_i y_i^2}{\sum_{i=1}^n y_i^2} \leq \lambda_{\max} \frac{\sum_{i=1}^n y_i^2}{\sum_{i=1}^n y_i^2} = \lambda_{\max}. \tag{7.15}$$

Assume that $\lambda_{\max}$ occurs at index (k,k). We have $Pe_k = x_k$, where x_k is column k of P and e_k is the kth standard basis vector, so $e_k = P^\mathrm{T}x_k$. By choosing $y = e_k$, the inequality in Equation 7.15 is an equality, and $\|A\|_2 = \sqrt{\lambda_{\max}}$. $\square$

Remark 7.6.   For values of p other than 1, 2, and ∞, there is no simple formula for the induced matrix p-norm.

Example 7.11.   In Example 7.10, we computed the largest singular value of $A = \begin{bmatrix} 2 & 5 \\ 1 & 4 \\ -1 & 6 \\ 7 & 8 \end{bmatrix}$, $\sigma_1 = 13.233$. Using the MATLAB command `norm`, compute the 2-norm of A.

```
>> norm(A,2)

ans =
      13.233
```

■

7.5   PROPERTIES OF THE MATRIX 2-NORM

There are a number of important properties of the matrix 2-norm. Since we have the tools to develop the properties, we will do so now and then refer to them as needed.

The matrix 2-norm inherits *orthogonal invariance* from the vector 2-norm. This means that multiplying matrix A on the left and right by orthogonal matrices does not change its 2-norm.

Theorem 7.8.   *For any orthogonal matrices U and V, $\|UAV\|_2 = \|A\|_2$.*

Proof. We first consider the case of multiplication on the left by a single orthogonal matrix P, and then use this to prove that multiplication on the right by a single orthogonal matrix also preserves the 2-norm. The combination of these two results will allow us to prove the more general result.

$$\|PA\|_2^2 = \max_{x \neq 0} \frac{(PAx)^{\mathrm{T}}(PAx)}{\|x\|_2^2} = \max_{x \neq 0} \frac{\left(x^{\mathrm{T}}A^{\mathrm{T}}P^{\mathrm{T}}\right)(PAx)}{\|x\|_2^2}$$

$$= \max_{x \neq 0} \frac{x^{\mathrm{T}}A^{\mathrm{T}}IAx}{\|x\|_2^2} = \max_{x \neq 0} \frac{(Ax)^{\mathrm{T}}(Ax)}{\|x\|_2^2} = \max_{x \neq 0} \frac{\|Ax\|_2^2}{\|x\|_2^2} = \|A\|_2^2 .$$

Now, noting that $\|x\|_2 = \|Px\|_2$ for any vector x,

$$\|AP\|_2^2 = \max_{x \neq 0} \frac{\|APx\|_2^2}{\|x\|_2^2} = \max_{x \neq 0} \frac{\|A(Px)\|_2^2}{\|Px\|_2^2} = \max_{y \neq 0} \frac{\|Ay\|_2^2}{\|y\|_2^2} = \|A\|_2^2 .$$

Now consider multiplication by orthogonal matrices U on the left and V on the right.

$$\|UAV\|_2 = \|U(AV)\| = \|AV\|_2 = \|A\|_2 .$$

This section concludes with the development of four properties of the matrix 2-norm. One of the properties involves the spectral radius of a matrix. $\square$

Definition 7.6.   Let A be an $n \times n$ real matrix with eigenvalues, $\{\lambda_1, \lambda_2, \ldots, \lambda_n\}$. Then the *spectral radius*, $\rho(A)$, of A is

$$\rho(A) = \max_{1 \leq i \leq n} |\lambda_i| .$$

Before considering the five properties in Theorem 7.9, there is a fact we have not needed until now.

Lemma 7.5.   *If A is invertible, the eigenvalues of A^{-1} are inverses of the eigenvalues of A; in other words, the eigenvalues of A^{-1} are $\{1/\lambda_1, 1/\lambda_2, \ldots, 1/\lambda_n\}$, where the λ_i are the eigenvalues of A. Note that there is no problem with division by 0, since an invertible matrix cannot have a 0 eigenvalue (Proposition 5.1). Furthermore, A and A^{-1} have the same eigenvectors. Also, the maximum eigenvalue of A^{-1} in magnitude is $1/\lambda_{\min}$, where $\lambda_{\min}$ is the smallest eigenvalue of A in magnitude.*

Proof. Let λ be an eigenvalue A and x be a corresponding eigenvector. Then $Ax = \lambda x$, and $A^{-1}x = (1/\lambda)x$. Thus, $1/\lambda$ is an eigenvalue of A^{-1} and x is a corresponding eigenvector. Clearly, if $\lambda_{\min}$ is the smallest eigenvalue of A in magnitude, then $1/\lambda_{\min}$ is the largest eigenvalue of A^{-1}. $\square$

Theorem 7.9.   *The matrix 2-norm has the following properties:*

1. *If A is a symmetric matrix, $\|A\|_2 = \rho(A)$, where $\rho(A)$ is the spectral radius of A.*
2. *If A is a symmetric matrix, its singular values are the absolute value of its eigenvalues.*
3. $\|A\|_2 = \|A^{\mathrm{T}}\|_2$.
4. $\|A^{\mathrm{T}}A\|_2 = \|AA^{\mathrm{T}}\|_2 = \|A\|_2^2$.
5. $\|A^{-1}\|_2 = \frac{1}{\sigma_{\min}}$, *where $\sigma_{\min}$ is the minimum singular value of A.*

Proof. The proof of parts 1 and 5 are left to the exercises.

For part (2), assume that λ is an eigenvalue of A with associated eigenvector, so $Av = \lambda v$. To obtain singular values, we need to find the eigenvalues of $A^{\mathrm{T}}A = A^2$ since A is symmetric. Multiply both sides of $Av = \lambda v$ by A, and we have

$$A^2 v = \lambda(Av) = \lambda(\lambda v) = \lambda^2 v.$$

Assume the eigenvalues are sorted by decreasing absolute value. The singular values are nonnegative and are square roots of λ^2, so $\sigma_1 = |\lambda_1|, \sigma_2 = |\lambda_2|, \ldots, \sigma_n = |\lambda_n|$.

Begin the proof of part (3) with

$$\|Ax\|_2^2 = (Ax)^{\mathrm{T}}(Ax) = x^{\mathrm{T}}A^{\mathrm{T}}Ax = \langle x, A^{\mathrm{T}}Ax \rangle \leq \|x\|_2 \|A^{\mathrm{T}}Ax\|_2 \tag{7.16}$$

by the Cauchy-Schwarz inequality. From Equation 7.9

$$\|x\|_2 \|A^{\mathrm{T}}Ax\|_2 \leq \|x\|_2^2 \|A^{\mathrm{T}}A\|_2 , \tag{7.17}$$

and by putting Equations 7.16 and 7.17 together, we have

$$\|Ax\|_2 \leq \sqrt{\left\|A^{\mathrm{T}}A\right\|_2}\,\|x\|_2. \tag{7.18}$$

By Equation 7.18,

$$\|A\|_2 = \max_{x \neq 0}\frac{\|Ax\|_2}{\|x\|_2} \leq \max_{x \neq 0}\frac{\sqrt{\left\|A^{\mathrm{T}}A\right\|_2}\,\|x\|_2}{\|x\|_2} = \sqrt{\left\|A^{\mathrm{T}}A\right\|_2}. \tag{7.19}$$

The matrix 2-norm is submultiplicative so by squaring both sides of Equation 7.19,

$$\|A\|_2^2 \leq \left\|A^{\mathrm{T}}A\right\|_2 \leq \left\|A^{\mathrm{T}}\right\|_2\|A\|_2.$$

Divide by $\|A\|_2$ when $A \neq 0$ to obtain

$$\|A\|_2 \leq \left\|A^{\mathrm{T}}\right\|_2.$$

Clearly, this is also true if $A = 0$. In the last inequality, replace A by A^{T} and we now have the two inequalities, $\|A\|_2 \leq \left\|A^{\mathrm{T}}\right\|_2$ and $\left\|A^{\mathrm{T}}\right\| \leq \left\|\left(A^{\mathrm{T}}\right)^{\mathrm{T}}\right\|_2 = \|A\|_2$, so

$$\|A\|_2 = \left\|A^{\mathrm{T}}\right\|_2.$$

For part (4), from Theorem 7.7, $\|A\|_2 = \sqrt{\lambda_{\max}}$, where $\lambda_{\max} \geq 0$ is the maximum eigenvalue of $A^{\mathrm{T}}A$, so $\|A\|_2^2 = \lambda_{\max}$. $A^{\mathrm{T}}A$ is a symmetric matrix, its eigenvalues are positive real numbers, and from (1) its 2-norm is its spectral radius, $\lambda_{\max}$, so

$$\left\|A^{\mathrm{T}}A\right\|_2 = \lambda_{\max} = \|A\|_2^2.$$

Now replace A by A^{T} to obtain

$$\left\|\left(A^{\mathrm{T}}\right)^{\mathrm{T}}A^{\mathrm{T}}\right\|_2 = \left\|AA^{\mathrm{T}}\right\|_2 = \left\|A^{\mathrm{T}}\right\|_2^2 = \|A\|_2^2$$

using part (3). □

7.6 CHAPTER SUMMARY

Vector Norms

A vector norm measures length. The most commonly used norm is the 2-norm, or the Euclidean norm, where

$$\|u\|_2 = \sqrt{u_1^2 + u_2^2 + \cdots + u_n^2} = \sqrt{\langle u, u\rangle}.$$

Motivated by the properties of the 2-norm, we define a general vector norm as a function mapping $\mathbb{R}^n$ into the real numbers with the following properties:

a. $\|x\| \geq 0$ for all $x \in \mathbb{R}^n$; $\|x\| = 0$ if and only if $x = 0$ (*positivity*);
b. $\|\alpha x\| = |\alpha|\|x\|$ for all $\alpha \in \mathbb{R}$ (*scaling*);
c. $\|x + y\| \leq \|x\| + \|y\|$ for all $x, y \in \mathbb{R}^n$ (*triangle inequality*).

Aside from the 2-norm, other useful norms are the 1-norm,

$$\|u\|_1 = \sum_{i=1}^{n}|u_i|,$$

and the ∞-norm,

$$\|u\|_\infty = \max_{1 \leq i \leq n}|u_i|.$$

Properties of the 2-Norm

The 2-norm has unique properties that will be useful throughout the book. It satisfies

- the Cauchy-Schwarz inequality, $|\langle u, v \rangle| = |u^\mathsf{T} v| \le \|u\|_2 \|v\|_2$,
- the Pythagorean Theorem, $\|u + v\|_2 = \|u\|_2^2 + \|v\|_2^2$, when u and v are orthogonal,
- orthogonal invariance, $\|Px\|_2 = \|x\|_2$ when P is an orthogonal matrix.

Orthogonal invariance is of particular importance in computer graphics. If a rotation is applied to the elements in an object, the object's size and shape do not change.

A set of k orthogonal vectors $u_1, u_2, \ldots, u_k$ is linearly independent and forms a basis for a $k-$dimensional subspace of $\mathbb{R}^n$. If we normalize each vector to form, $v_i = u_i / \|u_i\|_2$, then the matrix $[v_1 v_2 \ldots v_k]$ is orthogonal.

Spherical Coordinates

Spherical coordinates are useful in computer graphics, vibrating membranes, and the Schrodinger equation for the hydrogen atom. With some work, we can build an orthogonal matrix that implements a change of coordinates from rectangular to spherical and spherical to rectangular.

Matrix Norms

The definition of a matrix norm is the same as that for a vector norm. It has the properties of positivity, scaling, and the triangle inequality. Essentially, it measure the size of a matrix. The oldest matrix norm is the Frobenius norm defined by

$$\|A\|_\mathrm{F} = \sqrt{\sum_{i=1}^{m}\sum_{j=1}^{n} a_{ij}^2},$$

which can be viewed as the norm of a vector in $\mathbb{R}^{mn}$.

Induced Matrix Norm

A vector norm, $\|.\|$, can be used to define a corresponding matrix norm as follows:

$$\|A\| = \max_{x \ne 0} \frac{\|Ax\|}{\|x\|} = \max_{\|x\|=1} \|Ax\|.$$

From the definition, there results a frequently used inequality, $\|Ax\| \le \|A\| \|x\|$.

The three most used induced matrix norms are the 2-norm, the 1-norm, and the ∞-norm. The definition can be used to develop simple formulas for the matrix ∞- and 1-norms:

$$\|A\|_\infty = \max_{1 \le i \le m} \sum_{j=1}^{n} |a_{ij}|,$$

$$\|A\|_1 = \max_{1 \le j \le n} \sum_{i=1}^{m} |a_{ij}|.$$

Submultiplicative Matrix Norms

The induced norms are submultiplicative, which means that

$$\|AB\| \le \|A\| \|B\|.$$

The Frobenius norm is submultiplicative, but is not an induced matrix norm.

Computation of $\|A\|_2$

The computation of $\|A\|_2$ is more complex. First, note that for any $m \times n$ matrix $A^\mathsf{T} A$ is a symmetric $n \times n$ matrix. As such, its eigenvalues are all real, but in addition all its eigenvalues are greater than or equal to zero. The singular values of A are the square root of the eigenvalues of $A^\mathsf{T} A$, and the 2-norm is the largest singular value. In addition, the 2-norm of A^{-1} is the reciprocal of the smallest singular value of A.

Properties of $\|A\|_2$

The matrix 2-norm has the following properties:

1. For any orthogonal matrices U and V, $\|UAV\|_2 = \|A\|_2$.
2. If A is a symmetric matrix, $\|A\|_2 = \rho(A)$, where $\rho(A)$ is the spectral radius of A, the magnitude of its largest eigenvalue.
3. If A is a symmetric matrix, its singular values are the absolute value of its eigenvalues.
4. $\|A\|_2 = \|A^T\|_2$.
5. $\|A^T A\|_2 = \|AA^T\|_2 = \|A\|_2^2$.

7.7 PROBLEMS

7.1 Compute the 1-norm, the ∞-norm, and the 2-norm for the following vectors. Do the calculations with paper, pencil, and a calculator.

a. $\begin{bmatrix} 1 \\ 5 \end{bmatrix}$

b. $\begin{bmatrix} 4 \\ 1 \\ 3 \end{bmatrix}$

c. $\begin{bmatrix} -1 \\ -9 \\ -6 \end{bmatrix}$

d. $\begin{bmatrix} 1 \\ 4 \\ -1 \\ 2 \end{bmatrix}$

7.2 Compute the 1-norm, the ∞-norm, the 2-norm, and the Frobenius norm for the following vectors. Except for the 2-norm, do the calculations with paper and pencil. To compute the 2-norm, you may use MATLAB to find the required eigenvalues.

a. $\begin{bmatrix} 1 & 9 \\ -1 & 5 \end{bmatrix}$

b. $\begin{bmatrix} 2 & 5 & 3 \\ 0 & 4 & 1 \end{bmatrix}$

7.3 Show that $\|x\|_1$ is a vector norm by verifying Definition 7.1.

7.4 Prove that
   a. $\|I\|_2 = 1$
   b. $\|I\|_F = \sqrt{n}$.

7.5 Prove that

$$\|A\|_1 = \max_{1 \le k \le n} \sum |a_{ik}|$$

using the proof of Theorem 7.4 as a guide.

7.6 Prove the following:
   a.

$$\|A\|_{\max} = \max_{1 \le i,j \le n} |a_{ij}|$$

   is a matrix norm.

   b.

$$\|A\|_{\max} = \max_{1 \le i,j \le n} |a_{ij}|$$

   is not sub-multiplicative. Hint: Let A be a matrix whose entries are all equal to 1.

7.7

  a. In Lemma 7.1, we proved that $\|x\|_2 \leq \sqrt{n}\,\|x\|_\infty$. Prove that $\|x\|_1 \leq \sqrt{n}\,\|x\|_2$ Hint: Consider using the Cauchy-Schwarz inequality involving vectors x and $\begin{bmatrix} 1 & \dots & 1 \end{bmatrix}^{\mathrm{T}}$.

  b. Find a vector $\begin{bmatrix} x_1 \\ \vdots \\ x_n \end{bmatrix}$ for which the inequalities in (a) are an equality.

7.8 Show that if $\|.\|$ is an induced matrix norm, then $|\lambda| \leq \|A\|$, where λ is an eigenvalue of A.

7.9 This problem demonstrates why $\|\cdot\|_\infty$ is a p-norm by showing that

$$\lim_{p \longrightarrow \infty} \left(\sum_{i=1}^{n} |x_i|^p \right)^{1/p} = \max_{1 \leq i \leq n} |x_i| = \|x\|_\infty\,.$$

Reorder the elements in $\sum_{i=1}^{n} |x_i|^p$ so that $|x_1| = |x_2| = \cdots = |x_k|$ are the largest elements in magnitude, and let $|x_{k+1}|, |x_{k+2}|, \dots, |x|_n$ be the remaining elements. Show that

$$\left(\sum_{i=1}^{n} |x_i|^p \right)^{1/p} = |x_1| \left(k + \left| \frac{x_{k+1}}{x_1} \right|^p + \cdots + \left| \frac{x_n}{x_1} \right|^p \right)^{1/p},$$

and complete the argument.

7.10

  a. If $\|.\|$ is an induced matrix norm, show that $\|I\| = 1$.

  b. Show that the Frobenius norm is not an induced norm.

7.11 Let A be an $n \times n$ matrix, $\rho(A)$ its spectral radius, and $\|\cdot\|$ be an induced matrix norm. Prove that for every $k \geq 1$

$$\rho(A) \leq \left\| A^k \right\|^{1/k}.$$

Hint: Begin by showing that $A^k v = \lambda^k v$ if λ is an eigenvalue of A with corresponding eigenvector v.

7.12 Show that the Frobenius norm can be computed as $\|A\|_{\mathrm{F}} = \left(\mathrm{trace}\left(A^{\mathrm{T}}A\right) \right)^{1/2}$ or $\|A\|_{\mathrm{F}} = \left(\mathrm{trace}\left(AA^{\mathrm{T}}\right) \right)^{1/2}$.

7.13 Verify that $\sum_{i=1}^{n} \sum_{j=1}^{n} \left[\left(\sum_{k=1}^{n} a_{ik}^2 \right) \left(\sum_{k=1}^{n} b_{kj}^2 \right) \right] = \|A\|_{\mathrm{F}}^2 \|B\|_{\mathrm{F}}^2$.

7.14 If A is an $m \times n$ matrix, prove that

$$\max_{1 \leq i \leq m,\, 1 \leq j \leq n} |a_{ij}| \leq \|A\|_{\mathrm{F}}\,.$$

7.15 Prove the *Pythagorean Theorem* for orthogonal x and y,

$$\|x + y\|_2^2 = \|x\|_2^2 + \|y\|_2^2\,.$$

7.16 Prove that if A is a symmetric matrix and $\rho(A) < 1$, then

$$\lim_{k \to \infty} A^k = 0.$$

Use the spectral theorem.

7.17 Prove that if A is an $n \times n$ matrix and

$$\lim_{n \to \infty} A^n = 0,$$

then $\rho(A) < 1$. Consult the hint for Problem 7.11.

7.18 A matrix norm and a vector norm are compatible if it is true for all vectors x and matrices A that $\|Ax\| \leq \|A\|\,\|x\|$. Show that the Euclidean vector norm is compatible with the Frobenius matrix norm; in other words, show that

$$\|Ax\|_2 \leq \|A\|_{\mathrm{F}}\,\|x\|_2\,.$$

7.19 A Schatten p-norm is the p-norm of the vector of singular values of a matrix. If the singular values are denoted by σ_i, then the Schatten p-norm is defined by

$$\|A\|_p = \left(\sum_{i=1}^{n} \sigma_i^p \right)^{1/p}.$$

    a. Show that the Schatten ∞-norm is the spectral norm, $\|A\|_2$.

    b. In Chapter 15, we will be able to use the SVD to show that trace $\left(A^{\mathrm{T}}A\right) = \sum_{i=1}^{n} \sigma_i^2$. Using this result, show that the Schatten 2-norm is the Frobenius norm.

7.20 Prove

    a. If A is a symmetric matrix, $\|A\|_2 = \rho(A)$, where $\rho(A)$ is the spectral radius of A.

    b. Assume that $A^{\mathrm{T}}A$ and AA^{T} have the same eigenvalues, a fact we will deal with in Chapter 15. Prove that $\left\|A^{-1}\right\|_2 = 1/\sigma_{\min}$, where $\sigma_{\min}$ is the minimum singular value of A.

7.21 Complete the proof of Lemma 7.1 by showing that

    a. $\|x\|_2 \le \|x\|_1 \le \sqrt{n}\,\|x\|_2$

    b. $\|x\|_\infty \le \|x\|_1 \le n\,\|x\|_\infty$

7.22 Prove that for any induced matrix norm, $\rho(A) \le \|A\|$.

7.23 If A is an $n \times n$ matrix, $\rho(A) \ge 0$. Show that the spectral radius is not a matrix norm by using the matrices

$$A = \begin{bmatrix} 2 & 0 \\ 5 & 0 \end{bmatrix}, \quad B = \begin{bmatrix} 0 & 5 \\ 0 & 2 \end{bmatrix}.$$

7.7.1 MATLAB Problems

7.24 It can be shown that if A is an $m \times n$ matrix, then

$$\frac{1}{\sqrt{n}}\,\|A\|_\infty \le \|A\|_2 \le \sqrt{m}\,\|A\|_\infty$$

and

$$\frac{1}{\sqrt{m}}\,\|A\|_1 \le \|A\|_2 \le \sqrt{n}\,\|A\|_1\,.$$

Using MATLAB, verify these relationships for the matrix $A = \begin{bmatrix} 1 & 0 & 1 \\ 1 & 1 & 1 \\ 0 & 0 & 1 \\ 1 & 1 & 1 \\ 0 & 1 & 1 \\ 1 & 1 & 0 \end{bmatrix}.$

7.25 Run an experiment like that in Example 7.6 to formulate a good guess for the value of the matrix 1-norm.

7.26 The software distribution contains a function `P =sphereorthog(theta, phi)` that returns the orthogonal matrix in Equation 7.2. Convert $\begin{bmatrix} 2 \\ \frac{3}{8}\pi \\ \pi/6 \end{bmatrix}$ from the spherical coordinate basis to Cartesian coordinates. Using the MATLAB function `quiver3`, draw the spherical basis vectors $\mathbf{e_r}$, $\mathbf{e}_\phi$, $\mathbf{e}_\theta$, and the vector

$$u = 2\mathbf{e_r} + \frac{3}{8}\pi\mathbf{e}_\theta + \frac{\pi}{6}\mathbf{e}_\phi.$$

7.27 Let A be the 2×2 matrix $A = \begin{bmatrix} 1 & -1 \\ 3 & 5 \end{bmatrix}$. The following function plots the unit circle and the range of Ax as x varies over the circle. Add code that computes vectors x_1 and x_2 such that $\|Ax_1\|_2$ and $\|Ax_2\|_2$ are largest and smallest, respectively, among the points generated on the unit circle. Using the MATLAB function `quiver`, draw the two vectors Ax_1, Ax_2. Also output $\|Ax_1\|_2$ and $1/\|Ax_2\|$. Compute $\|A\|_2$ and $\left\|A^{-1}\right\|_2$. What conclusion can you make?

```
function matimage(A)

% build the unit circle
t=0:0.01:2*pi;
x=cos(t)';y=sin(t)';
npts = length(t);
Ax = zeros(npts,1);
%Image of the unit circle under A
for i = 1:npts
```

```
      v=[x(i);y(i)];
      w=A*v;
      Ax(i)=w(1);
      Ay(i)=w(2);
   end
   % Plot of the circle and its image
   plot(x,y,Ax,Ay,'.','MarkerSize',10,'LineWidth',3);
   grid on; axis equal;
   title('Action of a linear transformation on the unit circle');
   xlabel('x'); ylabel('y');
```

7.28 Plot the surface $r = \theta^2$, $\theta = \phi$, $0 \le \theta \le \pi$.

7.29 A magic square is an $n \times n$ matrix whose rows, columns, and both diagonals add to the same number. For instance, M is a 4×4 magic square, whose sums are 34.

$$M = \begin{bmatrix} 16 & 2 & 3 & 13 \\ 5 & 11 & 10 & 8 \\ 9 & 7 & 6 & 12 \\ 4 & 14 & 15 & 1 \end{bmatrix}$$

a. What is the sum for an $n \times n$ magic square?

b. The time required to execute one or more MATLAB commands can be timed by entering a single line_using `tic/toc` as follows:

```
tic;command1;command2;...;last command;toc;
```

It is faster to compute the Frobenius norm, the 1-norm, and the ∞-norm of a matrix than to compute the 2-norm. You are to perform an experiment. Generate a 1000×1000 magic square using the MATLAB command "M = `magic(1000)`;", and do not omit the ";" or 1,000,000 integers will begin spewing onto your screen. Using `tic/toc`, time the execution of `norm(A)` and then the execution of the other norms. Comment on the results.

7.30 As we have noted, a computer does not perform exact floating point arithmetic, and errors occur. Norms play a role in determining if one can depend on the solution to a linear system obtained using Gaussian elimination. Assume that the entries of matrix A are precise. Let x be the true solution to the system $Ax = b$ and that x_a is the solution obtained using Gaussian elimination. If the product Ax_a is not exact, then $Ax_a = b + \Delta b$, $\Delta b \ne 0$.

a. Using $Ax = b$, show that $x_a - x = A^{-1}\Delta b$.

b. Noting that $\|b\|_2 = \|Ax\|_2$, show that $(\|x_a - x\|_2/\|x\|_2) \le \|A^{-1}\|_2 \|A\|_2 \|\Delta b\|_2/\|b\|_2$.

c. The product $\|A^{-1}\|_2 \|A\|_2$ is called the *condition number* of A. If it is large, errors relative to the correct values can be large. For each matrix, find the condition number.

$$A = \begin{bmatrix} 1 & 3 & -1 \\ 5 & -1 & 2 \\ 1 & 7 & 8 \end{bmatrix}, \quad B = \begin{bmatrix} -4.0000 & 0.5000 & 0.3333 & 0.2500 \\ -120.0000 & 20.0000 & 15.0000 & 12.0000 \\ 240.0000 & -45.0000 & -36.0000 & -30.0000 \\ -140.0000 & 28.0000 & 23.3333 & 20.0000 \end{bmatrix}.$$

d. For each matrix, let b be a vector consisting of all ones. Find the MATLAB solution x. Then multiply b by 0.999 and solve the system again to obtain xp. Compute $\|x - xp\|_2$. What are your conclusions?

7.31 Let $A = \begin{bmatrix} 0.6 & 1 & 6 & -1 & 5 \\ 0 & 0.6 & 1 & 1 & 0 \\ 0 & 0 & 0.6 & 1 & 3 \\ 0 & 0 & 0 & 0.6 & 1 \\ 0 & 0 & 0 & 0 & -0.7 \end{bmatrix}$.

a. Without using MATLAB, find the eigenvalues of A.

b. Is there a basis of eigenvectors for $\mathbb{R}^5$?

c. What is the spectral radius of A?

d. Plot $\|A^n\|_2$ for $n = 0, 1, \ldots, 50$.

e. Find the maximum value of $\|A^n\|_2$, $0 \le n \le 50$.

f. Build another nonsymmetric matrix with $\rho(A) < 1$. Do parts (d) and (e) for it.

    g. Perform the same actions with the symmetric matrix SYMMAT in the software distribution.

    h. Attempt to explain your results. For a symmetric matrix, use the spectral theorem. We will develop Schur's triangularization in Chapter 19. It states that very $n \times n$ real matrix A with real eigenvalues can be factored into $A = PTP^{\mathrm{T}}$, where P is an orthogonal matrix and T is an upper-triangular matrix. Apply this result to the nonsymmetric case. In particular, is there a relationship between the eigenvalues of A and T that can explain what happens to $\|A^n\|_2$ as n increases?

7.32 The inner product of two $n \times 1$ vectors u, v is the real number $\langle u, v \rangle = u^{\mathrm{T}} v$. Now let's investigate the $n \times n$ matrix $A = uv^{\mathrm{T}}$. For $n = 5, 15, 25$, generate vectors $u = rand\,(n, 1)$ and $v = \mathrm{rand}\,(n, 1)$. In each case, compute rank $\left(uv^{\mathrm{T}}\right)$, $\|u\|_2 \|v\|_2$, and $\left\|uv^{\mathrm{T}}\right\|_2$. What do you conclude from the experiment? Prove each assertion. It will help to recall that *rank + nullity = n* (Theorem 3.4).

7.33 This problem investigates how some norm values compare with the maximum of the absolute values of all matrix entries,

$$\max_{1 \le i,j \le n} \left| a_{ij} \right|,$$

that can be computed using the MATLAB command `max(max(abs(A)))`. For $n = 5, 15, 25$, build matrices

    $A_n = $ `randi([-100 100],n,n)`

and compute

$$\max_{1 \le i,j \le n} \left| a_{ij}^{(n)} \right|, \ \|A_n\|_\infty, \ \|A_n\|_1, \ \|A\|_2 \text{ and } \|A_n\|_F.$$

What is the apparent relationship between

$$\max_{1 \le i,j \le n} \left| a_{ij}^{(n)} \right|$$

and the matrix norms? Prove your assertion. Hint: For the induced norms, assume

$$m = \max_{1 \le i,j \le n} \left| a_{ij}^{(n)} \right|$$

occurs at indices $(i_{\max}, j_{\max})$. By definition

$$\|A\| = \max_{\|x\| \ne 0} \frac{\|Ax\|}{\|x\|}.$$

If e_k is a standard basis vector, then $(\|Ae_k\|/\|e_k\|) = \|Ae_k\| \le \|A\|$.

Chapter 8

Floating Point Arithmetic

You should be familiar with

- Number systems, primarily binary, hexadecimal, and decimal
- Geometric series
- Euclidean vector norm
- Quadratic equation

In this day and age when we rely on computers for so many things, it seems unreasonable to think they make errors. In fact, when performing arithmetic with real numbers such as 0.3 and 1.67×10^8, there is error when the number is placed in computer memory, and it is called *round-off error*. And worse, the error propagates with arithmetic operations such as addition and division. In engineering and scientific applications, it is often necessary to deal with large-scale matrix operations. These computations must be done with minimal error. The engineer or scientist must be aware that errors will occur, why they occur, and how to minimize them.

Engineering applications deal with the computation of functions like e^x, $\cos x$, and the error function $erf(x) = (2/\sqrt{\pi}) \int_0^x e^{-t^2} dt$. These functions are defined in terms of infinite series, and we must cut off summing terms at some point. This is called *truncation error*. Engineering and science applications often have to deal with very large, sparse, matrices. Such matrices have a small number of nonzero entries, and methods of dealing with them are primarily iterative. An iterative method computes a sequence of approximate solutions $\{x_i\}$ that approaches a solution. Such methods suffer from truncation error.

We discuss the representation of numbers in digital computers and its associated arithmetic. A digital computer stores values using the binary number system, where a number is represented by a string of 0's and 1's. Each binary digit is termed a *bit*. Since digital computers have a finite bit capacity (memory), integers and real numbers are represented by a fixed number of binary bits. We will see that integers can be represented exactly as long as the integer value falls within the fixed number of bits. Floating point numbers are another story. Most such values cannot be represented exactly. We will describe representation systems so that we will understand the problems involved when dealing with both integers and floating point numbers.

8.1 INTEGER REPRESENTATION

Suppose that p bits are available to represent an integer. Here is a simple way to do it. A positive integer has a zero in the last bit and the $p-1$ other bits contain the binary (base-2) representation of the integer. For example, for $p = 8$, the positive integer

$$21 = 1 \times 2^4 + 0 \times 2^3 + 1 \times 2^2 + 0 \times 2^1 + 1 \times 2^0 \text{ is encoded as}$$

0		0	0	1	0	1	0	1

and $58 = 1 \times 2^5 + 1 \times 2^4 + 1 \times 2^3 + 0 \times 2^2 + 1 \times 2^1 + 0 \times 2^0$ is encoded as

0		0	1	1	1	0	1	0

For negative integers, most computers use the *two's-complement representation*. The system works like an ideal odometer. If the odometer reads 000000 and the car backs up 1 mile, the odometer reads 999999. In binary with $p = 8$, zero is 00000000, so -1 becomes 11111111. Continuing in this fashion, -2 becomes 11111110, -3 is 11111101, and so forth. This might appear strange, but it is really very effective and simple. If this representation of -1 is to make sense, there must be a logical way to take the negative of 1 and obtain this representation for -1. Invert all the bits $(0 \rightarrow 1, 1 \rightarrow 0)$ and add 1.

Numerical Linear Algebra with Applications. http://dx.doi.org/10.1016/B978-0-12-394435-1.00008-9

$$-1 \rightarrow \text{invert}\,(00000001) + 1 = 11111110 + 1 = 11111111. \qquad (8.1)$$

The inversion of bits is called the *1s-complement*, which we indicate by $\text{1comp}(n)$, so we can write Equation 8.1 as

$$-1 \rightarrow \text{1comp}(0000001) + 1.$$

1		1	1	1	1	1	1	1

The negative of 3 should be 11111101. To verify this, compute

$$-3 \rightarrow \text{1comp}(00000011) + 1 = 11111100 + 1 = 11111101.$$

1		1	1	1	1	1	0	1

Negation works both ways. For instance, $-(-2) = 2$.

$$-(-2) = \text{1comp}(11111110) + 1 = 00000001 + 1 = 00000010.$$

0		0	0	0	0	0	1	0

Definition 8.1. We denote the process of taking the one's complement and adding 1 with the notation $\text{2comp}(n)$. Thus, the negative of a two's complement number n is $\text{2comp}(n) = \text{1comp}(n) + 1$.

Remark 8.1. The leftmost bit is called the *sign bit*. It is always 0 when the integer is positive and 1 when it is negative.

We still have not discussed how to add or subtract two's-complement numbers. If we add the representations for -1 and 1 together, we should get 0. Thus, form the sum using ordinary binary arithmetic:

$$11111111 + 00000001 = \underline{1}00000000,$$

where the underlined bit is the *carry*. Discard the carry, retaining 8 bits, and we have a result of 00000000, or zero. One more example will suggest a formula for addition of two's complement numbers.

Example 8.1. Form the sum of 95 and -43.

$$95 \rightarrow 01011111$$
$$43 \rightarrow 00101011, \quad \text{and} \quad -43 \rightarrow 11010100 + 1 = 11010101$$
$$95 + (-43) \rightarrow 01011111 + 11010101 = \underline{1}00110100$$

Discard the carry, and the result is $00110100 \rightarrow 52$.

0		0	1	1	0	1	0	0

■

Remark 8.2. The following rules show how to perform addition or subtraction of two's-complement numbers.

- Given two p-bit integers m and n, form $m + n$ by performing binary addition and discarding the carry.
- The subtraction $m - n$ is performed adding $(-n)$ to m, so $m - n = m + \text{2comp}(n)$.

The largest positive integer is $0\,\underbrace{111\ldots111} = 2^{p-1} - 1$. Finding the most negative integer is more interesting. The representation of -1 is

1		1	1	1	1	1	1	1

and the representation of -2 is

1		1	1	1	1	1	1	0

Continuing in this way, we eventually arrive at

1		0	0	0	0	0	0	0

Going from 1111111 to 10000001 comprises $127 = 2^7 = 2^{p-1} - 1$ integers, and the range is $-1 \geq n \geq -127$. We get one more negative integer, namely, $1\underbrace{000\ldots000}_{p-1}$ that represents -2^{p-1}, so the range of integers we can represent is $-2^{p-1} \leq n \leq 2^{p-1} - 1$.

Example 8.2. You have heard of 32-bit and 64-bit operating systems. In a 32-bit system, $p = 32$, so the range of integers that can be represented is

$-2^{31} \leq n \leq 2^{31} - 1$ or $-2,147,483,648 \leq n \leq 2,147,483,647$.

In a 64-bit system, $p = 64$, and the range of an integer is

$-18,446,744,073,709,551,616 \leq n \leq 18,446,744,073,709,551,615$. ∎

Integer arithmetic can cause *overflow* when the range of a positive or negative integer is exceeded. For instance, let's assume $p = 8$ and add 120 and 88. Then

| 0 | 1 | 1 | 1 | 1 | 0 | 0 | 0 | + | 0 | 1 | 0 | 1 | 1 | 0 | 0 | 0 | = | 1 | 1 | 0 | 1 | 0 | 0 | 0 | 0 |

which is the value -48.

Remark 8.3. When the sign of the result is the opposite of what it should be, overflow has occurred.

Example 8.3. Compute $-18 + (-112) = 11101110 + 10010000 = \underline{1}|01111110$, so $-18 + (-112) = 126$. This results is an overflow. ∎

During integer operations, all results must lie within the interval $-2^{p-1} \leq n \leq 2^{p-1} - 1$. This is a significant limitation. If large integer arithmetic must be performed, software is required, often called a BigInteger package. For example, data encryption usually requires operations with very large integers.

8.2 FLOATING-POINT REPRESENTATION

For given integers b, p, e_{min}, and e_{max}, we define the nonzero floating-point numbers as real numbers of the form

$$\pm \left(0.d_1d_2\ldots d_p\right) \times b^n$$

with $d_1 \neq 0$, $0 \leq d_i \leq b - 1$, and $-e_{min} \leq n \leq e_{max}$. We denote by F the (finite) set of all floating-point numbers. In this notation,

a. b is the base. The most common bases are $b = 2$ (binary base), $b = 10$ (decimal base), and $b = 16$ (hexadecimal base).
b. $-e_{min} \leq n \leq e_{max}$ is the exponent that defines the order of magnitude of the number to be encoded.
c. The integers $0 \leq d_i \leq b - 1$ are called the digits and p is the number of *significant digits*. The *mantissa* is the integer $m = d_1d_2\ldots d_p$. Note that

$$\left(0.d_1d_2\ldots d_p\right) \times b^n = \left(d_1 \times b^{-1} + d_2b^{-2} + \cdots + d_pb^{-p}\right)b^n = b^n \sum_{k=1}^{p} d_kb^{-k}. \tag{8.2}$$

The following bounds hold for floating-point numbers

$$f_{min} \leq |f| \leq f_{max} \quad \text{for every } f \in F,$$

where $f_{min} = b^{-(e_{min}+1)}$ is the smallest positive real number. We can see this by setting $n = -e_{min}, d_1 = 1, d_2 = d_3 = \cdots = d_p = 0$ and applying Equation 8.2. The largest floating point number f_{max} is found by applying Equation 8.2 with $d_i = (b-1), 1 \leq i \leq p$ and exponent $n = e_{max}$. Using the formula for the sum of a geometric series

$$f_{max} = b^{e_{max}}(b-1)\left(b^{-1} + b^{-2} + \cdots + b^{-p}\right) = b^{e_{max}}(b-1)\left(\frac{1 - \left(\frac{1}{b}\right)^p}{b-1}\right)$$

$$= b^{e_{max}}\left(1 - \left(\frac{1}{b}\right)^p\right) = b^{e_{max}}\left(1 - b^{-p}\right).$$

Summary

$$f_{\min} = b^{-(e_{\min}+1)}$$
$$f_{\max} = b^{e_{\max}}\left(1 - b^{-p}\right)$$

Smaller numbers produce an *underflow* and larger ones an *overflow*.

Computers use base $b = 2$ and usually support single precision (representation with $p = 32$) and double precision (representation with $p = 64$). In our single-precision representation, we use 1 bit for the sign (0 means positive, 1 means negative), 8 bits for the exponent, and 23 bits for the mantissa (for a total of 32 bits). In our double-precision representation, we use 1 bit for the sign, 11 bits for the exponent, and 52 bits for the mantissa (for a total of 64 bits). For example, represent 81.625 in single precision.

$$81.625 = (1)\,2^6 + (0)\,2^5 + (1)\,2^4 + (0)\,2^3 + (0)\,2^2 + (0)\,2^1 + (1)\,2^0 + (1)\,2^{-1} + (0)\,2^{-2} + (1)\,2^{-3}$$
$$= \left((1)\,2^{-1} + (0)\,2^{-2} + (1)\,2^{-3} + (0)\,2^{-4} + (0)\,2^{-5} + (0)\,2^{-6} + (1)\,2^{-7} + (1)\,2^{-8} + (0)\,2^{-9} + (1)\,2^{-10}\right)2^7$$

Note that the leading digit is $d_1 = 1$, as required, and the exponent is properly adjusted. To avoid dealing with a negative exponent, encode it as an unsigned integer by adding to it a "bias" (127 is the usual bias in single precision). For our example of 81.625, the exponent will be stored as $7 + 127 = 134$ using bias 127. To obtain the actual exponent, compute $134 - 127 = 7$. If the exponent is -57, it is stored as $-57 + 127 = 70$. A stored exponent of 0 reflects an actual exponent of -127. Here is the internal representation for 81.625, where sign $= 0$, exponent $= 134$ in a field of 8 bits, and the mantissa follows the exponent in a field of 23 bits.

$$\boxed{0|10000110|10100011010000000000000}$$

There still remains the issue of $x = 0.0$. Represent it by filling both the exponent and mantissa with zeros. The sign bit can still be 0 (+) or 1 (−), so you will sometimes see output like −0.0000. Note that this situation does not occur with the two's-complement integer representation.

Example 8.4. Let $x = \frac{1}{6}$, and assume that $b = 2$ and $p = 8$. The binary representation of 1/6 is the infinite repeating pattern

$$0.00101010101010101010101010101010\ldots$$

We only have eight binary digits to work with, so when 1/6 is entered into computer memory, the binary sequence is either rounded or truncated. When using *rounding*, the approximation is 0.00101011, but with *truncation* the approximation is 0.00101010. ∎

To indicate the error in converting a floating number, x, to its computer representation, we use the notation $fl(x)$.

Definition 8.2. Let $fl(x)$ be the floating-point number associated with the real number x. For instance, in Example 8.4, the number $x = \frac{1}{6}$ was approximated using rounding, so

$$fl(x) = 0.00101011.$$

8.2.1 Mapping from Real Numbers to Floating-Point Numbers

The most widely used standard for floating-point computation is the *IEEE Standard for Floating-Point Arithmetic*. The most frequently used IEEE formats are single and double precision. In each case, a nonzero number is assumed to have a hidden 1 prior to the first digit. As a result, single precision uses 24 binary digits, and double precision 53. Table 8.1 lists the attributes of the two formats.

TABLE 8.1 IEEE Formats

Name	Base	Digits	$e_{\min}$	$e_{\max}$	Approximate Decimal Range
Single precision	2	23+1	−126	+127	1.18×10^{-38} *to* 3.4×10^{38}
Double precision	2	52+1	−1022	+1023	2.23×10^{-308} *to* 1.80×10^{308}

Floating point numbers are *granular*, which means there are gaps between numbers. The granularity is caused by the fact that a finite number of bits are used to represent a floating point number. We represented 81.625 perfectly because 0.625 is $\frac{1}{2} + \frac{1}{8}$, but most real numbers must be approximated because there is no exact conversion into binary (Example 8.4).

The distance from 1.0 to the next largest double-precision number is 2^{-52} in IEEE double precision. If a number smaller than 2^{-52} is added to 1, the result will be 1. The floating point numbers between 1.0 and 2.0 are equally spaced:

$$\left\{1, 1 + 2^{-52}, 1 + 2 \times 2^{-52}, 1 + 3 \times 2^{-52}, \ldots, 2\right\}.$$

The gap increases as the length of intervals become larger by a factor of 2. The numbers between 2.0 and 4.0 are separated by a gap of 2^{-51}.

$$2\left\{1, 1 + 2^{-52}, 1 + 2 \times 2^{-52}, 1 + 3 \times 2^{-52}, \ldots, 2\right\}.$$

In general, an interval from 2^k to 2^{k+1} has a gap between values of $2^k\left(2^{-52}\right)$. As k increases, the gap relative to 2^k remains 2^{-52}. In single precision, the relative gap between numbers is 2^{-23}. There is a name associated with this gap. It is called the *machine precision*, or eps, and it plays a significant role in analysis of floating point operations. Remember that the value of eps varies with the precision.

There is a formula for eps for any b and p, and let's intuitively determine it. Let $b = 10$ and $p = 4$, and assume we round to p digits. Let $x = 1$. Then, $\text{fl}(1 + 0.0001) = 1, \text{fl}(1 + 0.0003) = 1, \text{fl}(1 + 0.0004) = 1, \text{fl}(1 + 0.0005) = 1.0001$. If we repeat the experiment with $b = 10, p = 5$, we will find that $\text{fl}(1 + 0.00005) = 1.00001$. Let $b = 2, p = 3$ and again assume rounding. Let $x = 1$. Now, $\text{fl}\left(1 + 2^{-5}\right) = \text{fl}(1 + 0.00001) = 1, \text{fl}\left(1 + 2^{-4}\right) = \text{fl}(1 + 0.0001) = 1,$ $\text{fl}\left(1 + 2^{-3}\right) = \text{fl}(1 + 0.001) = 1, \text{fl}(1 + 2^{-2}) = \text{fl}(1.01) = 1.01$. This is enough information for us to define eps.

Definition 8.3. Assume b is the base of the number system, p is the number of significant digits, and that rounding is used. The machine precision, $\text{eps} = \frac{1}{2}b^{1-p}$, is the distance from 1.0 to the next largest floating point number.

For IEEE double-precision floating-point, we specified the $\text{eps} = 2^{-52}$. Applying the formula with $b = 2$ and $p = 52$, we have $\text{eps} = \frac{1}{2}2^{1-52} = 2^{-52}$. In single precision, $\text{eps} = \frac{1}{2}2^{1-23} = 2^{-23}$. Figure 8.1 shows the distribution of a floating-point number system with $b = 2, p = 3, e_{\min} = -3, e_{\max} = 3$, so $\text{eps} = \frac{1}{2}2^{-2} = \frac{1}{8}$. Notice how the gaps between numbers grow, but remember the gap remains constant relative to the number size. The nonnegative floating point numbers shown in Figure 8.1 are:

```
0.0000    0.0625    0.078125    0.09375    0.109375
0.125     0.15625   0.1875      0.21875
0.25      0.3125    0.375       0.4375
0.5       0.625     0.75        0.875
1.0000    1.2500    1.5000      1.7500
2.0000    2.5000    3.0000      3.5000
4.0000    5.0000    6.0000      7.0000
```

The conversion of real numbers to floating-point numbers is called floating-point representation or *rounding*, and the error between the true value and the floating-point value is called *round-off error*. We expect $(\text{fl}(x) - x)/x = \epsilon$ not to exceed eps in magnitude. The following formula holds for all real numbers $f_{\min} \le x \le f_{\max}$:

$$\text{fl}(x) = x(1 + \varepsilon) \tag{8.3}$$

with $|\epsilon| \le \text{eps}$.

Remark 8.4. For single-precision IEEE floating point representation, $\text{eps} = 2^{-23} \approx 1.192 \times 10^{-7}$, and for double precision, $\text{eps} = 2^{-52} \approx 2.22 \times 10^{-16}$. These numbers explain the maximum accuracy of 7 or 16 significant digits for single or double-precision arithmetic. Figure 8.2 shows a map of double-precision IEEE floating-point numbers. Figures 8.1 and 8.2 together provide a good picture of a floating-point number system.

FIGURE 8.1 Floating-point number system.

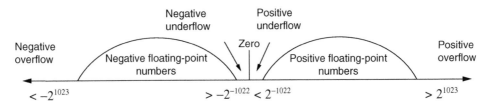

FIGURE 8.2 Map of IEEE double-precision floating-point.

8.3 FLOATING-POINT ARITHMETIC

Consider the operation $+$. The sum of two floating-point numbers is usually an approximation to the actual sum. We denote by $\oplus$ the computer result of the addition.

Definition 8.4.   For real numbers x and y,

$$x \oplus y = \mathrm{fl}\,(fl\,(x) + fl\,(y))\,. \tag{8.4}$$

Overflow occurs if the addition produces a number that is too large, $|x \oplus y| > f_{\max}$, and underflow occurs if it produces a number that is too small, $|x \oplus y| < f_{\min}$. Similar notation applies for the other operations: $\ominus, \otimes$, and $\oslash$.

Example 8.5.   Assume $b = 10, p = 4$, and that the true values of x and y are 0.34578×10^1 and 0.56891×10^1, respectively,

$$\mathrm{fl}\,(x) = 0.3458 \times 10^1, \quad \mathrm{fl}\,(y) = 0.5689 \times 10^1,$$

and

$$x \oplus y = 0.9147 \times 10^1\,. \qquad\blacksquare$$

When performing floating point operations on values with different exponents, a realignment must take place. For instance, let $b = 10, p = 5, x = 0.10002 \times 10^2$ and $y = 0.99982 \times 10^1$. Now compute $x \ominus y$.

0	.	1	0	0	0	2	**0**	×	10^2
				−					
0	.	0	9	9	9	8	2	×	10^2
				=					
0	.	3	8	0	0	0		×	10^{-2}

Without adding the extra 0 in x and sticking with 5 digits throughout the calculation, we will get

1	0	.	0	0	2
		−			
	9	.	9	9	8
		=			
0	.	0	0	4	0

an incorrect result. The additional zero is called a *guard digit*. Most computers use guard digits, so we will assume that all calculations are done using them.

Remark 8.5.   Floating point numbers have a fixed range so, as is the case with integers, dealing with floating point numbers with many significant digits and large exponents requires software.

8.3.1 Relative Error

There are two ways to measure error, using *absolute error* or *relative error*.

$$\text{Absolute error} = |\mathrm{fl}\,(x) - x|$$

$$\text{Relative error} = \frac{|\mathrm{fl}\,(x) - x|}{|x|}, \quad x \neq 0$$

Example 8.6.

a. In Example 8.5, $b = 10$ and $p = 4$. The value of eps for this representation is $\text{eps} = 0.0005 = 5 \times 10^{-4}$. The value $x = 0.34578 \times 10^1$ converts to floating point as $\text{fl}(x) = 0.3458 \times 10^1$. According to Equation 8.3, $\text{fl}\left(0.34578 \times 10^1\right) = 0.3458 \times 10^1 = 0.34578 \times 10^1(1 + \varepsilon)$, so $\epsilon = ((0.3458 \times 10^1)/(0.34578 \times 10^1)) - 1 = 0.5784 \times 10^{-4} < \text{eps}$, as expected. Also, $|\text{fl}(x) - x| = 0.0002$, as opposed to $(|\text{fl}(x) - x|)/|x| = 0.5784 \times 10^{-4}$.

b. Consider $x = 1.6553 \times 10^5$, $\text{fl}(x) = 1.6552 \times 10^5$. The absolute error is $|\text{fl}(x) - x| = 10$, while the relative error is $(|\text{fl}(x) - x|)/|x| = 6.04 \times 10^{-5}$. With large numbers, relative error is generally more meaningful, as we see here. This same type of example applies to small values.

c. Relative error gives an indication of how good a measurement is relative to the size of the thing being measured. Let's say that two students measure the distance to different objects using triangulation. One student obtains a value of $d_1 = 28.635$ m, and the true distance is $\overline{d_1} = 28.634$ m. The other student determines the distance is $d_2 = 67.986$ m, and the true distance is $\overline{d_2} = 67.987$ m. In each case, the absolute error is 0.001. The relative errors are $(|28.634 - 28.635|)/|28.634| = 3.49 \times 10^{-5}$ and $(|67.987 - 67.986|)/|67.987| = 1.47 \times 10^{-5}$. The relative error of measurement d_2 is about 237% better than that of measurement d_1, even though the amount of absolute error is the same in each case. ∎

Relative error provides a much better measure of change for almost any purpose. For instance, estimate the sum of the series

$$\sum_{i=1}^{\infty} \frac{1}{i^2 + \sqrt{i}}.$$

Compute the sequence of partial sums $s_n = \sum_{i=1}^{n} 1/(i^2 + \sqrt{i})$. until a given error tolerance, tol, is attained. The actual sum of the series is not known, so a comparison of the partial sum with the actual sum cannot be computed. There are two approaches commonly used:

a. Compute partial sums until $|s_{n+1} - s_n| < \text{tol}$.
b. Compute partial sums until $\frac{|s_{n+1} - s_n|}{|s_n|} < \text{tol}$.

Method 2 is preferable because it tells us how the new partial sum is changing relative to the previous sum.

Remark 8.6. Most computer implementations of addition (including the widely used IEEE arithmetic) satisfy the property that the relative error is less than the machine precision:

$$\left| \frac{(x \oplus y) - (x + y)}{x + y} \right| \leq \text{eps}$$

assuming $x + y \neq 0$.

The relative error for one operation is very small, but this is not always the case when a computation involves a sequence of many operations.

8.3.2 Rounding Error Bounds

It is important to understand how floating point errors propagate, since this leads to means of controlling them. We will do a mathematical analysis of rounding error for the addition of floating point numbers, but will not do so for $\ominus$, $\otimes$, or $\oslash$, or error propagation of vector and matrix operations. We will state results, and the interested reader can consult Refs. [9, 16, 17] for a rigorous analysis. In all cases, we will assume that the approximation of x by $\text{fl}(x)$ is done by rounding rather than truncation.

Remark 8.7. IEEE 754 requires that arithmetic operations produce results that are exactly rounded, i.e., the same as if the values were computed to infinite precision prior to rounding.

We will assume that once floating point numbers x and y are in computer memory that the basic arithmetic operations satisfy the following:

> For all floating point numbers x, y in a computer:
>
> $$\begin{aligned}
> x \oplus y &= (x + y)(1 + \epsilon), \\
> x \ominus y &= (x - y)(1 + \epsilon), \\
> x \otimes y &= (x \times y)(1 + \epsilon), \\
> x \oslash y &= (x/y)(1 + \epsilon), \quad (8.5)
> \end{aligned}$$
>
> where $|\epsilon| \leq$ eps.

If the reader is not interested in the technical details, the results for addition and multiplication of floating point numbers can be summarized as follows:

When adding n floating point numbers, the result is the exact sum of the n numbers, each perturbed by a small relative error. The errors are bounded by $(n - 1)$ eps, where eps is the unit roundoff error.

The relative error in computing the product of n floating point numbers is at most $1.06 (n - 1)$ eps, assuming that $(n - 1)$ eps < 0.1.

There are error bounds for matrix operations that depend on eps and the magnitude of the true values.

Addition

Lemma 8.1. *Let $x_1, x_2, \ldots, x_n$ be positive floating point numbers in a computer. Then,*

$$\begin{aligned}
x_1 \oplus x_2 \oplus x_3 \oplus \cdots \oplus x_n &= x_1 (1 + \varepsilon_1)(1 + \varepsilon_2) \ldots (1 + \epsilon_{n-1}) \\
&\quad + x_2 (1 + \varepsilon_1)(1 + \varepsilon_2) \ldots (1 + \epsilon_{n-1}) \\
&\quad + x_3 (1 + \varepsilon_2)(1 + \varepsilon_3) \ldots (1 + \epsilon_{n-1}) \\
&\quad \vdots \\
&\quad + x_{n-1} (1 + \epsilon_{n-2})(1 + \varepsilon_{n-1}) \\
&\quad + x_n (1 + \epsilon_{n-1}),
\end{aligned}$$

where $|\epsilon_i| \leq$ eps, $1 \leq i \leq n - 1$.

Proof. Assume the computation proceeds as follows:

$$s_2 = x_1 \oplus x_2, s_3 = s_2 \oplus x_3, \ldots, s_n = s_{n-1} \oplus x_n.$$

From Equation 8.5,

$$s_2 = \text{fl}(x_1 + x_2) = (x_1 + x_2)(1 + \epsilon_1) = x_1(1 + \epsilon_1) + x_2(1 + \epsilon_1),$$

where $|\epsilon_1| \leq$ eps. Now,

$$\begin{aligned}
s_3 &= \text{fl}(s_2 + x_3) = (s_2 + x_3)(1 + \epsilon_2) \\
&= s_2(1 + \epsilon_2) + x_3(1 + \epsilon_2) \\
&= x_1(1 + \epsilon_1)(1 + \epsilon_2) \\
&\quad + x_2(1 + \varepsilon_1)(1 + \epsilon_2) \qquad\qquad (8.6) \\
&\quad + x_3(1 + \varepsilon_2).
\end{aligned}$$

Equation 8.6 defines a pattern, and we have

$$
\begin{aligned}
s_n = \; & x_1 \left(1+\epsilon_1\right)\left(1+\epsilon_2\right)\ldots\left(1+\epsilon_{n-1}\right) \\
& +x_2 \left(1+\epsilon_1\right)\left(1+\epsilon_2\right)\ldots\left(1+\epsilon_{n-1}\right) \\
& +x_3 \left(1+\epsilon_2\right)\left(1+\epsilon_3\right)\ldots\left(1+\epsilon_{n-1}\right) \\
& \;\;\vdots \\
& +x_{n-1} \left(1+\varepsilon_{n-2}\right)\left(1+\varepsilon_{n-1}\right) \\
& +x_n \left(1+\varepsilon_{n-1}\right).
\end{aligned}
\tag{8.7}
$$

where $|\epsilon_i| \le \text{eps}, 1 \le i \le n-1$. $\qquad\square$

Following the analysis in Ref. [17, pp. 132-134], define

$$
\begin{aligned}
1+\eta_1 &= \left(1+\epsilon_1\right)\left(1+\epsilon_2\right)\ldots\left(1+\epsilon_{n-1}\right), \\
1+\eta_2 &= \left(1+\epsilon_1\right)\left(1+\epsilon_2\right)\ldots\left(1+\epsilon_{n-1}\right), \\
1+\eta_3 &= \left(1+\epsilon_2\right)\left(1+\epsilon_3\right)\ldots\left(1+\epsilon_{n-1}\right), \\
&\;\;\vdots \\
1+\eta_{n-1} &= \left(1+\varepsilon_{n-2}\right)\left(1+\varepsilon_{n-1}\right), \\
1+\eta_n &= \left(1+\varepsilon_{n-1}\right).
\end{aligned}
$$

We can now write Equation 8.7 as

$$
\begin{aligned}
s_n = \; & x_1 \left(1+\eta_1\right) + x_2 \left(1+\eta_2\right) + x_3 \left(1+\eta_3\right) + \cdots + \\
& x_{n-1} \left(1+\eta_{n-1}\right) + x_n \left(1+\eta_n\right).
\end{aligned}
\tag{8.8}
$$

In order for Equation 8.8 to be useful, we need bounds for the η_i. From $1+\eta_n = \left(1+\varepsilon_{n-1}\right)$, it follows that $|\eta_n| = |\epsilon_n| \le \text{eps}$. Now consider the term $1 + \eta_{n-1}$:

$$
1+\eta_{n-1} = 1 + \epsilon_{n-1} + \epsilon_{n-2} + \epsilon_{n-1}\epsilon_{n-2},
$$

so

$$
\eta_{n-1} = \epsilon_{n-1} + \epsilon_{n-2} + \epsilon_{n-1}\epsilon_{n-2},
$$

and

$$
|\eta_{n-1}| \le |\epsilon_{n-1} + \epsilon_{n-2}| + |\epsilon_{n-1}\epsilon_{n-2}| \le |\epsilon_{n-1}| + |\epsilon_{n-2}| + |\epsilon_{n-1}\epsilon_{n-2}|.
$$

The term $|\epsilon_{n-1}\epsilon_{n-2}|$ is bounded by 2eps^2. If we are using double-precision arithmetic, $\text{eps} = 2^{-52}$, so $2\text{eps}^2 = 2^{-103}$ and we can consider $|\epsilon_{n-1}\epsilon_{n-2}|$ negligible compared to $|\epsilon_{n-1}| + |\epsilon_{n-2}|$. Thus,

$$
|\eta_{n-1}| \le |\epsilon_{n-1}| + |\epsilon_{n-2}| \le 2 \, \text{eps}.
$$

Continuing in this fashion, we will obtain

$$
\begin{aligned}
|\eta_1| &\le (n-1)\,\text{eps}, \\
|\eta_i| &\le (n-i+1)\,\text{eps}, \quad 2 \le i \le n.
\end{aligned}
\tag{8.9}
$$

Equations 8.8 and 8.9 can be summarized as follows:

> When adding n floating point numbers, the result is the exact sum of the n numbers, each perturbed by a small relative error. The errors are bounded by $(n-1)\,\text{eps}$, where eps is the unit roundoff error.

It is useful to derive a bound for the relative error. Let

$$
s = \sum_{i=1}^{n} x_i
\tag{8.10}
$$

Subtract Equation 8.10 from Equation 8.7 to obtain

$$s_n - s = x_1 \eta_1 + x_2 \eta_2 + \cdots + x_n \eta_n. \tag{8.11}$$

Equation 8.9 implies that

$$|\eta_i| \le (n-1)\,\text{eps}, \quad 1 \le i \le n. \tag{8.12}$$

Taking the absolute value of both sides of Equation 8.11 and applying 8.12, we have

$$|s_n - s| \le (|x_1| + |x_2| + \cdots + |x_n|)\,(n-1)\,\text{eps},$$

This result leads to Theorem 8.1.

Theorem 8.1.   *If n floating point numbers, x_i, are added,*

$$\frac{|s_n - s|}{|x_1 + x_2 + \cdots + x_n|} \le K\,(n-1)\,\text{eps},$$

where

$$K = \frac{(|x_1| + |x_2| + \cdots + |x_n|)}{|x_1 + x_2 + \cdots + x_n|}.$$

Multiplication

Theorem 8.2.   *The relative error in computing the product of n floating point numbers is at most $1.06\,(n-1)\,\text{eps}$, assuming that $(n-1)\,\text{eps} < 0.1$.*

Matrix Operations

Theorem 8.3.   *For an $m \times n$ matrix M, define $|M| = (|m_{ij}|)$; that is $|M|$ is the matrix whose entries are the absolute value of those from M. Let A and B be two floating point matrices and let c be a floating point number. Then*

$$\text{fl}\,(cA) = cA + E, \quad |E| \le \text{eps}\,|cA|$$

$$\text{fl}\,(A + B) = (A + B) + E, \quad |E| \le \text{eps}\,|A + B|.$$

If the product of A and B is defined, then

$$\text{fl}\,(AB) = AB + |E|, \quad |E| \le n\,\text{eps}\,|A|\,|B| + K\text{eps}^2,$$

where K is a constant.

Example 8.7.   Assume $b = 10$ and $p = 5$. This example will test the error bound asserted for addition in Theorem 8.1. With this representation, $\text{eps} = 0.00005$. Table 8.2 gives the numbers, $\overline{x_i}$, before they are entered into the computer and the floating point approximations $\text{fl}\,(\overline{x_i}) = x_i$.

TABLE 8.2 Floating-Point Addition

$\overline{x}$	0.562937×10^0	0.129873×10^1	0.453219×10^1	0.765100×10^0	0.120055×10^1
x	0.56294×10^0	0.12987×10^1	0.45322×10^1	0.76510×10^0	0.12006×10^1

$$s = \sum_{i=1}^{5} x_i = 0.835954 \times 10^1$$

$$s_2 = \text{fl}\,(x_1 + x_2) = 0.18616 \times 10^1, \quad s_3 = \text{fl}\left(0.18616 \times 10^1 + x_3\right) = 0.63938 \times 10^1.$$

$$s_4 = \text{fl}\left(0.63938 \times 10^1 + x_4\right) = 0.71589 \times 10^1, \quad s_5 = \text{fl}\left(0.71589 \times 10^1 + x_5\right) = 0.83595 \times 10^1.$$

Now,

$$\frac{|s_5 - s|}{|s|} = \frac{\left|0.83640 \times 10^1 - 0.835954 \times 10^1\right|}{\left|0.835954 \times 10^1\right|} = \frac{.40000 \times 10^{-4}}{0.835954 \times 10^1} = 0.47850 \times 10^{-5}.$$

The value of K in Theorem 8.1

$$K = \frac{|x_1| + |x_2| + |x_3| + |x_4| + |x_5|}{|x_1 + x_2 + x_3 + x_4 + x_5|} = 1,$$

and

$$K(n-1)\,eps = (1)(4)(.00005) = 2.0000 \times 10^{-4},$$

validating the error bound of Theorem 8.1.

■

8.4 MINIMIZING ERRORS

The subject of error minimization is complex and sometimes involves clever tricks, but there are some general principles to follow. The article, *What Every Computer Scientist Should Know About Floating Point Arithmetic* [18], is an excellent summation of earlier sections of this chapter and contains a discussion of some techniques for minimizing errors.

8.4.1 Avoid Adding a Huge Number to a Small Number

Adding a very large number to a small number may eliminate any contribution of the small number to the final result.

Example 8.8. Assume $b = 10, p = 4$. Let $x = 0.267365 \times 10^3$ and $y = 0.45539 \times 10^{-3}$. Then,

$$\text{fl}(x) = 0.2674 \times 10^3, \quad \text{fl}(y) = 0.4554 \times 10^{-2},$$

and

$$\text{fl}(\text{fl}(x) + \text{fl}(y)) = 0.2674 \times 10^3.$$

■

Overflow can result from including very large numbers in calculations, and can sometimes be avoided by just modifying the order in which you do the computation. For instance, suppose you need to compute the Euclidean norm of a vector x:

$$\|x\|_2 = \sqrt{x_1^2 + x_2^2 + \cdots + x_n^2}$$

If some x_i are very large, overflow can occur. There is a way to avoid this. Compute $m = \max(|x_2|, |x_2|, \ldots, |x_n|)$, divide each x_i by m (called *normalizing* the vector), and then sum the squares of the normalized vector components and multiply by m. Overflow will be avoided. Here are the steps:

a. $m = \max(|x_2|, |x_2|, \ldots, |x_n|)$.
b. $y_i = x_i/m, \ 1 \le i \le n \left(y_i^2 \le 1\right)$.
c. $\|x\|_2 = m\sqrt{y_1^2 + y_2^2 + \cdots + y_n^2}$.

8.4.2 Avoid Subtracting Numbers That Are Close

Solving the quadratic equation

$$ax^2 + bx + c = 0, \quad a \ne 0$$

is a classic example where subtracting numbers that are close can be disastrous. We call this *cancellation* error. The usual way to finds its two roots is by using the *quadratic formula*:

$$x_1 = \frac{-b + \sqrt{b^2 - 4ac}}{2a}, \quad x_2 = \frac{-b - \sqrt{b^2 - 4ac}}{2a}.$$

If the product $4ac$ is small, then $\sqrt{b^2 - 4ac} \cong b$, and when computing one of x_1 or x_2 we will be subtracting two nearly equal numbers. This can be a serious problem, since we know a computer maintains only a fixed number of significant digits.

Example 8.9. Consider solving $ax^2 + bx + c = 0$ with $a = 1$, $b = 68.50$, $c = 0.1$ with the quadratic equation. Use rounded base-10 arithmetic with 4 significant digits. Now, $b^2 - 4ac = 4692 - 0.4000 = 4692$, and

$$x_1 = \frac{-68.50 + \sqrt{4692}}{2} = \frac{-68.50 + 68.50}{2} = 0,$$

$$x_2 = \frac{-68.50 - \sqrt{4692}}{2} = \frac{-68.50 - 68.50}{2} = -68.50.$$

The correct roots are

$$x_1 = -0.001460,$$

$$x_2 = -68.50.$$

The relative error in computing x_1 is $(|-0.001460 - 0.0000|)/|-0.0001460| = 1.0000$, which is quite awful; however, x_2 is correct. There are two causes of the problem. First, the contribution of $-4ac$ was lost during the subtraction from a much larger number, followed by the cancellation error. Cancellation can be avoided by writing the quadratic formula in a different way.

Multiply the two solutions:

$$\left(\frac{-b + \sqrt{b^2 - 4ac}}{2a}\right)\left(\frac{-b - \sqrt{b^2 - 4ac}}{2a}\right) = \frac{b^2 - (b^2 - 4ac)}{4a^2} = \frac{4ac}{4a^2} = \frac{c}{a}.$$

Thus, $x_1 x_2 = c/a$. Pick the one of the two solutions that does not cause subtraction and call it x_1.

$$x_1 = -\left(\frac{b + \text{sign}\,(b)\,\sqrt{b^2 - 4ac}}{2a}\right),$$

where sign (b) is $+1$ if $b > 0$ and -1 if $b < 0$. Then compute x_2 using

$$x_2 = \frac{c}{ax_1}.$$

Cancellation is avoided. For our example,

$$x_1 = -\left(\frac{68.5 + 68.5}{2}\right) = -68.5,$$

$$x_2 = \frac{0.1}{(1)\,(-68.5)} = -0.001460. \qquad \blacksquare$$

There are other classic examples where cancellation error causes serious problems, and some are included in the problems. See Ref. [19, pp. 42-44] for some interesting examples.

We will conclude this section by saying that underflow and overflow are not the only errors produced by a floating-point representation. The MATLAB constant `inf` returns the IEEE arithmetic representation for positive infinity, and in some situations its use is valid. Infinity is also produced by operations like dividing by zero (`1.0/0.0`), or from overflow (`exp(750)`). NaN is the IEEE arithmetic representation for Not-a-Number. A NaN results from mathematically undefined operations such as `0.0/0.0` and `inf-inf`. In practice, obtaining a NaN or an unexpected `inf` is a clear indication that something is wrong!

8.5 CHAPTER SUMMARY

Representation of Integers

Integers are represented using two's-complement notation. If the hardware uses n bits to store an integer, then the left-most bit is the sign bit, and is 0 if the integer is nonnegative and 1 if it is negative. The system functions like an ideal odometer. The representation $000\ldots00$ is the integer 0, and $111\ldots11$ represents -1. The two's-complement of a number is

$$2\text{comp}\,(k) = 1\text{comp}\,(k) + 1,$$

where 1comp (n) reverses the bits in n, and any remainder is discarded. It negates its argument. For instance,

$$2\text{comp}(1) = 1\text{comp}\left(\underbrace{00\ldots001}_{n\,\text{bits}}\right) + 1 =$$

$$\underbrace{11\ldots110}_{n\,\text{bits}} + 1 = \underbrace{11\ldots111}_{n\,\text{bits}} = -1$$

The range of integers that can be represented is

$$-2^{n-1} \le k \le 2^{n-1} - 1.$$

As an example, if an integer is stored using 32 bits, the range of integers is $2,147,483,648 \le k \le 2,147,483,647$.

To add, just perform binary addition using all n bits and discard any remainder. To subtract b from a, compute $a +$ 2comp (b). Addition or subtraction can overflow, meaning that the result cannot be represented using n bits. When this happens, the answer has the wrong sign.

Floating Point Format

A binary floating point number as described in this book has the form

$$\pm \left(0.d_1 d_2 \ldots d_p\right) \times b^n$$

with $d_1 \ne 0, d_i = 0, 1, -e_{\min} \le n \le e_{\max}$ is the exponent range, and p is the number of significant bits. Using this notation, the largest magnitude for a floating point number is $f_{\max} = 2^{e_{\max}}\left(1 - 2^{-p}\right)$, and smallest nonzero floating point number in magnitude is $f_{\min} = 2^{-(e_{\min}+1)}$.

Internally, the sign bit is the left-most bit, and 0 means nonnegative and 1 means negative. The exponent follows using e bits. To avoid having to represent negative exponents a bias of $2^{e-1} - 1$ is added to the true exponent. For instance, if 8 bits are used for the exponent, the bias is 127. If the true exponent is -18, then the stored exponent is $-18 + 127 = 109 = 01101101_2$. The true exponent of zero is stored as $127 = 01111111$. The first binary digit $d_1 = 1$, and is the coefficient of $2^{-1} = \frac{1}{2}$. The remaining digits can be 0 or 1, and represent coefficients of $2^{-2}, 2^{-3}, \ldots$.

Since numbers like $\frac{1}{7} = 0.001001001001001001001001001001 \ldots_2$ cannot be represented exactly using p digits, we round to p digits, and denote the stored number as fl (x). Doing this causes roundoff error, and this affects the accuracy of computations, sometimes causing serious problems.

Floating point numbers are granular, which means there are gaps between numbers. The gap is measured using the machine precision, eps, which is the distance between 1.0 and the next floating point number. In general, an interval from 2^k to 2^{k+1} has a gap between numbers of $2^k \times$ eps, and the gap relative to 2^k remains eps. If p binary digits are used, the value of eps is $\frac{1}{2} \times 2^{1-p}$.

IEEE single- and double-precision floating point arithmetic guarantees that

$$\text{fl}(x) = x(1 + \epsilon), \quad |\epsilon| \le \text{eps}.$$

This is a fundamental formula when analyzing errors in floating point arithmetic.

Floating Point Arithmetic

Represent floating point addition of the true numbers a and b as $a \oplus b$. After computation, what we actually get is

$$a \oplus b = \text{fl}(\text{fl}(a) + \text{fl}(b)),$$

and normally roundoff error is present.

Measurement of Error

There are two ways to measure error, using absolute error or relative error:

$$\text{Absolute error} = |\text{fl}(x) - x|,$$

$$\text{Relative error} = \frac{|\text{fl}(x) - x|}{|x|}, \quad x \ne 0.$$

Relative error is the most meaningful, as some examples in this chapter indicate. These types of error measurement apply to any calculation, not just measuring floating point error.

The analysis of roundoff error is complex, and we only do it for addition, presented in Theorem 8.1. Error bounds, as should be expected, involve eps.

Overflow and Underflow

Integer arithmetic can overflow, and the same is true for floating point arithmetic when the magnitude of a result exceeds the maximum allowable floating point number. In addition, underflow can occur, which means the magnitude of the result lies in the gap between 0 and the smallest floating point number.

Minimizing the Effects of Floating Point Error

Some computations are prone to floating point error, and should be replaced by an alternative. This chapter shows how to prevent overflow when computing

$$\|x\|_2 = \sqrt{x_1^2 + x_2^2 + \cdots + x_n^2}.$$

Another type of error is caused by cancellation. A classic example is evaluation of the quadratic formula. We will encounter more situations where we must be careful how we perform a computation.

8.6 PROBLEMS

8.1 Find the base 10 representation for each number.
   a. 1101101101 (base 2)
   b. 33671 (base 8)
   c. 8FB2 (base 16)
   d. 221341 (base 5)

8.2 Write each unsigned decimal number in binary, octal (base 8), and hexadecimal (base 16).
   a. 45
   b. 167
   c. 273
   d. 32763

8.3 Perform unsigned binary addition. Do not discard the carry.

a.	b.	c.	d.
11011	11110101	001111	10101010
+	+	+	+
11101	10001001	011111	10101010

8.4 Perform unsigned binary subtraction. Use borrowing.

a.	b.	c.	d.
11110	11110101	001110	11101110
−	−	−	−
11101	10001001	001101	10101011

8.5 Using $b = 2$ and $p = 8$, find the two's-complement integer representation for
   a. 25
   b. 127
   c. -1
   d. -127
   e. -37
   f. -101

8.6 If $b = 2$ and $p = 6$, indicate which two's-complement sums will cause overflow and compute the result in binary.
   a. $-9 + 30$
   b. $28 + 5$
   c. $-25 - 7$
   d. $-32 + 23$

8.7 Assume a two's-complement integer representation using 10 binary bits.

    a. What is the range of integers that can be represented?

    b. For each addition, determine if overflow occurs.

        i. $188 + 265$

        ii. $490 + 25$

        iii. $-400 + (-16)$

        iv. $-450 + (-70)$

8.8 What is the range of two's complement numbers if $b = 8$ and $p = 15$?

8.9 A computer that uses two's-complement arithmetic for integers has a subtract machine instruction. However, when the instruction executes the CPU does not have to perform the subtraction by borrowing or even worry about signs. Why?

8.10 Another integer representation system is one's-complement. If an integer is stored using p bits, the positive integers are ordinary binary values with the left-most bit set to 0. For instance, if $p = 8$, then 65 is stored as

$$0 \quad 1 \quad 0 \quad 0 \quad 0 \quad 0 \quad 0 \quad 1$$

Obtain the negative of a number by inverting bits; for instance, -65 is stored as

$$1 \quad 0 \quad 1 \quad 1 \quad 1 \quad 1 \quad 1 \quad 0$$

    a. Show how $1, -1, 25, 15, -21, 101, -120$ are stored in a one's-complement system with $b = 2, p = 8$.

    b. Show that there are two representations for 0 in a one's-complement system.

    c. Using base $b = 2$ and p digits, what is the range of one's-complement numbers?

8.11 Find the 32-bit single-precision representation for each number. Assume the format $\pm 0.d_1 d_2 \ldots d_{23} \times 2^e$, where $d_1 \neq 0$ unless the number is zero.

    a. 12.0625

    b. -18.1875

    c. 2.7

    Note: In binary $0.7 = 0.10110011001100110011001100...$

8.12 Assume you are performing decimal arithmetic with $p = 4$ significant digits. Using rounding, perform the following calculations:

    a. $26.8756 + 15.67883$

    b. $1.2567 * 14.77653$

    c. $12.98752 \times 10^3 * 23.47 \times 10^4$

8.13 Using four significant digits, compute the absolute and relative error for the following conversion to floating point form.

    a. $x_1 = 2.3566, \mathrm{fl}(x_1) = 2.357$

    b. $x_2 = 7.1434, \mathrm{fl}(x_2) = 7.143$

8.14 Verify the error bounds for addition and multiplication. Use $b = 10, p = 5$.

    a. $23.6643 + 45.6729 + 100.123$

    b. $8.25678*1.45729*5.35535$

8.15 Assume we are using IEEE double-precision arithmetic.

    a. What is the error bound in computing the sum of 20 positive floating point numbers?

    b. What is the error bound in computing the product of 20 floating point numbers?

8.16 If $b = 2, p = 12, e_{min} = 5, e_{max} = 8$, find f_{min} and f_{max}.

8.17 By constructing a counterexample, verify that floating-point addition and multiplication, $\oplus$ and $\otimes$, do not obey the distributive law $a \otimes (b \oplus c) = a \otimes b + a \otimes c$.

8.18 Show that, unlike the operation $+$, the operation $\oplus$ is not associative. In other words $(x \oplus y) \oplus z \neq x \oplus (y \oplus z)$ in general.

8.19 Let $b = 10$ and $p = 10$. Compute $\mathrm{fl}\left(A^2\right)$, where

$$A = \begin{bmatrix} 1 & 0 & 10^{-5} \\ 0 & 1 & 0 \\ 10^{-5} & 0 & 1 \end{bmatrix}.$$

Repeat your calculations with $p = 12$. Compare the results.

8.20 Assume you are using floating point arithmetic with $b = 10$, $p = 5$, and a maximum exponent of 8. Let x be the

$$\text{vector } x = \begin{bmatrix} 2500 \\ 6000 \\ 1000 \\ 8553 \end{bmatrix}.$$

a. What is f_{max}?

b. Directly compute $\|x\| = \sqrt{2500^2 + 6000^2 + 1000^2 + 8553^2}$.

c. Compute $\|x\|$ using the method discussed in Section 8.4.1 that is designed to prevent overflow.

8.21 For parts (a) and (b), use $b = 10, p = 4$.

a. What is the truncation error when approximating $\cos(0.5)$ by the first two terms of its McLaurin series?

b. Answer part (a) for $\tan(0.5)$.

8.22 There are problems evaluating $f(x) = \sqrt{x-1} - \sqrt{x}$ under some conditions. Give an example. Propose a means of accurately computing $f(x)$.

8.23 Using $b = 10, p = 4$, verify the bound

$$\text{fl}(cA) = cA + E, \quad |E| \le \text{eps}\,|cA|$$

for $c = 5.6797$ and

$$A = \begin{bmatrix} 2.34719 & -1.56219 & 5.89531 \\ -0.98431 & 23.764 & 102.35 \\ -77.543 & -0.87542 & 5.26743 \end{bmatrix}.$$

8.24 Find eps

a. for $b = 10, p = 8$.

b. for $b = 2, p = 128$.

8.25 Using floating point arithmetic, is it true that $a + b = a$ implies that $b = 0$? Justify your answer.

8.26 For parts (a) and (b), propose a method of computing each expression for small x. For part (c), propose a method for computing the expression for large x.

a. $\cos(x) - 1$

b. $\dfrac{\sin x - x}{x}$

c. $\dfrac{1}{x+1} - \dfrac{1}{x}$

8.27 Assume we have a floating point number system with base $b = 2$, p significant digits, minimum exponent e_{min} and maximum exponent e_{max}. A nonnegative number is either zero or of the form $0.1d_1 d_2 \ldots d_{p-1} \times 2^e$. Develop a formula for the number of nonnegative floating point values.

8.6.1 MATLAB Problems

8.28 Taking a double value x and developing MATLAB code that precisely rounds it to m significant digits is somewhat difficult. Consider the function

```
function y = roundtom(x,m)
%ROUNDTOM round to m significant decimal digits
%
%    Input: floating point number x.
%           number of decimal digits desirede
%    Output: x rounded to m decimal digits

pos = floor(log10(abs(x)))-m+1;
y = round(x/10^pos)*10^pos;
```

Explain how the code works and test it with several double values. It often works perfectly but can leave some trailing nonzero digits. For instance

```
>> n = 23.567927;
>> roundtom(n,5)

ans =
   23.568000000000001
```

Explain why this occurs.

8.29 It is possible to write a function, say `fmex`, in the programming language C in such a way that the function can be called from MATLAB. The function `fmex` must written so it conforms with what is termed the MEX interface and must be compiled using the MATLAB command `mex` (see Ref. [20]). This interface provides access to the input and output parameters when the function is called from MATLAB. The book software distribution contains the C program outputdouble.c in the subdirectory outputdouble of the software distribution as well as compiled versions. Since machine code is system dependent, there are multiple versions. The following table lists the names of the available compiled code for Windows, OS X, and Linux systems.

Windows 64-bit	OS X 64-bit	Linux-64 bit
outputdouble.mexw64	outputdouble.mexmaci64	outputdouble.mexa64

If you are using a 32-bit system, execute "mex outputdouble.c", and MATLAB should generate 32-bit code. On any system, the calling format is

```
>> oututdouble(x)
```

where x is a double variable or a constant. It prints the 64 binary bits of x in IEEE double format, marking the location of the sign bit, the exponent, and the mantissa. The binary bits represent a floating point number of the form $\pm 1.d_1 d_2 d_3 \ldots d_{52} \times 2^e$. The leading 1 is hidden; in other words it is considered present but is not stored, giving 53 bits for the mantissa. The exponent uses excess 1023 format.

a. Find the binary representation for each number, determine the exponent in decimal, and the mantissa in binary.
  i. 33
  ii. -35
  iii. -101
  iv. 0.000677
  v. 0

b. The MATLAB named constants `realmax` and `realmin` are the largest and smallest double values. Using `outputdouble`, determine each number in binary and then determine what each number is in decimal.

8.30 The infinite series $\sum_{n=1}^{\infty} (-1)^{n+1}/n$ converges to $\ln(2)$. MATLAB performs computations using IEEE double-precision floating point arithmetic. Sum the first 100,000 terms of the series using MATLAB and determine the truncation error by accepting as correct the MATLAB value `log(2)`. Explain your result.

8.31 Type the command "`format hex`" in MATLAB to see the hexadecimal representation of any number. For 64-bit double-precision floating-point numbers, the first three hexadecimal digits correspond to the sign bit followed by the 11 exponent bits. The mantissa is represented by the next 13 hexadecimal digits. We have seen that as the numbers grow larger, the gaps between numbers grows as well.

a. Find 2^{75} in hexadecimal.

b. Investigate the gap between 2^{75} and the next floating-point number by finding the first number of the form $u = 2^{75} + 2^i$ that has a different hexadecimal representation.

c. What is the result of the following MATLAB statements `x = 2^75; y = x + 2^(i-1); x == y`. i is the power found in part (b). Explain the result.

d. For $b = 2, p = 3, e_{\min} = -2, e_{\max} = 3$, draw the distribution of floating-point numbers as in Figure 8.1.

8.32 The matrix exponential e^A is defined by the McLaurin series

$$e^A = I + \frac{A}{1!} + \frac{A^2}{2!} + \frac{A^3}{3!} + \cdots + \frac{A^n}{n!} + \cdots$$

This computation can be quite difficult, as this problem illustrates.

a. Enter the function

```
function E = matexp(A)
% MATEXP Taylor series for exp(A)

E = zeros(size(A));
F = eye(size(A));
k = 1;
while norm(E+F-E,1) > 0
    E = E + F;
    F = A*F/k;
    k = k+1;
end
```

that estimates e^A using the McLaurin series.

b. Apply `matexp` to the matrix

$$A = \begin{bmatrix} 99 & -100 \\ 137 & -138 \end{bmatrix}$$

to obtain the matrix `A_McLaurin`.

c. `A_McLaurin` is far from the correct result. Compute the true value of e^A using the MATLAB function `expm` as follows:

```
>> A_true = expm(A);
```

The function uses the Padé approximation [21].

d. Explain why the result in part (b) is so far from the correct result. Use MATLAB help by typing

```
>> showdemo expmdemo
```

Remark 8.8.   The text by Laub [22, pp. 6-7] has a very interesting example dealing with the matrix exponential.

8.33 Write a MATLAB code segment that approximates eps. Start with epsest = 1.0 and halve epsest while 1.0+epsest > 1.0. Run it, and compare the result with the MATLAB's eps.

8.34 This problem was developed from material on MATLAB Central (http://www.mathworks.com/matlabcentral/), and is due to Loren Shure.

a. Describe the action of the MATLAB function `fix`.

b. Enter and run the following code

c.
```
format short
for ind = 0:.1:1
    f = fix(10*ind)/10;
    if ind ~= f
        disp(ind - f);
    end
end
```

d. Explain why there is output when, theoretically, there should be none.

8.35 This problem experiments with cancellation errors when solving the quadratic equation (Section 8.4.2). The equation for the problem is

$$0.0001x^2 + 10000x - 0.0001 = 0.$$

a. Find the smallest root in magnitude of the polynomial using the quadratic equation.

b. Use the MATLAB function `roots` to determine the same root.

c. Which value is most correct?

8.36 Using double-precision arithmetic, determine the largest value of x for which e^x does not overflow. Compute $|e^x - \text{realmax}|$ and $(|e^x - \text{realmax}|)/\text{realmax}$. Is your result acceptable? Why?

8.37 Enter and run the following MATLAB code.

```
x = 0.0;
y = exp(-x);
k = 0;
while y ~= 0.0
    x = x + 0.01;
    y = exp(-x);
    k = k + 1;
    if mod(k,500) == 0
        fprintf('x = %g   y = %.16e\n', x, y);
    end
end
```

Execute the MATLAB command `realmin` that determines the smallest positive normalized floating-point number. Does your output appear to contradict the value of `realmin`? If it does, use the MATLAB documentation to explain the results. Hint: What is an unnormalized floating point number?

Chapter 9

Algorithms

You should be familiar with

- The inner product
- The Euclidean vector norm
- The Frobenius matrix norm
- Matrix multiplication
- Truncation error
- Upper- and lower-triangular matrices
- Tridiagonal matrices

We have introduced some methods for solving problems in linear algebra; for instance, Gaussian elimination and the computation of eigenvalues using the characteristic equation. Each computation consisted of a series of steps leading to a solution of the problem. Such a series of steps is termed an *algorithm*. We will present algorithms of varying complexity throughout the remainder of this book.

Definition 9.1. Starting with input, if any, an *algorithm* is a set of instructions that describe a computation. When executed, the instructions eventually halt and may produce output.

To this point, our presentation of algorithms was done informally and supported by examples. Now we are beginning a rigorous presentation of algorithms in numerical linear algebra, and we need a more precise mechanism for describing how they work. A formal presentation will aid in understanding an algorithm and in implementing it in a programming language. In this book, you will use the MATLAB programming language or something similar, such as Octave, and perhaps you have also used C/C++, Java, or any of many other programming languages. Presenting an algorithm in MATLAB requires that we adhere to the strict syntax of the MATLAB programming language. We should be able to describe the instructions in a simple, less formal way, so they can be converted to statements in the MATLAB or any other programming language. We use *pseudocode* for this purpose. Pseudocode is a language for describing algorithms. It allows the algorithm designer to focus on the logic of the algorithm without being distracted by details of programming language syntax, such as variable declarations, the correct placement of semicolons and braces, and so forth. We provide pseudocode for all major algorithms and, in each case, there is a MATLAB implementation in the book software.

9.1 PSEUDOCODE EXAMPLES

Our pseudocode will use statements very similar to those of MATLAB such as

```
for i = 1:n do
    <statements>
end for
```

and

```
if abserr < tol
    <statements>
end if
```

Other pseudocode constructs include assignment statements, the while loop, a function, and so forth.

Numerical Linear Algebra with Applications. http://dx.doi.org/10.1016/B978-0-12-394435-1.00009-0

9.1.1   Inner Product of Two Vectors

As our first example, we present an algorithm for the computation of the inner product of $n \times 1$ vectors. A for loop forms the sum of the product of corresponding entries

---

Algorithm 9.1 Inner Product of Two Vectors

```
function INNERPROD(u,v)
   % Input: column vectors u and v
   % Output: ⟨u, v⟩
   inprod = 0.0
   for i = 1:n do
      inprod = inprod + u(i)v(i)
   end for
   return inprod
end function
```

9.1.2   Computing the Frobenius Norm

The algorithm for the computation of the inner product involves a single loop. The Frobenius norm requires that we cycle through all matrix entries, add their squares, and then take the square root. This involves an outer loop to traverse the rows and an inner loop that forms the sum of the squares of the entries of a row.

Algorithm 9.2 Frobenius Norm

```
function FROBENIUS(A)
   % Input: m × n matrix A.
   % Output: the Frobenius norm √(∑ᵢ₌₁ᵐ ∑ₖ₌₁ⁿ aᵢₖ²).
   fro = 0.0

   for i = 1:m do
      for j = 1:n do
         fro = fro + aᵢⱼ²
      end for
   end for
   return fro
end function
```

9.1.3   Matrix Multiplication

Matrix multiplication presents a more significant challenge. If A is an $m \times p$ matrix and B is a $p \times n$ matrix, the product is an $m \times n$ matrix whose elements are

$$c_{ij} = \sum_{k=1}^{p} a_{ik} b_{kj}.$$

Start with $i = 1$ and apply the formula for $j = 1, 2, \ldots n$. This gives the first row of the product. Follow this by letting $i = 2$ and applying the formula for $j = 1, 2, \ldots n$ to obtain the second row of the product. Continue in this fashion until computing the last row of AB. This requires three nested loops. The outer loop traverses the m rows of A. For each row i, another loop must cycle through the n columns of B. For each column, form the sum of the products of corresponding elements from row i of A and column j of B. (Figure 9.1).

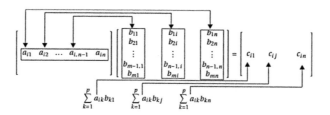

FIGURE 9.1 Matrix multiplication.

Algorithm 9.3 Product of Two Matrices

```
function MATMUL(A,B)
  % Input: m × p matrix A and p × m matrix B
  % Output A × B
  % for each row of A
  for i = 1:m do
    % for each column of B
    for j = 1:n do
      c(i, j) = 0
      % form the sum of the product of corresponding elements from row
      % i of A and column j of B
      for k = 1:p do
        c(i, j) = c(i, j) + a_{ik}b_{kj}
      end for
    end for
  end for

  return C
end function
```

9.1.4 Block Matrices

A block matrix is formed from sets of submatrices, and we briefly introduce the concept. In general, these matrices are useful for proving theorems and speeding up algorithms. We will use the idea only a few times in this book and refer the reader to Refs. [1, 2, 23] for an in-depth discussion.

We will confine the discussion to block matrices of order 2×2. Let

$$A = \begin{matrix} m_1 \\ m_2 \\ m = m_1 + m_2 \end{matrix} \begin{matrix} p_1 & p_2 \\ \begin{bmatrix} A_{11} & A_{12} \\ A_{21} & A_{22} \end{bmatrix} \\ p = p_1 + p_2 \end{matrix}$$

The submatrix block A_{11} has dimension $m_1 \times p_1$. In general A_{ij} has dimension $m_i \times p_j$.

Addition and scalar multiplication work as expected. If α is a scalar,

$$\alpha A = \begin{bmatrix} \alpha A_{11} & \alpha A_{12} \\ \alpha A_{21} & \alpha A_{22} \end{bmatrix}$$

and if $C = \begin{bmatrix} C_{11} & C_{12} \\ C_{21} & C_{22} \end{bmatrix}$ has the same dimensions as A,

$$A + C = \begin{bmatrix} A_{11} + C_{11} & A_{12} + C_{12} \\ A_{21} + C_{21} & A_{22} + C_{22} \end{bmatrix}.$$

Now we will discuss block matrix multiplication. Form matrix B as follows:

$$B = \begin{matrix} & \begin{matrix} n_1 & n_2 \end{matrix} \\ \begin{matrix} p_1 \\ p_2 \end{matrix} & \begin{bmatrix} B_{11} & B_{12} \\ B_{21} & B_{22} \end{bmatrix} \\ p = p_1 + p_2 & n = n_1 + n_2 \end{matrix}$$

To compute an ordinary matrix product AB, the number of columns of A must equal the number of rows in B. For block matrix multiplication, the number of columns of A is p, and the number of rows of B is p. Let's treat the blocks as individual scalar elements in an ordinary 2×2 matrix. Then,

$$AB = \begin{bmatrix} A_{11} & A_{12} \\ A_{21} & A_{22} \end{bmatrix} \begin{bmatrix} B_{11} & B_{12} \\ B_{21} & B_{22} \end{bmatrix} = \begin{bmatrix} A_{11}B_{11} + A_{12}B_{21} & A_{11}B_{12} + A_{12}B_{22} \\ A_{21}B_{11} + A_{22}B_{21} & A_{21}B_{12} + A_{22}B_{22} \end{bmatrix}.$$

Although we will not prove it, this is the product AB.

Example 9.1.   An example will clarify block matrix multiplication. Let

$$A = \left[\begin{array}{cc|cc} 1 & 4 & 6 & -1 \\ -1 & 3 & 1 & 4 \\ \hline 5 & 2 & -3 & 1 \end{array} \right] \qquad B = \left[\begin{array}{cc|c} 3 & 2 & 3 \\ 1 & 6 & 1 \\ \hline 4 & 1 & 3 \\ 2 & 0 & -1 \end{array} \right].$$

For these matrices $m_1 = 2, m_2 = 1, p_1 = 2, p_2 = 2, n_1 = 2,$ and $n_2 = 1,$ and

$$A_{11} = \begin{bmatrix} 1 & 4 \\ -1 & 3 \end{bmatrix}, \quad A_{12} = \begin{bmatrix} 6 & -1 \\ 1 & 4 \end{bmatrix}, \quad A_{21} = \begin{bmatrix} 5 & 2 \end{bmatrix}, \quad A_{22} = \begin{bmatrix} -3 & 1 \end{bmatrix}$$

$$B_{11} = \begin{bmatrix} 3 & 2 \\ 1 & 6 \end{bmatrix}, \quad B_{12} = \begin{bmatrix} 3 \\ 1 \end{bmatrix}, \quad B_{21} = \begin{bmatrix} 4 & 1 \\ 2 & 0 \end{bmatrix}, \quad B_{22} = \begin{bmatrix} 3 \\ -1 \end{bmatrix}.$$

Now,

$$AB = \left[\begin{array}{cc} \begin{bmatrix} 1 & 4 \\ -1 & 3 \end{bmatrix}\begin{bmatrix} 3 & 2 \\ 1 & 6 \end{bmatrix} + \begin{bmatrix} 6 & -1 \\ 1 & 4 \end{bmatrix}\begin{bmatrix} 4 & 1 \\ 2 & 0 \end{bmatrix} & \begin{bmatrix} 1 & 4 \\ -1 & 3 \end{bmatrix}\begin{bmatrix} 3 \\ 1 \end{bmatrix} + \begin{bmatrix} 6 & -1 \\ 1 & 4 \end{bmatrix}\begin{bmatrix} 3 \\ -1 \end{bmatrix} \\ \begin{bmatrix} 5 & 2 \end{bmatrix}\begin{bmatrix} 3 & 2 \\ 1 & 6 \end{bmatrix} + \begin{bmatrix} -3 & 1 \end{bmatrix}\begin{bmatrix} 4 & 1 \\ 2 & 0 \end{bmatrix} & \begin{bmatrix} 5 & 2 \end{bmatrix}\begin{bmatrix} 3 \\ 1 \end{bmatrix} + \begin{bmatrix} -3 & 1 \end{bmatrix}\begin{bmatrix} 3 \\ -1 \end{bmatrix} \end{array} \right]$$

$$= \left[\begin{array}{cc} \begin{bmatrix} 29 & 32 \\ 12 & 17 \\ 7 & 19 \end{bmatrix} & \begin{bmatrix} 26 \\ -1 \\ 7 \end{bmatrix} \end{array} \right]$$

If we ignore the blocks and consider A and B to have dimensions 3×4 and 4×3, respectively, the product is the 3×3 matrix

$$\begin{bmatrix} 29 & 32 & 26 \\ 12 & 17 & -1 \\ 7 & 19 & 7 \end{bmatrix}. \qquad \blacksquare$$

9.2   ALGORITHM EFFICIENCY

We all know that some algorithms take longer than others. Clearly, multiplying two $n \times n$ matrices takes longer than multiplying an $n \times 1$ column vector by a constant. If you have two algorithms that solve the same problem, one algorithm might be better than another under your current circumstances. For instance, consider the problem of computing eigenvalues. There are a number of algorithms; for instance, finding the roots of the characteristic polynomial, and using the power method (to be discussed later). What we need is a means of measuring the computational effort an algorithm requires. There are many factors that come into play, the number of floating-point arithmetic operations required, the amount of memory needed, the overhead of array subscripting, the maintenance of control variables in for loops, and so forth. Floating-point operations are slow compared to many other operations, so counting them exactly or approximately helps us compare algorithms.

Definition 9.2. A *flop* is a floating-point operation $\oplus$, $\ominus$, $\otimes$, and $\oslash$. The number of flops required to execute an algorithm is termed the *flop count* of the algorithm.

It is convenient to have a notation that gives us an idea of how much work the algorithm must perform, and we will use "Oh", or "Big-O" notation. For a floating point computation, we want to measure the number of flops required. Given an expression for the number of flops, we say the algorithm is $O\left(n^k\right)$ if the dominant term in the flop count is Cn^k, where C is a constant. We are saying that for large n, the other terms are negligible in comparison to n^k.

Example 9.2. If an algorithm requires $\frac{4}{3}n^3 + 9n^2 + 8n + 6$ flops, it is an $O\left(n^3\right)$ algorithm. When $n = 100$, the actual value of the expression is

$$V = \frac{4}{3}(100)^3 + 9(100)^2 + 8(100) + 6 = 1.424139 \times 10^6.$$

Discarding the lower order terms,

$$T = \frac{4}{3}(100)^3 = 1.333333 \times 10^6,$$

and $T/V = 0.9362$, so $\frac{4}{3}n^3$ contributes 93% of the expression's value. ∎

Suppose we want to form the sum of the components in a 3×1 vector. Use a variable sum, initialize sum to have value zero, and use a for loop as follows:

```
sum = 0
for i = 1:3 do
  sum = sum + x(i)
end for
```

The algorithm begins by adding $x(1)$ to $sum = 0$, so it executes three additions. However, looking at the sum abstractly as

$$x(1) + x(2) + x(3),$$

there are two additions. When counting flops, we will ignore the extra addition.

Example 9.3. Given two $n \times 1$ vectors u and v, the inner product $\langle u, v \rangle = u_1v_1 + u_2v_2 + \cdots + u_nv_n$ requires n multiplications and $n - 1$ additions, a total of $n + (n - 1) = 2n - 1$ flops. Computing the inner product is an $O(n)$ or a *linear algorithm*. ∎

Example 9.4. Computing the Frobenius norm requires one flop for each square (a_{ij}^2), and there are mn squares to compute, for a total of mn flops. The addition of the squares requires $mn - 1$ flops, so the flop count for the algorithm is $2mn - 1$, and computing the Frobenius norm is an $O(mn)$ algorithm. If the matrix A is square, the flop count is $O\left(n^2\right)$. We call this a quadratic algorithm. ∎

Example 9.5. You should consult Algorithm 9.3 as you read this flop count analysis. Consider the multiplication of an $m \times p$ matrix A by a $p \times n$ matrix B. The inner loop performs one multiplication and an addition, for a total of $2p$ flops per execution of the loop. This inner loop executes mn times, so the flop count is

$$2mnp. \tag{9.1}$$

This flop count is very useful. It tells you that matrix multiplication is an expensive operation; for instance, if $m = 10$, $n = 8, p = 12$, the multiplication requires $10(16)12 = 1920$ flops. Most of the matrices engineers and scientists deal with are $n \times n$, and matrix multiplication costs $2n^3$ flops. We say that square matrix multiplication is a *cubic algorithm*, or is $O\left(n^3\right)$. ∎

Throughout this book, we will do a detailed flop count analysis if it is instructional. In other cases, the flop count will be stated without proof.

9.2.1 Smaller Flop Count Is Not Always Better

In Chapter 14, we will introduce the QR decomposition of a matrix, which states that $A = QR$, where R is an upper-triangular matrix and Q has orthonormal columns. The decomposition is obtained using what is termed the Gram-Schmidt process. Chapter 17 presents two additional algorithms for finding the QR decomposition, using Givens rotations or Householder reflections. Although Gram-Schmidt has a lower flop count, both are preferable to Gram-Schmidt for a number of reasons that will be explained later. The flop count using Householder reflections for computing the QR decomposition of an $m \times n$ matrix, $m \geq n$, is

$$4 \left(m^2 n - mn^2 + \frac{n^3}{3} \right) + 2n^2 \left(m - \frac{n}{3} \right),$$

and the flop count using Givens rotations is

$$\frac{1}{2} (5 + 6n + 6n) \left(2mn - n^2 - n \right).$$

On the average, the Householder reflection method is superior in terms of flop count, but the Givens rotation method lends itself very well to parallelization. If you are using a machine with many cores, for instance, the Givens rotation method will likely be superior.

Another aspect that must be taken into account is memory requirements. An algorithm may have a smaller flop count but require much more memory. Depending on how much memory is on the system, an algorithm with a larger flop count but less memory use may run faster. There are many other things that influence speed. For instance, a better written implementation of algorithm A may run faster than algorithm B even if algorithm A has a larger flop count. For example, algorithm B may not reuse variables already allocated in memory, slowing it down.

Remark 9.1. We will use flop count as the primary means for comparing algorithms; in other words, we will ignore the evaluation of square roots, sine, cosine, etc. Different systems may implement these functions in different ways, some perhaps more efficient than others. However, the flop count will remain the same.

9.2.2 Measuring Truncation Error

When approximating a value using a finite sum of terms from a series, truncation error occurs.

Example 9.6. The McLaurin series for e^x is $e^x = \sum_{n=0}^{\infty} x^n/n!$. If x is small and we approximate e^x by

$$1 + x + \frac{x^2}{2},$$

we are leaving off $\frac{x^3}{3!} + \frac{x^4}{4!} + \cdots$, of which the largest term is $\frac{x^3}{3!}$, and we say the truncation error is $O\left(x^3\right)$. ∎

If you have studied numerical integration, you are familiar with fourth-order Runge-Kutta methods for estimating $\int_a^b f(x)\,dx$. If we divide an interval $a \leq x \leq b$ into n subintervals of length $h = (b - a)/n$ and apply a fourth-order Runge-Kutta method, the error is $O\left(h^4\right)$.

9.3 THE SOLUTION TO UPPER AND LOWER TRIANGULAR SYSTEMS

This section presents algorithms for solving upper- and lower-triangular systems of equations. In addition to providing additional algorithms for study, we will need to use both these algorithms throughout this book.

An *upper-triangular matrix* is an $n \times n$ matrix whose only nonzero entries are below the main diagonal; in other words

$$a_{ij} = 0, \quad j < i, \quad 1 \leq i, j \leq n.$$

If U is an $n \times n$ upper-triangular matrix, we know how to solve the linear system $Ux = b$ using back substitution. In fact, this is the final step in the Gaussian elimination algorithm that we discussed in Chapter 2. Compute the value of $x_n = b_n/u_{nn}$, and then insert this value into equation $(n - 1)$ to solve for x_{n-1}. Continue until you have found x_1. Algorithm 9.4 presents back substitution in pseudocode.

Algorithm 9.4 Solving an Upper Triangular System

```
function BACKSOLVE(U,b)
    % Find the solution to Ux = b, where U is an n × n upper-triangular matrix.
    xₙ = bₙ/uₙₙ
    for i = n-1:-1:1 do
        sum = 0.0
        for j = i+1:n do
            sum = sum + uᵢⱼxⱼ
        end for
        x(i) = (b(i) − sum)/uᵢᵢ
    end for
    return x
end function
```

NLALIB: The function `backsolve` implements Algorithm 9.4.

A *lower-triangular matrix* is a matrix all of whose elements above the main diagonal are 0; in other words

$$a_{ij} = 0, \quad j > i, \quad 1 \le i, j \le n.$$

A *lower-triangular system* is one with a lower-triangular coefficient matrix.

$$
\begin{bmatrix}
a_{11} & 0 & 0 & \cdots & 0 \\
a_{21} & a_{22} & 0 & \cdots & 0 \\
a_{31} & a_{32} & a_{33} & \cdots & 0 \\
\cdots & \cdots & \cdots & \ddots & 0 \\
a_{n1} & a_{n2} & \cdots & a_{n,n-1} & a_{nn}
\end{bmatrix}
\begin{bmatrix}
x_1 \\ x_2 \\ x_3 \\ \vdots \\ x_n
\end{bmatrix}
=
\begin{bmatrix}
b_1 \\ b_2 \\ b_3 \\ \vdots \\ b_n
\end{bmatrix}
$$

The solution to a lower-triangular system is just the reverse of the algorithm for solving an upper-triangular system—use *forward substitution*. Solve the first equation for $x_1 = \frac{b_1}{a_{11}}$, and insert this value into the second equation to find x_2, and so forth.

Example 9.7. Solve

$$
\begin{bmatrix}
2 & 0 & 0 \\
3 & 1 & 0 \\
1 & 4 & 5
\end{bmatrix}
\begin{bmatrix}
x_1 \\ x_2 \\ x_3
\end{bmatrix}
=
\begin{bmatrix}
2 \\ -1 \\ 8
\end{bmatrix}
$$

$x_1 = 2/2 = 1$
$3(1) + x_2 = -1, x_2 = -4$
$1(1) + 4(-4) + 5x_3 = 8, x_3 = 23/5$

SOLUTION: $x = \begin{bmatrix} 1 & -4 & 23/5 \end{bmatrix}^{\mathrm{T}}$. ■

Algorithm 9.5 Solving a Lower Triangular System

```
function FORSOLVE(L,b)
    % Find the solution to the system Lx = b, where L is an n × n lower-triangular matrix.
    x₁ = b₁/l₁₁
    for i = 2:n do
        sum = 0.0
        for j = 1:i-1 do
            sum = sum + lᵢⱼxⱼ
        end for
        x(i) = (b(i) − sum)/lᵢᵢ
    end for
    return x
end function
```

NLALIB: The function `forsolve` implements Algorithm 9.5.

Example 9.8.   Solve the systems

$$\begin{bmatrix} 1 & -1 & 3 \\ 0 & 2 & 9 \\ 0 & 0 & 1 \end{bmatrix} \begin{bmatrix} x_1 \\ x_2 \\ x_3 \end{bmatrix} = \begin{bmatrix} 1 \\ 9 \\ -2 \end{bmatrix}$$

and

$$\begin{bmatrix} 1 & 0 & 0 \\ -1 & 2 & 0 \\ 3 & 4 & 5 \end{bmatrix} \begin{bmatrix} y_1 \\ y_2 \\ y_3 \end{bmatrix} = \begin{bmatrix} 1 \\ 9 \\ -2 \end{bmatrix}.$$

```
>> U = [1 -1 3;0 2 9;0 0 1];
>> L = [1 0 0;-1 2 0;3 4 5];
>> b = [1 9 -2]';
>> x = backsolve(U,b)

x =
   20.5000
   13.5000
   -2.0000
>> U\b

ans =
   20.5000
   13.5000
   -2.0000
>> y = forsolve(L,b)

y =
    1
    5
   -5
>> L\b
ans =
    1
    5
   -5
```

∎

9.3.1 Efficiency Analysis

Algorithm 9.4 executes 1 division and then begins an outer loop having $n-1$ iterations. The inner loop executes $n-(i+1)+1 = n-i$ times, and each loop iteration performs 1 addition and 1 multiplication, for a total of $2(n-i)$ flops. After the inner loop finishes, 1 subtraction and 1 division execute. The total number of flops required is

$$\begin{aligned} 1 + \sum_{i=1}^{n-1} [2(n-i)+2] &= 1 + 2(n-1) + 2\sum_{i=1}^{n-1}(n-i) \\ &= 1 + 2(n-1) + 2[(n-1)+(n-2)+\cdots+1] \\ &= 1 + 2(n-1) + 2\left(\frac{n(n-1)}{2}\right) \\ &= n^2 + n - 1 \end{aligned}$$

Thus, back substitution is an $O(n^2)$ (quadratic) algorithm. It is left as an exercise to show that Algorithm 9.5 has exactly the same flop count.

9.4   THE THOMAS ALGORITHM

A tridiagonal matrix is square, and the only nonzero elements are those on the main diagonal, the first subdiagonal, and the first superdiagonal, as shown:

$$
\begin{bmatrix}
b_1 & c_1 & 0 & \dots & 0 & \dots & 0 \\
a_1 & b_2 & c_2 & & & & 0 \\
0 & a_2 & b_3 & c_3 & & \dots & \vdots \\
\vdots & & \ddots & \ddots & \ddots & & \vdots \\
0 & & & \ddots & \ddots & c_{n-2} & 0 \\
\vdots & & & & a_{n-2} & b_{n-1} & c_{n-1} \\
0 & \dots & 0 & \dots & 0 & a_{n-1} & b_n
\end{bmatrix}.
$$

A tridiagonal matrix is said to be *sparse*, since $n^2 - [n + 2(n-1)] = n^2 - 3n + 2$ entries are zero. Tridiagonal matrices are extremely important in applications; for instance, they occur in finite difference solutions to differential equations and in the computation of cubic splines.

A tridiagonal system $Ax = \text{rhs}$ can be solved by storing and using only the entries on the three diagonals

$$
\begin{aligned}
a &= \begin{bmatrix} a_1 & a_2 & \dots & a_{n-1} \end{bmatrix}^{\mathrm{T}}, \\
b &= \begin{bmatrix} b_1 & b_2 & \dots & b_{n-1} & b_n \end{bmatrix}^{\mathrm{T}}, \\
c &= \begin{bmatrix} c_1 & c_2 & \dots & c_{n-1} \end{bmatrix}^{\mathrm{T}}.
\end{aligned}
$$

The algorithm is similar to Gaussian elimination, in which the matrix is converted to upper-triangular form and then solved using back substitution, but the algorithm is much more efficient. For the purpose of explanation, we will display matrices. The first action is to divide the row 1 by b_1 to make the pivot in the first row and first column pivot 1. This gives

$$
\left[
\begin{array}{ccccccc|c}
1 & c_1' & 0 & 0 & \dots & \dots & 0 & \text{rhs}_1' \\
a_1 & b_2 & c_2 & & & & & \text{rhs}_2 \\
0 & a_2 & b_3 & c_3 & & & & \text{rhs}_3 \\
\vdots & & a_3 & \ddots & \ddots & & & \vdots \\
\vdots & \vdots & \vdots & \ddots & \ddots & \ddots & & \vdots \\
0 & & & & a_{n-2} & b_{n-1} & c_{n-1} & \text{rhs}_{n-1} \\
0 & 0 & 0 & \dots & 0 & a_{n-1} & b_n & \text{rhs}_n
\end{array}
\right],
$$

where $c_1' = c_1/b_1$ and $\text{rhs}_1' = \text{rhs}_1/b_1$.

Multiply row 1 by a_1 and subtract from row 2:

$$
\left[
\begin{array}{ccccccc|c}
1 & c_1' & 0 & 0 & \dots & \dots & 0 & \text{rhs}_1' \\
0 & b_2' & c_2 & 0 & \dots & \dots & & \text{rhs}_2' \\
0 & a_2 & b_3 & c_3 & & & & \text{rhs}_3 \\
\vdots & & a_3 & \ddots & \ddots & & & \vdots \\
\vdots & \vdots & \vdots & \ddots & \ddots & \ddots & & \vdots \\
0 & & & & a_{n-2} & b_{n-1} & c_{n-1} & \text{rhs}_{n-1} \\
0 & 0 & 0 & \dots & 0 & a_{n-1} & b_n & \text{rhs}_n
\end{array}
\right],
$$

where $b_2' = b_2 - a_1 c_1'$ and $\text{rhs}_2' = \text{rhs}_2 - a_1 \text{rhs}_1'$.

The steps we have performed define a process that will end in an upper-triangular matrix. To see this, continue by dividing row 2 by b_2', multiplying row 2 by a_2 and subtracting from row 3 to eliminate a_2 in row 3:

$$
\begin{bmatrix}
1 & c_1' & 0 & 0 & \cdots & \cdots & 0 & \Big| & \mathrm{rhs}_1' \\
0 & 1 & c_2' & 0 & \cdots & \cdots & & \Big| & \mathrm{rhs}_2'' \\
0 & 0 & b_3' & c_3 & & & & \Big| & \mathrm{rhs}_3' \\
\vdots & & a_3 & \ddots & \ddots & & & \Big| & \vdots \\
\vdots & \vdots & \vdots & \ddots & \ddots & \ddots & & \Big| & \vdots \\
0 & & & & a_{n-2} & b_{n-1} & c_{n-1} & \Big| & \mathrm{rhs}_{n-1} \\
0 & 0 & 0 & \cdots & 0 & a_{n-1} & b_n & \Big| & \mathrm{rhs}_n
\end{bmatrix},
$$

where

$$c_2' = c_2/b_2' = c_2/\left(b_2 - a_1 c_1'\right),$$

$$\mathrm{rhs}_2'' = \mathrm{rhs}_2'/b_2' = \left(\mathrm{rhs}_2 - a_1\mathrm{rhs}_1'\right)/\left(b_2 - a_1 c_1'\right),$$

and

$$b_3' = b_3 - a_2 c_2',$$

$$\mathrm{rhs}_3' = \mathrm{rhs}_3 - a_2\mathrm{rhs}_2''.$$

Note that b_3' and rhs_3' will be used to compute c_3' and rhs_3'' in the next elimination step. Continue the process row by row until the matrix is in upper-triangular form with ones on its diagonal:

$$
\begin{bmatrix}
1 & c_1' & 0 & 0 & \cdots & \cdots & 0 & \Big| & \mathrm{rhs}_1' \\
0 & 1 & c_2' & 0 & \cdots & \cdots & & \Big| & \mathrm{rhs}_2'' \\
0 & 0 & 1 & c_3' & & & & \Big| & \mathrm{rhs}_3'' \\
\vdots & & 0 & \ddots & \ddots & & & \Big| & \vdots \\
\vdots & \vdots & \vdots & \ddots & \ddots & \ddots & & \Big| & \vdots \\
0 & & & & 0 & 1 & c_{n-1}' & \Big| & \mathrm{rhs}_{n-1}'' \\
0 & 0 & 0 & \cdots & 0 & 0 & 1 & \Big| & \mathrm{rhs}_n''
\end{bmatrix}
$$

The resulting matrix is an *upper bidiagonal matrix*. It has nonzero entries only on the main diagonal and the one above. Back substitution is very fast for such a matrix, since the determination of x_i requires using only the computed value of x_{i+1} and rhs_i. Back substitution gives

$$x_n = \mathrm{rhs}_n'',$$

$$x_{n-1} = \mathrm{rhs}_{n-1}'' - c_{n-1}'x_n,$$

$$\vdots$$

$$x_1 = \mathrm{rhs}_1' - c_1'x_2.$$

Algorithm 9.5 formalizes the process.

The algorithm does involve division, so the Thomas algorithm can fail as a result of division by zero. A condition called *diagonal dominance* will guarantee that the algorithm will never encounter a zero divisor.

Algorithm 9.6 The Thomas Algorithm

```
1:    function THOMAS(a,b,c,rhs)
2:       % The function solves a tridiagonal system of linear equations Ax = rhs
3:       % using the linear Thomas algorithm. a is the lower diagonal, b the
4:       % diagonal, and c the upper diagonal.
5:
6:       % Begin elimination steps, resulting in a bidiagonal matrix
7:       % with 1s on its diagonal.
8:          c₁ = c₁/b₁
9:          rhs₁ = rhs₁/b₁
10:         for i = 2:n-1 do
11:            cᵢ = cᵢ/(bᵢ − aᵢ₋₁cᵢ₋₁)
12:            rhsᵢ = (rhsᵢ − aᵢ₋₁rhsᵢ₋₁)/(bᵢ − aᵢ₋₁cᵢ₋₁)
13:         end for
14:         rhsₙ = (rhsₙ − aₙ₋₁rhsₙ₋₁)/(bₙ − aₙ₋₁cₙ₋₁)
15:         % Now perform back substitution
16:         xₙ = rhsₙ
17:         for i = n-1:-1:1 do
18:            xᵢ = rhsᵢ − cᵢxᵢ₊₁
19:         end for
20:         return x
21:   end function
```

NLALIB: The function `thomas` implements Algorithm 9.6.

Definition 9.3. A square matrix is diagonally dominant if the absolute value of each diagonal element is greater than the sum of the absolute values of the other elements in its row, or

$$|a_{ii}| > \sum_{\substack{j=1 \\ j \neq i}}^{n} |a_{ij}|.$$

For instance, the tridiagonal matrix

$$A = \begin{bmatrix} 2 & -1 & 0 & 0 \\ 1 & 3 & -1 & 0 \\ 0 & 5 & -7 & 1 \\ 0 & 0 & 3 & 8 \end{bmatrix}$$

is diagonally dominant. This condition is easy to check and often occurs in problems.

Theorem 9.1. *If A is a diagonally dominant tridiagonal matrix with diagonals a, b, and c, the Thomas algorithm never encounters a division by zero.*

Remark 9.2.   If a matrix is not diagonally dominant, the Thomas algorithm may work. Diagonal dominance is only a sufficient condition.

9.4.1   Efficiency Analysis

To determine the flop count for the Thomas algorithm, note that lines 8 and 9 account for two divisions. The for loop beginning at line 10 executes $n - 2$ times, and each execution involves 2 divisions, 3 subtractions, and 3 multiplications, for a total of $8(n - 2)$ flops. The statement at line 14 involves 2 subtractions, 1 division, and 2 multiplications, a total of 5 flops. The for loop beginning at line 17 executes $n - 1$ times, and each execution performs 1 subtraction and 1 multiplication, for a total of $2(n - 1)$ flops. The total flop count is then

$$2 + 8(n - 2) + 5 + 2(n - 1) = 10n - 11.$$

The Thomas algorithm is linear ($O(n)$). As we will see in Chapter 11, the Gaussian elimination algorithm for a general $n \times n$ matrix requires approximately $\frac{2}{3}n^3$ flops. It is not uncommon when using finite difference methods for the solution of partial differential equations that tridiagonal systems of order 500×500 or higher must be solved. Standard Gaussian elimination will not take advantage of the sparsity of the tridiagonal system and will require approximately $\frac{2}{3}(500)^3 = 83333333$ flops. Using the Thomas algorithm requires $10(500) - 11 = 4989$ flops, quite a savings!

Example 9.9. Diagonals $a^{4999 \times 1}$, $b^{5000 \times 1}$, $c^{4999 \times 1}$, and right-hand side rhs$^{5000 \times 1}$ are generated randomly, and the example times the execution of function `thomas` when solving the 5000×5000 tridiagonal system formed from these vectors. The function `trid` in this book software distribution builds an $n \times n$ tridiagonal matrix from diagonals a, b, and c. The example computes the time required to solve the system using the MATLAB $' \backslash '$ operator. Again, we see the advantages of designing an algorithm that takes advantage of matrix structure.

```
>> a = randn(4999,1);
>> b =randn(5000,1);
>> c = randn(4999,1);
>> rhs = randn(5000,1);
>> tic;x1 = thomas(a,b,c,rhs);toc;
Elapsed time is 0.032754 seconds.
>> T = trid(a,b,c);
>> tic;x2 = T\rhs;toc;
Elapsed time is 0.386797 seconds.
```

■

9.5 CHAPTER SUMMARY

Stating an Algorithm Using Pseudocode

Starting with input, if any, an algorithm is a set of instructions that describe a computation. When executed, the instructions eventually halt and may produce output. This chapter begins a rigorous presentation of algorithms in numerical linear algebra using pseudocode. Pseudocode is a language for describing algorithms that allows the algorithm designer to focus on the logic of the algorithm without being distracted by details of programming language syntax. We provide pseudocode for all major algorithms and, in each case, there is a MATLAB implementation in the book software.

Algorithms are presented for computing the inner product, the Frobenius norm, and matrix multiplication. We also discuss block matrix formulation and operations with block matrices including multiplication. Using block matrices often simplifies the discussion of an algorithm.

Algorithm Efficiency

We measure the efficiency of an algorithm by explicitly counting or estimating the number of flops (floating point operations) it requires. Suppose an algorithm requires $n^3 + n^2 + 6n + 8$ flops. Using Big-O notation, we say it is an $O(n^3)$ algorithm, meaning that the dominant term is n^3. As n increases, the n^3 term accounts for almost all the value; for instance, if $n = 250$, $n^2 + 6n + 8 = 64,008$, and $n^3 = 1.5625 \times 10^7$. There is a mathematical description of this and similar notation for expression algorithm efficiency (see Ref. [24, pp. 52-61]). As examples, the inner product is an $O(n)$ algorithm, and computing the Frobenius norm is $O(n^2)$. Matrix multiplication is an interesting example. Multiplying an $m \times p$ by a $p \times n$ matrix requires $2mnp$ flops. If the matrix is $n \times n$, then the product requires $2n^3$ flops, and is an $O(n^3)$ algorithm. If two matrices are 500×500, their product requires 2.5×10^8 flops. Fortunately, when matrices in applications become that large, they are usually sparse, meaning there is a low percentage of nonzero entries. There are algorithms for rapid multiplication of sparse matrices, and we will deal with sparse matrices in Chapters 21 and 22.

It is possible that a lower flop count may not be better. This can occur when an algorithm with a higher flop count can be parallelized, but one with a lower flop count cannot. An algorithm with a lower flop count may require excessive amounts of memory and, as a result, perform more slowly.

We will have occasion to approximate a function by terms of a series. For instance, the McLaurin series for $\sin x$ is

$$\sin x = x - \frac{x^3}{3!} + \frac{x^5}{5!} - \frac{x^7}{7!} + \cdots$$

For small x, if we use $x - x^3/3!$ as an approximation, then the truncation error is $O(x^5)$. In order to approximate the solution to differential equations, we will use finite difference equations. If h is small, then

$$\frac{d^2 f}{dx^2} \approx \frac{f(x+h) - 2f(x) + f(x-h)}{h^2}$$

has a truncation error $O(h^2)$.

Solving Upper- and Lower-Triangular Systems

We have studied back substitution that solves a matrix equation of the form

$$\begin{bmatrix} a_{11} & a_{12} & \cdots & \cdots & a_{1n} \\ 0 & a_{22} & \cdots & \cdots & a_{2n} \\ 0 & 0 & \ddots & & \\ \vdots & \vdots & \ddots & \ddots & \\ 0 & 0 & \cdots & 0 & a_{nn} \end{bmatrix} x = b.$$

This is an $O(n^2)$ algorithm. If the matrix is in lower-triangular form, we use forward substitution, and it requires the same number of flops.

The Thomas Algorithm for Solving a Tridiagonal Linear System

Tridiagonal systems of equations occur often. When we approximate the solution to the one-dimensional heat equation in Chapter 12 and develop cubic splines in the same chapter, the solution involves solving a tridiagonal system. The Thomas algorithm is based on a clever use of Gaussian elimination and yields a solution in $O(n)$ flops. Solving a general linear system using Gaussian elimination requires approximately $\frac{2}{3}n^3$ flops, so the savings in using the Thomas algorithm is huge!

9.6   PROBLEMS

Whenever a problem asks for the development of an algorithm, do a flop count.

9.1   What is the flop count for each code segment?

a.
```
sum = 0.0;
for i = 1:n
    sum = sum + n²;
end
```

b.
```
x = 0.0:.01:2*pi;
n = length(x);
sum = 0.0;
for i = 1:length
    sum = sum + 1/(x(i)² + x(i) + 1);
end
```

c.
```
A = rand(5,8);
B = rand(8,6);
C = rand(6,12);
D = A*B*C;
```

d.
```
x = rand(10,1);
y = rand((10,1);
z = x*y';
```

9.2   Give the flop count for each matrix operation.
a. Multiplication of $m \times n$ matrix A by an $n \times 1$ vector x.
b. The product xy^T if x is an $m \times 1$ vector and y is a $p \times 1$ vector.
c. If u and v are $n \times 1$ vectors, the computation of $(\langle v, u \rangle / \|u\|^2) u$.
d. $\|A\|_\infty$ for $m \times n$ matrix A.

e. $\|A\|_1$ for $m \times n$ matrix A.

f. trace (A), where A is an $n \times n$ matrix.

9.3 What is the action of the following algorithm?

```
function PROBLEM(u,v)
   sum1 = 0.0
   sum2 = 0.0
   for i = 1:n do
      sum1 = sum1 + uᵢvᵢ
      sum2 = sum2 + uᵢ²
   end for
   k = sum1/sum2
   for i = 1:n do
      uᵢ = kuᵢ
   end for
   return u
end function
```

9.4 Determine the action of the following algorithm, and find the number of comparisons if the algorithm returns *true*. There is a name attached to this type of matrix. Determine what it is.

```
function ASYM(A)

   for i = 1:n do
      for j = i+1:n do
         if aᵢⱼ ≠ −aⱼᵢ then
            return false
         end if
      end for
   end for
   return true
end function
```

9.5 Let u be an $m \times 1$ column vector $\begin{bmatrix} u_1 & u_2 & \ldots & u_m \end{bmatrix}^{\mathrm{T}}$ and v be a $1 \times n$ row vector $\begin{bmatrix} v_1 & v_2 & \ldots & v_n \end{bmatrix}$. The *tensor product* of u and v, written $u \otimes v$, is the $m \times n$ matrix

$$
\begin{bmatrix}
u_1v_1 & u_1v_2 & \ldots & u_1v_n \\
u_2v_1 & u_2v_2 & \ldots & u_2v_n \\
\ldots & \ldots & \ldots & \ldots \\
u_mv_1 & u_mv_2 & \ldots & u_mv_n
\end{bmatrix}
$$

Write an algorithm, tensorprod, for the computation of $u \otimes v$.

9.6 Show that matrix multiplication can be implemented using tensor products as defined in Problem 9.5, and write an algorithm that does it.

9.7 What is the action of the following algorithm? Determine its flop count.

```
function PROBLEM(A,B)
   for i = 1:n do
      for j = i:n do
         sum = 0.0;
         for k = i:j do
            sum = sum + a(i, k) b(k, j)
         end for
         P(i, j) = sum
      end for
   end for
   return P
end function
```

Hint: To help determine its action, trace the algorithm using 3×3 matrices. These formulas are useful when determining the flop count:

$$\sum_{i=1}^{n} i = \frac{n(n+1)}{2},$$

$$\sum_{i=1}^{n} i^2 = \frac{n(n+1)(2n+1)}{6}.$$

9.8 Assume that the operation $A. * B$, where A and B are $n \times n$ matrices multiplies corresponding entries to form a new matrix. For instance,

$$\begin{bmatrix} 1 & 2 \\ 3 & 5 \end{bmatrix} . * \begin{bmatrix} 4 & 3 \\ -1 & 2 \end{bmatrix} = \begin{bmatrix} 4 & 6 \\ -3 & 10 \end{bmatrix}.$$

Write an algorithm to form $A. * B$.

9.9 Definition 6.4 within the problems of Chapter 6 defines the cross product of two vectors. Develop pseudocode for a function crossprod that computes the cross product of vectors u and v.

9.10 Write an efficient algorithm, addsym, that forms the sum of two $n \times n$ symmetric matrices.

9.11 Section 9.4 defined a tridiagonal matrix. Develop an algorithm trimul that forms the product of two $n \times n$ tridiagonal matrices.

9.12 An upper bidiagonal matrix is a matrix with a main diagonal and one upper diagonal:

$$\begin{bmatrix} a_{11} & a_{12} & & & 0 \\ & a_{22} & a_{23} & & \\ & & \ddots & \ddots & \\ & & & a_{n-1,n-1} & a_{n-1,n} \\ 0 & & & & a_{nn} \end{bmatrix}$$

Develop an algorithm, bisolve, to solve a system of equations $Ax = b$ that uses only the nonzero elements.

9.13 Develop an algorithm lowtrimul that forms the product of two lower triangular matrices.

9.14 Show that the flop count for Algorithm 9.5 is $n^2 + n - 1$.

9.15 Compute the product of the two block matrices.

$$A = \begin{bmatrix} \begin{bmatrix} 1 \\ 2 \\ 1 \end{bmatrix} & \begin{bmatrix} 1 & 0 & 1 \\ 5 & -1 & 3 \\ 2 & 7 & 5 \end{bmatrix} \\ \begin{bmatrix} 3 \\ -1 \end{bmatrix} & \begin{bmatrix} 1 & 1 & 0 \\ 2 & 1 & -1 \end{bmatrix} \end{bmatrix}, \quad B = \begin{bmatrix} \begin{bmatrix} 1 & 5 \end{bmatrix} & \begin{bmatrix} -1 & 5 & 7 \end{bmatrix} \\ \begin{bmatrix} 1 & -1 \\ 0 & 2 \\ 1 & 1 \end{bmatrix} & \begin{bmatrix} 1 & 0 & 2 \\ 1 & 1 & 1 \\ 4 & 0 & -1 \end{bmatrix} \end{bmatrix}.$$

9.6.1 MATLAB Problems

9.16

a. Implement the function matmul specified by Algorithm 9.3.

b. Run the following code, and explain the results by listing all the factors you can think of that determine the speed of matrix multiplication.

```
>> A = rand(1000,1000);
>> B = rand(1000,1000);
>> tic;C = A*B;toc;
>> tic;C = matmul(A,B);toc;
```

When a problem requires you to write a MATLAB function, always thoroughly test it with a variety of data. Some problems prescribe test data.

9.17 Implement Problem 9.4 in the MATLAB function asym, test it on the following matrices, and create one test case of your own.

$$A = \begin{bmatrix} 1 & -2 & 5 & 8 & 9 \\ 2 & -3 & 12 & 4 & 16 \\ -5 & -12 & 5 & -18 & 2 \\ -8 & -4 & 18 & 7 & 1 \\ -9 & -16 & -2 & -1 & 2 \end{bmatrix},$$

$$
B = \begin{bmatrix} 1 & 2 & 3 & -9 \\ -2 & 3 & 1 & 6 \\ -3 & -1 & -2 & 4 \\ 9 & 6 & -4 & 5 \end{bmatrix}.
$$

9.18 Implement Problem 9.9 with a MATLAB function `crossprod`. After implementation, create vectors u and v, and compute $u \times v$, $v \times u$, $\langle u \times v, u \rangle$, and $\langle v \times u, v \rangle$.

9.19 This problem deals with both upper- and lower-triangular matrix multiplication.

    a. Write the algorithm in Problem 9.7 as a MATLAB function, `updtrimul` and test it with random matrices of size 3×3, 8×8, and 25×25. For your test, compute $\|\text{updtrimul}\,(A,\ B) - A * B\|_2$. NOTE: The following statements generate a random, integer, upper-triangular matrix whose entries are in the range $-100 \le a_{ij} \le 100$:

```
>> n = value;
>> A = randi([-100 100],n,n);
>> A = triu(A);
```

    b. Implement the function `lowtrimul` from Problem 9.13 and test it as in part (a).

9.20 Write a MATLAB function, `addsym`, that implements the algorithm of Problem 9.10 and test it with matrices

```
A = rand(8,8);
A = A + A';
B = rand(8,8);
B = B + B';
```

9.21 Write a MATLAB function `bisolve` that implements the algorithm of Problem 9.12. The following statements generate a random 5×5 bidiagonal matrix and a random right-hand side. Use these statements as your first test of `bisolve`. Modify the statements and test your function with a 25×25 bidiagonal matrix.

```
>> d = rand(5,1);
>> ud = rand(4,1);
>> A = diag(d) + diag(ud,1);
>> b = rand(5,1);
```

9.22 Problem 9.5 defines the tensor product of two vectors. Implement a function, `tensorprod`, that computes the tensor product of an $m \times 1$ column vector, u, with a $1 \times n$ row vector, v.

9.23 Write a function, tenmatprd, that implements the multiplication of matrices using the tensor product.

9.24 Write a function, `trimul`, that computes the product of two tridiagonal matrices. Test your function using the following statements:

```
>> n = value;
>> a = rand(n-1,1);
>> b = rand(n,1);
>> c = rand(n-1,1);
>> A = trid(a,b,c)
```

9.25 For n odd, consider the "X-matrix"

$$
C = \begin{bmatrix}
a_1 & & & & & & b_n \\
 & a_2 & & & & \cdot^{\cdot^{\cdot}} & \\
 & & \ddots & & b_{k+1} & & \\
 & & & a_k & & & \\
 & & \cdot^{\cdot^{\cdot}} & & \ddots & & \\
 & b_2 & & & & a_{n-1} & \\
b_1 & & & & & & a_n
\end{bmatrix},
$$

where the diagonals contain no zero values.

a. The center element is $a_k = b_k$. Find a formula for k.

b. Develop a function x = xmatsolve(a,b,rhs) that solves the system $Cx = b$. The function must only use the nonzero values $\{a_i\}, \{b_i\}$. Output an error message and terminate under the following conditions:

- n is even.
- a and b do not both have n elements.
- The diagonals do not share a common center.

c. Develop a function X = buildxmat(a,b) that builds an X-matrix. Let $n = length\,(a)$. Generate an error message if n is even or $length\,(b) \neq n$.

d. Develop a function testxmatsolve(a,b,rhs) that times the execution of x = xmatsolve(a,b,rhs), builds the matrix X = buildxmat(a,b), and times the execution of the MATLAB command X\rhs.

e. Let a = randn(5001,1); b = randn(5001,1); and call buildxmat. Comment on the results.

9.26 Develop a MATLAB function

```
C = blockmul(A,B,m1,m2,p1,p2,n1,n2)
```

that computes the product of the block matrices A and B and returns the result into block matrix C. The scalars m1, m2, p1, p2, n1, and n2 are as described in Section 9.1.4. Test your function using the block matrices of Example 9.1 and Problem 9.15.

Chapter 10

Conditioning of Problems and Stability of Algorithms

You should be familiar with

- Solution of a linear system $Ax = b$ using Gaussian elimination
- Matrix inverse
- Vector and matrix norms
- Eigenvalues
- Singular values

This chapter begins with the question "Why do we need numerical linear algebra?" and answers it by presenting six examples, and there will be more later in this book. Some algorithms can be very sensitive to their input, meaning that the algorithm can produce poor results with perfectly good data. These algorithms are unstable. On the other hand, a stable algorithm can have input that cause the output to be poor. Such input is termed ill-conditioned. The chapter presents concrete examples of these issues and develops tools that are useful in detecting and dealing with them.

10.1 WHY DO WE NEED NUMERICAL LINEAR ALGEBRA?

We have studied fundamental concepts in linear algebra. We know how to solve linear systems using Gaussian elimination, how to find eigenvalues and eigenvectors, and that orthogonal matrices are important. We are familiar with subspaces, linear independence, and matrix rank. So why is it necessary for us to study numerical linear algebra, and what is it? This chapter provides some answers. It is not adequate to run an algorithm and accept the results, and we know this from our study of floating point arithmetic in Chapter 8. A computer does not perform exact floating point arithmetic, and this can cause even a time-honored algorithm to produce bad results. A good example is using the quadratic equation in its standard form and suffering very serious cancellation errors. Chapter 9 introduces the concept of an algorithm and algorithm efficiency. Suppose you have a problem to solve that involves matrices, and a computer must be used to obtain a solution. You may be faced with a choice from among a number of competing algorithms. In this case, you must consider efficiency, one aspect of which is flop count. Understanding issues of algorithm efficiency is one aspect of numerical linear algebra that sets it aside from theoretical linear algebra. We will see that certain algorithms are more prone to bad behavior from roundoff error than others, and we must avoid using them. Also, an algorithm that normally is very effective may not give good results for certain data, and this leads to the subject of conditioning, particularly as involves matrices. In short, numerical linear algebra is the study of how to accurately and efficiently solve linear algebra problems on a computer. Here are some classic examples that illustrate the issues.

a. *Using Gaussian elimination to solve a nonsingular $n \times n$ system $Ax = b$.* Chapter 2 discusses Gaussian elimination. During the process, if a 0 is encountered in the pivot position, a row exchange solved the problem. As we will see in Chapter 11, Gaussian elimination can perform very poorly unless we incorporate row exchange into the algorithm so that the pivot a_{ii} is the element of largest absolute value among the elements $\{a_{ki}\}, k \leq i \leq n$.

b. *Dealing with $m \times n$ systems, $m \neq n$.* A theoretical linear algebra course shows that systems $Ax = b$, where x is an $n \times 1$ vector and b is $m \times 1$, have an infinite number of solutions or none. This is done by transforming A to what is called reduced row echelon form. In numerical linear algebra, systems such as these arise in least-squares problems. Under the right conditions, there is a unique solution satisfying the requirement that $\|b - Ax\|_2$ is a minimum. Very seldom does $Ax = b$.

c. *Solving a linear algebraic system using Cramer's Rule.* We presented Cramer's Rule in Theorem 4.6 and mentioned that is was intended primarily for theoretical purposes. In practice, it is frequently necessary to solve square systems of

Numerical Linear Algebra with Applications. http://dx.doi.org/10.1016/B978-0-12-394435-1.00010-7

size greater than or equal to 50×50. Using Cramer's Rule for a 50×50 matrix involves computing 51 determinants of 50×50 matrices. If we use expansion by minors for each determinant, it will require evaluating $51\,(50!) \simeq 1.6 \times 10^{66}$ permutations to solve the system. Each permutation requires 49 multiplications. Assume a supercomputer can execute 10^{15} flops/s. The number of seconds required to do the multiplications is

$$\frac{51\,(50!)\,49}{10^{15}}\text{s} \simeq 2.4 \times 10^{45}$$

years!

d. *Computing the solution to a linear system $Ax = b$ by first finding A^{-1} and then computing $x = A^{-1}b$.* Computing A^{-1} takes more operations than using Gaussian elimination if you are solving one system $Ax = b$. But, is it effective to find A^{-1} if the solution to multiple systems $Ax_i = b_i, 1 \le i \le k$ is required? The solutions are $x_i = A^{-1}b_i$, so only k matrix-vector products need be computed. It can be shown that computing A^{-1} to solve the problem requires four (4) times as many flops as using Gaussian elimination to factor A into a product of a lower- and an upper-triangular matrix and then using forward and back substitution for each of the k systems. It is also the case that some inverses are very hard to compute accurately, and using A^{-1} gives very poor results.

e. *Computing the eigenvalues of a matrix by finding the roots of its characteristic polynomial.* There are long-standing algorithms for finding the roots of a polynomial, but remember that the coefficients of the characteristic polynomial will likely be corrupted by roundoff error. Section 10.3.1 demonstrates that even a slight change to one or more coefficients of a polynomial can cause large changes in its roots. If such a polynomial is the characteristic polynomial for the matrix, the eigenvalue computation can be disastrous. This method for computing eigenvalues should not be used.

f. *Finding the singular values of a matrix A by computing the eigenvalues of $A^T A$.* In computing the singular values by finding the eigenvalues of $A^T A$, errors introduced by matrix multiplication, followed by errors in computing the eigenvalues may be significant. We will show in Chapter 15 that the rank of a matrix is equal to the number of its nonzero singular values. Consider the example in Ref. [25]

$$A = \begin{bmatrix} 1 & 1 \\ \mu & 0 \\ 0 & \mu \end{bmatrix},$$

where $\mu < \sqrt{\text{eps}}$, so $1 + \mu^2 < 1 + \text{eps}$, and thus $\text{fl}\left(1 + \mu^2\right) = 1$. As a result,

$$\text{fl}\left(A^T A\right) = \begin{bmatrix} 1 & 1 \\ 1 & 1 \end{bmatrix},$$

its singular values are $\sigma_1 = \sqrt{2}$, $\sigma_2 = 0$, and the computed rank of A is 1. Using exact arithmetic,

$$A^T A = \begin{bmatrix} 1 + \mu^2 & 1 \\ 1 & 1 + \mu^2 \end{bmatrix},$$

and the singular values of A are

$$\sigma_1 = \sqrt{2 + \mu^2}, \quad \sigma_2 = |\mu|.$$

The true rank of A is 2. By computing $A^T A$, the term μ^2 entered the computations, giving rise to the roundoff error $\text{fl}\left(1 + \mu^2\right) = 1$. MATLAB does floating point arithmetic using 64-bits. The following MATLAB statements demonstrate the problem.

```
>> mu = sqrt(eps);
>> A = [1 1;mu 0;0 mu];
>> rank(A)
ans =
    2
```

```
>> B = A'*A;
>> eig(B)

ans =
    0.000000000000000
    2.000000000000000

>> rank(B)

ans =
    1
```

10.2 COMPUTATION ERROR

Our aim is to define criteria that help us decide what algorithm to use for a particular problem, or when the data for a problem may cause computational problems. In order to do this, we need to understand the two types of errors, forward error and backward error. Consider the problem to be solved by an algorithm as a function f mapping the input data x to the solution $y = f(x)$. However, due to inaccuracies during floating point computation, the computed result is $\hat{y} = \hat{f}(x)$. Before defining the types of errors and providing examples, it is necessary to introduce a new notation.

Definition 10.1. If A is an $m \times n$ matrix, $|A| = (|a_{ij}|)$; in other words, $|A|$ is the matrix consisting of the absolute values of a_{ij}, $1 \le i \le m$, $1 \le j \le n$.

Example 10.1.
a. If

$$A = \begin{bmatrix} -1 & 2 & 1.8 \\ -0.45 & 1.23 & 14.5 \\ -1.89 & 0.0 & -12.45 \end{bmatrix},$$

$$|A| = \begin{bmatrix} 1 & 2 & 1.8 \\ 0.45 & 1.23 & 14.5 \\ 1.89 & 0.0 & 12.45 \end{bmatrix}.$$

b. Let $x = \begin{bmatrix} -1 \\ 3 \\ -4 \end{bmatrix}$, $y = \begin{bmatrix} -8 \\ -1 \\ 2 \end{bmatrix}$. Then,

$$|x|^{\mathrm{T}}|y| = \begin{bmatrix} 1 & 3 & 4 \end{bmatrix} \begin{bmatrix} 8 \\ 1 \\ 2 \end{bmatrix} = 19.$$

∎

10.2.1 Forward Error

Forward error deals with rounding errors in the solution of a problem.

Definition 10.2. The *forward error* in computing $f(x)$ is $\left| \hat{f}(x) - f(x) \right|$. This measures errors in computation for input x.

The forward error is a natural quantity to measure but, in general, we do not know the true value $f(x)$ so we can only get an upper bound on this error.

Example 10.2.   The *outer product* of vectors x and y in $\mathbb{R}^n$ is the $n \times n$ matrix

$$A = xy^{\mathrm{T}} = \begin{bmatrix} x_1y_1 & x_1y_2 & \cdots & x_1y_n \\ x_2y_1 & x_2y_2 & \cdots & x_2y_n \\ \vdots & \vdots & \vdots & \vdots \\ x_ny_1 & x_ny_2 & \cdots & x_ny_n \end{bmatrix}. \tag{10.1}$$

Each entry of the computed result, $\hat{a}_{ij} = x_i y_j$ satisfies

$$\hat{a}_{ij} = x_i y_j \left(1 + \epsilon_{ij}\right), \quad \left|\epsilon_{ij}\right| \leq \text{eps},$$

and so

$$\hat{A} = \text{fl}\left(xy^{\mathrm{T}}\right) = xy^{\mathrm{T}} + \Delta,$$

where

$$\Delta = \begin{bmatrix} x_1y_1\epsilon_{11} & x_1y_2\epsilon_{12} & & x_1y_n\epsilon_{1n} \\ x_2y_1\epsilon_{21} & x_2y_2\epsilon_{22} & & x_2y_n\epsilon_{2n} \\ \vdots & \vdots & \vdots & \vdots \\ x_ny_1\epsilon_{n1} & x_ny_2\epsilon_{n2} & \cdots & x_ny_n\epsilon_{nn} \end{bmatrix}.$$

Since $\left|\epsilon_{ij}\right| \leq \text{eps}$, it follows that

$$\left|\Delta\right| \leq \text{eps} \left|xy^{\mathrm{T}}\right|.$$

This tells us that we can compute the outer product with forward error Δ, and that the error is bounded by $\text{eps} \left|xy^{\mathrm{T}}\right|$. ∎

As another example, consider the important problem of computing the inner product of vectors, u, v.

Example 10.3.   If x and y are $n \times 1$ vectors and $n \times \text{eps} \leq 0.01$, then,

$$\left|\text{fl}\left(x^{\mathrm{T}}y\right) - x^{\mathrm{T}}y\right| \leq 1.01\, n\, \text{eps}\, |x|^{\mathrm{T}}\, |y|.$$

The result shows that the forward error for the inner product is small. For a proof see Ref. [2, p. 99]. ∎

10.2.2 Backward Error

Backward error relates the rounding errors in the computation to the errors in the data rather than its solution. Generally, a backward error analysis is preferable to a forward analysis for this reason.

Consider the addition of n floating point numbers. In Section 8.3.2, we determined that

$$s_n = x_1 \left(1 + \eta_1\right) + x_2 \left(1 + \eta_2\right) + x_3 \left(1 + \eta_3\right) + \cdots + x_{n-1} \left(1 + \eta_{n-1}\right) + x_n \left(1 + \eta_n\right). \tag{10.2}$$

This says that s_n is the exact sum of the data perturbed by small errors, η_i. In other words, the result shows how errors in the data affect errors in the result.

Definition 10.3.   Roundoff or other errors in the data have produced the result $\hat{y}$. The *backward error* is the smallest Δx for which $\hat{y} = f(x + \Delta x)$; in other words, backward error tells us what problem we actually solved.

Figure 10.1 is an adaptation of Figure 1.1 from Ref. [16, p. 7] and illustrates the concepts of forward and backward error.

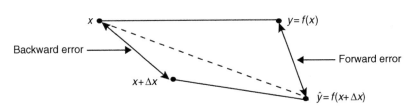

FIGURE 10.1 Forward and backward errors.

Example 10.4.   The McLaurin series for $f(x) = 1/(1-x)$ is

$$\frac{1}{1-x} = 1 + x + x^2 + \cdots = \sum_{n=0}^{\infty} x^n, \quad |x| < 1.$$

For $x = 0.57000$, approximate $f(x)$ by

$$\hat{f}(x) = 1 + x + x^2 + \cdots + x^7.$$

Then,

$$f(0.57000) = 2.3256, \quad \hat{f}(0.5700) = 2.2997.$$

The forward error is $|2.2997 - 2.3256| = 0.025900$. To find the backward error, we must find $\hat{x}$ such that

$$\frac{1}{1-\hat{x}} = 2.2997.$$

Solving for $\hat{x}$, we obtain $\hat{x} = 0.56516$, and the backward error is $|\hat{x} - x| = 0.0048400$.   ■

For another example of a backward error analysis, consider the computation of $\langle u, v \rangle$, where u and v are $n \times 1$ vectors. A proof of the following theorem can be found in Ref. [16, pp. 62-63].

Theorem 10.1.   *In the computation of the inner product, $\langle u, v \rangle$, where u and v are $n \times 1$ vectors,*

$$\text{fl}(\langle u, v \rangle) = x_1 y_1 (1 + \eta_n) + x_2 y_2 (1 + \eta_n') + x_3 y_3 (1 + \eta_{n-1}) + \cdots + x_n y_n (1 + \eta_2),$$

where $|\eta_i| \leq \frac{n \, \text{eps}}{1 - n \, \text{eps}}$ is very small.

This backward error result has an interpretation as follows:

The computed inner product is the exact inner product for a perturbed set of data $x_1, x_2, \ldots, x_n, y_1 (1 + \eta_n)$, $y_2 (1 + \eta_n'), \ldots, y_n (1 + \eta_2)$.

10.3   ALGORITHM STABILITY

Now that we are familiar with backward and forward error analysis, we are in a position to define algorithm stability. Intuitively, an algorithm is stable if it performs well in general, and an algorithm is unstable if it performs badly in significant cases. In particular, an algorithm should not be unduly sensitive to errors in its input or errors during its execution. In Section 8.4.2, we saw that using the quadratic equation in its classical form

$$x = \frac{-b \pm \sqrt{b^2 - 4ac}}{2a}$$

can produce poor results when $\sqrt{b^2 - 4ac} \approx b$. This is an *unstable algorithm*. Section 8.4.2 provided another algorithm that produces satisfactory results with the same values of a, b, and c. We will see in Chapter 11 that Gaussian elimination as we know it is unstable, but when a technique known as partial pivoting is added, the algorithm is stable in all but pathological cases. Given the same data, an unstable algorithm may produce poor results, while another, stable, algorithm produces good results. There are two types of stable algorithms, backward stable and forward stable.

Definition 10.4.   An algorithm is *backward stable* if for any x, it computes $\hat{f}(x)$ with small backward error, Δx. In other words, it computes the exact solution to a nearby problem,

$$f(x + \Delta x) = \hat{f}(x),$$

so that the solution is not sensitive to small perturbations in x.

By virtue of Equation 10.2, the addition of floating point numbers is backward stable, and by Example 10.3 so is the inner product of two vectors.

We use the process of back substitution as the final step of Gaussian elimination, so we need to know its stability properties. A proof that back substitution is backward stable can be found in Ref. [26, pp. 122-127].

Theorem 10.2. *Let back substitution be applied to the system, whose entries are floating point numbers. For any matrix norm, the computed solution $\hat{x}$ satisfies*

$$(R + \delta R)\,\hat{x} = b$$

for some upper-triangular matrix δR with

$$\frac{\|\delta R\|}{\|R\|} = O\,(\text{eps})\,.$$

Specifically, for each i, j,

$$\frac{|\delta r_{ij}|}{|r_{ij}|} \leq n\,\text{eps} + O\left(\text{eps}^2\right).$$

Definition 10.5. An algorithm is *forward stable* if whenever $f(x)$ is the true solution, the difference between the computed and true solutions is small. In other words,

$$\left|\hat{f}(x) - f(x)\right|$$

is small.

We have seen that the inner and outer product of two vectors is forward stable.

We know from our discussion in Chapter 8 that floating-point arithmetic does not follow the laws of real arithmetic. This can make forward error analysis difficult. In backward error analysis, however, real arithmetic is employed, since it is assumed that the computed result is the exact solution to a nearby problem. This is one reason why backward error analysis is often preferred, so we will refer to *stability* to mean backward stability. We have seen that floating point addition and inner product are stable, but we should provide examples of unstable algorithms.

10.3.1 Examples of Unstable Algorithms

The matrix xy^{T} has rank 1 (see Problem 10.3). For the computation of the outer product to be backward stable,

$$\hat{A} = (x + \Delta x)\,(y + \Delta y)^{\text{T}}\,,$$

must have rank 1. However,

$$\hat{A} = xy^{\text{T}} + \Delta,$$

and Δ in general does not have rank 1. The outer product is not backward stable, leading to the remark 10.1.

Remark 10.1. It is possible for an algorithm to be forward stable but not backward stable. In other words, it is possible for $\left|\hat{f}(x) - f(x)\right|$ to be small but the perturbation in the data, Δx, may be large.

We commented that one should not compute eigenvalues by finding the roots of the characteristic equation. Another example of an unstable algorithm is polynomial root finding.

Example 10.5. The matrix

$$A = \begin{bmatrix} 2 & 5 & 0 \\ 0 & 2 & 0 \\ -1 & 3 & 2 \end{bmatrix}$$

has characteristic polynomial

$$p\,(\lambda) = \lambda^3 - 6\lambda^2 + 12\lambda - 8 = (\lambda - 2)^3$$

with three equal roots. Suppose roundoff error causes the coefficient of λ^2 to become 5.99999, a perturbation of 10^{-5}. The roots of the characteristic equation are now

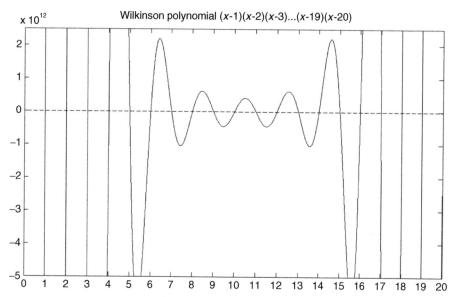

FIGURE 10.2 The Wilkinson polynomial.

```
2.016901481104199 + 0.029955270323774i
2.016901481104199 - 0.029955270323774i
1.966187037791599
```
■

Equal roots or roots very close together normally make the root finding problem unstable. The problem is actually more serious. If the roots are not close together, the root finding problem can still be unstable. A very clever example of this is the *Wilkinson polynomial*

$$p(x) = (x-1)(x-2)(x-3)\ldots(x-20).$$

Of course, all the roots are distinct and separated by 1. Expanded, $p(x)$ is

$$x^{20} - 210x^{19} + 20,615x^{18} - 1,256,850x^{17} + 53,327,946x^{16} - 1,672,280,820x^{15}$$
$$+ \cdots - 8,752,948,036,761,600,000x + 2,432,902,008,176,640,000.$$

Figure 10.2 shows a graph of the Wilkinson polynomial. Wilkinson found that when -210, the coefficient of x^{19}, was perturbed to $-210 - 2^{-23}$ the roots at $x = 16$ and $x = 17$ became approximately $16.73 \pm 2.81i$, quite a change for a perturbation of $2^{-23} \approx 1.192 \times 10^{-7}$!

This discussion definitely verifies that computing eigenvalues by finding the roots of the characteristic equation is a bad idea. Round-off errors when computing the coefficients of the characteristic polynomial may cause large errors in the determination of the roots.

10.4 CONDITIONING OF A PROBLEM

Even when using a stable algorithm to solve a problem, the problem may be sensitive to small changes (perturbations) in the data. The perturbations can come from roundoff error, small measurement errors when collecting experimental data, noise that is not filtered out of a signal, or truncation error when approximating the sum of an infinite series. If a problem falls in this category, we say it is *ill-conditioned*.

Definition 10.6.   A problem is *ill-conditioned* if a small relative change in its data can cause a large relative error in its computed solution, regardless of the algorithm used to solve the problem. If small perturbations in problem data lead to small relative errors in the solution, a problem is said to be *well-conditioned*.

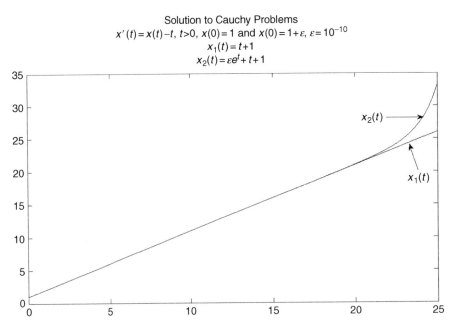

FIGURE 10.3 Ill-conditioned Cauchy problem.

Example 10.6.   The initial-value problem

$$\frac{dx}{dt} = x(t) - t, \quad t > 0, \quad x(0) = 1$$

is an example of a *Cauchy problem*. The unique solution to this problem is

$$x_1(t) = t + 1.$$

Now perturb the initial condition by a small amount ϵ, to obtain the Cauchy problem

$$\frac{dx}{dt} = x(t) - t, \quad t > 0, \quad x(0) = 1 + \epsilon.$$

The unique solution to this problem is

$$x_2(t) = \epsilon e^t + t + 1,$$

as is easily verified. Figure 10.3 is a graph of both solutions over the interval $0 \le t \le 25$ with $\epsilon = 10^{-10}$. A very small change in the initial condition created a very different solution for $x > 20$. This Cauchy problem is ill-conditioned. The L^2 function norm will give us the relative error over the interval:

$$\sqrt{\frac{\int_0^{25} (x_1(t) - x_2(t))^2 \, dt}{\int_0^{25} x_1(t)^2 \, dt}} = \sqrt{\frac{3}{2} (10^{-10})^2 \left(\frac{e^{50} - 1}{26^3 - 1} \right)} = 0.06652,$$

a very poor result.   ■

It is easy to confuse the concepts of stability and conditioning. Remark 10.2 provides a summary of the differences.

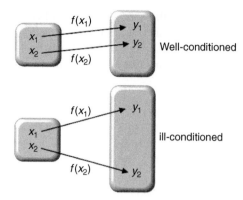

FIGURE 10.4 Conditioning of a problem.

Remark 10.2.   Summary

- Stable or unstable refers to an algorithm.
- Well or ill-conditioned refers to the particular problem, not the algorithm used.

Clearly, mixing roundoff error with an unstable algorithm is asking for disaster.

As mentioned in the introduction this section, a stable algorithm for the solution of a problem can produce poor results if the data are ill-conditioned. Assume that $f(x)$ represent an algorithm that takes data x and produces a solution $f(x)$. Here are some examples:

- In Example 10.6, f is an initial-value problem solver, and $x(0)$ is the initial condition.
- f is Gaussian elimination applied to solve a linear algebraic system $Ax = b$. The data are A and b.
- f is a function that takes a real number x and returns a real number y.

We can define ill and well-conditioning in terms of $f(x)$. Use the notation $\|\cdot\|$ to indicate a measure of size, such as the absolute value of a real number or a vector or matrix norm.

Definition 10.7.   Let x and $\bar{x}$ be the original and slightly perturbed data, and let $f(x)$ and $f(\bar{x})$ be the respective solutions. Then,

The problem is well-conditioned with respect to x if whenever $|x - \bar{x}|$ is small, $|f(x) - f(\bar{x})|$ is small.
The problem is ill-conditioned with respect to x if whenever $|x - \bar{x}|$ is small, $|f(x) - f(\bar{x})|$ can be large (Figure 10.4).

The sensitivity of a problem to data perturbations is measured by defining the *condition number*. The larger the condition number, the more sensitive a problem is to changes in data. For a particular x, assume there is a small error in the data so that the input to the problem is $\bar{x} = x + \Delta x$, and the computed value is $f(\bar{x})$ instead of $f(x)$. Form the relative error of the result divided by the relative error of the input:

$$\frac{\dfrac{|f(x) - f(\bar{x})|}{|f(x)|}}{\dfrac{|x - \bar{x}|}{|x|}} \tag{10.3}$$

The ratio measures how sensitive a function is to changes or errors in the input.

Remark 10.3.   In mathematics, the *supremum* (sup) of a subset S of an ordered set X is the least element of X that is greater than or equal to all elements of S. It differs from the maximum, in that it does not have to be a member of subset S.

This leads us to a mathematical definition of the condition number for a problem f.

Definition 10.8.   The condition number of problem f with input x is

$$C_f(x) = \lim_{\epsilon \to 0^+} \sup_{\|\delta x\| \le \epsilon} \frac{\dfrac{\|f(x + \delta x) - f(x)\|}{\|f(x)\|}}{\dfrac{\|\delta x\|}{\|x\|}}.$$

It is sufficient to think of the condition number as the limiting behavior of Equation 10.3 as the error δx becomes small.

As a first example of computing a condition number, let $f(x)$ be a function from $\mathbb{R}$ to $\mathbb{R}$. The question is whether evaluating the function is well-conditioned. The function $f(x) = x^2$ is clearly well-conditioned, but $f(x) = (1/(1-x))$ has serious problems near $x = 1$. In the general case, assume that f is differentiable so that the mean value theorem, proved in any calculus text, applies to f.

Theorem 10.3. *(Mean value theorem) If a function $f(x)$ is continuous on the closed interval $a \le x \le b$ and differentiable on the open interval $a < x < b$, then there exists a point ξ in $a < x < b$ such that*

$$f'(\xi) = \frac{f(b) - f(a)}{b - a}.$$

Fix x and let $\overline{x} = x + \delta x$. Applying the mean value theorem

$$\frac{\frac{|f(x) - f(x+\delta x)|}{|f(x)|}}{\frac{|\delta x|}{|x|}} = \frac{|x|}{|\delta x|} \frac{|f(x) - f(x+\delta x)|}{|f(x)|} = \frac{|x|}{|f(x)|} \frac{|f(x) - f(x+\delta x)|}{|\delta x|} = |x| \frac{|f'(\xi)|}{|f(x)|},$$

where ξ is between x and $x + \delta x$. As a result, $C_f(x) = |x| \, |f'(x)| / |f(x)|$.

Example 10.7. The example finds the condition number for three functions

a. $f(x) = x^2$, $C_f(x) = |x| \, |2x| / |x^2| = 2$, so f is well-conditioned.
b. $f(x) = 1/(1-x)$, $C_f(x) = |x| \, |1/(1-x)^2| / |1/(1-x)| = |x/(1-x)|$. $f(x)$ is ill-conditioned near $x = 1$ and well-conditioned everywhere else.
c. $f(x) = e^x$, $C_f(x) = |x| \, |e^x/e^x| = |x|$. $f(x)$ is ill-conditioned large x.

Parts (b) and (c) indicate that whenever x is within a certain range, a small relative error in x can cause a large relative error in the computation of $f(x)$. ∎

10.5 PERTURBATION ANALYSIS FOR SOLVING A LINEAR SYSTEM

For $A = \begin{bmatrix} 1.0001 & 1 \\ 1 & 1 \end{bmatrix}$, the problem $Ax = b$ is ill-conditioned. A good reason to suspect this is that $\det(A) = 0.0001$, so the matrix is almost singular. You will note that the exact solution to the following system is $x_1 = 1$ and $x_2 = 1$.

$$\begin{bmatrix} 1.0001 & 1 \\ 1 & 1 \end{bmatrix} \begin{bmatrix} x_1 \\ x_2 \end{bmatrix} = \begin{bmatrix} 2.0001 \\ 2 \end{bmatrix}$$

Now replace the right-hand side value 2.0001 by 2, and solve the system

$$\begin{bmatrix} 1.0001 & 1 \\ 1 & 1 \end{bmatrix} \begin{bmatrix} x_1 \\ x_2 \end{bmatrix} = \begin{bmatrix} 2 \\ 2 \end{bmatrix}.$$

The result is $x_1 = 0.0000$, $x_2 = 2.0000$. A very small change in the right-hand side caused a very large change in the solution.

In the remainder of this section, we will study the effect on the solution x if the elements of the linear system $Ax = b$ are slightly perturbed. This can occur in three ways:

a. One or more elements of b are perturbed, but the elements of A are exact.
b. One or more entries in A are perturbed, but the elements of b are exact.
c. There are perturbations in both A and b.

Theorem 10.4 specifies bounds for the errors involved in each case. If the reader does not desire to read through the proof, carefully look at Equations 10.4–10.6, and note the presence of the factor $\|A\| \, \|A^{-1}\|$. Our definition of the very important matrix condition is motivated by these results.

Remark 10.4. Recall that if $\|\cdot\|$ is a subordinate matrix norm, then $\|Ax\| \le \|A\| \, \|x\|$ (Equation 7.9)

Theorem 10.4. *Assume A is a nonsingular matrix, $b \ne 0$ is a vector, x is the solution to the system $Ax = b$, and $\|\cdot\|$ is a subordinate norm.*

1. *If $x + \delta x$ is the solution to the perturbed system $A (x + \delta x) = b + \delta b$, then*

$$\frac{\|\delta x\|}{\|x\|} \leq \|A\| \left\|A^{-1}\right\| \frac{\|\delta b\|}{\|b\|}. \tag{10.4}$$

2. *If $x + \delta x$ is the solution to the perturbed system $(A + \delta A) (x + \delta x) = b$, then*

$$\frac{\|\delta x\|}{\|x + \delta x\|} \leq \|A\| \left\|A^{-1}\right\| \frac{\|\delta A\|}{\|A\|}. \tag{10.5}$$

3. *If $x + \delta x$ is the solution to the perturbed system $(A + \delta A) (x + \delta x) = b + \delta b$, then*

$$\frac{\|\delta x\|}{\|x\|} \leq \|A\| \left\|A^{-1}\right\| \left(\frac{\|\delta A\|}{\|A\|} + \frac{\|\delta b\|}{\|A\| \, \|x + \delta x\|} \right). \tag{10.6}$$

Proof. To prove part 1, note that $A (x + \delta x) = Ax + A (\delta x) = b + A (\delta x) = b + \delta b$, so

$$A (\delta x) = \delta b.$$

It follows that $\delta x = A^{-1} (\delta b)$, so

$$\|\delta x\| \leq \left\|A^{-1}\right\| \|\delta b\|. \tag{10.7}$$

$Ax = b$, and thus

$$\|b\| \leq \|A\| \, \|x\|. \tag{10.8}$$

Multiply inequalities 10.7 and 10.8 together, and $\|\delta x\| \, \|b\| \leq \|A\| \left\|A^{-1}\right\| \|\delta b\| \, \|x\|$, from which we obtain the result

$$\frac{\|\delta x\|}{\|x\|} \leq \|A\| \left\|A^{-1}\right\| \frac{\|\delta b\|}{\|b\|}.$$

For part 2, $(A + \delta A) (x + \delta x) = Ax + A (\delta x) + \delta A (x + \delta x) = b$, and $b = Ax$, so

$$A (\delta x) + \delta A (x + \delta x) = 0. \tag{10.9}$$

Now, multiply Equation 10.9 by A^{-1} to obtain

$$\delta x = -A^{-1} \delta A (x + \delta x),$$

and so $\|\delta x\| \leq \left\|A^{-1}\right\| \|\delta A\| \, \|x + \delta x\|$, and

$$\frac{\|\delta x\|}{\|x + \delta x\|} \leq \left\|A^{-1}\right\| \|\delta A\|. \tag{10.10}$$

Multiply the right-hand side of Equation 10.10 by $\|A\| / \|A\|$ to obtain

$$\frac{\|\delta x\|}{\|x + \delta x\|} \leq \|A\| \left\|A^{-1}\right\| \left(\frac{\|\delta A\|}{\|A\|} \right).$$

For part 3, since $Ax = b$ and $(A + \delta A) (x + \delta x) = b + \delta b$ it follows that

$$A (\delta x) + \delta A (x + \delta x) = \delta b. \tag{10.11}$$

Multiply Equation 10.11 by A^{-1} to get

$$\delta x = A^{-1} (\delta b - \delta A (x + \delta x)),$$

so

$$\|\delta x\| \leq \left\|A^{-1}\right\| \|\delta b - \delta A (x + \delta x)\|. \tag{10.12}$$

Apply the triangle inequality as well as Equation 7.9 to $\|\delta b - \delta A (x + \delta x)\|$, and Equation 10.12 becomes

$$\|\delta x\| \leq \left\|A^{-1}\right\| (\|\delta b\| + \|\delta A\| \, \|x + \delta x\|). \tag{10.13}$$

Divide both sides of Equation 10.13 by $\|x + \delta x\|$ and multiply the right-hand side by $\|A\| / \|A\|$, and we obtain the required result

$$\frac{\|\delta x\|}{\|x + \delta x\|} \le \|A\| \, \left\|A^{-1}\right\| \left(\frac{\|\delta A\|}{\|A\|} + \frac{\|\delta b\|}{\|A\| \, \|x + \delta x\|} \right). \qquad \square$$

A question that naturally arises is whether these bounds are realistic. Is the upper bound far larger than any possible value for the actual relative error? As it turns out, the three inequalities are what is termed *optimal*.

Definition 10.9.   An upper bound in an inequality is *optimal* if there exist parameters for the inequality that attain the upper bound. In other words, if *expr* depends on a number of parameters and *expr* $\le$ *upperBound*, then there are parameters such that *expr* = *upperBound*.

It can be shown that each of the inequalities Equations 10.4–10.6 is optimal [27, pp. 80-81]. For instance, for any matrix A, there exists a perturbation δA, a right-hand side b and a vector x such that $\|\delta x\| / \|x + \delta x\| = \|A\| \, \left\|A^{-1}\right\| \, \|\delta A\| / \|A\|$.

Remark 10.5.   Even though the inequalities in Theorem 10.2 are optimal, they are *pessimistic*. This means that under most circumstances the upper bound is considerably larger than the relative error [27, p. 82].

Notice that Equations 10.4–10.6 involve the factor $\|A\| \, \left\|A^{-1}\right\|$, which has particular significance. It is the condition number of matrix A.

Definition 10.10.   The number $\|A\| \, \left\|A^{-1}\right\|$ is called the *condition number* of A and we denote it by $\kappa(A)$.

Theorem 10.4 says that relative change in the solution is bounded above by the product of $\kappa(A)$ and another factor that will be small if $\|\delta A\|$ and $\|\delta b\|$ are small. If $\kappa(A)$ is small, the relative change in the solution will be small, but if $\kappa(A)$ is large, then even small changes in A or b might drastically change the solution.

Remark 10.6.   Any matrix norm can be used to compute a condition number. We will assume that $\kappa(A)$ refers to the condition number relative to the 2-norm. The notation $\kappa_\infty(A)$, $\kappa_1(A)$, and $\kappa_F(A)$ refer to the ∞-, 1-, and Frobenius-norms, respectively.

In the following examples, $\|\cdot\|$ will refer to the 2-norm.

Example 10.8.   Let $A = \begin{bmatrix} 1 & 3 & 8 \\ -1 & 2 & 6 \\ 2 & -1 & 7 \end{bmatrix}$, $b = \begin{bmatrix} 1 \\ 1 \\ 1 \end{bmatrix}$. The solution to $Ax = b$ is $x = \begin{bmatrix} -0.1320755 \\ -0.075471698113208 \\ 0.169811320754717 \end{bmatrix}$. Let

$\delta b = \begin{bmatrix} 0.0001 \\ -0.0001 \\ 0.0005 \end{bmatrix}$, so $b + \delta b = \begin{bmatrix} 1.0001 \\ 0.9999 \\ 1.0005 \end{bmatrix}$. The solution to the perturbed system is $x + \delta x = \begin{bmatrix} -0.131964150943396 \\ -0.075550943396226 \\ 0.169839622641509 \end{bmatrix}$.

The relative perturbation is $\|\delta b\| / \|b\| = 3.0000 \times 10^{-4}$, which is quite small. The relative error of the solution is $\|\delta x\| / \|x\| = 6.1209 \times 10^{-4}$. The condition number of the matrix A is 9.6978, so good behavior should be expected.

Verify inequality 10.4:

$$\frac{\|\delta x\|}{\|x\|} = 6.1209 \times 10^{-4} \le \|A\| \, \left\|A^{-1}\right\| \frac{\|\delta b\|}{\|b\|} = 9.6978 \left(3.0000 \times 10^{-4} \right) = 2.9093 \times 10^{-3} \qquad \blacksquare$$

Example 10.9.   Let A be the matrix obtained in MATLAB using the command $A = $ gallery(3).

$$A = \begin{bmatrix} -149 & -50 & -154 \\ 537 & 180 & 546 \\ -27 & -9 & -25 \end{bmatrix}$$

If $b = \begin{bmatrix} 1 \\ 2 \\ 3 \end{bmatrix}$, the solution is $x = \begin{bmatrix} 324.3333 \\ -1035.8333 \\ 22.5000 \end{bmatrix}$. Perturb A by $\delta A = \begin{bmatrix} 0.00001 & 0 & -0.00001 \\ 0 & 0 & 0 \\ 0 & 0.00003 & 0 \end{bmatrix}$ so that $A + \delta A =$

$\begin{bmatrix} -148.99999 & -50 & -154.00001 \\ 537 & 180 & 546 \\ -27 & -8.99997 & -25 \end{bmatrix}$. The solution to the perturbed system is

$$x + \delta x = \begin{bmatrix} 326.31236 \\ -1042.17016 \\ 22.64266 \end{bmatrix},$$

which is very different from the solution to the unperturbed system relative to the small change in A. This is not surprising, since $\kappa(A) = 275848.6$.

Verify inequality 10.5:

$$\frac{\|\delta x\|}{\|x + \delta x\|} = \frac{6.64020}{1085.65605} = 0.0061163 \leq \|A\| \left\|A^{-1}\right\| \frac{\|\delta A\|}{\|A\|} = 275848.64261 \left(3.66856 \times 10^{-8}\right) = 0.010120. \quad \blacksquare$$

Example 10.10. If $A = \begin{bmatrix} -3 & \frac{1}{2} & \frac{1}{3} \\ -36 & 8 & 6 \\ 30 & -7.5 & -6 \end{bmatrix}$ and $b = \begin{bmatrix} 3 \\ 3 \\ 3 \end{bmatrix}$, the solution is $x = \begin{bmatrix} -6.5000 \\ -66.0000 \\ 49.5000 \end{bmatrix}$. Perturb A by $\delta A =$

$\begin{bmatrix} 0.0005 & 0 & 0.00001 \\ 0 & 0.0003 & 0 \\ 0 & 0 & -0.0007 \end{bmatrix}$ and b by $\delta b = \begin{bmatrix} -0.0001 \\ 0.0005 \\ -0.00002 \end{bmatrix}$, so $A + \delta A = \begin{bmatrix} -2.99950 & 0.50000 & 0.33334 \\ -36.00000 & 8.00030 & 6.00000 \\ 30.00000 & -7.50000 & -6.00070 \end{bmatrix}$ and

$b + \delta b = \begin{bmatrix} 2.99990 \\ 3.00050 \\ 2.99998 \end{bmatrix}$. The solution to the perturbed system $x + \delta x = \begin{bmatrix} -6.48637 \\ -65.72708 \\ 49.22128 \end{bmatrix}$, so $\delta x = \begin{bmatrix} 0.013630 \\ 0.272920 \\ -0.27872 \end{bmatrix}$. The relative error is

$$\frac{\|\delta x\|}{\|x\|} = 0.0047166.$$

Considering that $\|\delta A\|/\|A\| = 1.43018 \times 10^{-5}$ and $\|\delta b\|/\|b\| = 9.82061 \times 10^{-5}$, this relative error is poor, indicating the A is ill-conditioned. Indeed, $\kappa(A) = 2.39661 \times 10^3$.

Verify inequality 10.6:

$$\frac{\|\delta x\|}{\|x\|} = 0.0047166 \leq \|A\| \left\|A^{-1}\right\| \left(\frac{\|\delta A\|}{\|A\|} + \frac{\|\delta b\|}{\|A\| \|x + \delta x\|}\right)$$

$$= 2.39661 \times 10^3 \left(1.43018 \times 10^{-5} + \frac{5.10294 \times 10^{-4}}{48.95519 \, (82.37024)}\right) = 0.034579 \quad \blacksquare$$

Definition 10.11. If the condition number of matrix A is large, we say A is *ill-conditioned*; otherwise, A is well-conditioned.

The term "large" is vague. Condition numbers in the range of 10^4 or more definitely indicate ill-conditioning. For some matrices, a smaller condition number can indicate ill-conditioning, so determining ill-conditioning is not an exact science.

10.6 PROPERTIES OF THE MATRIX CONDITION NUMBER

The matrix condition number has importance in various applications and in proving some highly useful results. This section develops properties of the condition number and provides examples that illustrate its properties.

Theorem 10.5. *Let A be a nonsingular matrix.*

1. $\kappa_p(A) \geq 1$ *for any p-norm.*

2. $\kappa_G(\alpha A) = \kappa_G(A)$, *where $\alpha \neq 0$ is a constant. Here $\kappa_G(A)$ refers to any matrix norm.*

3. *Let A be an orthogonal matrix. Then, $\kappa\left(A\right) = 1$ if and only if $A^{\mathrm{T}}A = \alpha I$, where $\alpha \neq 0$.*
4. $\kappa\left(A^{\mathrm{T}}A\right) = \left(\kappa\left(A\right)\right)^2$.
5. $\kappa\left(A\right) = \kappa\left(A^{\mathrm{T}}\right)$; $\kappa_1\left(A\right) = \kappa_\infty\left(A^{\mathrm{T}}\right)$.
6. *For any matrix norm, $\kappa\left(AB\right) \leq \kappa\left(A\right)\kappa\left(B\right)$.*
7. $\kappa\left(A\right) = \sigma_{\max}/\sigma_{\min}$ *where $\sigma_{\max}$ and $\sigma_{\min}$ are, respectively, the largest and smallest singular values of A.*

We will prove properties (2), (4), and (7). The remaining properties are left as exercises.

Proof. To prove (2), note that since $(\alpha A)\left(\frac{1}{\alpha}A^{-1}\right) = I$, $\frac{1}{\alpha}A^{-1}$ must be the inverse of αA, and so

$$\kappa_G\left(\alpha A\right) = \|\alpha A\| \left\|\left(\alpha A\right)^{-1}\right\| = |\alpha| \, \|A\| \left\|\frac{1}{\alpha}A^{-1}\right\| = \|A\| \left\|A^{-1}\right\| = \kappa_G\left(A\right).$$

To prove (4), the condition number of $A^{\mathrm{T}}A$ is $\kappa\left(A^{\mathrm{T}}A\right) = \left\|A^{\mathrm{T}}A\right\|_2 \left\|\left(A^{\mathrm{T}}A\right)^{-1}\right\|_2$, so we have to deal with the two factors $\left\|A^{\mathrm{T}}A\right\|_2$ and $\left\|\left(A^{\mathrm{T}}A\right)^{-1}\right\|_2$. In Theorem 7.9, part (4) we proved that

$$\left\|A^{\mathrm{T}}A\right\|_2 = \|A\|_2^2. \tag{10.14}$$

Recall the property $\left(A^{\mathrm{T}}\right)^{-1} = \left(A^{-1}\right)^{\mathrm{T}}$ from Theorem 1.6, part (5), so

$$\left\|\left(A^{\mathrm{T}}A\right)^{-1}\right\|_2 = \left\|A^{-1}\left(A^{\mathrm{T}}\right)^{-1}\right\|_2 = \left\|A^{-1}\left(A^{-1}\right)^{\mathrm{T}}\right\|_2 = \left\|A^{-1}\right\|_2^2. \tag{10.15}$$

From Equations 10.14 and 10.15, we have

$$\kappa\left(A^{\mathrm{T}}A\right) = \|A\|_2^2 \left\|A^{-1}\right\|_2^2 = \left(\|A\|_2 \left\|A^{-1}\right\|_2\right)^2 = \left(\kappa\left(A\right)\right)^2.$$

Now consider property 7. By Theorem 7.7, $\|A\|_2 = \sqrt{\lambda_{\max}}$, where $\lambda_{\max}$ is the largest eigenvalue of $A^{\mathrm{T}}A$. From Theorem 7.9, part (5), $\left\|A^{-1}\right\|_2 = 1/\sqrt{\lambda_{\min}}$, where $\lambda_{\min}$ is the smallest eigenvalue of $A^{\mathrm{T}}A$. Then,

$$\|A\|_2 \left\|A^{-1}\right\|_2 = \frac{\sqrt{\lambda_{\max}}}{\sqrt{\lambda_{\min}}} = \frac{\sigma_{\max}}{\sigma_{\min}}. \qquad \square$$

Lemma 7.1 states that all norms on the vector space $\mathbb{R}^n$ are equivalent. That is, given any two norms $\|\cdot\|_a$ and $\|\cdot\|_b$, there exist constants C_l and C_h such that

$$C_l \|x\|_a \leq \|x\|_b \leq C_h \|x\|_a.$$

This same type of result also applies to the condition number of matrix norms. It answers the question "Can a condition number based on one norm be large and a condition number based on another norm be small?" The answer is "No." We state without proof the following relationships among norms [2, p. 88].

Theorem 10.6. *If A is an $n \times n$ matrix, any two condition numbers $\kappa_\alpha\left(A\right)$ and $\kappa_\beta\left(A\right)$ are equivalent in that there are constants c_1 and c_2 such that*

$$c_1\kappa\left(A\right) \leq \kappa_\beta\left(A\right) \leq c_2\kappa\left(A\right).$$

For $\mathbb{R}^{n \times n}$,

$$\frac{1}{n}\kappa_2\left(A\right) \leq \kappa_1\left(A\right) \leq n\kappa_2\left(A\right),$$

$$\frac{1}{n}\kappa_\infty\left(A\right) \leq \kappa_2\left(A\right) \leq n\kappa_\infty\left(A\right),$$

$$\frac{1}{n^2}\kappa_1\left(A\right) \leq \kappa_\infty\left(A\right) \leq n^2\kappa_1\left(A\right).$$

Thus, if a matrix is ill-conditioned in one norm, it is ill-conditioned in another, taking into account the constants c_1 and c_2.

We stated in Section 6.3 that orthogonal matrices are the most beautiful of all matrices. Now we can add another reason for supporting this claim.

Lemma 10.1. *In the 2-norm, an orthogonal matrix, P, is perfectly conditioned, in that $\kappa(P) = 1$.*

Proof. By Theorem 10.5, part 7, $\kappa(P) = \sigma_{\max}/\sigma_{\min} = 1$, since the singular values of P are the square roots of the eigenvalues of $P^T P = I$.   □

10.7 MATLAB COMPUTATION OF A MATRIX CONDITION NUMBER

To determine the condition number of a matrix, use the function cond. Its arguments are the matrix and one of the following:

- None (default is the 2-norm)
- 1 (the 1-norm)
- 2 (the 2-norm)
- 'inf' (the ∞-norm)
- 'fro' (the Frobenius-norm)

Here are the results when each of these norms are applied to the 10×10 bidiagonal matrix $A = \begin{bmatrix} 1 & 1 & 0 & 0 & \ldots & 0 \\ 0 & 1 & 1 & 0 & \ldots & 0 \\ 0 & \vdots & 1 & 1 & \ldots & 0 \\ \vdots & \vdots & \vdots & \ddots & \ldots & 0 \\ \vdots & \vdots & \vdots & \ldots & 1 & 1 \\ 0 & 0 & 0 & \ldots & \ldots & 1 \end{bmatrix}$

```
>>  A = diag(ones(10,1)) + diag(ones(9,1),1);
>> cond(A)

ans =
   13.2320

>> cond(A,'inf')

ans =
   20

>> cond(A,1)

ans =
   20

>> cond(A, 'fro')

ans =
   32.3265
```

10.8 ESTIMATING THE CONDITION NUMBER

Since the condition number is so important, we must either compute its value or have a good approximation. To compute $\kappa(A) = \left\|A^{-1}\right\|_2 \|A\|_2$ requires that we compute the maximum and minimum singular values of A. Recall from Theorem 10.6 that

$$\frac{1}{n}\kappa_2(A) \le \kappa_1(A) \le n\kappa_2(A),$$

and so it is reasonable to use $\kappa_1(A)$, since the matrix 1-norm is much easier to compute.

$$\|A\|_1 = \max_{1 \le k \le n} \sum_{i=1}^{m} |a_{ik}|.$$

However, the problem of accounting for A^{-1} still remains. It can be shown that

$$\kappa_1(A) \geq \frac{\|A\|_1 \|A^{-1}w\|_1}{\|w\|_1} \tag{10.16}$$

for any nonzero $w \in \mathbb{R}^n$ [23, pp. 131-133]. If we choose w so that $\|A^{-1}w\|_1 / \|w\|_1$ is close to its maximum, Equation 10.16 will give a sharp lower bound for $\kappa_1(A)$. The Hager algorithm is most frequently used to estimate w (see Refs. [1, pp. 50-54], [19, pp. 141-143], and [28]).

Example 10.11. The MATLAB function `condest(A)` estimates the condition number of a square matrix.

```
>> H = hilb(35);   % the Hilbert matrices are notoriously ill-conditioned
>> condest(H)
ans =
  4.7538e+019
>> A = [1 2 3;3 4 5;6 7 8.00001];
>> condest(A)
ans =
  1.6e+007
>> cond(A)
ans =
  1.0991e+007
```
∎

10.9 INTRODUCTION TO PERTURBATION ANALYSIS OF EIGENVALUE PROBLEMS

The 10×10 bidiagonal matrix in Section 10.7 is well-conditioned, since its condition number is 13.232. As it turns out, a matrix can be well-conditioned and yet its eigenvalues are sensitive to perturbations. As a result, the conditioning problem for eigenvalues must be considered separately from matrix conditioning. Perturbation analysis of eigenvalue problems will be discussed in Chapter 18, but at this point it is instructive to present some examples.

Example 10.12. The bidiagonal matrix, A, of Section 10.7 is an upper-triangular matrix, so its eigenvalues lie on the diagonal, and are all 1. After computing the eigenvalues of A, perturb $A(5,1)$ by 10^{-10} and compute the eigenvalues.

```
>> A = diag(ones(10,1)) + diag(ones(9,1),1);
>> eig(A)
ans =
     1
     1
    ...
     1
>> A(5,1) = 1.0e-10;
>> eig(A)
ans =
  0.9919 + 0.0059i
  0.9919 - 0.0059i
  1.0031 + 0.0095i
  1.0031 - 0.0095i
  1.0100
  1.0000
  1.0000
  1.0000
  1.0000
  1.0000
```

A perturbation of 10^{-10} in one entry caused four of the eigenvalues to become complex. ∎

10.10   CHAPTER SUMMARY

Reasons Why the Study of Numerical Linear Algebra Is Necessary

Floating point roundoff and truncation error cause many problems. We have learned how to perform Gaussian elimination in order to row reduce a matrix to upper-triangular form. Unfortunately, if the pivot element is small, this can lead to serious errors in the solution. We will solve this problem in Chapter 11 by using partial pivoting. Sometimes an algorithm is simply far too slow, and Cramer's Rule is an excellent example. It is useful for theoretical purposes but, as a method of solving a linear system, should not be used for systems greater than 2×2. Solving $Ax = b$ by finding A^{-1} and then computing $x = A^{-1}b$ is a poor approach. If the solution to a single system is required, one step of Gaussian elimination, properly performed, requires far fewer flops and results in less roundoff error. Even if the solution is required for many right-hand sides, we will show in Chapter 11 that first factoring A into a product of a lower- and an upper-triangular matrix and then performing forward and back substitution is much more effective. A classical mistake is to compute eigenvalues by finding the roots of the characteristic polynomial. Polynomial root finding can be very sensitive to roundoff error and give extraordinarily poor results. There are excellent algorithms for computing eigenvalues that we will study in Chapters 18 and 19. Singular values should not be found by computing the eigenvalues of $A^{T}A$. There are excellent algorithms for that purpose that are not subject to as much roundoff error. Lastly, if $m \neq n$ a theoretical linear algebra course deals with the system using a reduction to what is called reduced row echelon form. This will tell you whether the system has infinitely many solutions or no solution. These types of systems occur in least-squares problems, and we want a single meaningful solution. We will find one by requiring that x be such that $\|b - Ax\|_2$ is minimum.

Forward and Backward Error Analysis

Forward error deals with rounding errors in the solution of the problem. If the input is x, the solution is $f(x)$, and the result obtained is $\hat{f}(x)$, the forward error is $\left|f(x) - \hat{f}(x)\right|$. Normally we do not know the true solution $f(x)$, so we must resort to bounding the forward error. For instance, the forward error for the computation of the outer product of vectors x and y, xy^{T}, is bounded by eps $\times \left|xy^{T}\right|$, where $|.|$ returns the matrix with each entry replaced by its absolute value.

Backward error relates the rounding errors in the computation to the errors in the data rather than its solution. Generally, a backward error analysis is preferable to a forward analysis for this reason. Roundoff or other errors in the data have produced the result $\hat{y}$. The backward error is the smallest Δx for which $\hat{y} = f(x + \Delta x)$; in other words, backward error tells us what problem we actually solved. For instance, it can be shown that the inner product, $\langle x, y \rangle$, is the exact inner product for a perturbed set of data, where the perturbations are very small.

Algorithm Stability

Intuitively, an algorithm is stable if it performs well in general, and an algorithm is unstable if it performs badly in significant cases. In particular, an algorithm should not be unduly sensitive to errors in its input or errors during its execution. We saw in Chapter 8 that using the quadratic equation in its natural form is subject to serious cancellation error. There are two types of stable algorithms, backward stable, and forward stable.

An algorithm is backward stable if for any x, it computes $f(x)$ with small backward error, Δx. In other words, it computes the exact solution to a nearby problem,

$$f(x + \Delta x) = \hat{f}(x)$$

so that the solution is not sensitive to small perturbations in x. The addition of floating point numbers is backward stable, as is the computation of the inner product, and back substitution.

An algorithm is forward stable if whenever $f(x)$ is the true solution, the difference between the computed and true solutions is small. In other words,

$$\left|\hat{f}(x) - f(x)\right|$$

is small. The computation of the inner and outer product of two vectors is forward stable. We know from our discussion in Chapter 8 that floating-point arithmetic does not follow the laws of real arithmetic. This can make forward error analysis difficult. In backward error analysis, however, real arithmetic is employed, since it is assumed that the computed result is the exact solution to a nearby problem. This is one reason why backward error analysis is often preferred, so we will refer to stability to mean backward stability. We have seen that floating point addition and inner product are stable.

Computing the outer product is not backward stable. The root finding problem can be unstable, particularly in the presence of multiple roots. The Wilkinson polynomial is an example where all the roots are distinct, but their computation is unstable.

Conditioning of a Problem

A problem is ill-conditioned if a small relative change in its data can cause a large relative error in its computed solution, regardless of the algorithm used to solve the problem. If small perturbations in problem data lead to small relative errors in the solution, a problem is said to be well-conditioned. Let x and $\overline{x}$ be the original and slightly perturbed data, and let $f(x)$ and $f(\overline{x})$ be the respective solutions. Then,

- The problem is well-conditioned with respect to x if whenever $|x - \overline{x}|$ is small, $|f(x) - f(\overline{x})|$ is small.
- The problem is ill-conditioned with respect to x if whenever $|x - \overline{x}|$ is small, $|f(x) - f(\overline{x})|$ can be large.

The condition number defines the degree of conditioning for a particular problem. The chapter gives a mathematical definition of the condition number as the limiting ratio of the relative rate of change of the function divided by the relative change in the input. The larger the condition number, the more sensitive the problem is to errors.

Remark 10.7.

- Stable or unstable refers to an algorithm.
- Well or ill-conditioned refers to the particular problem, not the algorithm used.

Perturbation Analysis for Solving a Linear System

When solving a linear system, there are three situations we must consider, errors in the right-hand side, errors in the coefficient matrix, and errors in both. The text discusses each of these cases and, in each case, the expression

$$\|A\| \left\|A^{-1}\right\|$$

appears, where $\|.\|$ is any subordinate norm. This value is called the condition number of the matrix and is denoted by $\kappa(A)$. If the condition number is large, solving a linear system accurately is difficult.

Properties of the Matrix Condition Number

There are a number of important properties of the 2-norm matrix condition number. We list some of the most useful:

- If P is orthogonal, $\kappa(A) = 1$.
- $\kappa(A) = \sigma_{\max}/\sigma_{r\min}$

Using MATLAB to Compute the Matrix Condition Number

In MATLAB, compute the condition number using the function cond in one of the forms

`cond(A)`	2-norm
`cond(A,1)`	1-norm
`cond(A,'inf')`	∞-norm
`cond(A,'fro')`	Frobenius-norm

Approximating the Matrix Condition Number

The MATLAB function `condest` approximates the condition number of a matrix using the matrix 1-norm. We normally use this approximation with large sparse matrices, since the computation of the condition number is too costly.

Perturbation Analysis of Eigenvalue Problems

A matrix can be well-conditioned and yet its eigenvalues can be sensitive to perturbations. As a result, the conditioning problem for eigenvalues must be considered separately from matrix conditioning. Perturbation analysis of eigenvalue problems will be discussed in Chapter 18,

10.11 PROBLEMS

10.1 Show that the floating point multiplication of two numbers is backward stable.

10.2 Prove that a small residual for the solution of $Ax = b$ implies backward stability.

10.3 Prove that if x and y are in $\mathbb{R}^n$, the $n \times n$ matrix xy^T has rank 1.

10.4 If A and B are matrices and α is a floating point number, prove the following forward error results.

  a. $\text{fl}(\alpha A) = \alpha A + E$, $|E| \leq \text{eps } |\alpha A|$.

  b. $\text{fl}(A + B) = (A + B) + E$, $|E| \leq \text{eps } |A + B|$.

10.5 Show that the roots of the polynomial $x^3 - 12x^2 + 48x - 64$ are ill-conditioned and explain why.

10.6 Let $f(x) = \ln x$.

  a. Show that the condition number of f at x is $c(x) = 1/|\ln x|$.

  b. Using the above result, show that $\ln x$ is ill-conditioned near $x = 1$.

10.7 Show that computing $\sqrt{x}$ for $x > 0$ is well-conditioned.

10.8 What is the condition number for $f(x) = x/(x - 1)$ at x? Where is it ill-conditioned?

10.9 Let A be a nonsingular square matrix.

  a. Prove that if A is a scalar multiple of an orthogonal matrix, then $\kappa_2(A) = 1$.

  b. Prove that if $\kappa_2(A) = 1$, then A is a scalar multiple of an orthogonal matrix. You may assume the singular value decomposition that says

$$A = U\Sigma V^T,$$

  where U and V are orthogonal and Σ is a diagonal matrix of singular values.

10.10 If A is an $m \times n$ matrix, and x is an $n \times 1$ vector, then the linear transformation $y = Ax$ maps $\mathbb{R}^n$ to $\mathbb{R}^m$, so the linear transformation should have a condition number, $cond_{Ax}(x)$. Assume that $\|\cdot\|$ is a subordinate norm.

  a. Show that we can define $cond_{Ax}(x) = \|A\| \, \|x\|/\|Ax\|$ for every $x \neq 0$.

  b. Find the condition number of the linear transformation at $x = \begin{bmatrix} 1 & -1 & 2 \end{bmatrix}^T$ using the ∞-norm.

$$T = \begin{bmatrix} 1 & 7 & -1 \\ 3 & 2 & 1 \\ 5 & -9 & 3 \end{bmatrix} \begin{bmatrix} x_1 \\ x_2 \\ x_3 \end{bmatrix}.$$

  c. Show that $cond_{Ax}(x) \leq \|A\| \, \|A^{-1}\|$ for all x.

  d. Verify the result of part (c) for the matrix and vector of part (b).

10.11 Prove the following properties of the condition number.

  a. $\kappa_p(A) \geq 1$ for any p-norm.

  b. $\kappa_2(A) = \kappa_2(A^T)$; $\kappa_1(A) = \kappa_\infty(A^T)$.

  c. For any sub-multiplicative matrix norm, $\kappa(AB) \leq \kappa(A)\kappa(B)$.

10.12 Let $A = \begin{bmatrix} 1 & a \\ a & 1 \end{bmatrix}$. For what values of a is A ill-conditioned? What happens as $a \to \infty$.

10.13 Prove that for any orthogonal matrix P, $\kappa(PA) = \kappa(AP) = \kappa(A)$ for any $n \times n$ matrix A.

10.14 There is an equivalent definition of $\kappa(A)$ [27, pp. 84-85].

  The condition number $\kappa(A)$ of a nonsingular matrix A is

$$\frac{1}{\kappa(A)} = \min_{B \in S_n} \left\{ \frac{\|A - B\|_2}{\|A\|_2} \right\},$$

  where S_n is the set of singular $n \times n$ matrices.

  a. If A is ill-conditioned, what does this say about the relative distance of A from the subset of $n \times n$ singular matrices

  b. If A is a well-conditioned matrix, answer the question of part (a).

10.11.1 MATLAB Problems

10.15

    a. Write a MATLAB function that builds the matrix

$$
A = \begin{bmatrix}
n & n-1 & n-2 & \cdots & 3 & 2 & 1 \\
n-1 & n-1 & n-2 & \cdots & 3 & 2 & 1 \\
0 & n-2 & n-2 & \ddots & & \vdots & \vdots \\
\vdots & & & \ddots & \ddots & \ddots & \vdots & \vdots \\
\vdots & & & & \ddots & \ddots & 2 & \vdots \\
\vdots & & & & & 2 & 2 & 1 \\
0 & \cdots & & \cdots & \cdots & 0 & 1 & 1
\end{bmatrix}.
$$

    b. For $n = 5, 10, 15, 20$:

        i. Compute the vector of eigenvalues and assign it to the variable E1.

        ii. Perturb $A(1, n-1)$ by 10^{-8} and assign the eigenvalues of the perturbed matrix to the variable E2.

        iii. Compute $\|E1 - E2\|_2$.

    c. Which eigenvalues appear to be perturbed the most?

    d. The function `eigcond` in the software distribution with calling format

```
[c lambda] = eigcond(A)
```

computes the condition number c_i of eigenvalue λ_i. We will develop the function code in Section 18.11. Run `eigcond` for each of the matrices in part (b). Do the results confirm your conclusions of part (c)?

10.16 **a.** Let $x = \begin{bmatrix} 1 & 3 & 5 & 7 & 9 & 11 & 13 & 15 \end{bmatrix}$ and execute

```
A = gallery('sampling', x)
```

    Is A ill-conditioned?

    b. Execute the following code:

```
>> E = 0.0001*rand(8);
>> B = A+E;
>> eig(A)
>> eig(B)
```

    c. What do you conclude from the code? Justify your answer using `eigcond` introduced in Problem 10.15.

10.17 Let A = gallery(3).

    a. Is A ill-conditioned?

    b. Find the eigenvalues of A. Perturb A by a random matrix E whose values do not exceed 10^{-5}. Does it appear the eigenvalues of A are ill-conditioned?

    c. Find eigenvectors for A using the command [V E] = eig(A). Perturb the elements of A by 0.0001*rand(3,3) and compute the eigenvectors. What conclusion can you make?

10.18 Let A = gallery(5). Answer questions (a)-(c) in Problem 10.17 for this matrix.

10.19 A magic square of order n is an $n \times n$ matrix with entries $1 \ldots n^2$, such that the n numbers in all rows, all columns, and both diagonals sum to the same constant. MATLAB will create a magic square with the command A = magic(n).

    a. What is the largest eigenvalue of magic(n) for $n = 5, 8, 15$? In general, what is the largest eigenvalue of a magic square of order n? Prove it. Hint: Look at the Perron-Frobenius theorem, Theorem 5.6 in this book.

    b. What is the largest singular value of magic(n) for $n = 5, 8, 15$? In general, what is the largest singular value of a magic square of order n?

10.20 The symmetric Pei matrix is defined by

```
alpha*eye(n) + ones(n)
```

Enter the anonymous function

```
p = @(alpha,n) alpha*eye(n) + ones(n);
```

of two variables that creates a Pei matrix.

    Fix $n = 25$, and draw a graph of alpha = 0.5:-.01:.01 vs. cond(p(alpha(i),25)). What do you conclude?

10.21 The *Wilkinson bidiagonal matrix* has the general form

$$
A = \begin{bmatrix}
n & n & & & \\
 & n-1 & n & & \\
 & & \ddots & \ddots & \\
 & & & 2 & n \\
 & & & & 1
\end{bmatrix},
$$

and is often used for testing purposes.

a. Create an anonymous function using `diag` that builds an $n \times n$ Wilkinson bidiagonal matrix.

b. Graph the condition number of the Wilkinson bidiagonal matrices of orders 1 through 15. What are your conclusions?

c. Eigenvalues are tricky to compute accurately. The 20×20 *Wilkinson-bidiagonal matrix*, A, has eigenvalues $\lambda = 20, 19, \ldots, 1$. This matrix illustrates that even though the eigenvalues of a matrix are not equal or even close to each other, an eigenvalue problem can be ill-conditioned. Compute the eigenvalues of A, perturb $A(20, 1)$ by 10^{-10} and compute the eigenvalues. Comment on the results.

10.22 The symmetric rosser matrix is particularly nasty, intentionally. Investigate the conditioning of the matrix and its eigenvalues. If the eigenvalues appear well-conditioned, run an experiment that demonstrates this. Generate the Rosser matrix with the command

```
A = rosser;
```

10.23 The $n \times n$ Hilbert matrices have elements $h_{ij} = 1/(i+j-1)$ and are notoriously ill-conditioned.

a. Create `A = hilb(8)`. Perturb A(1,8) by 10^{-5}, and call the perturbed matrix B. Let `b = rand(8,1)`. Compute `x = A\b`, `y = B\b`, and follow by computing $\|x - y\|$ and $\|x - y\|/\|x\|$. Is ill-conditioning evident?

b. Compute the condition number of `hilb(8)` using the MATLAB command `cond(A)` and then use it to compute the theoretical upper bound given in Theorem 10.4, part (2). Do your calculations verify the inequality?

10.24 Let x be a known value, such as `ones(n,1)` and compute b = Ax. Theoretically, $\bar{x} = A\backslash b$ should be the known solution x. However, due to the nature of floating point computation the relative error $\|x - \bar{x}\|/\|x\|$ is not expected to be 0, but it should be small if A is not ill-conditioned. The idea for this problem is to observe ill-conditioning using the matrix `A = gallery('lotkin', n)`. Perform the steps presented in the following algorithm and comment on the results. The notation $\langle \ldots \rangle$ in a print statement indicates the value to print. For instance,

$$\text{print}\left('The\ condition\ number\ of\ A\ is', \langle condition\ number\ with\ 8\ significant\ digits\rangle\right)$$

```
procedure ERRORTEST
   for n = 2:11 do
      A = gallery('lotkin',n)
      x = [ 1  1  ...  1  1 ]ᵀ
      b = Ax
      x̂ = A\b
      Print "The condition number of A is ⟨η (A) with 15 significant digits⟩"
      Print "The relative error is ⟨ ‖x−x̂‖/‖x‖ with 15 significant digits⟩"
   end for
end procedure
```

10.25 The goal of this exercise is to observe the behavior of $\kappa(H_n)$ as n goes to ∞, where H_n is the $n \times n$ Hilbert matrix, defined by $(H_n)_{ij} = 1/(i+j-1)$. In MATLAB, generate H_n using the command `H = hilb(n)`. Let n vary from 2 to 500, and make a plot of n versus $log_{10}(\eta(H_n))$. Draw conclusions about the experimental behavior. Hint: Use the MATLAB plotting function `semilogy`.

Problems 10.26 and 10.27 deal with non-square systems.

The technique of linear least-squares solves a system $Ax = b$ by minimizing the residual $\|b - Ax\|_2$. Chapter 12 introduces the topic and Chapter 16 discusses it in depth. Least-squares problems usually involve dealing with $m \times n$ systems, $m \neq n$. Since solutions are obtained using floating point arithmetic, they are subject to errors, and it certainly is reasonable to ask the questions

"Is there a definition of ill-conditioning for non-square systems?".

"Can perturbations in entries of an $m \times n$ system, $m \neq n$ cause large fluctuations in the solution?"

The answer to both of these questions is "yes." Define the condition number of an $m \times n$ matrix as σ_1/σ_k, where σ_1 is the largest and σ_k the smallest nonzero singular values of A. Of course if A is square, this is $\eta(A)$. As we have noted, never compute singular values by finding the eigenvalues of $A^T A$. The MATLAB command

```
S = svd(A)
```

returns a vector S containing the singular values of A:

$$\sigma_1 = S(1) \geq \sigma_2 = S(2) \geq \ldots \sigma_k = S(k).$$

10.26 For each matrix, use svd to compute the condition number and verify your result by using MATLAB's cond. Note that you must find the smallest nonzero singular value, since is possible that there will be zero singular values. Explain the differences between $\kappa(A)$, $\kappa(B)$ and $\kappa(C)$.

a. $A = \begin{bmatrix} 1 & 3 & -1 \\ 8 & -4 & 12 \\ 1 & 9 & 0 \\ -1 & 7 & -8 \\ 5 & 6 & 1 \end{bmatrix}$

b. $B = \begin{bmatrix} 1 & 9 & 0 & 1 & -1 & 15 & 7 \\ 1 & -5 & 3 & -2 & 1 & -18 & 0 \\ 27 & 1 & 7 & -1 & 1 & 9 & -2 \\ 1 & 5 & -6 & 20 & 33 & 55 & 98 \end{bmatrix}$

c. $C = \begin{bmatrix} 1 & 2 & 8 & 19 \\ 5 & -1 & 29 & 62 \\ -1 & 3 & -3 & -4 \\ 3 & 5 & 23 & 54 \\ 1 & 6 & 12 & 31 \end{bmatrix}$

10.27 The QR decomposition of an $m \times n$ matrix is a decomposition of a matrix A into a product $A = QR$, where Q is an orthogonal matrix and R is an upper-triangular matrix. We will extensively discuss this decomposition in Chapters 14 and 17. There are two types of QR decomposition, the full and the reduced:

Full $\quad A^{m \times n} = Q^{m \times m} R^{m \times n}$
Reduced $\quad A^{m \times n} = Q^{m \times n} R^{n \times n}$

For $m > n$, the reduced decomposition saves space. To obtain a reduced QR decomposition, use the MATLAB function qr as follows:

```
[Q R] = qr(A,0).
```

For reasons we will explain in Chapter 16, if $m > n$, and $rank(A) = n$, one method of solving a least-squares problem is to follow these steps:

```
procedure SOLVELQ(A,b)
   Compute the reduced QR decomposition of A
   c = Q^T b
   Solve the upper-triangular system Rx = c.
   return x
end procedure
```

a. Implement solveq as a MATLAB function. Use backsolve from Chapter 9 to solve the upper-triangular system $Rx = c$.

b. Solve the system $\begin{bmatrix} 2 & 7 & 1 \\ -1 & 6 & 2 \\ 0 & 1 & 8 \\ 5 & 9 & 10 \\ 4 & 3 & 5 \end{bmatrix} \begin{bmatrix} x_1 \\ x_2 \\ x_3 \end{bmatrix} = \begin{bmatrix} 3 \\ 8 \\ 0 \\ -1 \\ 5 \end{bmatrix}$.

10.28 A *Toeplitz matrix* is a matrix whose entries are constant along each diagonal. The MATLAB command

```
T = toeplitz(c,r)
```

returns a Toeplitz matrix T having c as its first column and r as its first row. If the first elements of c and r are different, a message is printed and the column element is used.

a. Build the 100×100 pentadiagonal matrix

$$
A = \begin{bmatrix}
6 & -4 & 1 & 0 & \dots & 0 \\
-4 & 6 & -4 & 1 & \dots & 0 \\
1 & -4 & 6 & -4 & \ddots & \vdots \\
0 & 1 & -4 & \ddots & \ddots & 1 \\
\vdots & \vdots & \ddots & -4 & 6 & -4 \\
0 & 0 & \dots & 1 & -4 & 6
\end{bmatrix}.
$$

b. Verify that the matrix is ill-conditioned.

c. Run the following experiment to observe the behavior of the eigenvalues of A as it is slightly perturbed. The MATLAB program computes the eigenvalues of a randomly perturbed A and outputs the norm of the difference between the eigenvalues of A and those of the perturbed matrix. The perturbations to the elements of A are in the range $0 < a_{ij} < 10^{-8}$.

```
E1 = eig(T);
E1 = sort(E1);
for i = 1:3
    deltaT = 1.0e-8*rand(100,100);
    B = T + deltaT;
    E2 = eig(B);
    E2 = sort(E2);
    norm(E1-E2)
end
```

d. Are the eigenvalues of A ill-conditioned? For more information about this problem, see Ref. [29].

10.29 Use MATLAB's function `condest` to estimate the condition number of the 12×12 Hilbert matrix. Find the true 1-norm condition number and compare the results. Explain the warning message you receive when computing the true 1-norm condition number.

10.30 The MATLAB statement `A = rand(1000,1000)` creates a 1000×1000 random matrix. In order to observe the computational time difference between finding the condition number of A and estimating the condition number, execute

```
>> A = rand(1000,1000);
>> tic;cond(A);toc
>> tic;condest(A);toc
```

Comment on the results.

10.31 The $n \times (m+1)$ Vandermonde matrix $V = \begin{bmatrix} 1 & t_1 & \dots & t_1^m \\ 1 & t_2 & \dots & t_2^m \\ \vdots & \vdots & \ddots & \vdots \\ 1 & t_n & \dots & t_n^m \end{bmatrix}$ is created from the vector $t = \begin{bmatrix} t_1 & t_2 & \dots & t_{n-1} & t_n \end{bmatrix}^T$

and integer m. It has uses in polynomial approximation and other areas, and we will formally introduce it in Chapter 12. The function `vandermonde(t,m)` in the software distribution constructs the *Vandermonde matrix*, which is square if $m = n - 1$.

a. Construct the 11×11 Vandermonde matrix using $t = 1.0 : 0.1 : 2.0$. Compute `cond(A)` and `condest(A)`.

b. Construct the 21×21 Vandermonde matrix using $t = 1.0 : .05 : 2.0$. Compute `cond(A)` and `condest(A)`.

c. Are these matrices ill-conditioned? Is it safe to use these matrices in an application involving Gaussian elimination?

10.32 This problem deals with the instability of polynomial root finding. First, note how MATLAB handles polynomial operations by doing some simple computations. A polynomial, p, in MATLAB is an $n+1$-dimensional vector, where $p(1)$ is the coefficient of x^n, $p(2)$ is the coefficient of x^{n-2}, ..., $p(n+1)$ is the coefficient of x^0.

a. Show how to represent the polynomial $p(x) = x^4 - x^3 + x - 1$ in MATLAB.

b. To multiply two polynomials, use the MATLAB function `conv`. For instance, to compute the coefficients of $(x^2 + 1)(x + 3)$, proceed as follows:

```
>> factor1 = [1 0 1];
>> factor2 = [1 3];
>> p = conv(factor1,factor2)

p =
    1    3    1    3
% analytical result is x^3 + 3x^2 + x + 3
```

Using `conv` find the coefficients of $(x - 1)^4 (x - 2)(x + 1)$ and assign them to the vector p.

c. Using the MATLAB function `roots`, compute the roots of $(x - 1)^4 (x - 2)(x + 1)$. Explain the results.

Let a_i be the coefficient of x^i in polynomial $p(x)$ and r be a simple root of p. Suppose roundoff error perturbs a_i by an amount δa_i, so the root now becomes $r + \delta r$. Does a small δa_i result in a small δr? We can answer that question if we have a formula for the condition number,

$C_{a_i}(r)$, for the computation. By applying Theorem 2.1 in Ref. [30],

$$C_{a_i}(r) = \frac{|a_i r^{i-1}|}{|p'(r)|}.$$

Use this result in Problems 10.33 and 10.34.

10.33 If p is the MATLAB representation of a polynomial, the function `polyder` computes the derivative of p.

a. Using `polyder`, compute the derivative of the function in part (b) of Problem 10.32.

b. Compute the condition number for the roots $x_2 = 2$ and $x_3 = -1$ of the polynomial in part (b) of Problem 10.32 at the coefficient of x^5. Do your results confirm your explanation in part (c) of Problem 10.32?

10.34 The function `wilkpoly` in the software distribution computes the MATLAB representation of the Wilkinson polynomial

$$p(x) = (x - 1)(x - 2)(x - 3)\ldots(x - 19)(x - 20).$$

a. Compute the roots of the polynomial after a perturbation of -2^{-23} in the coefficient of x^{19}.

b. Using the definition of $C_{a_i}(x_k)$, compute the sensitivity of each root $x_k = k$, $1 \le k \le 20$ relative to a change in the coefficient of x^{19}. Use the function `polyder` discussed in Problem 10.33. Do your results agree with the errors in the roots you computed in part (a)?

Chapter 11

Gaussian Elimination and the *LU* Decomposition

You should be familiar with

- Matrix arithmetic
- Elementary row operations
- Gaussian elimination using an augmented matrix
- Upper- and lower-triangular matrices

In Chapter 2, we presented the process of solving a nonsingular linear system $Ax = b$ using Gaussian elimination. We formed the augmented matrix $A \mid b$ and applied the elementary row operations

1. Multiplying a row by a scalar.
2. Subtracting a multiple of one row from another
3. Exchanging two rows

to reduce A to upper-triangular form. Following this step, back substitution computed the solution. In many applications where linear systems appear, one needs to solve $Ax = b$ for many different vectors b. For instance, suppose a truss must be analyzed under several different loads. The matrix remains the same, but the right-hand side changes with each new load. Most of the work in Gaussian elimination is applying row operations to arrive at the upper-triangular matrix. If we need to solve several different systems with the same A, then we would like to avoid repeating the steps of Gaussian elimination on A for every different b. This can be accomplished by the *LU decomposition*, which in effect records the steps of Gaussian elimination.

Since Gaussian elimination is used so often, the algorithm must be stable. Unfortunately, this is not true, and we must add an operation termed partial pivoting. There are very rare cases when even the enhanced algorithm is still not stable. The solution to a linear system is actually much more difficult and interesting than it appeared to be in earlier chapters.

While the results of Gaussian elimination are normally very good, it is possible that a simple technique termed iterative improvement can help produce even better results.

11.1 *LU* DECOMPOSITION

The main idea of the *LU* decomposition is to record the steps used in Gaussian elimination with A in the places that would normally become zero. Consider the matrix:

$$A = \begin{bmatrix} 1 & -1 & 3 \\ 2 & -3 & 1 \\ 3 & 2 & 1 \end{bmatrix}.$$

The first step of Gaussian elimination is to use $a_{11} = 1$ as the pivot and subtract 2 times the first row from the second and 3 times the first row from the third. Record these actions by placing the multipliers 2 and 3 into the entries they made zero. In order to make it clear that we are recording multipliers and not elements of A, put the entries in parentheses. This leads to:

$$\begin{bmatrix} 1 & -1 & 3 \\ (2) & -1 & -5 \\ (3) & 5 & -8 \end{bmatrix}.$$

Numerical Linear Algebra with Applications. http://dx.doi.org/10.1016/B978-0-12-394435-1.00011-9

$$
\begin{bmatrix} * & * & * & * \\ * & * & * & * \\ * & * & * & * \\ * & * & * & * \end{bmatrix} = \begin{bmatrix} 1 & & & 0 \\ & 1 & & \\ & & \ddots & \\ * & & & 1 \end{bmatrix} \begin{bmatrix} * & & & * \\ & * & & \\ & & \ddots & \\ 0 & & & * \end{bmatrix}
$$
$$
\quad A \qquad\qquad L \qquad\qquad U
$$

FIGURE 11.1 *LU* decomposition of a matrix.

To zero-out the element in the third row, second column, the pivot is -1, and we need to subtract -5 times the second row from the third row. Record the -5 in the spot made zero.

$$
\begin{bmatrix} 1 & -1 & 3 \\ (2) & -1 & -5 \\ (3) & (-5) & -33 \end{bmatrix}.
$$

Let U be the upper-triangular matrix produced by Gaussian elimination and L be the lower-triangular matrix with the multipliers and ones on the diagonal, i.e.,

$$
L = \begin{bmatrix} 1 & 0 & 0 \\ 2 & 1 & 0 \\ 3 & -5 & 1 \end{bmatrix}, \quad U = \begin{bmatrix} 1 & -1 & 3 \\ 0 & -1 & -5 \\ 0 & 0 & -33 \end{bmatrix}.
$$

Now form the product of L and U:

$$
LU = \begin{bmatrix} 1 & 0 & 0 \\ 2 & 1 & 0 \\ 3 & -5 & 1 \end{bmatrix} \begin{bmatrix} 1 & -1 & 3 \\ 0 & -1 & -5 \\ 0 & 0 & -33 \end{bmatrix} = A.
$$

Thus, we see that A is the product of the lower triangular L and the upper triangular U. When a matrix can be written as a product of simpler matrices, we call that a *decomposition* and this one we call the *LU decomposition* (Figure 11.1). We will explain why this works in Section 11.4.

Remark 11.1. As the elimination process continues, the pivots are on the diagonal of U.

11.2 USING *LU* TO SOLVE EQUATIONS

Factor A into the product of L and U:

$$
Ax = b,
$$
$$
(LU)x = b,
$$
$$
L(Ux) = b.
$$

First solve $Ly = b$. This finds $y = Ux$. Now solve $Ux = y$ to find x. Each of these solution steps is simple. First, the system Ly is lower triangular.

$$
\begin{bmatrix} 1 & 0 & \dots & 0 \\ l_{21} & 1 & \dots & 0 \\ \vdots & \vdots & \dots & 0 \\ \vdots & \vdots & 1 & 0 \\ l_{n1} & l_{n2} & \dots & 1 \end{bmatrix} \begin{bmatrix} y_1 \\ y_2 \\ \vdots \\ y_{n-1} \\ y_n \end{bmatrix} = \begin{bmatrix} b_1 \\ b_2 \\ \vdots \\ b_{n-1} \\ b_n \end{bmatrix}.
$$

Solve for y using forward substitution.

$$
y_i = \frac{1}{l_{ii}} \left(b_i - \sum_{j=1}^{i-1} l_{ij} y_j \right).
$$

Next, the system $Ux = y$ is upper triangular.

$$
\begin{bmatrix}
u_{11} & u_{12} & \cdots & u_{1,n-1} & u_{1n} \\
0 & u_{22} & \cdots & u_{2,n-1} & u_{2n} \\
\vdots & \vdots & \ddots & \vdots & \vdots \\
0 & 0 & \cdots & u_{n-1,n-1} & u_{n-1,n} \\
0 & 0 & \cdots & 0 & u_{nn}
\end{bmatrix}
\begin{bmatrix}
x_1 \\ x_2 \\ \vdots \\ x_{n-1} \\ x_n
\end{bmatrix}
=
\begin{bmatrix}
y_1 \\ y_2 \\ \vdots \\ y_{n-1} \\ y_n
\end{bmatrix}.
$$

Solve for x using back substitution.

$$
x_i = \frac{1}{u_{ii}} \left(y_i - \sum_{j=i+1}^{n} u_{ij} x_j \right).
$$

Example 11.1.   Let A be the matrix $A = \begin{bmatrix} 1 & -1 & 3 \\ 2 & -3 & 1 \\ 3 & 2 & 1 \end{bmatrix}$ of Section 11.1, with $b = \begin{bmatrix} 1 \\ 3 \\ 1 \end{bmatrix}$. We determined that $A = LU$,

where

$$
L = \begin{bmatrix} 1 & 0 & 0 \\ 2 & 1 & 0 \\ 3 & -5 & 1 \end{bmatrix}, \quad U = \begin{bmatrix} 1 & -1 & 3 \\ 0 & -1 & -5 \\ 0 & 0 & -33 \end{bmatrix}.
$$

Execute forward substitution to solve $Ly = b$:

$$
\begin{aligned}
(1)\, y_1 &= 1, & y_1 &= 1, \\
2\,(1) + (1)\, y_2 &= 3, & y_2 &= 1, \\
(3)\,(1) - 5\,(1) + (1)\, y_3 &= 1, & y_3 &= 3.
\end{aligned}
$$

Execute back substitution to solve $Ux = y$:

$$
\begin{aligned}
-33x_3 &= 3, & x_3 &= -1/11, \\
-x_2 - 5\,(-1/11) &= 1, & x_2 &= -6/11, \\
x_1 - (-6/11) + 3\,(-1/11) &= 1, & x_1 &= 8/11.
\end{aligned}
$$

Solution: $x_1 = 8/11$, $x_2 = -6/11$, $x_3 = -1/11$ ∎

Example 11.2.   Solve $\begin{bmatrix} 1 & 2 & -1 \\ 2 & 3 & 2 \\ 5 & 1 & 4 \end{bmatrix} x = \begin{bmatrix} 1 \\ -1 \\ 2 \end{bmatrix}$. First factor A into the product LU.

$$
\begin{bmatrix} 1 & 2 & -1 \\ 2 & 3 & 2 \\ 5 & 1 & 4 \end{bmatrix}
\xrightarrow[R_3 = R_3 - (5)\,R_1]{R_2 = R_2 - (2)\,R_1}
\begin{bmatrix} 1 & 2 & -1 \\ (2) & -1 & 4 \\ (5) & -9 & 9 \end{bmatrix}
\xrightarrow{R_3 = R_3 - (9)R_2}
\begin{bmatrix} 1 & 2 & -1 \\ (2) & -1 & 4 \\ (5) & (9) & -27 \end{bmatrix}
$$

$$
L = \begin{bmatrix} 1 & 0 & 0 \\ 2 & 1 & 0 \\ 5 & 9 & 1 \end{bmatrix}, \quad U = \begin{bmatrix} 1 & 2 & -1 \\ 0 & -1 & 4 \\ 0 & 0 & -27 \end{bmatrix}
$$

Forward substitution:

$$
\begin{aligned}
y_1 &= 1, \\
y_2 &= -1 - 2\,(1) = -3, \\
y_3 &= 2 - 5\,(1) - 9\,(-3) = 24.
\end{aligned}
$$

Back substitution:

$$
\begin{aligned}
x_3 &= -24/27 = -8/9, \\
x_2 &= 3 - 4\,(8/9) = -5/9, \\
x_1 &= 1 - 2\,(-5/9) - 8/9 = 11/9.
\end{aligned}
$$
                                                                                                                       ∎

11.3 ELEMENTARY ROW MATRICES

Sections 11.1 and 11.2 describe the *LU* decomposition using examples. If a mathematical analysis of why the *LU* decomposition works is not required, the reader can skip this section and most of Section 11.4. However, it is recommended that Example 11.8 and Sections 11.4.1–11.4.3 be read.

The *elementary row matrices* are an important class of nonsingular matrices. Multiplication by one of these matrices performs an elementary row operation, and these matrices help us understand why the *LU* decomposition works.

Definition 11.1. To each of the three elementary row operations, there corresponds an *elementary row matrix* E_{ij}, E_i, and $E_{ij}(t)$:

a. E_{ij}, $i \neq j$, is obtained from the identity matrix I by exchanging rows i and j.
b. $E_i(t)$, $t \neq 0$ is obtained by multiplying the ith row of I by t.
c. $E_{ij}(t)$ $i \neq j$, is obtained from I by subtracting t times the jth row of I from the ith row of I.

Example 11.3. $E_{23} = \begin{bmatrix} 1 & 0 & 0 \\ 0 & 0 & 1 \\ 0 & 1 & 0 \end{bmatrix}$, $E_2(-1) = \begin{bmatrix} 1 & 0 & 0 \\ 0 & -1 & 0 \\ 0 & 0 & 1 \end{bmatrix}$, $E_{23}(-2) = \begin{bmatrix} 1 & 0 & 0 \\ 0 & 1 & 2 \\ 0 & 0 & 1 \end{bmatrix}$ ∎

The elementary row matrices have the following property, and it is this property that will allow us to explain why the *LU* decomposition works.

Theorem 11.1. *If an $n \times n$ matrix is premultiplied by an $n \times n$ elementary row matrix, the resulting $n \times n$ matrix is the one obtained by performing the corresponding elementary row-operation on A.*

Proof. We will prove that forming $C = E_{ij}A$ is equivalent to interchanging rows i and j of A. The remaining properties of the elementary row matrices is left to the exercises.

By the definition of matrix multiplication, row i of C has components

$$c_{ip} = \sum_{k=1}^{n} e_{ik}a_{kp}, \quad 1 \leq p \leq n.$$

Among the elements $\{e_{i1}, e_{i2}, \ldots, e_{in}\}$ only $e_{ij} = 1$ is nonzero. Thus, $c_{ip} = e_{ij}a_{jp} = a_{jp}$, $1 \leq p \leq n$, and elements of row i are those of row j. Similarly, the elements of row j are those of row i, and the other rows are unaffected. □

Example 11.4.

$$E_{23} \begin{bmatrix} a & b \\ c & d \\ e & f \end{bmatrix} = \begin{bmatrix} 1 & 0 & 0 \\ 0 & 0 & 1 \\ 0 & 1 & 0 \end{bmatrix} \begin{bmatrix} a & b \\ c & d \\ e & f \end{bmatrix} = \begin{bmatrix} a & b \\ e & f \\ c & d \end{bmatrix}.$$ ∎

Theorem 11.1 implies that premultiplying a matrix by a sequence of elementary row matrices gives the same result as performing the sequence of elementary row operations to A.

Example 11.5. Let $A = \begin{bmatrix} 1 & 2 & 8 & 2 \\ 3 & 9 & -1 & 2 \\ -1 & 2 & 6 & 3 \\ 1 & 5 & 3 & 2 \end{bmatrix}$. Multiply A first by E_{24} and then by $E_{43}(-3)$.

$$E_{43}(-3)E_{24}A = \begin{bmatrix} 1 & 0 & 0 & 0 \\ 0 & 1 & 0 & 0 \\ 0 & 0 & 1 & 0 \\ 0 & 0 & 3 & 1 \end{bmatrix} \begin{bmatrix} 1 & 0 & 0 & 0 \\ 0 & 0 & 0 & 1 \\ 0 & 0 & 1 & 0 \\ 0 & 1 & 0 & 0 \end{bmatrix} \begin{bmatrix} 1 & 2 & 8 & 2 \\ 3 & 9 & -1 & 2 \\ -1 & 2 & 6 & 3 \\ 1 & 5 & 3 & 2 \end{bmatrix}$$

$$= \begin{bmatrix} 1 & 0 & 0 & 0 \\ 0 & 1 & 0 & 0 \\ 0 & 0 & 1 & 0 \\ 0 & 0 & 3 & 1 \end{bmatrix} \begin{bmatrix} 1 & 2 & 8 & 2 \\ 1 & 5 & 3 & 2 \\ -1 & 2 & 6 & 3 \\ 3 & 9 & -1 & 2 \end{bmatrix}$$

$$\begin{bmatrix} 1 & 2 & 8 & 2 \\ 1 & 5 & 3 & 2 \\ -1 & 2 & 6 & 3 \\ 0 & 15 & 17 & 11 \end{bmatrix}.$$

Verify that if you perform elementary row operations by interchanging rows 2 and 4 and then subtracting -3 times row 3 from row 4 you obtain the same result. ∎

Theorem 11.2. *Elementary row matrices are nonsingular; in fact*

a. $E_{ij}^{-1} = E_{ij}$
b. $(E_i(t))^{-1} = E_i(t^{-1})$, $t \neq 0$
c. $(E_{ij}(t))^{-1} = E_{ij}(-t)$

Proof. $E_{ij}E_{ij} = I$. Swapping rows i and j of I and then swapping again gives I.
$E_i(t) E_i(1/t) = I$, if $t \neq 0$.
$E_{ij}(-t) E_{ij}(t) = I$. Multiply row j by t and subtract from row i, then reverse the operation by multiplying row j by $-t$ and subtracting from row i. □

Example 11.6. Find the 3×3 matrix $A = E_3(5) E_{23}(2) E_{12}$ and then find A^{-1}.

$$\begin{aligned} A &= E_3(5) E_{23}(2) E_{12} \\ &= \begin{bmatrix} 1 & 0 & 0 \\ 0 & 1 & 0 \\ 0 & 0 & 5 \end{bmatrix} \begin{bmatrix} 1 & 0 & 0 \\ 0 & 1 & -2 \\ 0 & 0 & 1 \end{bmatrix} \begin{bmatrix} 0 & 1 & 0 \\ 1 & 0 & 0 \\ 0 & 0 & 1 \end{bmatrix} \\ &= \begin{bmatrix} 0 & 1 & 0 \\ 1 & 0 & -2 \\ 0 & 0 & 5 \end{bmatrix}, \end{aligned}$$

$$\begin{aligned} A^{-1} &= (E_3(5) E_{23}(2) E_{12})^{-1} = E_{12}^{-1} (E_{23}(2))^{-1} (E_3(5))^{-1} \\ &= \begin{bmatrix} 0 & 1 & 0 \\ 1 & 0 & 0 \\ 0 & 0 & 1 \end{bmatrix} \begin{bmatrix} 1 & 0 & 0 \\ 0 & 1 & 2 \\ 0 & 0 & 1 \end{bmatrix} \begin{bmatrix} 1 & 0 & 0 \\ 0 & 1 & 0 \\ 0 & 0 & \frac{1}{5} \end{bmatrix} \\ &= \begin{bmatrix} 0 & 1 & \frac{2}{5} \\ 1 & 0 & 0 \\ 0 & 0 & \frac{1}{5} \end{bmatrix}. \end{aligned}$$
∎

If B is row equivalent to A, it seems reasonable that we can invert row operations and row reduce B to A.

Theorem 11.3. *If B is row-equivalent to A, then A is row equivalent to B.*

Proof. If $B = E_k E_{k-1} \ldots E_2 E_1 A$, then

$$A = (E_k E_{k-1} \ldots E_2 E_1)^{-1} B.$$

Since

$$E = E_k E_{k-1} \ldots E_2 E_1$$

is nonsingular by Theorem 11.2, $A = E^{-1}B$. Now, $E^{-1} = E_1^{-1} E_2^{-1} \ldots E_{k-1}^{-1} E_k^{-1}$ is a product of elementary row matrices, so forming $E^{-1}B$ is equivalent to performing elementary row operations on B to obtain A. □

Theorems 11.4 and 11.5 tell us how elementary row matrices and nonsingular matrices are related.

Theorem 11.4. *Let A be a nonsingular $n \times n$ matrix. Then*

a. *A is row-equivalent to I.*
b. *A is a product of elementary row matrices.*

Proof. A sequence of elementary row operations will reduce A to I; otherwise, the system $Ax = 0$ would have a non-trivial solution.

Assume $E_k, E_{k-1}, \ldots E_2, E_1$ is the sequence of elementary row operations that reduces A to I so that $E_k E_{k-1} \ldots E_2 E_1 A = I$. It follows that $A = E_1^{-1} E_2^{-1} \cdots E_{k-1}^{-1} E_k^{-1}$ is a product of elementary row matrices. $\qquad\square$

Theorem 11.5.   *Let A be an $n \times n$ matrix and suppose that A is row-equivalent to I. Then A is nonsingular, and A^{-1} can be found by performing the same sequence of elementary row operations on I as were used to convert A to I.*

Proof. Suppose that $E_k E_{k-1} \ldots E_2 E_1 A = I$. Thus $BA = I$, where $B = E_k E_{k-1} \ldots E_2 E_1$ is nonsingular, and $A^{-1} = B = (E_k E_{k-1} \ldots E_2 E_1) I$, which shows that A^{-1} is obtained by performing the same sequence of elementary row operations on I that were used to transform A to I. $\qquad\square$

Remark 11.2.   Theorems 11.4 and 11.5 together imply that A is nonsingular if and only if it is row equivalent to I. This means that a singular matrix is row-equivalent to a matrix that has a zero row.

Example 11.7.   Theorem 11.5 justifies the method we used in Chapter 2 for the computation of A^{-1}. If $A = \begin{bmatrix} 1 & 2 \\ -1 & 3 \end{bmatrix}$, find A^{-1} and express A as a product of elementary row matrices.

Attach I to A as a series of augmented columns, and apply a sequence of elementary row operations to A that reduce it to I. The attached matrix is A^{-1}.

$$\left[\begin{array}{cc} 1 & 2 \\ -1 & 3 \end{array}\middle|\begin{array}{cc} 1 & 0 \\ 0 & 1 \end{array}\right] \xrightarrow{R_2 = R_2 - (-1)R_1} \left[\begin{array}{cc} 1 & 2 \\ 0 & 5 \end{array}\middle|\begin{array}{cc} 1 & 0 \\ 1 & 1 \end{array}\right] \xrightarrow{R_1 = R_1 - \frac{2}{5}R_2}$$

$$\left[\begin{array}{cc} 1 & 0 \\ 0 & 5 \end{array}\middle|\begin{array}{cc} 3/5 & -2/5 \\ 1 & 1 \end{array}\right] \xrightarrow{R_2 = \frac{1}{5}R_2} \left[\begin{array}{cc} 1 & 0 \\ 0 & 1 \end{array}\middle|\begin{array}{cc} 3/5 & -2/5 \\ 1/5 & 1/5 \end{array}\right]$$

A is row-equivalent to I, so A is nonsingular, and

$$A^{-1} = \begin{bmatrix} 3/5 & -2/5 \\ 1/5 & 1/5 \end{bmatrix}.$$

The sequence of elementary row matrices that correspond to the row reduction from A to A^{-1} is

$$E_2\left(\frac{1}{5}\right) E_{12}\left(\frac{2}{5}\right) E_{21}(-1).$$

Thus,

$$A^{-1} = E_2\left(\frac{1}{5}\right) E_{12}\left(\frac{2}{5}\right) E_{21}(-1),$$

so

$$A = E_{21}(1) E_{12}\left(-\frac{2}{5}\right) E_2(5) = \begin{bmatrix} 1 & 0 \\ -1 & 1 \end{bmatrix}\begin{bmatrix} 1 & 2/5 \\ 0 & 1 \end{bmatrix}\begin{bmatrix} 1 & 0 \\ 0 & 5 \end{bmatrix} = \begin{bmatrix} 1 & 2 \\ -1 & 3 \end{bmatrix}. \qquad\blacksquare$$

11.4   DERIVATION OF THE *LU* DECOMPOSITION

We can use the elementary row matrices to explain the *LU* decomposition. For now we will assume there are no row exchanges, but will add row exchanges later when we discuss Gaussian elimination with partial pivoting. We will also assume no multiplication of a row by a scalar, so all we will use is the elementary row matrix $E_{ij}(t)$ that adds a multiple of row j to row i.

As we perform row operations, matrix elements change, but we will not use a notation for it, such as $\overline{a_{ij}}$ and simply maintain the notation a_{ij}. Let's look at the row reduction to an upper-triangular matrix column by column. First, start with column 1, multiply row 1 by a_{21}/a_{11}, and subtract from row 2. This eliminates the element in row 2, column 1:

$$
\begin{bmatrix}
a_{11} & a_{12} & \cdots & \cdots & a_{1n} \\
a_{21} & a_{22} & \cdots & \cdots & a_{2n} \\
a_{31} & \vdots & \ddots & \vdots & \vdots \\
\vdots & \vdots & \vdots & \ddots & \vdots \\
a_{n1} & a_{n2} & \cdots & \cdots & a_{nn}
\end{bmatrix}
\xrightarrow{\; R2 = R2 - \left(\dfrac{a_{21}}{a_{11}}\right) R1 \;}
\begin{bmatrix}
a_{11} & a_{12} & \cdots & \cdots & a_{1n} \\
0 & a_{22} & \cdots & \cdots & a_{2n} \\
a_{31} & \vdots & \ddots & \vdots & \vdots \\
\vdots & \vdots & \vdots & \ddots & \vdots \\
a_{n1} & a_{n2} & \cdots & \cdots & a_{nn}
\end{bmatrix}.
$$

Recall that a_{11} is called the pivot element. Now, multiply row 1 by a_{31}/a_{11} and subtract from row 3, eliminating the element in row 3, column 1:

$$
\begin{bmatrix}
a_{11} & a_{12} & \cdots & \cdots & a_{1n} \\
0 & a_{22} & \cdots & \cdots & a_{2n} \\
a_{31} & \vdots & \ddots & \vdots & \vdots \\
\vdots & \vdots & \vdots & \ddots & \vdots \\
a_{n1} & a_{n2} & \cdots & \cdots & a_{nn}
\end{bmatrix}
\xrightarrow{\; R3 = R3 - \left(\dfrac{a_{31}}{a_{11}}\right) R1 \;}
\begin{bmatrix}
a_{11} & a_{12} & \cdots & \cdots & a_{1n} \\
0 & a_{22} & \cdots & \cdots & a_{2n} \\
0 & \vdots & \ddots & \vdots & \vdots \\
\vdots & \vdots & \vdots & \ddots & \vdots \\
a_{n1} & a_{n2} & \cdots & \cdots & a_{nn}
\end{bmatrix}.
$$

Continue in this way until subtracting the multiple a_{n1}/a_{11} of row 1 from row n. These actions correspond to multiplication on the left by the series of elementary matrices

$$
E_{n1}\left(\frac{a_{n1}}{a_{11}}\right)\ldots E_{31}\left(\frac{a_{31}}{a_{11}}\right) E_{21}\left(\frac{a_{21}}{a_{11}}\right),
$$

resulting in the row reduced matrix

$$
E_{n1}\left(\frac{a_{n1}}{a_{11}}\right)\ldots E_{31}\left(\frac{a_{31}}{a_{11}}\right) E_{21}\left(\frac{a_{21}}{a_{11}}\right) A =
\begin{bmatrix}
a_{11} & a_{12} & \cdots & \cdots & a_{1n} \\
0 & a_{22} & \cdots & \cdots & a_{2n} \\
0 & a_{32} & \ddots & \vdots & \vdots \\
\vdots & \vdots & \vdots & \ddots & \vdots \\
0 & a_{n2} & \cdots & \cdots & a_{nn}
\end{bmatrix}.
$$

Continue this same process in column 2 with pivot element a_{22} to obtain

$$
\left[E_{n2}\left(\frac{a_{n2}}{a_{22}}\right)\ldots E_{42}\left(\frac{a_{42}}{a_{22}}\right) E_{32}\left(\frac{a_{32}}{a_{22}}\right)\right]\left[E_{n1}\left(\frac{a_{n1}}{a_{11}}\right)\ldots E_{31}\left(\frac{a_{31}}{a_{11}}\right) E_{21}\left(\frac{a_{21}}{a_{11}}\right)\right] A
$$

$$
=
\begin{bmatrix}
a_{11} & a_{12} & \cdots & \cdots & a_{1n} \\
0 & a_{22} & \cdots & \cdots & a_{2n} \\
0 & 0 & a_{33} & \vdots & \vdots \\
\vdots & \vdots & \vdots & \ddots & \vdots \\
0 & 0 & a_{n3} & \cdots & a_{nn}
\end{bmatrix}.
$$

In general, elimination in column i corresponds to the elementary matrix product

$$
E_{ni}\left(\frac{a_{ni}}{a_{ii}}\right)\ldots E_{i+2,i}\left(\frac{a_{i+2,i}}{a_{ii}}\right) E_{i+1,i}\left(\frac{a_{i+1,i}}{a_{ii}}\right).
$$

Putting this all together, we have

$$
E_{n,n-1}\left(\frac{a_{n,n-1}}{a_{n-1,n-1}}\right)
$$
$$
\times \cdots \times E_{ni}\left(\frac{a_{ni}}{a_{ii}}\right)\ldots E_{i+2,i}\left(\frac{a_{i+2,i}}{a_{ii}}\right) E_{i+1,i}\left(\frac{a_{i+1,i}}{a_{ii}}\right)
$$
$$
\times \cdots \times E_{n2}\left(\frac{a_{n2}}{a_{22}}\right)\ldots E_{42}\left(\frac{a_{42}}{a_{22}}\right) E_{32}\left(\frac{a_{32}}{a_{22}}\right)
$$
$$
\times E_{n1}\left(\frac{a_{n1}}{a_{11}}\right)\ldots E_{31}\left(\frac{a_{31}}{a_{11}}\right) E_{21}\left(\frac{a_{21}}{a_{11}}\right) A = U.
$$

We can simplify this expression by combining factors. Let E_i be the product of the elementary row matrices that perform row elimination in column i:

$$E_i = E_{ni}\left(\frac{a_{ni}}{a_{ii}}\right)\ldots E_{i+2,i}\left(\frac{a_{i+2,i}}{a_{ii}}\right)E_{i+1,i}\left(\frac{a_{i+1,i}}{a_{ii}}\right).$$

If A is a 3×3 matrix

$$\begin{bmatrix} a_{11} & a_{12} & a_{13} \\ a_{21} & a_{22} & a_{23} \\ a_{31} & a_{32} & a_{33} \end{bmatrix},$$

$$\begin{aligned} E_1 &= E_{31}\left(\frac{a_{31}}{a_{11}}\right)E_{21}\left(\frac{a_{21}}{a_{11}}\right) \\ &= \begin{bmatrix} 1 & 0 & 0 \\ 0 & 1 & 0 \\ -\dfrac{a_{31}}{a_{11}} & 0 & 1 \end{bmatrix}\begin{bmatrix} 1 & 0 & 0 \\ -\dfrac{a_{21}}{a_{11}} & 1 & 0 \\ 0 & 0 & 1 \end{bmatrix}, \\ &= \begin{bmatrix} 1 & 0 & 0 \\ -\dfrac{a_{21}}{a_{11}} & 1 & 0 \\ -\dfrac{a_{31}}{a_{11}} & 0 & 1 \end{bmatrix}. \end{aligned} \tag{11.1}$$

In general,

$$E_i = \begin{bmatrix} 1 & 0 & \ldots & 0 & 0 & 0 & \ldots & 0 \\ 0 & 1 & \ldots & 0 & 0 & 0 & \ldots & 0 \\ 0 & 0 & \ddots & \vdots & \vdots & \vdots & \ldots & 0 \\ & 0 & & 0 & 0 & \ldots & \ldots & 0 \\ & & & 1 & 0 & \ldots & \ldots & 0 \\ \vdots & \vdots & & -\dfrac{a_{i+1,i}}{a_{ii}} & 1 & \ldots & \ldots & 0 \\ & & & -\dfrac{a_{i+2,i}}{a_{ii}} & 0 & \ddots & \ldots & 0 \\ & & & \vdots & & \ldots & \ddots & \vdots \\ 0 & 0 & & -\dfrac{a_{ni}}{a_{ii}} & 0 & \ldots & 0 & 1 \end{bmatrix}. \tag{11.2}$$

We now have $E_{n-1}E_{n-2}\ldots E_2 E_1 A = U$, and

$$A = (E_{n-1}E_{n-2}\ldots E_2 E_1)^{-1}U = \left(E_1^{-1}E_2^{-1}E_3^{-1}\ldots E_{n-2}^{-1}E_{n-1}^{-1}\right)U.$$

In the case of a 3×3 matrix, from Equation 11.1 we see that

$$\begin{aligned} E_1^{-1} &= \left[E_{21}\left(\frac{a_{21}}{a_{11}}\right)\right]^{-1}\left[E_{31}\left(\frac{a_{31}}{a_{11}}\right)\right]^{-1} \\ &= E_{21}\left(-\frac{a_{21}}{a_{11}}\right)E_{31}\left(-\frac{a_{31}}{a_{11}}\right) \\ &\quad \begin{bmatrix} 1 & 0 & 0 \\ \dfrac{a_{21}}{a_{11}} & 1 & 0 \\ 0 & 0 & 1 \end{bmatrix}\begin{bmatrix} 1 & 0 & 0 \\ 0 & 1 & 0 \\ \dfrac{a_{31}}{a_{11}} & 0 & 1 \end{bmatrix} \\ &= \begin{bmatrix} 1 & 0 & 0 \\ \dfrac{a_{21}}{a_{11}} & 1 & 0 \\ \dfrac{a_{31}}{a_{11}} & 0 & 1 \end{bmatrix} \end{aligned}$$

In general,

$$
E_i^{-1} = \begin{bmatrix}
1 & 0 & \cdots & & 0 & 0 & 0 & \cdots & 0 \\
0 & 1 & \cdots & & 0 & 0 & 0 & \cdots & 0 \\
0 & 0 & \ddots & & \vdots & \vdots & \vdots & \cdots & 0 \\
& 0 & & & 0 & 0 & \cdots & \cdots & 0 \\
& & & & 1 & 0 & \cdots & \cdots & 0 \\
\vdots & \vdots & & \dfrac{a_{i+1,i}}{a_{ii}} & 1 & \cdots & \cdots & 0 \\
& & & \dfrac{a_{i+2,i}}{a_{ii}} & 0 & \ddots & \cdots & 0 \\
& & & \vdots & & \cdots & \cdots & \ddots & \vdots \\
0 & 0 & & \dfrac{a_{ni}}{a_{ii}} & 0 & \cdots & 0 & 1
\end{bmatrix}.
\tag{11.3}
$$

The product of the lower-triangular matrices $L = E_1^{-1}E_2^{-1}E_3^{-1}\ldots E_{n-2}^{-1}E_{n-1}^{-1}$ in Equation 11.3 is a lower-triangular matrix. From Equation 11.3, in the case of a 3×3 matrix,

$$
\begin{aligned}
L &= E_1^{-1}E_2^{-1} \\
&= \begin{bmatrix} 1 & 0 & 0 \\ \dfrac{a_{21}}{a_{11}} & 1 & 0 \\ \dfrac{a_{31}}{a_{11}} & 0 & 1 \end{bmatrix}
\begin{bmatrix} 1 & 0 & 0 \\ 0 & 1 & 0 \\ 0 & \dfrac{a_{32}}{a_{22}} & 1 \end{bmatrix} \\
&= \begin{bmatrix} 1 & 0 & 0 \\ \dfrac{a_{21}}{a_{11}} & 1 & 0 \\ \dfrac{a_{31}}{a_{11}} & \dfrac{a_{32}}{a_{22}} & 1 \end{bmatrix}
\end{aligned}
$$

For the $n \times n$ problem, $L = E_1^{-1}E_2^{-1}E_3^{-1}\ldots E_{n-2}^{-1}E_{n-1}^{-1}$ is a lower-diagonal matrix with ones on the main diagonal. Below each 1 are the multipliers used to perform row elimination. This is precisely what we described in Section 11.1, and we have shown that if the pivot is never zero, we can factor A as $A = LU$.

Remark 11.3.   For our decomposition algorithm to work, a_{ii} cannot equal 0. If during elimination $a_{ii} = 0$, requiring a row exchange, the *LU* decomposition as we have developed fails.

Example 11.8. $A = \begin{bmatrix} 2 & 1 & 1 \\ 2 & 1 & 3 \\ 4 & -1 & 1 \end{bmatrix}$. Let's carry out the *LU* decomposition.

$$
\begin{bmatrix} 2 & 1 & 1 \\ 2 & 1 & 3 \\ 4 & -1 & 1 \end{bmatrix}
\xrightarrow[\;R3 = R3 - (2)\,R1\;]{R2 = R2 - (1)\,R1}
\begin{bmatrix} 2 & 1 & 1 \\ (1) & 0 & 2 \\ (2) & -3 & -1 \end{bmatrix}
$$

$$
\xrightarrow{R2 \leftrightarrow R3}
\begin{bmatrix} 2 & 1 & 1 \\ (2) & -3 & -1 \\ (1) & 0 & 2 \end{bmatrix}
\xrightarrow{R_3 = R_3 - (0)\,R_2}
\begin{bmatrix} 2 & 1 & 1 \\ (2) & -3 & -1 \\ (1) & (0) & 2 \end{bmatrix}
$$

$$
L = \begin{bmatrix} 1 & 0 & 0 \\ 2 & 1 & 0 \\ 1 & 0 & 1 \end{bmatrix}, \; U = \begin{bmatrix} 2 & 1 & 1 \\ 0 & -3 & -1 \\ 0 & 0 & 2 \end{bmatrix}
$$

and

$$
LU = \begin{bmatrix} 2 & 1 & 1 \\ 4 & -1 & 1 \\ 2 & 1 & 3 \end{bmatrix}.
$$

The result is the original matrix with rows two and three swapped. Somehow, we need to account for the fact that we swapped rows two and three during row reduction. ∎

The process we have described results in $A = LU$. Perhaps we could have done the decomposition differently and arrived at another L and U; in other words, how do we know that there are no other LU factorizations of an $n \times n$ matrix A?

Theorem 11.6. *The LU decomposition of a nonsingular matrix A is unique.*

Proof. Assume there are two factorizations $L_1 U_1 = L_2 U_2 = A$. We need to show that $L_1 = L_2$ and $U_1 = U_2$. Since $L_1 U_1 = L_2 U_2$,

$$U_1 U_2^{-1} = L_1^{-1} L_2. \tag{11.4}$$

In Equation 11.4, $U_1 U_2^{-1}$ is an upper-triangular matrix, and $L_1^{-1} L_2$ is a lower-triangular matrix with 1s on its diagonal. Here is what we must have:

$$U_1 U_2^{-1} = \begin{bmatrix} \overline{a_{11}} & \overline{a_{12}} & \dots & \overline{a_{1n}} \\ 0 & \overline{a_{22}} & \dots & \overline{a_{2n}} \\ \vdots & \vdots & \ddots & \vdots \\ 0 & 0 & \dots & \overline{a_{nn}} \end{bmatrix} = L_1^{-1} L_2 = \begin{bmatrix} 1 & 0 & \dots & 0 \\ \overline{b_{21}} & 1 & \dots & 0 \\ \vdots & \vdots & \ddots & \vdots \\ \overline{b_{n1}} & \overline{b_{n2}} & \dots & 1 \end{bmatrix}.$$

The diagonals of both matrices must be equal, so $\overline{a_{ii}} = \overline{b_{ii}} = 1$. Since $U_1 U_2^{-1}$ is an upper-triangular matrix it must have 0s below the diagonal, and since $L_1^{-1} L_2$ is a lower-triangular matrix it must have 0s above the diagonal. Thus, both matrices must be the identity, and $U_1 U_2^{-1} = I$, so $U_1 = U_2$. Likewise, $L_1 = L_2$. □

With the exception of forming L, the algorithm for the LU decomposition is a straightforward expression of what we did by hand in Chapter 2. To maintain L, after performing a row elimination step, store the multiplier in the entry of A that becomes zero. We can now describe the algorithm simply:

Beginning with row 1:

Multiply row 1 by $\frac{a_{j1}}{a_{11}}$ and subtract from row j, $2 \le j \le n$, in order to eliminate all the elements in column 1 below a_{11}. For each j, store $\frac{a_{1j}}{a_{11}}$ in location $(j, 1)$.

Multiply row 2 by $\frac{a_{j2}}{a_{22}}$ and subtract from row j, $3 \le j \le n$, in order to eliminate all the elements in column 2 below a_{22}. For each j, store $\frac{a_{j2}}{a_{22}}$ in location $(j, 2)$.

$\dots$

Multiply row n-1 by $\frac{a_{n,n-1}}{a_{n-1,n-1}}$ and subtract from row n, in order to eliminate the element in row n, column n-1. Store $\frac{a_{n,n-1}}{a_{n-1,n-1}}$ in location $(n, n-1)$.

For the sake of clarity and brevity, we will begin using the colon notation in algorithms. It is the same format as the corresponding notation in MATLAB.

11.4.1 Colon Notation

When our pseudocode deals with an individual element in row i, column j, of an $m \times n$ matrix A, we use the notation a_{ij}. We adopt the MATLAB *colon notation* that allows us to access blocks of elements from matrices and perform operations. The notation $A(:, j)$ references column j of A:

$$A(:, j) = \begin{bmatrix} a_{1j} \\ a_{2j} \\ \vdots \\ a_{m-1,j} \\ a_{mj} \end{bmatrix}.$$

$$\begin{bmatrix} a_{11} & a_{12} & \cdots & a_{1k} & \cdots & a_{1n} \\ a_{21} & a_{22} & \cdots & a_{2k} & \cdots & a_{2n} \\ \vdots & \vdots & \cdots & \vdots & \vdots & \vdots \\ a_{k1} & a_{k2} & \cdots & a_{kk} & \cdots & a_{kn} \\ \vdots & \vdots & \vdots & \vdots & \vdots & \vdots \\ a_{n1} & a_{n2} & \cdots & a_{nk} & \cdots & a_{nn} \end{bmatrix}$$

FIGURE 11.2 $k \times k$ submatrix.

For row i, the notation is $A(i, :)$:

$$A(i, :) = \begin{bmatrix} a_{i1} & a_{i2} & \cdots & a_{i,n-1} & a_{in} \end{bmatrix}.$$

If A is an $n \times n$ matrix and we want to multiply row i by 3, use the following statement:

$$A(i, :) = 3 * A(i, :).$$

If we need to multiply row i by 7 and subtract it from row j, we can write

$$A(j, :) = A(j, :) - 7 * A(i, :).$$

The colon notation can reference both rows and columns. For instance, if A is an $n \times n$ matrix, $A(1 : k, 1 : k)$ is the $k \times k$ submatrix beginning in the upper left-hand corner of A (Figure 11.2).

The following statement replaces the upper $k \times k$ submatrix by the $k \times k$ matrix B:

$$A(1 : k, 1 : k) = B.$$

The next statement subtracts the elements in row i, columns i through n from the same portion of each row below row i:

$$A(i+1 : n, i : n) = A(i+1 : n, i : n) - \begin{bmatrix} 1 \\ \vdots \\ 1 \end{bmatrix}^{(n-i) \times 1} A(i, i : n).$$

For instance, if $A = \begin{bmatrix} 1 & 2 & 5 \\ 8 & 3 & 7 \\ 1 & 1 & 4 \end{bmatrix}$,

$$\begin{aligned} A(2 : 3, 1 : 3) &= A(2 : 3, 1 : 3) - \begin{bmatrix} 1 \\ 1 \end{bmatrix} A(1, 1 : 3) \\ &= \begin{bmatrix} 8 & 3 & 7 \\ 1 & 1 & 4 \end{bmatrix} - \begin{bmatrix} 1 \\ 1 \end{bmatrix} \begin{bmatrix} 1 & 2 & 5 \end{bmatrix} \\ &= \begin{bmatrix} 8 & 3 & 7 \\ 1 & 1 & 4 \end{bmatrix} - \begin{bmatrix} 1 & 2 & 5 \\ 1 & 2 & 5 \end{bmatrix} \\ &= \begin{bmatrix} 7 & 1 & 2 \\ 0 & -1 & -1 \end{bmatrix}, \end{aligned}$$

and

$$A = \begin{bmatrix} 1 & 2 & 5 \\ 7 & 1 & 2 \\ 0 & -1 & -1 \end{bmatrix}.$$

Remark 11.4.   Using the colon notation frees us to think at the vector and matrix level and concentrate on more important computational issues.

11.4.2 The *LU* Decomposition Algorithm

Algorithm 11.1 describes the *LU* factorization, assuming the pivot element is nonzero. The algorithm makes use of the colon notation and includes use of the functions triu and tril. A call to triu(A) returns the upper-triangular portion of A, and tril(A,-1) returns the portion of A below the main diagonal.

Algorithm 11.1 *LU* Decomposition Without a Zero on the Diagonal

```
function LUGAUSS(A)
   % Input: n × n matrix A.
   % Output: lower-triangular matrix L and upper-triangular matrix U such that A = LU.
   for i = 1:n-1 do
      if a_ii = 0 then
         print 'The algorithm has encountered a zero pivot.'
         exit
      end if
      % Replace the elements in column i, rows i+1 to n by the multipliers a_ji/a_ii
      A(i + 1 : n, i) = A(i + 1 : n, i)/a_ii
      % Modify the elements in rows i+1 to n, columns i+1 to n by subtracting
      % the multiplier for the row times the elements in row i.
      A(i + 1 : n, i + 1 : n) = A(i + 1 : n, i + 1 : n) − A(i + 1 : n, i)A(i, i + 1 : n)
   end for
   % Assign U the upper-triangular portion of A.
   U = triu(A)
   % Initialize L as the identity matrix.
   L = I % Add into L the portion of A below the main diagonal.
   L = L + tril(A,-1);
   return [ L, U ]
end function
```

NLALIB: The function lugauss implements Algorithm 11.1.

Example 11.9. In this example, we use the function lugauss to factor a 4 × 4 matrix.

```
A =
   1   -3    5    2
   1    0    1   -1
   6    1   -9    2
   1    0   -6    3
>> [L, U] = lugauss(A)

L =
           1           0           0           0
           1           1           0           0
           6      6.3333           1           0
           1           1      0.5122           1

U =
           1          -3           5           2
           0           3          -4          -3
           0           0     -13.667           9
           0           0           0    -0.60976
>> L*U

ans =
           1          -3           5           2
           1           0           1          -1
           6           1          -9           2
           1           0          -6           3
```

■

11.4.3 *LU* Decomposition Flop Count

Gaussian elimination is a relatively slow algorithm. Developing a flop count will tell how much work is actually involved in computing L and U. We will count first for $i = 1$, then $i = 2$, and so forth until $i = n - 1$ and form the sum of the counts. The annotated Figure 11.3 will aid in understanding the computation.

Flop Count

$i = 1$:

The $n - 1$ quotients $a_{21}/a_{11}, a_{31}/a_{11}, \ldots, a_{n1}/a_{11}$ are the row multipliers. The elements at indices $(2, 1), (3, 1), \ldots, (n, 1)$ are replaced by the multipliers, so row elimination operations occur in the $(n - 1) \times (n - 1)$ submatrix $A(2 : n, 2 : n)$. There are $(n - 1)^2$ elements that must be modified. To modify an element requires a multiplication and a subtraction, or 2 flops. The total flops required for row elimination is $2(n - 1)^2$. For $i = 1$, the flop count is $(n - 1) + 2(n - 1)^2$.

$i = 2$:

The $n - 2$ quotients $a_{32}/a_{22}, a_{42}/a_{22}, \ldots, a_{n2}/a_{22}$ are the row multipliers. There are $(n - 2)^2$ elements that must be modified in rows 3 through n, and the total flops required for the modification is $2(n - 2)^2$. For $i = 2$, the flop count is $(n - 2) + 2(n - 2)^2$.

$\ldots$

$i = n - 1$:

Compute 1 quotient $\frac{a_{n,n-1}}{a_{n-1,n-1}}$ and execute 1 multiplication and 1 subtraction, for a total of $1 + 2(1)$ flops.

The total count is

$$\sum_{i=1}^{n-1}(n - i) + \sum_{i=1}^{n-1}2(n - i)^2 = \sum_{i=1}^{n-1}i + 2\sum_{i=1}^{n-1}i^2. \tag{11.5}$$

To evaluate Equation 11.5, we need two summation formulas:

$$\begin{aligned}\sum_{i=1}^{k}i &= \frac{k(k+1)}{2}\\ \sum_{i=1}^{k}i^2 &= \frac{k(k+1)(2k+1)}{6}\end{aligned}$$

By applying these formulas to Equation 11.5, we have

$$\sum_{i=1}^{n-1}i + 2\sum_{i=1}^{n-1}i^2 = \frac{n(n-1)}{2} + 2\frac{(n-1)n(2n-1)}{6},$$

and after combining terms the flop count is

$$\frac{2}{3}n^3 - \frac{1}{2}n^2 - \frac{1}{6}n. \tag{11.6}$$

The dominant term in Equation 11.6 is $\frac{2}{3}n^3$, so the flop count is $\frac{2}{3}n^3 + O(n^2)$, and *LU* decomposition is a cubic algorithm.

A good reason for computing L and U is that the cubic *LU* decomposition algorithm allows us to repeatedly solve $Ax = b$ for many $b's$ without having to recompute either L or U, as we will show in Section 11.6.

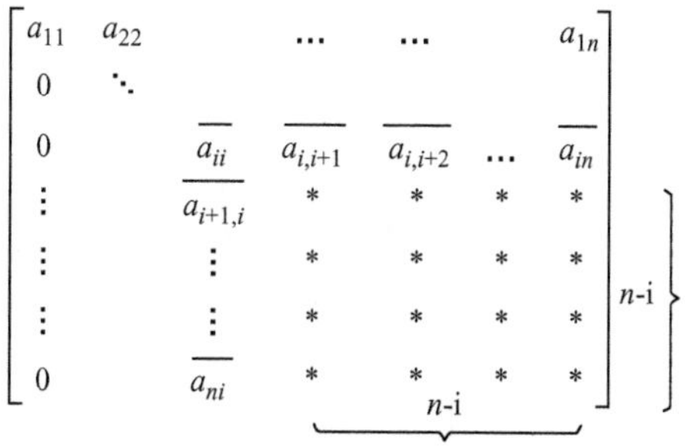

FIGURE 11.3 Gaussian elimination flop count.

11.5 GAUSSIAN ELIMINATION WITH PARTIAL PIVOTING

Gaussian elimination can produce extremely bad results under certain circumstances; in fact, the results can be completely wrong. Consider the matrix

$$A = \begin{bmatrix} 0.00001 & 3 \\ 2 & 1 \end{bmatrix},$$

and use three-digit arithmetic. There is only one step required to produce the LU decomposition. Use the multiplier $\frac{2}{0.00001} = 2 \times 10^5$. In exact arithmetic, the elimination step gives

$$\begin{bmatrix} 0.00001 & 1 \\ 1 & 1 \end{bmatrix} \rightarrow \begin{bmatrix} 0.00001 & 3 \\ 0 & 1 - 3\left(\frac{2}{0.00001}\right) \end{bmatrix} = \begin{bmatrix} 0.00001 & 3 \\ 0 & -599999 \end{bmatrix},$$

and the LU decomposition is $L = \begin{bmatrix} 1 & 0 \\ 200000 & 1 \end{bmatrix}$, $U = \begin{bmatrix} 0.00001 & 3 \\ 0 & -599999 \end{bmatrix}$.

In our three-digit arithmetic, the 1 is lost, and the result is

$$L = \begin{bmatrix} 1 & 0 \\ 200000 & 1 \end{bmatrix}, \ U = \begin{bmatrix} 0.00001 & 3 \\ 0 & -600000 \end{bmatrix}$$

Now compute LU to get

$$\begin{bmatrix} 0.00001 & 3 \\ 2 & 0 \end{bmatrix},$$

which is disastrously different from A! The problem arose when we divided by a small pivot element and obtained a large multiplier. The large multiplier resulted in the addition of a very large number to a much smaller number. The result was loss of any contribution from the small number. There is a good solution to this problem. When choosing the pivot element on the diagonal at position a_{ii}, locate the element in column i at or below the diagonal that is largest in magnitude, say a_{ji}, $i \leq j \leq n$. If $j \neq i$, interchange row j with row i, and then the multipliers, $\frac{a_{ji}}{a_{ii}}$, satisfy $\left| \frac{a_{ji}}{a_{ii}} \right| \leq 1$, $i+1 \leq j \leq n$, and we avoid multiplying a row by a large number and losing precision. We call this *Gaussian elimination with partial pivoting* (GEPP). Apply this strategy to our matrix, rounding to three-digit precision.

$$\begin{bmatrix} 0.00001 & 3 \\ 2 & 1 \end{bmatrix} \xrightarrow{R_1 \longleftrightarrow R_2} \begin{bmatrix} 2 & 1 \\ 0.00001 & 3 \end{bmatrix} \xrightarrow{R_2 = R_2 - \frac{0.00001}{2} R_1} \begin{bmatrix} 2 & 1 \\ 0 & 3 \end{bmatrix}$$

We now have
$L = \begin{bmatrix} 1 & 0 \\ \frac{0.00001}{2} & 1 \end{bmatrix}$, $U = \begin{bmatrix} 2 & 1 \\ 0 & 3 \end{bmatrix}$, and

$$LU = \begin{bmatrix} 2 & 1 \\ 0.00001 & 3 \end{bmatrix}.$$

Of course, this is not the original matrix A, but A with its two rows swapped (permuted). If we use GEPP, then an LU decomposition for A consists of three matrices P, L, and U such that

$$PA = LU.$$

P is a *permutation matrix*, also called the *pivot matrix*. Start with $P = I$, and swap rows i and j of the permutation matrix whenever rows i and j are swapped during GEPP. For instance,

$$P = \begin{pmatrix} 1 & 0 & 0 \\ 0 & 0 & 1 \\ 0 & 1 & 0 \end{pmatrix}$$

would be the permutation matrix if the second and third rows of A are interchanged during pivoting.

Example 11.10.   Factor the matrix $A = \begin{bmatrix} 3 & 8 & 1 \\ 5 & 2 & 0 \\ 6 & 1 & 12 \end{bmatrix}$.

$$L = \begin{bmatrix} 1 & 0 & 0 \\ 0 & 1 & 0 \\ 0 & 0 & 1 \end{bmatrix}, \quad P = \begin{bmatrix} 1 & 0 & 0 \\ 0 & 1 & 0 \\ 0 & 0 & 1 \end{bmatrix}, \quad A = \begin{bmatrix} 3 & 8 & 1 \\ 5 & 2 & 0 \\ 6 & 1 & 12 \end{bmatrix}.$$

Pivot row = 1. Swap rows 1 and 3, and permute *P*. Do not interchange rows of *L* until arriving at the pivot in row 2, column 2 (Remark 11.5).

$$L = \begin{bmatrix} 1 & 0 & 0 \\ 0 & 1 & 0 \\ 0 & 0 & 1 \end{bmatrix}, \quad P = \begin{bmatrix} 0 & 0 & 1 \\ 0 & 1 & 0 \\ 1 & 0 & 0 \end{bmatrix}, \quad A = \begin{bmatrix} 6 & 1 & 12 \\ 5 & 2 & 0 \\ 3 & 8 & 1 \end{bmatrix}.$$

Apply the pivot element, and add multipliers to *L*.

$$L = \begin{bmatrix} 1 & 0 & 0 \\ 5/6 & 1 & 0 \\ 1/2 & 0 & 1 \end{bmatrix}, \quad P = \begin{bmatrix} 0 & 0 & 1 \\ 0 & 1 & 0 \\ 1 & 0 & 0 \end{bmatrix}, \quad A = \begin{bmatrix} 6 & 1 & 12 \\ 0 & 7/6 & -10 \\ 0 & 15/2 & -5 \end{bmatrix}.$$

Pivot row = 2. Swap rows 2 and 3. Permute *P* and *L*.

$$L = \begin{bmatrix} 1 & 0 & 0 \\ 1/2 & 1 & 0 \\ 5/6 & 0 & 1 \end{bmatrix}, \quad P = \begin{bmatrix} 0 & 0 & 1 \\ 1 & 0 & 0 \\ 0 & 1 & 0 \end{bmatrix}, \quad A = \begin{bmatrix} 6 & 1 & 12 \\ 0 & 15/2 & -5 \\ 0 & 7/6 & -10 \end{bmatrix}.$$

Apply the pivot element and update *L*.

$$L = \begin{bmatrix} 1 & 0 & 0 \\ 1/2 & 1 & 0 \\ 5/6 & 7/45 & 1 \end{bmatrix}, \quad P = \begin{bmatrix} 0 & 0 & 1 \\ 1 & 0 & 0 \\ 0 & 1 & 0 \end{bmatrix}, \quad A = \begin{bmatrix} 6 & 1 & 12 \\ 0 & 15/2 & -5 \\ 0 & 0 & -83/9 \end{bmatrix}.$$

The final results are

$$L = \begin{bmatrix} 1 & 0 & 0 \\ 1/2 & 1 & 0 \\ 5/6 & 7/45 & 1 \end{bmatrix},$$

$$U = \begin{bmatrix} 6 & 1 & 12 \\ 0 & 15/2 & -5 \\ 0 & 0 & -83/9 \end{bmatrix},$$

$$P = \begin{bmatrix} 0 & 0 & 1 \\ 1 & 0 & 0 \\ 0 & 1 & 0 \end{bmatrix}.$$

You should verify that $PA = LU$. ∎

Remark 11.5.   Even if a row interchange occurs when dealing with column 1, do not interchange the corresponding rows of *L* until moving to column 2. Think of it this way. The matrix *A* after a row swap defines the starting configuration. The elements in the first column of *L* correspond to the multipliers after the row interchange involving pivot position $(1, 1)$, if any.

Summary

At step *i* of Gaussian elimination, the pivot element is a_{ii}. To eliminate each element a_{ji}, $i + 1 \le j \le n$, multiply row *i* by a_{ji}/a_{ii} and subtract from row *j*. These multipliers should not be large or precision can be lost. Find the largest of the elements $\{|a_{ii}|, |a_{i+1,i}|, |a_{i+2,i}|, \ldots, |a_{ni}|\}$. If the row index of the largest absolute value is $j \ne i$, exchange rows *i* and *j*. Now all the multipliers a_{ji}/a_{ii} have absolute value less than or equal to 1.

11.5.1   Derivation of *PA=LU*

For a reader primarily interested in applications, this subsection may be skipped; however, it is important to read Section 11.5.2.

For the explanation, we will assume that P_i is the permutation matrix corresponding a possible interchange when dealing with column *i*. The matrix E_i (Equation 11.2) is the product of the elementary row matrices that perform row elimination in column i. In the sequence of steps that follow, "exchange rows" means multiply by a permutation matrix or the identity if no row exchanges are required.

$$A = \begin{bmatrix} x & x & x & \dots & x & x & x \\ x & x & x & \dots & x & x & x \\ x & x & x & \dots & x & x & x \\ \vdots & \vdots & \vdots & \ddots & \vdots & \vdots & \vdots \\ x & x & x & \dots & x & x & x \\ x & x & x & \dots & x & x & x \\ x & x & x & \dots & x & x & x \end{bmatrix}.$$

Exchange rows, then perform elimination in column 1.

$$i = 1: \quad E_1 P_1 A = \begin{bmatrix} x & x & x & \dots & x & x & x \\ 0 & x & x & \dots & x & x & x \\ 0 & x & x & \dots & x & x & x \\ \vdots & \vdots & \vdots & \ddots & \vdots & \vdots & \vdots \\ 0 & x & x & \dots & x & x & x \\ 0 & x & x & \dots & x & x & x \\ 0 & x & x & \dots & x & x & x \end{bmatrix}.$$

Beginning with the matrix after step 1, exchange rows, then perform elimination in column 2.

$$i = 2: \quad E_2 P_2 E_1 P_1 A = \begin{bmatrix} x & x & x & \dots & x & x & x \\ 0 & x & x & \dots & x & x & x \\ 0 & 0 & x & \dots & x & x & x \\ \vdots & \vdots & \vdots & \ddots & \vdots & \vdots & \vdots \\ 0 & 0 & x & \dots & x & x & x \\ 0 & 0 & x & \dots & x & x & x \\ 0 & 0 & x & \dots & x & x & x \end{bmatrix}.$$

Beginning with the matrix after step 2, exchange rows, then perform elimination in column 3.

$$i = 3: \quad E_3 P_3 E_2 P_2 E_1 P_1 A = \begin{bmatrix} x & x & x & \dots & x & x & x \\ 0 & x & x & \dots & x & x & x \\ 0 & 0 & x & \dots & x & x & x \\ \vdots & \vdots & 0 & \ddots & \vdots & \vdots & \vdots \\ 0 & 0 & \vdots & \dots & x & x & x \\ 0 & 0 & 0 & \dots & x & x & x \\ 0 & 0 & 0 & \dots & x & x & x \end{bmatrix}$$

Beginning with the matrix after step $n - 2$, exchange rows, then eliminate the element in row n, column $n - 1$.

$$i = n - 1: \quad E_{n-1} P_{n-1} \dots E_3 P_3 E_2 P_2 E_1 P_1 A = \begin{bmatrix} x & x & x & \dots & x & x & x \\ 0 & x & x & \dots & x & x & x \\ 0 & 0 & x & \dots & x & x & x \\ \vdots & \vdots & 0 & \ddots & \vdots & \vdots & \vdots \\ 0 & 0 & \vdots & \dots & x & x & x \\ 0 & 0 & 0 & \dots & 0 & x & x \\ 0 & 0 & 0 & \dots & 0 & 0 & x \end{bmatrix} = U$$

The product $E_{n-1} P_{n-1} \dots E_3 P_3 E_2 P_2 E_1 P_1$ must be rewritten in order to isolate the permutation matrices in the order $P_{n-1} P_{n-2} \dots P_2 P_1$. First, note that if P_i is a permutation matrix, then $P_i^2 = I$, since $P_i P_i$ permutes I and then permutes the permutation back to I. Equivalently, $P_i^{-1} = P_i$. Let's look at a 3×3 example:

$$U = E_2 P_2 E_1 P_1 A = E_2 \left(P_2 E_1 P_2^{-1} \right) (P_2 P_1) A.$$

P_2P_1 is the product of permutation matrices, and so is a permutation matrix. Now, E_2 is an elementary matrix obtained by subtracting a multiple of row 2($\begin{bmatrix} 0 & 1 & 0 \end{bmatrix}$) from row 3 ($\begin{bmatrix} 0 & 0 & 1 \end{bmatrix}$) of I, and so it is a lower-triangular matrix. The factor $P_2E_1P_2^{-1}$ is interesting. P_2E_1 permutes rows of E_1, and E_1 is a lower-triangular matrix. Multiplying on the right by a permutation matrix exchanges columns (Problem 11.5). Thus, $P_2E_1P_2^{-1}$ is a lower-triangular matrix, and is E_1 with its subdiagonal elements permuted. The product of lower-triangular matrices is lower triangular, and $\overline{L}PA = U$, where $\overline{L} = E_2\left(P_2E_1P_2^{-1}\right)$ is a lower-triangular matrix. Since it is built from permutations of elementary matrices, it is invertible, and $PA = \left(\overline{L}\right)^{-1}U = LU$, where $L = \left(\overline{L}\right)^{-1}$. The inverse of a lower-triangular matrix is lower triangular, so L is lower triangular. As was the case without pivoting, L has ones on its diagonal.

Now consider a 4×4 example:

$$U = E_3P_3E_2P_2E_1P_1 = E_3\left(P_3E_2P_3^{-1}\right)\left(P_3P_2E_1P_2^{-1}P_3^{-1}\right)(P_3P_2P_1)A.$$

Each of $\{E_1, E_2, E_3\}$ is a lower-triangular matrix, and the permutation matrices are arranged such that each factor remains a lower-triangular matrix.

This manipulation can be carried out for any $n \times n$ matrix. Each factor before the final $(P_{n-1}P_{n-2}\ldots P_2P_1)$ is an invertible lower-triangular matrix, say $\hat{T}_i$, so we have

$$U = \hat{T}_{n-1}\hat{T}_{n-2}\ldots\hat{T}_2\hat{T}_1\,(P_{n-1}P_{n-2}\ldots P_2P_1)\,A,$$

and

$$(P_{n-1}P_{n-2}\ldots P_2P_1)\,A = \left(\hat{T}_{n-1}\hat{T}_{n-2}\ldots\hat{T}_2\hat{T}_1\right)^{-1}U,$$

or $PA = LU$.

Example 11.11. Let $A = \begin{bmatrix} -1 & -1 & 0 & 1 \\ -1 & 1 & 1 & 0 \\ 1 & 1 & 1 & 1 \\ 2 & 0 & 1 & 0 \end{bmatrix}$. This example illustrates the process just described that derives GEPP.

Start: $L = \begin{bmatrix} 1 & 0 & 0 & 0 \\ 0 & 1 & 0 & 0 \\ 0 & 0 & 1 & 0 \\ 0 & 0 & 0 & 1 \end{bmatrix}$, $P = \begin{bmatrix} 1 & 0 & 0 & 0 \\ 0 & 1 & 0 & 0 \\ 0 & 0 & 1 & 0 \\ 0 & 0 & 0 & 1 \end{bmatrix}$

$i = 1$: Exchange rows 1 and 4. Eliminate entries at indices $(2, 1) - (4, 1)$:

$$E_1 = E_{41}\left(-\tfrac{1}{2}\right)E_{31}\left(\tfrac{1}{2}\right)E_{21}\left(-\tfrac{1}{2}\right) = \begin{bmatrix} 1 & 0 & 0 & 0 \\ 0 & 1 & 0 & 0 \\ 0 & 0 & 1 & 0 \\ 0.5 & 0 & 0 & 1 \end{bmatrix}\begin{bmatrix} 1 & 0 & 0 & 0 \\ 0 & 1 & 0 & 0 \\ -0.5 & 0 & 1 & 0 \\ 0 & 0 & 0 & 1 \end{bmatrix}\begin{bmatrix} 1 & 0 & 0 & 0 \\ 0.5 & 1 & 0 & 0 \\ 0 & 0 & 1 & 0 \\ 0 & 0 & 0 & 1 \end{bmatrix} = \begin{bmatrix} 1 & 0 & 0 & 0 \\ 0.5 & 1 & 0 & 0 \\ -0.5 & 0 & 1 & 0 \\ 0.5 & 0 & 0 & 1 \end{bmatrix}$$

$$P_1 = \begin{bmatrix} 0 & 0 & 0 & 1 \\ 0 & 1 & 0 & 0 \\ 0 & 0 & 1 & 0 \\ 1 & 0 & 0 & 0 \end{bmatrix}$$

$$E_1P_1A = \begin{bmatrix} 2 & 0 & 1 & 0 \\ 0 & 1 & 1.5 & 0 \\ 0 & 1 & 0.5 & 1 \\ 0 & -1 & 0.5 & 1 \end{bmatrix}$$

$i = 2$: A row exchange is not necessary. Eliminate the entries at indices $(3, 2)$ and $(4, 2)$.

$$E_2 = E_{42}(-1) E_{32}(1) = \begin{bmatrix} 1 & 0 & 0 & 0 \\ 0 & 1 & 0 & 0 \\ 0 & 0 & 1 & 0 \\ 0 & 1 & 0 & 1 \end{bmatrix} \begin{bmatrix} 1 & 0 & 0 & 0 \\ 0 & 1 & 0 & 0 \\ 0 & -1 & 1 & 0 \\ 0 & 0 & 0 & 1 \end{bmatrix} = \begin{bmatrix} 1 & 0 & 0 & 0 \\ 0 & 1 & 0 & 0 \\ 0 & -1 & 1 & 0 \\ 0 & 1 & 0 & 1 \end{bmatrix}$$

$$P_2 = \begin{bmatrix} 1 & 0 & 0 & 0 \\ 0 & 1 & 0 & 0 \\ 0 & 0 & 1 & 0 \\ 0 & 0 & 0 & 1 \end{bmatrix}$$

$$E_2 P_2 E_1 P_1 A = \begin{bmatrix} 2 & 0 & 1 & 0 \\ 0 & 1 & 1.5 & 0 \\ 0 & 0 & -1 & 1 \\ 0 & 0 & 2 & 1 \end{bmatrix}$$

$i = 3$: Interchange rows 3 and 4. Eliminate the entry at $(4, 3)$.

$$E_3 = E_{43}\left(-\frac{1}{2}\right) = \begin{bmatrix} 1 & 0 & 0 & 0 \\ 0 & 1 & 0 & 0 \\ 0 & 0 & 1 & 0 \\ 0 & 0 & 0.5 & 1 \end{bmatrix}$$

$$P_3 = \begin{bmatrix} 1 & 0 & 0 & 0 \\ 0 & 1 & 0 & 0 \\ 0 & 0 & 0 & 1 \\ 0 & 0 & 1 & 0 \end{bmatrix}$$

$$E_3 P_3 E_2 P_2 E_1 P_1 A = \begin{bmatrix} 2 & 0 & 1 & 0 \\ 0 & 1 & 1.5 & 0 \\ 0 & 0 & 2 & 1 \\ 0 & 0 & 0 & 1.5 \end{bmatrix} = U$$

Form

$E_3 (P_3 E_2 P_3)(P_3 P_2 E_1 P_2 P_3)(P_3 P_2 P_1) A$

$$= \begin{bmatrix} 1 & 0 & 0 & 0 \\ 0 & 1 & 0 & 0 \\ 0 & 0 & 1 & 0 \\ 0 & 0 & 0.5 & 1 \end{bmatrix} \begin{bmatrix} 1 & 0 & 0 & 0 \\ 0 & 1 & 0 & 0 \\ 0 & 1 & 1 & 0 \\ 0 & -1 & 0 & 1 \end{bmatrix} \begin{bmatrix} 1 & 0 & 0 & 0 \\ .5 & 1 & 0 & 0 \\ .5 & 0 & 1 & 0 \\ -.5 & 0 & 0 & 1 \end{bmatrix} \begin{bmatrix} 0 & 0 & 0 & 1 \\ 0 & 1 & 0 & 0 \\ 1 & 0 & 0 & 0 \\ 0 & 0 & 1 & 0 \end{bmatrix} \begin{bmatrix} -1 & -1 & 0 & 1 \\ -1 & 1 & 1 & 0 \\ 1 & 1 & 1 & 1 \\ 2 & 0 & 1 & 0 \end{bmatrix} = U$$

$$P = \begin{bmatrix} 0 & 0 & 0 & 1 \\ 0 & 1 & 0 & 0 \\ 1 & 0 & 0 & 0 \\ 0 & 0 & 1 & 0 \end{bmatrix}, L = \left(\begin{bmatrix} 1 & 0 & 0 & 0 \\ 0 & 1 & 0 & 0 \\ 0 & 0 & 1 & 0 \\ 0 & 0 & 0.5 & 1 \end{bmatrix} \begin{bmatrix} 1 & 0 & 0 & 0 \\ 0 & 1 & 0 & 0 \\ 0 & 1 & 1 & 0 \\ 0 & -1 & 0 & 1 \end{bmatrix} \begin{bmatrix} 1 & 0 & 0 & 0 \\ 0.5 & 1 & 0 & 0 \\ 0.5 & 0 & 1 & 0 \\ -0.5 & 0 & 0 & 1 \end{bmatrix} \right)^{-1} = \begin{bmatrix} 1 & 0 & 0 & 0 \\ -0.5 & 1 & 0 & 0 \\ -0.5 & -1 & 1 & 0 \\ 0.5 & 1 & -0.5 & 1 \end{bmatrix}$$

Now,

$$PA = \begin{bmatrix} 0 & 0 & 0 & 1 \\ 0 & 1 & 0 & 0 \\ 1 & 0 & 0 & 0 \\ 0 & 0 & 1 & 0 \end{bmatrix} \begin{bmatrix} -1 & -1 & 0 & 1 \\ -1 & 1 & 1 & 0 \\ 1 & 1 & 1 & 1 \\ 2 & 0 & 1 & 0 \end{bmatrix} = \begin{bmatrix} 2 & 0 & 1 & 0 \\ -1 & 1 & 1 & 0 \\ -1 & -1 & 0 & 1 \\ 1 & 1 & 1 & 1 \end{bmatrix}$$

$$LU = \begin{bmatrix} 1 & 0 & 0 & 0 \\ -0.5 & 1 & 0 & 0 \\ -0.5 & -1 & 1 & 0 \\ 0.5 & 1 & -0.5 & 1 \end{bmatrix} \begin{bmatrix} 2 & 0 & 1 & 0 \\ 0 & 1 & 1.5 & 0 \\ 0 & 0 & 2 & 1 \\ 0 & 0 & 0 & 1.5 \end{bmatrix} = \begin{bmatrix} 2 & 0 & 1 & 0 \\ -1 & 1 & 1 & 0 \\ -1 & -1 & 0 & 1 \\ 1 & 1 & 1 & 1 \end{bmatrix}$$

∎

11.5.2 Algorithm for Gaussian Elimination with Partial Pivoting

Algorithm 11.2 specifies GEPP, making use of submatrix operations.

Algorithm 11.2 Gaussian Elimination with Partial Pivoting

```
function LUDECOMP(A)
   % LU decomposition using Gaussian elimination with partial pivoting.
   % [P U P interchanges] = ludecomp(A) factors a square
   % matrix so that PA = LU. U is an upper-triangular matrix,
   % L is a lower-triangular matrix, and P is a permutation
   % matrix that reflects the row exchanges required by
   % partial pivoting used to reduce round-off error.
   % In the event that is useful, interchanges is the number
   % of row interchanges required.
   L = I
   P = I
   for i = 1:n-1 do
      k = index of largest matrix entry in column i, rows i through n
```
$$pivotindex = i + k - 1$$
```
      if pivotindex ≠ i then
         % Exchange rows i and k, ignoring columns 1 through i-1 in each row.
         tmp = A(i,i:n)
         A(i,i:n) = A(pivotindex,i:n)
         A(pivotindex,i:n) = tmp
         % Swap whole rows in P.
         tmp = P(i,1:n)
         P(i,1:n) = A(pivotindex,1:n)
         P(pivotindex,1:n) = tmp
         % Swap rows of L also, but only in columns 1 through i-1.
         tmp = L(i,1:i-1)
         L(i,1:i-1) = A(pivotindex,1:i-1)
         P(pivotindex,1:i-1) = tmp
      end if
      % Compute the multipliers.
      multipliers = A(i+1:n,i)/A(i,i)
      % Use submatrix calculations instead of a loop to perform
      % the row operations on the submatrix A(i+1:n, i+1:n)
      A(i+1:n,i+1:n) = A(i+1:n,i+1:n) - multipliers*A(i,i+1:n);
      % Set entries in column i, rows i+1:n to 0.
```
$$A(i+1:n, i) = \begin{bmatrix} 0 & 0 & \dots & 0 & 0 \end{bmatrix}^{\mathrm{T}}$$
```
      L(i+1:n,i) = multipliers
   end for
   U = A
```
$$return \begin{bmatrix} L, & U, & P \end{bmatrix}$$
```
end function
```

NLALIB: The function `ludecomp` implements Algorithm 11.2.

Example 11.12.

Factor the 4 × 4 matrix of Example 11.11 using ludecomp.

```
>> [L, U, P] = ludecomp(A)

L =
    1.0000         0         0         0
   -0.5000    1.0000         0         0
   -0.5000   -1.0000    1.0000         0
    0.5000    1.0000   -0.5000    1.0000

U =
    2.0000         0    1.0000         0
         0    1.0000    1.5000         0
         0         0    2.0000    1.0000
         0         0         0    1.5000

P =
    0    0    0    1
    0    1    0    0
    1    0    0    0
    0    0    1    0
```

During the execution of ludecomp, if entries a_{ki}, $i \leq k \leq n$ are all less than or equal to eps, the algorithm simply moves to the next pivot location $(i + 1, i + 1)$. As a result, GEPP applies to any matrix, even one that is singular. Factor the singular matrix

$$A = \begin{bmatrix} 2 & 1 & 3 & 5 \\ 1 & 6 & -1 & 2 \\ 3 & 7 & 2 & 7 \\ 5 & 19 & 0 & 11 \end{bmatrix}.$$

```
>> [L, U, P] = ludecomp(A)

L =
    1.0000         0         0         0
    0.4000    1.0000         0         0
    0.6000    0.6667    1.0000         0
    0.2000   -0.3333         0    1.0000

U =
    5.0000   19.0000         0   11.0000
         0   -6.6000    3.0000    0.6000
         0         0         0    0.0000
         0         0   -0.0000   -0.0000

P =
    0    0    0    1
    1    0    0    0
    0    0    1    0
    0    1    0    0

>> P*A

ans =
    5   19    0   11
    2    1    3    5
    3    7    2    7
    1    6   -1    2
```

```
>> L*U
ans =
     5    19     0    11
     2     1     3     5
     3     7     2     7
     1     6    -1     2
>> P'*L*U
ans =
     2     1     3     5
     1     6    -1     2
     3     7     2     7
     5    19     0    11
```

■

11.6   USING THE *LU* DECOMPOSITION TO SOLVE $Ax_i = b_i$, $1 \leq i \leq k$

To use this decomposition to solve a single system $Ax = b$, first multiply both sides by the permutation matrix:

$$PAx = Pb.$$

Let $\bar{p} = Pb$, and substitute LU for PA:

$$LUx = \bar{b}.$$

Execute forward and back substitution:

$$Ly = \bar{b}$$

and

$$Ux = y.$$

Remark 11.6.   The MATLAB function lu also has a calling sequence [L U P] = lu(A). The *LU* decomposition functions ludecomp and lu produce the same results, but lu is faster because it is implemented in machine code.

Example 11.13.   Let $A = \begin{bmatrix} 1 & 4 & 9 \\ -1 & 5 & 1 \\ 3 & 1 & 5 \end{bmatrix}$, $b = \begin{bmatrix} 1 \\ 6 \\ 2 \end{bmatrix}$. We use the functions forsolve(L,b) and backsolve(U,b) from the book software that perform forward and back substitution, respectively.

```
>> [L,U,P] = ludecomp(A);y = forsolve(L,P*b);x = backsolve(U,y);
>> norm(b-A*x)
ans = 1.2561e-15
```

■

Solving systems $Ax = b$ for many $b's$ only requires that A be factored once, at a cost of $O\left(n^3\right)$ flops. The combined steps of forward and back substitution require $O\left(n^2\right)$ flops. If there are k right-hand sides, the flop count is $O\left(n^3\right) + kO\left(n^2\right)$. It would be extremely inefficient to perform Gaussian elimination for each right-hand side, since that would require $kO\left(n^3\right)$ flops. Algorithm lusolve takes the decomposition of A, a matrix of right-hand sides, B, and computes a matrix of solutions X.

Algorithm 11.3 Solve $Ax = b$ for Multiple Right-Hand Sides

```
function LUSOLVE(L,U,P,B)
  % Solve multiple equations Ax = b using
  % the result of the LU factorization.
  % X = lusolve(L,U,P,B), where B is an n × k matrix containing
  % k right-hand sides for which a solution to the linear system
  % Ax = b is required. L, U, P are the result of the LU factorization
  % P*A = L*U. The solutions are in the k columns of X.
  pb = P*B
  for i = 1:k do
    y_i = forsolve(L, pb(:, i))
    x_i = backsolve(U, y_i)
    X(:, i) = x_i
  end for
  return X
end function
```

NLALIB: The function `lusolve` implements Algorithm 11.3.
We now use `ludecomp` to factor a 5×5 randomly generated matrix A, and follow this by using `lusolve` to solve $Ax = b$ for three randomly generated right-hand sides. For each solution, the code outputs the 2-norm of the residual.

Example 11.14.

```
>> A = rand(5,5);
>> [L, U, P] = ludecomp(A);
>> B = [rand(5,1) rand(5,1) rand(5,1)];
X = zeros(5,3);
X = lusolve(L, U, P, B);
% check the results
for i=1:3
    norm(A*X(:,i)-B(:,i))
end

ans =
  3.5975e-016
ans =
  1.5701e-016
ans =
  2.0015e-016
```

■

Remark 11.7. There is an algorithm termed *Gaussian elimination with complete pivoting (GECP)*, in which the pivoting strategy exchanges both rows and columns. Note that column exchanges require renumbering the unknowns. Complete pivoting is more complex and is not often used in practice, since GEPP gives good results with less computational effort.

11.7 FINDING A^{-1}

As stated earlier, it is considerably expensive to compute the inverse of a general nonsingular matrix A, and its computation should be avoided whenever possible. We have previously mentioned that solving $Ax = b$ using $b = A^{-1}b$ is a poor choice.

In Chapter 18, will present an algorithm, called the inverse power method, for finding the smallest eigenvalue and a corresponding eigenvector of a real nonsingular matrix A. Under the correct circumstances, the iteration $x_{i+1} = A^{-1}x_i$ converges to the largest eigenvector of A^{-1}. To avoid computing A^{-1}, just repeatedly solve $Ax_{i+1} = x_i$ after factoring A.

If A^{-1} must be computed, Section 2.5 and Theorem 11.5 say that to find it we solve the systems $Ax_i = e_i$, $1 \leq i \leq n$, where e_i is the vector with a 1 in component i and zeros in all other entries. We are prepared for this computation. Use Algorithm 11.3 with $B = I$.

Example 11.15. Find the inverse of the matrix $A = \begin{bmatrix} -9 & 1 & 3 \\ 1 & 5 & 2 \\ -6 & 12 & 3 \end{bmatrix}$.

```
>> A = [-9 1 3;1 5 2;-6 12 3]
>> [L U P] = ludecomp(A);
>> A_inverse = lusolve(L, U, P, eye(3))
A_inverse =
   -0.0469    0.1719   -0.0677
   -0.0781   -0.0469    0.1094
    0.2187    0.5313   -0.2396

>> norm(A_inverse - inv(A))

ans =
   2.9394e-017
```

∎

Remark 11.8.   The flop count for finding the inverse is found by forming the flop count for the *LU* factorization plus n instances of forward and backward substitution, so we have
$\frac{2}{3}n^3 - \frac{1}{2}n^2 - \frac{1}{6}n + n\left[2\left(n^2 + n\right)\right] = \frac{8}{3}n^3 + \frac{3}{2}n^2 - \frac{1}{6}n \approx \frac{8}{3}n^3$. Suppose we simply solve n equations $Ax_i = e_i$ without using the *LU* decomposition. Each solution will cost $O\left(n^3\right)$ flops, so our efforts will take $O\left(n^4\right)$ flops. Of course, nobody would compute the inverse this way.

11.8 STABILITY AND EFFICIENCY OF GAUSSIAN ELIMINATION

Gaussian elimination without partial pivoting is not stable in general, as we showed by using the matrix $A = \begin{bmatrix} 0.00001 & 3 \\ 2 & 1 \end{bmatrix}$.

It is theoretically possible for Gaussian elimination with partial pivoting to be explosively unstable [31] on certain "cooked-up" matrices; however, if we consider performance in practice, it is stable. The unusual matrices that produce poor results have never been encountered in applications.

Theorem 11.7 provides a criteria for stability, and its proof can be found in Ref. [26, pp. 163-165]. Prior to stating the theorem, we need to define the *growth factor* of a matrix. During the *LU* factorization, the norm of the matrix L has an upper bound we can compute for the norms we use (Problem 11.8). However, the elements of U can grow very large relative to those of A. If this happens, we expect Gaussian elimination to produce poor results.

Definition 11.2. Apply the *LU* factorization to matrix A. During the elimination steps, we have matrices $A = A^{(0)}, A^{(1)}, A^{(2)}, A^{(k)}, \ldots, A^{(n-1)} = U$. The growth factor is the ratio

$$\rho = \frac{\max_{i,j,k} \left| a_{ij}^{(k)} \right|}{\max_{ij} \left| a_{ij} \right|};$$

in other words, find the ratio of the largest element in magnitude during Gaussian elimination to the largest element of A in magnitude.

Example 11.16. Compute the growth factor for

$$A = \begin{bmatrix} 0.0001 & 3 \\ -1 & 1 \end{bmatrix}.$$

Without partial pivoting:

$$\begin{bmatrix} 0.0001 & 3 \\ -1 & 1 \end{bmatrix} \xrightarrow{R2 = R2 - \left(\frac{-1}{0.0001}\right)R1} \begin{bmatrix} 0.0001 & 3 \\ 0 & 30001 \end{bmatrix},$$

and the growth factor is $\rho = \frac{30,001}{3} = 10000.33$.

With partial pivoting:

$$
\begin{bmatrix} 0.0001 & 3 \\ -1 & 1 \end{bmatrix} \xrightarrow{R1 \Leftrightarrow R_2} \begin{bmatrix} -1 & 1 \\ 0.0001 & 3 \end{bmatrix} \xrightarrow{R2 = R2 - (-0.0001)\,R1} \begin{bmatrix} -1 & 1 \\ 0 & 3.0001 \end{bmatrix},
$$

and $\rho = \frac{3.0001}{3} = 1.0000$. Partial pivoting, as expected, improves the growth factor. In this example, the improvement was huge. ∎

Theorem 11.7. *Let the factorization PA = LU of a matrix A be computed using Gaussian elimination with partial pivoting on a computer that satisfies Equations 8.3 and 8.5. Then the computed matrices $\tilde{L}$, $\tilde{U}$, and $\tilde{P}$ satisfy*

$$
\tilde{L}\tilde{U} = \tilde{P}A + \delta A, \qquad \frac{\|\delta A\|_2}{\|A\|} = O\left(\rho\ eps\right)
$$

for some $n \times n$ matrix δA, where ρ is the growth factor for A. If $|l_{ij}| < 1$ for each $i > j$, implying there are no ties in the selection of pivots in exact arithmetic, then $\tilde{P} = P$ for all sufficiently small eps.

Theorem 11.7 [26, pp. 163–165] says that GEPP is backward stable if the size of ρ does not vary with n for all $n \times n$ matrices. Using O notation, this means that ρ is $O(1)$ for all such matrices. Unfortunately, there are matrices for which ρ gets large (Problem 11.36).

We showed in Section 11.4.3 that Gaussian elimination without pivoting requires $O\left(n^2\right)$ comparisons. At step i, partial pivoting requires a search of elements $A(i:n, i)$ to locate the pivot element, and this requires $(n-1) + (n-2) + \cdots + 2 + 1 = n(n-1)/2$ comparisons, so Gaussian elimination with pivoting also requires $\frac{2}{3}n^3 + O\left(n^2\right)$ flops. When performing complete pivoting, step i requires a search of the submatrix $A(i:n, i:n)$ and may perform column interchanges. It requires $\frac{2}{3}n^3 + O\left(n^2\right)$ flops, but requires $O\left(n^3\right)$ comparisons, so is a slower algorithm. Incidentally, complete pivoting creates a factorization of the form $PAQ = LU$, where both P and Q are permutation matrices. See Refs. [2, pp. 131-133] and [19, pp. 104-109] for an explanation of complete pivoting.

11.9 ITERATIVE REFINEMENT

If a solution to $Ax = b$ is not accurate enough, it is possible to improve the solution using *iterative refinement*. Let $\bar{x}$ be the computed solution of the system $Ax = b$. If $x = \bar{x} + \delta x$ is the exact solution, then $Ax = A\left(\bar{x} + \delta x\right) = A\bar{x} + A\left(\delta x\right) = b$, and $A\left(\delta x\right) = b - A\bar{x} = r$, the residual. If we solve the system $A\left(\delta x\right) = r$ for δx, then $Ax = A\bar{x} + A\left(\delta x\right) = A\bar{x} + r = A\bar{x} + b - A\bar{x} = b$. It is unlikely that we will obtain an exact solution to $A\left(\delta x\right) = r$; however, $\bar{x} + \delta x$ might be better approximation to the true solution than $\bar{x}$. For this to be true, it is necessary to compute the residual r using twice the precision of the original computations; for instance, if the computation of $\bar{x}$ was done using 32-bit floating point precision, then the residual should be computed using 64-bit precision. This process provides a basis for an iteration that continues until we reach a desired relative accuracy or fail to do so. Unless the matrix is very poorly conditioned, the computed solution x is already close to the true solution, so only a few iterations are required. If the matrix has a large condition number, it is not reasonable to expect huge improvements. Algorithm 11.4 describes the iterative improvement algorithm. Note that the algorithm returns the number of iterations performed in an attempt to reach the tolerance or -1 if it is not attained.

Algorithm 11.4 Iterative Improvement

```
function ITERIMP(A,L,U,P,b,x₁,tol,numiter)
   % Perform iterative improvement of a solution x₁
   % to Ax = b, where L, U, P is the LU factorization of A.
   % tol is the error tolerance, and numiter the maximum number
   % of iterations to perform.
   % Returns the improved solution and the number of iterations
   % required, or -1 if the tolerance is not obtained.
   for k = 1:numiter do
      iter = k % Compute the residual.
      r = b − Axₖ
      % Compute the correction.
      δx = lusolve(L, U, P, r)
      % Add the correction to form a new approximate solution.
      xₖ₊₁ = xₖ + δx
      if  ‖xₖ₊₁ − xₖ‖ / ‖xₖ‖  < tol then
         x = xₖ₊₁
         return [ x iter ]
      end if
   end for
   % Tolerance not obtained.
   x = xₖ₊₁
   iter = -1
end function
```

NLALIB: The function `iterimp` implements Algorithm 11.4.

Example 11.17.

$$A = \begin{bmatrix} 1 & -6 & 3 \\ 2 & 4 & 1 \\ 3 & -9 & 0 \end{bmatrix}, \quad b = \begin{bmatrix} 1 \\ 2 \\ 5 \end{bmatrix}.$$

The solution accurate to four decimal places is $x = \begin{bmatrix} 1.3636 \\ -0.1010 \\ -0.3232 \end{bmatrix}$. Assume we have computed a solution $\bar{x} = \begin{bmatrix} 1 \\ 0 \\ -0.5 \end{bmatrix}$,

so let $x_1 = \begin{bmatrix} 1 \\ 0 \\ -0.5 \end{bmatrix}$.

$k = 1$:

$$r_1 = b - Ax_1 = \begin{bmatrix} 1.5000 \\ 0.5000 \\ 2.000 \end{bmatrix}.$$

The solution of $A(\delta x) = r_1$ is

$$\delta x_1 = \begin{bmatrix} 0.3636 \\ -0.1010 \\ 0.1768 \end{bmatrix}$$

and then

$$x_2 = x_1 + \delta x_1 = \begin{bmatrix} 1 \\ 0 \\ -0.5 \end{bmatrix} + \begin{bmatrix} 0.3636 \\ -0.1010 \\ 0.1768 \end{bmatrix} = \begin{bmatrix} 1.3636 \\ -0.1010 \\ -0.3232 \end{bmatrix}.$$

We obtained the answer correct to four places in one iteration. Note that $\kappa(A) = 4.6357$, so A is well-conditioned. ■

The flop count for Algorithm 11.4 is $O(n^2)$, since each solution of the equation $A\delta x = r_k$ using lusolve has flop count $O(n^2)$ and there a bounded number of iterations.

Example 11.18. Solve the 15×15 pentadiagonal system
$$
\begin{bmatrix}
6 & -4 & 1 & & & \\
-4 & 6 & -4 & \ddots & & \\
1 & -4 & \ddots & \ddots & 1 & \\
 & & \ddots & \ddots & 6 & -4 \\
 & & & 1 & -4 & 6
\end{bmatrix}
\begin{bmatrix}
x_1 \\ x_2 \\ \vdots \\ x_{14} \\ x_{15}
\end{bmatrix}
=
\begin{bmatrix}
1 \\ 1 \\ \vdots \\ 1 \\ 1
\end{bmatrix}
$$
and perform

iterative improvement. Assume the solution obtained using the LU factorization is

$x_1 = [\ 20.00 \quad 52.49 \quad 91.00 \quad 130.00 \quad 165.00 \quad 192.51 \quad 210.00 \quad 216.00 \quad 210.00 \quad 192.51 \quad 165.00 \quad 130.00 \quad 91.00 \quad 52.49 \quad 20.00\ ]^{\mathrm{T}}$,

whereas the true solution correct to two decimal places is

$x = [\ 20.00 \quad 52.50 \quad 91.00 \quad 130.00 \quad 165.00 \quad 192.50 \quad 210.00 \quad 216.00 \quad 210.00 \quad 192.50 \quad 165.00 \quad 130.00 \quad 91.00 \quad 52.50 \quad 20.00\ ]^{\mathrm{T}}$.

The true residual is $\|b - Ax\|_2 = 0$, and residual for the approximate solution is $\|b - Ax_1\|_2 = 0.166$. Apply iterative improvement using x_1, $tol = 1 \times 10^{-5}$ and $numiter = 2$.

```
>> [xnew, iter] = iterimp(A,L,U,P,b,x1,1.0e-5,5);
>> norm(b - A*xnew)

ans =
    3.2132e-13
```

The matrix has a condition number of approximately 2611, so with an ill-conditioned matrix we obtained improvement. ■

11.10 CHAPTER SUMMARY

The *LU* Decomposition

Without doing row exchanges, the actions involved in factoring a square matrix A into a product of a lower-triangular matrix, L, and an upper-triangular matrix, U, is simple. Assign L to be the identity matrix. Perform Gaussian elimination on A in order to reduce it to upper-triangular form. Assume we are ready to eliminate elements below the pivot element $a_{ii}, 1 \le i \le n - 1$. The multipliers used are

$$
\frac{a_{i+1,i}}{a_{ii}}, \frac{a_{i+2,i}}{a_{ii}}, \ldots, \frac{a_{ni}}{a_{ii}}.
$$

Place these multipliers in L at locations $(i + 1, i), (i + 2, i), \ldots, (n, i)$. When the row reduction is complete, A is matrix U, and $A = LU$.

Using the *LU* to Solve Equations

After performing the decomposition $A = LU$, consider solving the system $Ax = b$. Substitute LU for A to obtain

$$
\begin{aligned}
LUx &= b, \\
L(Ux) &= b.
\end{aligned}
$$

Consider $y = Ux$ to be the unknown and solve

$$
Ly = b
$$

using forward substitution. Now solve

$$
Ux = y
$$

using back substitution.

Elementary Row Matrices

Let A be an $n \times n$ matrix. An elementary row matrix, E, is an alteration of the identity matrix such that EA performs one of the three elementary row operations. For instance, if

$$E_{31}(2) = \begin{bmatrix} 1 & 0 & 0 \\ 0 & 1 & 0 \\ -2 & 0 & 1 \end{bmatrix}$$

then $E_{31}A$ subtracts (2) times row 1 from row 3. The primary purpose of these matrices is to show why the *LU* decomposition works.

Derivation of the *LU* Decomposition

Use products of elementary row matrices to row reduce A to upper-triangular form to arrive at a product

$$E_k E_{k-1} \ldots E_2 E_1 A = U,$$

and so

$$A = (E_k E_{k-1} \ldots E_2)^{-1} U.$$

$(E_k E_{k-1} \ldots E_2)^{-1}$ is precisely the matrix L.

An analysis shows that the flop count for the *LU* decomposition is $\approx \frac{2}{3}n^3$, so it is an expensive process.

Gaussian Elimination with Partial Pivoting

We use the pivot to eliminate elements $a_{i+1,i}, a_{i+2,i}, \ldots, a_{ni}$. If the pivot, a_{ii}, is small the multipliers $a_{k,i}/a_{ii}, i+1 \le k \le n$, will likely be large. As we saw in Chapter 8, adding or subtracting large numbers from smaller ones can cause loss of any contribution from the smaller numbers. For this reason, begin find the maximum element in absolute value from the set $a_{ii}, a_{i+1,i}, a_{i+2,i}, \ldots, a_{ni}$ and swap rows so the largest magnitude element is at position (i, i). Proceed with elimination in column i. The end result is a decomposition of the form $PA = LU$, where P is a permutation matrix that accounts for any row exchanges that occurred. This can be justified by an analysis using elementary row matrices.

Computing the *LU* Decomposition Once and Solving $Ax_i = b_i$, $1 \le i \le k$

A great advantage of performing the *LU* decomposition is that if the system must be solved for multiple right-hand sides, the $O(n^3)$ *LU* decomposition need only be performed once, as follows:

$$\begin{aligned} Ax &= b_i, \quad 1 \le i \le k, \\ PAx &= Pb_i, \\ LUx &= Pb_i, \\ L(Ux) &= Pb_i. \end{aligned}$$

Now solve $L(Ux_i) = Pb_i, 1 \le i \le k$ using forward and back substitution. The cost of the decomposition is $O(n^3)$, and the cost of the solutions using forward and back substitution is $O(kn^2)$. If we solved each system using Gaussian elimination, the cost would be $O(kn^3)$.

Computing A^{-1}

Conceptually, computing A^{-1} is simple. Apply the *LU* decomposition to obtain $PA = LU$, and use it to solve systems having as right-hand sides the standard basis vectors

$$e_1 = \begin{bmatrix} 1 & 0 & \ldots & 0 & 0 \end{bmatrix}^{\mathrm{T}}, e_2 = \begin{bmatrix} 0 & 1 & \ldots & 0 & 0 \end{bmatrix}^{\mathrm{T}}, \ldots, \begin{bmatrix} 0 & 0 & \ldots & 0 & 1 \end{bmatrix}^{\mathrm{T}}.$$

The solutions form the columns of A^{-1}. It should be emphasized that computing A^{-1} is expensive and roundoff error builds up.

Stability and Efficiency of Gaussian Elimination

There are instances where GEPP fails (see Problem 11.36), but these examples are pathological. None of these situations has occurred in 50 years of computation using GEPP. There is a method known as complete pivoting that involves exchanging both rows and columns. It is more expensive than GEPP and is not used often.

Iterative Refinement

If a solution to $Ax = b$ is not accurate enough, it is possible to improve the solution using iterative refinement. Let $\bar{x}$ be the computed solution of the system $Ax = b$. If $x = \bar{x} + \delta x$ is the exact solution, then $Ax = A\bar{x} + A(\delta x) = b$, and $A(\delta x) = b - A\bar{x} = r$, the residual. If we solve the system $A(\delta x) = r$ for δx, then $Ax = A\bar{x} + A(\delta x) = A\bar{x} + r = A\bar{x} + b - A\bar{x} = b$. It is unlikely that we will obtain an exact solution to $A(\delta x) = r$; however, $\bar{x} + \delta x$ might be better approximation to the true solution than $\bar{x}$. For this to be true, it is necessary to compute the residual r using twice the precision of the original computations; for instance, if the computation of $\bar{x}$ was done using 32-bit floating point precision, then the residual should be computed using 64-bit precision. This process provides a basis for an iteration that continues until we reach a desired relative accuracy or fail to do so. Unless the matrix is very poorly conditioned, the computed solution x is already close to the true solution, so only a few iterations are required. If the matrix has a large condition number, it is not reasonable to expect huge improvement

11.11 PROBLEMS

Note: The term "step-by-step" can involve computational assistance. For instance, suppose you want to subtract a multiple, t, of row i from row j of matrix A. The following statement will do this for you.

```
>> A(j,:) = A(j,:) - t*A(i,:)
```

This type of computation is essentially "by pencil and paper," except you do not have to perform the row elimination by hand or with a calculator.

11.1 Given $A = \begin{bmatrix} 1 & 2 & 3 \\ 2 & 5 & 4 \\ 3 & 5 & 4 \end{bmatrix}$, find the LU factorization of A step-by-step without pivoting.

11.2 Show that $A = \begin{bmatrix} 6 & 1 \\ -6 & 2 \end{bmatrix}$ is nonsingular, find A^{-1} and express A as a product of elementary row matrices.

11.3 What two elementary row matrices $E_{21}(t_1)$ and $E_{32}(t_2)$ put $A = \begin{bmatrix} 2 & 1 & 0 \\ 6 & 4 & 2 \\ 0 & 3 & 5 \end{bmatrix}$ into upper-triangular form

$E_{21}(t_2)E_{32}(t_1)A = U$? Multiply by $E_{21}^{-1}(t_2)E_{32}^{-1}(t_1) = L$ to factor A into $LU = E_{21}^{-1}(t_1)E_{32}^{-1}(t_2)U$.

11.4 Find the 3×3 matrix $A = E_2(5)E_{31}(2)E_{13}$? Also find A^{-1}.

11.5 Prove that multiplying a matrix A on the right by E_{ij} exchanges columns i and j of A.

11.6 This problem is taken from [8, p. 106]. Using matrix A, suppose you eliminate upwards (almost unheard of). Use the last row to produce zeros in the last column (the pivot is 1). Then use the second row to produce zero above the second pivot. Find the factors in the unusual order $A = UL$.

$$A = \begin{bmatrix} 5 & 3 & 1 \\ 3 & 3 & 1 \\ 1 & 1 & 1 \end{bmatrix}.$$

11.7 If A is singular, and $PA = LU$ is the LU factorization, prove there must be at least one zero on the diagonal of U.

11.8 In the LU factorization of a square matrix, show that

$$\|L\|_F \leq \frac{n(n+1)}{2}, \quad \|L\|_\infty \leq n,$$
$$\|L\|_1 \leq n \quad \text{and} \quad \|L\|_2 \leq n.$$

11.9 Prove Theorem 11.1 for the elementary row matrices $E_{ij}(t)$ and $E_i(t)$.

11.10 Show that during GECP, search for the new pivot element requires $O\left(n^3\right)$ comparisons. After the new pivot element is selected, explain why it may be necessary to exchange both rows and columns.

11.11 Solve the system

$$\begin{bmatrix} 0.00005 & 1 & 1 \\ 2 & -1 & 1 \\ 1 & 2 & 4 \end{bmatrix} \begin{bmatrix} x_1 \\ x_2 \\ x_3 \end{bmatrix} = \begin{bmatrix} 2 \\ 3 \\ 3 \end{bmatrix}$$

step-by-step with and without partial pivoting using 4 decimal place accuracy. Compare the results.

An alternative algorithm for *LU* factorization is *Crout's Method*. There is a variation of the algorithm with pivoting, but we will not include it here. The elements of *L* and *U* are determined using formulas that are easily programmed. In Crout's method, *U* is the matrix with ones on the diagonal, as indicated in Equation 11.7. Problems 11.12–11.15 develop the method.

$$\begin{bmatrix} a_{11} & a_{12} & a_{13} & a_{14} \\ a_{21} & a_{22} & a_{23} & a_{24} \\ a_{31} & a_{32} & a_{33} & a_{34} \\ a_{41} & a_{42} & a_{43} & a_{44} \end{bmatrix} = \begin{bmatrix} L_{11} & 0 & 0 & 0 \\ L_{21} & L_{22} & 0 & 0 \\ L_{31} & L_{32} & L_{33} & 0 \\ L_{41} & L_{42} & L_{43} & L_{44} \end{bmatrix} \begin{bmatrix} 1 & U_{12} & U_{13} & U_{14} \\ 0 & 1 & U_{23} & U_{24} \\ 0 & 0 & 1 & U_{34} \\ 0 & 0 & 0 & 1 \end{bmatrix}. \tag{11.7}$$

11.12 Form the product of the two matrices on the right-hand side of Equation 11.7.

11.13 The entries of the matrices *L* and *U* can be determined by equating the two sides of the result from Problem 11.12 and computing the L_{ij} and U_{ij} entries row by row; for instance, $L_{11} = a_{11}$. Once L_{11} is known, the values of U_{12}, U_{13}, and U_{14} are calculated by

$$U_{12} = \frac{a_{12}}{L_{11}}, \quad U_{13} = \frac{a_{13}}{L_{11}}, \quad U_{14} = \frac{a_{14}}{L_{11}}.$$

Continue and as far as necessary until you can state a general formula for the elements of *L* and *U* for an $n \times n$ matrix *A*.

11.14 Using Crout's Method, factor the following 3×3 matrix step-by-step.

a. $A = \begin{bmatrix} 1 & -2 & -1 \\ 2 & 0 & 3 \\ 1 & 5 & 0 \end{bmatrix}$

b. Use the factorization from part to solve the system $Ax = \begin{bmatrix} 1 \\ 1 \\ 0 \end{bmatrix}$.

11.15 In this problem, you will use the results of Problem 11.13 to develop an algorithm for the Crout Method. The algorithm involves four steps. Fill-in the missing elements of each step.

```
function CROUT(A)
   Calculate the first column of L.
   Place 1s on the diagonal of U.
   for i = 1:n do



   end for
   Calculate the elements in the first row of U. U₁₁ already calculated.
   for j = 2:n do



   end for
   Calculate the rest of the elements row after row.
   The entries of L are calculated first because they are used for
   calculating the elements of U.
   for i = 2:n do
      for j = 2:i do



      end for
      for j = i+1:n do
```

```
       end for
     end for
     return [ L  U ]
   end function
```

11.16 It is possible to determine that a matrix is singular during GEPP. Specify a condition that occurs while executing the algorithm that will indicate the matrix is singular.

11.17 There are situations where the LU decomposition must be done using pivoting.

a. For the matrix $\begin{bmatrix} 1 & 7 & 1 \\ 9 & -1 & 2 \\ 3 & 5 & 1 \end{bmatrix}$, compute the determinant of the submatrices $\begin{bmatrix} 1 \end{bmatrix}$, $\begin{bmatrix} 1 & 7 \\ 9 & -1 \end{bmatrix}$, and the whole matrix.

Can you do an LU factorization without pivoting?

b. Perform part (a) with the matrix $\begin{bmatrix} 1 & 3 & 1 \\ -1 & -3 & 5 \\ 6 & 3 & 9 \end{bmatrix}$.

c. For the matrix $\begin{bmatrix} 1 & 2 & -1 & 1 \\ 3 & 2 & -7 & 2 \\ 1 & 8 & 5 & 3 \\ 1 & -9 & 2 & 5 \end{bmatrix}$, calculate all determinants of submatrices with upper left-hand corner at a_{11}.

Can you do an LU decomposition without pivoting?

d. Propose a theorem that specifies when an LU decomposition without pivoting is possible. You do not have to prove it.

11.18 Prove that a permutation matrix is orthogonal.

11.19 If A is an $n \times n$ matrix whose LU decomposition is $PA = LU$, prove that

$$\det (A) = (-1)^r \, u_{11} u_{22} \ldots u_{nn},$$

where r is the number of row interchanges performed during row reduction, and u_{ii} are the diagonal elements of U.

11.20 A is *strictly column diagonally dominant* if $|a_{ii}| > \sum_{k=1, k \neq i}^{n} |a_{ki}|$. This problem investigates the LU decomposition of such a matrix.

a. Determine which matrices are strictly column diagonally dominant:

i. $\begin{bmatrix} 3 & 2 & 2 \\ -2 & 6 & 2 \\ 1 & -1 & 5 \end{bmatrix}$

ii. $\begin{bmatrix} -1 & 1 & 7 & -0.05 \\ 0.3 & 2.5 & -1 & -0.7 \\ 0.2 & 0.75 & -9 & 0.04 \\ 0.4 & 0.70 & 0.9 & 0.8 \end{bmatrix}$

iii. $\begin{bmatrix} 2 & 0.5 & & & \\ 0.5 & 2 & 0.5 & & \\ & & \ddots & & \\ & & 0.5 & 2 & 0.5 \\ & & & 0.5 & 2 \end{bmatrix}$

b. The matrix $A = \begin{bmatrix} 3 & 3 & 1 \\ 1 & 5 & 2 \\ 1 & 1 & -4 \end{bmatrix}$ is strictly column diagonally dominant. Using pencil and paper, form the LU decomposition with partial pivoting to show that no row interchanges are necessary.

c. Prove that if A is strictly column diagonally dominant, the LU decomposition with partial pivoting requires no row interchanges. Hint: Argue that row elimination in column 1 requires no row exchanges. Now consider the matrix $A^{(1)}$ that results from action with column 1. Show it is strictly column diagonally dominant by using the diagonal dominance of A. The result follows by induction.

11.21 The LDU decomposition is a variation of the LU decomposition. L is a lower diagonal matrix all of whose diagonal entries are 1, D is a diagonal matrix whose elements are the pivots, and U is a unit upper-triangular matrix. Without pivoting, $A = LDU$, and with pivoting $PA = LDU$. In this problem, assume that the *LU* decomposition uses no pivoting.

a. Show that the LDU decomposition exists when A is nonsingular.

b. Show that if A is symmetric and nonsingular, then $A = LDU$, where $L = U^T$ and $U = L^T$.

c. Is (b) true for the standard *LU* decomposition?

11.22 There are some matrices whose inverses can be computed easily.

a. Prove that the inverse of a diagonal matrix, D, with nonzero diagonal entries $d_1, d_2, \ldots, d_n$ is the diagonal matrix whose diagonal elements are $1/d_1, 1/d_2, \ldots, 1/d_n$.

b. We will have occasion to use Householder reflections later in the text when we discuss efficient algorithms for computing the *QR* decomposition. A Householder reflection is formed from a nonzero vector u as follows:

$$H_u = I - \frac{2uu^T}{u^T u}, \quad u \neq 0.$$

Show that $H_u^2 = I$

c. An atomic lower-triangular matrix is a special form of a unit lower-triangular matrix, also called a *Gauss transformation matrix*. All of the off-diagonal entries are zero, except for the entries in a single column. It has the form

$$L = \begin{bmatrix} 1 & & & & & & & \\ 0 & \ddots & & & & & & \\ 0 & & 1 & & & & & \\ 0 & & & 1 & & & & \\ & & & l_{i+1,i} & 1 & & & \\ \vdots & & & l_{i+2,i} & & \ddots & & \\ & & & \vdots & & & 1 & \\ 0 & \ldots & 0 & l_{ni} & 0 & \ldots & 0 & 1 \end{bmatrix}.$$

Show that

$$L^{-1} = \begin{bmatrix} 1 & & & & & & & \\ 0 & \ddots & & & & & & \\ 0 & & 1 & & & & & \\ 0 & & & 1 & & & & \\ & & & -l_{i+1,i} & 1 & & & \\ \vdots & & & -l_{i+2,i} & & \ddots & & \\ & & & \vdots & & & 1 & \\ 0 & \ldots & 0 & -l_{ni} & 0 & \ldots & 0 & 1 \end{bmatrix}.$$

11.23 Consider the system

$$\begin{bmatrix} 1 & 1 & 3 \\ 2 & 1 & 2 \\ 3 & 1 & 5 \end{bmatrix} \begin{bmatrix} x_1 \\ x_2 \\ x_3 \end{bmatrix} = \begin{bmatrix} -1 \\ 4 \\ 2 \end{bmatrix}.$$

Assuming an initial approximation $\bar{x} = \begin{bmatrix} 2 \\ 0 \\ -3 \end{bmatrix}$, perform one iterative improvement iteration using step-by-step, retaining four significant digits. Compare your result to the actual solution.

11.24 Let $A = \begin{bmatrix} 1 & 7 & -1 \\ 2 & 4 & 5 \\ 1 & 2 & -1 \end{bmatrix}$, $b = \begin{bmatrix} 1 \\ 2 \\ 3 \end{bmatrix}$. Compute the solution x using MATLAB. Let $\bar{x} = \begin{bmatrix} 3 \\ -1 \\ -1 \end{bmatrix}$. Perform one step of iterative improvement with four significant digits, obtain approximate solution $\bar{x}$, and compare it to x.

11.25 If in the *LU* decomposition, $|l_{ij}| < 1$ for each $i > j$, show there are no ties in the selection of pivots in exact arithmetic.

11.26 Compute the growth factor for $A = \begin{bmatrix} 0.0006 & -8 \\ 1 & 1 \end{bmatrix}$ with and without partial pivoting.

11.27 When using GEPP, show that $\rho \leq 2^{n-1}$, where ρ is the growth factor.

11.11.1 MATLAB Problems

11.28 Let $A = \begin{bmatrix} 1 & 0 & 1 \\ -1 & 1 & 1 \\ -1 & -1 & 1 \end{bmatrix}$.

   a. Find the decomposition $PA = LU$ for the matrix.
   b. Solve the system $Ax = \begin{bmatrix} 1 & 1 & 1 \end{bmatrix}^T$ using the results of part (a).

   c. Find the decomposition $PA = LU$ for the matrix $B = \begin{bmatrix} 1 & \frac{1}{2} & \frac{1}{3} \\ \frac{1}{2} & \frac{1}{3} & \frac{1}{4} \\ \frac{1}{3} & \frac{1}{4} & \frac{1}{5} \end{bmatrix}$.

   d. Solve the system $Bx = \begin{bmatrix} 1 & 1 & 1 \end{bmatrix}^T$ using the results of part (c).

11.29 Using GEPP, factor each matrix as $PA = LU$.

   a. $A = \begin{bmatrix} 1 & 2 & -1 \\ 2 & 1 & 3 \\ 5 & 8 & 2 \end{bmatrix}$

   b. $A = \begin{bmatrix} 0 & 1 & 0 & 0 \\ 0 & 0 & 1 & 0 \\ 0 & 0 & 0 & 1 \\ 2 & 3 & 4 & 5 \end{bmatrix}$

11.30 Use `ludecomp` and `lusolve` to find the solution to the following three systems. Compute $\|B - A * X\|_2$, where B is the matrix of right-hand sides, and the solutions are the columns of X.

$$A = \begin{bmatrix} 1 & 6 & 2 & 1 \\ 2 & 2 & 8 & 9 \\ 12 & 5 & 1 & 9 \\ -1 & -7 & 1 & 5 \end{bmatrix} \begin{bmatrix} x_1 \\ x_2 \\ x_3 \\ x_4 \end{bmatrix} = b_i, \text{ where } b_1 = \begin{bmatrix} 1 \\ 8 \\ 0 \\ -1 \end{bmatrix}, b_2 = \begin{bmatrix} 5 \\ 12 \\ 1 \\ -12 \end{bmatrix}, \text{ and } b_3 = \begin{bmatrix} 5 \\ 88 \\ 15 \\ 3 \end{bmatrix}$$

11.31 When a matrix T is tridiagonal, its L and U factors have only two nonzero diagonals. Factor each matrix into $A = LU$ without pivoting.

   a. $\begin{bmatrix} 1 & 1 & 0 \\ 1 & 2 & 1 \\ 0 & 1 & 2 \end{bmatrix}$

   b. $\begin{bmatrix} 1 & -1 & 0 \\ 2 & 3 & 1 \\ 0 & 4 & 5 \end{bmatrix}$

   c. $\begin{bmatrix} 1 & 2 & 0 & 0 \\ 2 & 3 & 1 & 0 \\ 0 & 1 & 2 & 3 \\ 0 & 0 & 3 & 4 \end{bmatrix}$

11.32 Write MATLAB functions to create the elementary row matrices, E_{ij}, $E_i(t)$, and $E_{ij}(t)$. Use the functions to verify your results for Problem 11.4.

11.33 **a.** Write a MATLAB function

   ```
   function [L U] = crout(A)
   ```
   that implements Crout's method developed in Problem 11.15.

b. Factor the matrix in Example 11.9. Apply your function to random 3×3, 5×5, and 25×25 matrices. Test the result in each case by computing $\|A - LU\|_2$.

11.34 a. Write

```
function x = gauss(A,b)
```

that uses Gaussian elimination without pivoting to solve the $n \times n$ system $Ax = b$. Note that the function does not perform the LU decomposition but just applies straightforward row operations as explained in Chapter 2. If a pivot element is zero, print an error message and exit.

b. Use your function to solve three random systems of size 3×3, 4×4, and 10×10.

11.35 Make a copy of the function ludecomp, name it `matgr`, and make modifications so it computes the growth factor of a square matrix using partial pivoting. Test it with the matrix in Problem 11.26 and the matrix

$$W = \begin{bmatrix} 1 & 0 & 0 & 0 & 1 \\ -1 & 1 & 0 & 0 & 1 \\ -1 & -1 & 1 & 0 & 1 \\ -1 & -1 & -1 & 1 & 1 \\ -1 & -1 & -1 & -1 & 1 \end{bmatrix}.$$

11.36 We have stated that Gaussian elimination with partial pivoting is stable in practice, but that contrived examples can cause it to become unstable. This problem uses a matrix pattern devised by *Wilkinson* in his definitive work [9].

a. Write a MATLAB function `function W = wilkpivot(n)` that returns an $n \times n$ matrix of the form

$$\begin{bmatrix} 1 & & & & & 1 \\ -1 & 1 & & & & 1 \\ -1 & -1 & 1 & & & 1 \\ \vdots & \vdots & \vdots & \ddots & & \vdots \\ -1 & -1 & -1 & \dots & 1 & 1 \\ -1 & -1 & -1 & \dots & -1 & 1 \end{bmatrix}$$

Your function requires no more than four MATLAB commands.

b. Will any row interchanges occur when solving a system with this coefficient matrix?

c. Using your function `wilkpivot`, show that a Wilkinson 5×5 matrix has the *LU* decomposition

$$\begin{bmatrix} 1 & & & & 1 \\ -1 & 1 & & & 1 \\ -1 & -1 & 1 & & 1 \\ -1 & -1 & -1 & 1 & 1 \\ -1 & -1 & -1 & -1 & 1 \end{bmatrix} = \begin{bmatrix} 1 & & & & \\ -1 & 1 & & & \\ -1 & -1 & 1 & & \\ -1 & -1 & -1 & 1 & \\ -1 & -1 & -1 & -1 & 1 \end{bmatrix} \begin{bmatrix} 1 & & & & 1 \\ & 1 & & & 2 \\ & & 1 & & 4 \\ & & & 1 & 8 \\ & & & & 16 \end{bmatrix}.$$

d. What is the value of $U(50, 50)$ in the *LU* decomposition of a 50×50 Wilkinson matrix, W50?

e. Using the matrix from part (d), form the vector b = [1:25 -1:-1:-25]' and compute x = A\b. Perturb b slightly using the statements

```
b(25) = b(25) - 0.00001;
b(30) = b(30) + 0.00001
```

and compute xbar = A\b.

f. Evaluate $\|x - \bar{x}\|_\infty / \|x\|_\infty$

g. What is the matrix growth factor? Explain the result of part (f) by considering the matrix growth factor.

11.37 Using row elimination with partial pivoting, write and test a MATLAB function `determinant(A)` that computes the determinant of matrix A and test it with matrices up to size 20×20.

11.38 Problem 11.20(c) required a proof that a strictly column diagonally dominant matrix requires no row interchanges during the *LU* decomposition. Use the MATLAB function `trid` in the software distribution to build the matrix

$$A = \begin{bmatrix} 1 & -0.25 & & & \\ -0.25 & 1 & -0.25 & & \\ & -0.25 & \ddots & \ddots & \\ & & \ddots & 1 & -0.25 \\ & & & -0.25 & 1 \end{bmatrix}$$

for $n = 5, 10, 25, 100$. For each n, demonstrate that no row interchanges are necessary when forming the LU decomposition. The function ludecomp returns the number of interchanges required as well as L and U:

```
[L, U, P, interchanges] = ludecomp(A)
```

11.39 Write and test a MATLAB function ldu(A) that computes the LDU decomposition of A as described in Problem 11.21.

11.40 Write and test the function inverse(A) that computes A^{-1} using the LU decomposition with partial pivoting. Include hilb(10), the 10×10 Hilbert matrix, in your tests.

11.41 A *band matrix* is a sparse matrix whose nonzero entries are confined to a diagonal band, consisting of the main diagonal and zero or more diagonals on either side. A tri- or bidiagonal matrix is a band matrix. Band matrices occur in applications, particularly when finding approximate solutions to partial differential equations. In many such applications, the banded system to be solved is very large. As we have emphasized earlier, it pays to take advantage of matrix structure, particularly when dealing with very large matrices. We will see in Chapter 13 that the LU decomposition of a tridiagonal matrix, T, requires $O(n)$ flops, where L is unit lower bidiagonal, and U is upper bidiagonal with its superdiagonal the same as that of T. Since $T^{-1} = U^{-1}L^{-1}$, and there are efficient algorithms for the inverse of a bidiagonal matrix (see Refs. [32, pp. 151-152] and [39, pp.248-249]), it would seem to be more efficient to compute $U^{-1}L^{-1}$ rather than executing a straightforward evaluation of A^{-1}. Although we will not discuss it here, there are formulas for the direct calculation of T^{-1} [39, 40].

The book software contains a function tridiagLU you call as follows:

```
% a, b, c are the subdiagonal, diagonal, and superdigonal of a tridiagonal matrix.

% L is the lower diagonal of the unit bidiagonal matrix, and U is the diagonal of the
upper bidiagonal matrix.

[L,U] = tridiagLU(a,b,c);
```

a. Let a, b, c be the subdiagonal, diagonal, and superdiagonal of the matrix in Problem 11.38 with $n = 5$. Use tridiagLU to compute L and U. Verify that tridiagLU returned the proper diagonals.

b. The book software distribution contains functions upbinv and lobinv that compute the inverse of an upper bidiagonal and a unit lower bidiagonal matrix, respectively. Scan through the source code and determine how to call the functions. Write a function, trinv, with arguments a, b, c that computes the inverse of a tridiagonal matrix using upbinv and lbinv. Test it thoroughly.

c. Build vectors

```
>> a = randn(499,1);
>> b = randn(500,1);
>> c = randn(499,1);
```

and execute the statements

```
>> A = trid(a,b,c);
>> tic;P1 = inv(A);toc;
>> tic;P2 = trinv(a,b,c);toc;
```

Why is inv faster than trinv?

d. Using the function inverse developed in Problem 11.40, execute the statements

```
tic;P1 = inverse(A);toc;
tic;P2 = trinv(a,b,c);toc;
```

Explain the results.

11.42 The MATLAB function `single(A)` converts each element of the array A from 64-bit to 32-bit precision. Create the following 20×20 pentadiagonal system

$$
T =
\begin{bmatrix}
6 & -4 & 1 & & & & & 0 \\
-4 & 6 & -4 & 1 & \ddots & & & \\
1 & -4 & 6 & -4 & 1 & \ddots & & \\
\vdots & \ddots & & & & \ddots & & \\
\cdots & 1 & -4 & 6 & -4 & 1 & & \\
\vdots & & & & & \ddots & & \\
0 & & & \cdots & & 1 & -4 & 6
\end{bmatrix}
\times
\begin{bmatrix}
x_1 \\ x_2 \\ \vdots \\ x_{19} \\ x_{20}
\end{bmatrix}
=
\begin{bmatrix}
1 \\ 1 \\ \vdots \\ 1 \\ 1
\end{bmatrix}
$$

using the function `pentd` in the book software distribution. Run the following MATLAB program. Explain what the program does and the results.

```
[L,U,P] = lu(T);
b1 = ones(20,1);
x = lusolve(L,U,P,b1);

C = single(T);
b2 = single(b1);
[L1,U1,P] = lu(C);
x1 = lusolve(L1,U1,P,b2);
fprintf('norm(x - single precision solution)/norm(x) = %g\n',...
        norm(x-x1)/norm(x));

[x2,iter] = iterimp(T,double(L1),double(U1),P,b1,...
                    double(x1),1.0e-12,10);
fprintf('norm(x - refined solution)/norm(x) = %g, requiring %d iterations\n',...
        norm(x-x2)/norm(x),iter);
```

11.43

a. Write a function, `makebidiag`, that takes a vector a with n elements and a vector b with $n - 1$ elements and builds an upper bidiagonal matrix.

b. Build a random 5×5 bidiagonal matrix with no zeros on either diagonal. Compute its singular values in vector S using the MATLAB function `svd`. Show that the singular values are unique using the MATLAB command "`numunique = length(unique(S));`." Repeat this experiment for random 10×10, 25×25, and 50×50 bidiagonal matrices. Formally state a theorem that describes the behavior.

c. It is true that a symmetric tridiagonal matrix for which $a_{i,i+1} = a_{i,i-1} \neq 0$ has n distinct eigenvalues. Using this fact, prove the theorem you stated in part (b).

Chapter 12

Linear System Applications

You should be familiar with

- Trigonometric functions and infinite series (Section 12.1)
- Evaluating an integral (Section 12.1)
- Derivatives of a function of one variable
- Partial derivatives (Sections 12.2 and 12.3)
- Acquaintance with finite difference equations (Section 12.2)
- Computing a local minima or maxima (Section 12.3)

We have discussed bases, dimensions, inner products, norms, and the LU decomposition with and without partial pivoting for the solution of $n \times n$ linear algebraic systems of equations. Linear algebra is used in almost all applied engineering and science applications. In many cases, its use is hidden from direct view, but is a critical component of an algorithm. Until this point, we have presented problems involving the solution to a truss, an electrical circuit, data encryption, as well as other applications. In this chapter, we discuss four additional applications that make use of linear algebra, Fourier series, finite difference techniques for solving partial differential equations, an introduction to least squares, and cubic spline interpolation.

12.1 FOURIER SERIES

To this point, we have dealt with finite-dimensional vector spaces, but there are many applications for vector spaces with infinitely many dimensions. A particularly important example is *Fourier series*. When we introduced the inner product of vectors in Chapter 6, we also presented the L^2 inner product of functions. If f and g are functions over the interval $a \leq x \leq b$, then

$$\langle f, g \rangle = \int_a^b f(x)\, g(x)\, dx, \quad \text{and} \quad \|f\|_{L^2}^2 = \int_a^b f^2(x)\, dx.$$

Consider the infinite sequence of trigonometric functions

$$\left\{ \frac{1}{\sqrt{2\pi}}, \frac{\cos x}{\sqrt{\pi}}, \frac{\sin x}{\sqrt{\pi}}, \frac{\cos 2x}{\sqrt{\pi}}, \frac{\sin 2x}{\sqrt{\pi}}, \ldots, \frac{\cos nx}{\sqrt{\pi}}, \frac{\sin nx}{\sqrt{\pi}}, \ldots \right\}$$

A straightforward computation shows that this is an orthonormal sequence over the interval $-\pi \leq x \leq \pi$.

$$\left\langle \frac{\cos ix}{\sqrt{\pi}}, \frac{\cos jx}{\sqrt{\pi}} \right\rangle = \frac{1}{\pi} \int_{-\pi}^{\pi} \cos ix \, \cos jx \, dx = 0, \quad i \neq j,$$

$$\left\langle \frac{\sin ix}{\sqrt{\pi}}, \frac{\sin jx}{\sqrt{\pi}} \right\rangle = \frac{1}{\pi} \int_{-\pi}^{\pi} \sin ix \, \sin jx \, dx = 0, \quad i \neq j,$$

$$\left\langle \frac{\cos ix}{\sqrt{\pi}}, \frac{\sin jx}{\sqrt{\pi}} \right\rangle = \frac{1}{\pi} \int_{-\pi}^{\pi} \cos ix \, \sin jx \, dx = 0, \quad i \neq j,$$

and

$$\left\langle \frac{\cos ix}{\sqrt{\pi}}, \frac{\cos ix}{\sqrt{\pi}} \right\rangle = \frac{1}{\pi} \int_{-\pi}^{\pi} \cos ix \cos ix \, dx = \frac{1}{\pi} \int_{-\pi}^{\pi} \cos^2(ix)\, dx = 1,$$

$$\left\langle \frac{\sin ix}{\sqrt{\pi}}, \frac{\sin ix}{\sqrt{\pi}} \right\rangle = \frac{1}{\pi} \int_{-\pi}^{\pi} \sin^2(ix)\, dx = 1.$$

Numerical Linear Algebra with Applications. http://dx.doi.org/10.1016/B978-0-12-394435-1.00012-0

We now consider this infinite sequence to be an orthonormal basis for what we term a *function space*, a set of linear combinations of the basis functions.

$$f(x) = a_0 \left(\frac{1}{\sqrt{2\pi}} \right) + a_1 \left(\frac{\cos x}{\sqrt{\pi}} \right) + b_1 \left(\frac{\sin x}{\sqrt{\pi}} \right) + a_2 \left(\frac{\cos 2x}{\sqrt{\pi}} \right) + b_2 \left(\frac{\sin 2x}{\sqrt{\pi}} \right) + \cdots =$$

$$a_0 \left(\frac{1}{\sqrt{2\pi}} \right) + \frac{1}{\sqrt{\pi}} \sum_{i=1}^{\infty} (a_i \cos ix + b_i \sin ix) \tag{12.1}$$

In general, such a linear combination is an infinite series. Not all functions belong to this function space; for instance, let $a_i = 1, b_i = 0, i \geq 0$. In this case, the series is

$$f(x) = \frac{1}{\sqrt{2\pi}} + \frac{1}{\sqrt{\pi}} (\cos x + \cos 2x + \cos 3x + \cdots).$$

If $x = 0$, then $f(0) = \frac{1}{\sqrt{2\pi}} + \frac{1}{\sqrt{\pi}} (1 + 1 + 1 + \cdots)$ is infinite. For f to belong to the function space, we require that $\|f\|_{L^2}^2$ be finite. Since

$$\left\{ \frac{1}{\sqrt{2\pi}}, \frac{\cos x}{\sqrt{\pi}}, \frac{\sin x}{\sqrt{\pi}}, \frac{\cos 2x}{\sqrt{\pi}}, \frac{\sin 2x}{\sqrt{\pi}}, \ldots, \frac{\cos nx}{\sqrt{\pi}}, \frac{\sin nx}{\sqrt{\pi}}, \ldots \right\}$$

is an orthonormal sequence

$$\langle f, f \rangle = \int_{-\pi}^{\pi} f^2(x) \, dx = a_0^2 + a_1^2 + b_1^2 + a_2^2 + b_2^2 + \cdots + a_n^2 + b_n^2 + \cdots. \tag{12.2}$$

Thus, a function is in this function space if the series 12.2 converges, and we call it the *Fourier series* for f. Since the function is composed of periodic functions, f is periodic. It can be shown that this function space obeys the rules for a vector space, and the Cauchy-Schwarz inequality, $\langle f, g \rangle \leq \|f\|_{L^2} \|g\|_{L^2}$ holds. This function space is an example of a *Hilbert space*, a mathematical concept that has many applications in such areas as electrical engineering, vibration analysis, optics, acoustics, signal and image processing, econometrics, and quantum mechanics.

The a_i, b_i in Equation 12.1 are called the *Fourier coefficients* for f. So far, we have said that if the series 12.2 converges, the function it defines is in the function space. Assume for the moment that the function f has a Fourier series. How do we compute the Fourier coefficients? The answer lies in the fact that the basis is orthonormal relative to the inner product. We'll compute the coefficients of the terms $\cos ix, 0 \leq i < \infty$. Start with

$$f(x) = a_0 \left(\frac{1}{\sqrt{2\pi}} \right) + \frac{1}{\sqrt{\pi}} \sum_{i=1}^{\infty} (a_i \cos ix + b_i \sin ix). \tag{12.3}$$

Multiply both sides of Equation 12.3 by $\cos kx$ for any $k \geq 1$ and integrate from $-\pi$ to π.

$$\int_{-\pi}^{\pi} f(x) \cos kx = a_0 \left(\frac{1}{\sqrt{2\pi}} \right) \int_{-\pi}^{\pi} \cos kx \, dx + \frac{1}{\sqrt{\pi}} \int_{-\pi}^{\pi} \left(\sum_{i=1}^{\infty} (a_i \cos kx \cos ix + b_i \cos kx \sin ix) \right) dx$$

$$= 0 + \frac{1}{\sqrt{\pi}} \sum_{i=1}^{\infty} \left(\int_{-\pi}^{\pi} (a_i \cos kx \cos ix) \, dx + \int_{-\pi}^{\pi} (b_i \sin kx \sin ix) \, dx \right) = a_k \frac{1}{\sqrt{\pi}} \int_{-\pi}^{\pi} \cos^2 kx = \sqrt{\pi} a_k,$$

and so

$$a_k = \frac{1}{\sqrt{\pi}} \int_{-\pi}^{\pi} f(x) \cos kx, \quad k \geq 1.$$

In a similar fashion,

$$b_k = \frac{1}{\sqrt{\pi}} \int_{-\pi}^{\pi} f(x) \sin kx, \quad k \geq 1.$$

It only remains to compute a_0:

$$\int_{-\pi}^{\pi} f(x) \, dx = a_0 \left(\frac{1}{\sqrt{2\pi}} \right) \int_{-\pi}^{\pi} dx + \frac{1}{\sqrt{\pi}} \sum_{i=1}^{\infty} \left(\int_{-\pi}^{\pi} (a_i \cos ix) \, dx + \int_{-\pi}^{\pi} (b_i \sin ix) \, dx \right) = \sqrt{2\pi} a_0 + 0 = \sqrt{2\pi} a_0.$$

Thus,

$$a_0 = \frac{1}{\sqrt{2\pi}} \int_{-\pi}^{\pi} f(x)\, dx.$$

Summary: The Fourier coefficients for the function f, $-\pi \leq x \leq \pi$, are

$$a_0 = \frac{1}{\sqrt{2\pi}} \int_{-\pi}^{\pi} f(x)\, dx.$$

$$a_k = \frac{1}{\sqrt{\pi}} \int_{-\pi}^{\pi} f(x) \cos kx\, dx, \ k \geq 1.$$

$$b_k = \frac{1}{\pi} \int_{-\pi}^{\pi} f(x) \sin kx\, dx, \ k \geq 1.$$

Now, if we have a periodic function f, is there a Fourier series that converges to f? The following theorem can be found in the literature on Fourier series.

Theorem 12.1. *Assume that f has period 2π and is piecewise continuously differentiable on $-\pi \leq x \leq \pi$. Then the Fourier series*

$$a_0 \left(\frac{1}{\sqrt{2\pi}} \right) + \frac{1}{\sqrt{\pi}} \sum_{i=1}^{\infty} (a_i \cos ix + b_i \sin ix)$$

converges at every point at which f is continuous and otherwise to

$$\frac{f(x+) + f(x-)}{2},$$

where

$$f(l+) = \lim_{x \to l+} f(x)$$

and

$$f(l-) = \lim_{x \to l-} f(x)$$

are the right and left-side limits of f at x.

Remark 12.1.   Our discussion of Fourier series deals with functions having period 2π. If the function is periodic over another interval, $-L \leq x \leq L$, let $t = \pi x / L$ and

$$g(t) = f(x) = f\left(\frac{Lt}{\pi} \right).$$

g has period 2π, and $x = -L, x = L$ correspond to $t = -\pi$, $t = \pi$, respectively. Find the Fourier series for $g(t)$ and, using substitution, find the Fourier series for $f(x)$.

12.1.1   The Square Wave

The *square wave* is a very useful function in many engineering applications. It is periodic and, in this example, has period 2π, and is $+1$ in one-half the period and -1 in the other half (Figure 12.1). The Fourier series for the square wave is not difficult to compute. All the cosine terms in its Fourier series are 0, since the function is odd ($f(-x) = -f(x)$). The series is

$$\frac{4}{\pi} \left[\frac{\sin x}{1} + \frac{\sin 3x}{3} + \frac{\sin 5x}{5} + \cdots \right].$$

It is interesting to see how the series converges (Figure 12.2). As $n \to \infty$, the partial sums of sine terms "wiggle less and less" and converge to the line segments comprising the wave. The book software contains the sound file square.mp3. Play it to hear this waveform.

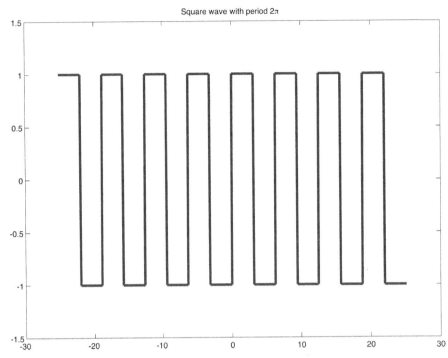

FIGURE 12.1 Square wave with period 2π.

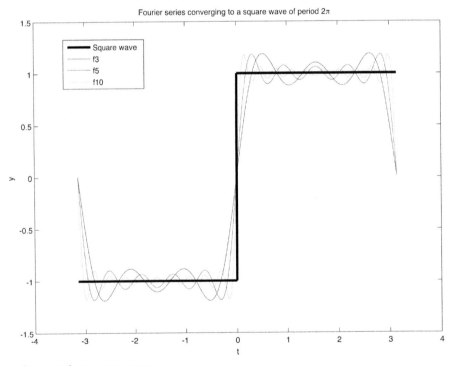

FIGURE 12.2 Fourier series converging to a square wave.

12.2 FINITE DIFFERENCE APPROXIMATIONS

Differential equations are at the heart of many engineering problems. They appear in fluid mechanics, thermodynamics, vibration analysis, and many other areas. For most of the problems that appear in practice, we cannot find an analytical solution and must rely on numerical techniques. Applications of linear algebra appear particularly in what are termed *initial-boundary value problems*. In problems like these, an initial condition is known at $t = 0$, and the value of the solution is known on the boundary of an object. We must use these known values to approximate the solution in the interior of the object.

12.2.1   Steady-State Heat and Diffusion

Suppose a thin rod is given an initial temperature distribution, then insulated on the sides. The ends of the rod are kept at the same fixed temperature; e.g., suppose at the start of the experiment, both ends are immediately plunged into ice water. We are interested in how the temperature along the rod varies with time. Suppose that the rod has a length L (in meters), and we establish a coordinate system along the rod as illustrated in Figure 12.3. Let $u(x, t)$ represent the temperature at the point x meters along the rod at time t (in seconds). We start with an initial temperature distribution $u(x, 0) = f(x)$. This problem is modeled using a partial differential equation called the *heat equation*:

$$\frac{\partial u}{\partial t} = c\frac{\partial^2 u}{\partial x^2}, \quad 0 \le x \le L, \quad 0 \le t \le T,$$
$$u(0, t) = u(L, t) = 0,$$
$$u(x, 0) = f(x).$$

The constant c is the *thermal diffusivity*. For most functions $f(x)$, we cannot obtain an exact solution to the problem, so we must resort to numerical methods. So where does linear algebra come in? We divide the space interval $0 \le x \le L$ into small subintervals of length $h = L/m$, and the time interval into small subintervals of length $k = T/n$, where m and n are integers (Figure 12.4). We wish to approximate $u(x, t)$ at the grid points (x_i, t_j), where $(x_1, x_2, x_3, \ldots, x_m, x_{m+1}) = (0, h, 2h, \ldots, L - h, L)$ and $(t_1, t_2, t_3, \ldots, t_n, t_{n+1}) = (0, k, 2k, \ldots, T - k, T)$. We denote this approximate solution by

$$u_{ij} \approx u(x_i, t_j).$$

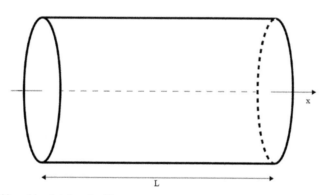

FIGURE 12.3 The heat equation: a thin rod insulated on its sides.

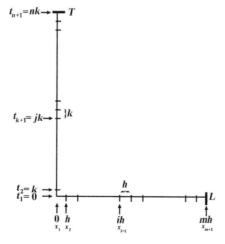

FIGURE 12.4 Numerical solution of the heat equation: subdivisions of the x and t axes.

If m and n are large, the values of h and k are small, and we can approximate each derivative at a point (x_i, t_j) using *finite difference equations* as follows:

$$\frac{\partial u}{\partial t}\left(x_i, t_{j+1}\right) \approx \frac{u_{i,j+1} - u_{ij}}{k}, \quad 1 \le j \le n, \tag{12.4}$$

$$\frac{\partial^2 u}{\partial x^2}\left(x_i, t_{j+1}\right) \approx \frac{1}{h^2}\left(u_{i-1,j+1} - 2u_{i,j+1} + u_{i+1,j+1}\right), \quad 2 \le i \le m. \tag{12.5}$$

These approximations are developed using Taylor series for a function of two variables, and the interested reader can refer to Ref. [33, Chapter 6], for an explanation. By equating these approximations, we have

$$\frac{u_{i,j+1} - u_{ij}}{k} = \frac{c}{h^2}\left(u_{i-1,j+1} - 2u_{i,j+1} + u_{i+1,j+1}\right).$$

All but one term in the finite difference formula involves $j + 1$, so isolate it to obtain

$$u_{i,j} = -ru_{i-1,j+1} + (1 + 2r)\, u_{i,j+1} - ru_{i+1,j+1},\ 2 \le i \le m,\ 1 \le j \le n, \tag{12.6}$$

where $r = ck/h^2$. Notice that Equations 12.4 and 12.5 specify relationships between points in the grid pattern of Figure 12.5. Figure 12.6 provides a view of the entire grid. In Figure 12.6, the black circles represent boundary and initial values and the open circles represent a general pattern of the four points used in each difference equation.

Using matrix notation, Equation 12.6 becomes

$$u_j = Bu_{j+1} - rb_{j+1}, \tag{12.7}$$

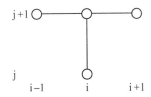

FIGURE 12.5 Numerical solution of the heat equation:locally related points in the grid.

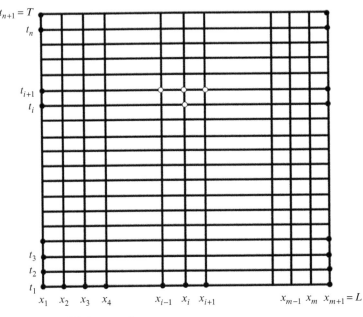

FIGURE 12.6 Grid for the numerical solution of the heat equation.

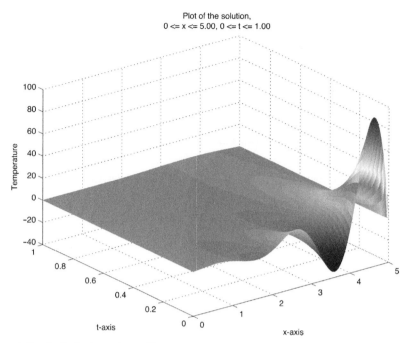

Plot of the solution,
0 <= x <= 5.00, 0 <= t <= 1.00

FIGURE 12.7 Graph of the solution for the heat equation problem.

where

$$B = \begin{bmatrix} 1+2r & -r & 0 & \dots & 0 \\ -r & 1+2r & -r & \dots & 0 \\ 0 & \ddots & \ddots & \ddots & \vdots \\ \vdots & \vdots & -r & 1+2r & -r \\ 0 & 0 & 0 & -r & 1+2r \end{bmatrix}$$

and b_{j+1} accounts for the boundary or initial conditions. Now, we can write Equation 12.7 as

$$Bu_{j+1} = u_j + rb_{j+1}. \tag{12.8}$$

This is a linear algebraic system to solve for u_{j+1}. Most of the elements of B are 0, so it is a *sparse matrix*; furthermore, it is *tridiagonal*. The Thomas algorithm presented in Section 9.4 takes advantage of the tridiagonal structure and solves system 12.7 in $O(n)$ flops as opposed to $O(n^3)$. Think of Equation 12.8 this way. Values of u_j and rb_{j+1} give us values for u_{j+1}, and we work our way up the grid. This is not a "chicken and an egg" problem, since the right-hand side of Equation 12.8 is known at time $t = 0$. We now give an example of our finite difference equations in action by approximating and graphing the solution to the heat flow problem

$$\begin{aligned} \frac{\partial u}{\partial t} &= 0.875 \frac{\partial^2 u}{\partial x^2}, \quad 0 \le x \le 5.0, \quad 0 \le t \le 1.0, \\ u(0,t) &= u(5,t) = 0, \\ u(x,0) &= e^x \sin(\pi x). \end{aligned} \tag{12.9}$$

The required MATLAB code, `heateq.m`, is located in the software distribution. We present a graph of the approximate solution obtained in Figure 12.7.

We will deal with more problems like this in Chapters 20 and 21, where we discuss the Poisson and biharmonic equations.

12.3 LEAST-SQUARES POLYNOMIAL FITTING

We now consider a problem in data analysis. Assume that during an experiment, we collect m measurements of a quantity y that depends on a parameter t; in other words, we have the set of experimental points $(t_1, y_1), (t_2, y_2), \dots, (t_m, y_m)$. We want to use those m measurements to approximate other y values inside and outside the range t_1 to t_m, processes called *interpolation* and *extrapolation*, respectively. One approach to the problem is to fit a polynomial to the data in the "least-squares" sense. Assume the polynomial is

$$p(x) = a_n x^n + a_{n-1} x^{n-1} + \cdots + a_2 x^2 + a_1 x + a_0.$$

Fit the polynomial by minimizing the sum of the squares of the deviation of each data point from the corresponding polynomial value; in other words, find the polynomial that will minimize the *residual*

$$E = \sum_{i=1}^{m} \left(y_i - \left(a_0 + a_1 t_i + a_2 t_i^2 + \cdots + a_n t_i^n \right) \right)^2.$$

We know from multivariable calculus that we do this by requiring the partial derivatives of E with respect to $a_0, a_1, \ldots, a_n$ be 0, or

$$\frac{\partial E}{\partial a_i} = 0, \quad 0 \le i \le n.$$

The partial derivatives are

$$
\left.
\begin{aligned}
\frac{\partial E}{\partial a_0} &= -2 \sum_{i=1}^{m} \left(y_i - a_0 - a_1 t_i - a_2 t_i^2 - \cdots - a_n t_i^n \right) \\
\frac{\partial E}{\partial a_1} &= -2 \sum_{i=1}^{m} t_i \left(y_i - a_0 - a_1 t_i - a_2 t_i^2 - \cdots - a_n t_i^n \right) \\
\frac{\partial E}{\partial a_2} &= -2 \sum_{i=1}^{m} t_i^2 \left(y_i - a_0 - a_1 t_i - a_2 t_i^2 - \cdots - a_n t_i^n \right) \\
&\;\;\vdots \\
\frac{\partial E}{\partial a_m} &= -2 \sum_{i=1}^{m} t_i^n \left(y_i - a_0 - a_1 t_i - a_2 t_i^2 - \cdots - a_n t_i^n \right)
\end{aligned}
\right\}.
$$

Setting these equations to zero, we have

$$a_0 m + a_1 \sum_{i=1}^{m} t_i + a_2 \sum_{i=1}^{m} t_i^2 + \cdots + a_n \sum_{i=1}^{m} t_i^n = \sum_{i=1}^{m} y_i$$

$$a_0 \sum_{i=1}^{m} t_i + a_1 \sum_{i=1}^{m} t_i^2 + \cdots + a_n \sum_{i=1}^{m} t_i^{n+1} = \sum_{i=1}^{m} t_i y_i$$

$$a_0 \sum_{i=1}^{m} t_i^2 + a_1 \sum_{i=1}^{m} t_i^3 + \cdots + a_n \sum_{i=1}^{m} t_i^{n+2} = \sum_{i=1}^{m} t_i^2 y_i$$

$$\vdots$$

$$a_0 \sum_{i=1}^{m} t_i^n + a_1 \sum_{i=1}^{m} t_i^{n+1} + \cdots + a_n \sum_{i=1}^{m} t_i^{2n} = \sum_{i=1}^{m} t_i^n y_i \tag{12.10}$$

Let $S_k = \sum_{i=1}^{m} t_i^k$, $k = 0, 1, \ldots, 2n$, $b_k = \sum_{i=1}^{m} t_i^k y_i, k = 0, 1, \ldots, n$, and write Equation 12.10 as a matrix equation.

$$
\begin{bmatrix}
S_0 & S_1 & \cdots & S_n \\
S_1 & S_2 & \cdots & S_{n+1} \\
S_2 & S_3 & \cdots & S_{n+2} \\
\vdots & \vdots & \ddots & \vdots \\
S_n & S_{n+1} & \cdots & S_{2n}
\end{bmatrix}
\begin{bmatrix}
a_0 \\
a_1 \\
a_2 \\
\vdots \\
a_n
\end{bmatrix}
=
\begin{bmatrix}
b_0 \\
b_1 \\
b_2 \\
\vdots \\
b_n
\end{bmatrix}.
\tag{12.11}
$$

(Note that $S_0 = m$.) This is a system of $(n + 1)$ equations in $n + 1$ unknowns $a_0, a_1, \ldots, a_n$.

We can write system 12.11 in a different form. Define the $m \times (n + 1)$ *Vandermonde matrix and the vector y*

$$
V =
\begin{bmatrix}
1 & t_1 & \cdots & t_1^n \\
1 & t_2 & \cdots & t_2^n \\
\vdots & \vdots & \ddots & \vdots \\
1 & t_m & \cdots & t_m^n
\end{bmatrix},
\quad
y =
\begin{bmatrix}
y_1 \\
y_2 \\
\vdots \\
y_m
\end{bmatrix}.
$$

Using Equation 12.11, it follows that $V^{\mathrm{T}}y = b$:

$$V^{\mathrm{T}}y = \begin{bmatrix} 1 & 1 & \cdots & 1 \\ t_1 & t_2 & \cdots & t_m \\ \vdots & \vdots & \ddots & \vdots \\ t_1^n & t_2^n & \cdots & t_m^n \end{bmatrix} \begin{bmatrix} y_1 \\ y_2 \\ \vdots \\ y_m \end{bmatrix} = \begin{bmatrix} \sum_{i=1}^{m} y_i \\ \sum_{i=1}^{m} t_i y_i \\ \vdots \\ \sum_{i=1}^{m} t_i^n y_i \end{bmatrix} = \begin{bmatrix} b_0 \\ b_1 \\ b_2 \\ \vdots \\ b_n \end{bmatrix} = b.$$

Now note that $V^{\mathrm{T}}Va = b$:

$$V^{\mathrm{T}}Va = \begin{bmatrix} 1 & 1 & \cdots & 1 \\ t_1 & t_2 & \cdots & t_m \\ \vdots & \vdots & \ddots & \vdots \\ t_1^n & t_2^n & \cdots & t_m^n \end{bmatrix} \begin{bmatrix} 1 & t_1 & \cdots & t_1^n \\ 1 & t_2 & \cdots & t_2^n \\ \vdots & \vdots & \ddots & \vdots \\ 1 & t_m & \cdots & t_m^n \end{bmatrix} \begin{bmatrix} a_0 \\ a_1 \\ \vdots \\ a_n \end{bmatrix}$$

$$= \begin{bmatrix} S_0 & S_1 & \cdots & S_n \\ S_1 & S_2 & \cdots & S_{n+1} \\ \vdots & \vdots & \ddots & \vdots \\ S_n & S_{n+1} & \cdots & S_{2n} \end{bmatrix} \begin{bmatrix} a_0 \\ a_1 \\ \vdots \\ a_n \end{bmatrix} = \begin{bmatrix} b_0 \\ b_1 \\ \vdots \\ b_n \end{bmatrix} = b$$

by referencing Equation 12.11. Thus,

$$V^{\mathrm{T}}Va = V^{\mathrm{T}}y = b.$$

A system of equations of this form is called the normal equations.

12.3.1 Normal Equations

Definition 12.1. Let $A \in R^{m \times n}$. The system of n equations and n unknowns

$$A^{\mathrm{T}}Ax = A^{\mathrm{T}}y$$

is called the *normal equations*.

Normal equations will become very important when we discuss linear least-squares problems in Chapter 16. Usually in these types of problems, either $m > n$ (*overdetermined system*) or $m < n$ (*underdetermined system*). If we have 25 data points and want to fit a straight line in the least-squares sense, we have $m = 25$ and $n = 2$.
To solve our least-squares problem, compute $V^{\mathrm{T}}y$ to obtain b. Then compute $V^{\mathrm{T}}V$ and solve the square system $\left(V^{\mathrm{T}}V\right)a = b$.

Example 12.1. The following eight data points show the relationship between the number of fishermen and the amount of fish (in thousand pounds) they can catch a day.

Number of Fishermen	Fish Caught
4	7
5	8
9	9
10	12
12	15
14	20
18	26
22	35

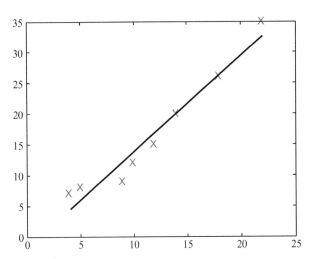

FIGURE 12.8 Linear least-squares approximation.

Case 1: First, we will fit a straight line ($n = 1$) to the data. The Vandermonde matrix V is of dimension 8×2, and its transpose has dimension 2×8, so $V^T V$ is a 2×2 matrix. Compute

$$V^T y = b = \begin{bmatrix} 1 & 1 & 1 & 1 & 1 & 1 & 1 & 1 \\ 4 & 5 & 9 & 10 & 12 & 14 & 18 & 22 \end{bmatrix} \begin{bmatrix} 7 \\ 8 \\ 9 \\ 12 \\ 15 \\ 20 \\ 26 \\ 35 \end{bmatrix} = \begin{bmatrix} 132 \\ 1967 \end{bmatrix}.$$

Now solve $V^T Va = \begin{bmatrix} 132 \\ 1967 \end{bmatrix}$ to obtain the linear least-squares solution

$$a_0 = -1.9105, \quad a_1 = 1.5669.$$

Thus, the line that fits the data in the least-squares sense is $y = 1.5669x - 1.9105$. Figure 12.8 shows the data and the least-squares line (often called the *regression line*).

To estimate the value for 16 fishermen, compute $1.5669\,(16) - 1.9105 = 23.1591$.

Case 2: We will do a quadratic fit ($n = 2$).

$$V^T y = \begin{bmatrix} 132 \\ 1967 \\ 33,685 \end{bmatrix} = b$$

Now solve $V^T Va = \begin{bmatrix} 132 \\ 1967 \\ 33,685 \end{bmatrix}$ to obtain $a = \begin{bmatrix} 5.2060 \\ 0.1539 \\ 0.0554 \end{bmatrix}$. The best quadratic polynomial in the least-squares sense is

$5.2060 + 0.1539x + 0.0554x^2$. Figure 12.9 shows the data and the least-squares quadratic polynomial.

To estimate the value for 16 fishermen, compute $5.2060 + 0.1539\,(16) + 0.0554(16)^2 = 21.8485$. ■

Remark 12.2.   In Example 12.1, the quadratic polynomial appears to be more accurate; however, using a higher degree polynomial does not guarantee better results. As the dimensions of the Vandermonde matrices get larger, they become ill-conditioned.

Example 12.2.   Estimating absolute zero.

Charles's Law for ideal gas states that at constant volume a linear relationship exists between the pressure p and the temperature t. An experiment takes gas in a sealed container that is initially submerged in ice water ($t = 0\,°C$). The temperature is increased in $10°$ increments with the pressure measured each $10°$.

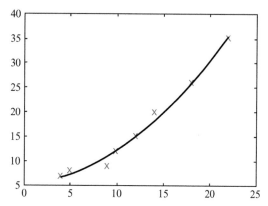

FIGURE 12.9 Quadratic least-squares approximation.

t	0	10	20	30	40	50	60	70	80	90	100
p	0.94	0.96	1.00	1.05	1.07	1.09	1.14	1.17	1.21	1.24	1.28

After finding a linear least-squares estimate, extrapolate the function to determine absolute zero, i.e., the temperature where the pressure is 0.

$$V^{\mathrm{T}}y = \begin{bmatrix} 1 & 1 & 1 & 1 & 1 & 1 & 1 & 1 & 1 & 1 & 1 \\ 0 & 10 & 20 & 30 & 40 & 50 & 60 & 70 & 80 & 90 & 100 \end{bmatrix} \begin{bmatrix} 0.94 \\ 0.96 \\ 1.00 \\ 1.05 \\ 1.07 \\ 1.09 \\ 1.14 \\ 1.17 \\ 1.21 \\ 1.24 \\ 1.28 \end{bmatrix} = \begin{bmatrix} 12.1500 \\ 645.1000 \end{bmatrix}.$$

Solve $V^{\mathrm{T}}Va = \begin{bmatrix} 12.1500 \\ 645.100 \end{bmatrix}$ to obtain the regression line $p = 0.0034t + 0.9336$. The temperature when $p = 0$ is $t_{\text{abs zero}} = -\frac{0.9336}{0.0034} = -273.1383$. The true value of absolute zero in Celsius is $-273.15\,°C$.

The function vandermonde(t,m) in the software distribution constructs the Vandermonde matrix. The following MATLAB program finds the least-squares approximation to absolute zero and draws a graph (Figure 12.10) showing the regression line and the data points.

```
>> t = 0:10:100;
>> V = vandermonde(t,1);
>> p = [.94 .96 1 1.05 1.07 1.09 1.14 1.17 1.21 1.24 1.28]';
>> b = V'*p;
>> a = (V'*V)\b

a =

    0.9336
    0.0034

>> abszero_approx = -a(1)/a(2)

abszero_approx =

 -273.1383

>> temp = [-300 100];
>> plot(t, p, 'o', temp, a(2)*temp + a(1));
```

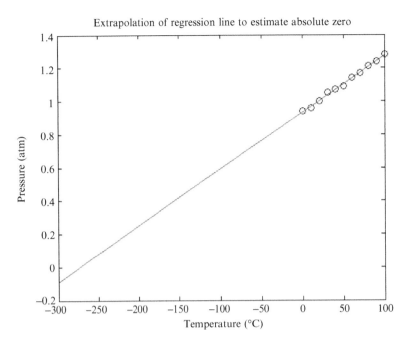

FIGURE 12.10 Estimating absolute zero.

```
>> xlabel('temperature (C)');
>> ylabel('pressure (atm)');
>> title('Extrapolation of Regression Line to Estimate Absolute Zero');
```                                                                          ∎

An alternative to using least-squares is *Lagrange interpolation*. This process takes n distinct data points and finds the unique polynomial of degree $n - 1$ that passes through the points. Lagrange interpolation is discussed in the problems.

Both least-squares and Lagrange interpolation fit one polynomial to the data. An alternative is to fit a piecewise polynomial, and a premier method that uses this approach is cubic splines.

12.4 CUBIC SPLINE INTERPOLATION

Least squares can be used for either interpolation or extrapolation. Example 12.2 used extrapolation. When applying interpolation in a range $a \leq t \leq b$, estimates can be computed only for values in that range. Most engineers and scientists are familiar with *linear interpolation* (also called *linear splines*), in which the data points are joined by line segments, and a value between two data points is estimated using a point on the line segment.

Example 12.3. An experiment yields measurements

$$\{(1.3, 2.8), (1.7, 3.2), (1.9, 3.1), (2.3, 3.5), (2.7, 4.8), (3.1, 4.2), (3.6, 5.3), (4.0, 4.8)\}.$$

Given any value t in the range $t_i \leq t \leq t_{i+1}$

$$f(t) = y_i + \frac{(t - t_i)}{(t_{i+1} - t_i)}(y_{i+1} - y_i) \tag{12.12}$$

is a piecewise linear function defined for all points in the data set. For instance, approximate the measurement at $t = 2.5$. The value of t lies between $t_4 = 2.3$ and $t_5 = 2.7$. Using Equation 12.12, the approximate value is $f(2.5) = 3.5 + \frac{(2.5-2.3)}{(2.7-2.3)}(4.8 - 3.5) = 4.15$ (Figure 12.11)   ∎

Cubic splines use the same idea, but in a more sophisticated fashion. Instead of using a line segment between two points, the algorithm uses a cubic polynomial (Figure 12.12) to form a piecewise cubic function. The data points t_{i-1}, t_i where two polynomials from adjacent intervals meet are called *knots*. Following the development of cubic splines in Ref. [34] with assistance from Ref. [35], assume the data points are

$$\{(t_1, y_1), (t_2, y_2), (t_3, y_3), \ldots, (t_n, y_n), (t_{n+1}, y_{n+1})\},$$

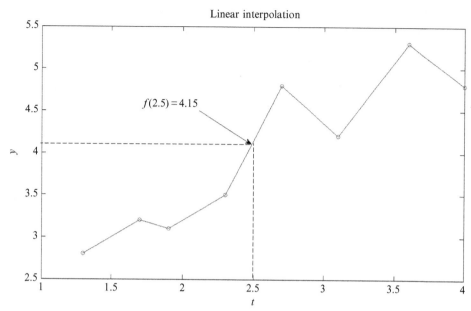

FIGURE 12.11 Linear interpolation.

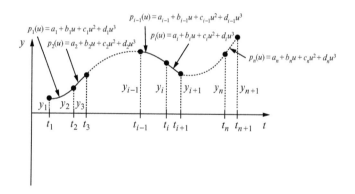

FIGURE 12.12 Cubic splines.

and that the cubic polynomial p_i, $1 \le i \le n$ between t_i and t_{i+1} is parameterized by u

$$p_i(u) = a_i + b_i u + c_i u^2 + d_i u^3, \quad 0 \le u \le 1. \tag{12.13}$$

As it turns out, we will never need to deal directly with the t_i. The algorithm deals with them implicitly. The polynomials must agree at the points t_i, $2 \le i \le n$, which leads to

$$\begin{aligned} p_i(0) &= a_i = y_i, & (12.14) \\ p_i(1) &= a_i + b_i + c_i + d_i = y_{i+1}. & (12.15) \end{aligned}$$

These conditions make the piecewise polynomial continuous. Let D_i, $1 \le i \le n+1$, be the value of the first derivative of the p_i at the knots, and require that at the interior points the first derivatives must agree. This assures that the function defined by the piecewise polynomials is continuous and differentiable. Thus, for $i = 2, \ldots, n$

$$\begin{aligned} p_i'(0) &= b_i = D_i, & (12.16) \\ p_i'(1) &= b_i + 2c_i + 3d_i = D_{i+1}. & (12.17) \end{aligned}$$

Solve Equations 12.14–12.17 for a_i, b_i, c_i, and d_i to obtain

$$\begin{aligned} a_i &= y_i, & (12.18) \\ b_i &= D_i, & (12.19) \\ c_i &= 3(y_{i+1} - y_i) - 2D_i - D_{i+1}, & (12.20) \\ d_i &= 2(y_i - y_{i+1}) + D_i + D_{i+1}. & (12.21) \end{aligned}$$

Using Equations 12.18–12.21, the a_i, b_i, c_i, and d_i follow from the D_i and y_i.

Now require that the second derivatives match at the $n - 1$ interior points $\{t_2, t_3, \ldots, t_n\}$, so

$$p''_{i-1}(1) = p''_i(0),$$

and

$$2c_{i-1} + 6d_{i-1} = 2c_i, \quad 2 \leq i \leq n. \tag{12.22}$$

Also require that $p_1(0) = y_1$ and $p_n(1) = y_{n+1}$, which gives

$$a_1 = y_1, \tag{12.23}$$

$$a_n + b_n + c_n + d_n = y_{n+1}. \tag{12.24}$$

Each cubic polynomial p_i, $1 \leq i \leq n$ has four unknown coefficients $\{a_i, b_i, c_i, d_i\}$, so there are a total of $4n$ unknowns. Here are the equations we have:

- Equating the polynomial values at the interior points: $2(n - 1)$ equations
- Equating the first derivatives at the interior points: $(n - 1)$ equations (substitute 12.14 into 12.15 to obtain only one equation)
- Equating the second derivatives at the interior points: $(n - 1)$ equations
- Require that $p_1(0) = y_1$ and $p_n(1) = y_{n+1}$: 2 equations

TOTAL = 4n − 2 equations.

We still require two more equations. The choice of the two additional equations determines the type of cubic spline. Require

$$p''_1(0) = 0,$$
$$p''_n(1) = 0, \tag{12.25}$$

which implies

$$c_1 = 0, \tag{12.26}$$

$$2c_n + 6d_n = 0. \tag{12.27}$$

With some manipulation, we can reduce the problem to the solution of an $(n + 1) \times (n + 1)$ system of equations that we can solve much faster. Using Equations 12.20 and 12.22, we have

$$2c_{i-1} + 6d_{i-1} = 2\left[3(y_{i+1} - y_i) - 2D_i - D_{i+1}\right].$$

From Equations 12.20 and 12.21, it follows that

$$c_{i-1} = 3(y_i - y_{i-1}) - 2D_{i-1} - D_i$$
$$d_{i-1} = 2(y_{i-1} - y_i) + D_{i-1} + D_i,$$

so

$$2\left[3(y_i - y_{i-1}) - 2D_{i-1} - D_i\right] + 6\left[2(y_{i-1} - y_i) + D_{i-1} + D_i\right] = 2\left[3(y_{i+1} - y_i) - 2D_i - D_{i+1}\right] \tag{12.28}$$

Simplify Equation 12.28 and move the unknowns to the left-hand side, and we have the $n-1$ equations in $n+1$ unknowns.

$$D_{i+1} + 4D_i + D_{i-1} = 3(y_{i+1} - y_{i-1}), \quad 2 \leq i \leq n. \tag{12.29}$$

We must add two more equations to Equation 12.29. From Equation 12.20, $c_1 = 3(y_2 - y_1) - 2D_1 - D_2$, and by using Equation 12.26

$$2D_1 + D_2 = 3(y_2 - y_1). \tag{12.30}$$

From Equation 12.27 $c_n = -3d_n$, and by substituting it into Equation 12.24, there results

$$a_n + b_n - 2d_n = y_{n+1}. \tag{12.31}$$

Apply Equations 12.18, 12.19, and 12.21 with $i = n$ in Equation 12.31, and perform some algebra, to obtain

$$D_n + 2D_{n+1} = 3(y_{n+1} - y_n). \tag{12.32}$$

After adding Equations 12.30 and 12.32 to Equation 12.29, we have the symmetric tridiagonal system

$$
\begin{bmatrix}
2 & 1 & & & & & & \\
1 & 4 & 1 & & & & & \\
 & 1 & 4 & 1 & & & & \\
 & & 1 & 4 & 1 & & & \\
\vdots & \ddots & \ddots & \ddots & \ddots & \ddots & & \vdots \\
 & & & & 1 & 4 & 1 \\
 & & & & & 1 & 2
\end{bmatrix}
\begin{bmatrix}
D_1 \\ D_2 \\ D_3 \\ D_4 \\ \vdots \\ \vdots \\ D_n \\ D_{n+1}
\end{bmatrix}
=
\begin{bmatrix}
3\,(y_2 - y_1) \\
3\,(y_3 - y_1) \\
3\,(y_4 - y_2) \\
\vdots \\
3\,(y_n - y_{n-2}) \\
3\,(y_{n+1} - y_{n-1}) \\
3\,(y_{n+1} - y_n)
\end{bmatrix}.
\tag{12.33}
$$

Solve this $(n + 1) \times (n + 1)$ tridiagonal system using the linear $[O(n)]$ Thomas algorithm presented in Section 9.4, and apply Equations 12.18–12.21 to determine the $\{a_i, b_i, c_i, d_i\}$.

We have defined *natural cubic splines*. There are other ways to obtain the two additional equations. The *not-a-knot* condition requires that the third derivatives are continuous at t_2, and t_n. In a *clamped cubic spline*, the first derivative of the spline is specified at the end points, so that $p'_1(0) = f'(0)$ and $p'_n(1) = f'(x_{n+1})$. This requires that the derivatives are known or can be estimated at the endpoints.

Algorithm 12.1 builds natural cubic splines. It also graphs the data and the spline approximation, but it does not perform interpolation at a specified point or points. That is left to the exercises.

Algorithm 12.1 Cubic Spline Approximation

```
function CUBICSPLINEB(t,y)
    % CUBSPLINEB Natural cubic spline interpolation
    % CUBSPLINEB(t,y) computes a natural cubic spline approximation
    % to the data t, y. It then plots the data and the cubic spline.
    % Input: n+1 data points (t_1 y_i).
    % Output: n × 4 matrix. Row 1 is the cubic polynomial fit to
    % t_1 ≤ t ≤ t_2,..., Row n is the cubic polynomial fit to
    % t_n ≤ t ≤ t_n+1. The function makes a plot of the data
    % and the cubic spline approximation.
    b_1 = 2
    b_n+1 = 2
    % build the right-hand side of the system.
    rhs_1 = 3 (y_2 − y_1)
    rhs_n+1 = 3 (y_n+1 − y_n)

    for i = 2:n do
        rhs_i = 3 (y_i+1 − y_i-1)
    end for
    % solve the system using the linear Thomas algorithm.
    D = thomas (a, b, c, rhs)
    % construct the cubic polynomials in the rows of S.
    for i = 1:n do
        S_i1 = 2 (y_i − y_i+1) + D_i + D_i+1
        S_i2 = 3 (y_+1 − y_i) − 2D_i − D_i+1
        S_i3 = D_i
        S_i4 = y_i
    end for
    % plot the data and the cubic spline approximation.
    for i = 1:n do
        p_i = spline S_i
        plot p_i over the interval t_i ≤ t ≤ t_i+1
    end for
    plot (t, y), marking each point with '*'
end function
```

NLALIB: The function `cubicsplineb` implements Algorithm 12.1.

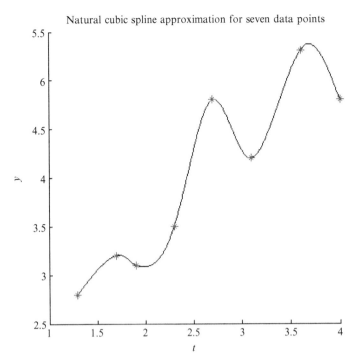

FIGURE 12.13 Cubic spline approximation.

Example 12.4. Using the data from Example 12.3, fit the data using cubic spline interpolation and graph the results (Figure 12.13). Approximate the value at *tval* = 2.5.

```
>> S = cubicsplineb(t,y);
>> p = S(4,:);
>> tval = 2.5;
>> % compute value of parameter u by interpolation;
>> u = (tval - t(4))/(t(5) - t(4));
>> polyval(p, u)

ans =

    4.2868
```
■

12.5 CHAPTER SUMMARY

Fourier Series

A Fourier series is an infinite series of trigonometric functions that, under the correct conditions, converges to a periodic function. Fourier series expose frequencies in waves and can serve as an exact solution to some partial differential equations. Computing Fourier series for square, sawtooth, and triangle waves is useful in the analysis of musical sounds.

Finite Difference Approximations

A finite difference approximation is an expression involving the function at various points that approximates an ordinary or a partial derivative. To approximate the solution to an ordinary or partial differential equation, approximate the derivatives at a series of grid points, normally close together. Using boundary and initial conditions that determine the right-hand side, create a linear system of equations. The solution provides approximations to the actual solution at the grid points. In this chapter, we use finite differences to approximate the solution to the one-dimensional heat equation with initial and boundary conditions. This requires solving a large tridiagonal system, for which we use the Thomas algorithm from Section 9.4.

Least-Squares Polynomial Fitting

Given experimental data points $(t_1, y_1), (t_2, y_2), \ldots (t_m, y_m)$, find a unique polynomial $p(t) = a_n t^n + a_{n-1} t^{n-1} + \cdots + a_2 t^2 + a_1 t + a_0$ such that

$$E = \sum_{i=1}^{m} \left(y_i - \left(a_0 + a_1 t_i + a_2 t_i^2 \cdots + a_n t_i^n \right) \right)^2$$

is a minimum. The minimum occurs when

$$\frac{\partial E}{\partial a_0} = 0, \frac{\partial E}{\partial a_1} = 0, \ldots, \frac{\partial E}{\partial a_n} = 0.$$

After some manipulation, we determine that this results in an $(n+1) \times (n+1)$ system of equations. If we define the $m \times (n+1)$ Vandermonde matrix as

$$V = \begin{bmatrix} 1 & t_1 & \cdots & t_1^n \\ 1 & t_2 & \cdots & t_2^n \\ \vdots & \vdots & \ddots & \vdots \\ 1 & t_m & \cdots & t_m^n \end{bmatrix},$$

then it follows that the system of equations we have to solve is

$$V^{\mathrm{T}} V \begin{bmatrix} a_0 \\ a_1 \\ \vdots \\ a_n \end{bmatrix} = V^{\mathrm{T}} \begin{bmatrix} y_1 \\ y_2 \\ \vdots \\ y_m \end{bmatrix}.$$

This is an example of the normal equations, the primary equations for linear least-squares problems that we will study in Chapter 16.

Cubic Spline Interpolation

A polynomial least-squares approximation can be used for interpolation or extrapolation. If high-accuracy interpolation is required, cubic splines are a superior choice. Given the experimental data points

$$\{(t_1, y_1), (t_2, y_2), (t_3, y_3), \ldots, (t_n, y_n), (t_{n+1}, y_{n+1})\},$$

fit a cubic polynomial p_i, $1 \le i \le n$ between t_i and t_{i+1} parameterized by u

$$p_i(u) = a_i + b_i u + c_i u^2 + d_i u^3, \quad 0 \le u \le 1.$$

Require that the cubics agree with the y-data at $t_1, t_2, \ldots, t_{n+1}$; in addition, require that the first and second derivatives of the splines match at the interior points

$$(t_2, y_2), (t_3, y_3), \ldots, (t_n, y_n),$$

and that the second derivative of p_1 is zero at t_1, and the second derivative of p_n is zero at t_{n+1}. After some manipulation, there results an $(n+1) \times (n+1)$ tridiagonal system of equations that must be solved. This is quickly done using the Thomas algorithm.

Cubic splines have many applications, including computer graphics, image interpolation and digital filtering, and modeling airplane drag as a function of mach number, the speed of the airplane with respect to the free stream airflow [36].

12.6   PROBLEMS

12.1 The following function S is called a *sawtooth wave* and has period 2π:

$$S(t) = \begin{cases} t & -\pi < t < \pi \\ S(t + 2\pi k) = S(t) & -\infty < t < \infty, \quad k \in \mathbb{Z} \end{cases}$$

The file sawtooth.mp3 in the book software distribution plays the sawtooth wave sound. Figure 12.14 is a graph of the function over the interval $-4\pi \le t \le 4\pi$. Find its Fourier series. Hint: The sawtooth wave is an odd function.

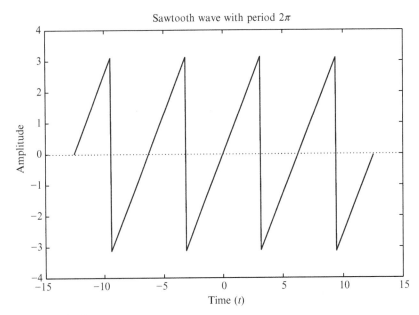

FIGURE 12.14 Sawtooth wave with period 2π.

12.2 Find the Fourier series for the triangle wave defined by

$$f(t) = \begin{cases} \frac{\pi}{2} + t, & -\pi \leq t \leq 0 \\ \frac{\pi}{2} - t, & 0 < t \leq \pi \end{cases}.$$

12.3 Find the Fourier series for the function

$$f(t) = \begin{cases} -1, & -\pi \leq t \leq -\frac{\pi}{2} \\ 0, & -\frac{\pi}{2} < t \leq \frac{\pi}{2} \\ 1, & \frac{\pi}{2} < t \leq \pi \end{cases}.$$

12.4 Orthogonal functions play an important role in mathematics, science, and engineering, and the Fourier series is an example. Another example is the *Chebyshev polynomials* that play an important role in proving the convergence of the conjugate gradient method for solving large, sparse, linear systems, among other uses. We will discuss this method in Chapter 21. One approach to define the Chebyshev polynomials is through the use of trigonometric functions.

a. By using the trigonometric identities

$$\cos(\alpha + \beta) = \cos\alpha\cos\beta - \sin\alpha\sin\beta,$$

$$\sin(\alpha + \beta) = \sin(\alpha)\cos(\beta) + \cos(\alpha)\sin(\beta),$$

$$\cos^2(\theta) + \sin^2(\theta) = 1,$$

show that

$$\cos((n+1)\theta) = 2\cos\theta\cos n\theta - \cos((n-1)\theta). \tag{12.34}$$

b. The Chebyshev polynomial, $T_n(x)$ of degree n is a polynomial with integer coefficients such that

$$\cos n\theta = T_n(\cos\theta);$$

in other words there is a polynomial

$$T_n(x) = a_n x^n + a_{n-1}x^{n-1} + \cdots + a_2 x^2 + a_1 x + a_0$$

such that

$$\cos n\theta = a_n \cos^n\theta + a_{n-1}\cos^{n-1}\theta + \cdots + a_2\cos^2\theta + a_1\cos\theta + a_0. \tag{12.35}$$

i. Show that Equation 12.35 is true for $n = 0$ and $n = 1$.

ii. By applying Equation 12.34, use mathematical induction to prove Equation 12.35 for all n.

iii. Let $x = \cos \theta$ so that θ

$$T_n(x) = \cos\left(n \cos^{-1}(x)\right), \quad -1 \le x \le 1.$$

iv. Show that

$$\max_{-1 \le x \le 1} T_n(x) = 1.$$

v. Prove that the n roots of $T_n(x)$ are

$$x_i = \cos\left(\frac{(2i-1)\pi}{2n}\right).$$

vi. Prove that the Chebyshev polynomials satisfy the recurrence relation

$$T_{n+1}(x) = 2x T_n(x) - T_{n-1}(x). \tag{12.36}$$

vii. Using Equation 12.36, find the first five Chebyshev polynomials.

c. If f and g are continuous functions over the interval $a \le x \le b$, then f and g are orthogonal with respect to the weight function w if

$$\int_a^b f(x) g(x) w(x) \, dx = 0.$$

Show that the Chebyshev polynomials are orthogonal over $-1 \le x \le 1$ with respect to the weight function

$$\frac{1}{\sqrt{1 - x^2}}.$$

by proceeding as follows.

i. Investigate the possible values of

$$\int_0^\pi \cos m\theta \, \cos n\theta \, d\theta.$$

ii. Make a change of variable $x = \cos \theta$ and show that

$$\int_{-1}^1 T_m(x) \, T_n(x) \, \frac{dx}{\sqrt{1 - x^2}} = 0, \quad m \ne n$$

12.5 Finite difference methods are also used to approximate the solution to ordinary differential equations. Consider the boundary value problem for the general second-order equation with constant coefficients

$$\frac{d^2 y}{dx^2} + p \frac{dy}{dx} + qy = r(x), \quad a \le x \le b,$$
$$y(a) = YA, \quad y(b) = YB.$$

Let the interval $a \le x \le b$ be divided into n subintervals of width $h = (b - a)/n$. Using the central difference approximations

$$\frac{d^2 y}{dx^2} \approx \frac{y_{i+1} - 2y_i + y_{i-1}}{h^2}, \quad \frac{dy}{dx} \approx \frac{y_{i+1} - y_{i-1}}{2h},$$

find the linear system that must be solved to approximate $y_2, y_3, \ldots, y_n$.

12.6 Although not used as often as cubic splines, *quadratic splines* use a quadratic equation $p_i(x) = a_i x^2 + b_i x + c_i$ between knots. Assume that the data points are

$$(x_1, y_1), (x_2, y_2), \ldots, (x_n, y_n), (x_{n+1}, y_{n+1})$$

and that the piecewise quadratic function is continuous and differentiable. Determine the set of linear equations that determine the coefficients a_i, b_i, and c_i. You will need one additional equation so that the system is square. Do this by assuming that

$$p_1''(x_1) = 0.$$

12.6.1 MATLAB Problems

12.7 Recall Figure 12.2 in which partial sums of the Fourier series for the square wave were graphed on the same set of axes, demonstrating convergence. Write a MATLAB program that graphs the sawtooth wave defined in Problem 12.1 over $-2\pi \leq t \leq 2\pi$, along with the partial sums for $n = 3, 7, 10, 12$.

12.8 Repeat Problem 12.7 for the triangle wave defined in Problem 12.2.

12.9 Repeat Problem 12.7 for the periodic function defined in Problem 12.3.

12.10 The *Gibbs phenomenon* is the odd way in which the Fourier series of a piecewise continuously differentiable periodic function behaves at a jump discontinuity, such as that in a square or triangle wave [37]. The nth partial sum of the Fourier series exhibits large oscillations near the jump, which might cause the partial sum's value to rise above the function value. The overshoot does not die out as the number of terms in the partial sum increases, but approaches a finite limit. Experiment with the sawtooth wave defined in Problem 12.1 and present evidence of the phenomenon.

12.11 Use `heateq.m` to approximate and graph the solution of the following heat conduction problem:

$$\frac{\partial u}{\partial t} = 1.25 \frac{\partial^2 u}{\partial x^2}, \quad 0 \leq x \leq 5.0, \quad 0 \leq t \leq 1.0,$$
$$u(0, t) = u(L, t) = 0,$$
$$u(x, 0) = x \sin x.$$

12.12

a. Write a function

```
[x,y] = ode2ndconst(a,b,n,YA,YB,p,q,r)
% The function solves a general second order linear ODE of the form:
%    d2y/dx2 + p(dy/dx) + qy = r(x)
% using the finite difference method.
% The boundary conditions are assumed to be of the form:
%    y(a) = YA and y(b) = YB
% Second order central differences are used.
%Input arguments:
% a First value of x (first point in the domain).
% b Last value of x (last point in the domain).
% n Number of subintervals.
% YA Boundary condition at x=a.
% YB Boundary condition at x=b.
% p The constant p.
% q The constant q.
% r Right-hand side function r(x).
%Output arguments:
% x A vector with the x coordinate of the solution points.
% y A vector with the y coordinate of the solution points.
```

that uses the result of Problem 12.5 to solve a boundary value problem for a general linear ordinary differential equation with constant coefficients.

b. Graph the solution to the problem

$$\frac{d^2 y}{dx^2} + 8 \frac{dy}{dx} + 5y = \sin(x), \quad 0 \leq x \leq 2,$$
$$y(0) = 1, \quad y(2) = 2.$$

12.13 **a.** Modify your function in Problem 12.12 to approximate the solution of a boundary value problem for a linear ordinary differential equation of the form

$$\frac{dy^2}{dx^2} + p(x) \frac{dy}{dx} + q(x) y = r(x), \quad a \leq x \leq b$$
$$y(a) = YA, \quad y(b) = YB$$

Name your function `ode2nd`.

b. Plot the solution to each boundary value problem.

    i. $\frac{d^2y}{dx^2} + \left(\frac{1}{1+x^2}\right)\frac{dy}{dx} + (x^2 - 1)y = 1$, $y(1) = 1$, $y(5) = 2$

    ii. $\frac{d^2y}{dx^2} + x\sin(x)\frac{dy}{dx} + x^2 y = \coth(x)$, $y(-1) = -1$, $y(6) = -2$

12.14

a. Plot the data, the linear regression line, and the quadratic least-squares curve on the same set of axes.

t	y
1.3	-12.3
1.8	-8.5
2.3	-5.6
3.6	-2.3
4.9	-0.5
5.3	1.3
6.5	1.5

b. Compute both linear and quadratic least-squares estimates for $t = 2.0$ and $t = 6.8$.

12.15 The data in the table give the approximate population of the United States for selected years from 1845 until 2000.

Year	1845	1871	1915	1954	2000
Populations (millions)	20	30	100	160	280

Assume that the population growth can be modeled with an exponential function $p = be^{mx}$, where x is the year and p is the population in millions. Use linear least squares to determine the constants b and m for which the function best fits the data. Use the equation to estimate the population in the year 1970. Hint: Take the logarithm of the function.

12.16

a. Develop a MATLAB function

```
function p = lq(t, y, n)
```

that finds the least-squares fit for the data (t, y). The return value is the polynomial in MATLAB format; in other words, if the polynomial is

$$a_n t^n + a_{n-1} t^{n-1} + \cdots + a_2 t^2 + a_1 t + a_0,$$

then $p = \begin{bmatrix} a_n & a_{n-1} & \cdots & a_2 & a_1 & a_0 \end{bmatrix}^{\mathsf{T}}$.

b.

Use your function to verify your calculations in Problem 12.14. Also, fit a cubic polynomial and compare the results with the lower-order approximations.

12.17 Find out how MATLAB performs least-squares fitting, and solve Problem 12.14.

12.18 Lagrange interpolation is an alternative to least squares. Given n points

$$\{(t_1, y_1), (t_2, y_2), \ldots, (t_n, y_n)\},$$

there is a unique polynomial of degree $n - 1$ that passes through all n points. One approach is to find the polynomial

$$p(t) = a_{n-1}t^{n-1} + a_{n-2}t^{n-2} + \cdots + a_2 t^2 + a_1 t + a_0$$

by solving the system of equations

$$
\begin{aligned}
a_{n-1}t_1^{n-1} + a_{n-2}t_1^{n-2} + \cdots + a_2 t_1^2 + a_1 t_1 + a_0 &= y_1 \\
a_{n-1}t_2^{n-1} + a_{n-2}t_2^{n-2} + \cdots + a_2 t_2^2 + a_1 t_2 + a_0 &= y_2 \\
&\vdots \\
a_{n-1}t_n^{n-1} + a_{n-2}t_n^{n-2} + \cdots + a_2 t_n^2 + a_1 t_n + a_0 &= y_n
\end{aligned}
$$

To approximate a value $\bar{y}$ at $\bar{t}$ between t_1 and t_n, compute $\bar{y} = p(\bar{t})$.

a. Write a function `p = lagrange1(t,y)` that returns the interpolating polynomial for the data (t, y).

b. Find the fourth-order interpolating polynomial for the points $\{(1, 2), (1.5, 2.5), (2.3, 3.5), (2.9, 4.5), (3.3, 5.0)\}$ and graph the points and the polynomial on the same set of axes.

c. Graph the points and the seventh-order polynomial that passes through
t = (0:7)', y = [0 2 0 2 0 2 0 2]'.

d. Graph the points and the cubic spline approximation.

e. Comment on the results.

12.19 Sometimes it is not necessary to employ linear algebra to solve a problem. It may be more efficient to approach the problem a different way. This problem presents an alternate approach to Lagrange interpolation as presented in Problem 12.18. Given n points

$$\{(t_1, y_1), (t_2, y_2), \ldots, (t_n, y_n)\},$$

there is a unique polynomial $p(t)$ of degree $n - 1$ that passes through all n points.

a. Show that

$$p(t) = \frac{(t - t_2)(t - t_3)\ldots(t - t_n)}{(t_1 - t_2)(t_1 - t_3)\ldots(t_1 - t_n)}y_1 + \frac{(t - t_1)(t - t_3)\ldots(t - t_n)}{(t_2 - t_1)(t_2 - t_3)\ldots(t_2 - t_n)}y_2 + \cdots + \frac{(t - t_1)(t - t_2)\ldots(t - t_{n-1})}{(t_n - t_1)(t_n - t_2)\ldots(t_n - t_{n-1})}y_n.$$

b. Repeat Problem 12.18, part (b), by developing a function p = lagrange2(t,y) that uses the polynomial of part (a). You will find the MATLAB function conv useful in constructing the product of polynomial factors.

12.20 There is a famous example, known as Runge's phenomenon, showing that a Lagrange polynomial (discussed in Problems 12.18 and 12.19) may oscillate at the edges of an interval when using Lagrange interpolation with polynomials of high degree. Let

$$f(x) = \frac{1}{1 + 25x^2},$$

and consider interpolation at the points $(x_i, f(x_i))$, where

$$x_i = -1 + \frac{2(i - 1)}{n}, \quad 1 \le i \le n + 1.$$

Plot $f(x)$ and the Lagrange interpolate for $n = \{5, 10, 15, 20\}$ on the same axes. Comment on the results.

12.21 Write a function lininterp(t,y,tval) that performs linear interpolation to approximate the data value at tval. Test it with the data from Example 12.3.

12.22

a. Remove the plotting code from cubsplineb and name the function cubspline. It returns only the matrix of cubic polynomials.

b. Write a function ival = cubsplineinterp(S, t, tval).

S is the matrix of cubic polynomials that defines a cubic spline.

t is the vector of independent variables.

tval is the value at which to interpolate.

ival is the interpolated value at tval.

c. Using the data from Example 12.3, approximate the values at $t = 1.5, 1.85, 2.55, 3.0, 3.8$.

12.23 Determine how MATLAB performs not-a-knot cubic spline interpolation and use MATLAB to solve Problem 12.22(c).

Chapter 13

Important Special Systems

You should be familiar with

- Tridiagonal matrices
- Gaussian elimination
- Pivoting
- Symmetric matrices
- Eigenvalues
- Expansion by minors

In Chapter 12, we used finite difference methods to approximate the solution to the heat equation

$$\frac{\partial u}{\partial t} = c\frac{\partial^2 u}{\partial x^2}, \quad 0 \le x \le L, \quad 0 \le t \le T,$$
$$u(0,t) = u(L,t) = 0,$$
$$u(x,0) = f(x).$$

The technique involved successively solving a system of equations with the following matrix:

$$B = \begin{bmatrix} 1+2r & -r & 0 & \dots & & 0 \\ -r & 1+2r & -r & \dots & & 0 \\ 0 & \ddots & \ddots & \ddots & & \vdots \\ \vdots & & \vdots & -r & 1+2r & -r \\ 0 & & 0 & 0 & -r & 1+2r \end{bmatrix}.$$

There are three things we should notice about the matrix:

- It is symmetric
- It is tridiagonal
- It is *diagonally dominant*; in other words, each diagonal element a_{ii} has larger absolute value than the sum of the other entries in its row ($|a_{ii}| > \sum_{j=1,\dots,n, j \ne i} |a_{ij}|$).

When discussing cubic splines in Chapter 12, we encountered another symmetric, diagonally dominant, tridiagonal coefficient matrix:

$$\begin{bmatrix} 2 & 1 & & & & & \\ 1 & 4 & 1 & & & & \\ & 1 & 4 & 1 & & & \\ & & 1 & 4 & 1 & & \\ \vdots & \ddots & \ddots & \ddots & \ddots & \ddots & \vdots \\ & & & & 1 & 4 & 1 \\ & & & & & 1 & 2 \end{bmatrix}$$

Both of these matrices have yet another feature in common; they are positive definite. Matrices with these special features frequently occur in engineering and science applications, and it is appropriate that we devote a chapter to them.

13.1 TRIDIAGONAL SYSTEMS

When a matrix T is tridiagonal and nonsingular, its LU decomposition without pivoting yields *bidiagonal matrices L* and U. L has 1's on the main diagonal as usual, but the *superdiagonal entries* of U are the same as those of T.

Numerical Linear Algebra with Applications. http://dx.doi.org/10.1016/B978-0-12-394435-1.00013-2

Example 13.1. Let $A = \begin{bmatrix} 1 & 4 & 0 & 0 \\ -1 & 5 & 1 & 0 \\ 0 & 2 & -1 & -9 \\ 0 & 0 & 3 & 7 \end{bmatrix}$. The function lugauss developed in Chapter 11 performs the LU

decomposition without partial pivoting, and the MATLAB segment factors A using lugauss.

```
>> [L U] = lugauss(A)
L =
    1.0000         0         0         0
   -1.0000    1.0000         0         0
         0    0.2222    1.0000         0
         0         0   -2.4545    1.0000

U =
    1.0000    4.0000         0         0
         0    9.0000    1.0000         0
         0         0   -1.2222   -9.0000
         0         0         0  -15.0909
```
■

We will investigate the general problem using a 4×4 matrix. Doing this will make it clear how to factor a general tridiagonal matrix. Consider the equation

$$\begin{bmatrix} b_1 & c_1 & 0 & 0 \\ a_1 & b_2 & c_2 & 0 \\ 0 & a_2 & b_3 & c_3 \\ 0 & 0 & a_3 & b_4 \end{bmatrix} = \begin{bmatrix} 1 & 0 & 0 & 0 \\ l_1 & 1 & 0 & 0 \\ 0 & l_2 & 1 & 0 \\ 0 & 0 & l_3 & 1 \end{bmatrix} \begin{bmatrix} u_1 & c_1 & 0 & 0 \\ 0 & u_2 & c_2 & 0 \\ 0 & 0 & u_3 & c_3 \\ 0 & 0 & 0 & u_4 \end{bmatrix} \tag{13.1}$$
$$= \begin{bmatrix} u_1 & c_1 & 0 & 0 \\ l_1 u_1 & l_1 c_1 + u_2 & c_2 & 0 \\ 0 & l_2 u_2 & l_2 c_2 + u_3 & c_3 \\ 0 & 0 & l_3 u_3 & l_3 c_3 + u_4 \end{bmatrix}.$$

Equate both sides of Equation 13.1 to obtain

$$u_1 = b_1, \quad l_1 u_1 = a_1, \quad l_1 c_1 + u_2 = b_2,$$
$$l_2 u_2 = a_2, \quad l_2 c_2 + u_3 = b_3, \quad l_3 u_3 = a_3,$$
$$l_3 c_3 + u_4 = b_4,$$

from which follows

$$u_1 = b_1, \quad l_1 = a_1/u_1, \quad u_2 = b_2 - l_1 c_1, \tag{13.2}$$
$$l_2 = a_2/u_2, \quad u_3 = b_3 - l_2 c_2, \quad l_3 = a_3/u_3, \tag{13.3}$$
$$u_4 = b_4 - l_3 c_3. \tag{13.4}$$

Since $u_1 = b_1$, we can compute $l_1 = a_1/u_1$. Knowing l_1, we have $u_2 = b_2 - l_1 c_1$. Similarly, we can compute l_2, l_3 and u_3, u_4. Using the 4×4 case as a model (Equation 13.3), the l_i and u_i are computed as follows:

$$u_1 = b_1,$$
$$l_i = a_i/u_i, \quad 1 \le i \le n-1,$$
$$u_{i+1} = b_{i+1} - l_i c_i, \quad 1 \le i \le n-1.$$

Algorithm 13.1 Computing the *LU* Decomposition of a Tridiagonal Matrix

```
function TRIDIAGLU(a,b,c)
   % Factor the tridiagonal matrix defined by subdiagonal a,
   % main diagonal b and superdiagonal c
   % into a product of two bidiagonal matrices
   % Input: vectors a, b, c.
   % Output: L is the subdiagonal of the left bidiagonal factor.
   % U is the diagonal of the right bidiagonal factor.
   U₁ = b₁
   for i =1:n-1 do
       Lᵢ = aᵢ/Uᵢ
       Uᵢ₊₁ = bᵢ₊₁ − LᵢCᵢ
   end for
end function
```

NLALIB: The function `tridiagLU` implements Algorithm 13.1.

Remark 13.1.

- After running Algorithm 13.1, the bidiagonal systems must be solved in the order (i) $Ly = b$ and (ii) $Ux = y$.
- For efficiency, Algorithm 13.1 accepts the three vector diagonals and returns the two vectors that must be computed by the decomposition.
- It should be noted that this algorithm does not work if any $u_i = 0$, but this occurs very seldom in practice. Unfortunately, the stability of the algorithm cannot be guaranteed.

This is an $O(n)$ algorithm and thus very fast compared to the general LU decomposition. The for loop executes $n - 1$ times, each execution involves one division, one subtraction, and one multiplication, so the flop count for the decomposition is $3(n - 1)$. Forward substitution requires $2(n - 1)$ flops (verify), and back substitution requires $1 + 3(n - 1)$ (verify) flops. The total number of flops to solve a system $Tx = b$ is thus $3(n - 1) + 2(n - 1) + 3(n - 1) + 1 = 8n - 7$ flops. The Thomas algorithm presented in Chapter 9 requires $10n - 3$ flops, so decomposition followed by forward and back substitution is more efficient. Furthermore, if multiple systems $Tx_i = b_i$, $1 \leq i \leq k$ need to be solved, the Thomas algorithm will cost $k(10n - 3)$, but the cost of the decomposition approach is $3(n - 1) + k(5n - 4)$ flops, a significant improvement.

Example 13.2. Factor the matrix $A = \begin{bmatrix} 1 & 2 & 0 \\ 5 & 7 & 1 \\ 0 & 1 & 3 \end{bmatrix}$

$u_1 = 1$

i = 1:

$l_1 = \frac{a_1}{u_1} = \frac{5}{1} = 5;$ $u = b_2 - l_1 c_1 = 7 - (5)(2) = -3$

i = 2:

$l_2 = \frac{a_2}{u_2} = \frac{1}{-3} = -\frac{1}{3}$ $u_3 = b_3 - l_2 c_2 = 3 - \left(-\frac{1}{3}\right)(1) = \frac{10}{3}$

The factorization is

$$L = \begin{bmatrix} 1 & 0 & 0 \\ 5 & 1 & 0 \\ 0 & -\frac{1}{3} & 1 \end{bmatrix}, \quad U = \begin{bmatrix} 1 & 2 & 0 \\ 0 & -3 & 1 \\ 0 & 0 & \frac{10}{3} \end{bmatrix}. \qquad \blacksquare$$

Algorithm 13.2 solves $Tx = LUx = b$. Since the superdiagonal of U is the same as T, the superdiagonal will need to be passed as an argument to trisolve.

NLALIB: The function `trisolve` implements Algorithm 13.2.

Algorithm 13.2 Solve a Factored Tridiagonal System

```
function TRISOLVE(L,U,c,rhs)
    % Solve the equation Tx = b, where T is a tridiagonal matrix.
    % T has been factored into a unit lower-bidiagonal matrix and an upper
    % bidiagonal matrix.
    % Input: L is the subdiagonal of the lower diagonal matrix, U is the diagonal
    % of the upper diagonal matrix, c is the superdiagonal
    % of the original tridiagonal matrix,
    % and rhs is the right-hand side of the system Tx = rhs.
    % Output: The solution x.

    % forward substitution
    y₁ = rhs₁
    for i = 2:n do
        yᵢ = rhsᵢ − Lᵢ₋₁yᵢ₋₁
    end for
    % back substitution
    xₙ = yₙ/Uₙ
    for i = n-1:-1:1 do
        xᵢ = (yᵢ − cᵢxᵢ₊₁)/Uᵢ
    end for
end function
```

Example 13.3. Let

$$T = \begin{bmatrix} 1 & 15 & 0 & 0 & 0 \\ 2 & -1 & 3 & 0 & 0 \\ 0 & -8 & 5 & 7 & 0 \\ 0 & 0 & 4 & 6 & 12 \\ 0 & 0 & 0 & -18 & 7 \end{bmatrix},$$

and solve

$$Tx = \begin{bmatrix} 1 & -1 & 5 & 0 & 3 \end{bmatrix}^{\mathrm{T}}.$$

The MATLAB code factors T using trifactLU and then solves the system $Tx = b$ using trisolve. The results are verified by performing the calculation using the MATLAB operator "\".

```
>> a = [2 -8 4 -18]';
>> b = [1 -1 5 6 7]';
>> c = [15 3 7 12]';
>> rhs = [1 -1 5 0 3]';
>> [L U] = tridiagLU(a,b,c);
>> x = trisolve(L,U,c,rhs)

x =

  -3.2789
   0.2853
   1.9477
  -0.3509
  -0.4738

>> A = trid(a,b,c);
>> A\rhs
```

```
ans =
    -3.2789
     0.2853
     1.9477
    -0.3509
    -0.4738
```

■

13.2   SYMMETRIC POSITIVE DEFINITE MATRICES

Let $A = \begin{bmatrix} 1 & 0 \\ 0 & 1 \end{bmatrix}$ and form $x^T A x$:

$$x^T A x = \begin{bmatrix} x_1 & x_2 \end{bmatrix} \begin{bmatrix} 1 & 0 \\ 0 & 1 \end{bmatrix} \begin{bmatrix} x_1 \\ x_2 \end{bmatrix} = x_1^2 + x_2^2.$$

Note that $x^T A x > 0$ for all $x = \begin{bmatrix} x_1 \\ x_2 \end{bmatrix} \neq \begin{bmatrix} 0 \\ 0 \end{bmatrix}$. This expression is an example of a symmetric positive definite matrix, and $x^T A x$ is a quadratic form.

Definition 13.1.   A symmetric matrix A is *positive definite* if for every nonzero vector $x = \begin{bmatrix} x_1 \\ x_2 \\ x_3 \\ \vdots \\ x_n \end{bmatrix}$, $x^T A x > 0$. The

expression $x^T A x = \sum_{i=1}^{n} \sum_{j=1}^{n} a_{ij} x_i x_j$ is called the *quadratic form* associated with A. If $x^T A x \geq 0$ for all $x \neq 0$, then the symmetric matrix A is called *positive semidefinite*.

Remark 13.2.   In this book, all positive definite matrices will also be symmetric, so we simply use the term positive definite. It is possible for a nonsymmetric matrix to satisfy

$$x^T A x > 0, \quad x \neq 0$$

(Problem 13.3), but there is no general agreement on what positive definite means for nonsymmetric matrices.

Positive definite matrices are important in a wide variety of applications. Many large matrices used in finite difference approximations to the solution of partial differential equations are positive definite, and positive definite matrices are important in electrical engineering problems, optimization algorithms, and least squares.

Example 13.4.   **a.** Show that the symmetric matrix $A = \begin{bmatrix} 2 & -1 \\ -1 & 2 \end{bmatrix}$ is positive definite.

$$\begin{bmatrix} x_1 & x_2 \end{bmatrix} \begin{bmatrix} 2 & -1 \\ -1 & 2 \end{bmatrix} \begin{bmatrix} x_1 \\ x_2 \end{bmatrix} = \begin{bmatrix} 2x_1 - x_2 & -x_1 + 2x_2 \end{bmatrix} \begin{bmatrix} x_1 \\ x_2 \end{bmatrix} =$$

$$2x_1^2 - 2x_1 x_2 + 2x_2^2 = x_1^2 + (x_1 - x_2)^2 + x_2^2 > 0, \quad \begin{bmatrix} x_1 \\ x_2 \end{bmatrix} \neq 0$$

b. Suppose we try using the definition to show the matrix $C = \begin{bmatrix} 1 & 1 & 1 \\ 1 & 2 & 1 \\ 1 & 1 & 3 \end{bmatrix}$ is positive definite. Compute $x^T A x$ to obtain

$$x_1 (x_1 + x_2 + x_3) + x_2 (x_1 + 2x_2 + x_3) + x_3 (x_1 + x_2 + 3x_3).$$

Showing that this expression is greater than zero for all $x > 0$ is messy. Imagine the problem with a 50×50 matrix. There must be a better way than using the definition. Theorem 13.1 begins to address this issue.

■

We will prove properties 1 and 3 of Theorem 13.1 and leave the proofs of 4 and 5 to the exercises. For a proof of property 2, see Ref. [38].

Theorem 13.1. **1.** *A symmetric $n \times n$ matrix A is positive definite if and only if all its eigenvalues are positive.*
2. *A symmetric matrix A is positive definite if and only if all its leading principle minors are positive; that is* $\det A(1:i, 1:i) > 0$, $1 \le i \le n$. *This called Sylvester's criterion.*
3. *If $A = (a_{ij})$ is positive definite, then $a_{ii} > 0$ for all i.*
4. *If $A = (a_{ij})$ is positive definite, then the largest element in magnitude of all matrix entries must lie on the diagonal.*
5. *The sum of two positive definite matrices is positive definite.*

Proof. In Chapter 19 we will prove that any $n \times n$ real symmetric matrix has orthonormal eigenvectors that form a basis for $\mathbb{R}^n$. To prove (1), assume that x_i is an eigenvector of A with corresponding eigenvalue λ_i, so $Ax_i = \lambda x_i$. Let $x \ne 0$ be a vector in $\mathbb{R}^n$. Then, $x = c_1 x_1 + c_2 x_2 + \cdots + c_n x_n$, and

$$x^T Ax = (c_1 x_1 + c_2 x_2 + \cdots + c_n x_n)^T A (c_1 x_1 + c_2 x_2 + \cdots + c_n x_n) \tag{13.5}$$

$$= (c_1 x_1 + c_2 x_2 + \cdots + c_n x_n)^T (c_1 \lambda_1 x_1 + c_2 \lambda_2 x_2 + \cdots + c_n \lambda_n x_n) \tag{13.6}$$

$$= c_1^2 \lambda_1 \|x_1\|_2^2 + c_2^2 \lambda_2 \|x_2\|_2^2 + \cdots + c_n^2 \lambda_n \|x_n\|_2^2 > 0 \tag{13.7}$$

If all $\{\lambda_i\}$ are positive, Equation 13.7 guarantees that A is positive definite. Now assume that A is positive definite and that x is an eigenvector of A corresponding to eigenvalue λ. Then,

$$x^T Ax = x^T \lambda x = \lambda \|x\|_2^2 > 0, \quad x \ne 0$$

so $\lambda > 0$.
To prove (3), let e_i be the ith standard basis vector $e_i = \begin{bmatrix} 0 & 0 & \dots & 1 & 0 & \dots & 0 \end{bmatrix}^T$. Then,

$$e_i^T A e_i = a_{ii} > 0. \qquad \square$$

Remark 13.3. To show that a matrix is positive definite, one can compute its eigenvalues and verify that they are all positive, although we know that computing eigenvalues is a tricky process. Sylvester's criterion can be made practical (Problem 13.29), although it requires some care. Note that items 3 and 4 are necessary conditions only; in other words, if a matrix A does not satisfy either item 3 or 4, then it cannot be positive definite. You can use the items only to show that A is not positive definite.

Remark 13.4. A matrix is negative definite if $x^T Ax < 0$ for all $x \ne 0$. In this case, $-A$ is positive definite. A matrix is *symmetric indefinite* if it has both positive and negative eigenvalues or, put another way, if $x^T Ax$ assumes both positive and negative values.

Example 13.5.
a. The matrices

$$A = \begin{bmatrix} 1 & 4 & 0 & 3 & 9 \\ 23 & 8 & 1 & -1 & 4 \\ 0 & 4 & 7 & -8 & 7 \\ 2 & -13 & 12 & 0 & 5 \\ 1 & 4 & 2 & 8 & 1 \end{bmatrix}, \quad B = \begin{bmatrix} 1 & -1 & 0 & 9 \\ 8 & 45 & 3 & 19 \\ 0 & 15 & 16 & 35 \\ 3 & -55 & 2 & 22 \end{bmatrix}$$

cannot be positive definite because A has a diagonal element of 0, and the largest element in magnitude (-55) is not on the diagonal of B.
b. The eigenvalues of $C = \begin{bmatrix} 1 & 1 & 1 \\ 1 & 2 & 1 \\ 1 & 1 & 3 \end{bmatrix}$ are $\lambda_1 = 0.3249$, $\lambda_2 = 1.4608$, $\lambda_3 = 4.2143$, all positive, so C is positive definite. ∎

Tridiagonal matrices appear so frequently in engineering and science applications that we state Theorem 13.2 without proof, since it provides a simple way to test if a tridiagonal matrix is positive definite.

Theorem 13.2.

Suppose that a real symmetric tridiagonal matrix $A = \begin{bmatrix} b_1 & a_1 & & & \\ a_1 & b_2 & a_2 & & \\ & a_2 & \ddots & \ddots & \\ & & \ddots & b_{n-1} & a_{n-1} \\ & & & a_{n-1} & b_n \end{bmatrix}$ *with diagonal entries all positive is*

strictly diagonally dominant, i.e., $b_i > |a_{i-1}| + |a_i|$, $1 \le i \le n$. *Then A is positive definite.*

13.2.1 Applications

In Section 12.3.1, we defined the normal equations, $A^\mathrm{T}Ax = A^\mathrm{T}y$, and observed their connection with least-squares polynomial approximation. It is important to note that if A is nonsingular

$$x^\mathrm{T}\left(A^\mathrm{T}A\right)x = (Ax)^\mathrm{T}(Ax) = \langle Ax, Ax \rangle > 0, \quad x \neq 0,$$

so $A^\mathrm{T}A$ is positive definite.

Positive definite matrices frequently occur when using finite difference or finite element methods to approximate the solution to partial differential equations. For instance, in Section 12.2.1, we used a finite difference technique to approximate the solution to the heat equation and needed to solve a tridiagonal system with the coefficient matrix

$$B = \begin{bmatrix} 1+2r & -r & 0 & \dots & 0 \\ -r & 1+2r & -r & \dots & 0 \\ 0 & \ddots & \ddots & \ddots & \vdots \\ \vdots & \vdots & -r & 1+2r & -r \\ 0 & 0 & 0 & -r & 1+2r \end{bmatrix}.$$

The matrix B satisfies the conditions of Theorem 13.2, so it is positive definite.

The topic of Section 12.4 is cubic splines. In order to compute a cubic spline for n data points, a system with the matrix

$$S = \begin{bmatrix} 2 & 1 & & & & & \\ 1 & 4 & 1 & & & & \\ & 1 & 4 & 1 & & & \\ & & 1 & 4 & 1 & & \\ \vdots & \ddots & \ddots & \ddots & \ddots & \ddots & \vdots \\ & & & & 1 & 4 & 1 \\ & & & & & 1 & 2 \end{bmatrix}$$

must be solved. By Theorem 13.2, S is positive definite.

Positive definite matrices play a role in electrical engineering. As an example, consider the circuit in Figure 13.1. The matrix equation for the determination of V_1 and V_2 is

$$\begin{bmatrix} \left(\frac{1}{R_1} + \frac{1}{R_2} + \frac{1}{R_3}\right) & -\frac{1}{R_3} \\ -\frac{1}{R_3} & \left(\frac{1}{R_3} + \frac{1}{R_4}\right) \end{bmatrix} \begin{bmatrix} V_1 \\ V_2 \end{bmatrix} = \begin{bmatrix} \frac{V_S}{R_1} \\ 0 \end{bmatrix}.$$

Note that the coefficient matrix is symmetric. Now, $\det \left(\frac{1}{R_1} + \frac{1}{R_2} + \frac{1}{R_3}\right) = \frac{1}{R_1} + \frac{1}{R_2} + \frac{1}{R_3} > 0$, and

$$\begin{vmatrix} \left(\frac{1}{R_1} + \frac{1}{R_2} + \frac{1}{R_3}\right) & -\frac{1}{R_3} \\ -\frac{1}{R_3} & \left(\frac{1}{R_3} + \frac{1}{R_4}\right) \end{vmatrix} = \frac{1}{R_1 R_3} + \frac{1}{R_1 R_4} + \frac{1}{R_2 R_3} + \frac{1}{R_2 R_4} + \frac{1}{R_3 R_4} > 0.$$

By property 2 in Theorem 13.1, the matrix is positive definite.

13.3 THE CHOLESKY DECOMPOSITION

Determining whether a symmetric matrix is positive definite by showing its eigenvalues are positive is computationally intensive. Showing that all its leading principle minors are positive (Theorem 13.2, part 1) can be made to work, but is tricky.

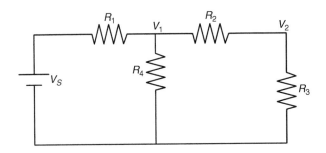

FIGURE 13.1 Conductance matrix.

The French engineer Andre-Louis Cholesky discovered the *Cholesky decomposition*, a result very important in computation with a positive definite matrix and in demonstrating that a matrix is positive definite.

The Cholesky decomposition is based on following theorem, and we will prove existence of the decomposition by showing how to construct the upper-triangular factor R. A proof that the decomposition is unique can be found in Ref. [26, Lecture 23].

Theorem 13.3. *Let A be a real positive definite $n \times n$ matrix. Then there is exactly one upper-triangular matrix $R = (r_{ij})$ with $r_{ii} > 0$, $1 \le i \le n$ such that*

$$A = R^{\mathrm{T}} R. \tag{13.8}$$

13.3.1 Computing the Cholesky Decomposition

We will find the matrix R by equating both sides of Equation 13.8 and demonstrating how to compute the entries of R. It is sufficient to develop the algorithm by considering the 3×3 case. The general case follows precisely the same pattern. Require

$$
\underbrace{\begin{bmatrix} a_{11} & a_{12} & a_{13} \\ a_{12} & a_{22} & a_{23} \\ a_{13} & a_{23} & a_{33} \end{bmatrix}}_{A} = \underbrace{\begin{bmatrix} r_{11} & 0 & 0 \\ r_{12} & r_{22} & 0 \\ r_{13} & r_{23} & r_{33} \end{bmatrix}}_{R^{\mathrm{T}}} \underbrace{\begin{bmatrix} r_{11} & r_{12} & r_{13} \\ 0 & r_{22} & r_{23} \\ 0 & 0 & r_{33} \end{bmatrix}}_{R} = \begin{bmatrix} r_{11}^2 & r_{11}r_{12} & r_{11}r_{13} \\ r_{11}r_{12} & r_{12}^2 + r_{22}^2 & r_{12}r_{13} + r_{22}r_{23} \\ r_{13}r_{11} & r_{13}r_{12} + r_{23}r_{22} & r_{13}^2 + r_{23}^2 + r_{33}^2 \end{bmatrix}
$$

1. Find the first column of R by equating the elements in the first column of A with those in the first column of $R^{\mathrm{T}}R$:

$$a_{11} = r_{11}^2 \Longrightarrow r_{11} = \sqrt{a_{11}}, \tag{13.9}$$

$$a_{12} = r_{11}r_{12} \Longrightarrow r_{12} = \frac{a_{12}}{r_{11}}, \tag{13.10}$$

$$a_{13} = r_{13}r_{11} \Longrightarrow r_{13} = \frac{a_{13}}{r_{11}}. \tag{13.11}$$

2. Find the second column of R. We already know r_{11} and r_{12}, so we only need to equate the second and third entries of the second column of both sides.

$$a_{22} = r_{12}^2 + r_{22}^2 \Longrightarrow r_{22} = \sqrt{a_{22} - r_{12}^2}, \tag{13.12}$$

$$a_{23} = r_{13}r_{12} + r_{23}r_{22} \Longrightarrow r_{23} = \frac{a_{23} - r_{12}r_{13}}{r_{22}}.$$

3. Find the third column of R. We have computed all entries of R except r_{33}. Equate the third entry of the third column of both sides.

$$a_{33} = r_{13}^2 + r_{23}^2 + r_{33}^2 \Longrightarrow r_{33} = \sqrt{a_{33} - r_{13}^2 - r_{23}^2} \tag{13.13}$$

Before formally giving the algorithm, we will investigate how the Cholesky decomposition can not only factor A but also tell us if A is positive definite. The discussion assumes that the diagonal entries of A are greater than zero, as required by Theorem 13.1, part 3. Equations 13.9–13.13 are in agreement. However, even with all the diagonal entries of A greater than 0, Equations 13.12 and 13.13 must not involve the square root of a negative number. During the algorithm, if the argument of a square root is negative, the matrix is not positive definite. This is a much less computationally intensive method than using either property 1 or 2 stated in Theorem 3.1.

Algorithm 13.3 The Cholesky Decomposition

```
function CHOLESKY(A)
    % Factor the positive definite matrix A
    % using the Cholesky decomposition algorithm.
    % If the algorithm fails, A is not positive definite.
    % Output an error message and return and empty array R.
    for i =1:n do
        tmp=a_ii − ∑_{j=1}^{i−1} r_{ji}^2
        if tmp ≤ 0 then
            Output error message.
            % Return an empty array.
            return []
        end if
        r_ii = √tmp
        for j = i+1:n do
            r_ij = (a_ij − ∑_{k=1}^{i−1} r_ki r_kj) / r_ii
        end for
    end for
end function
```

NLALIB: The function `cholesky` implements Algorithm 13.3.

Example 13.6.   The MATLAB code finds the Cholesky decomposition of $A = \begin{bmatrix} 1 & 1 & 4 & -1 \\ 1 & 5 & 0 & -1 \\ 4 & 0 & 21 & -4 \\ -1 & -1 & -4 & 10 \end{bmatrix}$ and shows that

$B = \begin{bmatrix} 1 & 5 & 6 \\ -7 & 12 & 5 \\ 2 & 1 & 10 \end{bmatrix}$ is not positive definite.

```
>> R1 = cholesky(A)

R1 =

        1       1       4      -1
        0       2      -2       0
        0       0       1       0
        0       0       0       3

>> R1'*R1

ans =

        1       1       4      -1
        1       5       0      -1
        4       0      21      -4
       -1      -1      -4      10

>> R2 = cholesky(B);
The matrix is not positive definite
```

■

Remark 13.5.   The MATLAB function `chol` computes the Cholesky decomposition.

13.3.2 Efficiency

The computation $tmp = a_{ii} - \sum_{j=1}^{i-1} r_{ji}^2$ requires $1 + 2(i-1)$ flops. Most of the work takes place in the inner loop,

```
for  j = i+1:n  do
    r_ij = (a_ij - Σ^{i-1}_{k=1} h_ki h_kj) / r_ii
end for
```

It requires $(n-i)(2 + 2(i-1)) = 2i(n-i)$ flops, so the total flop count for iteration i is $1 + 2(i-1) + 2i(n-i)$. Now, the algorithm flop count is $\sum_{i=1}^{n} [1 + 2(i-1) + 2i(n-i)]$. Using the formulas $\sum_{i=1}^{n} i = n(n+1)/2$ and $\sum_{i=1}^{n} i^2 = (n(n+1)(2n+1))/6$ along with some simplification, the flop count for the algorithm is

$$\frac{n^3}{3} + n^2 + \frac{5n}{3}.$$

Like the LU decomposition, Cholesky decomposition is $O(n^3)$, but the leading term of the LU decomposition flop count is $2n^3/3$. Cholesky decomposition requires computing n square roots, but the flop count for those computations is not significant compared to $n^3/3$ flops. Thus, Cholesky decomposition is approximately twice as fast as the LU decomposition for a positive definite matrix.

13.3.3 Solving $Ax = b$ If A Is Positive Definite

If a matrix A is positive definite, it is very straightforward to solve $Ax = b$.

a. Use the Cholesky decomposition to obtain $A = R^{\mathrm{T}}R$.
b. Solve the lower-triangular system $R^{\mathrm{T}}y = b$.
c. Solve the upper-triangular system $Rx = y$.

This a very efficient means of solving $Ax = b$. Recall that we showed in Section 9.3.1 that each of forward and back substitution requires approximately n^2 flops, so the solution to $Ax = b$ using the Cholesky decomposition requires approximately $(n^3/3) + 2n^2$ flops. The standard LU decomposition requires $(2n^3/3) + 2n^2$ flops. Because of its increased speed, Cholesky decomposition is preferred for a large positive definite matrix.

The MATLAB function cholsolve in the software distribution solves the linear system $Ax = b$, where A is a positive definite matrix.

Example 13.7.

Let $A = \begin{bmatrix} 1 & 3 & 7 \\ -1 & -1 & 3 \\ 5 & 4 & 2 \end{bmatrix}$ and compute $B = A^{\mathrm{T}}A$. Show that B is positive definite, and solve $Bx = \begin{bmatrix} 25 & 3 & 35 \end{bmatrix}^{\mathrm{T}}$ using cholsolve.

```
>> B = A'*A

B =

    27    24    14
    24    26    26
    14    26    62

>> R = cholesky(B); % no complaint. R is positive definite
>> b = [25 3 35]';
>> cholsolve(R,b)

ans =
    15.0455
   -18.8409
     5.0682

>> B\b

ans =
    15.0455
   -18.8409
     5.0682
```

■

Remark 13.6.   If matrix A is tridiagonal and positive definite, it is more efficient to use the algorithm `tridiagLU` to factor the matrix.

13.3.4   Stability

Theorem 13.4 shows that the Cholesky algorithm is backward stable [26, pp. 176-177].

Theorem 13.4.   *Let A be a positive definite matrix. Compute a Cholesky decomposition of A on a computer satisfying Equations 8.3 and 8.7. For all sufficiently small eps, this process is guaranteed to run to completion (no square roots of a negative number) generating a computed factor $\hat{R}$ that satisfies*

$$\hat{R}^{\mathrm{T}}\hat{R} = A + \delta A$$

with

$$\frac{\|\delta A\|_2}{\|A\|_2} = O\,(\text{eps})$$

for some $\delta A \in R^{n \times n}$.

In the discussion following the theorem, Trefethen and Bau [26] make a very good point. A forward error analysis would involve $\kappa\,(A)$, and if A is ill-conditioned the forward error bound would look unfavorable. However, a backward error analysis looks at $\hat{R}^{\mathrm{T}}\hat{R}$, and the errors in the two factors interact to remove error, or as stated in Ref. [26], the errors in $\hat{R}^{\mathrm{T}}$ and $\hat{R}$ must be "diabolically correlated." It can also be shown that solving $Ax = b$ when A is positive definite is also backward stable and that pivoting is not necessary for stability.

13.4   CHAPTER SUMMARY

Factoring a Tridiagonal Matrix

A tridiagonal matrix, T, can be factored into the product of two bidiagonal matrices, L and U. L has ones on its diagonal, and the superdiagonal entries of U are the same as T. To solve $Tx = b$, once T is factored, solve the lower-bidiagonal system

$$Ly = b,$$

followed by the upper-bidiagonal system

$$Ux = y.$$

The flop count for the decomposition is $3\,(n - 1)$, forward substitution requires $2\,(n - 1)$ flops, and back substitution costs $1 + 3\,(n - 1)$ flops, for a total of

$$8n - 7$$

flops. The Thomas algorithm presented in Chapter 9 requires $10n - 3$ flops, so decomposition followed by forward and back substitution is more efficient. Furthermore, if multiple systems $Tx_i = b_i$, $1 \le i \le k$ need to be solved, the Thomas algorithm will cost $k\,(10n - 3)$, but the cost of the decomposition approach is $3\,(n - 1) + k\,(5n - 4)$ flops, a significant improvement.

Symmetric Positive Definite Matrices

Symmetric positive definite matrices occur frequently in engineering and science applications. For instance, the coefficient matrix for the solution of the heat equation in Section 12.2 is symmetric positive definite. We will see other important matrices of this type, including the Poisson and biharmonic matrices used in many applications.

A symmetric matrix is positive definite if $x^{\mathrm{T}}Ax > 0$ for all $n \times 1$ vectors $x \neq 0$. This is nearly impossible to verify for most matrices, so there are other criteria that assures a matrix is positive definite.

- A is positive definite if and only if its eigenvalues are all greater than zero.
- A symmetric matrix A is positive definite if and only if all its leading principle minors are positive; that is det $A\,(1 : i,\ 1 : i) > 0$, $1 \le i \le n$. This called Sylvester's criterion.

There are criteria that allow us to reject a matrix as positive definite.

- If $a_{ii} \leq 0, 1 \leq i \leq n$, A is not positive definite.
- If the largest element in magnitude is not on the diagonal, A is not positive definite.

It is important to note that $A^{T}A$ is positive definite for any $n \times n$ nonsingular matrix A.

The Cholesky Decomposition

Let A be a real positive definite matrix $n \times n$ matrix. Then there is exactly one upper-triangular matrix $R = (r_{ij})$ with $r_{ii} > 0, 1 \leq i \leq n$ such that $A = R^{T}R$. This is called the Cholesky decomposition of A. The flop count for the algorithm is

$$\frac{n^3}{3} + n^2 + \frac{5n}{3}.$$

Like the LU decomposition, Cholesky decomposition is $O(n^3)$, but the leading term of the LU decomposition flop count is $2n^3/3$. The Cholesky decomposition requires computing n square roots, but the flop count for those computations are not significant compared to $n^3/3$ flops. Thus, Cholesky decomposition is approximately twice as fast as the LU decomposition for a positive definite matrix.

Unless the matrix is tridiagonal, it is faster to solve a large positive definite system by first applying the Cholesky decomposition. Execute the following steps:

a. Solve $R^{T}y = b$ using forward substitution.
b. Solve $Rx = y$ using back substitution.

Checking for positive eigenvalues or that all leading principle minors are positive is very time consuming. The standard technique is to apply the Cholesky decomposition and see if it fails due to an attempt to take the square root of a negative number. If not, then the matrix is positive definite.

Note that the Cholesky algorithm is backward stable. It can also be shown that solving $Ax = b$ when A is positive definite is also backward stable and that pivoting is not necessary for stability.

13.5 PROBLEMS

13.1 Using pencil and paper, find the LU decomposition of the tridiagonal matrix

$$A = \begin{bmatrix} 1 & 1 & 0 \\ 2 & 1 & 5 \\ 0 & 3 & 4 \end{bmatrix}.$$

13.2 Show that $A = \begin{bmatrix} 1 & 0 \\ 0 & 2 \end{bmatrix}$ is positive definite by verifying that $x^{T}Ax > 0$ for all vectors $x = \begin{bmatrix} x_1 \\ x_2 \end{bmatrix}$.

13.3 We have only dealt with symmetric positive define matrices. It is possible for a matrix to be positive definite and not symmetric. Show this is the case for $A = \begin{bmatrix} 1 & 1 \\ -1 & 1 \end{bmatrix}$.

13.4 In Example 13.5(b), we showed that the matrix $C = \begin{bmatrix} 1 & 1 & 1 \\ 1 & 2 & 1 \\ 1 & 1 & 3 \end{bmatrix}$ is positive definite because all its eigenvalues are positive. Show it is positive definite by computing all its principle minors and showing that they are all positive.

13.5 By inspection, which matrices cannot be positive definite? For the remaining matrices, determine if each is positive definite.

a. $\begin{bmatrix} 2 & 1 & 0 \\ 2 & 5 & 1 \\ -1 & 1 & -1 \end{bmatrix}$

b. $\begin{bmatrix} 2 & 1 & 1 \\ 1 & 2 & 1 \\ 1 & 1 & 2 \end{bmatrix}$

c. $\begin{bmatrix} 2 & -1 & -9 \\ 3 & 3 & 1 \\ 1 & -1 & 8 \end{bmatrix}$

d. $\begin{bmatrix} 1 & 0 & 0 & 0 \\ 1 & 2 & 0 & 0 \\ -5 & 3 & -6 & 0 \\ 1 & 7 & 1 & 3 \end{bmatrix}$

e. $\begin{bmatrix} 1 & 1 & 1 & 1 \\ 1 & 2 & 1 & 2 \\ 1 & 1 & 3 & 1 \\ 1 & 2 & 1 & 4 \end{bmatrix}$

13.6 Using pencil and paper, find the Cholesky decomposition of $A = \begin{bmatrix} 25 & 15 & -5 \\ 15 & 18 & 0 \\ -5 & 0 & 11 \end{bmatrix}$.

13.7 Prove that if A is positive definite so is A^{-1}. Hint: Note property 1 in Theorem 13.1.

13.8 Show that if A is positive definite, then $\kappa(A) = (\kappa(R))^2$, where R is the Cholesky factor. Hint: See Theorem 7.9, part 4.

13.9 Let R be the Cholesky factor for a positive definite matrix A, and $T = R^{\mathrm{T}}$. Prove that

$$\sum_{k=1}^{i} t_{ik}^2 = a_{ii}.$$

13.10 Without performing any computation whatsoever, state why $A = \begin{bmatrix} 3 & -1 & 0 \\ -1 & 5 & 1.7 \\ 0 & 1.7 & 2 \end{bmatrix}$ is positive definite.

13.11 Explain why the following MATLAB statement determines if matrix M is positive definite.

```
all(all(M == M')) & min(eig( M )) > 0
```

13.12 Assume M, N, and $M - N$ are positive definite matrices.
   a. Show that $N^{-1} - M^{-1} = M^{-1}(M - N)M^{-1} + M^{-1}(M - N)N^{-1}(M - N)M^{-1}$.
   b. Show that $N^{-1} - M^{-1}$ is positive definite.

13.13 Using the following steps, prove that if $A = (a_{ij})$ is positive definite, then the largest element in magnitude lies on the diagonal.
   a. Recall that if A is symmetric, then Equation 7.12 follows; in other words, $\langle Au, v \rangle = \langle u, Av \rangle$. Assume that A is positive definite and u, v are $n \times 1$ vectors. Show that $\langle u, v \rangle_A = \langle Au, v \rangle = \langle u, Av \rangle$ is an inner product by verifying the following properties.
   i. $\langle u + v, w \rangle_A = \langle u, w \rangle_A + \langle v, w \rangle_A$
   ii. $\langle \alpha u, v \rangle_A = \alpha \langle u, v \rangle_A$
   iii. $\langle u, v \rangle_A = \langle v, u \rangle$
   iv. $\langle u, u \rangle_A > 0$ if and only if $u \neq 0$.

   This inner product defines the *A-norm* or the *energy norm*, and it will become very important when we present the conjugate gradient method in Chapter 21.
   b. Prove $2a_{ij} < a_{ii} + a_{jj}$, $i \neq j$ Hint: If e_i and e_j are standard basis vectors $\langle e_i - e_j, e_i - e_j \rangle_A > 0$.
   c. Let i and j be distinct indices and define a vector w such that

$$w_k = \begin{cases} 0 & k \neq i \quad \text{and} \quad k \neq j \\ 1 & k = i \quad \text{or} \quad k = j \end{cases}, \ 1 \leq k \leq n$$

   Use the fact that $0 < \langle w, w \rangle_A$ to complete the proof that the element of largest magnitude lies on the diagonal of A.

13.14 Prove that the sum of two positive definite matrices is positive definite.

13.15 Prove that a positive definite matrix is nonsingular.

Remark 13.7. There is a modification to the Cholesky decomposition so it applies to a positive semidefinite matrix [15, pp. 438-442].

13.16 The square root of a real number $y \geq 0$ is a real number x such that $y = x^2$, and we write $x = \sqrt{y}$. The square root is not unique since $(-x)^2 = y$ as well; however, we refer to $x \geq 0$ as the square root of y. There is an analogous definition in matrix theory.

> If A is a positive semidefinite matrix, there exists a matrix X such that $A = X^2$, where X is positive semidefinite.

We will discuss the singular value decomposition in Chapter 15. A special case states that if A is an $n \times n$ matrix, then $A = U\Sigma V^T$, where U and V are $n \times n$ orthogonal matrices, and Σ is an $n \times n$ diagonal matrix containing the singular values of A. Let $A = R^T R$ be the Cholesky decomposition of A, $R = U\Sigma V^T$ be the singular value decomposition of R, and define $X = V\Sigma V^T$.

a. Show that X is positive semidefinite.

b. Show that $A = X^2$.

It can be shown that X is unique.

13.17 The inverse of an upper-bidiagonal matrix can be computed using a simple formula. Assume

$$U = \begin{bmatrix} u_1 & c_1 & & & \\ & u_2 & c_2 & & \\ & & \ddots & \ddots & \\ & & & u_{n-1} & c_{n-1} \\ & & & & u_n \end{bmatrix} \quad u_i \neq 0, \quad 1 \leq i \leq n.$$

a. Let $\beta_i = \begin{bmatrix} \beta_{1i} & \beta_{2i} & \cdots & \beta_{n-1,i} & \beta_{ni} \end{bmatrix}^T$ be column i of U^{-1}. By letting

$$U \begin{bmatrix} \beta_1 & \beta_2 & \cdots & \beta_{n-1} & \beta_n \end{bmatrix} = I,$$

develop the equations

$$\begin{aligned} \beta_{ij} &= 0, \quad i > j \\ \beta_{ii} &= \frac{1}{u_i} \\ \beta_{ij} &= -\frac{c_i \beta_{i+1,j}}{u_i}, \quad i < j \end{aligned}$$

b. From part (a), show that

$$\beta_{ij} = \begin{cases} 0, & i > j \\ \frac{1}{u_j} \prod_{k=i}^{j-1} \left(-\frac{c_k}{u_k} \right), & i \leq j \end{cases}, \tag{13.14}$$

where $\prod_{k=i}^{0} = 1$.

c. Show that the elements, α_{ij}, of the inverse for a bidiagonal matrix

$$L = \begin{bmatrix} 1 & & & & \\ l_1 & 1 & & & \\ & l_2 & \ddots & & \\ & & \ddots & 1 & \\ & & & l_{n-1} & 1 \end{bmatrix}$$

are given by

$$\alpha_{ij} = \begin{cases} 0, & i < j \\ \prod_{k=j}^{i-1} (-l_k), & i \geq j \end{cases}. \tag{13.15}$$

Remark 13.8. These results imply that the inverse of an upper-bidiagonal matrix is upper triangular and that the inverse of a unit lower-bidiagonal matrix is lower triangular. The article by Higham [32] discusses these results and then continues to develop very efficient algorithms for computing the condition number of a tridiagonal matrix.

13.5.1   MATLAB Problems

13.18 Let $A = \begin{bmatrix} 5 & -2 & 0 & \ldots & 0 \\ -2 & 5 & -2 & \ldots & 0 \\ 0 & \ddots & \ddots & \ddots & \vdots \\ \vdots & \vdots & -2 & 5 & -2 \\ 0 & 0 & 0 & -2 & 5 \end{bmatrix}$

be a 15×15 matrix. Factor A into a unit lower-bidiagonal matrix L and an upper-bidiagonal matrix U. Then solve the equation

$$Ax = \begin{bmatrix} 0.1 \\ 0.2 \\ 0.3 \\ \vdots \\ 1.5 \end{bmatrix}$$

13.19 Matrix $A = \begin{bmatrix} 1 & 1 & 1 \\ 1 & 2 & 1 \\ 1 & 1 & 3 \end{bmatrix}$ is positive definite.

Factor A using
a. Gaussian elimination without pivoting
b. The Cholesky algorithm
c. For each case, solve the system $Ax = b$, where

$$b = \begin{bmatrix} -1 \\ 3 \\ 4 \end{bmatrix}$$

13.20 Show that the matrix

$$A = \begin{bmatrix} 1.0000 & 0.2500 & 0.0625 & 0.0156 \\ 0.2500 & 1.0000 & 0.2500 & 0.0625 \\ 0.0625 & 0.2500 & 1.0000 & 0.2500 \\ 0.0156 & 0.0625 & 0.2500 & 1.0000 \end{bmatrix}$$

is positive definite
a. by showing all its eigenvalues are > 0
b. by showing that all the principle minors are > 0
c. using the Cholesky decomposition

13.21 The MATLAB command `hilb(n)` creates the Hilbert matrix of order n. Compute cholesky(A), $A = \{$hilb (5) , hilb (6) , hilb (7) , . . .$\}$ until the decomposition fails. All Hilbert matrices are symmetric positive definite. Explain why failure occurred.

13.22 Find the Cholesky decomposition $R^{T}R$ for

$$A = \begin{bmatrix} 0.2090 & 0.0488 & -0.0642 & -0.0219 \\ 0.0488 & 0.1859 & -0.0921 & -0.0475 \\ -0.0642 & -0.0921 & 0.4257 & 0.0364 \\ -0.0219 & -0.0475 & 0.0364 & 0.1973 \end{bmatrix}.$$

Compute the residual $\left\| A - R^{T}R \right\|_{2}$.

13.23 Write a function `choldet` that uses the Cholesky decomposition to compute the determinant of a positive definite matrix. Test your function with random positive definite matrices of orders $3 \times 3, 5 \times 5, 10 \times 10$, and 50×50. In each case compute the relative error

$$\frac{|\det (A) - \text{choldet} (A)|}{|\det (A)|}.$$

Here, we assume that the MATLAB `det` computes the correct value. Why is relative error a better measure of error for this problem?

Here is one way to generate a random positive definite matrix:

```
A = randn(n,n);
while rank(A) ~= n
    A = randn(n,n);
end
A = A'*A;
```

13.24 The Hilbert matrices, H_n, are symmetric positive definite.

a. Compute the condition number of the Hilbert matrices of orders 5, 25, 50, and 100.

b. Repeat part (a), except compute the 1-norm and the 2-norm of each Hilbert matrix.

c. Give a formula for $\|H_n\|_1$. What can you say about $\lim_{n\to\infty} \|H_n\|_1$? Why does $\lim_{n\to\infty} \|H_n\|_2$ behave in the same way?

13.25 Recall that in Section 11.5, we showed that with three-digit arithmetic when Gaussian elimination is applied to the matrix $A = \begin{bmatrix} 0.00001 & 3 \\ 2 & 1 \end{bmatrix}$ without a row exchange, the factors L and U are

$$L = \begin{bmatrix} 1 & 0 \\ 200000 & 1 \end{bmatrix}, \quad U = \begin{bmatrix} 0.00001 & 3 \\ 0 & -599999 \end{bmatrix}.$$

Large entries appear in L and U, and LU is very far from A. Gaussian elimination without pivoting is not a stable algorithm. However, if a matrix is positive definite, it can be shown that Gaussian elimination without pivoting is stable [9].

a. Let $A = \begin{bmatrix} 0.0006 & 0.01 \\ 0.01 & 0.50 \end{bmatrix}$. Find the LU decomposition without pivoting and note that the entries of L and U are not large.

b. Construct three random positive definite matrices (see Problem 13.23). Perform the LU decomposition without pivoting to three matrices and observe that the entries in L and U do not grow to be large relative to the entries of A.

13.26 a. Let A be a positive definite matrix. Develop an algorithm for computing the lower-triangular matrix H such that $A = HH^{\mathrm{T}}$.

b. Develop a function cholH that finds H and prints an error message if A is not positive definite.

c. Test your function using the matrices A and B from Example 13.6.

13.27 Perform the following numerical experiment and explain the results.

```
>> R = triu(randi([-10 10],5,5));
>> A = R'*R;
>> while rank(A) ~= 5
        R = triu(randi([-10 10],5,5));
        A = R'*R;
   end
>> Rhat = cholesky(A);
>> norm(Rhat - R)/norm(R)
>> norm(A - Rhat'*Rhat)
```

13.28 This problem examines the properties of a positive definite matrix. The MATLAB statement

```
A = gallery('gcdmat',n);
```

creates an $n \times n$ positive definite matrix, where $a_{ij} = \gcd(i, j)$. The function gcd computes the greatest common divisor of i and j, the largest integer that divides both i and j. For instance,

$$\gcd(540, 252) = 36$$

a. Create a 4×4 gcd matrix. Verify that the principle minors are all positive by computing
det(A(1,1)), det(A(1:2,1:2)), det(A(1:3,1:3)), det(A)

b. Verify that all the eigenvalues of A are positive.

i. The function lugauss in the software distribution performs the LU decomposition without pivoting. Modify lugauss so it returns [L, U, pivot], where pivot is an $n \times 1$ vector containing the $n - 1$ pivots. Name the function `lupiv`. Run `lupiv` on a gcd matrix of dimensions 4, 10, 15, 25, and 50. In each case execute `min(pivot) > 0`

ii. If there is a pattern to your experiment, state a theorem. You need not prove it.

13.29 Sylvester's criterion can be used to check the positive definiteness of any symmetric matrix A without using expansion by minors. Row reduce A to an upper-triangular matrix using Gaussian elimination without pivoting. Check Sylvester's criterion each time a leading principle submatrix is in upper-triangular form by forming the product of its diagonal elements. If the criterion is true for all the leading principle submatrices, then the matrix is positive definite. Write a function

> `isposdef = sylvester(A)`

that determines if A is positive definite. Test your function with matrices of dimension 3×3, 5×5, 10×10, and 25×25. For each dimension, use two matrices, one positive definite, and the other not. *Note:* This algorithm can be done using pivoting, but we elect not to consider it here.

13.30 This problem develops a means for computing the inverse of a tridiagonal matrix

$$
A = \begin{bmatrix} b_1 & c_1 & & & \\ a_1 & b_2 & c_2 & & \\ & \ddots & \ddots & \ddots & \\ & & a_{n-12} & b_{n-1} & c_{n-1} \\ & & & a_{n-1} & b_n \end{bmatrix}.
$$

Write a function `trinv` with the calling format

> `C = trinv(a,b,c)`

that uses the function `tridiagLU` from the book software distribution to compute the inverse of A. Create a 500×500 tridiagonal matrix with a, b, and c consisting of random numbers. Time the execution using `trinv` and the MATLAB `inv` function.

Remark 13.9. We have stated that computing the inverse is generally not a good idea; however, tridiagonal matrices have many applications in the numerical solution of boundary value problems, in the solution of second order difference equations, and so forth, so a reliable means for computing the inverse can be useful. There are a number of papers on the subject [39–43].

13.31 We know a tridiagonal matrix, A, can be factored into a product of a unit lower-bidiagonal matrix, L, and an upper-triangular matrix, U. For the purpose of this problem, assume that A, L, and U have the following form:

$$
A = \begin{bmatrix} b_1 & c_1 & & & \\ a_2 & b_2 & c_2 & & \\ & \ddots & \ddots & \ddots & \\ & & a_{n-1} & b_{n-1} & c_{n-1} \\ & & & a_n & b_n \end{bmatrix}, \quad L = \begin{bmatrix} 1 & & & & \\ \gamma_2 & 1 & & & \\ & \gamma_3 & \ddots & & \\ & & \ddots & 1 & \\ & & & \gamma_n & 1 \end{bmatrix}, \quad U = \begin{bmatrix} \alpha_1 & c_1 & & & \\ & \alpha_2 & c_2 & & \\ & & \ddots & \ddots & \\ & & & \alpha_{n-1} & c_{n-1} \\ & & & & \alpha_n \end{bmatrix}.
$$

We also assume that $\alpha_i \neq 0$, $1 \leq i \leq n$. Equations 13.14 and 13.15 given in Problem 13.17 are put together in Ref. [39] to develop an explicit formula for the inverse of a general tridiagonal matrix with the $\alpha_i \neq 0$ restriction. Let $\tau_i = \frac{c_i}{\alpha_i}, 1 \leq i \leq n - 1$ and c_{ij}, $1 \leq i, j \leq n$ be the entries of the inverse. The following is the algorithm for computing A^{-1}.

```
function TRIDIAGINV(a,b,c)
    % Find the inverse of the tridiagonal matrix whose diagonals are a, b, c.

    % Compute the LU decomposition.
    α₁ = b₁
    for i = 2:n do
        τᵢ₋₁ = cᵢ₋₁/αᵢ₋₁
        αᵢ = bᵢ - aᵢτᵢ₋₁
        γᵢ = aᵢ/αᵢ₋₁
    end for
```

```
% Perform error checking.
if αᵢ = 0 for some i then
   if αₙ = 0 then
      Output that A is singular and terminate.
   else
      Output that some αᵢ is zero and terminate.
   end if
end if

% Compute the main diagonal entries.
cₙₙ = 1/αₙ
for i = n-1:-1:1 do
   cᵢᵢ = 1/αᵢ + τᵢγᵢ₊₁cᵢ₊₁, ᵢ₊₁
end for

% Compute the ith row elements cᵢⱼ, j < i.
for i = n:-1:2 do
   for j = i-1:-1:1 do
      cᵢⱼ = -γⱼ₊₁cᵢ, ⱼ₊₁
   end for
end for

% Compute the ith column elements cⱼᵢ, j < i.
for i = n:-1:2 do
   for j = i-1:-1:1 do
      cⱼᵢ = -τⱼcⱼ₊₁, ᵢ
   end for
end for

   return C
end function
```

The computational cost of this algorithm is shown in Ref. [39] to be $O\left(n^2\right)$.

a. In general, explain why this algorithm is superior to directly computing A^{-1}.

b. Write a function, tridiaginv, that implements the algorithm. Test your function on the following matrices:

    i. 5×5 tridiagonal matrix with $a = \begin{bmatrix} 0 & 1 & 1 & 1 & 1 \end{bmatrix}^{\mathrm{T}}$, $b = \begin{bmatrix} 3 & 3 & 3 & 3 & 3 \end{bmatrix}^{\mathrm{T}}$, and $c = \begin{bmatrix} 5 & 5 & 5 & 5 \end{bmatrix}^{\mathrm{T}}$.

    ii. 500×500 tridiagonal matrix with a, b, and c consisting of random numbers. Time the execution using tridiaginv and the MATLAB inv function. Also include trinv if you did Problem 13.30 or have access to the source code. Note that inv is compiled into machine code. Can you beat MATLAB?

Chapter 14

Gram-Schmidt Orthonormalization

You should be familiar with

- The inner product
- Orthogonal vectors and matrices
- Rank
- Algorithm stability
- The determinant

Orthogonal vectors and matrices are of great importance in many fields of science and engineering; for instance, they play an important role in least-squares problems, analysis of waves, the finite element method for the numerical solution of partial differential equations, financial engineering, and quantum mechanics. We determined in Chapter 7 that orthogonal matrices preserve length, a property called orthogonal invariance; for example, in computer graphics, orthogonal matrices are used to rotate an image without altering its size. Also, an orthogonal matrix is used in a change of basis, such as constructing a basis for spherical or cylindrical coordinates. Many problems in dynamics involve a change of basis. We begin this chapter by reviewing some of properties of orthogonal vectors and matrices.

- If $x_1, x_2, \ldots, x_k$ are orthogonal, they are linearly independent.
- If A is a real symmetric matrix, then any two eigenvectors corresponding to different eigenvalues are orthogonal.
- If P is an orthogonal matrix, then $P^{-1} = P^{\mathrm{T}}$.
- If P is an orthogonal matrix, then $\|Px\|_2 = \|x\|_2$.
- Let P be a $n \times n$ real matrix. Then P is an orthogonal matrix if and only if the columns of P are orthogonal and have unit length.
- For any orthogonal matrices U and V, $\|UAV\|_2 = \|A\|_2$.

If we have a basis $v_1, v_2, \ldots, v_n$ for a subspace, it would be valuable to have the means of constructing an orthonormal basis $e_1, e_2, \ldots, e_n$ that spans the same subspace. We can then use the orthonormal basis to build an orthogonal matrix. The Gram-Schmidt process does exactly that, and leads to the QR decomposition of a general real $m \times n$ matrix.

14.1 THE GRAM-SCHMIDT PROCESS

The *Gram-Schmidt process* takes a set of linearly independent vectors $S = \{v_1, v_2, \ldots, v_n\} \in \mathbb{R}^m$, and transforms them into a set of orthonormal vectors $S' = \{e_1, e_2, \ldots, e_n\}$. The orthonormal set $S' = \{e_1 e_2, \ldots, e_n\}$ spans the same n-dimensional subspace as S. It follows that $m \geq n$ or the vectors would be linearly dependent.

Remark 14.1.   Although we will deal with the inner product of vectors, v_i, $1 \leq i \leq n$, we can just as well have a set of functions that we transform into an orthonormal set relative to the L^2 norm

$$\langle v_i, v_j \rangle = \int_a^b v_i(t)\, v_j(t)\, \mathrm{d}t.$$

The Gram-Schmidt process works by successively subtracting orthogonal projections from vectors. The projection operator we now define is one of the most important uses of the inner product.

Definition 14.1.   The orthogonal projection of vector v onto vector u is done by the projection operator $\mathrm{proj}_u(v) = \left(\langle v, u \rangle / \|u\|_2^2\right) u$. Figure 14.1 depicts the projection.

Numerical Linear Algebra with Applications. http://dx.doi.org/10.1016/B978-0-12-394435-1.00014-4

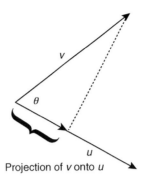

FIGURE 14.1 Vector orthogonal projection.

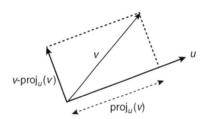

FIGURE 14.2 Removing the orthogonal projection.

The use of some geometry will help understand the projection operator. The length of the projection, d, of v onto u has the value

$$d = \|v\|_2 \cos\theta,$$

which can be negative if $\theta > \pi/2$. The inner product of v and u is $\langle v, u \rangle = \langle v, u \rangle = \|v\|_2 \|u\|_2 \cos\theta$. As a result

$$d = \frac{\langle v, u \rangle}{\|u\|_2}.$$

Now create a vector with magnitude $|d|$ by multiplying the unit vector $u/\|u\|$ by d to obtain the vector

$$\text{proj}_u (v) = \left(\frac{\langle v, u \rangle}{\|u\|_2} \right) \frac{u}{\|u\|_2} = \left(\frac{\langle v, u \rangle}{\|u\|_2^2} \right) u.$$

Now comes the most critical point. If we remove the orthogonal projection of v onto u from v by computing $v - \text{proj}_u (v)$ we obtain a vector orthogonal to u, as depicted in Figure 14.2. We can verify this algebraically as follows:

$$u^T \left(v - \text{proj}_u (v) \right) = u^T v - \left(\frac{u^T v}{\|u\|_2^2} \right) u^T u = u^T v - u^T v = 0$$

The Gram-Schmidt process works by successively removing the orthogonal projection of v_i from the orthonormal vectors $e_1, e_2, \ldots, e_{i-1}$ already built from $v_1, v_2, \ldots, v_{i-1}$. This will produce a vector orthogonal to $\{e_1, e_2, \ldots, e_{i-1}\}$ (verify).

Step 1: Begin by taking the first vector, v_1, normalize it to obtain a vector e_1, and define $r_{11} = \|v_1\|_2$. As our description of the process progresses, we will define a set, $\{r_{ij}, 1 \le i, j \le n, j \ge i\}$. While not strictly necessary for building an orthonormal basis from $v_1, v_2, \cdots, v_k$, we will use these values to form an upper triangular matrix when we discuss the QR decomposition later in this chapter. Let $u_1 = v_1$, and then

$$e_1 = \frac{v_1}{\|v_1\|_2} = \frac{v_1}{r_{11}}, \quad r_{11} = \|v_1\|_2.$$

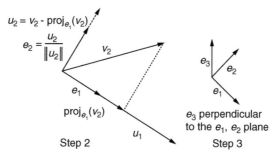

FIGURE 14.3 Result of the first three steps of Gram-Schmidt.

Step 2: Let $r_{12} = \langle v_2, e_1 \rangle$. Define vector, u_2, normal to e_1, by removing the orthogonal projection of v_2 onto e_1. Then normalize it to obtain e_2.

$$u_2 = v_2 - \text{proj}_{e_1}(v_2) = v_2 - \frac{\langle v_2, e_1 \rangle}{\|e_1\|_2^2} e_1 = v_2 - \langle v_2, e_1 \rangle e_1 = v_2 - r_{12}e_1.$$

Normalize the vector u_2, let $r_{22} = \|u\|_2$ and define

$$e_2 = \frac{u_2}{\|u_2\|_2} = \frac{u_2}{r_{22}}.$$

Step 3: Now obtain a vector u_3 normal to both e_1 and e_2 by removing the orthogonal projections of v_3 onto e_1 and e_2. Define $r_{13} = \langle v_3, e_1 \rangle$ and $r_{23} = \langle v_3, e_2 \rangle$.

$$u_3 = v_3 - \text{proj}_{e_1}(v_3) - \text{proj}_{e_2}(v_3) = v_3 - \langle v_3, e_1 \rangle e_1 - \langle v_3, e_2 \rangle e_2 = v_3 - r_{13}e_1 - r_{23}e_2.$$

Normalize u_3, and let $r_{33} = \|u_3\|_2$ to obtain

$$e_3 = \frac{u_3}{\|u_3\|_2} = \frac{u_3}{r_{33}}.$$

See Figure 14.3 showing the results of steps 1-3.

Continue in this fashion until all the vectors up to e_n are determined. The formula for computing a general vector, e_i, is:

Step i: Find

$$u_i = v_i - \sum_{j=1}^{i-1} r_{ji}e_j, \quad 1 \leq i \leq n, \tag{14.1}$$

where

$$r_{ji} = \langle v_i, e_j \rangle.$$

Let $r_{ii} = \|u_i\|_2$. Normalize u_i to obtain

$$e_i = \frac{u_i}{\|u_i\|_2} = \frac{u_i}{r_{ii}}.$$

The sequence $e_1, \ldots, e_k$ is the required set of orthonormal vectors, and the process is known as *Gram-Schmidt orthonormalization*.

This process can be explained geometrically. The vectors e_1 and e_2 are orthogonal (Figure 14.3). The vector u_3 is orthogonal to e_1 and e_2, since it is formed by removing their orthogonal projections from v_3. Continuing in this fashion, vector e_i is orthogonal to the vectors $e_1, e_2, \ldots, e_{i-1}$; furthermore, the vectors $e_1, e_2, \ldots, e_i$ span the same subspace as $v_1, v_2, \ldots, v_i$. To see this, let vector x be a linear combination of the vectors $v_1, v_2, \ldots, v_i$.

$$x = c_1v_1 + c_2v_2 + c_3v_3 + \cdots + c_iv_i.$$

Each vector $v_i = u_i + \sum_{j=1}^{i-1} \text{proj}_{e_j}(v_i)$. Now, each $\text{proj}_{e_j}(v_i)$ is a scalar multiple of e_j, and $u_i = \|u_i\| e_i$, so x can be written as a linear combination of $e_1, e_2, \ldots, e_i$. Since $e_i, e_2, \ldots, e_i$ are orthonormal, they are linearly independent and form a basis for the same subspace as $v_1, v_2, \ldots, v_i$.

If the Gram-Schmidt process is applied to a linearly dependent sequence, then at least one v_i is a linear combination of the remaining vectors,

$$v_i = c_1 v_1 + c_2 v_2 + \cdots + c_{i-1} v_{i-1} + c_{i+1} v_{i+1} + \cdots + c_k v_k.$$

In this situation, on the ith step the result is $u_i = 0$. Discard u_i and v_i and continue the computation. If this happens again, do the same. The number of vectors output by the process will then be the dimension of the space spanned by the original vectors, less the dependent vectors. For the sake of simplicity, we will assume throughout this chapter that all the vectors we deal with are linearly independent.

Example 14.1. Consider the following set of three linearly independent vectors in $\mathbb{R}^3$.

$$S = \left\{ v_1 = \begin{bmatrix} 1 \\ -1 \\ 3 \end{bmatrix}, v_2 = \begin{bmatrix} 3 \\ 1 \\ 4 \end{bmatrix}, v_3 = \begin{bmatrix} 3 \\ 2 \\ 5 \end{bmatrix} \right\}$$

Execute Gram-Schmidt to obtain an orthonormal set of vectors.

$$r_{11} = \sqrt{11} = 3.3166, \quad e_1 = \frac{\begin{bmatrix} 1 \\ -1 \\ 3 \end{bmatrix}}{r_{11}} = \begin{bmatrix} 0.30151 \\ -0.30151 \\ 0.90453 \end{bmatrix},$$

$$r_{12} = \langle v_2, \ e_1 \rangle = 4.2212, u_2 = v_2 - r_{12} e_1 = \begin{bmatrix} 1.7273 \\ 2.2727 \\ 0.18182 \end{bmatrix}, \quad e_2 = \frac{u_2}{\|u_2\|_2} = \begin{bmatrix} 0.60386 \\ 0.79455 \\ 0.063564 \end{bmatrix},$$

$$r_{22} = \|u_2\| = 2.8604,$$

$$r_{13} = \langle v_3, e_1 \rangle = 4.8242, \quad r_{23} = \langle v_3, e_2 \rangle = 3.7185, \quad u_3 = v_3 - r_{13} e_1 - r_{23} e_2 = \begin{bmatrix} -0.7 \\ 0.5 \\ 0.4 \end{bmatrix},$$

$$e_3 = \frac{u_3}{\|u_3\|} = \begin{bmatrix} -0.737861 \\ 0.52705 \\ 0.42164 \end{bmatrix}, \quad r_{33} = \|u_3\| = 0.94868.$$

Summary

$$e_1 = \begin{bmatrix} 0.30151 \\ -0.30151 \\ 0.90453 \end{bmatrix}, \quad e_2 = \begin{bmatrix} 0.60386 \\ 0.79455 \\ 0.063564 \end{bmatrix}, \quad e_3 = \begin{bmatrix} -0.737861 \\ 0.52705 \\ 0.42164 \end{bmatrix},$$

$$R = \begin{bmatrix} r_{11} & r_{12} & r_{13} \\ 0 & r_{22} & r_{23} \\ 0 & 0 & r_{33} \end{bmatrix} = \begin{bmatrix} 3.3166 & 4.2212 & 4.8242 \\ 0 & 2.8604 & 3.7185 \\ 0 & 0 & 0.94868 \end{bmatrix}. \quad \blacksquare$$

Algorithm 14.1 describes the Gram-Schmidt process. The algorithm name, clgrsch, reflects the fact that the process we have described is called *classical Gram-Schmidt*. As we will see in Section 14.2, classical Gram-Schmidt has numerical stability problems, and we normally use a simple modification of the classical process.

Remark 14.2. The function clgrsch in Algorithm 14.1 takes an $m \times n$ matrix as input. It is important to understand that there are n vectors to be orthonormalized, and the dimension of each vector is m; however, the function does not compute the values r_{ij}. We will use them when developing the QR decomposition of a matrix.

NLALIB: The function clgrsch implements Algorithm 14.1.

14.2 NUMERICAL STABILITY OF THE GRAM-SCHMIDT PROCESS

During the execution of the Gram-Schmidt process, the vectors u_i are often not quite orthogonal, due to rounding errors. For the classical Gram-Schmidt process we have described, this loss of orthogonality is particularly bad. The computation also

---

Algorithm 14.1 Classical Gram-Schmidt

```
function CLGRSCH(V)
   % Converts a set of linearly independent vectors to a set
   % of orthonormal vectors spanning the same subspace
   % Input: An m × n matrix V whose columns are the vectors to be normalized.
   % Output: An m × n matrix E whose columns are an orthonormal set of
   % vectors spanning the same subspace as the columns of V.
   for i = 1:n do
      sumproj = 0
      for j = 1:i-1 do
         sumproj = sumproj + E(:, j)ᵀ V(:, i) E(:, j)
      end for
      E(:, i) = V(:, i) − sumproj
      E(:, i) = E(:, i) / ‖E(:, i)‖₂
   end for
   return E
end function
```

---

yields poor results when some of the vectors are almost linearly dependent. For these reasons, it is said that the classical Gram-Schmidt process is numerically unstable.

The Gram-Schmidt process can be improved by a small modification. The computation of u_i using the formula

$$u_i = v_i - \sum_{j=1}^{i-1} r_{ji} e_j$$

removes the projections all at once. Split the computation into smaller parts by removing the projections one at a time.

$u_i^{(1)} = v_i - \langle v_i, e_1 \rangle e_1$ remove the projection of v_i onto e_1.

$u_i^{(2)} = u_i^{(1)} - \langle v_i, e_2 \rangle e_2$ remove the projection of v_i onto e_2.

$u_i^{(3)} = u_i^{(2)} - \langle v_i, e_3 \rangle e_3$ remove the projection of v_i onto e_3.

$$\vdots$$

$u_i^{i-1} = u_i^{(i-2)} - \langle v_i, e_{i-2} \rangle e_{i-2}$ remove the projection of v_i onto e_{i-2}

$u_i = u_i^{(i-1)} - \langle v_i, e_{i-1} \rangle e_{i-1}$ remove the projection of v_i onto e_{i-1}.

This approach (sometimes referred to as *modified Gram-Schmidt (MGS) process*) gives the same result as the original formula in exact arithmetic, but it introduces smaller roundoff errors when executed on a computer.

The following algorithm implements the MGS process. The input and output format is the same as for clgrsch.

NLALIB: The function modgrsch implements Algorithm 14.2.

Remark 14.3. The orthonormal basis obtained by modgrsch are the columns of the output matrix E, so E is an orthogonal matrix.

Example 14.2. Apply the function modgrsch to the vectors of Example 14.1, obtain an orthonormal basis. Verify that the columns in matrix E are orthonormal by computing $E^T E$. ∎

```
>> E = modgrsch(V)

E =

      0.30151       0.60386      -0.73786
     -0.30151       0.79455       0.52705
```

Algorithm 14.2 Modified Gram-Schmidt

```
function MODGRSCH(V)
   % Modified Gram-Schmidt process for converting a set of linearly independent vectors to a set
   % of orthonormal vectors spanning the same subspace
   % Input: An m × n matrix V whose columns are the vectors to be normalized.
   % Output: An m × n matrix E whose columns are an orthonormal set of
   % vectors spanning the same subspace as the columns of V
   for i = 1:n do
      E(:,i) = V(:,i)
      for j = 1:i-1 do
         E(:, i) = E(:, i) − E(:, j)ᵀ E(:, i) E(:, j)
      end for
      E(:, i) = E(:, i)/‖E(:, i)‖₂
   end for
   return E
end function
```

```
      0.90453        0.063564        0.42164

>> E'*E

ans =
            1 -1.3878e-017  -4.996e-016
  -1.3878e-017           1   8.6736e-016
  -4.996e-016   8.6736e-016            1
```

The difference in results between the classical and MGS methods can be startling. For Example 14.3, the author is indebted to an example found on the MIT OpenCourseWare site for the course [44] 18.335J, Introduction to Numerical Methods.

Example 14.3. Choose $\epsilon = 10^{-8}$, and form the 4×3 matrix $A = \begin{bmatrix} 1 & 1 & 1 \\ \epsilon & 0 & 0 \\ 0 & \epsilon & 0 \\ 0 & 0 & \epsilon \end{bmatrix}$. Apply the classical and MGS methods to

A and test for the orthogonality of columns.

```
>> epsilon = 1.0e-8;
>> A = [1 1 1;epsilon 0 0;0 epsilon 0;0 0 epsilon];
>> E1 = clgrsch(A);
>> E2 = modgrsch(A);
>> E1(:,2)'*E1(:,3)

ans =

   0.500000000000000

>> E2(:,2)'*E2(:,3)

ans =

   1.1102e-016
```

The huge difference between the results is caused by the fact that the columns are almost equal, and cancellation errors occur when using the classical method. ∎

The concept of rank is very important in linear algebra. The best situation is when an $m \times n$ matrix has full rank.

Definition 14.2. The $m \times n$ real matrix is said to have *full rank* if $rank = \min(m, n)$. If $A \in R^{m \times n}$ and $m > n$ then to have full rank the n columns of A must be linearly independent. If $m < n$, then to be of full rank, the m rows must be linearly independent.

If $m > n$, the application of the Gram-Schmidt process to the column vectors of an $m \times n$ full rank matrix A while recording the values r_{ij} yields the QR decomposition, one of the major achievements in linear algebra. The QR decomposition is very important in the accurate, efficient, computation of eigenvalues and is very useful in least-squares problems.

14.3 THE QR DECOMPOSITION

Assume $A = \begin{bmatrix} v_1 & v_2 & \cdots & v_{n-1} & v_n \end{bmatrix}$ is an $m \times n$ matrix with columns $v_1, v_2, \ldots, v_n$. The Gram-Schmidt process can be used to factor A into a product $A = QR$, where $Q^{m \times n}$ has orthonormal columns, and $R^{n \times n}$ is an upper-triangular matrix. The decomposition comes directly from the Gram-Schmidt process by using the r_{ij} values we defined in the description of Gram-Schmidt. Arrange Equation 14.1 so the v_i are on the left-hand side.

Equation 1: $v_1 = e_1 r_{11}$

Equation 2: $v_2 = u_2 + r_{12}e_1$. Note that $e_2 = u_2/\|u_2\|_2$, so $u_2 = e_2\|u_2\|_2 = e_2 r_{22}$, and we have

$$v_2 = e_1 r_{12} + e_2 r_{22}.$$

Equation 3: $v_3 = u_3 + r_{13}e_1 + r_{23}e_2$. We have $e_3 = u_3/\|u_3\|_2$, so $u_3 = r_{33}e_3$, and then

$$v_3 = e_1 r_{13} + e_2 r_{23} + e_3 r_{33}.$$

The general formula for the v_k is

$$v_k = \sum_{j=1}^{k-1} r_{jk}e_j + e_k r_{kk}, \quad k = 1, 2, \ldots, n. \tag{14.2}$$

Let $Q = \begin{bmatrix} e_1 & e_2 & \cdots & e_{n-1} & e_n \end{bmatrix}$, the matrix whose columns are the orthonormal vectors

$$e_1 = \begin{bmatrix} e_{11} \\ e_{21} \\ \vdots \\ e_{m-1,1} \\ e_{m1} \end{bmatrix}, e_2 = \begin{bmatrix} e_{12} \\ e_{22} \\ \vdots \\ e_{m-1,2} \\ e_{m2} \end{bmatrix}, \ldots, e_n = \begin{bmatrix} e_{1n} \\ e_{2n} \\ \vdots \\ e_{m-1,n} \\ e_{mn} \end{bmatrix}$$

$$\cdots$$

and

$$R = \begin{bmatrix} r_{11} & r_{12} & r_{13} & r_{14} & \cdots & r_{1n} \\ 0 & r_{22} & r_{23} & r_{24} & \cdots & r_{2n} \\ 0 & 0 & r_{33} & r_{34} & \cdots & r_{3n} \\ 0 & 0 & 0 & \ddots & \cdots & \vdots \\ \vdots & \vdots & \vdots & \vdots & \ddots & \vdots \\ 0 & 0 & 0 & 0 & \cdots & r_{nn} \end{bmatrix}.$$

It is sufficient to show that $A = QR$ for a general 3×3 matrix. Let A be the 3×3 matrix

$$A = \begin{bmatrix} v_{11} & v_{12} & v_{13} \\ v_{21} & v_{22} & v_{23} \\ v_{31} & v_{32} & v_{33} \end{bmatrix},$$

and

$$v_1 = \begin{bmatrix} v_{11} \\ v_{21} \\ v_{31} \end{bmatrix}, \quad v_2 = \begin{bmatrix} v_{12} \\ v_{22} \\ v_{32} \end{bmatrix}, \quad v_3 = \begin{bmatrix} v_{13} \\ v_{23} \\ v_{33} \end{bmatrix}.$$

Then,

$$
\begin{bmatrix} e_{11} & e_{12} & e_{13} \\ e_{21} & e_{22} & e_{23} \\ e_{31} & e_{32} & e_{33} \end{bmatrix} \begin{bmatrix} r_{11} & r_{12} & r_{13} \\ 0 & r_{22} & r_{23} \\ 0 & 0 & r_{33} \end{bmatrix} = \begin{bmatrix} r_{11}e_{11} & r_{12}e_{11} + r_{22}e_{12} & r_{13}e_{11} + r_{23}e_{12} + r_{33}e_{13} \\ r_{11}e_{21} & r_{12}e_{21} + r_{22}e_{22} & r_{13}e_{21} + r_{23}e_{22} + r_{33}e_{23} \\ r_{11}e_{31} & r_{12}e_{31} + r_{22}e_{32} & r_{13}e_{31} + r_{23}e_{32} + r_{33}e_{33} \end{bmatrix}. \tag{14.3}
$$

From Equation 14.2,

$$
v_1 = r_{11}e_1, \quad v_2 = r_{12}e_1 + r_{22}e_2, \quad v_3 = r_{13}e_1 + r_{23}e_2 + r_{33}e_3. \tag{14.4}
$$

Comparing Equations 14.3 and 14.4, we see that

$$
A = QR,
$$

where

$$
Q = \begin{bmatrix} e_{11} & e_{12} & e_{13} \\ e_{21} & e_{22} & e_{23} \\ e_{31} & e_{32} & e_{33} \end{bmatrix}, \quad R = \begin{bmatrix} r_{11} & r_{12} & r_{13} \\ 0 & r_{22} & r_{23} \\ 0 & 0 & r_{33} \end{bmatrix}.
$$

Theorem 14.1. *(QR Decomposition) If A is a full rank $m \times n$ matrix, $m \geq n$, then there exists an $m \times n$ matrix Q with orthonormal columns and an $n \times n$ upper-triangular matrix R such that $A = QR$.*

Remark 14.4.   The decomposition is unique (Problem 14.10).

The MATLAB function clqrgrsch in the software distribution executes the classical Gram-Schmidt QR decomposition process. The application of the classical Gram-Schmidt process for finding the QR decomposition has the same problems as the classical Gram-Schmidt process for finding an orthonormal basis. Algorithm 14.3 specifies the MGS process to find the QR decomposition. The implementation is that of the MGS process with the added maintenance of the r_{ij}, $1 \leq i$, $j \leq n$, $j \geq i$, $r_{ij} = 0$, $j < i$.

Algorithm 14.3 Modified Gram-Schmidt QR Decomposition

```
function MODQRGRSCH(A)
    % Input: m × n matrix A.
    % Output: the QR decomposition A = QR, where
    % Q is an m × n matrix with orthonormal columns, and
    % R is an n × n upper-triangular matrix.
    for i = 1:n do
        Q(:, i) = A(:, i)
        for j = 1:i-1 do
            R(j, i) = Q(:, j)ᵀ Q(:, i)
            Q(:, i) = Q(:, i) − R(j, i) Q(:, j)
        end for
        R(i, i) = ‖Q(:, i)‖
        Q(:, i) = Q(:, i)/R(i, i)
    end for
    return [Q, R]
end function
```

NLALIB: The function modqrgrsch implements Algorithm 14.3.

Example 14.4.   Use MGS to compute the QR decomposition of the matrix

$$A = \begin{bmatrix} 1 & 6 & -1 & 4 & 7 \\ -7 & 0 & 12 & -8 & 2 \\ 14 & 4 & 5 & 3 & 35 \end{bmatrix}.$$

Check the results by computing $\|A - QR\|_2$.

```
>> [Q R] = modqrgrsch(A)

Q =
     0.0638    0.9531   -0.2960   -0.3841    0.2643
    -0.4463    0.2925    0.8457    0.5121   -0.7362
     0.8926    0.0782    0.4440    0.7682    0.6230

R =
    15.6844    3.9530   -0.9564    6.5033   30.7950
         0    6.0311    2.9481    1.7066    9.9929
         0         0   12.6647   -6.6178   15.1595
         0         0         0    0.0000    0.0000
         0         0         0         0    0.0000

>> norm(A - Q*R)

ans =
   9.9301e-016
```

∎

Remark 14.5.   The Gram-Schmidt QR algorithm produces an $m \times n$ matrix Q and an $n \times n$ matrix R. This is termed the *reduced QR decomposition*. MATLAB has a function qr that computes the QR decomposition of an $m \times n$ matrix. Normally, qr factors A into the product of an $m \times m$ orthogonal matrix Q and an $m \times n$ upper-triangular matrix R, called the *full QR decomposition*, and we will discuss this decomposition in Chapter 17. This decomposition gives useful information not provided by the reduced version. The term "reduced" derives from the fact that if $m > n$, the resulting matrices are smaller in the reduced decomposition. The MATLAB statement [Q R] = qr(A,0) gives the reduced QR decomposition.

14.3.1   Efficiency

We will determine the flop count for MGS. This analysis is somewhat more involved than earlier ones, so it is suggested that the reader refer to the steps in Algorithm 14.3 while reading the analysis.

The outer loop executes n times ($i = 1 : n$), and for each execution of the outer loop, an inner loop executes $i - 1$ times. Let's count the number of flops in the inner loop.

```
i = 1: inner loop executes (n-1) times
i = 2: inner loop executes (n-2) times
...
i = n-1: inner loop executes 1 time
i = n: inner loop does not execute
```

The statements in the inner loop thus execute $(n - 1) + (n - 2) + \cdots + 1 = n(n - 1)/2$ times. The inner loop computes an inner product, requiring $2m$ flops, followed by an expression that performs a scalar multiplication and a vector subtraction. This expression requires $2m$ flops. After the inner loop, a 2-norm is computed, which costs $2m$ flops. Following the computation of the 2-norm, a vector is normalized, requiring m divisions. Now add this all up:

Inner loop: $\frac{n(n-1)}{2}(2m+2m) = \frac{n(n-1)}{2}(4m)$

Statements after the inner loop: $n(2m+m) = n(3m)$ flops

TOTAL: $n(3m) + \frac{n(n-1)}{2}(4m) = 2mn^2 + mn$ flops.

Assume the term $2mn^2$ will dominate mn, and we obtain an approximation of $2mn^2$ flops.

14.3.2 Stability

Before we can discuss the stability of the Gram-Schmidt QR decomposition, we need a definition of for the condition number of a general an $m \times n$ matrix. For a nonsingular square matrix, we know that

$$\kappa_2(A) = \|A\|_2 \left\|A^{-1}\right\|_2 = \frac{\sigma_1}{\sigma_n},$$

where σ_1 and σ_n are the largest and smallest singular values of A. A non-square matrix has no inverse, but it does have a largest and a smallest nonzero singular value, leading to the following definition.

Definition 14.3. If A is an $m \times n$ matrix, the 2-norm condition number of A is

$$\kappa_2(A) = \frac{\sigma_1}{\sigma_r},$$

where σ_1 and σ_r are the largest and smallest nonzero singular values of A.

The Gram-Schmidt process for computing the reduced QR decomposition is conceptually simpler than methods we will discuss in Chapter 17, but it is not as stable as those methods. The stability of the algorithm is linked to the condition number of the matrix, as we see in Theorem 14.2, whose proof can be found in Ref. [16, pp. 372-373]. This is not the case with other algorithms we will discuss for computing the QR decomposition. In reading the theorem, note that if Q is an $m \times n$ matrix with orthonormal columns, $m > n$, Q^TQ is the $n \times n$ identity matrix (Problem 14.4).

Theorem 14.2. *Suppose the MGS process is applied to the $m \times n$ matrix $A = \begin{bmatrix} a_1 & a_2 & \dots & a_{n-1} & a_n \end{bmatrix}$ having rank n, yielding an $m \times n$ matrix $\hat{Q}$ and $n \times n$ matrix $\hat{R}$. Then there are constants c_i, $i = 1, 2, 3$, depending on m and n, such that*

$$A + \Delta A_1 = \hat{Q}\hat{R}, \quad \|\Delta A_1\|_2 \le c_1 \operatorname{eps} \|A\|_2,$$
$$\left\|\hat{Q}^T\hat{Q} - I\right\|_2 \le c_2 \operatorname{eps} \kappa_2(A) + O\left((\operatorname{eps}\kappa_2(A))^2\right),$$

and there exists a matrix Q with orthonormal columns such that

$$A + \Delta A_2 = Q\hat{R}, \quad \|\Delta A_2(:,j)\|_2 \le c_3 \operatorname{eps} \|a_j\|_2, \quad 1 \le j \le n.$$

There are three observations we can obtain from Theorem 14.2.

a. The residual $A - \hat{Q}\hat{R}$ is small.
b. How close $\hat{Q}$ is to having orthonormal columns depends on the condition number of A. If A is well conditioned, then $\hat{Q}$ has close to orthonormal columns.
c. $\hat{R}$ is the exact upper-triangular factor of a matrix near to A.

14.4 APPLICATIONS OF THE QR DECOMPOSITION

There are numerous applications for the QR decomposition. Chapters 18 and 19 discuss the use of the QR decomposition to find eigenvalues and eigenvectors, a problem we know should never be done by finding the roots of the characteristic polynomial. We discussed polynomial fitting using least squares in Section 12.3. The general linear least-squares problem involves finding a vector $x \in \mathbb{R}^n$ that minimizes $\|Ax - b\|_2$, where A is an $m \times n$ matrix, and b an $m \times 1$ vector. The QR decomposition can be effectively used to solve such problems, as you will see in Chapter 16. In this section, we discuss two useful but simpler applications, computing the determinant, and finding the range of a matrix.

14.4.1 Computing the Determinant

We can use the QR decomposition to find the absolute value of the determinant of a square matrix A. We will assume that the columns of A are linearly independent. Since Q has orthonormal columns,

$$Q^{\mathrm{T}}Q = I, \text{ and } \det\left(Q^{\mathrm{T}}Q\right) = \det\left(Q^{\mathrm{T}}\right)\det\left(Q\right) = (\det\left(Q\right))^2 = 1\,,$$

so $|\det\left(Q\right)| = 1$. It follows that

$$|\det\left(A\right)| = |\det\left(QR\right)| = |\det\left(Q\right)|\,|\det\left(R\right)| = |r_{11}r_{22}r_{33}\ldots r_{nn}|\,,$$

since the determinant of an upper-triangular matrix is the product of its diagonal elements.

Example 14.5. Find the absolute value of the determinant of the matrix $A = \begin{bmatrix} 8.0 & 2.6 & 4.0 & 9.8 \\ 4.2 & 6.3 & -1.2 & 5.0 \\ -2.0 & 0.0 & 9.1 & 8.5 \\ 18.7 & 25.0 & -1.0 & 23.5 \end{bmatrix}$.

```
>> [Q R] = modqrgrsch(A);
>> abs(prod(diag(R)))

ans =
  519.8238

>> det(A)

ans =
  -519.8238
```

■

14.4.2 Finding an Orthonormal Basis for the Range of a Matrix

Recall that the range, $R\left(A\right)$, of an $m \times n$ matrix A is defined by

$$R\left(A\right) = \left\{y \in \mathbb{R}^m \mid Ax = y \text{ for some vector } x \in \mathbb{R}^n\right\}.$$

In other words, the range of A is the set of all vectors y for which the equation $Ax = y$ has a solution. If $m \geq n$, A has rank n, and v_i, $1 \leq i \leq n$ are the columns of A, then

$$Ax = x_1v_1 + x_2v_2 + \cdots + x_{n-1}v_{n-1} + x_nv_n,$$

so the v_i are a basis for the range of A. Another way of putting this is that $R\left(A\right)$ is the column space of A. Now suppose that $A = QR$ is a reduced decomposition of A, $m \geq n$, and the diagonal entries r_{ii} of R are nonzero. We do not make the assumption that the decomposition was found using the Gram-Schmidt algorithm; in fact, we will discuss two other algorithms for computing the reduced decomposition. It is reasonable to require $r_{ii} \neq 0$, since we know the QR decomposition can be done using Gram-Schmidt with $r_{ii} > 0$. The following theorem connects Q and $R\left(A\right)$.

Theorem 14.3. *If A is a full rank $m \times n$ matrix, $m \geq n$, and $A = QR$ is a reduced QR decomposition of A with $r_{ii} \neq 0$, the columns of Q are an orthonormal basis for the range of A.*

Proof. If $x \in \mathbb{R}^n$, then $Ax = Q\left(Rx\right)$. Rx is a vector in $\mathbb{R}^n$, so range $\left(A\right) \subseteq$ range $\left(Q\right)$. If we can show that range $\left(Q\right) \subseteq$ range $\left(A\right)$, then range $\left(A\right) =$ range $\left(Q\right)$, and the columns of Q are an orthonormal basis for the range of A. Since R is upper diagonal with nonzero diagonal entries, it is invertible, and $AR^{-1} = Q$. Then, $Qx = A\left(R^{-1}x\right)$, and range $\left(Q\right) \subseteq$ range $\left(A\right)$, completing the proof. □

14.5 CHAPTER SUMMARY

The Gram-Schmidt Process

The Gram-Schmidt process takes a set of k linearly independent vectors, v_i, $1 \leq i \leq k$, and builds an orthonormal basis that spans the same subspace. Compute the projection of vector v onto vector u using

$$\text{proj}_u(v) = \left(\frac{\langle v, u \rangle}{\|u\|_2^2} \right) u.$$

The vector $v - \text{proj}_u(v)$ is orthogonal to u, and this forms the basis for the Gram-Schmidt process. Begin with the first vector, v_1, normalize it, name it e_1, and form $u_2 = e_1 - \text{proj}_{e_1}(v_2)$, and let $e_2 = u_2/\|u\|_2$. Continue this process by subtracting all projections of v_i onto e_j, $1 \leq j \leq i-1$ to obtain a vector u_i orthogonal to the e_j and normalize it to obtain e_i. Continue until you have an orthonormal set $e_1, e_2, \ldots, e_k$. To prepare for constructing the QR decomposition, maintain the upper-triangular matrix entries $r_{ji} = \langle v_i, e_j \rangle$, $1 \leq j \leq i-1$ and $r_{ii} = \|u_i\|_2$.

Numerical Stability of the Gram-Schmidt Process

During the execution of the Gram-Schmidt process, the vectors u_i are often not quite orthogonal, due to rounding errors. For the classical Gram-Schmidt process just described, this loss of orthogonality is particularly bad. The computation also yields poor results when some of the vectors are almost linearly dependent. For these reasons, it is said that the classical Gram-Schmidt process is numerically unstable.

Subtracting the projections of v_i onto the e_j all at once causes the problem. Split the computation into smaller parts by removing the projections one at a time. This approach (referred to as *MGS* process) gives the same result as the original formula in exact arithmetic, but it introduces smaller roundoff errors when executed on a computer.

The *QR* Decomposition

An $m \times n$ real matrix A is said to have full rank if $\text{rank}(A) = \min(m, n)$. If $A \in R^{m \times n}$ and $m > n$, then to have full rank the n columns of A must be linearly independent. If $m < n$, then to be of full rank, the m rows must be linearly independent. If $m \geq n$, the application of the Gram-Schmidt process to the column vectors of an $m \times n$ full rank matrix A while recording the values r_{ij} yields the QR decomposition, $A = QR$, where Q has orthonormal columns and R is an $n \times n$ upper-triangular matrix.

The decomposition requires approximately $2mn^2$ flops, which is better than other methods we will discuss. However, the algorithm is not as stable. Unlike the other methods, how close the computed Q is to having orthonormal columns depends on the condition number of A.

Applications of the *QR* Decomposition

We will see in Chapter 16 that the QR decomposition can be very effectively used to solve linear least-squares problems. In Chapters 18 and 19, we will see that the QR decomposition is very important in the accurate, efficient, computation of eigenvalues. This chapter presents two simpler applications.

The QR decomposition can be used to determine the absolute value of a determinant, namely,

$$|\det A| = |r_{11} r_{22} \ldots r_{nn}|.$$

The matrix R computed by Gram-Schmidt has positive diagonal elements, and Theorem 14.3 tells us that the columns of Q are an orthonormal basis for the range of A.

14.6 PROBLEMS

14.1 The vectors $\begin{bmatrix} 1 & -1 & 5 \end{bmatrix}^T$ and $\begin{bmatrix} -1 & -\frac{3}{5} & \frac{2}{25} \end{bmatrix}^T$ are orthogonal. Divide them by their norms to produce orthonormal vectors e_1 and e_2. Create a matrix Q with e_1 and e_2 as its columns and calculate $Q^T Q$ and QQ^T. Explain the results.

14.2 Compute the projection of v onto u and verify that $v - \text{proj}_u(v)$ is orthogonal to u.

    a. $v = \begin{bmatrix} 1 & 2 & 3 & 4 \end{bmatrix}^T$, $u = \begin{bmatrix} -1 & 3 & -1 & 7 \end{bmatrix}^T$

    b. $v = \begin{bmatrix} -1 & 6 & 2 & 0 & 1 \end{bmatrix}^T$, $u = \begin{bmatrix} 0 & 6 & -1 & -1 & 1 \end{bmatrix}^T$

14.3 Find an orthonormal basis e_1, e_2 for the subspace spanned by $v = \begin{bmatrix} 1 & -1 & 2 & 3 \end{bmatrix}^T$ and $w = \begin{bmatrix} 2 & 0 & 1 & 6 \end{bmatrix}^T$.

14.4 If Q is an $m \times n$ matrix with orthonormal columns, prove that $Q^T Q$ is the $n \times n$ identity matrix.

For Problems 14.5 and 14.6, assume that $A = QR$ is the QR decomposition of an $n \times n$ matrix A.

14.5 Show that $A^T A = R^T R$.

14.6 Show that $\det\left(A^T A\right) = (r_{11} r_{22} \ldots r_{nn})^2$.

14.7 **a.** Let $A = \begin{bmatrix} a & b \\ c & d \end{bmatrix}$, with $\det A = ad - bc > 0$. Find the QR decomposition of A.

   b. Gram-Schmidt breaks down if the column vectors are linearly dependent. Using the result of part (a), show how the breakdown occurs.

14.8 Assume A is an $m \times n$ matrix, $m < n$, and rank $(A) = m$. prove that A can be written in the form $A = LQ$, where L is a $m \times m$ lower-triangular matrix and Q is an $m \times n$ matrix with orthonormal rows.

14.9 This problem takes another approach to developing the QR decomposition.

   a. Show that $A^T A = R^T R$, where R is upper triangular with $r_{ii} > 0$, $1 \le i \le n$.

   b. Show that R nonsingular so we can define $Q = AR^{-1}$, and thus $A = QR$.

   c. Show that $Q^T Q = I$, so Q has orthonormal columns.

14.10 This problem is a proof that the reduced decomposition is unique.

   a. Assume that $A = \hat{Q}\hat{R}$ is another QR decomposition of A. Show that $A^T A = \hat{R}^T \hat{R}$.

   b. Using the fact that the Cholesky decomposition of a positive definite matrix is unique, show that $\hat{Q} = Q$ and $\hat{R} = R$.

14.11 Assume that $A = QR$ is the QR decomposition of the $m \times n$ matrix A. In the proof of Theorem 14.3, we showed that $R(A) \subseteq R(Q)$. This problem provides an alternative approach to developing this subset relationship. Let $C = QR$. Show that each element c_{ki}, $1 \le k \le m$, in column i of C has the value

$$c_{ki} = \sum_{p=1}^{i} q_{kp} r_{pi}.$$

Now show that

$$c_i = \sum_{p=1}^{i} q_p r_{pi},$$

where c_i is column i of QR and q_p is column p of Q. Explain why this shows that $R(A) \subseteq R(Q)$.

14.6.1 MATLAB Problems

14.12

   a. Show that the vectors $\begin{bmatrix} 1 \\ 5 \\ -1 \end{bmatrix}$, $\begin{bmatrix} 2 \\ 3 \\ 8 \end{bmatrix}$, $\begin{bmatrix} 0 \\ 1 \\ 3 \end{bmatrix}$ are linearly independent by showing that

$$\det \begin{bmatrix} 1 & 2 & 0 \\ 5 & 3 & 1 \\ -1 & 8 & 3 \end{bmatrix} \ne 0$$

   b. Using `modqrgrsch`, find the QR decomposition of $A = \begin{bmatrix} 1 & 2 & 0 \\ 5 & 3 & 1 \\ -1 & 8 & 3 \end{bmatrix}$.

14.13

   a. Find an orthonormal basis for the subspace spanned by the columns of $A = \begin{bmatrix} 1 & 4 & 7 \\ -1 & 2 & 3 \\ 9 & 1 & 0 \\ 4 & 1 & 8 \end{bmatrix}$.

   b. Find an orthonormal basis for the subspace spanned by the rows of $B = \begin{bmatrix} 5 & -1 & 3 & 6 \\ 0 & 5 & -7 & 1 \\ 1 & 2 & -1 & 0 \end{bmatrix}$.

c. Our implementation of Gram-Schmidt assumes that the columns of the matrix are linearly independent. Will `modgrsch` apply to the columns of B in part (b)? Why or why not?

14.14 Find an orthonormal basis for the column space of A.

$$A = \begin{bmatrix} 1 & -2 \\ 1 & 0 \\ 1 & 1 \\ 1 & 3 \end{bmatrix}$$

14.15 Use `modgrsch` to find an orthonormal basis for the columns of the matrix

$$A = \begin{bmatrix} 1 & 9 & 0 & 5 & 3 & 2 \\ -6 & 3 & 8 & 2 & -8 & 0 \\ 3 & 15 & 23 & 2 & 1 & 7 \\ 3 & 57 & 35 & 1 & 7 & 9 \\ 3 & 5 & 6 & 15 & 55 & 2 \\ 33 & 7 & 5 & 3 & 5 & 7 \end{bmatrix}$$

14.16 The QR decomposition using Gram-Schmidt gives a result if rank $(A) \neq \min (m, n)$.

a. Demonstrate this for the matrices

i. $A = \begin{bmatrix} 1 & -1 & 3 & 4 \\ 2 & 1 & 4 & 9 \\ 0 & 3 & 2 & 5 \\ 1 & 5 & -1 & 6 \\ 4 & -8 & 6 & 6 \end{bmatrix}$

ii. $B = \begin{bmatrix} 1 & 8 & -1 & 3 & 2 \\ 5 & 7 & -9 & 1 & 4 \\ 13 & 71 & -17 & 25 & 20 \end{bmatrix}$

b. Explain why the results of part (a) are actually not QR decompositions.

c. Use the MATLAB function `qr` with the two matrices in part (a). Are these actually QR decompositions? Why?

14.17 Compute the absolute value of the determinant of the matrix in Problem 14.15 using `modqrgrsch`. Compare the computed value to that obtained from the MATLAB function `det`.

14.18 The QR decomposition can be used to solve a linear system. Let A be an $n \times n$ matrix, with $A = QR$. Then, the linear system $Ax = b$ can be written as

$$QRx = b.$$

The process goes as follows:

Solve $Qy = b$ for y.

Solve $Rx = y$ for x.

a. It is very easy to solve for y without using Gaussian elimination. Why?

b. The solution to $Rx = y$ can be done quickly why?

c. Develop a function `qrsolve` that solves an $n \times n$ system using the MATLAB reduced QR decomposition. If the matrix is singular, the QR decomposition will complete with no error. Make the return values [x, resid], where resid is the residual $\|Ax - b\|_2$. A large residual will indicate that the matrix is singular or ill-conditioned.

d. Apply the function to solve each system.

i. $\begin{bmatrix} 1 & -1 & 0 \\ 2 & 4 & 5 \\ -7 & 1 & 3 \end{bmatrix} x = \begin{bmatrix} 1 \\ -1 \\ 8 \end{bmatrix}$

ii. $\begin{bmatrix} 21 & 3 & -4 & 8 \\ 1 & 3 & 59 & 0 \\ 1 & 2 & -22 & 35 \\ 3 & 78 & 100 & 3 \end{bmatrix} x = \begin{bmatrix} 1 \\ -1 \\ 1 \\ 2 \end{bmatrix}$

iii. The rosser matrix A = rosser.

We have three methods available for computing the *QR* decomposition of a matrix:

a. `[Q, R] = clqrgrsch(A);`

b. `[Q, R] = modqrgrsch(A);`

c. `[Q, R] = qr(A);`

Problems 14.19 and 14.20 involve running numerical experiments using these functions.

14.19 A Hilbert matrix is square and has entries $H(i,j) = 1/(i+j-1)$, $1 \le i,j \le n$. Hilbert matrices are notoriously ill-conditioned. An interesting fact about a Hilbert matrix is that every Hilbert matrix has an inverse consisting entirely of large integers. For $n \le 15$, the MATLAB command $H = \text{invhilb}(n)$ returns the exact inverse of the $n \times n$ Hilbert matrix. Use methods 1-3 for *QR* decomposition to produce Q_1R_1, Q_2R_2, and Q_3R_3 for the inverse Hilbert matrix of order 12. Each Q_i should be orthogonal. Test this by computing $\|Q_i\|_2$ and $\left\|Q_i^T Q_i - I\right\|$ for each method. Explain your results.

14.20 Let $A = \begin{bmatrix} -1 & 2 & 7 \\ -1 & 2 & 5 \\ 1 & -1 & 3 \end{bmatrix}$, and perform the following numerical experiment.

```
>> for i = 1:10
       [Q R] = modqrgrsch(A);
       A = R*Q;
   end
```

Examine *R*, and determine what it tells you about the original matrix *A*.

14.21 This problem tests the performance of MGS. Recall that Theorem 14.2 links the performance of MGS to the condition number of *A*.

a. Let *A* be the Rosser matrix by executing `A = rosser`. *A* is an 8×8 symmetric matrix with some very bad properties.

b. Find the condition number of `A`.

c. The function `hqr` in the software distribution computes the *QR* decomposition of an $m \times n$ matrix using what are termed *Householder reflections*. We will develop `hqr` in Chapter 17. Execute the following statements

```
>> [Q1, R1] = modqrgrsch(A);
>> [Q2, R2] = hqr(A);
>> norm(Q1'*Q1 - eye(8))
>> norm(Q2'*Q2 - eye(8))
```

and comment on the results.

d. We defined the Vandermonde matrix in Section 12.3, and it is implemented by the function `vandermonde` in the software distribution. Given a vector $x \in \mathbb{R}^n$ and an integer *m*, the Vandermonde matrix is of dimension $m \times (n+1)$. Normally, *m* is much larger than *n*. Explain the output of the following code:

```
>> x = rand(100,1);
>> V = vandermonde(x,25);
>> [Q1 R1] = modqrgrsch(V);
>> [Q2 R2] = qr(V,0);
>> norm(Q1'*Q1 - eye(26))
>> norm(Q2'*Q2 - eye(26))
```

14.22 Determining the rank of a matrix is a difficult problem, as will become evident in subsequent chapters. The *QR* decomposition can be used to determine the exact or approximate rank of a matrix. Column pivoting can be used in performing the *QR* decomposition, and the process is termed *rank revealing QR decomposition* [2, pp. 248-250]. A discussion of this method is beyond the scope of this book, but it is useful to know that it exists and to experiment with it. In MATLAB, there is a version of the function `qr` that has the calling sequence

```
[Q,R,E] = qr(A,0)
```

It produces a matrix `Q` with orthonormal columns, upper triangular `R`, and a permutation matrix `E` so that AE = QR. The rank of *A* is determined by counting the number of nonzero values on the diagonal of *R*. What is a nonzero value? Since roundoff error occurs, it is likely that an entry that should be exactly zero is small instead. We will assign a tolerance, tol, and any value less than or equal to tol will be considered zero. Portions of the following questions are taken from Ref. [23, Exercise 4.2.21, pp. 271-272].

a. The Kahan matrix $K_n(\theta)$ is an $n \times n$ upper triangular matrix whose entries are a function of θ. It provides an interesting problem in rank determination. Generate the 90×90 Kahan matrix with $\theta = 1.2$ using the MATLAB statement
```
K = gallery('kahan',90,1.2,0);
```
b. Enter
```
>> [Q R E] = qr(K,0);
>> tol = 1.0e-14;
>> rdiag = diag(R);
>> sum(rdiag > tol);
>> rank(K)
```

Is rank determination using the QR decomposition correct?

c. To ensure that the QR factorization with column pivoting does not interchange columns in the presence of rounding errors, the diagonal is perturbed by `pert*eps*diag([n:-1:1])`. Obtain a slightly perturbed version of K using the statement
```
>> K25 = gallery('kahan',90,1.2,25);
```

Enter the same sequence of statements as in part (b), replacing K by K25. Does the QR decomposition with column pivoting give the correct rank?

14.23 Example 14.3 provided evidence that the classical Gram-Schmidt method was error-prone. In this problem, you will repeat an expanded version of Example 14.3. Define $\epsilon = 0.5 \times 10^{-7}$ and build the matrix

$$A = \begin{bmatrix} 1 & 1 & 1 & 1 & 1 \\ \epsilon & 0 & 0 & 0 & 0 \\ 0 & \epsilon & 0 & 0 & 0 \\ 0 & 0 & \epsilon & 0 & 0 \\ 0 & 0 & 0 & \epsilon & 0 \\ 0 & 0 & 0 & 0 & \epsilon \end{bmatrix}.$$

Execute the following MATLAB command sequence. The function `givensqr` is in the software distribution and will be developed in Chapter 17.
```
>> [Q1 R1] = clqrgrsch(A);
>> [Q2 R2] = modqrgrsch(A);
>> [Q3 R3] = qr(A);
>> [Q4 R4] = givensqr(A);
```

For each decomposition, compute $\|Q_i^T Q_i - I\|$. Comment on the results.

14.24 This problem investigates what happens when the matrix A given as an argument to `modqrgrsch` is complex. The two primary components in the implementation of MGS are the inner product $\langle u, v \rangle$ and the 2-norm $\|u\|_2$. The remaining operations are arithmetic.

a. Enter the complex vectors

$$x = \begin{bmatrix} 1-i & 2+i & 3-2i \end{bmatrix}^T, \quad y = \begin{bmatrix} 8+i & i & 5+4i \end{bmatrix}^T$$

and compute the expression $\sum_{i=1}^{3} x_i y_i$. MATLAB will compute the inner product using the function `dot(x,y)`. Use `dot` with x and y. Do you obtain the same results?

b. Compute $\sum_{i=1}^{3} \overline{x_i} y_i$, where $\overline{x_i}$ is the complex conjugate of x_i. Use the MATLAB function `conj` to compute the complex conjugate. Does your result agree with `dot(x,y)`?

c. Based upon the your answers in parts (a) and (b), what is the definition for the inner product of two vectors $x, y \in \mathbb{C}^n$?

d. Define the norm of a vector $x \in \mathbb{C}^n$. Use your definition to find the norm of x and compare the result with the MATLAB statement `norm(x)`.

e. Run `modqrgrsch` with a 3×3 and a 5×3 complex matrix of your choosing. In each case, compare your result with that of the reduced QR decomposition computed using MATLAB's `qr`. If the results are very close, explain why, and if not explain why `modqrgrsch` fails with complex matrices.

14.25 We will see in Chapters 21 and 22 that there are important algorithms which are subject to problems when the vectors they are generating begin to lose orthogonality. A standard solution is to perform *reorthogonalization*. At point i in the algorithm, assume it has generated vectors $v_1, v_2, \ldots, v_i$ that should be orthogonal if exact arithmetic were used. Reorthogonalization performs operations on the vectors to improve orthogonality. This can be done with both the classical and MGS methods. In Ref. [45, p. 1071], the classical and MGS methods are discussed with and without reorthogonalization. Also see Ref. [23, pp. 231-233]. In this problem, we will only discuss the classical Gram-Schmidt method. The following listing is from the classical Gram-Schmidt method modified to compute the QR decomposition.

```
1:   for i = 1:n do
2:      Q(:,i) = A(:,i)
3:      sumproj = 0
4:      for j = 1:i-1 do
5:         R(j, i) = Q(:, j)ᵀ Q(:, i)
6:         sumproj = sumproj + R(j, i) Q(:, j)
7:      end for
8:      Q(:, i) = Q(:, i) − sumproj
9:      R(i, i) = ‖Q(:, i)‖₂
10:       Q(:, i) = Q(:, i)/R(i, i)
11:  end for
```

After steps 3-8, in exact arithmetic $Q(:, i)$ is orthogonal to $Q(:, 1 : i)$; however with roundoff and cancellation error present, it likely is not. We need to start with the already computed $Q(:, i)$ and repeat steps 3-8, hopefully improving orthogonality. The inner loop computed the terms $R(j, i) = Q(:, j)^T Q(:, i)$. When starting with $Q(:, i)$ and repeating steps 3-8, the values of $Q(:, j)^T Q(:, i)$ will be small and serve as corrections to the original values of R. They must be assigned to another matrix, S, and then added to R. The following statements perform reorthogonalization when placed between statements 8 and 9.

```
sumproj = 0
for j = 1:i-1 do
   S(j, i) = Q(:, j)ᵀ Q(:, i)
   sumproj = sumproj + S(j, i) Q(:, j)
end for
Q(:, i) = Q(:, i) − sumproj
R(1 : i − 1, i) = R(1 : i − 1, i) + S(1 : i − 1, i)
```

a. Modify clqrgrsch so it selectively performs reorthogonalization. Its calling format should be
`[Q,R] = clqrgrsch(A,reorthog)`. When reorthog = 1, perform reorthogonalization; otherwise, don't. If the argument is omitted, the default value should be 1. If you don't know how to deal with variable input arguments, use MATLAB help for `nargin`.

b. The matrix west0167 in the software distribution has dimension 167×167 and a condition number of 4.7852×10^{10}. It was involved in a chemical process simulation problem. Apply clqrgrsch to west0167 with and without reorthogonalization. In each case, compute $\|Q\|_2$ and $\|Q^TQ - I\|_2$. Explain the results.

c. Modify your `clqrgrsch` so it returns an array of values orthog $= \|Q(:, 1 : i)^T Q(:, 1 : i) - I^{i \times i}\|_2$, $1 \le i \le n$, in addition to Q and R. The array records how well orthogonality is maintained. Take the function modqrgrsch in the software distribution and do the same. Apply classical Gram-Schmidt, with and without reorthogonalization, and modqrgrsch to west0167. On the same set of axes, draw a `semilogy` graph of $m = 1, 2, \ldots, 167$ against orthog for each algorithm. Comment on the results.

Chapter 15

The Singular Value Decomposition

You should be familiar with

- Eigenvalues and eigenvectors
- Singular values
- Orthogonal matrices
- Matrix column space, row space, and null space
- The matrix 2-norm and Frobenius norm
- The matrix inverse

Matrix decompositions play a critical role in numerical linear algebra, and we have already see the *QR* decomposition, one of the great accomplishments in the field. The *singular value decomposition* (SVD) is also among the greatest results in linear algebra. Just like the *QR* decomposition, the SVD is a matrix decomposition that applies to any matrix, real, or complex. The SVD is a powerful tool for many matrix computations because it reveals a great deal about the structure of a matrix. A number of its many applications are listed in Section 15.7.2.

This chapter proves the SVD theorem but does not develop a useable algorithm for its computation. We will use the SVD from this point forward in the book and will see some of its powerful applications. Unfortunately, computing the SVD efficiently is quite difficult, and we will present two methods for its computation in Chapter 23. In the meantime, we will use the built-in MATLAB command `svd` to compute it.

15.1 THE SVD THEOREM

Recall that if A is an $m \times n$ matrix, then $A^{\mathrm{T}}A$ is an $n \times n$ symmetric matrix with nonnegative eigenvalues (Lemma 7.4). The singular values of an $m \times n$ matrix are the square roots of the eigenvalues of $A^{\mathrm{T}}A$, and the 2-norm of a matrix is the largest singular value. The SVD factors A into a product of two orthogonal matrices and a diagonal matrix of its singular values.

Theorem 15.1. *Let $A \in \mathbb{R}^{m \times n}$ be a matrix having r positive singular values, $m \geq n$. Then there exist orthogonal matrices $U \in R^{m \times m}, V \in \mathbb{R}^{n \times n}$, and a diagonal matrix $\widetilde{\Sigma} \in \mathbb{R}^{m \times n}$ such that*

$$A = U\tilde{\Sigma}V^{\mathrm{T}}$$
$$\tilde{\Sigma} = \begin{bmatrix} \Sigma & 0 \\ 0 & 0 \end{bmatrix},$$

where $\Sigma = \mathrm{diag}\,(\sigma_1, \sigma_2, \ldots, \sigma_r)$, and $\sigma_1 \geq \sigma_2 \geq \cdots \geq \sigma_r > 0$ are the positive singular values of A.

Overview:

The proof is by construction. Build the $m \times n$ matrix $\widetilde{\Sigma} = \begin{bmatrix} \Sigma & 0 \\ 0 & 0 \end{bmatrix}$ by placing the positive singular values on the diagonal of Σ, so

$$\Sigma = \begin{bmatrix} \sigma_1 & & & \\ & \sigma_2 & & \\ & & \ddots & \\ & & & \sigma_r \end{bmatrix}.$$

Find an orthonormal basis v_i, $1 \leq i \leq n$ of eigenvectors of $A^{\mathrm{T}}A$. This can be done because $A^{\mathrm{T}}A$ is symmetric (Theorem 7.6, the spectral theorem). Then, $V = \begin{bmatrix} v_1 & v_2 & \ldots & v_n \end{bmatrix}$. Build the orthogonal matrix U using A, v_i, and σ_i.

Numerical Linear Algebra with Applications. http://dx.doi.org/10.1016/B978-0-12-394435-1.00015-6

Proof. The matrix $A^T A$ is symmetric and by Lemma 7.4, its eigenvalues are real and nonnegative. Listing the eigenvalues in descending order we obtain

$$\sigma_1^2 \geq \sigma_2^2 \geq \cdots \geq \sigma_r^2 \geq \sigma_{r+1}^2 \geq \cdots \geq \sigma_n^2 \geq 0.$$

Assume that the first r eigenvalues are positive, and the eigenvalues $\sigma_{r+1}^2 = \sigma_{r+2}^2 = \cdots \sigma_n^2 = 0$. Let

$$\widetilde{\Sigma} = \begin{bmatrix} \Sigma & 0 \\ 0 & 0 \end{bmatrix},$$

where $\Sigma = \mathrm{diag}(\sigma_1, \sigma_2, \ldots, \sigma_r)$ is a diagonal matrix of singular values. The spectral theorem (Theorem 7.6) guarantees that $A^T A$ has an orthonormal basis for $\mathbb{R}^n$ of eigenvectors. Let

$$V = \begin{bmatrix} v_1 & v_2 & \ldots & v_{n-1} & v_n \end{bmatrix},$$

and it turns out this is the matrix we are looking for. We now need to find U. The matrix U must be orthogonal, so its columns will form a basis for $\mathbb{R}^m$. Let

$$u_i = \frac{A v_i}{\sigma_i}, \quad 1 \leq i \leq r.$$

These vectors are orthonormal since

$$\langle u_i, u_j \rangle = \frac{(A v_i)^T (A v_j)}{\sigma_i \sigma_j} = \frac{(\sigma_i v_i)^T (\sigma_j v_j)}{\sigma_i \sigma_j} = v_i^T v_j = \langle v_i, v_j \rangle = \begin{cases} 0 & i \neq j \\ 1 & i = j \end{cases}.$$

If $r < m$, we still need $m - r$ additional vectors $\{ u_{r+1} \ u_{r+1} \ \ldots \ u_{m-1} \ u_m \}$ so that $\{ u_1 \ u_2 \ \ldots \ u_{m-1} \ u_m \}$ forms an orthonormal set. Beginning with

$$u_1, \ u_2, \ \ldots \ u_{r-1}, \ u_r$$

use a Gram-Schmidt algorithm step to add the standard basis vector e_{r+1} to the set to obtain the orthonormal set

$$u_1, \ u_2, \ \ldots \ u_{r-1}, \ u_r, \ u_{r+1} .$$

Continue by adding in the same fashion $e_{r+2}, e_{r+3}, \ldots, e_m$ to obtain the basis

$$u_1, \ u_2, \ \ldots \ u_{m-1}, \ u_m$$

for $\mathbb{R}^m$ and the matrix $U = \begin{bmatrix} u_1 & u_2 & \ldots & u_{m-1} & u_m \end{bmatrix}$. Now we need to show that $A = U \widetilde{\Sigma} V^T$, or $U^T A V = \widetilde{\Sigma}$. Before we begin, note two things:

a. The vectors $v_1, \ v_2, \ \ldots \ , v_r$ are the nonzero eigenvectors of $A^T A$, so $A^T A v_i = 0$, $r + 1 \leq i \leq n$. Multiply by v_i^T to get

$$v_i^T A^T A v_i = 0,$$

so

$$(A v_i)^T (A v_i) = \| A v_i \|_2^2 = 0,$$

and $A v_i = 0$, $r + 1 \leq i \leq n$.

b. Write $U^T = \begin{bmatrix} u_1 \\ u_2 \\ \vdots \\ u_r \\ u_{r+1} \\ \vdots \\ u_m \end{bmatrix}$, where each u_i is a row of U^T. Since $u_1, u_2, \ldots, u_r$ are an orthonormal set, it follows that

$U^T u_i = e_i$, $1 \leq i \leq r$, and so $\sigma_i U^T u_i = \sigma_i e_i$, $1 \leq i \leq r$.

Now continue.

$$U^T A V = U^T A \begin{bmatrix} v_1 & v_2 & \ldots & v_{n-1} & v_n \end{bmatrix} = U^T \begin{bmatrix} A v_1 & A v_2 & \ldots & A v_{n-1} & A v_n \end{bmatrix} = \tag{15.1}$$

$$U^T \begin{bmatrix} \sigma_1 u_1 & \sigma_2 u_2 & \ldots & \sigma_r u_r & A v_{r+1} & \ldots & A v_n \end{bmatrix} = \tag{15.2}$$

$$U^T \begin{bmatrix} \sigma_1 u_1 & \sigma_2 u_2 & \dots & \sigma_r u_r & 0 & \dots & 0 \end{bmatrix} = \tag{15.3}$$

$$\begin{bmatrix} \sigma_1 U^T u_1 & \sigma_2 U^T u_2 & \dots & \sigma_r U^T u_r & 0 & \dots & 0 \end{bmatrix} = \tag{15.4}$$

$$\begin{bmatrix} \sigma_1 e_1 & \sigma_2 e_2 & \dots & \sigma_r e_r & 0 & \dots & 0 \end{bmatrix} = \tilde{\Sigma} \tag{15.5}$$

$\square$

Remark 15.1.

a. There is no loss of generality in assuming that $m \geq n$, for if $m < n$, find the SVD for A^T and transpose back. We have

$$A^T = U\tilde{\Sigma}V^T$$

so

$$A = V\tilde{\Sigma}U^T,$$

and we have an SVD for A.

The columns of U and V are called the *left* and *right singular vectors*, respectively. The largest and smallest singular values are denoted, respectively, as $\sigma_{\max}$ and $\sigma_{\min}$.

Example 15.1. The matrix $A = \begin{bmatrix} 2 & 2 \\ -1 & 1 \end{bmatrix}$ has SVD

$$\begin{bmatrix} 2 & 2 \\ -1 & 1 \end{bmatrix} = \begin{bmatrix} 1 & 0 \\ 0 & 1 \end{bmatrix} \begin{bmatrix} 2\sqrt{2} & 0 \\ 0 & \sqrt{2} \end{bmatrix} \begin{bmatrix} 1/\sqrt{2} & 1/\sqrt{2} \\ -1/\sqrt{2} & 1/\sqrt{2} \end{bmatrix}$$

We see that the columns of U and V have unit length since $U = I$, and

$$\left(\frac{1}{\sqrt{2}}\right)^2 + \left(\frac{1}{\sqrt{2}}\right)^2 = 1.$$

A simple calculation of inner products will show the columns of U and V are mutually orthogonal. ∎

Example 15.2. Let $A = \begin{bmatrix} 1 & -1 & 3 \\ 1 & 0 & 1 \\ 1 & 2 & 0 \end{bmatrix}$ and $B = \begin{bmatrix} 1 & 1 & -1 \\ 1 & 0 & 2 \\ 2 & 1 & 1 \end{bmatrix}$. Here are SVDs for each matrix:

$$A = \begin{bmatrix} -0.9348 & 0.0194 & 0.3546 \\ 0.3465 & -0.2684 & -0.8988 \\ 0.0778 & -0.9631 & 0.2577 \end{bmatrix} \begin{bmatrix} 3.5449 & 0 & 0 \\ 0 & 2.3019 & 0 \\ 0 & 0 & 0.3676 \end{bmatrix} \begin{bmatrix} -0.3395 & 0.3076 & -0.8889 \\ -0.5266 & -0.8452 & -0.0913 \\ -0.7794 & 0.4371 & 0.4489 \end{bmatrix},$$

$$r = 3, \sigma_1 = 3.5449, \sigma_2 = 2.3019, \sigma_3 = 0.3676$$

$$B = \begin{bmatrix} -0.1355 & 0.8052 & -0.5774 \\ -0.6295 & -0.5199 & -0.5774 \\ -0.7651 & 0.2852 & 0.5774 \end{bmatrix} \begin{bmatrix} 3.1058 & 0 & 0 \\ 0 & 2.0867 & 0 \\ 0 & 0 & 0.0000 \end{bmatrix} \begin{bmatrix} -0.7390 & -0.2900 & -0.6081 \\ 0.4101 & 0.5226 & -0.7475 \\ 0.5345 & -0.8018 & -0.2673 \end{bmatrix},$$

$$r = 2, \sigma_1 = 3.1058, \sigma_2 = 2.0867$$

Note: The rank of A is 3, and the rank of B is 2. ∎

Example 15.3. Consider the matrix

$$A = \begin{bmatrix} 1 & 1 & 1 & 1 & 1 \\ 0 & 0 & 0 & 0 & 0 \\ -1 & -1 & -1 & -1 & -1 \\ 0 & 0 & 0 & 0 & 0 \\ 1 & 1 & 1 & 1 & 1 \\ -1 & -1 & -1 & -1 & -1 \\ 0 & 0 & 0 & 0 & 0 \end{bmatrix}$$

An SVD is

$$
U = \begin{bmatrix}
-0.5000 & -0.8660 & 0 & 0.0000 & 0 & 0.0000 & 0 \\
0.0000 & 0.0000 & 1 & 0.0000 & 0 & 0.0000 & 0 \\
0.5000 & -0.2887 & 0 & 0.8165 & 0 & 0.0000 & 0 \\
0.0000 & 0.0000 & 0 & 0.0000 & -1 & 0.0000 & 0 \\
-0.5000 & 0.2887 & 0 & 0.4082 & 0 & 0.7071 & 0 \\
0.5000 & -0.2887 & 0 & -0.4082 & 0 & 0.7071 & 0 \\
0.0000 & 0.0000 & 0 & 0.0000 & 0 & 0.0000 & 1
\end{bmatrix},
$$

$$
\widetilde{\Sigma} = \begin{bmatrix}
4.4721 & 0 & 0 & 0 & 0 \\
0 & 0 & 0 & 0 & 0 \\
0 & 0 & 0 & 0 & 0 \\
0 & 0 & 0 & 0 & 0 \\
0 & 0 & 0 & 0 & 0 \\
0 & 0 & 0 & 0 & 0 \\
0 & 0 & 0 & 0 & 0
\end{bmatrix},
$$

$$
V = \begin{bmatrix}
-.4472 & -0.3651 & -0.6712 & 0.4614 & 0.0581 \\
-.4472 & -0.3651 & -0.0671 & -0.8074 & -0.1016 \\
-.4472 & 0.5477 & 0.0000 & -0.0883 & 0.7016 \\
-.4472 & -0.3651 & 0.7383 & 0.3460 & 0.0435 \\
-.4472 & 0.5477 & 0.0000 & 0.0883 & -0.7016
\end{bmatrix}
$$

$$
\Sigma = \begin{bmatrix} 4.4721 \end{bmatrix}, \sigma_1 = 4.4721, r = 1
$$

Note: The rank of A is 1. ∎

If the singular values of A are placed in descending order in $\widetilde{\Sigma}$, then $\widetilde{\Sigma}$ is unique; however, U and V are not in general. In the proof of Theorem 15.1, we extended the orthonormal basis

$$
\left\{ u_1 \quad u_2 \quad \ldots \quad u_{r-1} \quad u_r \right\}
$$

to an orthonormal basis for $\mathbb{R}^m$. This can be done in many ways. For the matrix of Example 15.3, the matrices

$$
U = \begin{bmatrix}
-0.5000 & 0.8660 & 0.0000 & 0 & 0.0000 & 0.0000 & 0 \\
0.0000 & 0.0000 & -1.0000 & 0 & 0.0000 & 0.0000 & 0 \\
0.5000 & 0.2887 & 0.0000 & 0 & 0.5774 & -0.5774 & 0 \\
0.0000 & 0.0000 & 0.0000 & 1 & 0.0000 & 0.0000 & 0 \\
-0.5000 & -0.2887 & 0.0000 & 0 & 0.7887 & 0.2113 & 0 \\
0.5000 & 0.2887 & 0.0000 & 0 & 0.2113 & 0.7887 & 0 \\
0.0000 & 0.0000 & 0.0000 & 0 & 0.0000 & 0.0000 & 1
\end{bmatrix}, V = \begin{bmatrix}
-0.4472 & 0.3651 & 0.0000 & -0.5774 & -0.5774 \\
-0.4472 & 0.3651 & 0.0000 & -0.2113 & -0.7887 \\
-0.4472 & -0.5477 & 0.7071 & 0.0000 & 0.0000 \\
-0.4472 & 0.3651 & 0.0000 & 0.7887 & -0.2113 \\
-0.4472 & -0.5477 & -0.7071 & 0.0000 & 0.0000
\end{bmatrix}
$$

also produce a valid SVD.

Remark 15.2. If A is square and the σ_i are distinct, then u_i and v_i are uniquely determined except for sign.

15.2 USING THE SVD TO DETERMINE PROPERTIES OF A MATRIX

The *rank* of a matrix is the number of linearly independent columns or rows. Notice that in Example 15.2, the matrix A has three nonzero singular values, and the matrix B has two. The matrix of Example 15.3 has only one nonzero singular value. The rank of the matrices is 3, 2, and 1, respectively. This is not an accident. The rank of a matrix is the number of nonzero singular values in $\widetilde{\Sigma}$.

We now present a theorem dealing with rank that we have not needed until now. The result will allow us to show the relationship between the rank of a matrix and its singular values.

Theorem 15.2. *If A is an $m \times n$ matrix, X is an invertible $m \times m$ matrix, and Y is an invertible $n \times n$ matrix, then* rank (XAY) = rank (A).

Proof. Since X is invertible, it can be written as a product of elementary row matrices, so $X = E_k^{(X)}E_{k-1}^{(X)}\ldots E_2^{(X)}E_1^{(X)}$. Similarly, Y is a product of elementary row matrices, $Y = E_p^{(Y)}E_{p-1}^{(Y)}\ldots E_2^{(Y)}E_1^{(Y)}$, and so

$$XAY = E_k^{(X)}E_{k-1}^{(X)}\ldots E_2^{(X)}E_1^{(X)}AE_p^{(Y)}E_{p-1}^{(Y)}\ldots E_2^{(Y)}E_1^{(Y)}.$$

The product of the elementary row matrices on the left performs elementary row operations on A, and this does not change the rank of A. The product of elementary row matrices on the left perform elementary column operations, which also do not alter rank. Thus, rank (XAY) = rank (A). $\qquad\square$

Theorem 15.3. *The rank of a matrix A is the number of nonzero singular values.*

Proof. Let $A = U\widetilde{\Sigma}V^{\mathrm{T}}$ be the SVD of A. Orthogonal matrices are invertible, so by Theorem 1.2,

$$\text{rank}\,(A) = \text{rank}\,\left(U\widetilde{\Sigma}V^{\mathrm{T}}\right) = \text{rank}\,\left(\widetilde{\Sigma}\right).$$

The rank of $\widetilde{\Sigma}$ is r, since

$$\begin{bmatrix}\sigma_1 & 0 & 0 & \ldots & 0\end{bmatrix}^{\mathrm{T}},\quad \begin{bmatrix}0 & \sigma_2 & 0 & \ldots & 0\end{bmatrix}^{\mathrm{T}},\quad \begin{bmatrix}0 & 0 & \sigma_3 & \ldots & 0\end{bmatrix}^{\mathrm{T}},\quad \begin{bmatrix}0 & 0 & \ldots & \sigma_r & 0 & \ldots & 0\end{bmatrix}^{\mathrm{T}}$$

is a basis for the column space of $\widetilde{\Sigma}$. $\qquad\square$

From the components of the SVD, we can determine other properties of the original matrix. Recall that the *null space* of a matrix A, *written* null (A), is the set of *vectors x* for which $Ax = 0$, and the *range* of A is the set all linear combinations of the columns of A (the column space of A). Let u_i, $1 \le i \le m$ and v_i, $1 \le i \le n$ be the column vectors of U and V, respectively. Then

$$Av_i = U\widetilde{\Sigma}V^{\mathrm{T}}v_i.$$

The matrix V^{T} can be written as $\begin{bmatrix}v_1^{\mathrm{T}}\\ \vdots\\ v_i^{\mathrm{T}}\\ \vdots\\ v_n^{\mathrm{T}}\end{bmatrix}$, where the v_i are the orthonormal columns of V. The product $V^{\mathrm{T}}v_i = \begin{bmatrix}v_1^{\mathrm{T}}\\ \vdots\\ v_i^{\mathrm{T}}\\ \vdots\\ v_n^{\mathrm{T}}\end{bmatrix}v_i = e_i$,

where $e_i = \begin{bmatrix}0\\ \vdots\\ 1\\ 0\\ \vdots\\ 0\end{bmatrix}$ is the *i*th standard basis vector in $\mathbb{R}^n$. Now,

$$\widetilde{\Sigma}e_i = \begin{bmatrix}\sigma_1 & 0 & \cdots & 0 & \cdots & 0 & 0 & 0\\ 0 & \sigma_2 & \cdots & 0 & \cdots & 0 & 0 & 0\\ 0 & 0 & \cdots & \sigma_i & \cdots & \vdots & \cdots & \vdots\\ 0 & 0 & \cdots & 0 & \ddots & \vdots & \cdots & \vdots\\ \vdots & \vdots & \vdots & \vdots & \cdots & \sigma_r & \cdots & 0\\ \vdots & \vdots & \vdots & \vdots & \cdots & \vdots & \ddots & \vdots\\ 0 & 0 & \cdots & 0 & \cdots & 0 & \cdots & 0\end{bmatrix}e_i = \begin{bmatrix}0\\ \vdots\\ \sigma_i\\ 0\\ \vdots\\ \vdots\\ 0\end{bmatrix},\ i \le r,$$

and

$$
U \begin{bmatrix} 0 \\ \vdots \\ \vdots \\ \sigma_i \\ \vdots \\ \vdots \\ 0 \end{bmatrix} = \begin{bmatrix} u_{11} & u_{12} & \cdots & u_{1r} & \cdots & u_{1m} \\ u_{21} & u_{22} & \cdots & u_{2r} & \cdots & u_{2m} \\ \vdots & \vdots & \ddots & \vdots & \cdots & \vdots \\ \vdots & \vdots & & u_{rr} & \cdots & u_{rm} \\ \vdots & \vdots & \cdots & \vdots & \ddots & \vdots \\ u_{m1} & u_{m2} & \cdots & u_{mr} & \cdots & u_{mm} \end{bmatrix} \begin{bmatrix} 0 \\ \vdots \\ 0 \\ \sigma_i \\ \vdots \\ 0 \end{bmatrix} = \sigma_i u_i, \quad 1 \le i \le r.
$$

For v_i, $r + 1 \le i \le m$, we have

$$
Av_i = U\widetilde{\Sigma}e_i = \begin{bmatrix} u_{11} & \cdots & u_{1i} & \cdots & u_{1m} \\ u_{21} & \cdots & u_{2i} & \cdots & u_{2m} \\ \vdots & \ddots & \vdots & \cdots & \vdots \\ \vdots & \cdots & u_{ii} & \cdots & u_{im} \\ \vdots & \cdots & \vdots & \cdots & \vdots \\ \vdots & \cdots & \vdots & \ddots & \vdots \\ \vdots & \vdots & \vdots & \vdots & \vdots \\ u_{m1} & \cdots & u_{mi} & \cdots & u_{mm} \end{bmatrix} \begin{bmatrix} \sigma_1 & \cdots & 0 & \cdots & \cdots & 0 & \cdots & 0 \\ \vdots & \ddots & \vdots & \cdots & \vdots & & \cdots & \vdots \\ 0 & \cdots & \sigma_r & \ddots & \vdots & & \cdots & 0 \\ \vdots & \vdots & \vdots & 0 & \cdots & \cdots & & \vdots \\ \vdots & \vdots & \vdots & \vdots & \ddots & \vdots & & \vdots \\ 0 & \cdots & 0 & 0 & 0 & 0 & \cdots & \vdots \\ \vdots & \cdots & \vdots & \vdots & \vdots & \vdots & \ddots & \vdots \\ 0 & \cdots & 0 & 0 & 0 & 0 & \cdots & 0 \end{bmatrix} \begin{bmatrix} 0 \\ \vdots \\ 0 \\ \vdots \\ \vdots \\ 1 \\ \vdots \\ 0 \end{bmatrix} = 0
$$

and $Av_i = 0$, $r + 1 \le i \le m$.

In summary, we have

$$
Av_i = \sigma_i u_i, \quad \sigma_i \neq 0, \ 1 \le i \le r
$$

$$
Av_i = 0, \quad r + 1 \le i \le n
$$

Since U and V are orthogonal matrices, all u_i and v_i are linearly independent. For $1 \le i \le r$, $Av_i = \sigma_i u_i$, $\sigma_i \neq 0$, and u_i, $1 \le i \le r$ is in the range of A. Since by Theorem 15.2 the rank of A is r, the u_i are a basis for the range of A. For $r + 1 \le i \le n$, $Av_i = 0$, so v_i is in null (A). Since rank (A) + nullity $(A) = n$, nullity $(A) = n - r$. There are $n - (r + 1) + 1 = n - r$ orthogonal vectors v_i, so the v_i, $r + 1 \le i \le n$, are a basis for the null space of A.

Example 15.4. Let $B = \begin{bmatrix} 1 & 1 & -1 \\ 1 & 0 & 2 \\ 2 & 1 & 1 \end{bmatrix}$ be the matrix in Example 15.2. From the SVD, the vectors $\begin{bmatrix} -0.1355 \\ -0.6295 \\ -0.7651 \end{bmatrix}$ and $\begin{bmatrix} 0.8052 \\ -0.5199 \\ 0.2852 \end{bmatrix}$ are a basis for the range of B, and the vector $\begin{bmatrix} 0.5345 \\ -0.8018 \\ -0.2673 \end{bmatrix}$ is a basis for the null space of B. Remember when looking at the decomposition of B, V^{T} appears, not V. ∎

15.2.1 The Four Fundamental Subspaces of a Matrix

There are four fundamental subspaces associated with an $m \times n$ matrix. We have seen two, the range and the null space, and have determined how compute a basis for each using the SVD. The other two subspaces are the range and the null space of

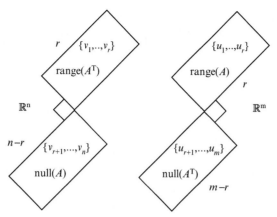

FIGURE 15.1 The four fundamental subspaces of a matrix.

A^T. If we take the transpose of the SVD for A, the result is

$$A^T = V \widetilde{\Sigma} U^T. \tag{15.6}$$

Applying the same procedure that we used to determine an orthonormal basis for the range of A to Equation 15.6, it follows that v_i, $1 \le i \le r$ is a basis for the range of A^T. Note that the range of A^T is the row space of A. We already know that v_i, $r+1 \le i \le n$, is a basis for null (A). Since all the v_i are orthogonal, it follows that the vectors in range (A^T) are orthogonal to the vectors in null (A). Again using Equation 15.6, we see that u_i, $r + 1 \le i \le m$ is an orthonormal basis for null (A^T). We have shown that u_i, $1 \le i \le r$ is a basis for range (A). Thus, range (A) is orthogonal to null (A^T). Table 15.1 summarizes the four fundamental subspaces, and Figure 15.1 provides a graphical depiction. In the figure, the symbol $\square$ indicates the subspaces are orthogonal.

   We have stated a number of times that the dimension of the column space and row space are equal, and now we can prove it.

Theorem 15.4.   *The dimension of the column space and the dimension of the row space of a matrix are equal and is called the rank of the matrix.*

Proof. Our discussion of the SVD has shown that if r is the number of nonzero singular values, u_i, $1 \le i \le r$ is a basis for the range of A, and v_i, $1 \le i \le r$ is a basis for the range of A^T, which is the row space of A. $\square$

Example 15.5.   Let $A = \begin{bmatrix} 1 & 4 & 2 \\ -1 & 0 & 2 \\ 5 & -1 & -11 \\ 0 & 2 & 2 \\ 1 & 1 & -1 \end{bmatrix}$. The SVD of A is

TABLE 15.1 The Four Fundamental Subspaces of a Matrix

	Range	Null Space
A	$u_i, 1 \le i \le r$	$v_i, r+1 \le i \le n$
A^T	$v_i, 1 \le i \le r$	$u_i, r+1 \le i \le m$

$$A = \begin{bmatrix} -0.1590 & -0.8589 & -0.3950 & 0.1461 & 0.2443 \\ -0.1740 & 0.0738 & 0.4983 & 0.6089 & 0.5876 \\ 0.9532 & -0.1726 & 0.0759 & 0.2364 & -0.0054 \\ -0.1665 & -0.3926 & 0.5709 & 0.1858 & -0.6766 \\ 0.0907 & -0.2701 & 0.5139 & -0.7194 & 0.3705 \end{bmatrix} \begin{bmatrix} 12.6906 & 0 & 0 \\ 0 & 4.7905 & 0 \\ 0 & 0 & 0.0000 \\ 0 & 0 & 0 \\ 0 & 0 & 0 \end{bmatrix} \begin{bmatrix} 0.3839 & -0.1443 & -0.9120 \\ -0.4312 & -0.9014 & -0.0389 \\ 0.8165 & -0.4082 & 0.4082 \end{bmatrix},$$

so $r = 2$. Using Table 15.1, the four fundamental subspaces have orthonormal bases as follows:

$$\text{range}\,(A) = \left\{ \begin{bmatrix} -0.1590 \\ -0.1740 \\ 0.9532 \\ -0.1665 \\ 0.0907 \end{bmatrix}, \begin{bmatrix} -0.8589 \\ 0.0738 \\ -0.1726 \\ -0.3926 \\ -0.2701 \end{bmatrix} \right\}, \quad \text{null}(A^T) = \left\{ \begin{bmatrix} -0.3950 \\ 0.4983 \\ 0.0759 \\ 0.5709 \\ 0.5139 \end{bmatrix}, \begin{bmatrix} 0.1461 \\ 0.6089 \\ 0.2364 \\ 0.1858 \\ -0.7194 \end{bmatrix}, \begin{bmatrix} 0.2443 \\ 0.5876 \\ -0.0054 \\ -0.6766 \\ 0.3705 \end{bmatrix} \right\}$$

$$\text{range}(A^T) = \left\{ \begin{bmatrix} 0.3839 \\ -0.1443 \\ -0.9120 \end{bmatrix}, \begin{bmatrix} -0.4312 \\ -0.9014 \\ -0.0389 \end{bmatrix} \right\}, \quad \text{null}\,(A) = \left\{ \begin{bmatrix} 0.8165 \\ -0.4082 \\ 0.4082 \end{bmatrix} \right\}$$

$\blacksquare$

15.3 SVD AND MATRIX NORMS

The SVD provides a means of computing the 2-norm of a matrix, since $\|A\|_2 = \sqrt{\sigma_1}$. If A is invertible, then $\|A^{-1}\|_2 = \sqrt{\frac{1}{\sigma_n}}$. The SVD can be computed accurately, so using it is an effective way to find the 2-norm. The SVD also provides a means of computing the Frobenius norm.

Remark 15.3. If M is a general matrix (Problem 7.12),

$$\|A\|_F^2 = \text{trace}\left(A^T A\right) = \text{trace}\left(A A^T\right).$$

There is means of computing the Frobenius norm using the singular values of matrix A. Before developing the formula, we need to prove the invariance of the Frobenius norm under multiplication by orthogonal matrices.

Lemma 15.1. *If U is an $m \times m$ orthogonal matrix, and V is an $n \times n$ orthogonal matrix, then $\|UAV\|_F^2 = \|A\|_F^2$.*

Proof.

$$\|UA\|_F^2 = \text{trace}\left((UA)^T (UA)\right) = \text{trace}\left((A^T U^T)(UA)\right) = \text{trace}\left(A^T IA\right) = \text{trace}\left(A^T A\right) = \|A\|_F^2,$$

showing that the Frobenius norm is invariant under left multiplication by an orthogonal matrix. Now,

$$\|AV\|_F^2 = \text{trace}\left((AV)(AV)^T\right) = \text{trace}\left((AV)(V^T A^T)\right) = \text{trace}\left(A A^T\right) = \|A\|_F^2,$$

so the Frobenius norm is invariant under right multiplication by an orthogonal matrix. Now form the complete product.

$$\|UAV\|_F^2 = \|U(AV)\|_F^2 = \|AV\|_F^2 = \|A\|_F^2. \qquad \square$$

Theorem 15.5. $\|A\|_F = \left(\sum_{i=1}^r \sigma_i^2\right)^{\frac{1}{2}}$

Proof. By the SVD, there exist orthogonal matrices U and V such that $A = U\widetilde{\Sigma}V^T$. Then, $\|A\|_F = \|U\widetilde{\Sigma}V^T\|_F = \|\widetilde{\Sigma}\|_F$ by Lemma 15.1. The only nonzero entries in $\widetilde{\Sigma}$ are the singular values $\sigma_1, \sigma_2, \ldots, \sigma_r$, so $\|A\|_F = \left(\sum_{i=1}^r \sigma_i^2\right)^{\frac{1}{2}}$. $\square$

15.4 GEOMETRIC INTERPRETATION OF THE SVD

Multiplying a vector x by a matrix A stretches or contracts the vector. The singular values of A significantly add to a geometric understanding of the linear transformation Ax. For $x = \begin{bmatrix} x_1 \\ \vdots \\ x_m \end{bmatrix} \in \mathbb{R}^m$, the m-dimensional *unit sphere* is defined by

$$\sum_{i=1}^m x_i^2 = 1$$

If A is an $m \times n$ matrix, the product Ax takes a vector $x \in \mathbb{R}^n$ and produces a vector in $\mathbb{R}^m$. In particular, we will look at the subset of vectors $y = Ax$ in $\mathbb{R}^m$ as x varies over the unit sphere, $\|x\|_2 = 1$. The matrix A has n singular values $\sigma_1 \geq \sigma_2 \geq \cdots \geq \sigma_r > 0, \sigma_i = 0, r + 1 \leq i \leq n$ and $A = U\widetilde{\Sigma}V^T$, so we will investigate

$$Ax = U\widetilde{\Sigma}V^T x, \quad \sum_{i=1}^m x_i^2 = 1.$$

If K is an $n \times n$ orthogonal matrix and x is on the unit sphere, then

$$\langle Kx, Kx \rangle = (Kx)^T Kx = x^T K^T Kx = x^T Ix = x^T x = 1$$

Thus, the orthogonal linear transformation Kx maps x to another vector on the unit sphere. Now choose any $x \in \mathbb{R}^n$ on the unit sphere. Since the columns $k_i, 1 \leq i \leq n$, of K are an orthonormal basis for $\mathbb{R}^n$,

$$x = c_1 k_1 + c_2 k_2 + \cdots + c_p k_p = K \begin{bmatrix} c_1 \\ \vdots \\ c_p \end{bmatrix}.$$

Then, $\|x\|_2^2 = \sum_{i=1}^p c_i^2 = 1$, so $\begin{bmatrix} c_1 \\ \vdots \\ c_p \end{bmatrix}$ is on the unit sphere. Thus, Kx maps out the unit sphere, $\|x\|_2 = 1$, in $\mathbb{R}^p$.

Assume $\|x\|_2 = 1$. Since the $n \times n$ matrix V^T in the SVD of A is an orthogonal matrix, $y = V^T x$ is on the unit sphere in $\mathbb{R}^n$, and

$$\widetilde{\Sigma}y = \begin{bmatrix} \sigma_1 & & & & 0 \\ & \ddots & & & \\ & & \sigma_r & & \\ & & & 0 & \\ & & & & \ddots \\ 0 & & & & 0 \end{bmatrix} \begin{bmatrix} y_1 \\ \vdots \\ y_n \end{bmatrix} = \begin{bmatrix} \sigma_1 y_1 \\ \vdots \\ \sigma_r y_r \\ 0 \\ \vdots \\ 0 \end{bmatrix} = \begin{bmatrix} x_1' \\ \vdots \\ x_r' \\ 0 \\ \vdots \\ 0 \end{bmatrix}$$

Now,

$$\left(\frac{x_1'}{\sigma_1}\right)^2 + \left(\frac{x_2'}{\sigma_2}\right)^2 + \cdots + \left(\frac{x_r'}{\sigma_r}\right)^2 = 1,$$

which is an r-dimensional *ellipsoid* in $\mathfrak{R}^m$ with semiaxes of length $\sigma_i, 1 \leq i \leq r$. For instance, if $m = 3$ and $r = 2$, then we have a circle in 3-dimensional space. If $m = 3$ and $r = 3$, then the surface is an ellipsoid. We still have to account for multiplication by U. Since U is an orthogonal matrix, U causes a change of orthonormal basis, or a rotation.

Summary If A is an $m \times n$ matrix, then Ax applied to the unit sphere $\|x\|_2 \leq 1$ in $\mathbb{R}^n$ is a rotated ellipsoid in $\mathbb{R}^m$ with semiaxes $\sigma_i, 1 \leq i \leq r$, where the σ_i are the nonzero singular values of A. '

Example 15.6. We will illustrate the geometric interpretation of the SVD with a MATLAB function `svdgeom` in the software distribution that draws the unit circle, $x_1^2 + x_2^2 = 1$, in the plane, computes Ax for values of x, and draws the resulting ellipse. The function then outputs the SVD.

The linear transformation used for the example is $A = \begin{bmatrix} 1.50 & 0.75 \\ -0.50 & -1.00 \end{bmatrix} \begin{bmatrix} x \\ y \end{bmatrix}$. Note that the semiaxes of the ellipse are 1.9294 and 0.5831 (Figure 15.2).

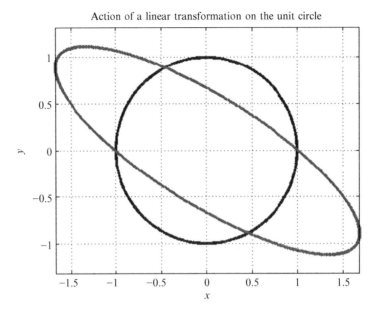

FIGURE 15.2 SVD rotation and distortion.

```
>> svdgeom(A)
The singular value decomposition for A is
U =
   -0.8550    0.5187
    0.5187    0.8550
S =
    1.9294         0
         0    0.5831
V =
   -0.7991    0.6012
   -0.6012   -0.7991
```

■

15.5   COMPUTING THE SVD USING MATLAB

Although it is possible to compute the SVD by using the construction in Theorem 15.1, this is far too slow and prone to roundoff error, so it should not be used. The MATLAB function svd computes the SVD:

a. [U S V] = svd(A)
b. S = svd(A)

Form 2 returns only the singular values in descending order in the vector S.

Example 15.7.   Find the SVD for the matrix of Example 15.5. Notice that $\sigma_3 = 4.8021 \times 10^{-16}$ and yet the rank is 2. In this case, the true value is 0, but roundoff error caused svd to return a very small singular value. MATLAB computes the rank using the SVD, so rank decided that σ_3 is actually 0.

```
>> [U,S,V] = svd(A)

U =
    0.15897      0.8589    -0.2112    -0.3642    -0.24447
    0.17398    -0.073783    -0.9063     0.22223    0.30581
   -0.95316      0.17263    -0.18552    0.14793   -0.073387
    0.16648      0.39256     0.21339    0.87898   -0.0063535
   -0.090746     0.27006     0.23251   -0.15324    0.91722
```

```
S =
        12.691              0              0
            0         4.7905              0
            0              0     4.8021e-16
            0              0              0
            0              0              0
V =
      -0.38387        0.43125         0.8165
       0.1443        0.90139       -0.40825
       0.91204       0.038895        0.40825

>> rank(A)

ans =
     2
```
■

The function svd applies equally well to a matrix of dimension $m \times n$, $m < n$. Of course, in this case the rank does not exceed m.

Example 15.8. Let $A = \begin{bmatrix} 7 & 9 & -5 & 10 & 10 & -8 \\ 9 & 3 & 1 & -7 & 0 & -2 \\ -8 & 8 & 10 & 10 & 6 & 9 \end{bmatrix}$

```
>> [U S V] = svd(A)

U =
      -0.42586       -0.89303       -0.14539
       0.26225       -0.27562        0.9248
      -0.86595        0.35571        0.35157

S =
        22.577              0              0              0              0              0
            0         20.176              0              0              0              0
            0              0         9.5513              0              0              0

V =
       0.27935       -0.57383         0.4704        0.60332       0.008204       0.085659
      -0.44176        -0.2983        0.44795       -0.37549       -0.51939       -0.32318
      -0.27763        0.38396        0.54102        0.10972        0.60466       -0.32426
      -0.65349        -0.1707       -0.46191        0.53138      0.0076852       -0.21915
      -0.41876       -0.33684       0.068636       -0.28352        0.38297         0.6924
      -0.21753         0.5401         0.2594        0.34673       -0.46673         0.5056

>> rank(A)

ans =
     3
```
Since $m = 3$ and the rank is 3, A has full rank.
■

15.6 COMPUTING A^{-1}

We know that the inverse is often difficult to compute accurately and that, under most circumstances, its computation should be avoided. When it is necessary to compute A^{-1}, the SVD can be used. Since A is invertible, the matrix $\widetilde{\Sigma}$ cannot have a 0 on its diagonal (rank would be $< n$), so $\widetilde{\Sigma} = \Sigma$. From $A = U\Sigma V^{\mathrm{T}}$, $A^{-1} = \left(V^{\mathrm{T}}\right)^{-1}\Sigma^{-1}U^{-1}$, where all the matrices have dimension $n \times n$. U and V are orthogonal, so

$$A^{-1} = V \begin{bmatrix} \frac{1}{\sigma_1} & & & & 0 \\ & \frac{1}{\sigma_2} & & & \\ & & \ddots & & \\ & & & \frac{1}{\sigma_{n-1}} & \\ 0 & & & & \frac{1}{\sigma_n} \end{bmatrix} U^{\mathrm{T}}.$$

Example 15.9. Let $A = \begin{bmatrix} 1 & -1 & 3 \\ 4 & 2 & 3 \\ 5 & 1 & -1 \end{bmatrix}$.

```
>> A = [1 -1 3;4 2 3;5 1 -1];
>> [U S V] = svd(A);
>> Ainv = V*diag(1./diag(S))*U'

Ainv =
    0.1190   -0.0476    0.2143
   -0.4524    0.3810   -0.2143
    0.1429    0.1429   -0.1429

>> inv(A)

ans =
    0.1190   -0.0476    0.2143
   -0.4524    0.3810   -0.2143
    0.1429    0.1429   -0.1429
```

■

15.7 IMAGE COMPRESSION USING THE SVD

Suppose you are given a fairly large image, at least 256×256 pixels. In any large image, some pixels will not be noticed by the human eye. By applying the SVD to a matrix representing the image, we can take advantage of this. The idea involves using only portions of the SVD that involve the larger singular values, since the smaller singular values do not contribute much to the image. Assume matrix A contains the image in some format. Then,

$$A = U\tilde{\Sigma}V^{\mathrm{T}} = \begin{bmatrix} u_{11} & u_{12} & \cdots & \cdots & u_{1m} \\ u_{21} & u_{22} & \cdots & \cdots & u_{2m} \\ \vdots & & \ddots & & \vdots \\ \vdots & & & \ddots & \vdots \\ u_{m1} & \cdots & \cdots & \cdots & u_{mm} \end{bmatrix} \begin{bmatrix} \sigma_1 & & & & \\ & \ddots & & 0 & \\ & & \sigma_r & & \\ & 0 & & \ddots & \\ & & & & 0 \end{bmatrix} \begin{bmatrix} v_{11} & v_{21} & \cdots & \cdots & v_{n1} \\ v_{12} & v_{22} & \cdots & \cdots & v_{n2} \\ \vdots & & \ddots & & \\ \vdots & & & \ddots & \\ v_{1n} & \cdots & \cdots & & v_{nn} \end{bmatrix}.$$

Definition 15.1. A *rank 1 matrix* is a matrix with only one linearly independent column or row.

The primary idea in using the SVD for image compression is that we can write a matrix A as a sum of rank 1 matrices.

Lemma 15.2. *After applying the SVD to an $m \times n$ matrix A, we can write A as follows:*

$$A = \sigma_1 u_1 v_1^{\mathrm{T}} + \sigma_2 u_2 v_2^{\mathrm{T}} + \cdots + \sigma_r u_r v_r^{\mathrm{T}} = \sum_{i=1}^{r} \sigma_i u_i v_i^{\mathrm{T}},$$

where each term in the sum is a rank 1 matrix.

Proof. For $1 \le i \le r$, let Σ_i be the $m \times n$ matrix

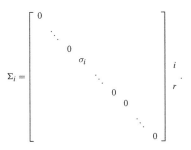

Then

$$\tilde{\Sigma} = \Sigma_1 + \Sigma_2 + \cdots + \Sigma_r,$$

and

$$A = U\Sigma_1 V^{\mathrm{T}} + U\Sigma_2 V^{\mathrm{T}} + \cdots + U\Sigma_r V^{\mathrm{T}} = \sigma_1 u_1 v_1^{\mathrm{T}} + \sigma_2 u_2 v_2^{\mathrm{T}} + \cdots + \sigma_r u_r v_r^{\mathrm{T}}.$$

Each product $\sigma_i u_i v_i^{\mathrm{T}}$ has dimension $(m \times 1)(1 \times n) = m \times n$, so the sum is an $m \times n$ matrix. Each portion of the sum, $u_i v_i^{\mathrm{T}}$ has the form

$$\sigma_i \begin{bmatrix} u_{1i} \\ u_{2i} \\ u_{3i} \\ \vdots \\ u_{mi} \end{bmatrix} \begin{bmatrix} v_{1i} & v_{2i} & v_{3i} & \cdots & v_{ni} \end{bmatrix} = \sigma_i \begin{bmatrix} u_{1i}v_{1i} & u_{1i}v_{2i} & u_{1i}v_{3i} & \cdots & u_{1i}v_{ni} \\ u_{2i}v_{1i} & u_{2i}v_{2i} & u_{2i}v_{3i} & \cdots & u_{2i}v_{ni} \\ u_{3i}v_{1i} & u_{31i}v_{2i} & u_{31i}v_{3i} & \cdots & u_{3i}v_{ni} \\ \vdots & \vdots & \vdots & \vdots & \vdots \\ u_{mi}v_{1i} & u_{m1i}v_{2i} & u_{m1i}v_{3i} & \cdots & u_{mi}v_{ni} \end{bmatrix}. \tag{15.7}$$

Each column in Equation 15.7 is a multiple of the vector $\begin{bmatrix} u_{1i} \\ u_{2i} \\ u_{3i} \\ \vdots \\ u_{mi} \end{bmatrix}$, so each matrix $\sigma_i u_i v_i^{\mathrm{T}}$ has rank 1.   □

Each term $\sigma_i u_i v_i^{\mathrm{T}}$ is called a *mode*, so we can view an image as a sum of modes. Because the singular values σ_i are ordered $\sigma_1 \geq \sigma_2 \geq \cdots \geq \sigma_r > 0$, significant compression of the image is possible if the set of singular values has only a few large modes. Form the sum of those modes, and it will be indistinguishable from the original image. For instance if the rank of the matrix containing the image is 350, it is possible that only modes 1-25 are necessary to cleanly represent the image. If the first k modes are summed, the rank of the matrix will be k, since the first k singular values are in the sum. In other words, if $\overline{A} = \sum_{i=1}^{k} \sigma_i u_i v_i^{\mathrm{T}}$, then $\mathrm{rank}(\overline{A}) = k$. If the first k modes dominate the set of r modes, then $\overline{A}$ will be a good approximation to A.

15.7.1 Image Compression Using MATLAB

Image processing with MATLAB is somewhat complex, so we will confine ourselves to simple operations with gray scale images. In order to use the SVD for working with images, you must input the file using the command `imread`.

```
>> A = imread('filename.ext');
```

The extension "ext" can be one of many possibilities, including "tif/tiff," "bmp," "gif," "jpg," and so forth. Display the image with the commands

```
>> imagesc(A);
>> colormap(gray);
```

The first function "imagesc" scales the image so it uses the full `colormap` and displays it. A color map is a matrix that may have any number of rows, but it must have exactly 3 columns. Each row is interpreted as a color, with the first element specifying the intensity of red light, the second green, and the third blue (RGB). For gray scale, the MATLAB colormap is `gray`. After reading the file into the array A, the data type of its elements must be converted from `uint8` (unsigned 8-bit integers) to `double`.

```
>> A = double(A);
```

At this point, the SVD can be applied to the array A.

```
>> [U S V] = svd(A);
```

Compute the sum of the first k modes using the following statement:

```
>> Aapprox = U(:,1:k)*S(1:k,1:k)*V(:,1:k)';
```

Now display the approximation using

```
>> imagesc(Aapprox);
```

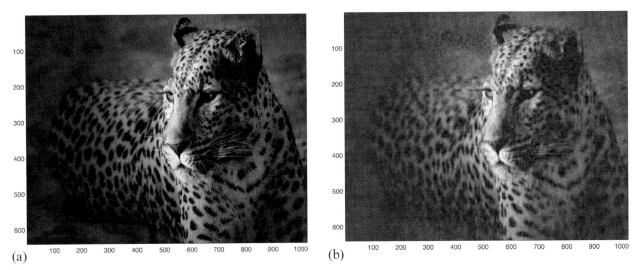

FIGURE 15.3 (a) jaguar (640 × 1024) and (b) jaguar using 35 modes.

FIGURE 15.4 Jaguar using 145 modes.

The following MATLAB sequence reads a picture of a jaguar, displays it, and then displays the image using 35 of the 640 modes (Figure 15.3). The 640 × 1024 image is in JPG format within the software distribution. Notice that the use of 35 modes is somewhat blurry but very distinguishable.

```
>> JAGUAR = imread('jaguar.jpg');
>> JAGUAR = double(JAGUAR);
>> imagesc(JAGUAR); colormap(gray);
>> [U S V] = svd(JAGUAR);
>> rank(JAGUAR)

ans = 640

>> figure(2);
>> JAGUAR35 = U(:,1:35)*S(1:35,1:35)*V(:,1:35)';
>> imagesc(JAGUAR35);colormap(gray);
```

Now let us use 145 modes and see the result in Figure 15.4.

```
>> JAGUAR145 = U(:,1:145)*S(1:145,1:145)*V(:,1:145)';
>> imagesc(JAGUAR145);colormap(gray);
```

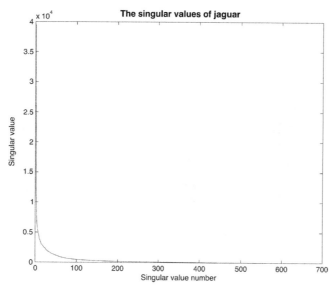

FIGURE 15.5 Singular value graph of jaguar.

Figure 15.5 is a graph of the 640 singular value numbers versus the singular values. Note that the initial singular values are very large, and modes larger than number 145 contribute almost nothing to the image.

In general, if an image is stored in an $m \times n$ matrix, we need to retain $m \times n$ numbers. Suppose that after converting the image to a matrix, we perform the SVD on this matrix and discover that only the largest k singular values capture the "important" information. Instead of keeping $m \times n$ numbers, we keep the k singular values, plus the k vectors $u_1, u_2, \ldots, u_k$ of dimension m, plus vectors $v_1, v_2, \ldots, v_k$ of dimension n, for a total of $k + km + kn$ numbers. For jaguar using $k = 145$, the ratio of the compressed image to the original image is 0.3684.

In addition to `jaguar.jpg`, the software distribution provides the files

black-hole.tif, horsehead-nebula.tif, planets.tif, saturn.tif, and whirlpool.tif

in the subdirectory SVD_compress. For each file, there is a corresponding file with the same name and extension ".mat." Each file name is in uppercase and contains the image converted to array format. By using the `load` command, you directly obtain the array in `double` format; for example,

```
>> load SATURN;
```

The software distribution contains a function `svdimage` that allows you to start with any mode of an image, add one mode per mouse click, and watch the image improve. Input to the function is the image in matrix format, the starting mode number, and the colormap, which should be `gray`. Here is an example using the image of Saturn with a starting mode of 50 (Figure 15.6). Be sure to terminate the function by typing "q," or you will receive a MATLAB error message.

```
>> load SATURN;
>> svdimage(SATURN, 50, gray);
```

15.7.2 Additional Uses

The SVD has many applications:

- Most accurate way to determine the rank of a matrix and the four fundamental subspaces.
- Determining the condition number of a matrix.
- Solve least-squares problems. Least-squares problems are discussed in Chapter 16.
- Used in computer graphics because its factors provide geometric information about the original matrix.
- Principal components analysis approximating a high-dimensional data set with a lower-dimensional subspace. Used in statistics.
- Image compression

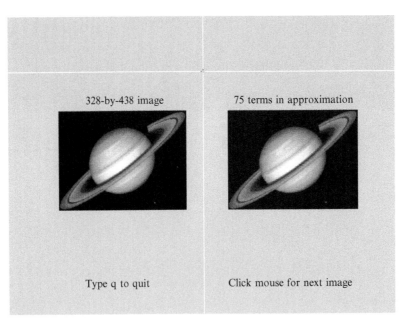

FIGURE 15.6 SVD image capture.

- Image restoration, in which "noise" has caused an image to become blurry. The noises are caused by small singular values, and the SVD can be used to remove them.
- Applications in digital signal processing. For example, as a method for noise reduction. Let a matrix A represents the noisy signal, compute the SVD, and then discard small singular values of A. The small singular values primarily represent the noise, and thus a rank-k matrix, $k <$ rank (A), represents a filtered signal with less noise.

15.8 FINAL COMMENTS

The SVD of a matrix is a powerful technique for matrix computations. Despite its power, however, there are some disadvantages. The SVD is computationally expensive. Many real world problems involve very large matrices. In these cases, applying simpler techniques, such as the QR decomposition or another of a number of matrix decompositions may be indicated. The SVD operates on a fixed matrix, and hence it is not useful in problems that require adaptive procedures. An *adaptive procedure* is a procedure that changes its behavior based on the information available. A good example is adaptive quadrature, a very accurate method for approximating $\int_a^b f(x)\,dx$. As an adaptive method runs, it estimates error, isolates regions where the error tolerance has not been met, and deals with those regions separately. Adaptive quadrature is particularly effective when $f(x)$ behaves badly near a point in the interval $a \leq x \leq b$.

15.9 CHAPTER SUMMARY

The SVD Theorem

The SDV theorem states that there exist orthogonal matrices $U \in R^{m \times m}$, $V \in \mathbb{R}^{n \times n}$, $m \geq n$, and a diagonal matrix $\tilde{\Sigma} \in \mathbb{R}^{m \times n}$ such that

$$A = U\tilde{\Sigma}V^{\mathrm{T}}$$
$$\tilde{\Sigma} = \begin{bmatrix} \Sigma & 0 \\ 0 & 0 \end{bmatrix},$$

where $\Sigma = \mathrm{diag}\,(\sigma_1, \sigma_2, \ldots, \sigma_r)$, and $\sigma_1 \geq \sigma_2 \geq \cdots \geq \sigma_r > 0$ are the positive singular values of A. The proof of the theorem is by construction. The columns of $V = \begin{bmatrix} v_1 & v_2 & \ldots & v_{n-1} & v_n \end{bmatrix}$ are the eigenvalues of $A^{\mathrm{T}}A$, and U is constructed from Av_i and the use of Gram-Schmidt to fill out the matrix, if necessary. If $m < n$, then find the SVD of

$$A^T = U\tilde{\Sigma}V^T,$$

and form

$$A = V\Sigma\tilde{U}^T.$$

Determining Matrix Properties Using the SVD

The SVD reveals substantial information about A:

The rank of A is the number of nonzero singular values, r. The following table lists the bases of four subspaces immediately available from the SVD:

$$A = U\tilde{\Sigma}V^T$$

	Range	Null Space
A	$u_i, 1 \leq i \leq r$	$v_i, r+1 \leq i \leq n$
A^T	$v_i, 1 \leq i \leq r$	$u_i, r+1 \leq i \leq m$

Note that the column space and the row space both have dimension r, proving that the row and column space of a matrix have the same dimension.

The SVD and Matrix Norms

The SVD provides a means of computing the 2-norm of a matrix, since $\|A\|_2 = \sqrt{\sigma_1}$. If A is invertible, then $\|A^{-1}\|_2 = \sqrt{\frac{1}{\sigma_r}}$. The SVD can be computed accurately, so using it is an effective way to find the 2-norm. The SVD also provides a means of computing the Frobenius norm, namely, $\|A\|_F = \left(\sum_{i=1}^{r} \sigma_i^2\right)^{\frac{1}{2}}$.

Geometric Interpretation of the SVD

If A is an $m \times n$ matrix, the SVD tells us that Ax applied to the unit sphere $\|x\|_2 \leq 1$ in $\mathbb{R}^n$ is a rotated ellipsoid in $\mathbb{R}^m$ with semiaxes $\sigma_i, 1 \leq i \leq r$, where the σ_i are the nonzero singular values of A. In $\mathbb{R}^2$, the image is a rotated ellipse centered at the origin and in $\mathbb{R}^3$, the image is a rotated ellipsoid centered at the origin.

Computation of the SVD Using MATLAB

In MATLAB, compute just the singular values with the statement

```
S = svd(S);
```

and the full SVD with

```
[U,S,V] = svd(A);
```

Using the SVD to Compute A^{-1}

Normally, we do not compute A^{-1}, but if it is required we can proceed as follows:

$$A = U\Sigma V^T$$
$$A^{-1} = V\Sigma^{-1}U^T$$

$$A^{-1} = V \begin{bmatrix} \frac{1}{\sigma_1} & & & & 0 \\ & \frac{1}{\sigma_2} & & & \\ & & \ddots & & \\ & & & \frac{1}{\sigma_{n-1}} & \\ 0 & & & & \frac{1}{\sigma_n} \end{bmatrix} U^T.$$

Since A is invertible, all the singular values are nonzero.

Image Compression Using the SVD

After applying the SVD to an $m \times n$ matrix A, we can write A as follows:

$$A = \sigma_1 u_1 v_1^T + \sigma_2 u_2 v_2^T + \cdots + \sigma_r u_r v_r^T = \sum_{i=1}^{r} \sigma_i u_i v_i^T,$$

where each term in the sum is a rank 1 matrix. To use the SVD for image compression, convert the image to a matrix and discard the terms involving small singular values. For instance, if the first k singular values dominate the remaining ones the approximation is

$$\tilde{A} = \sum_{i=1}^{k} \sigma_i u_i v_i^T.$$

After applying the MATLAB function svd and obtaining U, S, and V, the approximation can be written using colon notation as

```
U(:,1:k)*S(1:k,1:k)*V(:,1:k)'.
```

The reader should follow Section 15.7 and use the compression technique on the images supplied in the subdirectory SVD_compress with the book software distribution.

15.10   PROBLEMS

15.1 Prove that for all rank-one matrices, $\sigma_1^2 = \sum_{i=1}^{m} \sum_{j=1}^{n} a_{ij}^2$. Hint: Use Theorem 15.5.

15.2 Suppose $u_1, u_2, ..., u_n$ and $v_1, v_2, ..., v_n$ are orthonormal bases for $\mathbb{R}^n$. Construct the matrix A that transforms each v_i into u_i to give $Av_1 = u_1, Av_2 = u_2, ..., Av_n = u_n$.

15.3 Let $A = \begin{bmatrix} a_{11} & a_{12} \\ 0 & a_{22} \end{bmatrix}$. Determine value(s) for the a_{ij} so that A has distinct singular values.

15.4 Prove that rank $(A^T A)$ = rank (AA^T).

15.5 Find the SVD of
   a. $A^T A$.
   b. $(A^T A)^{-1}$

15.6 If S is a subspace of $\mathbb{R}^n$, the *orthogonal complement* of S, written $S^\perp$, is the set of all vectors x orthogonal to S; in other words

$$S^\perp = \left\{ x \in \mathbb{R}^n \mid x^T y = < x, \, y > = 0 \text{ for all } y \in S \right\}.$$

Prove that $S^\perp$ is a subspace. Find orthogonal complements among the four fundamental subspaces.

15.7 Let $A \in \mathbb{R}^n$ be a rank-one matrix. Show that there exist two vectors u and v such that $A = uv^T$.

15.8 If $B = kA$, where k is a positive integer, what is the SVD of B?

15.9 Find an SVD of a column vector and a row vector.

15.10
   a. What is the SVD for A^T?
   b. What is the SVD for A^{-1}?

15.11 Suppose P is an orthogonal matrix, and B is an $n \times n$ matrix. Show that $A = P^{-1}BP$ has the same singular values as B.

15.12 Prove that if A is nonsingular, all its singular values are greater than zero.

15.13 Prove that the null space of $A^T A$ and A are equal.

15.14 Show that if A is an $n \times n$ matrix and $A = U\tilde{\Sigma}V^T$ is its SVD, then

$$\left\| A^2 \right\|_2 = \left\| \tilde{\Sigma} V^T U \tilde{\Sigma} \right\|_2.$$

15.15
   a. Prove that the matrices $A^T A$ and AA^T have the same eigenvalues

$$\left\{ \sigma_1^2 \ \sigma_2^2 \ ... \ \sigma_r^2 \ 0 \ ... \ 0 \right\}.$$

b. Prove that the orthonormal column vectors of V are orthonormal eigenvectors of $A^{\mathrm{T}}A$ and that the column vectors of U are the orthonormal eigenvectors of AA^{T}.

c. Using the results of (a) and (b), describe an algorithm for the computation of the SVD.

d. Using your algorithm, compute the SVD for the matrix $A = \begin{bmatrix} 1 & 4 \\ 2 & 3 \end{bmatrix}$. Do the computations with pencil and paper.

15.16 Assume A is nonsingular with SVD

$$A = U\Sigma V^{\mathrm{T}}.$$

a. Prove that

$$\sigma_n \|x\|_2 \leq \|Ax\|_2 \leq \sigma_1 \|x\|_2.$$

Hint: To prove the left half of the inequality, use A^{-1}.

b. Show that $\frac{\|Ax\|_2}{\|x\|_2}$ attains its maximum value σ_1 at $x = v_1$.

c. Show that $\frac{\|A^{-1}x\|_2}{\|x\|_2}$ attains its maximum value $\frac{1}{\sigma_n}$ at $x = u_n$.

15.17 An $n \times n$ matrix X is said to be the *square root* of A if $A = X^2$.

a. Show that if A is positive definite and not diagonal, then the Cholesky factor R is not a square root.

b. Let A be positive definite and $A = S^{\mathrm{T}}S$ be the Cholesky decomposition of A. Let $S = U\Sigma V^{\mathrm{T}}$ be the SVD for S, and define $X = V\Sigma V^{\mathrm{T}}$. Show that X is positive definite and $X^2 = A$, so that a positive definite matrix has a positive definite square root.

15.10.1 MATLAB Problems

15.18 **a.** Find the SVD for the singular matrix

$$A = \begin{bmatrix} 2 & 2 \\ 1 & 1 \end{bmatrix}.$$

b. Find a basis for its range and null space.

15.19 **a.** Find the SVD for the matrix

$$A = \begin{bmatrix} 2 & 2 \\ -1 & 1 \end{bmatrix}.$$

b. What is its rank and the dimension of its null space?

15.20 Find the SVD for the matrices

a. $\begin{bmatrix} 2 & 5 & -1 & -7 \\ -4 & 1 & 5 & 0 \end{bmatrix}$

b. $\begin{bmatrix} -1 & 1 & 2 & 3 & 5 \\ 1 & 2 & 3 & 4 & 5 \\ -4 & -3 & -2 & -1 & 0 \end{bmatrix}$

In each case, find an orthonormal basis for the range, null space, row space, and the null space of its transpose.

15.21 **a.** Execute the MATLAB function svdgeom with the matrix $\begin{bmatrix} 6 & 1 \\ -7 & 3 \end{bmatrix}$.

b. Do part (a) using the matrix $\begin{bmatrix} 0.092091 & -0.0043853 \\ 0.035082 & 0.052623 \end{bmatrix}$.

15.22 Develop a method for building a 2×2 matrix with specified singular values σ_1, σ_2. Use your method to construct matrix A with singular values $\sigma_1 = 55.63$, $\sigma_2 = 25.7$, and matrix B with singular values $\sigma_1 = .2$, $\sigma_2 = .1$. In each case, use the function svdgeom to show how the linear transformation transforms the unit circle.

15.23 The software distribution contains a graphics file "black-hole.tif." Use MATLAB to read the image, convert it to a matrix, and use the SVD to display the graphic using only large modes. Compute the percentage of image storage you save.

15.24 The software distribution contains the files SATURN.mat and WHIRLPOOL.mat. Each of these is an image matrix stored in MATLAB format. Input each one using the load command and experiment with it using the SVD to compress the images. In each case, draw a graph of the singular value number vs. the singular value similar to Figure 15.5.

Remark 15.4.   There is another image, HORSEHEAD.mat, with which you might want to experiment.

15.25 The Hilbert matrices $H_{ij} = \frac{1}{i+j-1}$, $1 \leq i, j \leq n$ have notoriously bad properties. It can be shown that any Hilbert matrix is nonsingular and, as such, has rank n. MATLAB constructs a Hilbert matrix with the command H = hilb(n).
   a. Use MATLAB to verify that the rank of the 8×8 Hilbert matrix H is 8.
   b. Find the SVD of H.
   c. Comment on the singular values.
   d. Compute $\frac{\sigma_1}{\sigma_8}$, the condition number of H.

15.26
   a. Construct A = rand(m,n) for $m = 5$, $n = 4$. Find the eigenvalues and eigenvectors for $A^T A$ and $A A^T$. Do the experiment again with $m = 3$, $n = 5$.
   b. If you see a pattern of behavior, state a theorem a prove it.
      The following problem is adapted from material in Ref. [46, Section 10.7].

15.27 This problem is adapted from the material in Ref. [46], Section 10.7.
   a. MATLAB provides a matrix, gallery(5), for eigenvalue testing. Use the MATLAB poly function to determine the characteristic polynomial of A. Is the matrix singular? Would computation of the eigenvalues of A be ill-conditioned? Explain.

   The remainder of the problem deals with the singular values of gallery(5).
   b.
      i.   Show that $\widetilde{\Sigma} + \delta\widetilde{\Sigma} = U^T (A + \delta A) V$.
      ii. Using (b), part (i) show that $\|\delta A\|_2 = \|\delta\widetilde{\Sigma}\|_2$. This says that the size of the errors in computing the singular values is the same as the errors involved in forming A. This type of perturbation result is ideal.
      iii. The result of (b), part (ii) deals with all elements of the singular value problem. There can be very large and very small singular values, and the primary problem is with the small singular values. If A is singular, one or more singular values will be zero, but may not actually be reported as zero due to rounding errors. It is hard to distinguish between a very small singular value and one that is actually 0. gallery (5) will help in understanding the problems with small singular values. Execute the statements

```
A = gallery(5);
format long e;
svd(A)
```

Comment on the distribution of singular values. In Ref. [46], it is stated that the small singular values that should be 0 lie somewhere between *eps* and $\|A\|_2$ *eps*. Is this the case here?
      iv. Compute the SVD of a randomly perturbed matrix by running the following code. The function randn(5,5) creates a random 5×5 matrix, and randn(5,5).*A multiplies each entry a_{ij} by r_{ij}. The sum A+eps* randn(5,5).*A perturbs gallery(5) by a small amount. Run

```
format long e
clc

for i = 1:5
    svd(A + eps*randn(5,5).*A)
    fprintf('----------------------------------');
end
fprintf('\n');
```

The MATLAB output is in the format *digit.$d_1 d_3 d_3 \ldots d_{15} e \pm e_1 e_2 e_3$*. Write down one line of svd output; for example,

```
1.010353607103610e+005
1.679457384066240e+000
1.462838728085211e+000
```

```
1.080169069985495e+000
4.288402425161663e-014
```

Analyze the output of the for statement and place a star ('*') at every digit position in your written line of output that changes. The asterisks show the digits that change as as a result of random perturbations. Comment on the results.

15.28

a. Write a function, buildtaumat, that builds the matrix

$$T = \begin{bmatrix} 0 & \Gamma_n \\ \Gamma_n & 0 \end{bmatrix},$$

where Γ_n is an $n \times n$ submatrix of ones, and 0 represents an $n \times n$ zero submatrix. Thus, T is a $2n \times 2n$ matrix.

For instance, $T = \begin{bmatrix} 0 & 0 & 1 & 1 \\ 0 & 0 & 1 & 1 \\ 1 & 1 & 0 & 0 \\ 1 & 1 & 0 & 0 \end{bmatrix}$.

b. Construct the matrix for $n = 2, 4, 8$, and 16. In each case, compute the eigenvalues and the singular values. Propose a theorem that explains what you observe. Prove it.

15.29 If A is nonsingular, the SVD can be used to solve a linear system $Ax = b$.

a. Explain why $\widetilde{\Sigma}$ in the SVD of A is invertible. What is its inverse?

b. Show that

$$x = V\widetilde{\Sigma}^{-1}U^T b.$$

c. Develop a one line MATLAB command to compute $\widetilde{\Sigma}^{-1}$ that only uses the function diag.

d. Solve

$$\begin{bmatrix} 1 & -1 & 0 \\ 8 & 4 & 1 \\ -9 & 0 & 3 \end{bmatrix} x = \begin{bmatrix} 1 \\ 2 \\ 3 \end{bmatrix}$$

and

$$\begin{bmatrix} 1 & 3 & 0 & 1 \\ 1 & 5 & -3 & 8 \\ 12 & 5 & 7 & 0 \\ 6 & 77 & 15 & 35 \end{bmatrix} x = \begin{bmatrix} -1 \\ 0 \\ 1 \\ 2 \end{bmatrix}$$

using the results of parts (a)-(c).

e. Does using the SVD seems like a practical means of solving $Ax = b$? Explain your answer.

15.30 There is a geometric interpretation of the condition number. You are given the matrix sequence

$$\left\{ \begin{bmatrix} 1 & 2 \\ 1 & 1 \end{bmatrix}, \begin{bmatrix} 1 & 2 \\ 1 & 1.5 \end{bmatrix}, \begin{bmatrix} 1 & 2 \\ 1 & 1.9 \end{bmatrix}, \begin{bmatrix} 1 & 2 \\ 1 & 1.99 \end{bmatrix}, \begin{bmatrix} 1 & 2 \\ 1 & 1.999 \end{bmatrix} \right\}.$$

a. Remove textual output from the function svdgeom introduced in Section 9.4 so it only plots a graph, and name the function svdgeom1.

b. For each matrix in the sequence, do the following

(a) Compute the condition number.

(b) Call the function svdgeom1 and observe the graph.

c. Do you see a relationship between the condition number of a matrix and its action as a linear transformation?

15.31 In discussing image compression using the SVD, we approximated the graphical representation by using the first k singular values and discarding the rest. The image looked just as good as the original if we included sufficient singular values. This process is known as computing a rank k approximation. There is theoretical justification for this technique. A proof of the following theorem can be found in Refs. [1, pp. 110-113] and [47, pp. 83-84]. It says that using the SVD we obtain the optimal rank k approximation to a matrix.

Theorem. *Assume that the $m \times n$ matrix A, $m \geq n$, has the SVD $A = U\tilde{\Sigma}V^{\mathrm{T}}$. If $k <$ rank (A), a matrix of rank k closest to A as measured by the 2-norm is $A_k = \sum_{i=1}^{k} \sigma_i u_i v_i^{\mathrm{T}}$, and $\|A - A_k\|_2 = \sigma_{k+1}$. A_k can also be written as $A_k = U\tilde{\Sigma}_k V^{\mathrm{T}}$, where $\tilde{\Sigma}_k = \mathrm{diag}\,(\sigma_1, \sigma_2, \ldots, \sigma, 0, \ldots, 0)$. Another way of putting this is*

$$\sigma_{k+1} = \min_{B \in \mathbb{R}^{m \times n},\, \mathrm{rank}(B)=k} \|A - B\|_2 = \|A - A_k\|_2.$$

Run the following MATLAB program and explain the output.

```
load('HORSEHEAD.mat');
[U,S,V] = svd(HORSEHEAD);
HORSEHEAD250 = U(:,1:250)*S(1:250,1:250)*V(:,1:250)';
figure(2);
imagesc(HORSEHEAD250);
colormap(gray);
fprintf('sigma251 = %.12f\n\n', S(251,251));
deltaS = 0.1:-.005:0;
for i = 1:length(deltaS)
    B = HORSEHEAD;
    B = B + deltaS(i)*ones(566,500);
    B = makerank(B,250);
    fprintf('%.12f\n', abs(norm(HORSEHEAD - B)-S(251,251)));
end
```

15.32 The *polar decomposition* of an $n \times n$ matrix A is

$$A = UP,$$

where U is orthogonal and P is symmetric positive semidefinite ($x^{\mathrm{T}}Px \geq 0$ for all $x \neq 0$). Intuitively, the polar decomposition factors A into a component P that stretches Ax along a set of orthogonal axes followed by a rotation U. This is analogous to the polar form of a complex number $z = re^{i\theta}$. P plays the role of r, and U plays the role of $e^{i\theta}$. Applications of the polar decomposition include factor analysis and aerospace computations [48].
a. If $A = U\Sigma V^{\mathrm{T}}$ is the SVD for A, show that $A = \left(UV^{\mathrm{T}}\right)\left(V\Sigma V^{\mathrm{T}}\right)$ is a polar decomposition for A.
b. For what class of matrices can we guarantee that P is positive definite?
c. Write a function [U P] = polardecomp(A) that computes a polar decomposition for the square matrix A. Test your function with matrices of dimensions 3×3, 5×5, 10×10, and 50×50.
15.33 a. Using the result of Problem 15.17, write a function sqrroot that computes the square root of a positive definite matrix.
b. For $n = 5$, 10, 25, and 50, test your function using the matrices

```
gallery('moler',n)
```

15.34 Generate a random 2×2 matrix A = rand(2,2). Then type eigshow(A) at the MATLAB prompt. A window will open. Click on the svd button on the right side of the window. Your matrix A will appear (in MATLAB notation) in the menu bar above the graph. Underneath the graph the statement "Make $A*x$ perpendicular to $A*y$" should appear. The graph shows a pair of orthogonal unit vectors x and y, together with the image vectors Ax and Ay. Move the pointer onto the vector x, and then make the pair of vectors x, y go around in a circle. The transformed vectors Ax and Ay then move around an ellipse, as we expect from the discussion in Section 15.4. Generally Ax will not be perpendicular to Ay. Keep moving vector x until you find a position where Ax is perpendicular to Ay. When this happens, then the singular values σ_1 and σ_2 of A are the lengths of the vectors Ax and Ay. Estimate the lengths from the graph. Take note of the fact that $\|x\|_2 = \|y\|_2 = 1$. Confirm your estimates by using MATLAB's svd command.

Chapter 16

Least-Squares Problems

You should be familiar with

- Rank
- Proof by contradiction
- Cholesky decomposition
- *QR* decomposition
- SVD
- Residual
- Block matrix notation

We introduced the concept of least squares in Section 12.3 by developing the least-squares algorithm for fitting a polynomial to data. Even in that case, we found the unique polynomial by solving a square system of equations $Ax = b$, where A was nonsingular. In many areas such as curve fitting, statistics, and geodetic modeling, A is either singular or has dimension $m \times n$, $m \neq n$. If $m > n$, there are more equations than unknowns, and the system is said to be *overdetermined*. In most cases, overdetermined systems have no solution. In the case $m < n$, there are more unknowns than equations, and we say such systems are *underdetermined*. In this situation, there are usually an infinite number of solutions. It is clear that the Gaussian elimination techniques we have studied will not be useful in these cases.

Since singular, over- and underdetermined systems do not give us a solution in the exact sense, the solution is to find a vector x such that Ax is as close as possible to b. A way of doing this is to find a vector x such that the residual $r(x) = \|Ax - b\|_2$ is a minimum. Recall that the Euclidean norm, $\|\cdot\|_2$, of a vector in $\mathbb{R}^n$ is $\sqrt{x_1^2 + x_2^2 + \cdots + x_n^2}$, so if we want to minimize $\|Ax - b\|_2$, we call x a *least-squares* solution. Finding a least-squares solution to $Ax = b$ is known as the *linear least-squares problem*. We need to formally define the problem so we can develop rigorous techniques for solving it.

Definition 16.1 (The least-squares problem). Given a real $m \times n$ matrix A and a real vector b, find a real vector $x \in \mathbb{R}^n$ such that the function $r(x) = \|Ax - b\|_2$ is minimized. It is possible that the solution x will not be unique.

Assume that $m > n$. Since $x \in \mathbb{R}^n$, and A is an $m \times n$ matrix, Ax is a linear transformation from $\mathbb{R}^n$ to $\mathbb{R}^m$, and the range of the transformation, $R(A)$, is a subspace of $\mathbb{R}^m$. Given any $y \in R(A)$, there is an $x \in R^n$ such that $Ax = y$. If $b \in \mathbb{R}^m$ is in $R(A)$, we have a solution. If b is not in $R(A)$, consider the vector $Ax - b$ that joins the endpoints of the vectors Ax and b. Since b is not in $R(A)$, project b onto the plane $R(A)$ to obtain a vector $u \in R(A)$. There must be a vector $x \in \mathbb{R}^n$ such that $Ax = u$. The distance between the two points is $\|Ax - b\|_2$ is as small as possible, so x is the solution we want (Figure 16.1).

The vector $b - Ax$ is orthogonal to $R(A)$, and since every vector in $R(A)$ is a linear combination of the columns of A (vectors in R^m), it must be the case that $b - Ax$ is orthogonal to the every column of A. Mathematically this says that the inner product of $b - Ax$ with each column of A must be zero. If

$$a_i = \begin{bmatrix} a_{1i} \\ a_{2i} \\ \vdots \\ a_{m-1,i} \\ a_{mi} \end{bmatrix}$$

is column i, then $\langle a_i, b - Ax \rangle = a_i^{\mathrm{T}} (b - Ax) = 0$, $1 \leq i \leq n$, and

$$A^{\mathrm{T}} (b - Ax) = 0,$$

Numerical Linear Algebra with Applications. http://dx.doi.org/10.1016/B978-0-12-394435-1.00016-8

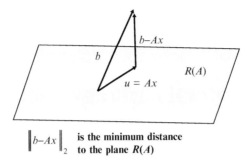

$$\left\| b-Ax \right\|_2 \quad \text{is the minimum distance to the plane } R(A)$$

FIGURE 16.1 Geometric interpretation of the least-squares solution.

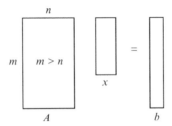

FIGURE 16.2 An overdetermined system.

and

$$A^{\mathrm{T}}Ax = A^{\mathrm{T}}b.$$

These are the normal equations specified in Definition 12.1.

16.1 EXISTENCE AND UNIQUENESS OF LEAST-SQUARES SOLUTIONS

The geometric argument we presented is not an mathematical proof. We should give some mathematical justification that if $m > n$, a solution exists and satisfies the normal equations. Figure 16.2 graphically represents an overdetermined system.

16.1.1 Existence and Uniqueness Theorem

In order to prove the existence and uniqueness to the solution of the least-squares problem, we first consider the case $m \geq n$.

Let A be an $m \times n$ matrix. Then, each of the n columns has m components and is a member of $\mathbb{R}^m$, and each of m rows has n components and is a member of $\mathbb{R}^n$. The columns of A span a subspace of $\mathbb{R}^m$, and the rows of A span a subspace of $\mathbb{R}^n$. The column rank of A is the number of linearly independent columns of A, and the row rank of A is the number of linear independent rows of A. Theorem 15.3 proves that the column rank and row rank of A are equal.

Definition 16.2. An $m \times n$ matrix A has full rank if rank $(A) = \min(m, n)$.

If $m \geq n$, and A has full rank, then rank$(A) = n$, and the columns of A are linearly independent.

Lemma 16.1. *Let A be an $m \times n$ matrix, $m \geq n$. A has full rank if and only if the $n \times n$ matrix $A^{\mathrm{T}}A$ is nonsingular.*

Proof. Use proof by contradiction. Assume A has full rank, but $A^{\mathrm{T}}A$ is singular. In this case, the $n \times n$ homogeneous system $A^{\mathrm{T}}Ax = 0$ has a nonzero solution x. As a result, $x^{\mathrm{T}}A^{\mathrm{T}}Ax = 0$, which says that $\langle Ax, Ax \rangle = \|Ax\|_2^2 = 0$, so $Ax = 0$, and A cannot have full rank.

Again, use proof by contradiction. Assume $A^{\mathrm{T}}A$ is nonsingular, but A does not have full rank. Since A is rank deficient, there is a nonzero vector x such that $Ax = 0$. As a result, $A^{\mathrm{T}}Ax = 0$, $x \neq 0$, so $A^{\mathrm{T}}A$ is singular. $\square$

Theorem 16.1 establishes the link between a least-squares solution and the normal equations and tells us the solution is unique if the matrix has full rank. The proof is somewhat technical and can be omitted if desired, but the results are very important to remember.

Theorem 16.1.   **a.**   *Given an $m \times n$ matrix A with $m \geq n$ and an $m \times 1$ column vector b, an $n \times 1$ column vector x exists such that x is a least-squares solution to $Ax = b$ if and only if x satisfies the normal equations*

$$A^{\mathrm{T}} A x = A^{\mathrm{T}} b.$$

b.   *The least-squares solution x is unique if and only if A has full rank.*

Proof.   To prove part (1), assume that $A^{\mathrm{T}} A \bar{x} = A^{\mathrm{T}} b$, so that $\bar{x}$ is a solution of the normal equations. Now, if x is any vector in $\mathbb{R}^n$,

$$
\begin{aligned}
\|Ax - b\|_2^2 &= \|Ax - A\bar{x} + A\bar{x} - b\|_2^2 = \langle [(A\bar{x} - b) + A(x - \bar{x})], [(A\bar{x} - b) + A(x - \bar{x})] \rangle \\
&= \|A\bar{x} - b\|_2^2 + 2\langle A(x - \bar{x}), (A\bar{x} - b) \rangle + \|A(x - \bar{x})\|_2^2 \\
&= \|A\bar{x} - b\|_2^2 + 2(A(x - \bar{x}))^{\mathrm{T}}(A\bar{x} - b) + \|A(x - \bar{x})\|_2^2 \\
&= \|A\bar{x} - b\|_2^2 + 2(x - \bar{x})^{\mathrm{T}}(A^{\mathrm{T}} A\bar{x} - A^{\mathrm{T}} b) + \|A(x - \bar{x})\|_2^2 \\
&= \|A\bar{x} - b\|_2^2 + \|A(x - \bar{x})\|_2^2 \\
&\geq \|A\bar{x} - b\|_2^2,
\end{aligned}
$$

and $\bar{x}$ is a solution to the least-squares problem.

Now assume that $\bar{x} \in \mathbb{R}^n$ is a solution to the least-squares problem, so that $\|A\bar{x} - b\|_2$ is minimum. Thus, $\|b - A\bar{x}\|_2^2 \leq \|b - Ay\|_2^2$ for any $y \in \mathbb{R}^n$. Given any vector $z \in \mathbb{R}^n$, let $y = \bar{x} + tz$, where t is a scalar. Then,

$$
\begin{aligned}
\|b - A\bar{x}\|_2^2 \leq \|b - A(\bar{x} + tz)\|_2^2 &= ([b - A\bar{x}] - tAz)^{\mathrm{T}}([b - A\bar{x}] - tAz) \\
&= \|b - A\bar{x}\|_2^2 - 2t(b - A\bar{x})^{\mathrm{T}} Az + t^2 \|Az\|_2^2.
\end{aligned}
$$

Thus,

$$0 \leq -2t(b - A\bar{x})^{\mathrm{T}} Az + t^2 \|Az\|_2^2.$$

If $t > 0$,

$$0 \leq -2(b - A\bar{x})^{\mathrm{T}} Az + t \|Az\|_2^2,$$

and

$$2(b - A\bar{x})^{\mathrm{T}} Az \leq t \|Az\|_2^2.$$

If $t < 0$,

$$0 \leq 2(b - A\bar{x})^{\mathrm{T}} Az + |t| \|Az\|_2^2.$$

As $t \to 0^+$ or $t \to 0^-$, we have $2(b - A\bar{x})^{\mathrm{T}} Az \leq 0$ and $0 \leq 2(b - A\bar{x})^{\mathrm{T}} Az$, so

$$(b - A\bar{x})^{\mathrm{T}} Az = 0$$

for all $z \in \mathbb{R}^n$. Thus,

$$(b - A\bar{x})^{\mathrm{T}} Az = (Az)^{\mathrm{T}}(b - A\bar{x}) = z^{\mathrm{T}} A^{\mathrm{T}}(b - A\bar{x}) = z^{\mathrm{T}}\left(A^{\mathrm{T}} b - A^{\mathrm{T}} A\bar{x}\right) = 0$$

for all $z \in \mathbb{R}^n$. Choose $z = \left(A^{\mathrm{T}} b - A^{\mathrm{T}} A\bar{x}\right)$, so $\left\|A^{\mathrm{T}} b - A^{\mathrm{T}} A\bar{x}\right\|_2^2 = 0$ and $A^{\mathrm{T}} A x = A^{\mathrm{T}} b$.

For part (2), if x is the unique solution to $A^{\mathrm{T}} A x = A^{\mathrm{T}} b$, then $A^{\mathrm{T}} A$ is nonsingular so A must have full rank according to Lemma 16.1. If A has full rank, $A^{\mathrm{T}} A$ is nonsingular by Lemma 16.1, and $A^{\mathrm{T}} A x = A^{\mathrm{T}} b$ has a unique solution. $\square$

16.1.2 Normal Equations and Least-Squares Solutions

The least-squares residual equation $r = b - Ax$ is what we defined as the residual for the solution of an ordinary $n \times n$ linear system. If we multiply the residual equation by A^T, the result is

$$A^T r = A^T b - A^T Ax,$$

which says that we do not expect r to be 0. We want $A^T r$ to be zero.

Theorem 16.2 (Least-squares residual equation). *Let* $r = b - Ax$. *Then* $A^T r = 0$ *if and only if* x *is a least-squares solution.*

Proof. If x is a least-squares solution, then it satisfies the normal equations by Theorem 16.1. Then $A^T Ax = A^T b$ and $A^T b - A^T Ax = A^T (b - Ax) = A^T r = 0$.

When $A^T r = 0$, $A^T (b - Ax) = 0$, and $A^T Ax = A^T b$, so by Theorem 16.1, x is a least-squares solution. $\qquad\square$

Example 16.1. Let $A = \begin{bmatrix} 1 & 3 \\ 2 & 4 \\ 3 & 8 \\ 2 & 9 \end{bmatrix}$, $b = \begin{bmatrix} 1 \\ 3 \\ 5 \\ 8 \end{bmatrix}$, which is a system of four equations in two unknowns. The rank of A is 2, so we know there is a unique least-squares solution. Formulate and solve the normal equations.

$A^T A = \begin{bmatrix} 18 & 53 \\ 53 & 170 \end{bmatrix}$ and $A^T b = \begin{bmatrix} 38 \\ 127 \end{bmatrix}$. Solve $\begin{bmatrix} 18 & 53 \\ 53 & 170 \end{bmatrix} x = \begin{bmatrix} 38 \\ 127 \end{bmatrix}$ to obtain the solution $x = \begin{bmatrix} -1.0797 \\ 1.0837 \end{bmatrix}$. ∎

16.1.3 The Pseudoinverse, $m \geq n$

Assume that A is a full rank $m \times n$ matrix, $m \geq n$. By Lemma 16.1, $A^T A$ is invertible, so the matrix product $\left(A^T A\right)^{-1} A^T$ is defined. From Theorem 16.1, if A has full rank and x is a solution to the least-squares problem, then $A^T Ax = A^T b$, and so

$$x = \left(A^T A\right)^{-1} A^T b.$$

Definition 16.3. The matrix

$$A^\dagger = \left(A^T A\right)^{-1} A^T$$

is called the *pseudoinverse* or the *Moore-Penrose generalized inverse*. The solution to the full-rank overdetermined least-squares problem $Ax = b$ is $x = A^\dagger b$.

Remark 16.1. There is an equivalent definition for the pseudoinverse that uses components from the singular value decomposition (SVD). This form can be used for rank-deficient matrices, and we will present the definition in Section 16.4.

Example 16.2. $A = \begin{bmatrix} 1 & 6 & -1 \\ 4 & 2 & 1 \\ 0 & 3 & 5 \\ 2 & 6 & 9 \\ -1 & 5 & -8 \end{bmatrix}$, $b = \begin{bmatrix} 2 \\ 12 \\ 5 \end{bmatrix}$. The rank of A is 3, so A has full rank. Solve the system $Ax = b$.

$$x = A^\dagger b = \left[\left(A^T A\right)^{-1} A^T\right] b = \begin{bmatrix} 0.0691 & -0.0109 & -0.0101 \\ -0.0109 & 0.0111 & 0.0002 \\ -0.0101 & 0.0002 & 0.0075 \end{bmatrix} \begin{bmatrix} 1 & 4 & 0 & 2 & -1 \\ 6 & 2 & 3 & 6 & 5 \\ -1 & 1 & 5 & 9 & -8 \end{bmatrix} b$$

$$= \begin{bmatrix} 0.0138 & 0.2447 & -0.0831 & -0.0178 & -0.0431 \\ 0.0556 & -0.0212 & 0.0345 & 0.0469 & 0.0647 \\ -0.0162 & 0.0324 & 0.038 & 0.0485 & -0.0487 \end{bmatrix} \begin{bmatrix} 2 \\ 12 \\ 5 \end{bmatrix} = \begin{bmatrix} 2.5662 \\ -0.0174 \\ -0.2790 \end{bmatrix}$$

The pseudoinverse, $A^\dagger$ generalizes the definition of the inverse of a square matrix. In particular, when A is a nonsingular $n \times n$ matrix

$$A^\ddagger = \left(A^{\mathrm{T}}A\right)^{-1} A^{\mathrm{T}} = A^{-1} \left(A^{\mathrm{T}}\right)^{-1} A^{\mathrm{T}} = A^{-1}I = A^{-1} \qquad \blacksquare$$

Recall that if B is a nonsingular matrix, $\kappa(B) = \|B\|_2 \|B^{-1}\|_2$. Given that $A^\dagger = A^{-1}$ for a nonsingular matrix, the following definition makes good sense for a full-rank matrix.

Definition 16.4. If A is a full rank $m \times n$ matrix, then the condition number of A is

$$\kappa(A) = \|A^\dagger\| \, \|A\|. \qquad (16.1)$$

Definition 14.3 specifies the condition number of a general matrix. For a full rank matrix, $m \ge n$, Definition 16.4 is equivalent (Problem 16.4).

Example 16.3. If A is the 4×2 matrix in Example 16.1, then

$$\|A\| \, \|A^\dagger\| = 13.6622 \left\|\left(A^{\mathrm{T}}A\right)^{-1} A^{\mathrm{T}}\right\| = 13.662184 \left\| \begin{bmatrix} 18 & 53 \\ 53 & 170 \end{bmatrix}^{-1} \begin{bmatrix} 1 & 2 & 3 & 2 \\ 3 & 4 & 8 & 9 \end{bmatrix} \right\|$$

$$= 13.6621841 \, (0.8623494) = 11.7815762$$

A is a well-conditioned matrix. On the other hand, let V be the 6×4 Vandermonde matrix formed from $x = \begin{bmatrix} 1.1 & 1.8 & 2.3 & 2.7 & 3.3 & 3.5 \end{bmatrix}^{\mathrm{T}}$, $m = 4$. The function cond in MATLAB computes the condition number of a nonsquare matrix, as you can see.

```
>> V = vandermonde(x,4);
>> norm(inv(V'*V)*V')*norm(V)

ans =
   3.0888e+004

>> cond(V)

ans =
   3.0888e+004
```

Clearly, V is an ill-conditioned matrix. $\qquad \blacksquare$

16.1.4 The Pseudoinverse, *m<n*

If $m < n$ and rank $(A) = m$, then A^{T} is of dimension $n \times m$, $n > m$, and the pseudoinverse is defined for A^{T}. We have

$$\left(A^{\mathrm{T}}\right)^\dagger = \left(\left(A^{\mathrm{T}}\right)^{\mathrm{T}} A^{\mathrm{T}}\right)^{-1} \left(A^{\mathrm{T}}\right)^{\mathrm{T}} = \left(AA^{\mathrm{T}}\right)^{-1} A$$

Now take the transpose to obtain

$$A^\dagger = A^{\mathrm{T}} \left[\left(AA^{\mathrm{T}}\right)^{-1}\right]^{\mathrm{T}} = A^{\mathrm{T}} \left(AA^{\mathrm{T}}\right)^{-1}.$$

Note that AA^{T} is an $m \times m$ matrix, and is nonsingular because A has full rank. This leads us to the definition.

Definition 16.5. Assume that A is an $m \times n$ matrix, $m < n$ and has full rank. The pseudoinverse is the well-defined product

$$A^\dagger = A^{\mathrm{T}} \left(AA^{\mathrm{T}}\right)^{-1}$$

16.2 SOLVING OVERDETERMINED LEAST-SQUARES PROBLEMS

There are three basic methods for solving overdetermined least-squares problems. The first of these, using the normal equations, was the standard method for many years. However, using the QR decomposition or the SVD gives much better results in most cases.

16.2.1   Using the Normal Equations

$A^T A$ is a symmetric matrix. Consider the product $x^T A^T A x = (Ax)^T (Ax) = \|Ax\|_2^2$. If we assume that A has full rank, $Ax \neq 0$ for any $x \neq 0$, and so $A^T A$ is positive definite. As a result, the Cholesky decomposition applies (Section 13.3). Recall that the Cholesky decomposition of a positive definite matrix M is of the form $M = R^T R$, where R is an upper triangular matrix. To solve the normal equations, proceed as follows:

Solve the Normal Equations Using the Cholesky Decomposition

a. Find the Cholesky decomposition $A^T A = R^T R$.
b. Solve the system $R^T y = A^T b$ using forward substitution.
c. Solve the system $Rx = y$ using back substitution.

Example 16.4.   There are three mountains m_1, m_2, m_3 that from one site have been measured as 2474 ft., 3882 ft., and 4834 ft. But from m_1, m_2 looks 1422 ft. taller and m_3 looks 2354 ft. taller, and from m_2, m_3 looks 950 ft. taller. This data gives rise to an overdetermined set of linear equations for the height of each mountain.

$$
\begin{array}{rrrcl}
m_1 & & & = & 2474 \\
& m_2 & & = & 3882 \\
& & m_3 & = & 4834 \\
-m_1 & +m_2 & & = & 1422 \\
-m_1 & & +m_3 & = & 2354 \\
& -m_2 & +m_3 & = & 950
\end{array}
$$

In matrix form, the least-squares problem is

$$
\begin{bmatrix}
1 & 0 & 0 \\
0 & 1 & 0 \\
0 & 0 & 1 \\
-1 & 1 & 0 \\
-1 & 0 & 1 \\
0 & -1 & 1
\end{bmatrix}
\begin{bmatrix} m_1 \\ m_2 \\ m_3 \end{bmatrix}
=
\begin{bmatrix}
2474 \\ 3882 \\ 4834 \\ 1422 \\ 2354 \\ 950
\end{bmatrix}.
$$

$$
A^T A =
\begin{bmatrix}
3 & -1 & -1 \\
-1 & 3 & -1 \\
-1 & -1 & 3
\end{bmatrix}
\text{ has the Cholesky decomposition}
$$

$$
\begin{bmatrix}
3 & -1 & -1 \\
-1 & 3 & -1 \\
-1 & -1 & 3
\end{bmatrix}
=
\begin{bmatrix}
1.7321 & 0 & 0 \\
-0.5774 & 1.6330 & 0 \\
-0.5774 & -0.8165 & 1.4142
\end{bmatrix}
\begin{bmatrix}
1.7321 & -0.5774 & -0.5774 \\
0 & 1.6330 & -0.8165 \\
0 & 0 & 1.4142
\end{bmatrix}
$$

Solve

$$
\begin{bmatrix}
1.7321 & 0 & 0 \\
-0.5774 & 1.6330 & 0 \\
-0.5774 & -0.8165 & 1.4142
\end{bmatrix}
y = A^T
\begin{bmatrix}
2474 \\ 3882 \\ 4834 \\ 1422 \\ 2354 \\ 950
\end{bmatrix}
=
\begin{bmatrix}
-1302 \\ 4354 \\ 8138
\end{bmatrix},
\quad
y =
\begin{bmatrix}
-751.7 \\ 2400.5 \\ 6833.5
\end{bmatrix}
$$

Solve

$$
\begin{bmatrix}
1.7321 & -0.5774 & -0.5774 \\
0 & 1.6330 & -0.8165 \\
0 & 0 & 1.4142
\end{bmatrix}
x = y
$$

$$
m =
\begin{bmatrix}
2472.0 \\ 3886.0 \\ 4832.0
\end{bmatrix}
\qquad \blacksquare
$$

Algorithm 16.1 uses the normal equations to solve the least-squares problem.

Algorithm 16.1 Least-Squares Solution Using the Normal Equations

---

```
function NORMALSOLVE(A,b)
   % Solve the overdetermined least-squares problem using the normal equations.
   % Input: m × n full-rank matrix A, m ≥ n and an m × 1 column vector b.
   % Output: the unique least-squares solution to Ax = b and the residual
   c = Aᵀb
   Use the Cholesky decomposition to obtain AᵀA = RᵀR
   Solve the lower triangular system Rᵀy = c
   Solve the upper triangular system Rx = y
   return[ x  ‖b − Ax‖₂ ]
end function
```

NLALIB: The function `normalsolve` implements Algorithm 16.1 by using the functions `cholesky` and `cholsolve`.

Efficiency

The efficiency analysis is straightforward.

- $c = A^{T}b : 2nm$ flops
- Form $A^{T}A : 2mn^2$ flops
- Cholesky decomposition of $A^{T}A : \frac{n^3}{3} + n^2 + \frac{5n}{3}$ flops
- Forward and back substitution: $2\left(n^2 + n - 1\right)$ flops

The total is $n^2 \left(2m + \frac{n}{3}\right) + 2mn + 3n^2 + \frac{11}{3}n - 2 \simeq n^2 \left(2m + \frac{n}{3}\right)$ flops.

Computational Note

It must be noted that, although relatively easy to understand and implement, it has serious problems in some cases.

- There may be some loss of significant digits when computing $A^{T}A$. It may even be singular to working precision.
- We will see in Section 16.3 that the accuracy of the solution using the normal equations depends on the square of condition number of the matrix. If $\kappa\left(A\right)$ is large, the results can be seriously in error.

16.2.2   Using the *QR* Decomposition

The *QR* decomposition can be used very effectively to solve the least-squares problem $Ax = b$, $m \geq n$, when A has full rank. Our implementation of the *QR* decomposition using Gram-Schmidt required A to have full rank, but the algorithm is not as stable as others. In Chapter 17, we will develop two better algorithms for computing the *QR* decomposition. In this chapter, we will use the MATLAB function `qr` to compute a unique solution to the full-rank least-squares problem.

Using the reduced *QR* decomposition, $A = Q^{m \times n}R^{n \times n}$, we have

$$A^{T}A = (QR)^{T} QR = R^{T}Q^{T}QR = R^{T}IR = R^{T}R,$$

so the normal equations become

$$R^{T}Rx = A^{T}b = (QR)^{T} b = R^{T}Q^{T}b. \tag{16.2}$$

Note that in reduced *QR* decomposition R is an $n \times n$ matrix. We can simplify Equation 16.2 by using an important property of the reduced *QR* decomposition.

Lemma 16.2.   *If $A^{T}A = QR$ is the reduced decomposition of $A^{T}A$, then $r_{ii} > 0$, $1 \leq i \leq n$, where the r_{ii} are the diagonal entries of R.*

Proof. $A^T A = (QR)^T (QR) = R^T Q^T QR = R^T R$. $A^T A$ is positive definite, and by Theorem 13.3, R^T has positive diagonal entries. $\square$

By Lemma 16.2, the diagonal entries of R^T are positive, so R^T is invertible, and from Equation 16.2 we have

$$Rx = Q^T b \qquad (16.3)$$

Since R is an upper triangular, system 16.3 can be solved simply by back substitution. This leads to Algorithm 16.2.

Algorithm 16.2 Solving the Least-Squares Problem Using the QR Decomposition

---

```
function QRLSTSQ(A,b)
    % Solve the least-squares problem using QR decomposition.
    % Input: m × n matrix A of full rank and m × 1 column vector b.
    % Output: The solution to the least-squares problem Ax=b and the residual
    Compute the reduced QR decomposition of A: A=QR
    c = QᵀB
    Solve the upper triangular system Rx=c
    return[ x ‖b − Ax‖₂ ]
end function
```

NLALIB: The function `qrlstsq` implements Algorithm 16.2.

Example 16.5. The table shows the length-weight relation for a species of salmon. Find a power function $W = \alpha L^\beta$ which best fits the data.

L	0.5	1.0	2.0
W	1.77	10	56.6

Take the natural logarithm of the power function to obtain $\ln(W) = \beta \ln(L) + \ln(\alpha)$. In problems of this type, it is often necessary or convenient to perform a change of variable. Let $y = \ln(W)$, $x = \ln(L)$, and $b = \ln(\alpha)$, and $m = \beta$. Create a table containing the logarithms of L and W.

x	−0.6931	0.0000	0.6931
y	0.5710	2.3026	4.0360

We want the line $y = mx + b$ which best fits the data, so we must find b and m such that

$$\begin{array}{rcrcl} -0.6931m & + & b & = & 0.5710 \\ 0m & + & b & = & 2.3026 \\ 0.6931m & + & b & = & 4.0360 \end{array},$$

an overdetermined system.

We must solve the least-squares problem $\begin{bmatrix} -0.6931 & 1 \\ 0 & 1 \\ 0.6931 & 1 \end{bmatrix} \begin{bmatrix} m \\ b \end{bmatrix} = \begin{bmatrix} 0.5710 \\ 2.3026 \\ 4.0360 \end{bmatrix}$ for b and m. The following MATLAB code does this for us.

```
>> A = [-0.6931 1;0 1;0.6931 1];
>> b = [0.5710 2.3026 4.0360]';
>> x = qrlstsq(A,b)

x =
    2.4996
    2.3032
```

The least-squares line is $y = 2.4996x + 2.3032$. Now we need to find the corresponding power function.

$W = e^y = e^{2.4996x + 2.3032} = e^{2.3032}(e^x)^{2.4996} = e^{2.3032}\left(e^{\ln(L)}\right)^{2.4996} = e^{2.3032}L^{2.4996}$. The power function that best fits this data is (Figure 16.3)

$$W = 10.0062 L^{2.4996}$$ ∎

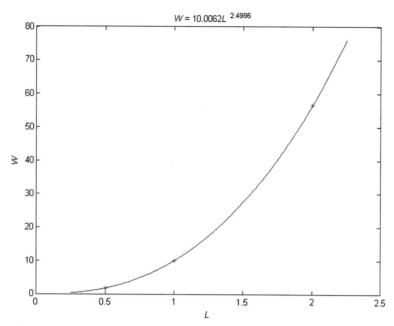

FIGURE 16.3 Least-squares estimate for the power function.

Efficiency

The decomposition is normally computed by the use of Householder reflections (Sections 17.8 and 17.9), and the cost of this algorithm is approximately $2n^2 \left(m - \frac{n}{3}\right)$ flops. The product $Q^{\mathrm{T}}b$ requires $2mn$, and back substitution costs $n^2 + n - 1$ flops, so the total cost of Algorithm 16.2 is

$$2mn^2 - \frac{2n^3}{3} + 2mn + n^2 + n + 1.$$

The cost of 16.2 is dominated by computing the reduced QR factorization, so we ignore $2mn + n^2 + n + 1$ and say that the QR solution to the full-rank least-squares problem, $m \geq n$ costs approximately

$$2mn^2 - \frac{2n^3}{3}$$

flops.

16.2.3 Using the SVD

Just like the QR decomposition, there is a *reduced SVD*, and it has the form (Figure 16.4)

$$A^{m \times n} = U^{m \times n} \Sigma^{n \times n} \left(V^{n \times n}\right)^{\mathrm{T}}.$$

In MATLAB, use the command $[U \ S \ V] = \mathrm{svd}(A, 0)$. The reduced SVD is also a powerful tool for computing full-rank least-squares solutions, $m \geq n$. The following manipulations show how to use it.

Apply the reduced SVD to A and obtain $A = U \Sigma V^{\mathrm{T}}$, where $U \in \mathbb{R}^{m \times n}$ has orthonormal columns, and $V \in \mathbb{R}^{n \times n}$ is orthogonal, and $\Sigma = \mathrm{diag}\,(\sigma_1, \sigma_2, \ldots, \sigma_n) \in \mathbb{R}^{n \times n}$, $\sigma_i > 0$, $1 \leq i \leq n$. Form the normal equations.

$$A^{\mathrm{T}}A = \left(U \Sigma V^{\mathrm{T}}\right)^{\mathrm{T}} U \Sigma V^{\mathrm{T}}$$

and so

$$A^{\mathrm{T}}A = (V \Sigma)\, \Sigma V^{\mathrm{T}}.$$

Now, $A^{\mathrm{T}}b = (V \Sigma)\, U^{\mathrm{T}}b$, and so the normal equations become

$$(V \Sigma)\, \Sigma V^{\mathrm{T}} x = (V \Sigma)\, U^{\mathrm{T}}b$$

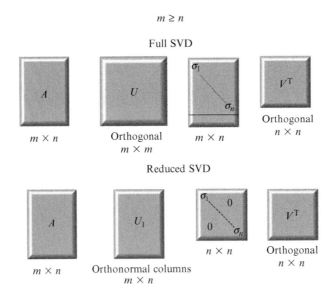

FIGURE 16.4 The reduced SVD for a full rank matrix.

Since V is orthogonal and Σ is a diagonal matrix with positive entries, $V\Sigma$ is invertible, and after multiplying the previous equation by $(V\Sigma)^{-1}$ we have

$$\Sigma V^\mathrm{T} x = U^\mathrm{T} b.$$

First solve $\Sigma y = U^\mathrm{T} b$, followed by $V^\mathrm{T} x = y$. Since Σ is a diagonal matrix, the solution to $\Sigma y = U^\mathrm{T} b$ is simple. Let
$U^\mathrm{T} b = \begin{bmatrix} c_1 \\ c_2 \\ \vdots \\ c_{n-1} \\ c_n \end{bmatrix}$. Then,

$$\begin{bmatrix} \sigma_1 & & & 0 \\ & \sigma_2 & 0 & \\ & 0 & \ddots & \\ 0 & & & \sigma_n \end{bmatrix}\begin{bmatrix} y_1 \\ y_2 \\ \vdots \\ y_n \end{bmatrix} = \begin{bmatrix} c_1 \\ c_2 \\ \vdots \\ c_n \end{bmatrix},$$

and $y_i = \frac{c_i}{\sigma_i}$, $1 \le i \le n$. The solution to $V^\mathrm{T} x = y$ is $x = Vy$. We summarize these steps in Algorithm 16.3.

Algorithm 16.3 Solving the Least-Squares Problem Using the SVD

```
function SVDLSTSQ(A,b)
  % Use the SVD to solve the least-squares problem.
  % Input: An m × n full-rank matrix A, m ≥ n, and an m × 1 vector b.
  % Output: The solution to the least-squares problem Ax=b and the residual
  Compute the reduced SVD of A: A = UΣVᵀ
  c = Uᵀb
  % Solve the system Σy = c
  for i =1:n do
    yᵢ = cᵢ/σᵢ
  end for
  x = Vy
  return[ x  ‖b − Ax‖₂ ]
end function
```

NLALIB: The function `svdlstsq` implements Algorithm 16.3. The implementation uses `svd(A,0)` to compute the reduced SVD for A, and computes y using

$$y = c./diag(S),$$

where S is the diagonal matrix.

Example 16.6.   The velocity of an enzymatic reaction with Michaelis-Menton kinetics is given by

$$v(s) = \frac{\alpha s}{1 + \beta s} \tag{16.4}$$

Find the Michaelis-Menton equation which best fits the data:

s	1	4	6	16
v	4	10	12	16

Inverting Equation 16.4 gives the Lineweaver-Burke equation:

$$\frac{1}{v} = \frac{1}{\alpha}\frac{1}{s} + \frac{\beta}{\alpha} \tag{16.5}$$

Perform the following change of variable: $y = \frac{1}{v}$ and $x = \frac{1}{s}$. Let $m = \frac{1}{\alpha}$ and $b = \frac{\beta}{\alpha}$. Equation 16.5 then becomes

$$y = mx + b$$

Recompute the table to reflect the change of variables.

x	1.0000	0.2500	0.1667	0.0625
y	0.2500	0.1000	0.0833	0.0625

Find the least-squares fit for $y = mx + b$ by solving the following 4×2 set of equations

$$1.0000m + b = 0.2500$$
$$0.2500m + b = 0.1000$$
$$0.1667m + b = 0.0833$$
$$0.0625m + b = 0.0625$$

that correspond to the matrix equation

$$\begin{bmatrix} 1.0000 & 1 \\ 0.2500 & 1 \\ 0.1667 & 1 \\ 0.0625 & 1 \end{bmatrix} \begin{bmatrix} m \\ b \end{bmatrix} = \begin{bmatrix} 0.2500 \\ 0.1000 \\ 0.0833 \\ 0.0625 \end{bmatrix}.$$

The following MATLAB code solves for m and b and then computes α, β.

```
>> A = [1.0000 1.0000;0.2500 1.0000;0.1667 1.0000;0.0625 1.0000];
>> b = [0.2500 0.1000 0.0833 0.0625]';
>> x = svdlstsq(A,b);
>> m = x(1);
>> b = x(2);
>> alpha = 1/m

alpha =

   4.9996

>> beta = alpha*b

beta =

   0.2499
```

The least-squares approximation is (Figure 16.5)

$$v(s) = \frac{4.9996s}{1 + 0.2499s}$$ ∎

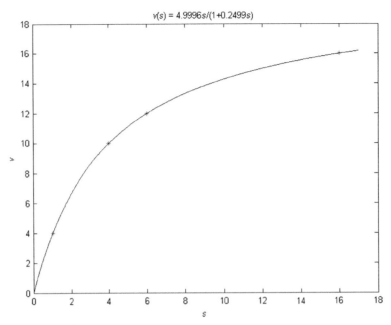

FIGURE 16.5 Velocity of an enzymatic reaction.

Efficiency

Algorithms for the computation of the SVD are complex, and we will develop two in Chapter 21. We will use the estimate given in Ref. [2, p. 493] of approximately $14mn^2 + 8n^3$ flops. Generally, using QR is faster for full rank least-squares problems. As we will see in Section 16.4, the SVD should be used for rank deficient problems.

16.2.4 Remark on Curve Fitting

To fit a polynomial of degree n to a set of points (x_i, y_i), you can avoid using the normal equations by computing the vandermonde matrix $V(x, n)$ and solving the least-squares problem $Vx = y$ using the QR or SVD approach.

16.3 CONDITIONING OF LEAST-SQUARES PROBLEMS

A perturbation analysis of least-squares problems is quite complex. This section will present a theorem without proof and provide some examples to illustrate what the theorem tells us. As you might expect, the perturbation result depends of the condition number for an $m \times n$ matrix, $m \geq n$. We will see that least-squares problems are more sensitive to perturbations than the solution of linear systems using Gaussian elimination. A sketch of the proof and a discussion of its consequences can be found in Ref. [1, pp. 117-118].

Theorem 16.3. *Assume that δA and δb are perturbations of the full-rank matrix $A \in \mathbb{R}^{m \times n} (m \geq n)$ and the vector $b \in \mathbb{R}^m$, respectively. Assume that x is the unique solution to the least-squares problem $Ax = b$ and that $\hat{x}$ is a solution to the least-squares problem $(A + \Delta A)\hat{x} = b + \Delta b$. Let r be the residual $r = b - Ax$, and assume that*

$$\epsilon = \max\left(\frac{\|\delta A\|_2}{\|A\|_2}, \frac{\|\delta b\|_2}{\|b\|_2}\right) < \frac{1}{\kappa(A)}.$$

Then,

$$\frac{\|\hat{x} - x\|_2}{\|x\|_2} \leq \epsilon \left(\frac{2\kappa_2(A)}{\cos\theta} + \tan\theta \ (\kappa_2(A))^2\right) + O\left(\epsilon^2\right),$$

where $\sin\theta = \frac{\|r\|_2}{\|b\|_2}$. In other words, θ, $0 < \theta < \frac{\pi}{2}$, is the angle between the vectors b and Ax (Figure 16.1).

There are a number of important consequences of Theorem 16.3.

a. If θ is small, the residual is small (Figure 16.1), and the sensitivity to perturbations depends on $\kappa\ (A)$.

b. If θ is not close to $\frac{\pi}{2}$, but $\kappa\ (A)$ is large, then the sensitivity depends on $(\kappa\ (A))^2$.

c. If θ is near $\frac{\pi}{2}$, the solution is nearly zero, but the solution to the least-squares problem, $\hat{x}$, produced by perturbations in A and b will not be zero, and $\frac{\|\hat{x}-x\|_2}{\|x\|_2}$ will be very large.

Example 16.7. In Ref. [49], Golub refers to the matrix

$$A = \begin{bmatrix} 1 & 1 & 1 & 1 & 1 \\ \epsilon & 0 & 0 & 0 & 0 \\ 0 & \epsilon & 0 & 0 & 0 \\ 0 & 0 & \epsilon & 0 & 0 \\ 0 & 0 & 0 & \epsilon & 0 \\ 0 & 0 & 0 & 0 & \epsilon \end{bmatrix}.$$

Choose $\epsilon = 0.01$. The following MATLAB script solves the least-squares problem $Ax = b$ with $b = \begin{bmatrix} 1 & 1 & 1 & 1 & 1 & 1 \end{bmatrix}^T$ using the function svdlstsq. Then, the nonzero elements of A are each perturbed by the same random value in the range $0 < \delta a_{ij} < 0.00001$, the elements of b perturbed by random values in the range $0 < \delta b_i < 0.001$, and svdlstsq finds the solution to the perturbed problem. Then, the code computes the norm of the difference between the two solutions. The result is interesting.

```
>> B = .00001*rand*A + A;
>> c = b + .001*rand(6,1);
>> x = svdlstsq(A,b);
>> y = svdlstsq(B,c);
>> norm(x-y)

ans =

    0.038215669993397
```

There is a very good reason for this. The condition number $\kappa\ (A) = \|A\|_2\ \|A^{\ddagger}\| = 223.609$. That does not seem too bad, but remember Theorem 16.3 says that the relative error can be bounded by the square of the condition number, and $(223.609)^2 = 50001$. ∎

16.3.1 Sensitivity when using the Normal Equations

[19], p. 118, states the following result that is an alternative to the bound in Theorem 16.3:

$$\frac{\|\tilde{x} - x\|_2}{\|x\|_2} \leq \frac{\epsilon \kappa_2\ (A)}{1 - \epsilon \kappa_2\ (A)} \left(2 + (\kappa_2\ (A) + 1)\ \frac{\|r\|_2}{\|A\|_2\ \|x\|_2} \right),$$

where r is the residual $r = \|b - Ax\|_2$. Using the normal equations requires solving $A^T Ax = A^T b$, and so the accuracy depends on the condition number $\kappa_2\ (A^T A)$. From Theorem 10.5, part (4)

$$\kappa_2\ (A^T A) = \kappa_2^2\ (A).$$

Thus the relative error, $\frac{\|\tilde{x}-x\|_2}{\|x\|_2}$, is always bounded by $\kappa_2^2\ (A)$ rather than $\kappa_2\ (A)$. As a result, use of the normal equations can lose twice as many digits of accuracy compared to the use of the QR or SVD method. Using the normal equations is only acceptable when the condition number of A is small, and in this case it is faster than either the QR or SVD approaches.

16.4 RANK-DEFICIENT LEAST-SQUARES PROBLEMS

To this point, we have dealt with least-squares problems $Ax = b$, $m \geq n$ and A having full rank. What happens if $m \geq n$ and rank $(A) < n$? Such problems are termed *rank deficient*, and arise in areas of science and engineering such as acoustics, tomography, electromagnetic scattering, image restoration, remote sensing, and the study of atmospheres, so there has to be a means of handling them.

Recall that a solution using the reduced QR decomposition requires solving $Rx = Q^T b$. In a rank-deficient problem, the reduced QR decomposition gives an upper triangular matrix R that is singular, and there will be infinitely many solutions to the least-squares problem; in other words there are infinitely many vectors x that minimize $\|b - Ax\|_2$ for $x \in \mathbb{R}^n$.

It is also possible that a problem is "almost rank deficient."

Example 16.8. Let $A = \begin{bmatrix} 1.0000 & -0.3499 \\ -2.0000 & 0.6998 \\ 8.0000 & -2.8001 \end{bmatrix}$. In the reduced QR decomposition,

$$R = \begin{bmatrix} 8.3066 & -2.9072 \\ 0 & 0.0002 \end{bmatrix},$$

and R is very close to singular and is singular to 3-digit accuracy. Using the QR method for computing the unique least-squares solution in situations like this will be highly sensitive to perturbations or will be impossible if R is singular. ∎

The solution to rank-deficient or nearly rank-deficient problems is to make a wise choice among the solutions, and the choice is to pick the solution with the smallest 2-norm, the *minimum norm solution*. There are two approaches to solving such a problem, using the QR method with column pivoting [2, pp. 288-298], or using the SVD. We will choose the latter.

Let $A \in \mathbb{R}^{m \times n}$ have rank r. The full SVD has the following form:

$$A = U \begin{bmatrix} \Sigma_r & 0 \\ 0 & 0 \end{bmatrix} V^T$$

$$\begin{array}{cccc} r & m-r & r & n-r \end{array}$$

$$U = [U_r, \quad U_{m-r}], \quad V = [V_r \quad V_{n-r}]$$

where U_r is the submatrix consisting of the first r columns, and so forth. The SVD is a very revealing orthogonal decomposition, as we saw in Section 15.2.1 when we discussed the four fundamental subspaces of a matrix. The SVD can be used to determine a formula for the minimum norm least-squares solution x_{LS} as well as the norm of the minimum residual r_{LS}. Theorem 16.4 shows how to compute x_{LS} and r_{LS}:

$$x_{LS} = \sum_{i=1}^{r} \left(\frac{u_i^T b}{\sigma_i} \right) v_i,$$

and the corresponding residual

$$r_{LS} = \sqrt{\sum_{i=r+1}^{m} \left(u_i^T b \right)^2}.$$

Computing the rank is a nontrivial problem. After computing the SVD, MATLAB determines the rank as follows:

```
tol = max(size(A))*eps(max(sigma))
r = sum(sigma > tol)
```

In the computation of `tol`, `max(size(A))` is the largest dimension of A, and `eps(max(sigma))` is the distance between σ_1 and the next largest floating point number in double precision arithmetic. To estimate the rank, r, find the number of singular values larger than tol. Theorem 16.4 provides a proof that x_{LS} and r_{LS} are correct, and can be omitted if desired.

Theorem 16.4. *Let $A = U\tilde{\Sigma}V^T$ be the SVD of $A^{m \times n}$, with $r = \text{rank}(A)$. If $U = \begin{bmatrix} u_1 & \cdots & u_m \end{bmatrix}$ and $V = \begin{bmatrix} v_1 & \cdots & v_n \end{bmatrix}$ are the columns of U and V, and $b \in \mathbb{R}^m$, then*

$$x_{LS} = \sum_{i=1}^{r} \left(\frac{u_i^T b}{\sigma_i} \right) v_i$$

minimizes $\|Ax - b\|_2$ and has the smallest 2-norm of all minimizers. In addition,

$$r_{LS}^2 = \|b - Ax_{LS}\|_2^2 = \sum_{i=r+1}^{m} \left(u_i^T b \right)^2.$$

Proof. Multiplication by an orthogonal matrix does not change the 2-norm, so

$$\|b - Ax\|_2^2 = \|U^T (b - Ax)\|_2^2. \tag{16.6}$$

Using the fact that $U^T A V = \tilde{\Sigma}$, Equation 16.6 gives

$$\|b - Ax\|_2^2 = \|U^T b - (U^T Ax)\|_2^2 = \|U^T b - (U^T AV)(V^T x)\|_2^2 = \|U^T b - \tilde{\Sigma}\alpha\|_2^2, \tag{16.7}$$

where $\alpha = V^T x = [\begin{array}{ccc} \alpha_1 & \cdots & \alpha_n \end{array}]^T$. Now,

$$\tilde{\Sigma}\alpha = \begin{bmatrix} \Sigma_r & 0 \\ 0 & 0 \end{bmatrix} \alpha = [\begin{array}{ccccccc} \overset{r}{\sigma_1 \alpha_1} & \cdots & \sigma_r \alpha_r & \overset{m-r}{0} & \cdots & 0 \end{array}]^T.$$

Write U^T as $\begin{bmatrix} u_1 \\ \vdots \\ u_m \end{bmatrix}$, where u_i are the rows of U^T, and then note that row i of the product $U^T b$ is the inner product of u_i and b, which is $u_i^T b$, so we obtain

$$U^T b = \begin{bmatrix} u_1 \\ \vdots \\ u_m \end{bmatrix} b = \begin{bmatrix} u_1^T b \\ \vdots \\ u_m^T b \end{bmatrix}. \tag{16.8}$$

Use Equations 16.8 and 16.9 in Equation 16.7, to obtain

$$\|b - Ax\|_2^2 = \left\| \begin{bmatrix} u_1^T b - \sigma_1 \alpha_1 \\ \vdots \\ u_r^T b - \sigma_r \alpha_r \\ u_{r+1}^T b \\ \vdots \\ u_m^T b \end{bmatrix} \right\|_2^2,$$

from which we get

$$\|b - Ax\|_2^2 = \sum_{i=1}^{r} \left(u_i^T b - \sigma_i \alpha_i\right)^2 + \sum_{i=r+1}^{m} \left(u_i^T b\right)^2. \tag{16.9}$$

Notice that the term $\sum_{i=r+1}^{m} \left(u_i^T b\right)^2$ in Equation 16.9 is independent of x, so the minimum value of $\|Ax - b\|_2^2$ occurs when

$$\alpha_i = \frac{u_i^T b}{\sigma_i}, \quad 1 \leq i \leq r. \tag{16.10}$$

Since $\alpha = V^T x$, $x = V\alpha$, from Equation 16.10, we have

$$x = \sum_{i=1}^{r} \left(\frac{u_i^T b}{\sigma_i}\right) v_i + \sum_{i=r+1}^{n} \alpha_i v_i \tag{16.11}$$

If we choose x such that $\alpha_i = 0$, $r+1 \leq i \leq n$, and then x will have the smallest possible 2-norm, and the minimum residual is

$$r_{LS} = \sqrt{\sum_{i=r+1}^{m} \left(u_i^T b\right)^2}. \tag{16.12}$$

$\square$

Remark 16.2. If $m \geq n$ and rank $(A) = n$, then the term $\sum_{i=r+1}^{n} \alpha_i v_i$ is not present in Equation 16.11, and the solution is unique. If A is rank deficient, then $r < n$, and there are infinitely many ways to choose α_i, $r+1 \leq i \leq n$, so the rank-deficient least-squares problem has infinitely many solutions. However, there is only one solution x_{LS} with minimum norm; that is, when we choose $\alpha_i = 0$, $r + 1 \le i \le n$ so that

$$x_{LS} = \sum_{i=1}^{r} \left(\frac{u_i^T b}{\sigma_i} \right) v_i. \tag{16.13}$$

Example 16.9. Let $A = \begin{bmatrix} 2 & 1 & 1 & 2 \\ 1 & 2 & 1 & 2 \\ 1 & 1 & 2 & 2 \\ 2 & 2 & 2 & 3 \end{bmatrix}$, $b = \begin{bmatrix} -1 \\ 5 \\ 3 \\ 2 \end{bmatrix}$. The rank of A is 3, so A is rank deficient and we will apply the results of Theorem 16.4. First, compute the SVD:

$$A = \begin{bmatrix} -0.4364 & 0.8165 & 0 & -0.3780 \\ -0.4364 & -0.4082 & 0.7071 & -0.3780 \\ -0.4364 & -0.4082 & 0.7071 & -0.3780 \\ -0.6547 & 0 & 0 & 0.7559 \end{bmatrix} \begin{bmatrix} 7 & 0 & 0 & 0 \\ 0 & 1 & 0 & 0 \\ 0 & 0 & 1 & 0 \\ 0 & 0 & 0 & 0 \end{bmatrix} \begin{bmatrix} -0.4364 & -0.4364 & -0.4364 & -0.6547 \\ 0.8165 & -0.4082 & -0.4082 & 0 \\ 0 & 0.7071 & -0.7071 & 0 \\ -0.3780 & -0.3780 & -0.3780 & 0.7559 \end{bmatrix}$$

Equation 16.13 gives the minimum norm solution.

$$x_{LS} = \sum_{i=1}^{3} \left(\frac{u_i^T b}{\sigma_i} \right) v_i$$

$$= \frac{\begin{bmatrix} -0.4364 & -0.4364 & -0.4364 & -0.6547 \end{bmatrix} \begin{bmatrix} -1 & 5 & 3 & 2 \end{bmatrix}^T}{7} \begin{bmatrix} -0.4364 \\ -0.4364 \\ -0.4364 \\ -0.6547 \end{bmatrix} + \cdots = \begin{bmatrix} -3.0612 \\ 2.9388 \\ 0.9388 \\ 0.4082 \end{bmatrix}$$

∎

Algorithm 16.4 finds the minimum norm solution to the rank-deficient least-squares problem, but it will work for a full-rank problem; however, with full-rank problems, use either svdlstsq or qrlstsq.

Algorithm 16.4 Minimum Norm Solution to the Least-Squares Problem

---

```
function RDLSTSQ(A,b)
   % Compute the minimum norm solution to the
   % linear least-squares problem using the SVD.
   % Input: m × n matrix A and m × 1 vector b.
   % Output: Solution x and the residual
   [U Σ̃ V] = svd(A)
   σ = diag(Σ̃)
   % Compute the rank of A.
   tol = max(size(A)) eps(max(σ))
   r = ∑(elements of σ > tol)
   x = 0
   for i=1:r do
      x = x + (U(:, i)ᵀb / σ(i)) V(:, i)
   end for
   residual = 0
   for i=r+1:m do
      residual = residual + (bᵀU(:, i))²
   end for
   return[ x √residual ]
end function
```

NLALIB: The function `rdlstsq` implements Algorithm 16.4.

Example 16.10. Let

$$A = \begin{bmatrix} 7 & 1 & 8 \\ -1 & 5 & 0 \\ -1 & 6 & 9 \\ 0 & 2 & 9 \\ 1 & 2 & 3 \\ 3 & 5 & 7 \end{bmatrix}, \quad b1 = \begin{bmatrix} 1 \\ 1 \\ 1 \\ 1 \\ 1 \\ 1 \end{bmatrix},$$

and

$$B = \begin{bmatrix} 1 & 1 & 1 \\ 1 & 1 & 1 \\ 1 & 1 & 1 \\ 1 & 1 & 1 \end{bmatrix}, \quad b2 = \begin{bmatrix} 1 \\ 2 \\ 3 \\ 4 \end{bmatrix}.$$

The rank of A is 3, so it is not rank deficient, and `qrlstsq` applies. However, the rank of B is 1 and we must use `rdlstsq`.

```
>> x1 = qrlstsq(A,b1)

x1 =
    0.0618
    0.1548
    0.0454
>> x = rdlstsq(B,b2)

x =
    0.8333
    0.8333
    0.8333
```

■

It is now appropriate to introduce an equivalent definition for the pseudoinverse that is a helpful theoretical tool.

Definition 16.6. Let A be an $m \times n$ matrix, $m \geq n$, having rank $r \leq n$, with SVD $A = U\tilde{\Sigma}V^{\mathrm{T}}$. The pseudoinverse is

$$A^{\dagger} = V\Sigma^{+}U^{\mathrm{T}},$$

where

$$\Sigma^{+} = \mathrm{diag}\left(\frac{1}{\sigma_1}, \frac{1}{\sigma_2}, \ldots, \frac{1}{\sigma_r}, 0, 0, \ldots, 0\right)$$

is an $n \times m$ diagonal matrix. If A has full rank, this definition is equivalent to Definition 16.3 (Problem 16.11).

This definition enables us to reformulate the results of Theorem 16.4 using the pseudoinverse, namely, that $x_{\mathrm{LS}} = A^{\dagger}b$. Let u_i, $1 \leq i \leq m$, be the columns of U, and v_i, $1 \leq i \leq n$, be the columns of V. Begin by computing $U^{\mathrm{T}}b$, which we can write as

$$U^{\mathrm{T}}b = \begin{bmatrix} u_1 \\ u_2 \\ \vdots \\ u_r \\ u_{r+1} \\ \vdots \\ u_m \end{bmatrix} b = \begin{bmatrix} \langle u_1, b \rangle \\ \langle u_2, b \rangle \\ \vdots \\ \langle u_r, b \rangle \\ \langle u_{r+1}, b \rangle \\ \vdots \\ \langle u_m, b \rangle \end{bmatrix},$$

and then

$$\Sigma^{+} U^{\mathrm{T}} b = \begin{bmatrix} \frac{1}{\sigma_1} \langle u_1, b \rangle \\ \frac{1}{\sigma_2} \langle u_2, b \rangle \\ \vdots \\ \frac{1}{\sigma_r} \langle u_r, b \rangle \\ 0 \\ \vdots \\ 0 \end{bmatrix}.$$

To finish up,

$$V \Sigma^{+} U^{\mathrm{T}} b = \begin{bmatrix} v_{11} & v_{12} & \dots & v_{1n} \\ v_{21} & v_{22} & \dots & v_{2n} \\ \vdots & \vdots & \ddots & \\ v_{n1} & v_{n2} & \dots & v_{nn} \end{bmatrix} \begin{bmatrix} \frac{1}{\sigma_1} \langle u_1, b \rangle \\ \frac{1}{\sigma_2} \langle u_2, b \rangle \\ \vdots \\ \frac{1}{\sigma_r} \langle u_r, b \rangle \\ 0 \\ \vdots \\ 0 \end{bmatrix}$$

$$= \begin{bmatrix} \frac{1}{\sigma_1} \langle u_1, b \rangle v_{11} + \frac{1}{\sigma_2} \langle u_2, b \rangle v_{12} + \dots + \frac{1}{\sigma_r} \langle u_r, b \rangle v_{1r} \\ \frac{1}{\sigma_1} \langle u_1, b \rangle v_{21} + \frac{1}{\sigma_2} \langle u_2, b \rangle v_{22} + \dots + \frac{1}{\sigma_r} \langle u_r, b \rangle v_{2r} \\ \vdots \\ \frac{1}{\sigma_1} \langle u_1, b \rangle v_{n1} + \frac{1}{\sigma_2} \langle u_2, b \rangle v_{n2} + \dots + \frac{1}{\sigma_r} \langle u_r, b \rangle v_{nr} \end{bmatrix}$$

$$= \frac{1}{\sigma_1} \langle u_1, b \rangle v_1 + \frac{1}{\sigma_2} \langle u_1, b \rangle v_2 + \dots \frac{1}{\sigma_r} \langle u_1, b \rangle v_r$$

$$= \sum_{i=1}^{r} \left(\frac{1}{\sigma_i} u_i^{\mathrm{T}} b \right) v_i = x_{\mathrm{LS}}$$

It is also the case that

$$\|b - A x_{\mathrm{LS}}\|_2 = \left\| \left(I^{m \times m} - A A^{\dagger} \right) b \right\|_2 \qquad (16.14)$$

(Problem 16.12). The algorithm rdlstsq is more efficiently stated as is, rather than using the pseudoinverse. However, the pseudoinverse plays an important role in developing theorems dealing with least squares.

16.4.1 Efficiency

The full SVD computation requires approximately $4m^2n + 8mn^2 + 9n^3$ flops [2, p. 493]. Once the SVD computation is complete, the rank computation requires negligible effort. The for loop executes r times. Each execution requires 1 division, 1 addition, $2m$ flops for computing $U(:,i)^{\mathrm{T}} b$, and n multiplications, so the cost of the for loop is $r(2m + n + 2)$. This is negligible, so the cost of rdlstsq is $\approx 4m^2n + 8mn^2 + 9n^3$.

Remark 16.3. If a matrix is almost rank deficient (has one or more small singular values), an SVD approach should be used. The algorithm rdlstsq is a good choice, but svdlstsq can also be used.

16.5 UNDERDETERMINED LINEAR SYSTEMS

Consider the least-squares problem $Ax = b$, where A is an $m \times n$ matrix, $m < n$, and b is an $m \times 1$ vector, and the solution x is has dimension $n \times 1$. Such a system is called *underdetermined* (Figure 16.6).

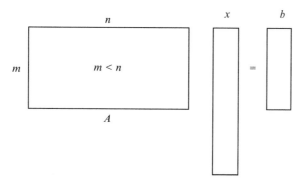

FIGURE 16.6 Underdetermined system.

If A has full rank (rank $(A) = m$), then the matrix A^{T} with dimension $n \times m$, $n > m$ has full rank. The QR decomposition can be used to obtain a solution to a full-rank underdetermined system by applying it to A^{T}. Compute the full QR decomposition $A^{\mathrm{T}} = QR$, where Q is $n \times n$ and R is $n \times m$. Then, using block matrix notation

$$
Q^{\mathrm{T}} A^{\mathrm{T}} = R = \begin{array}{c} \\ m \\ n-m \end{array} \overset{\displaystyle m}{\left[\begin{array}{c} R_1 \\ 0 \end{array} \right]}
$$

where R_1 is an $m \times m$ upper triangular matrix. Now, $A = R^{\mathrm{T}} Q^{\mathrm{T}}$, so $Ax = b$ can be written as $R^{\mathrm{T}} \left(Q^{\mathrm{T}} x \right) = b$. Again using block matrix notation the equation becomes

$$
m \overset{\displaystyle m \quad n-m}{\left[R_1^{\mathrm{T}} \ \ 0 \right]} \left[\begin{array}{c} y_1 \\ \vdots \\ y_n \end{array} \right] = \left[\begin{array}{c} b_1 \\ \vdots \\ b_m \end{array} \right],
$$

where $y = Q^{\mathrm{T}} x$. Solve $R_1^{\mathrm{T}} \left[\begin{array}{c} y_1 \\ \vdots \\ y_m \end{array} \right] = b$, and let $y = \left[\begin{array}{ccccc} y_1 & \cdots & y_m & 0 & \cdots & 0 \end{array} \right]^{\mathrm{T}}$. We must now compute $x = Qy$ to obtain the

solution. There is no need to perform the entire $(n \times n)\,(n \times 1)$ product due to the presence of the $n-m$ zeros in y. We have

$$
Qy = \left[\begin{array}{ccccccc} q_{11} & \cdots & q_{1m} & q_{1,m+1} & \cdots & q_{1n} \\ q_{21} & \cdots & q_{2m} & q_{2,m+1} & \cdots & q_{2n} \\ \vdots & \ddots & \vdots & \vdots & \vdots & \vdots \\ \vdots & \vdots & \vdots & \vdots & \vdots & \vdots \\ q_{n1} & \cdots & q_{nm} & q_{n,m+1} & \cdots & q_{nn} \end{array} \right] \left[\begin{array}{c} y_1 \\ \vdots \\ y_m \\ 0 \\ \vdots \\ 0 \end{array} \right] = Q_1 \left[\begin{array}{ccc} q_{1,m+1} & \cdots & q_{1n} \\ q_{2,m+1} & \cdots & q_{2n} \\ \vdots & \vdots & \vdots \\ \vdots & \vdots & \vdots \\ q_{n,m+1} & \cdots & q_{nn} \end{array} \right] \left[\begin{array}{c} y_1 \\ \vdots \\ y_m \\ 0 \\ \vdots \\ 0 \end{array} \right]
$$

$$
= Q_1^{n \times m} \left[\begin{array}{c} y_1 \\ \vdots \\ y_m \end{array} \right] = \left[\begin{array}{c} x_1 \\ x_2 \\ \vdots \\ \vdots \\ x_{n-1} \\ x_n \end{array} \right]
$$

Algorithm 16.5 implements this process.

Algorithm 16.5 Solution of Full-Rank Underdetermined System Using QR Decomposition

---

```
function UQRLSTSQ(A,b)
    % Solve the m × n full rank underdetermined system Ax=b
    % using the QR decomposition
    % Input: m × n matrix A, m < n, and m × 1 vector b.
    % Output: minimum norm solution n × 1 vector x and the residual
    [ Q R ] = qr (Aᵀ)
    R₁ = R(1 : m, 1 : m)
    Solve the m × m lower triangular system R₁ᵀy = b
    Q₁ = Q(1 : n, 1 : m)
    x = Q₁y
    return[ x  ‖b − Ax‖₂ ]
end function
```

NLALIB: The function `uqrlstsq` implements Algorithm 16.5.

Example 16.11. Find the minimum norm solution to the full-rank system $Ax = b$, where

$$A = \begin{bmatrix} 1 & 3 & 5 & 7 & 9 \\ -1 & -2 & -3 & -4 & -5 \\ 6 & 12 & 8 & 9 & 10 \end{bmatrix}$$

and

$$b = \begin{bmatrix} 1 & 5 & 8 \end{bmatrix}^{\mathrm{T}}.$$

```
>> rank(A)

ans =
     3

>> uqrlstsq(A,b)

ans =
  -18.4286
   13.6000
   -7.5143
   -2.0571
    3.4000
```

∎

The results of Theorem 16.4 apply to the underdetermined rank deficient problems as well. By changing a_{32} from 12 to 7, the matrix A has rank 2, and we must use the algorithm implemented by the function `rdlstsq`.

Example 16.12. $A = \begin{bmatrix} 1 & 3 & 5 & 7 & 9 \\ -1 & -2 & -3 & -4 & -5 \\ 6 & 7 & 8 & 9 & 10 \end{bmatrix}$ and $b = \begin{bmatrix} 1 & 5 & 8 \end{bmatrix}^{\mathrm{T}}.$

```
>> x = rdlstsq(A,b)

x =
    1.1741
    0.7361
    0.2980
   -0.1401
   -0.5782
```

∎

Remark 16.4. The MATLAB function `pinv` computes the pseudoinverse and can be used for overdetermined, underdetermined, or rank-deficient problems in the following way:

```
x = pinv(A)*b
```

16.5.1 Efficiency

The QR decomposition requires approximately $4\left(n^2m - nm^2 + \frac{m^3}{3}\right)$ flops when Q is required. The solution of the $m \times m$ lower triangular system requires $m^2 + m - 1$ flops, and forming the product Q_1y costs $2mn$ flops. The QR decomposition dominates, so the cost is approximately $4\left(n^2m - nm^2 + \frac{m^3}{3}\right)$ flops.

16.6 CHAPTER SUMMARY

The Least-Squares Problem

If A is an $m \times n$ matrix, a least-squares solution to the problem $Ax = b$, $b \in \mathbb{R}^m$, $x \in \mathbb{R}^n$ is a value of x for which $\|b - Ax\|_2$ is minimum. Figure 16.1 shows that the residual vector, $b - Ax$ must be orthogonal to the range of A, a subset of $\mathbb{R}^m$. This leads to the fact that the normal equations $A^TAx = A^Tb$ must be satisfied by any least-squares solution.

Existence, Uniqueness, and the Normal Equations

If A is an $m \times n$ matrix, $m \geq n$, a vector $x \in \mathbb{R}^n$ is a least-squares solution if and only if x satisfies the normal equations; furthermore, x is unique if and only if A has full rank.

The Pseudoinverse

The pseudoinverse, $A^{\dagger} = \left(A^TA\right)^{-1}A^T$, is a generalization of the square matrix inverse; in fact, if A is square and nonsingular, $A^{\dagger} = A^{-1}$. The pseudoinverse allows the definition of the condition number for a general $m \times n$ matrix as $\kappa(A) = \left\|A^{\dagger}\right\|_2 \|A\|_2$.

Overdetermined Problems

A least-squares problem is overdetermined if A has dimension $m \times n$, $m > n$; in other words, there are more equations than unknowns. This is the most common type of least-squares problem.

Solving Overdetermined Problems Using the Normal Equations

If A has full rank, solving the normal equations $A^TAx = A^Tb$ by applying the Cholesky decomposition to the symmetric positive definite matrix A^TA yields a unique least-squares solution. Despite its simplicity, this approach is almost never used, since

- There may be some loss of significant digits when computing A^TA. It may even be singular to working precision.
- The accuracy of the solution using the normal equations depends on the square of condition number of the matrix. If $\kappa(A)$ is large, the results can be seriously in error.

Solving Overdetermined Problems Using the QR Decomposition

This is a time-honored technique for solving full-rank overdetermined least-squares problems. Find the reduced QR decomposition of A, and the solution is $Rx = Q^Tb$, easily obtained using back substitution. The primary cost is the computation of the QR decomposition.

Solving overdetermined Problems Using the SVD

By first computing the reduced SVD of A, $A = U\Sigma V^T$, the solution to the full-rank overdetermined least-squares problem is found by first solving $\Sigma y = U^Tb$, followed by computing $x = Vy$.

Conditioning of Least-Squares Problems

Perturbation analysis for least-squares problems is complex. If θ is the angle between b and the range of Ax, then

a. if θ is small, the residual is small (Figure 16.1), and the sensitivity to perturbations depends on $\kappa(A)$.

b. if θ is not close to $\frac{\pi}{2}$, but $\kappa(A)$ is large, then the sensitivity depends on $(\kappa(A))^2$.

c. if θ is near $\frac{\pi}{2}$, the solution is nearly zero, but the solution to the least-squares problem, $\hat{x}$, produced by perturbations in A and b will not be zero, and $\frac{\|\hat{x}-x\|_2}{\|x\|_2}$ will be very large.

Thus, assuming that b is not almost orthogonal to $R(A)$, relative errors depend at best on $\kappa(A)$ and at worst on $(\kappa(A))^2$.

Rank Deficient Problems

An overdetermined problem is rank deficient if rank $(A) < n$. Such systems have infinitely many solutions, so the strategy is to determine the unique solution $\hat{x}$ with minimum norm among the solutions. After computing the SVD of A, there is a simple formula for the minimum norm solution. The section also develops an alternative but equivalent definition of the pseudoinverse. This formulation is primarily a theoretical tool.

Underdetermined Problems

A problem is underdetermined when A has dimension $m \times n$, $m < n$. The matrix A^T is of full rank, and a solution can be obtained using the QR decomposition and submatrix operations that avoid computations dealing with blocks of zeros.

16.7 PROBLEMS

16.1 Three variables x, y, and z are required to satisfy the equations

$$
\begin{aligned}
3x - y + 7z &= 0 \\
2x - y + 4z &= 1/2 \\
x - y + z &= 1 \\
6x - 4y + 10z &= 3
\end{aligned}
$$

Is there a solution in the normal sense, not a least-squares solution? If so, is it unique, or are there infinitely many solutions?

16.2 If A is an invertible matrix and assuming exact arithmetic, are the solutions using Gaussian elimination and least squares identical? Prove your assertion.

16.3 If A is an $m \times n$ matrix, $m \geq n$, and b is an $m \times 1$ vector, show that if $A^T b = 0$, b is orthogonal to the range of A.

16.4 If A is an $m \times n$ full-rank matrix, $m \geq n$, show that

$$
\kappa(A) = \left\| A^{\ddagger} \right\|_2 \|A\|_2 = \frac{\sigma_1}{\sigma_n},
$$

where σ_1 and σ_n are the largest and smallest singular values of A, respectively.

16.5

   a. What is the Moore-Penrose inverse of a nonzero column vector?

   b. Answer the same question for a row vector.

16.6

   a. If A is a full rank $m \times n$ matrix, with $m < n$, show that

$$
x = A^T \left(AA^T \right)^{-1} b + \left(I - A^T \left(AA^T \right)^{-1} A \right) y
$$

satisfies the normal equations, where y is any $n \times 1$ vector. If $y = 0$, $x = A^T \left(AA^T \right)^{-1} b$ is the minimum norm solution.

   b. Show how to evaluate x without computing $\left(AA^T \right)^{-1}$.

   c. Is using this formula a reliable way to compute x? Explain your answer.

16.7 There are many identities involving the pseudoinverse. Let $A \in \mathbb{R}^{m \times n}$. Show that

   a. $A^{\dagger} A A^{\dagger} = A^{\dagger}$

b. $\left(A^{\dagger}A\right)^{T} = I$

c. $AA^{\dagger}A = A$

d. $\left(AA^{\dagger}\right)^{T} = AA^{\dagger}$

16.8 Show that if $A \in \mathbb{R}^{m \times n}$,

  a. $\left(I - AA^{\dagger}\right)A = 0$

  b. $A\left(I - A^{\dagger}A\right) = 0$

16.9 Show that if A has orthonormal columns, then $A^{\dagger} = A^{T}$.

16.10 If $A \in \mathbb{R}^{m \times n}$, then $B^{n \times m}$ is a *right inverse* of A if $AB = I^{m \times m}$. $C^{n \times m}$ is a *left inverse* of A if $CA = I^{n \times n}$.

  a. Prove that if rank $(A) = n$, $A^{\dagger}$ is a left inverse of A.

  b. Prove that if rank $(A) = m$, $A^{\dagger}$ is a right inverse of A.

16.11 If $A^{m \times n}$, $m \geq n$, is of full rank, show that Definitions 16.3 and 16.6 for $A^{\dagger}$ are equivalent.

16.12 Show that $\|b - Ax_{LS}\|_{2} = \left\|\left(I - AA^{\dagger}\right)b\right\|_{2}$.

16.13 Prove that all solutions $\hat{x}$ of $\min_{x \in \mathbb{R}^{n}} \|Ax - b\|_{2}$ are of the form $\hat{x} = A^{\dagger}b + \eta$, where $\eta \in$ null (A).

16.14 If some components of $Ax - b$ are more important than others, a weight $w_{i} > 0$ can be attached to each component.

By forming the diagonal matrix $W = \begin{bmatrix} w_1 & & 0 \\ & \ddots & \\ 0 & & w_m \end{bmatrix}$, the problem becomes one of minimizing the norm of

$W(Ax - b)$. This called a *weighted least-squares* problem. The approach is to define an inner product and minimize the norm of $Ax - b$ relative to this inner product. Of course, this inner product must use W in some way.

  a. If S is a symmetric positive definite matrix, define $\langle x, y \rangle_{S} = x^{T}Sy$, and verify it is an inner product by showing that

   i. $\langle x, y + z \rangle_{S} = \langle x, y \rangle_{S} + \langle x, z \rangle_{S}$

   ii. $\langle cx, y \rangle_{S} = \langle x, cy \rangle_{S} = c \langle x, y \rangle_{S}$, where c is a scalar

   iii. $\langle x, y \rangle_{S} = \langle y, x \rangle_{S}$

   iv. $\langle x, 0 \rangle_{S} = 0$

   v. If $\langle x, x \rangle_{S} = 0$, then $x = 0$.

  b. Show that the weight matrix W is positive definite.

  c. Using an argument similar to the one used to derive the normal equations $A^{T}Ax = A^{T}b$, derive, $A^{T}WAx = A^{T}Wb$, the normal equations for the weighted least-squares problem.

  d. Let $W^{\frac{1}{2}} = \text{diag}\left(\sqrt{w_1}, \sqrt{w_2}, \ldots, \sqrt{w_m}\right)$, so that $\left(W^{\frac{1}{2}}\right)^{2} = W$. Rewrite the normal equations in part (c) so that the problem becomes an ordinary least-squares problem with the rescaled matrix $W^{\frac{1}{2}}A$.

  e. Develop an algorithm to solve the weighted least-squares problem, $m \geq n$, using the QR decomposition.

16.15 In Ref. [2, pp. 288-291], there is a discussion of rank deficient least-squares problem that includes some sensitivity issues, which are more complex than those for full-rank problems. It is stated that small changes in A and b can cause large changes to the solution, $x_{LS} = A^{\dagger}b$. The sensitivity is related to the behavior of the pseudoinverse. Using the example in Ref. [2], let $A = \begin{bmatrix} 1 & 0 \\ 0 & 0 \\ 0 & 0 \end{bmatrix}$ and $\delta A = \begin{bmatrix} 0 & 0 \\ 0 & \epsilon \\ 0 & 0 \end{bmatrix}$.

  a. Show that

$$\left\|A^{\dagger} - (A + \delta A)^{\dagger}\right\|_{2} = \frac{1}{\epsilon}.$$

  b. What does this say about

$$\lim_{\delta A \to 0} \left\|A^{\dagger} - (A + \delta A)^{\dagger}\right\|_{F}?$$

16.7.1 MATLAB Problems

16.16 To facilitate printing vectors, implement a function

```
printvec(msg,x)
```

that outputs the string `msg`, followed by the elements of x, eight per line. Place the file `printvec.m` in a directory of your choice, and add the directory to the MATLAB search path.

16.17 Using least squares, fit a line and a parabola to the data. Plot the two curves on the same set of axes.

x	0	2	3	5	8	11	12	15
y	50	56	60	72	85	100	110	125

Compute the condition number of the associated Vandermonde matrix in each case.

16.18 Let

$$A = \begin{bmatrix} 1 & 1 \\ 2 & 3 \\ 0 & 1 \end{bmatrix}, \quad b = \begin{bmatrix} 0 \\ 5 \\ 1 \end{bmatrix}.$$

In parts (a)–(d), find the unique least-squares solution x using
a. $x = A^{\dagger}b$
b. The normal equations
c. The QR method
d. The SVD method
e. Find $\kappa(A)$.

16.19 The data in the table give the approximate population of the United States every decade from 1900 to 1990.

Year	1900	1910	1920	1930	1940	1950	1960	1970	1980	1990
Populations (millions)	76.1	92.4	106.5	123.1	132.1	152.3	180.7	205.1	227.2	249.4

Assume that the population growth can be modeled with an exponential function $p = be^{mx}$, where x is the year and p is the population in millions. Use least squares to determine the constants b and m for which the function best fits the data, and graph the data and the exponential curve on the same set of axes. Use the equation to estimate the population in the years 1998, 2010, and 2030.

16.20 Consider the following data:

x	1.0	1.4	1.9	2.2	2.8	3.0
y	0.26667	0.23529	0.20513	0.19048	0.16667	0.16

Determine the coefficients α and β in the function $y = \frac{1}{\alpha x + \beta}$ that best fits the data using least squares, and graph the data and the curve $\frac{1}{\alpha x + \beta}$ on the same set of axes.

16.21 The following are values of a function of the form $y = \frac{x}{\alpha + \beta x}$. Using least squares, estimate α and β and approximate y at $x = \{\, 25.0 \quad 33.8 \quad 36.0 \,\}$. Graph the data and the curve $\frac{x}{\alpha x + \beta}$ on the same set of axes.

x	20	21.3	21.9	30.6	32.0	33.3
y	1.0152	1.027	1.032	1.0859	1.0922	1.0976

16.22 Radium-226 decays exponentially. The data shows the amount, y, of radium-226 in grams after t years. Find an exponential function $y = Ae^{rt}$ which best fits the data. Estimate the half-life of radium-226. Graph the data and the curve $y = Ae^{rt}$ on the same set of axes.

t	0	100	350	500	1000	1800	2000
y	10	9.5734	8.5847	8.0413	6.4662	4.5621	4.1811

16.23 The table shows the length-weight relation for Pacific halibut. Find a power function $W = \alpha L^{\beta}$ which best fits the data. Graph the data and the curve $W = \alpha L^{\beta}$ on the same set of axes.

L	0.5	1.0	1.5	2.0	2.5
W	1.3	10.4	35	82	163

16.24 The equation for a stretched beam is

$$y = l + Fs,$$

where l is the original length, F is the force applied, and s is the inverse coefficient of stiffness. The following measurements were taken:

F	20	25	27	30	33
y	22.3	22.8	23.2	23.5	25.5

Estimate the initial length and the inverse coefficient of stiffness.

16.25 A resistor in an electric circuit satisfies the equation $V = RI$, where V is the voltage and I is the current. The following measurements of voltage and current were taken:

I	2.00	1.90	1.85	2.06	1.95
V	4.95	5.05	5.10	4.92	5.02

Estimate the resistance.

16.26 The equation $z = Ax + By + D$ represents a plane, and the points in the table are on or close to a plane. Use least squares to determine the plane, and graph the data points and the plane on the same set of xyz-axes.

x	−1.5	−1.0	−1.9	0.0	0.5	1.0	1.3	1.95
y	1.0	1.6	1.5	−1.4	−1.3	−1.6	−1.8	1.9
z	1.0	1.4	−0.3	6.4	7.3	8.6	9.4	7.0

The following MATLAB statements plot the data points:

```
% plot the data points in black circles
H=plot3(x,y,z,'ko');
% double the circle size
set(H,'Markersize',2*get(H,'Markersize'));
% fill the circles with black
set(H,'Markerfacecolor','k');
hold on;
```

To plot the plane, use the functions `meshgrid` and `surf`. After calling `surf`, the statement `alpha(0.5)` will assure that the plane will not obscure portions of the circles.

16.27 Graph the following data, and then fit and graph polynomials of degree 1, 2, and 3. All graphs must be on the same set of axes. Which fit seems best?

x	−0.25	0.00	0.25	0.50	0.75	1.00	1.25	1.50	1.75	2.00
y	1.9047	1.3	0.63281	0.1375	0.048437	0.6	2.0266	4.5625	8.4422	13.9

16.28 The MATLAB command `pinv(A)` computes $A^\dagger$, the pseudoinverse. The MATLAB statement `pinv(A)*b` computes the solution to a least-squares problem. For each part, you are given a least-squares problem. Solve each problem two ways, using one of the functions developed in this chapter and by using `pinv`.

a. $\begin{bmatrix} 1 & -1 & 2 \\ 3 & 0 & 1 \\ 4 & 2 & -8 \\ 5 & 2 & 7 \end{bmatrix} x = \begin{bmatrix} 1 \\ -1 \\ 3 \\ 7 \end{bmatrix}$

b. $\begin{bmatrix} 2 & 3 & 1 \\ -1 & 2 & 3 \\ 3 & 1 & -2 \\ 6 & 9 & 3 \end{bmatrix} x = \begin{bmatrix} -1 \\ 3 \\ 5 \\ 0 \end{bmatrix}$

c. $\begin{bmatrix} 2 & 5 & 6 & 8 \\ -1 & 3 & 7 & 1 \\ 5 & 1 & 6 & 2 \end{bmatrix} x = \begin{bmatrix} 1 \\ -1 \\ 10 \end{bmatrix}$

d. $\begin{bmatrix} 1 & -6 & 2 & -4 \\ 3 & -18 & 6 & -12 \\ 9 & 0 & 12 & 0 \end{bmatrix} x = \begin{bmatrix} -2 \\ 8 \\ 12 \end{bmatrix}$

16.29 Let $A = \begin{bmatrix} 1 & 3 & -1 \\ 2 & 1 & 8 \\ 7 & 6 & 23 \end{bmatrix}$.

a. If $b = \begin{bmatrix} 1 & 1 & 3 \end{bmatrix}^{\mathrm{T}}$, use Gaussian elimination to determine if there is a unique solution, infinitely many solutions, or no solution to $Ax = b$.

b. If your answer to part (a) is infinitely many or no solutions, can you find a minimum norm solution?

c. For $b = \begin{bmatrix} 3 & 11 & 36 \end{bmatrix}^{\mathrm{T}}$, find the infinitely many solutions using Gaussian elimination.

d. Find the least-squares minimum norm solution to the rank deficient problem (c), and verify that the minimum norm solution is one of the infinitely many solutions obtained in part (c).

16.30 Let

$$A = \begin{bmatrix} 1 & -1 & 3 \\ 8 & 8 & 1 \\ 4 & 6 & -12 \\ 6 & -9 & 0 \\ 3 & 4 & 4 \end{bmatrix}, \quad b = \begin{bmatrix} 1 \\ 3 \\ -1 \\ 6 \\ 15 \end{bmatrix}.$$

a. Show that there is a unique solution.

b. Find the solution in the following ways:

    i.   Using the normal equations

    ii. Using the QR method

    iii. Using the SVD method

16.31 Let $A = \begin{bmatrix} 1 & 4 & 10 \\ -2 & 6 & -6 \\ 5 & -7 & 23 \\ 3 & 3 & 21 \end{bmatrix}$ and $b = \begin{bmatrix} 1 \\ 12 \\ -1 \\ 2 \end{bmatrix}$.

a. Try to solve the least-squares problem $Ax = b$ using the QR decomposition.

b. Try to solve the least-squares problem $Ax = b$ using the SVD.

c. Explain the results of parts (a) and (b).

d. Try to find a solution using `rdlstsq`. Explain your result.

16.32 The 50×40 matrix `RD.mat` in the software distribution is a rank deficient. Let `b = rand(50,1)`.

a. Find the minimum norm solution to `RD*x = b`.

b. Find two different solutions that produce the same residual, and show that their norms are greater than that of the minimum norm solution.

16.33 Build the matrix

$$A = \begin{bmatrix} -7 & -3 & 1 \\ -6 & -2 & 2 \\ -5 & -1 & 3 \\ -4 & 0 & 4 \end{bmatrix}$$

and vectors

$$b_1 = \begin{bmatrix} -5 & 2 & 9 & 15 \end{bmatrix}^{\mathrm{T}}, \quad b_2 = \begin{bmatrix} 7 & 1 & 3 & 6 \end{bmatrix}^{\mathrm{T}}.$$

The rank of A is 2, so A is rank deficient. Find the minimal norm solution to the problems

$$\min_{x \in \mathbb{R}^3} \|Ax - b_1\| \quad \text{and} \quad \min_{x \in \mathbb{R}^3} \|Ax - b_2\|,$$

and compute $\|Ax_i - b_i\|$, $i = 1, 2$. Explain the results.

16.34

    a. Show that the 50×35 matrix, ARD, in the software distribution is "almost rank deficient."

    b. Let $b = \begin{bmatrix} 1 & 1 & \dots & 1 & 1 \end{bmatrix}^T$ and solve the least-squares problem $ARDx = b$ using `normalsolve`, `qrlstsq`, `svdlstsq`, and `rdlstsq`. In each case, also compute the residual. Compare the results by filling-in the table.

Method	Residual
normalsolve	
qrlstsq	
svdlstsq	
rdlstsq	

16.35 Build the matrix A and vector b using the following statements:

```
A = rosser; A(9:10,:) = ones(2,8); b = ones(10,1).
```

    a. Compute x = `pinv(A)*b`, and accept x as the correct solution.

    b. Solve the least-squares problem $Ax = b$ using the normal equations to obtain solution x_1.

    c. Do part (a), except use the QR decomposition algorithm to obtain solution x_2.

    d. Do part (a), except use the SVD decomposition algorithm to obtain solution x_3.

    e. Compute $\frac{\|x - x_i\|_2}{\|x\|_2}$, $i = 1, 2, 3$.

    f. Explain the results, taking Remark 16.2 into account.

16.36 Build the Lauchli matrix A = `gallery('lauchli',50)`, and let b = `(1:51)'`.

    a. Does A have full rank?

    b. Compute the condition number of A.

    c. Compute the least-squares solution, x_1, using `qrlstsq`.

    d. Obtain solution x_2 by using the normal equations.

    e. Comment on the results.

16.37 Solve the underdetermined system and give the residual.

$$\begin{bmatrix} 2 & -4 & 4 & 0.077 \\ 0 & -2 & 2 & -0.056 \\ 2 & -2 & 0 & 0 \end{bmatrix} \begin{bmatrix} x_1 \\ x_2 \\ x_3 \\ x_4 \end{bmatrix} = \begin{bmatrix} 3.86 \\ -3.47 \\ 0 \end{bmatrix}$$

16.38 Build a 9×25 system using the following MATLAB code, find the minimum norm solution and the residual.

```
>> x = 1:25;
>> V = vandermonde(x,8);
>> V = V';
>> b = (1:9)';
```

16.39 Find the rank of the matrix.

$$\begin{bmatrix} 1 & 3 & 6 & 2 & 1 \\ 3 & -7 & 9 & 2 & 1 \\ 5 & -9 & 51 & 14 & 7 \\ -16 & 36 & -66 & -16 & -8 \end{bmatrix} x = \begin{bmatrix} 1 \\ -1 \\ 3 \\ 5 \end{bmatrix}.$$

Will `uqrlstsq` work? If it does, use it, and if not find the minimum norm solution using `rdlstsq`.

16.40 A Toeplitz matrix is a matrix in which each diagonal from left to right is constant. For instance,

$$T = \begin{bmatrix} 5 & 4 & 3.001 & 2 & 1 & 1 & 2 & 3 \\ 6 & 5 & 4 & 3.001 & 2 & 1 & 1 & 2 \\ 7 & 6 & 5 & 4 & 3.001 & 2 & 1 & 1 \\ 8 & 7 & 6 & 5 & 4 & 3.001 & 2 & 1 \\ 9 & 8 & 7 & 6 & 5 & 4 & 3.001 & 2 \\ 10 & 9 & 8 & 7 & 6 & 5 & 4 & 3.001 \\ 11 & 10 & 9 & 8 & 7 & 6 & 5 & 4 \\ 12 & 11 & 10 & 9 & 8 & 7 & 6 & 5 \\ 13 & 12 & 11 & 10 & 9 & 8 & 7 & 6 \\ 14 & 13 & 12 & 11 & 10 & 9 & 8 & 7 \\ 15 & 14 & 13 & 12 & 11 & 10 & 9 & 8 \end{bmatrix}$$ is a Toeplitz matrix. This interesting matrix was

found in Ref. [50]. We will use this matrix to investigate the conditioning of $m \times n$ least-squares problems, $m > n$.

a. Build T and vectors b and bp as follows:

```
>> r = [5 4 3.001 2 1 1 2 3];
>> c = [5 6 7 8 9 10 11 12 13 14 15];
>> T = toeplitz(c,r);
>> b = [21.001 24.001 29.001 36.001 44.001 52.001 60.0 68.0...
76.0 84.0 92.0]';
>> bp = [21.001 24.0002 29.0001 36.0003 44.00 52.0003 59.999...
68.0001 75.9999 84.0001 91.9999]';
```

b. Compute $\frac{\|b-bp\|_2}{\|b\|_2}$.

c. Use the function svdlstsq to solve the two problems

$$Tx = b, \quad T(xp) = bp,$$

d. Compute

$$\frac{\|x - xp\|_2}{\|x\|_2}.$$

e. Discuss your results.

16.41

a. Using the results of Problem 16.14, write a function

```
function [x, r] = wlstsq(A,b,w)
```

that solves the weighted least-squares problem. The argument w is a vector of weights, not a diagonal matrix.

b. Given weights $\begin{bmatrix} 2 & 4 & 5 & 1 & 6 \end{bmatrix}^T$, solve the weighted least-squares problem using wlstsq.

$$x_1 + 2x_2 + x_3 - x_4 = 1$$
$$2x_1 + 5x_2 - x_3 + x_4 = 2$$
$$4x_1 + x_2 - 3x_3 - x_4 = -1$$
$$-x_1 + x_2 + 3x_3 + 7x_4 = 0$$
$$5x_1 - x_2 + x_3 - 8x_4 = 3$$

16.42 Section 11.9 discussed iterative improvement for the solution of a linear system. Iterative improvement can also be done for least-squares solutions. Consider the following algorithm outline for the full-rank overdetermined problem:

```
function LSTSQIMP(A,b,x,numsteps)
   % Iterative refinement for least squares.
   % Execute numsteps of improvement.
   % Normally numsteps is small.
   for i=1:numsteps do
      r = b - Ax
      correction=qrlstsq(A,r)
      x=x+correction
```

```
   end for
   return [x,b-Ax]
end function
```

a. Implement the function lstsqimp in MATLAB.

b. Add iterative refinement to the following code, and see if you can improve the result.

```
A = gallery('lauchli',50);
b = (1:51)';

[x,r] = normalsolve(A,b);
fprintf('The residual using the normal equations = %g\n',r);
```

Remark 16.5.   This method is helpful only when the initial residual is small. The algorithm actually used in practice is described in Ref. [2, pp. 268-269].

Chapter 17

Implementing the *QR* Decomposition

You should be familiar with

- Rank, orthonormal basis
- Orthogonal matrices
- *QR* decomposition using Gram-Schmidt
- Computation with submatrices

In Chapter 14, we developed the *QR* decomposition of an arbitrary $m \times n$ matrix, $m \geq n$, into a product of an $m \times n$ matrix Q with orthonormal columns and an $n \times n$ upper triangular matrix R such that $A = QR$. We constructed the decomposition using the Gram-Schmidt classical and modified algorithms. Chapter 16 discussed the solution of least-squares problems using the *QR* decomposition, and the decomposition plays a very important role in eigenvalue computation. We must be able to compute it accurately. There are other approaches to the *QR* decomposition that are numerically superior to Gram-Schmidt, *Givens method* and *Householder's Method*, and this chapter presents both methods.

17.1 REVIEW OF THE *QR* DECOMPOSITION USING GRAM-SCHMIDT

We begin this section with a formal statement of the *QR* decomposition theorem from Chapter 14.

Theorem 17.1. *If A is an $m \times n$ matrix, $m \geq n$, with linearly independent columns, then A can be factored as $A = QR$ where Q is an $m \times n$ matrix with orthonormal columns and R is an $n \times n$ upper triangular matrix.*

In Chapter 14, we built the matrices Q and R by using the Gram-Schmidt process, and developed algorithms for both the classical (clqrgrsch) and modified Gram-Schmidt process (modqrgrsch). The modified Gram-Schmidt algorithm avoided possibly costly cancelation errors and is always to be preferred. For the sake of brevity, we will use the acronym MGS to refer to the algorithm. We mentioned in the introduction that the *QR* decomposition using MGS was not as good numerically compared to the Givens or Householder methods. The columns of Q tend to lose orthogonality in proportion to the $\kappa(A)$ [51, 52], and MGS can have problems when the matrix A is ill-conditioned. Example 17.1 clearly demonstrates this.

Example 17.1.
```
>> H = hilb(5);
>> [Q,R] = modqrgrsch(H);
>> norm(eye(5)-Q'*Q)
ans =
    1.1154e-011

>> [Q,R] = clqrgrsch(H);
>> norm(eye(5)-Q'*Q)
ans =
  5.7917e-008

>> H = hilb(15);
>> [Q,R] = modqrgrsch(H);
>> norm(eye(15)-Q'*Q)

ans =
    0.9817    % indicates severe effects from ill-conditioned H
```

The reason the MGS was so poor with the 15×15 Hilbert matrix H is that $\text{Cond}_2(H) = 2.5699 \times 10^{17}$. ∎

Numerical Linear Algebra with Applications. http://dx.doi.org/10.1016/B978-0-12-394435-1.00017-X

Remark 17.1.   The use of Gram-Schmidt can be improved by a process known as reorthogonalization. See Problem 14.25. This issue of losing orthogonality will become a major concern in Chapters 21 and 22 when we discuss methods for solving systems and computing eigenvalues for large sparse matrices.

Gram-Schmidt requires $2mn^2$ flops to find both Q and R. As we will see, this flop count is better than those for the Givens or Householder methods when both Q and R are desired. However, its possible instability usually dictates the use of other methods. This does not mean that Gram-Schmidt is unimportant. In fact, using it to orthonormalize a set of vectors is a basis for the Arnoldi and Lanczos methods we will study in conjunction with large sparse matrix problems in Chapters 21 and 22.

17.2   GIVENS ROTATIONS

Recall we developed the LU decomposition in Chapter 11 by applying a sequence of elementary matrices to the left side of A. In the resulting decomposition LU, L is the product of the elementary matrices and U is an upper triangular matrix. We will employ this same idea to the transformation of A into the product QR by applying on the left a sequence of orthogonal matrices called Givens matrices that transform A into an upper triangular matrix R. Each Givens matrix product zeros out a matrix element in the creation of R. Here is the idea:

Apply $n - 1$ Givens matrices, $J_{i1}, 2 \le i \le n$, on the left of A so that $a_{21}, a_{31}, \ldots, a_{n1}$ are zeroed out, and

$$(J_{n1}J_{n-1,1},\ldots,J_{31}J_{21})\,A \to \begin{bmatrix} X & X & \ldots & X \\ 0 & X & \cdots & X \\ 0 & X & \ldots & X \\ \vdots & \vdots & \vdots & \vdots \\ 0 & X & \ldots & X \end{bmatrix}.$$

Now use $n - 2$ Givens matrices to zero out the elements at indices $(3, 2)\,, (4, 2)\,,\ldots, (n, 2)$, and we have

$$(J_{n2},\ldots J_{42}J_{32})\,(J_{n1}J_{n-1,1},\ldots,J_{31}J_{21})\,A \to \begin{bmatrix} X & X & \ldots & X \\ 0 & X & \cdots & X \\ 0 & 0 & \ldots & X \\ \vdots & \vdots & \vdots & \vdots \\ 0 & 0 & \ldots & X \end{bmatrix}.$$

Let J_i be the product of Givens matrices acting on column i. Continue this process $n - 3$ more times until we have

$$(J_{n-1}J_{n-2}\ldots J_2 J_1)\,A \to \begin{bmatrix} X & X & \ldots & X & X \\ 0 & X & \ldots & X & X \\ 0 & 0 & X & \ldots & X \\ \vdots & \vdots & \vdots & \ddots & \vdots \\ 0 & 0 & 0 & \ldots & X \end{bmatrix} = R,$$

and

$$A = (J_{n-1}J_{n-2}\ldots J_2 J_1)^{\mathrm{T}}\, R = QR.$$

Q is a product of orthogonal matrices, and so it is orthogonal.

We start with the definition of the general $n \times n$ *Givens matrix*.

Definition 17.1.   A matrix of the form (Figure 17.1) is a Givens matrix. The value c is on the diagonal at indices (i, i) and (j, j), $i < j$. The value $-s$ is at index (j, i), and s is at index (i, j). The remaining diagonal entries are 1, and all off-diagonal elements at indices other than (j, i) and (i, j) are zero.

We want a Givens matrix to be orthogonal $(J\,(i, j, c, s)^{\mathrm{T}}\, J\,(i, j, c, s) = I)$, and clearly the columns in the definition are orthogonal; however, each column must have unit length. This requires that $c^2 + s^2 = 1$. For clarity, Example 17.2 illustrates the form of $J\,(i, j, c, s)^{\mathrm{T}}\, J\,(i, j, c, s)$ for a 3×3 matrix.

Example 17.2.   Let $J\,(1, 3, c, s) = \begin{bmatrix} c & 0 & s \\ 0 & 1 & 0 \\ -s & 0 & c \end{bmatrix}$. Then,

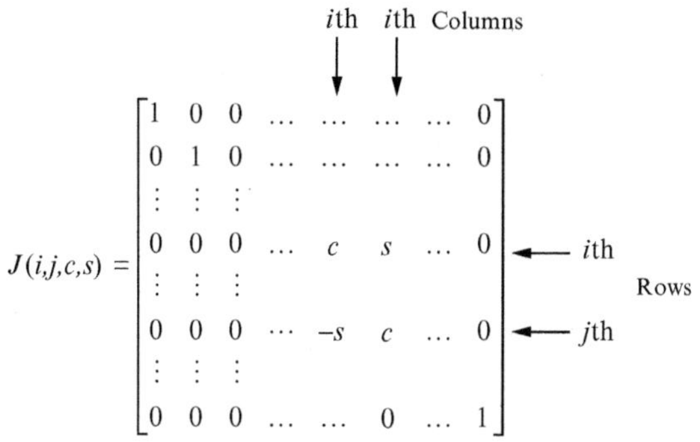

$$J(i,j,c,s) = \begin{bmatrix} 1 & 0 & 0 & \dots & \dots & \dots & \dots & 0 \\ 0 & 1 & 0 & \dots & \dots & \dots & \dots & 0 \\ \vdots & \vdots & \vdots & & & & & \\ 0 & 0 & 0 & \dots & c & s & \dots & 0 \\ \vdots & \vdots & \vdots & & & & & \\ 0 & 0 & 0 & \cdots & -s & c & \dots & 0 \\ \vdots & \vdots & \vdots & & & & & \\ 0 & 0 & 0 & \dots & \dots & 0 & \dots & 1 \end{bmatrix}$$

*i*th *i*th Columns

*i*th Rows

*j*th

FIGURE 17.1 Givens matrix.

$$J(1,3,c,s)^{\mathrm{T}} J(1,3,c,s) = \begin{bmatrix} c & 0 & -s \\ 0 & 1 & 0 \\ s & 0 & c \end{bmatrix} \begin{bmatrix} c & 0 & s \\ 0 & 1 & 0 \\ -s & 0 & c \end{bmatrix} = \begin{bmatrix} c^2 + s^2 & 0 & 0 \\ 0 & 1 & 0 \\ 0 & 0 & c^2 + s^2 \end{bmatrix},$$

so if $c^2 + s^2 = 1$, $J(1,3,c,s)$ is orthogonal. ∎

In the $n \times n$ case,

$$\begin{bmatrix} 1 & 0 & \dots & \dots & \dots & \dots & 0 \\ & \ddots & & & & & \\ & & c & & -s & & \\ & & \vdots & \ddots & \vdots & & \\ & & s & \dots & c & & \vdots \\ & & & & & \ddots & \\ 0 & 0 & \dots & \dots & \dots & \dots & 1 \end{bmatrix} \begin{bmatrix} 1 & 0 & \dots & \dots & \dots & \dots & 0 \\ & \ddots & & & & & \\ & & c & & s & & \\ & & \vdots & \ddots & \vdots & & \\ & & -s & \dots & c & & \vdots \\ & & & & & \ddots & \\ 0 & 0 & \dots & \dots & \dots & \dots & 1 \end{bmatrix} = \begin{bmatrix} 1 & 0 & \dots & \dots & \dots & \dots & 0 \\ & \ddots & & & & & \\ & & c^2+s^2 & & 0 & & \\ & & \vdots & \ddots & \vdots & & \\ & & 0 & \dots & c^2+s^2 & & \vdots \\ & & & & & \ddots & \\ 0 & 0 & \dots & \dots & \dots & \dots & 1 \end{bmatrix}.$$

Require that $c^2 + s^2 = 1$ and recall the identity $\sin^2\theta + \cos^2\theta = 1$. For instance, the matrix

$$J(4,6,c,s) = \begin{bmatrix} 1 & 0 & 0 & 0 & 0 & 0 \\ 0 & 1 & 0 & 0 & 0 & 0 \\ 0 & 0 & 1 & 0 & 0 & 0 \\ 0 & 0 & 0 & \cos\left(\frac{\pi}{6}\right) & 0 & \sin\left(\frac{\pi}{6}\right) \\ 0 & 0 & 0 & 0 & 1 & 0 \\ 0 & 0 & 0 & -\sin\left(\frac{\pi}{6}\right) & 0 & \cos\left(\frac{\pi}{6}\right) \end{bmatrix},$$

where $i = 4, j = 6, c = \cos\left(\frac{\pi}{6}\right), s = \sin\left(\frac{\pi}{6}\right)$ is orthogonal since $\cos^2\left(\frac{\pi}{6}\right) + \sin^2\left(\frac{\pi}{6}\right) = 1$. In general, we can choose any angle θ, let $c = \cos(\theta), s = \sin(\theta)$ and always obtain a Givens matrix. When we make such a choice, we will use the notation $J(i,j,\theta)$. Such a matrix is actually a rotation matrix that rotates a pair of coordinate axes through an angle θ in the (i,j) plane, so it is also known as a *Givens* rotation. In $\mathbb{R}^2$, the Givens matrix $J(1,2,\theta)$ is $\begin{bmatrix} \cos\theta & \sin\theta \\ -\sin\theta & \cos\theta \end{bmatrix}$, and you will recognize this as a rotation matrix, a topic we discussed in Chapter **??**. It rotates a vector clockwise through the angle θ.

Remark 17.2. Operations with Givens matrices can be done implicitly; in other words, it is not necessary to actually build the Givens matrix. However, to reinforce understanding of exactly how these matrices operate, we will explicitly build them until Section 17.4.

17.2.1 Zeroing a Particular Entry in a Vector

The Givens *QR* decomposition algorithm relies on being able to place zeros at specified locations in a matrix. Let $x = \begin{bmatrix} x_1 & x_2 & \dots & x_{n-1} & x_n \end{bmatrix}^{\mathrm{T}}$, and form the product $J(i,j,c,s)\,x$.

$$
J\,(i,j,c,s)\,x =
\begin{bmatrix}
1 & 0 & 0 & \cdots & \cdots & & \cdots & \cdots & 0 \\
0 & 1 & 0 & \cdots & \cdots & & \cdots & \cdots & 0 \\
\vdots & \vdots & \vdots & & & & & & \vdots \\
0 & 0 & 0 & \cdots & c & \cdots & s & \cdots & 0 \\
\vdots & \vdots & \vdots & & & & & & \vdots \\
0 & 0 & 0 & \cdots & -s & \cdots & c & \cdots & 0 \\
\vdots & \vdots & \vdots & & & & & \cdots & \vdots \\
0 & 0 & 0 & \cdots & \cdots & \cdots & 0 & \cdots & 1
\end{bmatrix}
\begin{bmatrix}
x_1 \\ \vdots \\ x_i \\ \vdots \\ x_j \\ \vdots \\ x_n
\end{bmatrix}
=
\begin{bmatrix}
x_1 \\ \vdots \\ cx_i + sx_j \\ \vdots \\ -sx_i + cx_j \\ \vdots \\ x_{n-1} \\ x_n
\end{bmatrix}
$$

Note that the product changes only the components i and j of x.

Example 17.3. Let $J\,(2,4,c,s) =
\begin{bmatrix}
1 & 0 & 0 & 0 \\
0 & c & 0 & s \\
0 & 0 & 1 & 0 \\
0 & -s & 0 & c
\end{bmatrix}
\begin{bmatrix}
x_1 \\ x_2 \\ x_3 \\ x_4
\end{bmatrix}
=
\begin{bmatrix}
x_1 \\ cx_2 + sx_4 \\ x_3 \\ -sx_2 + cx_4
\end{bmatrix}$ ■

Assume we have a vector $x \in \mathbb{R}^n$ and want to zero out $x_j, j > 1$, as illustrated in Equation 17.1.

$$
x = \begin{bmatrix} x_1 \\ x_2 \\ \vdots \\ x_j \\ \vdots \\ x_n \end{bmatrix},
\quad
J\,(i,j,c,s)\,x = \begin{bmatrix} x_1 \\ X \\ \vdots \\ 0 \\ \vdots \\ X \end{bmatrix}
\tag{17.1}
$$

We must determine the values of i, j, c, and s to use, and then form the Givens rotation by creating the identity matrix, inserting c in locations (i,i) and (j,j), $-s$ in location (j,i), and s into position (i,j). Since multiplication by a Givens matrix only affects components i and j of the vector, we can reduce finding c and s to a two-dimensional problem. Create a Givens matrix $\begin{bmatrix} \cos\theta & \sin\theta \\ -\sin\theta & \cos\theta \end{bmatrix}$ that rotates vector $\begin{bmatrix} x \\ y \end{bmatrix}$ in $\mathbb{R}^2$ onto the x-axis. In that way, we zero out the y-coordinate. From Figure 17.2, we have $\cos\theta = \frac{x}{\sqrt{x^2+y^2}}, \sin\theta = \frac{y}{\sqrt{x^2+y^2}}$, so

$$
\begin{bmatrix} \cos\theta & \sin\theta \\ -\sin\theta & \cos\theta \end{bmatrix}
\begin{bmatrix} x \\ y \end{bmatrix}
=
\begin{bmatrix} \frac{x}{\sqrt{x^2+y^2}} & \frac{y}{\sqrt{x^2+y^2}} \\ -\frac{y}{\sqrt{x^2+y^2}} & \frac{x}{\sqrt{x^2+y^2}} \end{bmatrix}
\begin{bmatrix} x \\ y \end{bmatrix}
=
\begin{bmatrix} \sqrt{x^2+y^2} \\ 0 \end{bmatrix}.
$$

By looking at Figure 17.2, this is exactly what we should get. The vector $\begin{bmatrix} x \\ y \end{bmatrix}$ after rotation will have an x-coordinate of $\sqrt{x^2+y^2}$ and a y-coordinate of 0.

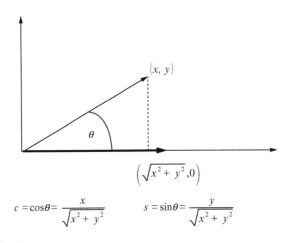

$$
c = \cos\theta = \frac{x}{\sqrt{x^2+y^2}} \qquad s = \sin\theta = \frac{y}{\sqrt{x^2+y^2}}
$$

FIGURE 17.2 Givens rotation.

Summary

To zero out entry j of vector x, choose index $i < j$ and compute the values

$$c = \frac{x_i}{\sqrt{x_i^2 + x_j^2}}, \quad s = \frac{x_j}{\sqrt{x_i^2 + x_j^2}}.$$

Let $J(i, j, c, s) = I$, followed by

$$J(i, i, c, s) = J(j, j, c, s) = c$$
$$J(j, i, c, s) = -s$$
$$J(i, j, c, s) = s$$

and compute $J(i, j, c, s)\, x$.

Example 17.4.   Suppose we want to zero out component 2 of the 3×1 vector $x = \begin{bmatrix} -1 \\ 2 \\ 3 \end{bmatrix}$.

Choose $i = 1, j = 2$ and compute $c = \frac{-1}{\sqrt{5}}, s = \frac{2}{\sqrt{5}}$. The Givens rotation matrix is

$$J(1, 2, c, s) = \begin{bmatrix} c & s & 0 \\ -s & c & 0 \\ 0 & 0 & 1 \end{bmatrix} = \begin{bmatrix} \frac{-1}{\sqrt{5}} & \frac{2}{\sqrt{5}} & 0 \\ -\frac{2}{\sqrt{5}} & \frac{-1}{\sqrt{5}} & 0 \\ 0 & 0 & 1 \end{bmatrix}.$$

Now form the product $J(1, 2, c, s)\, x$ to obtain

$$\begin{bmatrix} -\frac{1}{\sqrt{5}} & \frac{2}{\sqrt{5}} & 0 \\ -\frac{2}{\sqrt{5}} & -\frac{1}{\sqrt{5}} & 0 \\ 0 & 0 & 1 \end{bmatrix} \begin{bmatrix} -1 \\ 2 \\ 3 \end{bmatrix} = \begin{bmatrix} \sqrt{5} \\ 0 \\ 3 \end{bmatrix}.$$

Let us look at a more complex example.

We want to zero out component 3 of the 4×1 vector $x = \begin{bmatrix} 3 & -7 & -1 & 5 \end{bmatrix}^{\mathrm{T}}$. Choose $i = 1, j = 3$, and then

$$c = \frac{3}{\sqrt{10}}, \quad s = \frac{-1}{\sqrt{10}},$$

and form

$$J(1, 3, c, s)\, x = \begin{bmatrix} \frac{3}{\sqrt{10}} & 0 & \frac{-1}{\sqrt{10}} & 0 \\ 0 & 1 & 0 & 0 \\ \frac{1}{\sqrt{10}} & 0 & \frac{3}{\sqrt{10}} & 0 \\ 0 & 0 & 0 & 1 \end{bmatrix} \begin{bmatrix} 3 \\ -7 \\ -1 \\ 5 \end{bmatrix} = \begin{bmatrix} \sqrt{10} \\ -7 \\ 0 \\ 5 \end{bmatrix}.$$

It is necessary to choose $i < j$, and any index $i < j$ can be used. Choose $i = 2$ and $j = 3$, form $J(2, 3, c, s)$, and verify that $J(2, 3, c, s)\, x$ zeros out x_3. ∎

17.3   CREATING A SEQUENCE OF ZEROS IN A VECTOR USING GIVENS ROTATIONS

Given a vector $x = \begin{bmatrix} x_1 & x_2 & \dots & x_n \end{bmatrix}^{\mathrm{T}}$, we can use a product of Givens matrices to zero out the $n - i$ elements of x below entry x_i. To zero out x_{i+1}, compute $J(i, i+1, c_{i+1}, s_{i+1})\, x = \overline{x_{i+1}}$. To zero out x_{i+2}, compute $J(i, i+2, c_{i+2}, s_{i+2})\, \overline{x_{i+1}} = \overline{x_{i+2}}$, and continue the process until computing $J(i, n, c_n, s_n)\, \overline{x_{n-1}} = \overline{x_n}$. In summary, the product

$$J(i, n, c_n, s_n) \dots J(i, i+2, c_{i+2}, s_{i+2})\, J(i, i+1, c_{i+1}, s_{i+1})\, x$$

transforms x into a vector of the form $\begin{bmatrix} x_1 & x_2 & \dots & x_{i-1} & * & 0 & \dots & 0 \end{bmatrix}^{\mathrm{T}}$.

Example 17.5. Let $x = \begin{bmatrix} 5 \\ -1 \\ 3 \end{bmatrix}$ and zero out the second and third components of x using Givens rotations.

$$\overline{x_2} = J(1,2,c_2,s_2)\,x = \begin{bmatrix} \frac{5}{\sqrt{26}} & -\frac{1}{\sqrt{26}} & 0 \\ \frac{1}{\sqrt{26}} & \frac{5}{\sqrt{26}} & 0 \\ 0 & 0 & 1 \end{bmatrix} \begin{bmatrix} 5 \\ -1 \\ 3 \end{bmatrix} = \begin{bmatrix} \sqrt{26} \\ 0 \\ 3 \end{bmatrix},$$

$$\overline{x_3} = J(1,3,c_3,s_3)\,\overline{x_2} = \begin{bmatrix} \frac{\sqrt{26}}{\sqrt{35}} & 0 & \frac{3}{\sqrt{35}} \\ 0 & 1 & 0 \\ -\frac{3}{\sqrt{35}} & 0 & \frac{\sqrt{26}}{\sqrt{35}} \end{bmatrix} \begin{bmatrix} \sqrt{26} \\ 0 \\ 3 \end{bmatrix} = \begin{bmatrix} \sqrt{35} \\ 0 \\ 0 \end{bmatrix} \qquad ■$$

17.4 PRODUCT OF A GIVENS MATRIX WITH A GENERAL MATRIX

If A is an $m \times n$ matrix, a huge advantage when dealing with Givens matrices is the fact that you can compute a product $J(i,j,c,s)\,A$ without ever constructing $J(i,j,c,s)$.

Example 17.6. If $A = \begin{bmatrix} a_{11} & a_{12} & a_{13} \\ a_{21} & a_{22} & a_{23} \\ a_{31} & a_{32} & a_{33} \\ a_{41} & a_{42} & a_{43} \end{bmatrix}$, then

$$J(1,3,c,s)\,A = \begin{bmatrix} c & 0 & s & 0 \\ 0 & 1 & 0 & 0 \\ -s & 0 & c & 0 \\ 0 & 0 & 0 & 1 \end{bmatrix} \begin{bmatrix} a_{11} & a_{12} & a_{13} \\ a_{21} & a_{22} & a_{23} \\ a_{31} & a_{32} & a_{33} \\ a_{41} & a_{42} & a_{43} \end{bmatrix} = \begin{bmatrix} ca_{11}+sa_{31} & ca_{12}+sa_{32} & ca_{13}+sa_{33} \\ a_{21} & a_{22} & a_{23} \\ ca_{31}-sa_{11} & ca_{32}-sa_{12} & ca_{33}-sa_{13} \\ a_{41} & a_{42} & a_{43} \end{bmatrix}$$

The product only affects rows 1 and 3. $\qquad ■$

In the general case where $J(i,j,c,s)$ is $m \times m$ and A is $m \times n$ the product $J(i,j,c,s)\,A$ only affects rows i and j. Using Example 17.6 as a guide, convince yourself that rows i and j, $i < j$, have the following form:

$$\text{Row } i: \qquad ca_{i1}+sa_{j1},\, ca_{i2}+sa_{j2},\dots,\, ca_{in}+sa_{jn}, \qquad (17.2)$$

$$\text{Row } j: \qquad ca_{j1}-sa_{i1},\, ca_{j2}-sa_{i2},\dots,\, ca_{jn}-sa_{in}. \qquad (17.3)$$

This makes the product $J(i,j,c,s)\,A$ very efficient to compute, requiring only $6n$ flops. Just change rows i and j according to Equations 17.2 and 17.3.

Algorithm 17.1 Product of a Givens Matrix J with a General Matrix A

```
function GIVENSMUL(A,i,j,c,s)
   % Multiplication by a Givens matrix.
   % Input: An m × n matrix A, matrix indices i, j,
   % and Givens parameters c, s.
   % Output: The m × n matrix J(i,j,c,s)A.
   a = A(i,:)
   b = A(j,:)
   A(i,:)=ca+sb
   A(:,j)=-sa+cb
end function
```

NLALIB: The function `givensmul` implements Algorithm 17.1.

17.5 ZEROING-OUT COLUMN ENTRIES IN A MATRIX USING GIVENS ROTATIONS

We know how to zero out all the entries of a vector x below an entry x_i, and we know how to very simply form the product of a Givens matrix with another matrix. Now we will learn how to zero out all the entries in column i below a matrix entry $A(i, i)$. This is what we must do to create an upper triangular matrix.

First, consider the problem of computing the product $J(i, j, c, s)A, j > i$, so that the result will have a zero at index (j, i) and only rows i and j of A will change. Think of column i, $\begin{bmatrix} a_{1i} & a_{2i} & \dots & a_{ii} & \dots & a_{ji} & \dots & a_{mi} \end{bmatrix}^T$, as a vector and find c and s using a_{ii} and a_{ji} so the product will zero out a_{ji}. Using `givensmul`, implicitly form a Givens matrix with c at indices (i, i) and (j, j), $-s$ at index (j, i) and s at index (i, j) and compute $J(i, j, c, s)A$ using givensmul. Entries in both rows i and j will change. This makes no difference, since we are only interested in zeroing out a_{ji}.

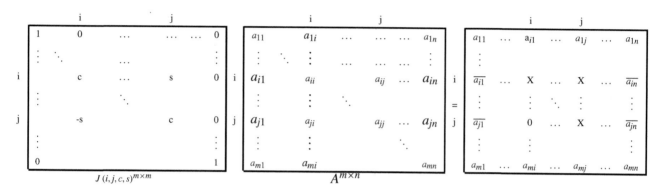

To zero out every element in column i below a_{ii}, compute the sequence

$$\overline{A_{i+1}} = J(i, i+1, c_{i+1}, s_{i+1})A, \overline{A_{i+2}} = J(i, i+2, c_{i+2}, s_{i+2})\overline{A_{i+1}}, \dots, \overline{A_m} = J(i, m, c_m, d_m)\overline{A_{m-1}}.$$

Example 17.7. Let $A = \begin{bmatrix} 1 & 3 & -6 & -1 \\ 4 & 8 & 7 & 3 \\ 2 & 3 & 4 & 5 \\ -9 & 6 & 3 & 2 \end{bmatrix}$

Zero out all entries in column 1 below a_{11}. To form $\overline{A_2}$, implicitly multiply by

$$J(1, 2, c_2, s_2) = \begin{bmatrix} 0.2425 & 0.9701 & 0 & 0 \\ -0.9701 & 0.2425 & 0 & 0 \\ 0 & 0 & 1 & 0 \\ 0 & 0 & 0 & 1 \end{bmatrix}.$$

$\overline{A_2} = J(1, 2, c_2, s_2)A$

$$= \begin{bmatrix} 0.2425\,(1) + 0.9701\,(4) & 0.2425\,(3) + 0.9701\,(8) & 0.2425\,(-6) + 0.9701\,(7) & 0.2425\,(-1) + 0.9701\,(3) \\ -0.9701\,(1) + 0.2425\,(4) & -0.9701\,(3) + 0.2425\,(8) & (-0.9701)\,(-6) + 0.2425\,(7) & (-0.9701)\,(-1) + 0.2425\,(3) \\ 2 & 3 & 4 & 5 \\ -9 & 6 & 3 & 2 \end{bmatrix}$$

$$= \begin{bmatrix} 4.1231 & 8.4887 & 5.3358 & 2.6679 \\ 0 & -0.9701 & 7.5186 & 1.6977 \\ 2 & 3 & 4 & 5 \\ -9 & 6 & 3 & 2 \end{bmatrix}$$

Implicitly multiply by

$$J(1, 3, c_3, c_4) = \begin{bmatrix} 0.8997 & 0 & 0.4364 & 0 \\ 0 & 1 & 0 & 0 \\ -0.4364 & 0 & 0.8997 & 0 \\ 0 & 0 & 0 & 1 \end{bmatrix}$$

$$\overline{A_3} = J(1,3,c_3,s_3)\overline{A_2}$$

$$= \begin{bmatrix} 4.5826 & 8.9469 & 6.5465 & 4.5826 \\ 0 & -0.9701 & 7.5186 & 1.6977 \\ 0 & -1.0056 & 1.2702 & 3.3343 \\ -9 & 6 & 3 & 2 \end{bmatrix}$$

Implicitly multiply by

$$J(1,4,c_4,s_4) = \begin{bmatrix} -0.4537 & 0 & 0 & 0.8911 \\ 0 & 1 & 0 & 0 \\ 0 & 0 & 1 & 0 \\ -0.8911 & 0 & 0 & -0.4537 \end{bmatrix}$$

$$\overline{A_4} = J(1,4,c_4,s_4)\overline{A_3}$$

$$= \begin{bmatrix} -10.0995 & 1.2872 & -0.2970 & -0.2970 \\ 0 & -0.9701 & 7.5186 & 1.6977 \\ 0 & -1.0056 & 1.2702 & 3.3343 \\ 0 & -10.6954 & -7.1951 & -4.9912 \end{bmatrix}$$

Form

$$P = J(1,4,c_4,s_4)J(1,3,c_3,s_3)J(1,2,c_2,s_2) = \begin{bmatrix} -0.0990 & -0.3961 & -0.1980 & 0.8911 \\ -0.9701 & 0.2425 & 0 & 0 \\ -0.1059 & -0.4234 & 0.8997 & 0 \\ -0.1945 & -0.7778 & -0.3889 & -0.4537 \end{bmatrix},$$

and $PA = \overline{A_4}$. ∎

The matrix P in Example 17.7 is the product of orthogonal matrices, and so P is orthogonal, as proved in Lemma 17.1.

Lemma 17.1. *If matrix $P = P_k P_{k-1} \ldots P_2 P_1$, where each matrix P_i is orthogonal, so is P.*

Proof. $P^{-1} = P_1^{-1} P_2^{-1} \ldots P_{k-1}^{-1} P_k^{-1} = P_1^{\mathrm{T}} P_2^{\mathrm{T}} \ldots P_{k-1}^{\mathrm{T}} P_k^{\mathrm{T}} = (P_k P_{k-1} \ldots P_2 P_1)^{\mathrm{T}}$, so $P^{-1} = P^{\mathrm{T}}$, and P is orthogonal. □

Note that we can continue with $\overline{A_4}$ in Example 17.8 and use Givens rotations to zero out the elements in column 2 below $\overline{A_4}(2,2)$. By zeroing out the element at index $(4,3)$, we will have an upper triangular matrix.

It is critical that c and s be computed as accurately as possible when implicitly multiplying by $J(i,j,c,s)$. As it turns out, this is not a simple matter.

17.6 ACCURATE COMPUTATION OF THE GIVENS PARAMETERS

The computation of c and s can have problems with overflow or underflow. We saw in Sections 8.4.1 and 8.4.2 that in order to minimize computation errors, it may be necessary to rearrange the way we perform a computation. The following algorithm provides improvement in overall accuracy [19, pp. 195-196] by cleverly employing the normalization procedure described in Section 8.4.1. The algorithm takes care of the case where $x_j = 0$ by assigning $c = 1$ and $s = 0$ so that the Givens rotation is the identity matrix. The signs of c and s may be different from those obtained from $c = \frac{x_i}{\sqrt{x_i^2 + x_j^2}}$, $s = \frac{x_j}{\sqrt{x_i^2 + x_j^2}}$, but that does not change the rotation's effect.

NLALIB: The function `givensparms` implements Algorithm 17.2.

Algorithm 17.2 requires five flops and a square root.

Algorithm 17.2 Computing the Givens Parameters

```
function GIVENSPARMS(xᵢ, xⱼ)
   % Input: value xᵢ at index i and xⱼ at index j>i of a vector x.
   % Output: the Givens parameters for x.

   if xⱼ = 0 then
      c = 1
      s = 0
   else if |xⱼ| > |xᵢ| then
      t = xᵢ/xⱼ
      s = 1/√(1+t²)
      c = st
   else
      t = xⱼ/xᵢ
      c = 1/√(1+t²)
      s = ct;
   end if
      return [c, s]
end function
```

17.7 THE GIVENS ALGORITHM FOR THE *QR* DECOMPOSITION

All the necessary tools are now in place to construct Q and R using Givens rotations. First, we formalize our understanding of the term *upper triangular matrix* A.

Definition 17.2.   An $m \times n$ matrix $A = [a_{ij}]$ is upper triangular if $a_{ij} = 0$ for $i > j$. Another way of putting it is that all entries below a_{ii} are 0.

Example 17.8.   The matrices A and B are upper triangular.

$$A = \begin{bmatrix} 1 & -1 & 7 & 12 & 1 & 3 \\ 0 & 2 & 8 & 2 & 4 & 1 \\ 0 & 0 & 3 & -9 & 10 & 6 \\ 0 & 0 & 0 & 4 & 2 & 1 \\ 0 & 0 & 0 & 0 & 5 & 18 \end{bmatrix}, \quad B = \begin{bmatrix} 1 & 2 & 6 & -1 \\ 0 & 4 & -2 & 8 \\ 0 & 0 & -1 & -9 \\ 0 & 0 & 0 & -4 \\ 0 & 0 & 0 & 0 \\ 0 & 0 & 0 & 0 \end{bmatrix}$$ ∎

The decomposition algorithm uses Givens rotations to zero out elements below the diagonal element until arriving at an upper triangular matrix. The question is "How many steps, k, should we execute?" This depends on the dimension of A. Clearly, if A is an $n \times n$ matrix, we need to execute the process $k = n - 1$ times. Now consider two examples where $m \neq n$.

---

m > n: Look at a 5×3 matrix $A_1 = \begin{bmatrix} X & X & X \\ X & X & X \\ X & X & X \\ X & X & X \\ X & X & X \end{bmatrix}$. Here are the transformations that occur to A_1.

$$A_1 = \begin{bmatrix} X & X & X \\ X & X & X \\ X & X & X \\ X & X & X \\ X & X & X \end{bmatrix} \implies \begin{bmatrix} X & X & X \\ 0 & X & X \\ 0 & X & X \\ 0 & X & X \\ 0 & X & X \end{bmatrix} \implies \begin{bmatrix} X & X & X \\ 0 & X & X \\ 0 & 0 & X \\ 0 & 0 & X \\ 0 & 0 & X \end{bmatrix} \implies \begin{bmatrix} X & X & X \\ 0 & X & X \\ 0 & 0 & X \\ 0 & 0 & 0 \\ 0 & 0 & 0 \end{bmatrix}.$$

We executed three steps until we could not continue to zero out elements below the diagonal, so $k = 3 = n$.

m < n: Look at a 4×6 matrix

$$A_2 = \begin{bmatrix} X & X & X & X & X & X \\ X & X & X & X & X & X \\ X & X & X & X & X & X \\ X & X & X & X & X & X \end{bmatrix} \implies \begin{bmatrix} X & X & X & X & X & X \\ 0 & X & X & X & X & X \\ 0 & X & X & X & X & X \\ 0 & X & X & X & X & X \end{bmatrix} \implies \begin{bmatrix} X & X & X & X & X & X \\ 0 & X & X & X & X & X \\ 0 & 0 & X & X & X & X \\ 0 & 0 & X & X & X & X \end{bmatrix}$$

$$\implies \begin{bmatrix} X & X & X & X & X & X \\ 0 & X & X & X & X & X \\ 0 & 0 & X & X & X & X \\ 0 & 0 & 0 & X & X & X \end{bmatrix}.$$

We executed three steps until we could not continue to zero out elements below the diagonal, so $k = 3 = m - 1$.

From these examples, we see that the sequence of steps in the Givens QR decomposition algorithm is $k = \min(m - 1, n)$. We are now ready to provide the algorithm for the Givens QR decomposition. To compute Q, start with $Q = I$. As the algorithm progresses, build Q by successive uses of givensmul. Note also that when the rotations have produced R, $QA = R$, and so $A = Q^{T}R$. Thus, the Q returned is the transpose of the one built during the construction of R.

Algorithm 17.3 Givens QR Decomposition

```
function GIVENSQR(A)
   % Computes the QR decomposition of A
   % Input: m × n matrix A
   % Output: m × n orthogonal matrix Q
   % and m × n upper triangular matrix R
   Q = I

   for i=1:min(m-1,n) do
      for j=i+1:m do
         [c s]=givensparms(a_ii, a_ji)
         A=givensmul(A,i,j,c,s)
         Q=givensmul(Q,i,j,c,s)
      end for
   end for

   R=A;
   Q= Q^T
end function
```

NLALIB: The function givensqr implements Algorithm 17.3.

The construction we developed that is realized in Algorithm 17.3 proves the following theorem.

Theorem 17.2 (QR decomposition). *If A is an $m \times n$ matrix, it can be expressed in the form $A = QR$, where Q is an $m \times m$ orthogonal matrix and R is an $m \times n$ upper triangular matrix.*

Now let us do Example 17.1 again, this time using the Givens QR decomposition. Compare the results to those of produced by the modified Gram-Schmidt algorithm.

Example 17.9.
```
>> H = hilb(5);
>> [Q R] = givensqr(H);
>> norm(eye(5) - Q'*Q)

ans =
```

```
   5.6595e-016
>> H = hilb(15);
>> [Q R] = givensqr(H);
>> norm(eye(15) - Q'*Q)

ans =

   1.0601e-015
```

The columns of the matrix Q in the Givens algorithm do not lose orthogonality like the columns of Q do when using MGS. ∎

17.7.1   The Reduced *QR* Decomposition

Gram-Schmidt produces the reduced *QR* decomposition. Without knowledge of Gram-Schmidt, we can prove the existence of this decomposition using Theorem 17.2 and, at the same time, determine how to compute it using Givens rotations.

If A is has full rank, $m \geq n$, let $A^{m \times n} = Q^{m \times m} R^{m \times n}$ be the full *QR* decomposition of A. Let the columns of Q be denoted by $q_i, 1 \leq i \leq m$ so that we can write

$$Q = \left[\underbrace{q_1 \; q_2 \; \cdots \; q_n} \; \underbrace{q_{n+1} \; q_{n+2} \; \cdots \; q_m} \right].$$

Let $q_i, 1 \leq i \leq n$ form the matrix Q_1, and $q_i, n+1 \leq i \leq m$ form Q_2 so that

$$Q = \left[\; Q_1^{m \times n} \;\; Q_2^{m \times (m-n)} \; \right]$$

The upper triangular matrix

$$R = \begin{bmatrix} r_{11} & r_{12} & \cdots & r_{1,n} \\ 0 & r_{22} & \cdots & r_{2,n} \\ \vdots & & \ddots & \ddots \\ 0 & & & r_{n,n} \\ 0 & & & 0 \\ \vdots & & & \vdots \\ 0 & & & 0 \end{bmatrix}$$

can be written as

$$R = \begin{bmatrix} R_1^{n \times n} \\ 0^{(m-n) \times n} \end{bmatrix},$$

and so

$$A = \left[\; Q_1^{m \times n} \;\; Q_2^{m \times (m-n)} \; \right] \begin{bmatrix} R_1^{n \times n} \\ 0^{(m-n) \times n} \end{bmatrix}.$$

This implies that $A = Q_1 R_1$.

It is actually not necessary that A have full rank. We will still have $A = Q_1 R_1$, and the matrix R_1 has the form

$$\begin{bmatrix} r_{11} & r_{12} & \cdots & & \cdots & & r_{1n} \\ 0 & r_{22} & \cdots & & \cdots & & r_{2n} \\ 0 & 0 & \ddots & & & & \vdots \\ \vdots & \vdots & 0 & r_{kk} & r_{k,k+1} & \cdots & r_{kn} \\ \vdots & \vdots & \vdots & \vdots & 0 & & \\ & & & & & \vdots & \\ 0 & 0 & \cdots & 0 & 0 & \cdots & 0 \end{bmatrix}.$$

Remark 17.3.   If m is much larger than n, the reduced *QR* decomposition saves significant memory.

17.7.2 Efficiency

We will assume $m \geq n$. If $m > n$, the outer loop executes n times, and if $m = n$, it executes $n - 1$ times. Since $m > n$ is the more general case, assume the outer loop executes n times. Each call to givensparms requires five flops, and the call to givensmul(A,i,j,c,s) requires $6n$ flops. The matrix Q is of size $m \times m$, so givensmul(Q,i,j,c,s) executes $6m$ flops. The total flop count is

$$(5 + 6n + 6m) \sum_{i=1}^{n} (m - i). \tag{17.4}$$

After some work, Equation 17.4 expands to

$$3n \left(mn + 2m^2 - n^2\right) + \text{lower order terms}.$$

Disregarding the lower-order terms, we have

$$\text{Flop count} \cong 3n \left(mn + 2m^2 - n^2\right). \tag{17.5}$$

If $m = n$, Equation 17.5 gives $6n^3$ flops, so in this case the algorithm is $O\left(n^3\right)$.

The Givens QR algorithm is stable [16, pp. 365-368].

17.8 HOUSEHOLDER REFLECTIONS

The use of Householder reflections is an alternative to Givens rotations for computing the QR decomposition of a matrix. A Givens rotation zeros out one element at a time, and a sequence of rotations is required to transform the matrix into upper triangular form. The Householder algorithm for the QR decomposition requires about two-thirds the flop count of the Givens algorithm because it zeros out all the elements under a diagonal element a_{ii} in one multiplication. However, it should be noted that Givens rotations lend themselves to parallelization. Also, if a matrix has a particular structure, it may be efficient to zero out one element at a time, as we will see in Chapters 18, 19, and 21. Givens rotations are perfect for that purpose.

A Householder reflection is an $n \times n$ orthogonal matrix formed from a vector in $\mathbb{R}^n$.

Definition 17.3. A *Householder reflection* (or Householder transformation) H_u is a transformation that takes a vector u and reflects it about a plane in $\mathbb{R}^n$. The transformation has the form

$$H_u = I - \frac{2uu^\mathrm{T}}{u^\mathrm{T}u}, \quad u \neq 0.$$

Clearly, H_u is an $n \times n$ matrix, since uu^T is a matrix of dimension $n \times n$. The Householder transformation has a geometric interpretation (Figure 17.3).

A Householder reflection applied to u gives $-u$. As a result, $H_u(cu) = -cu$, where c is a scalar. For all other vectors w, let $v = w - \mathrm{proj}_u(w)$. We know from our work with Gram-Schmidt that v is orthogonal to u. The vector w is a linear combination of the vectors u and v (Figure 17.4).

$$w = c_1 u + c_2 v$$
$$c_1 = \langle w, v \rangle / \|v\|_2^2$$
$$c_2 = \langle w, v \rangle / \|v\|_2^2$$

$H_u(w)$ is a reflection of w in the plane through 0 perpendicular to u. We will show that these claims are true.

$$H_u(u) = \left(I - \frac{2uu^\mathrm{T}}{u^\mathrm{T}u}\right) u = u - \frac{2uu^\mathrm{T}u}{u^\mathrm{T}u} = u - \frac{2u\|u\|_2^2}{\|u\|_2^2} = u - 2u = -u.$$

Now,

$$H_u(c_1 u + c_2 v) = \left(I - 2\frac{uu^\mathrm{T}}{u^\mathrm{T}u}\right)(c_1 u + c_2 v)$$

$$= c_1 u + c_2 v - 2\left(\frac{uu^\mathrm{T}}{u^\mathrm{T}u}\right)(c_1 u + c_2 v) = c_1 u + c_2 v - 2c_1 u - 2c_2 \left(\frac{uu^\mathrm{T}}{u^\mathrm{T}u}\right) v$$

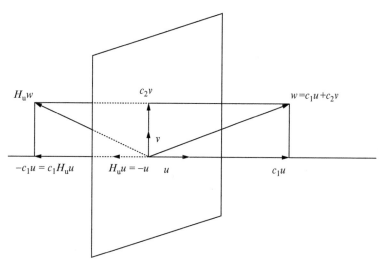

FIGURE 17.3 Householder reflection.

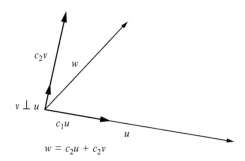

FIGURE 17.4 Linear combination associated with Householder reflection.

$$= -c_1 u + c_2 v - 2c_2 \left(\frac{u}{u^\mathrm{T} u} \right) \langle u, v \rangle = -c_1 u + c_2 v.$$

The Householder transformation also has a number of other interesting and useful properties.

Theorem 17.3. *Let H_u be a Householder reflection with vector $u \in \mathbb{R}^n$. Then*

1. *H_u is symmetric.*
2. *H_u is orthogonal.*
3. *$H_u^2 = I$*
4. *$H_u v = v$ if $\langle v, u \rangle = 0$.*

Proof. 1. $H_u^\mathrm{T} = \left(I - \frac{2uu^\mathrm{T}}{u^\mathrm{T} u} \right)^\mathrm{T} = I - \left(\frac{2uu^\mathrm{T}}{u^\mathrm{T} u} \right)^\mathrm{T} = I - \frac{2uu^\mathrm{T}}{u^\mathrm{T} u} = H_u.$

2. This is the most important property of H_u. Let $\beta = \frac{2}{u^\mathrm{T} u} = \frac{2}{\|u\|_2^2}$. Now, $H_u^\mathrm{T} H_u = H_u H_u = \left(I - \beta uu^\mathrm{T} \right) \left(I - \beta uu^\mathrm{T} \right)$, since H_u is symmetric.

$$\left(I - \beta uu^\mathrm{T} \right) \left(I - \beta uu^\mathrm{T} \right) = I - 2\beta uu^\mathrm{T} + \beta^2 \left(uu^\mathrm{T} uu^\mathrm{T} \right) = I - 2\beta uu^\mathrm{T} + \beta^2 \|u\|_2^2 uu^\mathrm{T} = I - 2 \left(\frac{2}{\|u\|_2^2} \right) uu^\mathrm{T}$$

$$+ \left(\frac{2}{\|u\|_2^2} \right)^2 \|u\|_2^2 uu^\mathrm{T} = I,$$

so H_u is orthogonal.

The proofs of properties 3 and 4 are left to the problems. $\qquad \square$

Example 17.10. Let $u = \begin{bmatrix} 3 & -1 & 2 \end{bmatrix}^{\mathrm{T}}$.

$$H_u = I - 2\frac{uu^{\mathrm{T}}}{u^{\mathrm{T}}u} = \begin{bmatrix} 1 & 0 & 0 \\ 0 & 1 & 0 \\ 0 & 0 & 1 \end{bmatrix} - \frac{2}{3^2 + (-1)^2 + 2^2}\begin{bmatrix} 3 \\ -1 \\ 2 \end{bmatrix}\begin{bmatrix} 3 & -1 & 2 \end{bmatrix}$$

$$= \begin{bmatrix} 1 & 0 & 0 \\ 0 & 1 & 0 \\ 0 & 0 & 1 \end{bmatrix} - \frac{1}{7}\begin{bmatrix} 9 & -3 & 6 \\ -3 & 1 & -2 \\ 6 & -2 & 4 \end{bmatrix} = \begin{bmatrix} -0.2857 & 0.4286 & -0.8571 \\ 0.4286 & 0.8571 & 0.2857 \\ -0.8571 & 0.2857 & 0.4286 \end{bmatrix}$$

Note that H_u is symmetric. Other properties of H_u are easier to illustrate using MATLAB.

```
>> u = [3 -1 2]';
>> Hu = eye(3) - 2*(u*u')/(u'*u);
Hu*u % Hu*u = -u

ans =
   -3.0000
    1.0000
   -2.0000

>> w = [-1 5 2]';
>> v = w - ((w'*u)/(u'*u))*u;
v'*u % v orthogonal to u

ans =
  -8.8818e-16

>> c1 = (w'*u)/(u'*u);
>> c2 = (w'*v)/(v'*v);
>> norm((c1*u + c2*v)-w) % w = c1*u + c2*v

ans =
   2.2204e-16

>> Hu*w % should be -c1*u + c2*v

ans =
    0.7143
    4.4286
    3.1429

>> -c1*u + c2*v

ans =
    0.7143
    4.4286
    3.1429

>> Hu'*Hu   % Hu is orthogonal

ans =
    1.0000    0.0000    0.0000
    0.0000    1.0000    0.0000
    0.0000    0.0000    1.0000

>> Hu^2   % Hu*Hu = I

ans =
    1.0000    0.0000    0.0000
    0.0000    1.0000    0.0000
    0.0000    0.0000    1.0000

>> z = [-1 -1 1]';
z'*u % z is orthogonal to u
```

```
ans =
     0
>> Hu*z   % should be z
ans =
    -1.0000
    -1.0000
     1.0000
```

■

17.8.1 Matrix Column Zeroing Using Householder Reflections

A single Householder reflection can zero out all the elements $a_{i+1,i}a_{i+2,i}, \ldots, a_{mi}$ below a diagonal element a_{ii}. By applying a sequence of Householder reflections, we can transform a matrix A into an upper triangular matrix. For simplicity, we will begin by developing a Householder matrix H_u that will zero out the elements $a_{21}, a_{31}, \ldots, a_{m1}$ below a_{11}. Thus, our goal is to transform

$$A = \begin{bmatrix} X & X & X & X & X & X & X & \cdots & X \\ X & X & X & X & X & X & X & \cdots & X \\ \vdots & \vdots & \vdots & \vdots & \vdots & \vdots & \vdots & \cdots & \vdots \\ X & X & X & X & X & X & X & \cdots & X \\ X & X & X & X & X & X & X & \cdots & X \end{bmatrix}$$

into the form

$$H_u A = \begin{bmatrix} X & X & X & X & X & X & X & \cdots & X \\ 0 & X & X & X & X & X & X & \cdots & X \\ \vdots & \vdots & \vdots & \vdots & \vdots & \vdots & \vdots & \cdots & \vdots \\ 0 & X & X & X & X & X & X & \cdots & X \\ 0 & X & X & X & X & X & X & \cdots & X \end{bmatrix}.$$

Let x be the first column of A. Then, $H_u x$ will be the first column of $H_u A$ (Equation 2.3). If $x \neq k e_1 = \begin{bmatrix} k & 0 & \ldots & 0 \end{bmatrix}^T$, we want to choose u so that $H_u x$ has zeros everywhere in column 1 except at location $(1, 1)$. Elements in the remaining columns will also be affected, but that is of no importance.

The process of choosing u can be viewed geometrically. We know that $H_u x$ reflects x through the plane perpendicular to u, and that $\|H_u x\|_2 = \|x\|$ because H_u is an orthogonal matrix. We want to determine u in such a way that $H_u x$ reflects x to a vector $\pm \|x\|_2 e_1$, i.e., $H_u x = \pm \|x\|_2 e_1$. Figure 17.5 helps in developing an approach. Reflect x through a hyperplane that bisects the angle between x and e_1. This can be done by choosing $u = x - \|x\|_2 e_1$, as Figure 17.5 indicates. A direct computation verifies the result. Begin with

$$H_u x = \left(I - 2\frac{uu^T}{u^T u} \right) x = x - 2\frac{(x - \|x\|_2 e_1)\left(x^T - \|x\|_2 e_1^T\right)x}{\|u\|_2^2}. \tag{17.6}$$

$$\|u\|_2^2 = \|x - \|x\|_2 e_1\|_2^2 = 2 \|x\|_2 \left(\|x\|_2 - x^T e_1\right) = 2 \|x\|_2 \left(\|x\|_2 - x_1\right) \tag{17.7}$$

Evaluating the numerator of Equation 17.6 using the results $x e_1^T x = x_1 x$ and $e_1 e_1^T x = x_1 e_1$, we have

$$(x - \|x\|_2 e_1)\left(x^T - \|x\|_2 e_1^T\right)x = \|x\|_2 \left[(\|x\|_2 - x_1) x - \|x\|_2 (\|x\|_2 - x_1) e_1\right]. \tag{17.8}$$

Using Equations 17.6–17.8, there results

$$H_u x = x - \frac{2 \|x\|_2 \left[(\|x\|_2 - x_1) x - \|x\|_2 (\|x\|_2 - x_1) e_1\right]}{2 \|x\|_2 (\|x\|_2 - x_1)} = x - (x - \|x\|_2 e_1) = \|x\|_2 e_1.$$

A similar calculation shows that if $u = x + \|x\|_2 e_1$, $H_u x = -\|x\|_2 e_1$. Thus $H_u x$ will eliminate all entries of x except that at index 1 by choosing either $u = x - \|x\|_2 e_1$ or $u = x + \|x\|_2 e_1$.

The sign in $u = x \pm \|x\|_2 e_1$ must be chosen carefully to avoid cancelation error. Now,

$$\begin{bmatrix} x_1 & x_2 & \ldots & x_{m-1} & x_m \end{bmatrix}^T \pm \|x\|_2 e_1 = \begin{bmatrix} x_1 \pm \|x\|_2 & x_2 & \ldots & x_{m-1} & x_m \end{bmatrix}^T$$

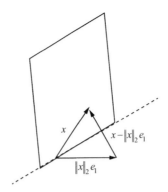

FIGURE 17.5 Householder reflection to a multiple of e_1.

so the only component different from x is the first component of u. To avoid subtraction and possible cancelation error, choose the sign to be that of x_1 so an addition is done instead of a subtraction.

$$u = \begin{cases} x + \|x\|_2\, e_1, & x_1 > 0 \\ x - \|x\|_2\, e_1, & x_1 < 0 \\ x + \|x\|_2\, e_1, & x_1 = 0 \end{cases} \quad . \tag{17.9}$$

Another possible problem is overflow or underflow when computing $\|x\|_2$. Section 8.4.1 presents a strategy to avoid overflow during the calculation.

1. colmax $= \max\left(|x_2|, |x_2|, \ldots, |x_m|\right)$.
2. $\bar{x} = x/\text{colmax}$.
3. $\|x\|_2 = \text{colmax}\,\|\bar{x}\|_2$

One of the components of $\bar{x}$ will have absolute value 1, and the remaining components will have absolute value less than or equal to 1, so there is no possibility of overflow. For computing u from Equation 17.9, we only execute steps 1 and 2. Then, $1 \le \|\bar{x}\| \le \sqrt{m}$, and there can be no overflow or underflow in computing $\bar{x}$. Let

$$\bar{u} = \left(\frac{x}{\text{colmax}}\right) + \left\|\frac{x}{\text{colmax}}\right\|_2 e_1$$

and use the Householder reflection $H_{\bar{u}}$. It follows that $\bar{u} = \left(\frac{1}{\text{colmax}}\right) u$ (Problem 17.4), and as Lemma 17.2 shows, multiplying u by a constant does not change the value of H_u.

Lemma 17.2. *If H_u is a Householder reflection and k is a constant, then $H_{ku} = H_u$.*

Proof.

$$H_{ku} = \left(I - \frac{2\,(ku)\,(ku)^{\mathrm{T}}}{\|ku\|_2^2}\right)$$

$$= I - \left(\frac{2k^2 uu^{\mathrm{T}}}{k^2\,\|u\|_2^2}\right) = H_u \qquad \qquad \square$$

Example 17.11. For $A = \begin{bmatrix} 1 & 2 & 0 \\ -1 & 4 & 1 \\ -3 & 1 & 2 \end{bmatrix}$, find and apply a Householder reflection to zero out entries a_{21} and a_{31}. We will do this in four steps, retaining four decimal places.

1. *Determine x.* The maximum element in magnitude in column 1 is -3, so

$$x = \frac{1}{3}\begin{bmatrix} 1 & -1 & -3 \end{bmatrix} = \begin{bmatrix} 0.3333 & -0.3333 & -1.0000 \end{bmatrix}^{\mathrm{T}}.$$

2. *Compute u and β. Let*

$$u = x \pm \|x\|_2\, e_1 = \begin{bmatrix} 0.3333 \\ -0.3333 \\ -1.0000 \end{bmatrix} \pm \left\| \begin{bmatrix} 0.3333 \\ -0.3333 \\ -1.0000 \end{bmatrix} \right\| \begin{bmatrix} 1 \\ 0 \\ 0 \end{bmatrix} = \begin{bmatrix} 0.3333 \\ -0.3333 \\ -1.0000 \end{bmatrix} \pm 1.1055 \begin{bmatrix} 1 \\ 0 \\ 0 \end{bmatrix}.$$

Choose the sign + because $x_1 = 0.3333 \geq 0$, and

$$u = \begin{bmatrix} 0.3333 \\ -0.3333 \\ -1.0000 \end{bmatrix} + \begin{bmatrix} 1.1055 \\ 0 \\ 0 \end{bmatrix} = \begin{bmatrix} 1.4388 \\ -0.3333 \\ -1.0000 \end{bmatrix}, \quad \beta = \frac{2}{u^T u} = 0.6287.$$

3. *The Householder reflection is*

$$H_u = \begin{bmatrix} 1 & 0 & 0 \\ 0 & 1 & 0 \\ 0 & 0 & 1 \end{bmatrix} - \beta \begin{bmatrix} 1.4388 \\ -0.3333 \\ -1.0000 \end{bmatrix} \begin{bmatrix} 1.4388 \\ -0.3333 \\ -1.0000 \end{bmatrix}^T = \begin{bmatrix} -0.3015 & 0.3015 & 0.9045 \\ 0.3015 & 0.9301 & -0.2095 \\ 0.9045 & -0.2095 & 0.3713 \end{bmatrix}.$$

4. *Form*

$$H_u A = \begin{bmatrix} 1 & 2 & 0 \\ -1 & 4 & 1 \\ -3 & 1 & 2 \end{bmatrix} - \beta u u^T \begin{bmatrix} 1 & 2 & 0 \\ -1 & 4 & 1 \\ -3 & 1 & 2 \end{bmatrix} = \begin{bmatrix} -3.3166 & 1.5076 & 2.1106 \\ 0.0000 & 4.1141 & 0.5111 \\ 0.0000 & 1.3422 & 0.5332 \end{bmatrix}. \qquad \blacksquare$$

17.8.2 Implicit Computation with Householder Reflections

As was the case with Givens rotations, multiplication by a Householder reflection H_u does not require construction of the matrix. If H_u is an $m \times m$ Householder matrix, x is a vector in $\mathbb{R}^m$, and A is an $m \times n$ matrix, then the products $H_u x$ and $H_u A$ can be computed by a simple formula. If H_u is an $n \times n$ matrix, then the same is true for $A H_u$.

> Let $\beta = \frac{2}{u^T u}$.
>
> $$H_u v = \left(I - \beta u u^T\right) v = v - \beta u \left(u^T v\right) \qquad (17.10)$$
>
> $$H_u A = \left(I - \beta u u^T\right) A = A - \beta u u^T A \qquad (17.11)$$
>
> $$A H_u = A \left(I - \beta u u^T\right) = A - \beta A u u^T \qquad (17.12)$$

We are now prepared to formally state the algorithm, hzero1, for zeroing out all elements below a_{11} using a Householder reflection. The function returns the vector u because it will be needed to form Q during the Householder *QR* decomposition algorithm.

Algorithm 17.4 Zero Out Entries in the First Column of a Matrix using a Householder Reflection

---

```
function HZERO1(A)
    % Zero out all elements a₂₁...aₘ₁ in the
    % m × n matrix A using a Householder reflection Hᵤ
    % [A u] - hzero1(A) returns u and a new matrix A implicitly
    % premultiplied by the Householder matrix Hu.

    x = A(:,1)
    colmax = max([ |x₁|, |x₂|, ..., |xₘ₋₁|, |xₘ| ])
    x = x/colmax
    colnorm = ‖x‖₂
    u = x
    if u₁ ≥ 0 then
        u₁ = u₁ + colnorm
    else
        u₁ = u₁ - colnorm
    end if

    % implicitly form HᵤA.
    unorm = ‖u‖₂
    if unorm ≠ 0 then
        β = 2/unorm
    else
        β = 0
    end if
    A = A - (βu)(uᵀA)
    return [ A, u ]
end function
```

NLALIB: The function `hzero1` implements Algorithm 17.4.

17.9 COMPUTING THE *QR* DECOMPOSITION USING HOUSEHOLDER REFLECTIONS

To transform an $m \times n$ matrix into upper triangular form, we must zero out all the elements below the diagonal entries $a_{11}, a_{22}, \ldots, a_{kk}$, where $k = \min(m - 1, n)$. We know how to do this for a_{11} (Algorithm 17.4), and now we will demonstrate how to zero out the elements below the remaining diagonal entries. This is done by implicitly creating a sequence of Householder matrices that deal with submatrix blocks, as illustrated in Figure 17.6.

Zeroing out all the elements below a_{11} using a Householder matrix gives matrix A_1. Now we must deal with the submatrices having a diagonal element in their upper left-hand corner. Assume we have zeroed out all the elements below diagonal indices $(1, 1)$ through $(i - 1, i - 1)$ (Figure 17.7) by computing a sequence of matrices $A_1, A_2, \ldots, A_{i-1}$. We must find a Householder reflection that zeros out $\overline{a_{i+1,i}}, \ldots, \overline{a_{mi}}$ and only modifies elements in the submatrix denoted in Figure 17.7. Consider this $(m - i + 1) \times (n - i + 1)$ matrix as the matrix whose first column must be transformed to $\begin{bmatrix} X & 0 & \ldots & 0 \end{bmatrix}^{\mathrm{T}}$ by using hzero1. The vector x used for the formation of H_u is $\begin{bmatrix} \overline{a_{ii}} & \overline{a_{i+1,i}} & \ldots & \overline{a_{mi}} \end{bmatrix}^{\mathrm{T}}$. Imagine we build the matrix in Figure 17.8, say $\tilde{H}_u$, and compute $\tilde{H}_u A_{i-1}$. $\tilde{H}_u$ is orthogonal and is structured so it only affects the elements of the submatrix shown in Figure 17.7. To compute R we only have to carry out the calculations that modify the $(m - i + 1) \times (n - i + 1)$ matrix. Do this using Equation 17.13.

$$\begin{bmatrix} A(i:m, i:n) & u \end{bmatrix} = \text{hzero1}(A(i:m, i:n)). \tag{17.13}$$

The application of Equation 17.13 for $i = 1, 2, k = \min(m - 1, n)$ determines R in a highly efficient fashion, since it only deals with the submatrices that must be modified, and the submatrices become smaller with each step.

$$A = \begin{bmatrix} X & X & X & X & X & X & X & X \\ X & X & X & X & X & X & X & X \\ X & X & X & X & X & X & X & X \\ X & X & X & X & X & X & X & X \\ X & X & X & X & X & X & X & X \\ X & X & X & X & X & X & X & X \\ X & X & X & X & X & X & X & X \\ X & X & X & X & X & X & X & X \end{bmatrix} \xrightarrow[H_{u_1}]{A_1} \begin{bmatrix} X & X & X & X & X & X & X & X \\ 0 & X & X & X & X & X & X & X \\ 0 & X & X & X & X & X & X & X \\ 0 & X & X & X & X & X & X & X \\ 0 & X & X & X & X & X & X & X \\ 0 & X & X & X & X & X & X & X \\ 0 & X & X & X & X & X & X & X \\ 0 & X & X & X & X & X & X & X \end{bmatrix} \rightarrow$$

$$H_{u_2}H_{u_1}A = \begin{bmatrix} X & X & X & X & X & X & X & X \\ 0 & X & X & X & X & X & X & X \\ 0 & 0 & X & X & X & X & X & X \\ 0 & 0 & X & X & X & X & X & X \\ 0 & 0 & X & X & X & X & X & X \\ 0 & 0 & X & X & X & X & X & X \\ 0 & 0 & X & X & X & X & X & X \\ 0 & 0 & X & X & X & X & X & X \end{bmatrix} \xrightarrow[H_{u_3}H_{u_2}H_{u_1}]{A_3} \begin{bmatrix} X & X & X & X & X & X & X & X \\ 0 & X & X & X & X & X & X & X \\ 0 & 0 & X & X & X & X & X & X \\ 0 & 0 & 0 & X & X & X & X & X \\ 0 & 0 & 0 & X & X & X & X & X \\ 0 & 0 & 0 & X & X & X & X & X \\ 0 & 0 & 0 & X & X & X & X & X \\ 0 & 0 & 0 & X & X & X & X & X \end{bmatrix} \rightarrow$$

$$H_{u_4}H_{u_3}H_{u_2}H_{u_1} = \begin{bmatrix} X & X & X & X & X & X & X & X \\ 0 & X & X & X & X & X & X & X \\ 0 & 0 & X & X & X & X & X & X \\ 0 & 0 & 0 & X & X & X & X & X \\ 0 & 0 & 0 & 0 & X & X & X & X \\ 0 & 0 & 0 & 0 & X & X & X & X \\ 0 & 0 & 0 & 0 & X & X & X & X \\ 0 & 0 & 0 & 0 & X & X & X & X \end{bmatrix} \quad A_4$$

FIGURE 17.6 Transforming an $m \times n$ matrix to upper triangular form using householder reflections.

$$A_{i+1} = \begin{bmatrix} X & X & \cdots & X & X & X & X & \cdots & X \\ 0 & X & \cdots & X & X & X & X & \cdots & X \\ 0 & 0 & \ddots & X & X & X & X & \cdots & X \\ & & & X & X & X & X & \cdots & X \\ 0 & & & & \overline{a_{ii}} & \overline{a_{i,i+1}} & \overline{a_{i,i+2}} & \cdots & \overline{a_{in}} \\ \vdots & \vdots & \vdots & \vdots & \overline{a_{i+1,i}} & \overline{a_{i+1,i+1}} & \overline{a_{i+1,i+2}} & \cdots & \overline{a_{i+1,n}} \\ & & & & \overline{a_{i+2,i}} & \overline{a_{i+2,i+1}} & \overline{a_{i+2,i+2}} & \cdots & \overline{a_{i+2,n}} \\ & & & & \vdots & \vdots & \vdots & \ddots & \vdots \\ 0 & 0 & \cdots & 0 & \overline{a_{mi}} & \overline{a_{m,i+1}} & \overline{a_{m,i+2}} & \cdots & \overline{a_{mn}} \end{bmatrix}$$

FIGURE 17.7 Householder reflections and submatrices.

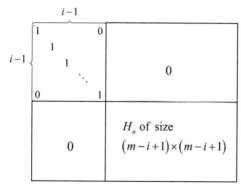

FIGURE 17.8 Householder reflection for a submatrix.

Example 17.12. Let $A = \begin{bmatrix} 1 & 5 & -1 & 8 & 3 \\ -1 & 4 & 12 & 6 & -9 \\ 0 & 3 & 16 & -1 & -6 \\ 8 & 1 & 4 & 9 & -2 \\ 1 & 2 & 7 & 8 & 0 \\ 15 & 22 & 17 & -1 & 5 \\ 23 & -7 & 1 & 7 & 9 \end{bmatrix}$. After zeroing out elements below the entries at indices (1,1) and

(2,2), we have the matrix

$$H_{u_2}H_{u_1}A = \begin{bmatrix} -1.2458 & -6.2820 & -10.6097 & -7.9573 & -9.7023 \\ 0 & -1.4375 & -17.4780 & 1.4939 & 2.3462 \\ 0 & 0 & 12.7841 & -1.5443 & -4.8219 \\ 0 & 0 & 3.5983 & 5.0658 & -6.2295 \\ 0 & 0 & 4.9398 & 7.1680 & 0.2076 \\ 0 & 0 & -5.3266 & -12.0282 & 4.9730 \\ 0 & 0 & 10.4307 & -2.5192 & -7.0378 \end{bmatrix} = A_2.$$

The task is to zero out all the elements in column 3 below 12.7841. Start with

$$x = \begin{bmatrix} 12.7841 \\ 3.5983 \\ 4.9398 \\ -5.3266 \\ 10.4307 \end{bmatrix}$$

and apply the steps we have described.

$$x = x/12.7841 = \begin{bmatrix} 1.0000 \\ 0.2815 \\ 0.3864 \\ -0.4167 \\ 0.8159 \end{bmatrix}, \quad u = x + \|x\| \begin{bmatrix} 1 \\ 0 \\ 0 \\ 0 \\ 0 \end{bmatrix} = \begin{bmatrix} 2.4380 \\ 0.2815 \\ 0.3864 \\ -0.4167 \\ 0.8159 \end{bmatrix}$$

$$\beta = \frac{2}{u^T u} = 0.2852, \quad A_3\,(3:7,3:5) = A_2\,(3:7,3:5) - \beta u u^T A_2\,(3:7,3:5),$$

$$A_3 = \begin{bmatrix} -1.2458 & -6.2820 & -10.6097 & -7.9573 & -9.7023 \\ 0 & -1.4375 & -17.4780 & 1.4939 & 2.3462 \\ 0 & 0 & -1.4380 & -3.8995 & 9.9508 \\ 0 & 0 & 0 & 4.7939 & -4.5240 \\ 0 & 0 & 0 & 6.7947 & 2.5490 \\ 0 & 0 & 0 & -11.6257 & 2.4483 \\ 0 & 0 & 0 & -3.3074 & -2.0939 \end{bmatrix}$$

We can now develop the QR decomposition using Householder reflections. It may be that we are only interested in computing R. For instance,

$$A^T A = (QR)^T (QR) = R^T Q^T QR = R^T R,$$

and R is the factor in the Cholesky decomposition of $A^T A$. Not forming Q saves computing time. If we require Q, it will be necessary to implicitly compute the $\tilde{H}_i$ (Figure 17.8) as we transform A into the upper triangular matrix R. Then we have

$$\tilde{H}_k \tilde{H}_{k-1} \tilde{H}_{k-2} \ldots \tilde{H}_2 \tilde{H}_1 A = R$$

$$A = \left(\tilde{H}_k \tilde{H}_{k-1} \tilde{H}_{k-2} \ldots \tilde{H}_2 \tilde{H}_1 \right)^{\mathrm{T}} R$$

$$= \left(\tilde{H}_1 \tilde{H}_2 \tilde{H}_3 \ldots \tilde{H}_{k-1} \tilde{H}_k \right) R$$

and

$$Q = \tilde{H}_1 \tilde{H}_2 \tilde{H}_3 \ldots \tilde{H}_{k-1} \tilde{H}_k.$$

Start with the $m \times m$ identity matrix $Q = I$. Implicitly compute $Q(1:m, 1:m) = I H_{u_1}$, which changes entries in all rows and columns of I. Now compute $Q(1:m, 2:m) = Q(1:m, 2:m) H_{u_2}$, which only affects columns 2 through m. Continue this k times. Algorithm 17.5 implements the full *QR* decomposition of an $m \times n$ matrix.

Algorithm 17.5 Computation of *QR* Decomposition Using Householder Reflections

---

```
function HQR(A)
    % Compute the QR decomposition of matrix A using Householder reflections.
    % Input: m × n matrix A. There are no restrictions on the values of m and n.
    % Output: m × m matrix Q and m × n upper triangular matrix R
    % such that A=QR.
    R = A
    Q = I
    k = min (m − 1, n)
    for i=1:k do
        [ R(i : m, i : n) , u ] = hzero1 (R(i : m, i : n))
        Q(1 : m, i : m) = Q(1 : m, i : m) − (2/ ‖u‖²₂) Q(1 : m, i : m) (uuᵀ)
    end for
end function
```

NLALIB: The function `hqr` implements Algorithm 17.5.

This is generally the preferred method for computing the *QR* decomposition, since it is faster than using Givens rotations. However, as we have stated, the *QR* decomposition using Givens rotations can be parallelized, and this is a advantage in an era of multicore processors and GPU computing. In addition, we will require the use of Givens rotations when computing eigenvalues and the singular value decomposition (SVD).

Example 17.13.

Let x be the vector of values beginning at -1.0 and ending at 1.0 in steps of 0.01. Use x to build a Vandermonde matrix, V, using $m = 20$, and apply both `hqr` and `qr` to V.

```
>> x = -1.0:0.01:1.0;
>> V = vandermonde(x,20);
>> cond(V)

ans =
  1.7067e+007

>> [Q R] = hqr(V);
>> [QM RM] = qr(V);
>> norm(V - Q*R)

ans =
  5.9967e-014

>> norm(V - QM*RM)

ans =
  9.5622e-015
```

```
>> norm(Q'*Q - eye(201))
ans =
  2.6553e-015
>> norm(QM'*QM - eye(201))
ans =
  1.7922e-015
```

■

17.9.1 Efficiency and Stability

Our analysis will not include the computation of Q. Including this calculation more than doubles the flop count. Assume $m > n$ so $k = n$. If we determine the flop count $\text{flop}_R(i)$ for the statement

$$\begin{bmatrix} R(i:m, i:n) & u \end{bmatrix} = \text{hzero1}(R(i:m, i:n))$$

then $\sum_{i=1}^{n} \text{flop}_R(i)$ is the flop count for building R.

Before determining the flop count, we need to consider the computation $A - \beta u u^T A$, since the order of evaluation drastically effects its flop count (Problem 17.5). We will assume that the expression is computed in the following order:

1. $v = \beta u^T$: $(m - i + 1)$ flops
2. $v = vA$: $(1 \times (m - i + 1))$ matrix times $(m - i + 1) \times (n - i + 1)$ matrix requires $2(m - i + 1)(n - i + 1)$ flops.
3. $A - uv$: A is an $(m - i + 1) \times (n - i + 1)$ and uv is an $(m - i + 1) \times (n - i + 1)$ matrix. Using nested loops, this calculation can be done in $2(m - i + 1) \times (n - i + 1)$ flops, so we can compute $A - \beta u u^T A$ in $(m - i + 1) + 4(m - i + 1)(n - i + 1)$ flops.

Each instance of Algorithm 17.4 works with a matrix A of dimension $(m - i + 1) \times (n - i + 1)$. For each i, list the computations required, the flop count for each, and then form the total sum, $\text{flop}_R(i)$.

Expression	Flop Count
$x = x/\text{colmax}$	$m - i + 1$
$\text{colnorm} = \|x\|_2$	$2(m - i + 1)$
$u_1 \pm \text{colnorm}$	1
$\text{unorm} = \|u\|_2$	$2(m - i + 1)$
$\beta = 2/\text{unorm}^2$	2
$A - u((\beta u^T)A)$	$(m - i + 1) + 4(m - i + 1)(n - i + 1)$

$$\text{flop}_R(i) = 6(m - i + 1) + 4(m - i + 1)(n - i + 1) + 3$$

The total flop count for computing R is

$$\sum_{i=1}^{n} [6(m - i + 1) + 4(m - i + 1)(n - i + 1) + 3]$$

Discard $\sum_{i=1}^{n} [6(m - i + 1) + 3]$, since it contributes only low-order terms. Using standard summation formulas and some algebra, we obtain

$$4\sum_{i=1}^{n} (m - i + 1)(n - i + 1) = \frac{2}{3}n(n + 1)(3m - n + 1)$$

$$= 2\left(mn^2 - \frac{n^3}{3}\right) + 2mn + \frac{2}{3}n$$

Again, exclude low-order terms, and we have

$$\text{flop count} \approx 2\left(mn^2 - \frac{n^3}{3}\right).$$

For the Givens algorithm, we computed the flop count for determining both R and Q. If we do the analysis and exclude Q, we have

$$\text{Givens flop count} = (5 + 6n) \sum_{i=1}^{n} (m - i) \approx 3n^2 (2m - n)$$

If we consider an $n \times n$ matrix, the Givens algorithm is approximately twice as costly as the Householder algorithm. Of course, this is because hqr zeros out all the elements in column i below a_{ii} in one reflection, whereas the Givens algorithm requires $(m - i)$ rotations.

The Householder algorithm for computing the *QR* decomposition is stable [16, pp. 357-360].

17.10   CHAPTER SUMMARY

Gram-Schmidt *QR* Decomposition

The modified Gram-Schmidt process (never use classical Gram-Schmidt unless you perform reorthogonalization) gives a reduced *QR* decomposition, and its algorithm for orthonormalization of set of linearly independent vectors has other applications. Its flop count of $2mn^2$ is superior to the Givens and Householder when both Q and R are required. However, its stability depends on the condition number of the matrix, so it is not as stable as the other methods.

Givens *QR* Decomposition

Assume A is an $m \times n$ matrix. If c and s are constants, an $m \times m$ Givens matrix $J(i, j, c, s)\, i < j$, also called a Givens rotation, places c at indices (i, i) and (j, j), $-s$ at (j, i), and s at (i, j) in the identify matrix. $J(i, j, c, s)$ is orthogonal, and by a careful choice of constants (Algorithm 17.2), $J(i, j, c, s)\, A$ affects only rows i and j of A and zeros out a_{ji}. The product is performed implicitly by changing just rows i and j, so it is rarely necessary to build a Givens matrix. By zeroing out all the elements below the main diagonal, the Givens *QR* algorithm produces the upper triangular matrix R in the decomposition. By maintaining a product of Givens matrices, Q can also be found. The algorithm is stable, and its perturbation analysis does not involve the condition number of A. While the Givens *QR* decomposition is efficient and stable, Householder reflections are normally used for the *QR* decomposition. However, because premultiplication by a Givens matrix can zero out a particular element, these matrices are very useful when A has a structure that lends itself to zeroing out one element at a time (Problem 17.12).

Householder Decomposition

The use of Householder matrices, also termed a Householder reflections, is the most commonly used method for performing the *QR* decomposition. If u is an $m \times 1$ vector, the Householder matrix defined by

$$H_u = I - \left(\frac{2}{u^T u} \right) uu^T$$

is orthogonal and symmetric. Products $H_u v$, $H_u A$, and AH_u, where A is an $m \times n$ matrix and v is an $m \times 1$ vector can be computed implicitly without the need to build H_u. By a proper choice of u (Equation 17.9), $H_u A$ zeros out all the elements below a diagonal element a_{ii}, and so it is an ideal tool for the *QR* decomposition. The Householder *QR* decomposition is stable and, like the Givens *QR* process, its perturbation analysis does not depend on the condition number of A. It is this "all at once" feature of Householder matrices that makes them so useful for matrix decompositions. They will be very important in our study of eigenvalue computation in Chapters 18 and 19.

17.11   PROBLEMS

17.1 Using pencil and paper, compute the *QR* decomposition of $A = \begin{bmatrix} 1 & -1 \\ -2 & 3 \end{bmatrix}$ using a Givens rotation that you explicitly build.

17.2 Using pencil and paper, compute the *QR* decomposition of $A = \begin{bmatrix} 1 & -1 \\ -2 & 3 \end{bmatrix}$ using a Householder reflection that you explicitly build.

17.3 Show that in the full QR decomposition of the full rank $m \times n$ matrix $A, m \geq n$, the vectors $q_{n+1}, \ldots, q_m$ are an orthonormal basis for the null space of A^T. Hint: Using block matrix notation, write

$$Q = \begin{bmatrix} Q_1^{m \times n} & Q_2^{(m-n) \times n} \end{bmatrix},$$

where Q_1 consists of the first n columns of Q, and the columns of Q_2 are the remaining $m - n$ columns. Write R in block matrix notation also.

17.4 If $u = x + \|x\|_2 e_1$, show that if $\bar{u} = kx + \|kx\|_2 e_1$, where $k \geq 0$ is a constant, then $\bar{u} = ku$.

17.5 If A is an $m \times n$ matrix, and $u \in \mathbb{R}^m$, $v \in \mathbb{R}^m$, the amount of work to evaluate $A - uv^TA$ depends dramatically on the order in which the operations are performed.

  a. How many flops are required to compute it in the order

$$
\begin{aligned}
T_1 &= uv^T \\
T_2 &= T_1 A \\
T_3 &= A - T_2
\end{aligned}
$$

  b. Show that computing $A - \left(uv^T\right) A$ can be done in approximately $2m^2n + 2mn$ flops. Hint: Determine the structure of each row of uv^T, and consider each row of $\left(uv^T\right) A$ as the inner product of a row of uv^T with all columns of A.

  c. Show that the $A - uv^TA$ can be done in $4mn$ flops by first computing v^TA and then $A - u\left(v^TA\right)$.

17.6 The QR decomposition can be used to solve an $n \times n$ linear system $Ax = b$ using the following steps:

> Find the QR decomposition of A: $A = QR$.
> Form $b' = Q^Tb$.
> Solve $Rx = b'$.

Assume that a full QR decomposition of an $n \times n$ matrix requires $4\left(m^2n - mn^2 + \frac{n^3}{3}\right)$ using Householder reflections.

  a. Determine the flop count for the solution.

  b. Compare your flop count with that of Gaussian elimination. Which method is generally preferable?

17.7 Prove these properties of Householder matrices

  a. $H_u^2 = I$

  b. $H_u v = v$ if $\langle v, u \rangle = 0$.

17.8 Prove that if u is chosen to be parallel to vector $x - y$, where $x \neq y$ but $\|x\|_2 = \|y\|_2$, then $H_u x = y$. Hint: Let $u = k(x - y)$, where k is a constant, and note that

$$x = \frac{1}{2}(x + y) + \frac{1}{2}(x - y).$$

Compute Hx and apply the relation $H_u(u) = -u$. Show that $\langle x + y, x - y \rangle = 0$ and apply Theorem 17.3, part 4.

17.9 Let $A \in \mathbb{R}^{m \times n}$, $m \geq n$, have reduced QR decomposition $A = QR$. Show that $\|A\|_2 = \|R\|_2$.

17.10 If u and v are $n \times 1$ vectors, the $n \times n$ matrix uv^T has rank 1 (Problem 10.3). If A is an $n \times n$ matrix, we say that the matrix $B = A + uv^T$ is a *rank 1 update* of A. Let $A = QR$ be the QR decomposition of A. Show that

$$A + uv^T = Q\left(R + wv^T\right),$$

where $w = Q^Tu$.

17.11 Problem 17.10 defines a rank 1 update of a matrix. A Householder reflection, $H_u = I - \left(\frac{2}{\|u\|_2^2}\right)uu^T$, is a rank 1 update of the identity matrix. We know a Householder matrix is symmetric, orthogonal, and is its own inverse ($H_u^2 = I$). This problem investigates a more general rank 1 update of the identity,

$$R_1 = I - uv^T.$$

  a. Prove that R_1 is nonsingular if and only if $\langle v, u \rangle \neq 1$.

  b. If R_1 is nonsingular, show that $R_1^{-1} = I - \beta uv^T$. Do this by finding a formula for β.

17.12 In this problem, you will investigate a special type of square matrix called an upper Hessenberg matrix. Such a matrix has the property $a_{ij} = 0$, $i > j + 1$, and is often called almost upper triangular.

a. Give an example of a 4×4 and a 5×5 upper Hessenberg matrix.

b. Develop an algorithm for computing the *QR* decomposition of an upper Hessenberg matrix. Hint: Use Givens rotations. How many will be necessary?

c. Show that the flop count is $O\left(n^2\right)$.

17.13 Suppose that $A \in \mathbb{R}^{m \times n}$ has full column rank. Prove that the reduced decomposition

$$A = QR$$

is unique where $Q \in \mathbb{R}^{m \times n}$ has orthonormal columns and $R^{n \times n}$ is upper triangular with positive diagonal entries. Do this in steps.

a. Show that $A^{\mathrm{T}}A = R^{\mathrm{T}}R$.

b. Prove that $A^{\mathrm{T}}A$ is symmetric positive definite, and apply Theorem 13.3 to show that R is the unique Cholesky factor of $A^{\mathrm{T}}A$.

c. Show that Q must be unique.

d. Give an example to show that there is no guarantee of uniqueness if A does not have full column rank.

17.14 We have studied the *LU*, *QR*, and the SVDs. There are many more, and the book will construct additional ones in later chapters. This problem develops two variants of the *QR* decomposition, the *QL* and the *RQ*.

An $m \times m$ matrix of the form

$$K_m = \begin{bmatrix} & & & & 1 \\ & & & 1 & \\ & & \iddots & & \\ & 1 & & & \\ 1 & & & & \end{bmatrix} = \left[k_{ij}\right],$$

where $k_{i,m-i+1} = 1$, $1 \le i \le m$, and all other entries are zero is termed a *reversal matrix* and sometimes the *reverse identity matrix*.

a. Show that $K_m^2 = I$ (a very handy feature).

b. If A is an $m \times n$ matrix, $m \ge n$, what is the action of $K_m A$? What about AK_n?

c. If R is upper triangular $n \times n$ matrix, what is the form of the product $K_n R K_n$?

d. Let $AK_n = \hat{Q}\hat{R}$ be the reduced *QR* decomposition of AK_n, $m \ge n$. Show that $A = \left(\hat{Q}K_n\right)\left(K_n\hat{R}K_n\right)$, and from that deduce the decomposition

$$A = QL,$$

where Q is an $m \times n$ matrix with orthogonal columns, and L is an $n \times n$ lower triangular matrix. This is a *reduced QL decomposition*.

e. If $m < n$, show to form an *RQ* decomposition, $A = RQ$, where R is $m \times m$ and Q is $m \times n$.

17.15 If $A \in \mathbb{R}^{m \times n}$, there exists an $m \times n$ lower triangular matrix and an $n \times n$ orthogonal matrix Q such that $A = LQ$.

a. Given the 1×2 vector $\begin{bmatrix} x & y \end{bmatrix}$, show there is a Givens rotation, J, such that $\begin{bmatrix} x & y \end{bmatrix}J = \begin{bmatrix} * & 0 \end{bmatrix}$. This type of rotation eliminates elements from columns.

b. Develop an algorithm using Givens rotations that computes an *LQ* decomposition of A for any $m \times n$ matrix. Hint: Write a function givensmulpost(A,i,j,c,s) that affects only columns i and j and zeros out A(i,j) in row i. The function givensparms does not change.

17.16 In a series of steps, this problem develops the result:

Assume $J\left(i,j,c,s\right)$ is a Givens rotation, A is a symmetric matrix, and define $B = J^{\mathrm{T}}\left(i,j,c,s\right)AJ\left(i,j,c,s\right)$. Then,

$$\sum_{i=1}^{n}\sum_{j=1}^{n}b_{ij}^2 = \sum_{i=1}^{n}\sum_{j=1}^{n}a_{ij}^2.$$

Recall the following relationships:

- $\|X\|_F^2 = \sum_{i=1}^{n}\sum_{j=1}^{n}x_{ij}^2 = \text{trace}\left(X^{\mathrm{T}}X\right)$
- $\text{trace}\left(XY\right) = \text{trace}\left(YX\right)$

a. Show that $\|B\|_F^2 = \text{trace}\left(J^T(i,j,c,s)A^TAJ(i,j,c,s)\right)$.

b. Show that $\text{trace}\left(J^T(i,j,c,s)A^TAJ(i,j,c,s)\right) = \text{trace}\left(AA^T\right)$

c. Conclude that $\sum_{i=1}^{n}\sum_{j=1}^{n}b_{ij}^2 = \sum_{i=1}^{n}\sum_{j=1}^{n}a_{ij}^2$.

17.17 Find the eigenvalues of a Householder reflection. Hint: Run some numerical experiments to look for a pattern. If H_u is a Householder reflection, $H_u u = -u$. Starting with u, build a basis $u, v_1, v_2, \ldots, v_n$, and take note of part 4 in Theorem 17.3.

17.11.1   MATLAB Problems

17.18 Given $v = \begin{bmatrix} -1 & 3 & 7 \end{bmatrix}^T$, build the Givens matrix $J(1,2,c,s)$ such that the second component of $J(1,2,c,s)v$ is zero. Build the Givens matrix $J(1,3,c,s)$ such that $J(1,3)J(1,2,c,s)v$ has the form $\begin{bmatrix} * & 0 & 0 \end{bmatrix}^T$. Use a sequence like the following to explicitly build a Givens matrix.

```
>> [c s] = givensparms(xi,xj);
>> J = eye(n);
>> J(i,i) = c;
>> J(j,j) = c;
>> J(j,i) = -s;
>> J(i,j) = s;
```

17.19 Explicitly build the Givens rotations that transform A to upper triangular form. Use the product of the Givens matrices to compute Q. You might want to use the MATLAB statements given in Problem 17.18. Verify that the decomposition is correct by computing $\|A - QR\|_2$. Also compute the decomposition using qr.

$$A = \begin{bmatrix} 1 & -1 & 2 \\ 9 & 3 & 4 \\ 3 & -8 & 1 \\ 12 & 10 & 5 \end{bmatrix}$$

17.20 To reinforce your understanding of using Householder reflections to find the QR decomposition, it is helpful to execute the algorithm step-by-step. Here are MATLAB statements that explicitly perform the QR decomposition of the matrix

$$M = \begin{bmatrix} 1 & -1 & 1 \\ 2 & 1 & 0 \\ 3 & -1 & 1 \\ 4 & 5 & 3 \end{bmatrix}.$$

The function $houseparms$ in the book software distribution computes u and β for the Householder reflection $H_u(x)$ that zeros out all the elements of x except $x(1)$.

```
A = [1 -1 1;2 1 0;3 -1 1;4 5 3];
R = A;
[m,n] = size(A);

Q = eye(m);

[u, beta] = houseparms(R(:,1));
Hu1 = eye(m) - beta*u*u';
R = Hu1*R;
Q(1:m,1:m) = Q(1:m,1:m) - beta*Q(1:m,1:m)*(u*u');

[u,beta] = houseparms(R(2:m,2));
Hu2 = eye(m-1) - beta*u*u';
R(2:m,2:n) = Hu2*R(2:m,2:n);
Q(1:m,2:m) = Q(1:m,2:m) - beta*Q(1:m,2:m)*(u*u');

[u,beta] = houseparms(R(3:m,3));
Hu3 = eye(m-2) - beta*u*u';
R(3:m,3:n) = Hu3*R(3:m,3:n);
Q(1:m,3:m) = Q(1:m,3:m) - beta*Q(1:m,3:m)*(u*u');
```

Study the code, and the explicitly construct the *QR* decomposition of the matrix

$$A = \begin{bmatrix} 1 & 4 & 6 & 2 & -1 \\ 3 & 6 & 1 & 9 & 10 \\ -6 & 7 & 8 & 1 & 0 \\ 3 & -4 & 1 & -1 & 2 \\ 9 & 12 & 15 & 1 & 5 \\ 35 & 1 & 2 & 3 & 4 \end{bmatrix}.$$

Compute $\|A - QR\|_2$.

17.21 Let

$$A = \begin{bmatrix} 7 & 1 & 6 \\ 9 & 10 & -8 \\ -8 & 10 & -2 \\ 9 & -7 & 9 \\ 3 & 10 & 6 \\ -8 & 10 & 10 \\ -5 & 0 & 3 \end{bmatrix}$$

have *QR* decomposition $A = QR$. Note Problem 17.3 and Theorem 14.3.

a. Use the *QR* decomposition to find the rank of *A*, and verify that rank $(A) = n$. Find an orthonormal basis for the range of *A*.

b. Find an orthonormal basis for the null space of A^T.

17.22 Find the *QR* decomposition of each matrix using the modified Gram-Schmidt process, Givens rotations, and Householder reflections. In each case compute $\left\| Q^T Q - I \right\|_2$.

a. $A = \begin{bmatrix} 3 & 2 & 1 & -1 \\ 1 & 3 & 1 & -1 \\ 4 & 1 & 3 & 1 \\ -1 & 1 & 1 & 3 \end{bmatrix}$

b. $B = \begin{bmatrix} 0 & 1 & 0 & 1 \\ 1 & 0 & 1 & 0 \\ 0 & 1 & 0 & 1 \\ -1 & 0 & 1 & 0 \end{bmatrix}$. Explain the results.

c. The Frank matrix of order 5 using the statement "C = gallery('frank', 5);". Explain the results.

d. The 30 × 20 Chebyshev Vandermonde matrix using the statement "D = gallery('chebvand',30,20);". Explain the results.

17.23 Given the four vectors $\begin{bmatrix} 1 & -1 & 4 & 3 & 8 \end{bmatrix}^T, \begin{bmatrix} 9 & -1 & 3 & 0 & 12 \end{bmatrix}^T, \begin{bmatrix} 2 & 1 & -1 & 3 & 5 \end{bmatrix}^T,$ and $\begin{bmatrix} 0 & 6 & 7 & -1 & 5 \end{bmatrix}^T$, find an orthonormal basis that spans the same subspace using the Givens *QR* decomposition.

17.24 Develop a MATLAB function

```
function [Q R] = myhqr(A)
```

using Householder reflections that returns one of a set of possibilities according to the format of the calling sequence.

```
%MYHQR Executes the full or reduced QR decomposition to return
%both Q and R or, optionally, just R.
%
%    Full decomposition:
%       [Q, R] = myhqr(A) returns an orthogonal m x m matrix Q
%       and an upper triangular m x n matrix R such that A = QR.
%
%    Reduced decomposition
%       If m > n, [Q, R] = myhqr(A,0) returns an m x n matrix Q with
%       orthonormal columns and an upper triangular n x n matrix R
%       such that A = QR. If m <= n, returns the full QR decomposition.
%
%    R = myhqr(A) returns the m x n upper triangular matrix
```

```
%     from the full QR decomposition.
%
%     R = myhqr(A,0) returns the n x n upper triangular matrix
%     from the reduced QR decomposition as long as m > n; otherwise,
%     it returns the m x n upper triangular matrix of the full
%     QR decomposition
%
%     myhqr(A) returns the m x n upper triangular matrix
%     from the full QR decomposition.
%
%     myhqr(A,0) returns the n x n upper triangular matrix
%     from the reduced QR decomposition as long as m > n; otherwise,
%     it returns the m x n upper triangular matrix of the full
%     QR decomposition
```

Use the MATLAB constructs

```
varargout, nargout, varargin, nargin.
```

If you are not familiar with these, consult the MATLAB documentation. Test your function by constructing a 3×2 and a 2×3 matrix and using all the options on each.

17.25 Problem 17.14 presented the QL decomposition of an $m \times n$ matrix $A, m \geq n$. Implement the algorithm in a function ql and test it on at two matrices of different sizes. Hint: If I is the $m \times m$ identity matrix,

```
rot90(I)
```

gives the reversal matrix K_m by rotating I $90°$ counterclockwise.

17.26 Use the result of Problem 17.15 to develop a function lq that performs the LQ decomposition and test it with random matrices of sizes 10×7 and 50×75.

17.27 In this problem, we will use computation to motivate a theoretical result.

   a. Build a series of Givens rotations, and find the eigenvalues and corresponding eigenvectors for each. A pattern emerges.

   b. Propose a formula for the eigenvalues and corresponding eigenvectors of a Givens rotation, and prove you are correct.

Chapter 18

The Algebraic Eigenvalue Problem

You should be familiar with

- Ordinary differential equations for Sections 18.1.1 and 18.1.3.
- Eigenvalues, eigenvectors, and their basic properties covered in Chapter 5.
- Vector and matrix norms and the matrix condition number $\kappa\,(A)$.
- QR decomposition, both full and reduced.
- Householder reflections and Givens rotations.
- Reduction to upper triangular form using orthogonal matrices.

This chapter discusses the computation of eigenvalues and eigenvectors of nonsymmetric matrices. As we discussed in Chapter 10, finding the roots of the characteristic polynomial is not acceptable, since the problem of finding roots of polynomials is unstable. Since the eigenvalue problem is extremely important in areas such as engineering, physics, chemistry, statistics, and economics, we need to know how to solve the problem accurately. The chapter provides three examples where the eigenvalue problem arises, vibration analysis, the Leslie model for population ecology, and buckling of a column.

Some applications only need the largest or smallest eigenvalue of a matrix, so the iterative power and inverse power methods are often used.

Application of the QR decomposition is the most commonly used method for computing eigenvalues. Multiple versions of the QR algorithm are discussed. The basic QR algorithm for computing eigenvalues and eigenvectors is very simple to implement; however, is not efficient. Recall that similar matrices have the same eigenvalues. The basic QR algorithm can be greatly improved by first reducing the matrix to what is termed upper Hessenberg form using a transformation $H = P^{\mathrm{T}}AP$, where H is upper Hessenberg and P is orthogonal. The application of the basic algorithm finds the eigenvalues of H, which are the same as those of A (Theorem 5.1). In practice, the basic QR algorithm is improved by shifting the problem to computing the eigenvalue of a nearby matrix and, as a result, better isolate an eigenvalue from nearby eigenvalues.

For computing the eigenvalues and eigenvectors of a small matrix ($n < 1000$), the algorithm of choice for many years has been the implicit QR iteration, often called the Francis algorithm. This method involves a transformation to upper Hessenberg form, followed by a series of transformations that reduce the Hessenberg matrix to upper triangular form using what are termed single and double shifts. These shifts are done implicitly to save significant computation. In this chapter, we will present the algorithm that uses the double shift to compute the eigenvalues for a general matrix. The double shift is necessary to find complex conjugate eigenvalues. Chapter 19 discusses the algorithm for a symmetric matrix, in which only single shifts are used.

It may be that a particular eigenvalue is known and a corresponding eigenvector is required. In this case, the Hessenberg inverse iteration can be used to compute the eigenvector. In Section 18.10, we build a general eigenvalue/eigenvector solver that uses transformation to upper Hessenberg form, followed by the Francis algorithm to compute the eigenvalues, and the Hessenberg inverse iteration to compute corresponding eigenvectors.

We discussed perturbation theory for solving the linear system $Ax = b$, and there are similar results for the computation of eigenvalues. In particular, we can define a condition number for a particular eigenvalue and use it to estimate the difficulty of computing the eigenvalue.

18.1 APPLICATIONS OF THE EIGENVALUE PROBLEM

The applications of the eigenvalue problem in engineering and science are vast, including such areas as the theory of vibration, analysis of buckling beams, principle component analysis in statistics, economic models, and quantum physics. In this section, we will present three applications.

Numerical Linear Algebra with Applications. http://dx.doi.org/10.1016/B978-0-12-394435-1.00018-1

FIGURE 18.1 Tacoma Narrows Bridge collapse. *Source: University of Washington Libraries, Special Collections, UW 21422.*

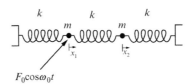

FIGURE 18.2 Mass-spring system.

18.1.1   Vibrations and Resonance

One application of eigenvalues and eigenvectors is in the analysis of vibration problems. You have probably heard of the collapse of the Tacoma Narrows Bridge (Figure 18.1) in the state of Washington. For a period of time, the bridge would move in small waves and became a tourist attraction. One day, the wind was approximately 40 miles/hr, and the oscillations of the bridge increased to the point where the bridge tore apart and crashed into the water. There is still debate as to the cause of the collapse, but one explanation is that the frequency of the wind was close to the fundamental frequency of the bridge. The fundamental frequency of the bridge is the magnitude of the smallest eigenvalue of a system that mathematically models the bridge. The lowest frequency is the most dangerous for a structure or machine because that mode corresponds to the largest displacement.

Consider the vibration of two objects of mass m attached to each other and the walls by three springs with spring constant k. There is no damping, and the springs cannot move vertically. A driving force $F_0 \cos(\beta t)$ acts on the left mass (Figure 18.2). The equations of motion for the displacement of the masses are:

$$m\frac{\mathrm{d}^2 x_1}{\mathrm{d}t^2} = -2kx_1 + kx_2 + F_0 \cos \omega_0 t$$

$$m\frac{\mathrm{d}^2 x_2}{\mathrm{d}t^2} = kx_1 - 2kx_2$$

In matrix form, we have

$$\frac{\mathrm{d}^2 x}{\mathrm{d}t^2} = \frac{1}{m}Kx + \frac{F}{m}, \tag{18.1}$$

where

$$K = \begin{bmatrix} -2k & k \\ k & -2k \end{bmatrix}, \quad F = \begin{bmatrix} F_0 \cos \omega_0 t \\ 0 \end{bmatrix}.$$

where K is the *stiffness* matrix and F is the *force vector*. The general solution to Equation 18.1 is a linear combination of the solution to the homogeneous system

$$\frac{d^2 x}{dt^2} - \frac{K}{m}x = 0 \tag{18.2}$$

and a particular solution to Equation 18.1. Try a solution to the homogeneous equation of the form $x = ve^{\omega t}$, where ω is a frequency. Substituting $x = ve^{\omega t}$ into Equation 18.2 gives

$$\omega^2 v e^{\omega t} = \frac{K}{m}v e^{\omega t},$$

and we have the eigenvalue problem

$$Kv = \left(m\omega^2\right)v. \tag{18.3}$$

The coefficient matrix is symmetric, so it has real eigenvalues and corresponding real linearly independent eigenvectors (Lemma 7.3 and Theorem 7.6). The characteristic equation is

$$\lambda^2 + 4k\lambda + 3k^2 = (\lambda + 3k)(\lambda + k) = 0.$$

The eigenvalues and corresponding normalized eigenvectors are

$$\lambda_1 = -3k \quad v_1 = \frac{1}{\sqrt{2}}\begin{bmatrix} -1 \\ 1 \end{bmatrix}$$

$$\lambda_2 = -k \quad v_2 = \frac{1}{\sqrt{2}}\begin{bmatrix} 1 \\ 1 \end{bmatrix}$$

Now, using Equation 18.3,

$$\omega_1^2 = -3\frac{k}{m}, \quad \omega_2^2 = -\frac{k}{m},$$

and

$$\omega_1 = i\sqrt{\frac{3k}{m}}, \quad \omega_2 = i\sqrt{\frac{k}{m}}.$$

The corresponding solutions are

$$v_1 e^{i\sqrt{\frac{3k}{m}}t}, \quad v_2 e^{i\sqrt{\frac{k}{m}}t}.$$

Applying Euler's formula (Equation A.2) and taking the real and imaginary parts, the general solution to the homogeneous equation is

$$x_h(t) = \frac{1}{\sqrt{2}}\begin{bmatrix} -1 \\ 1 \end{bmatrix}\left(c_1 \cos\sqrt{\frac{3k}{m}}t + c_2 \sin\sqrt{\frac{3k}{m}}t\right) + \frac{1}{\sqrt{2}}\begin{bmatrix} 1 \\ 1 \end{bmatrix}\left(c_3 \cos\sqrt{\frac{k}{m}}t + c_4 \sin\sqrt{\frac{k}{m}}t\right).$$

The frequencies $\sqrt{\frac{3k}{m}}$ and $\sqrt{\frac{k}{m}}$ are known as the *natural frequencies* of the system. Without a driving force, the system vibrates with these frequencies.

We now determine a particular solution, $x_p(t)$, by using complex variables and replacing the driving force $F = \begin{bmatrix} F_0 \cos \omega_0 t \\ 0 \end{bmatrix}$ by $\begin{bmatrix} F_0 \\ 0 \end{bmatrix}e^{i\omega_0 t}$. After finding a imaginary particular solution, the real part is the particular solution we are looking for.

Using the method of undetermined coefficients, the trial solution is $x_p(t) = De^{i\omega_0 t}$. Substitute it into Equation 18.1 to obtain

$$\left(\frac{K}{m} + \omega_0^2 I\right)D = -\frac{1}{m}\begin{bmatrix} F_0 \\ 0 \end{bmatrix}.$$

If $\left(\frac{K}{m} + \omega_0^2 I\right)$ is invertible

$$D = -\frac{1}{m}\left(\frac{K}{m} + \omega_0^2 I\right)^{-1}\begin{bmatrix} F_0 \\ 0 \end{bmatrix},$$

and

$$x_{\text{p}}(t) = D \cos \omega_0 t,$$

giving the general solution $x(t) = x_{\text{h}}(t) + x_{\text{p}}(t)$.

Now, let $\det\left(\frac{K}{m} + \omega_0^2 I\right) = 0$.

$$\det\left(\frac{K}{m} + \omega_0^2 I\right) = \det\left(\begin{bmatrix} \omega_0^2 - \frac{2k}{m} & \frac{k}{m} \\ \frac{k}{m} & \omega_0^2 - \frac{2k}{m} \end{bmatrix}\right)$$

$$= \left(\omega_0^2 - \frac{3k}{m}\right)\left(\omega_0^2 - \frac{k}{m}\right).$$

When

$$\omega_0 = \sqrt{\frac{k}{m}}, \quad \omega_0 = \sqrt{\frac{3k}{m}},$$

$\left(\frac{K}{m} + \omega_0^2 I\right)$ is not invertible. Thus, as ω_0 approaches either natural frequency $\sqrt{\frac{k}{m}}$ or $\sqrt{\frac{3k}{m}}$, $D\left(\frac{K}{m} + \omega_0^2 I\right)$ is close to singular. Example 18.1 investigates what happens when ω_0 approaches these frequencies.

Example 18.1. The initial conditions determine the constants c_i, $1 \leq i \leq 4$. Assume at time $t = 0$ the spring system is at rest, so that $x(0) = x'(0) = 0$, and that the driving force starts motion. We will leave ω_0 as a variable and investigate what happens as ω_0 varies. The constants are determined by solving the system of equations

$$\begin{bmatrix} -\frac{1}{\sqrt{2}} & 0 & \frac{1}{\sqrt{2}} & 0 \\ \frac{1}{\sqrt{2}} & 0 & \frac{1}{\sqrt{2}} & 0 \\ 0 & -\sqrt{\frac{k}{2m}} & 0 & \sqrt{\frac{3k}{2m}} \\ 0 & \sqrt{\frac{k}{2m}} & 0 & \sqrt{\frac{3k}{2m}} \end{bmatrix}\begin{bmatrix} c_1 \\ c_2 \\ c_3 \\ c_4 \end{bmatrix} = \begin{bmatrix} -d_1 \\ -d_2 \\ 0 \\ 0 \end{bmatrix},$$

where d_1 and d_2 are the components of D. The solution $x(t) = x_{\text{h}}(t) + x_{\text{p}}(t)$ can be computed when we have values for m, k, and F_0, and we assume $m = 1$, $k = 1$, and $F_0 = 5$. Figure 18.3 shows the behavior of $x_1(t)$, $x_2(t)$, $0 \leq t \leq 20$ for $\omega_0 = 2.0$ and $\omega_0 = \sqrt{3} - 0.001$. Note the large oscillations for the latter value of ω_0. This caused by the fact that ω_0 is close to the natural frequency $\sqrt{\frac{3k}{m}}$, and the system approaches *resonance*. Resonance is the tendency of a system to oscillate at a greater amplitude at some frequencies than at others. Note that resonance occurs in our example when the frequency of the driving force approaches $\sqrt{\frac{3k}{m}}$, one of the natural frequencies of the system. ∎

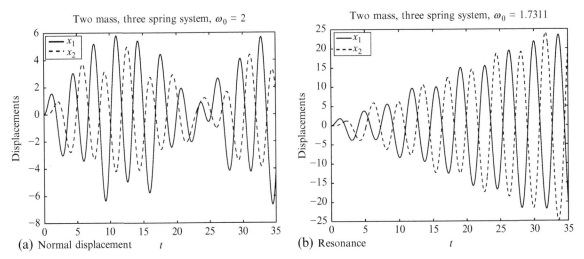

FIGURE 18.3 Solution to a system of ordinary differential equations.

18.1.2   The Leslie Model in Population Ecology

The model of Leslie is a heavily used tool in population ecology. It is a model of an age-structured population which predicts how distinct populations change over time. The model runs for n units of time, beginning at 0. For simplicity of modeling, we sort individuals into discrete age classes, denoted $n_i(t)$. We further simplify our model by assuming that there is an approximately 50:50 male to female ratio, and that the number of offspring per year depends primarily on the number of females. Therefore, we only consider the females present in the population. There are m age classes, $1, \ldots, m$, where m is the maximum reproductive age of an individual. The model uses the following parameters for the groups:

p_i probability of surviving from age i to $i+1$

f_i average number of offspring surviving to age 1 in class i (fertility function)

$n_i(t)$ number of females of age i at time t (the number of i-year olds at time t)

In specifying f_i, the mortality rate of offspring and parents is included.

The number of individuals in age class i at time $t+1$ depends on the number of individuals surviving from age class $i-1$ at time t. The age class $i=1$ at time t consists of new borns. First, consider how many individuals in age class i survive from t to $t+1$, and do not consider births. This is given by the equation

$$n_i(t+1) = p_{i-1}n_{i-1}(t), \quad 2 \le i \le m$$

Now consider age class 1, the newly born individuals. The number individuals in age class 1 at time $t+1$ is the number of offspring born to existing individuals in the population at time t:

$$n_1(t+1) = \sum_{i=1}^{m} f_i n_i(t)$$

Now place all of the $n_i(t)$ values into a column vector $N(t)$ of dimension m and build the $m \times m$ *Leslie matrix L* as follows:

The first row of L consists of the fertility values f_i, $1 \le i \le m$, and the subdiagonal contains the probabilities of survival. The remaining entries are zero.

$$L = \begin{bmatrix} f_1 & f_2 & \cdots & f_{m-1} & f_m \\ p_1 & 0 & 0 & \cdots & 0 \\ 0 & p_2 & 0 & \cdots & 0 \\ \vdots & \vdots & \ddots & \vdots & \vdots \\ 0 & 0 & & p_{m-1} & 0 \end{bmatrix}$$

The product of L and $N(t)$ is

$$L(N(t)) = \begin{bmatrix} f_1 & f_2 & \cdots & f_{m-1} & f_m \\ p_1 & 0 & 0 & \cdots & 0 \\ 0 & p_2 & 0 & \cdots & 0 \\ \vdots & \vdots & \ddots & \vdots & \vdots \\ 0 & 0 & & p_{m-1} & 0 \end{bmatrix} \begin{bmatrix} n_1(t) \\ n_2(t) \\ \vdots \\ n_{m-1}(t) \\ n_m(t) \end{bmatrix}$$

$$= \begin{bmatrix} \sum_{i=1}^{m} f_i n_i(t) \\ p_1 n_1(t) \\ p_2 n_2(t) \\ \vdots \\ p_{m-1} n_{m-1}(t) \end{bmatrix} = N(t+1),$$

so

$$N(t+1) = L(N(t)).$$

Thus,

$$N(t+2) = L(N(t+1)) = L^2 N(t),$$

and in general for each time $t+k$

$$N(t+k) = L^k N(t).$$ (18.4)

Equation 18.4 says that we can project into the future by computing matrix powers.

The digraph produced by L is strongly connected, so L is irreducible. By Theorem 5.6, L has an eigenvector with strictly positive entries. Thus, we can assume that v is an eigenvector of L, with $v_1 \neq 0$. Divide by v_1 so that

$$v = \begin{bmatrix} 1 \\ v_2 \\ \vdots \\ v_{m-1} \\ v_m \end{bmatrix}.$$

Our aim is to determine the characteristic equation for L, from which we will obtain valuable information. Let $Lv = \lambda v$ to obtain

$$\begin{bmatrix} f_1 + f_2 v_2 + \cdots + f_m v_m \\ p_1 \\ p_2 v_2 \\ \vdots \\ \vdots \\ p_{m-1} v_{m-1} \end{bmatrix} = \begin{bmatrix} \lambda \\ \lambda v_2 \\ \lambda v_3 \\ \vdots \\ \lambda v_{m-1} \\ \lambda v_m \end{bmatrix}.$$ (18.5)

Equate entries 2 through m to get

$$\begin{aligned} p_1 &= \lambda v_2 \\ p_2 v_2 &= \lambda v_3 \\ &\vdots \\ p_{m-1} v_{m-1} &= \lambda v_m. \end{aligned}$$

From these equations, we have $v_2 = \left(\frac{1}{\lambda}\right) p_1$ and $v_3 = \left(\frac{1}{\lambda}\right) p_2 v_2$, so $v_3 = \left(\frac{1}{\lambda}\right)^2 p_1 p_2$. In general,

$$v_k = \left(\frac{1}{\lambda}\right)^{k-1} p_1 p_2 \ldots p_{k-1}, \quad 2 \leq k \leq m.$$

The first components of Equation 18.5 must be equal, so

$$f_1 + f_2 v_2 + \cdots + f_m v_m = \lambda,$$

and

$$f_1 + f_2 \left[\left(\frac{1}{\lambda}\right) p_1\right] + f_3 \left[\left(\frac{1}{\lambda}\right)^2 p_1 p_2\right] + \cdots + f_k \left[\left(\frac{1}{\lambda}\right)^k p_1 p_2 \ldots p_k\right] + \cdots + f_m \left[\left(\frac{1}{\lambda}\right)^{m-1} p_1 p_2 \ldots p_{m-1}\right] = \lambda.$$

Divide by λ to obtain

$$\left(\frac{1}{\lambda}\right) f_1 + f_2 \left[\left(\frac{1}{\lambda}\right)^2 p_1\right] + f_3 \left[\left(\frac{1}{\lambda}\right)^3 p_1 p_2\right] + \cdots + f_k \left[\left(\frac{1}{\lambda}\right)^{k+1} p_1 p_2 \ldots p_k\right] + \cdots + f_m \left[\left(\frac{1}{\lambda}\right)^m p_1 p_2 \ldots p_{m-1}\right] = 1.$$ (18.6)

Let $l_1 = 1$ and for $k = 2, 3, \ldots, m$, define $l_k = p_1 p_2 \ldots p_{k-1}$. Then Equation 18.6 becomes

$$f_1 l_1 \lambda^{-1} + f_2 l_2 \lambda^{-2} + f_3 l_3 \lambda^{-3} + \cdots + \lambda^{-m} f_m l_m = 1.$$

Using summation notation, we have

$$\sum_{k=1}^{m} f_k l_k \lambda^{-k} - 1 = 0,$$ (18.7)

which is known as the *Euler-Lotka equation*. The value l_k is the fraction of 1-year olds that survive to age k. The characteristic equation can be derived by evaluating $det\ (L - \lambda I)$ using expansion by minors or the elementary row operation of multiplying a row by a constant and subtracting from another row, and the result is

$$\lambda^m \left(1 - \left[\frac{f_1 l_1}{\lambda} + \frac{f_2 l_2}{\lambda^2} + \ldots + \frac{f_m l_m}{\lambda^m} \right] \right) = 0, \lambda \neq 0.$$

Both the Euler-Lotka equation and the characteristic equation have the same roots. Let

$$f(\lambda) = \sum_{k=1}^{m} f_k l_k \lambda^{-k} - 1 = \frac{f_1 l_1}{\lambda} + \frac{f_2 l_2}{\lambda^2} + \ldots + \frac{f_m l_m}{\lambda^m} - 1,$$

and we have

$$f'(\lambda) = -\frac{f_1 l_1}{\lambda^2} - \frac{2 f_2 l_2}{\lambda^3} - \ldots - \frac{m f_m l_m}{\lambda^{m+1}} < 0, \lambda > 0.$$

The function f is decreasing for $\lambda > 0$, and we know there is a value of λ such that $f(\lambda) = 0$. The second derivative

$$f''(\lambda) = \frac{2 f_1 l_1}{\lambda^3} + \frac{6 f_2 l_2}{\lambda^4} + \ldots + \frac{m(m+1) f_m l_m}{\lambda^{m+2}} > 0, \lambda > 0,$$

so f is concave upward, and there is one positive real root, λ_1, the dominant eigenvalue. All the other eigenvalues are negative or imaginary.

If we assume that L has m distinct eigenvalues, then there is a basis of eigenvectors $u_1, u_2, \ldots, u_m$ such that

$$L = UDU^{-1},$$

where $U = (u_1, u_2, \ldots, u_m)$ and $D = \begin{bmatrix} \lambda_1 & 0 & \cdots & 0 \\ 0 & \lambda_2 & \ldots & 0 \\ \vdots & \vdots & \ddots & \vdots \\ 0 & 0 & \ldots & \lambda_m \end{bmatrix}$ is the matrix of eigenvalues. As a result (Section 5.3.1),

$$L^k = U \begin{bmatrix} \lambda_1^k & 0 & \ldots & 0 \\ 0 & \lambda_2^k & \ldots & 0 \\ \vdots & \vdots & \ddots & \vdots \\ 0 & 0 & \ldots & \lambda_n^k \end{bmatrix} U^{-1}$$

From Equation 18.4,

$$N(t+k) = U \begin{bmatrix} \lambda_1^k & 0 & \ldots & 0 \\ 0 & \lambda_2^k & \ldots & 0 \\ \vdots & \vdots & \ddots & \vdots \\ 0 & 0 & \ldots & \lambda_n^k \end{bmatrix} U^{-1} N(t).$$

If we let $U^{-1} N(t) = c(t)$, then

$$N(t+k) = U \begin{bmatrix} \lambda_1^k & 0 & \ldots & 0 \\ 0 & \lambda_2^k & \ldots & 0 \\ \vdots & \vdots & \ddots & \vdots \\ 0 & 0 & \ldots & \lambda_m^k \end{bmatrix} c(t),$$

and

$$N(t+k) = c_1(t) \lambda_1^k u_1 + c_2(t) \lambda_2^k u_2 + \cdots + c_m(t) \lambda_m^k u_m.$$

Since λ_1 is the largest eigenvalue in magnitude,

$$N(t+k) \approx c_1(t) \lambda_1^k u_1.$$

This means that as time increases, the age distribution vector tends to a scalar multiple of the eigenvector associated with the largest eigenvalue of the Leslie matrix. In other words, at equilibrium the proportion of individuals belonging to each age class will remain constant, and the number of individuals will increase by λ_1 times each period. Divide the eigenvector, u_1, with all positive entries, associated with λ_1 by the sum of its components $\left(u_1 = \left(\frac{1}{\sum_{i=1}^{m} u_{1i}} \right) u_1 \right)$. The components of the new vector have the same relative proportions as those in the original eigenvector, and they determine the percentage of females in each age class after an extended period of time.

Example 18.2. Suppose a population has four age classes, and that the following table specifies the data for the population.

Age Class	$n_i(0)$	f_i	p_i
1	8	0	0.60
2	10	6	0.45
3	12	3	0.25
4	7	2	

The Leslie matrix for this population is

$$
L = \begin{bmatrix} 0 & 6 & 3 & 2 \\ 0.60 & 0 & 0 & 0 \\ 0 & 0.45 & 0 & 0 \\ 0 & 0 & 0.25 & 0 \end{bmatrix}
$$

Using MATLAB, find the eigenvalues and eigenvectors of L.

```
>> [U D] = eig(L);
>> diag(D)

ans =
      2.0091
     -1.7857
     -0.11171 +    0.15858i
     -0.11171 -    0.15858i

>> u1 = U(:,1);
>> u1 = (u1/sum(u1))*100

u1 =
      72.788
      21.737
      4.8687
      0.60582
```

The dominant eigenvalue is 2.0091, so the number of individuals at equilibrium will increase by 2.0091 each time period. The percentage of females in each age class after an extended period of time is given by the components of u1. Now, we will look at the situation graphically. Let $n_0 = \begin{bmatrix} 8 & 10 & 12 & 7 \end{bmatrix}^T$, and compute the age distribution vector over a 10-year period. Store the initial distribution in column 1 of a 4×11 matrix N, and use Equation 18.5 to compute vectors of the age distribution for years 1-10. The vectors grow exponentially, so graph the results for each class using a logarithmic scale of the vertical (population) axis (Figure 18.4).

```
>> format shortg
>> n0 =[8 10 12 7]';
>> N = zeros(4,11);
>> N(:,1) = n0;
>> for k = 2:11
       N(:,k) = L*N(:,k-1);
   end
>> t = 0:10;
>> semilogy(t,N);
>> xlabel('Time');
>> ylabel('log_{10}(population)');
>> legend('Age class 1', 'Age class 2', 'Age class 3', 'Age class 4',...
          'Location','NorthWest');
```

■

18.1.3 Buckling of a Column

This example shows that eigenvalues can be associated with functions as well as matrices.

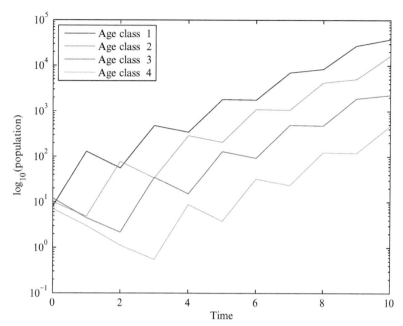

FIGURE 18.4 Populations using the Leslie matrix.

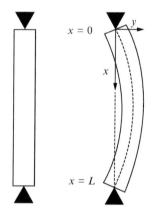

FIGURE 18.5 Column buckling.

Apply a compressive axial force, or load, P, to the top of a vertical thin elastic column of uniform cross-section having length L (Figure 18.5). The column will buckle, and the deflection $y(x)$ satisfies the differential equation

$$EI\frac{d^2y}{dx^2} = -Py,$$

where E is Young's modulus of elasticity and I is the area moment of inertia of column cross-section. Assume that the column is hinged at both ends so that $y(0) = y(L) = 0$, and we have the boundary value problem

$$EI\frac{d^2y}{dx^2} + Py = 0, \quad y(0) = 0, \ y(L) = 0.$$

Note that $y = 0$ is a solution to the problem, and this corresponds to the situation where the load P is not large enough to cause deflection. We wish to determine values of P that cause the column to buckle; in other words, for what values of P does the boundary value problem have nontrivial solutions?

Let $\lambda = \frac{P}{EI}$, and we have the problem

$$\frac{d^2y}{dx^2} + \lambda y = 0, \quad y(0) = 0, \ y(L) = 0.$$

Let $\lambda = \beta^2$, $\beta > 0$. The equation is homogeneous, so try a solution of the form $y = e^{px}$. After substitution into the equation, we have

$$p^2 + \beta^2 = 0,$$

and $p = \pm\beta i$, yielding complex solutions $e^{\pm\beta ix}$. These give rise to the real solutions $y = c_1 \cos \beta x + c_2 \sin \beta x$. Apply the boundary conditions:

$$y(0) = c_1 = 0,$$
$$y(L) = c_2 \sin \beta L = 0.$$

If $\sin \beta L = 0$, we can choose c_2 to be any nonzero value, so let $c_2 = 1$. We have $\sin \beta L = 0$ when $\beta L = n\pi$, $n = 1, 2, 3, \ldots$, and so our values of β must be

$$\beta_n = \frac{n\pi}{L}, \quad n = 1, 2, 3, \ldots.$$

Since $\lambda = \beta^2$, it follows that:

$$\lambda_1 = \frac{\pi^2}{L^2}, \quad \lambda_2 = \frac{4\pi^2}{L^2}, \quad \lambda_3 = \frac{9\pi^2}{L^2}, \quad \ldots, \quad \lambda_n = \frac{n^2\pi^2}{L^2}, \ldots$$

and the sequence of functions

$$y_n(x) = k \sin\left(\frac{n\pi}{L}x\right), \quad n = 1, 2, 3, \ldots$$

where k is a constant are nontrivial solutions to the boundary value problem. These functions are called *eigenfunctions* with corresponding eigenvalues λ_n. For our column buckling equation, we have eigenvalues $\lambda_n = \frac{P_n}{EI} = \frac{n^2\pi^2}{L^2}$, $n = 1, 2, 3, \ldots$, and loads

$$P_n = \frac{EI\pi^2 n^2}{L^2}, \quad n = 1, 2, 3, \ldots$$

The column will buckle only when the compressive force is one of these values. This sequence of forces are called *critical loads*. The deflection function corresponding to the smallest critical load $P_1 = \frac{EI\pi^2}{L^2}$ is termed the *first buckling mode*.

Example 18.3. For a thin 2.13 m column of aluminum, $E = 69 \times 10^9 \, \text{N/m}^2$, and $I = 3.26 \times 10^{-4} \, \text{m}^4$. Compute the critical loads and graph the deflection curves for $n = 1, 2, 3$ (Figure 18.6).

```
The first buckling mode = 4.89336e+07
The second critical load = 1.95734e+08
The third critical load = 4.40402e+08
```

If the column has a physical restraint on it at $x = L/2$, then the smallest critical load will be $P_2 = 1.69 \times 10^7$ and the deflection curve is $\sin\left(\frac{2\pi x}{L}\right)$. If restraints are placed on the column at $x = L/3$ and at $2L/3$, then the column will not buckle until the column is subjected to the critical load $P_3 = 3.81 \times 10^7$, and the deflection curve is $\sin\left(\frac{3\pi x}{L}\right)$ [53, pp. 167-169]. ∎

Remark 18.1.   See Problem 18.43 for another approach to the buckling problem.

18.2   COMPUTATION OF SELECTED EIGENVALUES AND EIGENVECTORS

In some applications, it is only necessary to compute a few of the largest or the smallest eigenvalues and their corresponding eigenvectors. Examples include:

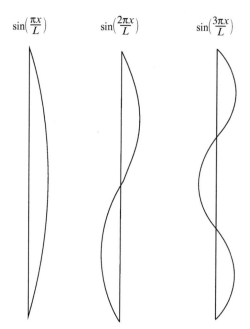

$$\sin\left(\frac{\pi x}{L}\right) \qquad \sin\left(\frac{2\pi x}{L}\right) \qquad \sin\left(\frac{3\pi x}{L}\right)$$

FIGURE 18.6 Deflection curves for critical loads P_1, P_2, and P_3.

- The Leslie matrix.
- The buckling problem. The most important eigenvalue is the smallest.
- Vibration of structures. The most important eigenvalues are a few of the smallest ones.
- Statistical applications. Only the first few of the largest eigenvalues need to be computed.

The power and inverse power methods compute the largest and smallest eigenvector, respectively. If v is an eigenvector of A, then $Av = \lambda v$, so $\langle Av, v \rangle = \lambda \langle v, v \rangle$, and

$$\lambda = \frac{v^{\mathrm{T}}(Av)}{v^{\mathrm{T}}v}. \tag{18.8}$$

Equation 18.8 is called the *Rayleigh quotient*. Given any eigenvector of A, Equation 18.8 computes the corresponding eigenvalue. This will be useful to us, since the power and inverse power methods compute an eigenvector, and the Rayleigh quotient finds the corresponding eigenvalue.

If the eigenvalues of an $n \times n$ matrix A are such that

$$|\lambda_1| > |\lambda_2| \geq |\lambda_3| > \cdots \geq |\lambda_n|, \tag{18.9}$$

the eigenvalue λ_1 is said to be the *dominant eigenvalue* of A. Thus, if we know that v_{d} is an eigenvector of the dominant eigenvalue, the eigenvalue is $\lambda_1 = \frac{v_{\mathrm{d}}^{\mathrm{T}}(Av_{\mathrm{d}})}{v_{\mathrm{d}}^{\mathrm{T}}v_{\mathrm{d}}}$. Not all matrices have a dominant eigenvalue. For instance, let $A = \begin{bmatrix} 1 & 0 \\ 0 & -1 \end{bmatrix}$. Its eigenvalues are $\lambda_1 = 1$ and $\lambda_2 = -1$. The matrix $B = \begin{bmatrix} 2 & 0 & 0 \\ 0 & 2 & 0 \\ 0 & 0 & 1 \end{bmatrix}$ has eigenvalues $\lambda_1 = \lambda_2 = 2, \lambda_3 = 1$.

18.2.1 Additional Property of a Diagonalizable Matrix

Theorem 5.3 states that if the $n \times n$ matrix A has n linearly independent eigenvectors $v_1, v_2, \ldots, v_n$, then A can be diagonalized by the matrix the *eigenvector matrix* $X = (v_1 v_2 \ldots v_n)$. The converse of Theorem 5.3 is also true; that is, if a matrix can be diagonalized, it must have n linearly independent eigenvectors. We need this result for the purposes of developing the power method in Section 18.2.2.

Theorem 18.1. *If A is a real $n \times n$ matrix that is diagonalizable, it must have n linearly independent eigenvectors.*

Proof. We know there is an invertible matrix V such that $V^{-1}AV = D$, where $D = \begin{bmatrix} \lambda_1 & & & \\ & \lambda_2 & & \\ & & \ddots & \\ & & & \lambda_n \end{bmatrix}$ is a diagonal matrix,

and let $v_1, v_2, \ldots, v_n$ be the columns of V. Since V is invertible, the v_i are linearly independent. The relationship $V^{-1}AV = D$ gives $AV = VD$, and using matrix column notation we have

$$A \begin{bmatrix} v_1 & v_2 & \ldots & v_n \end{bmatrix} = \begin{bmatrix} v_1 & v_2 & \ldots & v_n \end{bmatrix} \begin{bmatrix} \lambda_1 & & & \\ & \lambda_2 & & \\ & & \ddots & \\ & & & \lambda_n \end{bmatrix}.$$

Column i of $A \begin{bmatrix} v_1 & v_2 & \ldots & v_n \end{bmatrix}$ is Av_i, and column i of $\begin{bmatrix} v_1 & v_2 & \ldots & v_n \end{bmatrix} \begin{bmatrix} \lambda_1 & & & \\ & \lambda_2 & & \\ & & \ddots & \\ & & & \lambda_n \end{bmatrix}$ is $\lambda_i v_i$, so $Av_i = \lambda_i v_i$.

Thus, the linearly independent set $v_1, v_2, \ldots, v_n$ are eigenvectors of A corresponding to eigenvalues $\lambda_1, \lambda_2, \ldots, \lambda_n$. $\qquad \square$

18.2.2 The Power Method for Computing the Dominant Eigenvalue

We now develop the *power method*, a simple iteration, for computing the dominant eigenvalue of a matrix, if it has one. Make an initial guess for the eigenvector, usually $v_0 = \begin{bmatrix} 1 & \ldots & 1 \end{bmatrix}^{\mathrm{T}}$, and normalize it by assigning $v_0 = \frac{v_0}{\|v_0\|_2}$. Compute $v_1 = Av_0$, and then normalize v_1. Repeat this process until satisfying a convergence criterion. Example 18.4 demonstrates this algorithm.

Example 18.4. Let $A = \begin{bmatrix} 1 & 3 & 0 \\ 2 & 5 & 1 \\ -1 & 2 & 3 \end{bmatrix}$, $x_0 = \begin{bmatrix} 1 \\ 1 \\ 1 \end{bmatrix}$.

$k = 1$

$$x_1 = Ax_0 = \begin{bmatrix} 4 \\ 8 \\ 4 \end{bmatrix}, \quad x_1 = x_1 / \|x_1\|_2 = \begin{bmatrix} 0.4082 \\ 0.8165 \\ 0.4082 \end{bmatrix}$$

$k = 2$

$$x_2 = Ax_1 = \begin{bmatrix} 2.8577 \\ 5.3072 \\ 2.4495 \end{bmatrix}, \quad x_2 = x_2 / \|x_2\|_2 = \begin{bmatrix} 0.4392 \\ 0.8157 \\ 0.3765 \end{bmatrix}$$

$k = 3$

$$x_3 = Ax_2 = \begin{bmatrix} 2.8863 \\ 5.3334 \\ 2.3216 \end{bmatrix}, \quad x_3 = x_3 / \|x_3\|_2 = \begin{bmatrix} 0.4445 \\ 0.8213 \\ 0.3575 \end{bmatrix}$$

$k = 4$

$$x_4 = Ax_3 = \begin{bmatrix} 2.9085 \\ 5.3532 \\ 2.2708 \end{bmatrix}, \quad x_4 = x_4 / \|x_4\|_2 = \begin{bmatrix} 0.4473 \\ 0.8233 \\ 0.3493 \end{bmatrix}$$

$$\lambda_1 = \frac{(Ax_4) \cdot x_4}{x_4 \cdot x_4} = 6.5036$$

The eigenvalues of A are 6.5050, -0.4217, 2.9166, so the relative error in computing the largest eigenvalue is 2.1856×10^{-4}. $\qquad \blacksquare$

In Algorithm 18.1 (the power method), we assume that the real matrix A is diagonalizable. Theorem 18.1 guarantees there exists a basis of eigenvectors for $\mathbb{R}^n$. This assumption will allow us to prove the power method converges under condition 18.9. The convergence test is to compute the Rayleigh quotient after normalization ($\lambda_k = x_k^T (Ax_k)$) at each iteration and determine if $\|Ax_k - \lambda x_k\|_2 < $ tol; in other words, is λx_k sufficiently close to Ax_k?

Algorithm 18.1 The Power Method

```
function LARGEEIG(A,x₀,tol,numiter)
   % Use the power method to find the dominant eigenvalue and
   % the corresponding eigenvector of real diagonalizable matrix A.
   % [lambda x iter] = largeeig(A,x0,n,tol) computes the largest eigenvalue
   % lambda in magnitude and corresponding eigenvector x of real matrix A.
   % x0 is the initial approximation, tol is the desired error tolerance,
   % and maxiter is the maximum number of iterations to perform.
   % If the algorithm converges, iter contains the number of iterations required.
   % If the method does not converge, iter=-1.

   x₀ = x₀/‖x₀‖₂
   for k=1:numiter do
      xₖ = Axₖ₋₁
      xₖ = xₖ/‖xₖ‖₂
      λ = xₖᵀAxₖ
      error=‖Axₖ − λxₖ‖₂
      if error<tol then
         x = xₖ
         iter=k
         return [λ, x, iter]
      end if
   end for
   x = xₖ
   iter=-1
   return [λ, x, iter]
end function
```

NLALIB: The function `largeeig` implements Algorithm 18.1.

Theorem 18.2 specifies conditions under which the power iteration is guaranteed to converge.

Theorem 18.2.   *Assume that A is diagonalizable, with real eigenvalues $\lambda_1, \lambda_2, \ldots, \lambda_n$ and associated real eigenvectors $v_1, v_2, \ldots, v_n$, and that λ_1 is a simple eigenvalue with the largest magnitude, i.e.,*

$$|\lambda_1| > |\lambda_2| \geq |\lambda_3| \geq \cdots \geq |\lambda_n|. \tag{18.10}$$

Then, the power method converges to a normalized eigenvector corresponding to eigenvalue λ_1.

Proof. Let x_0 be the initial approximation. By Theorem 18.1, the eigenvectors are linearly independent, and so $x_0 = c_1 v_1 + c_2 v_2 + \cdots + c_n v_n$. We can assume that $c_1 \neq 0$, or otherwise rearrange the linear combination so it is. Now,

$$x_1 = Ax_0, \quad x_1 = \frac{Ax_0}{\|Ax_0\|_2}$$

$$x_2 = A\left(\frac{Ax_0}{\|Ax_0\|_2}\right) = \frac{A^2x_0}{\|Ax_0\|_2}, \quad x_2 = \frac{A^2x_0}{\|Ax_0\|_2}\left(\frac{\|Ax_0\|_2}{\|A^2x_0\|_2}\right) = \frac{A^2x_0}{\|A^2x_0\|_2}$$

$$x_k = \frac{A^kx_0}{\|A^kx_0\|_2}, \quad \cdots \tag{18.11}$$

Since $x_0 = c_1 v_1 + c_2 v_2 + \cdots + c_n v_n$,

$$
\begin{aligned}
A^k x_0 &= A^k (c_1 v_1 + c_2 v_2 + \cdots + c_n v_n) = c_1 \lambda_1^k v_1 + c_2 \lambda_2^k v_2 + \cdots + c_n \lambda_n^k v_n \\
&= \lambda_1^k \left(c_1 v_1 + c_2 \left(\frac{\lambda_2}{\lambda_1}\right)^k v_2 + c_3 \left(\frac{\lambda_3}{\lambda_1}\right)^k v_3 + \cdots + c_n \left(\frac{\lambda_n}{\lambda_1}\right)^k v_n \right).
\end{aligned} \tag{18.12}
$$

By substituting Equation 18.12 into Equation 18.11, we obtain

$$
\begin{aligned}
x_k &= \frac{\lambda_1^k \left(c_1 v_1 + c_2 \left(\frac{\lambda_2}{\lambda_1}\right)^k v_2 + c_3 \left(\frac{\lambda_3}{\lambda_1}\right)^k v_3 + \cdots + c_n \left(\frac{\lambda_n}{\lambda_1}\right)^k v_n \right).}{\left\| \lambda_1^k \left(c_1 v_1 + c_2 \left(\frac{\lambda_2}{\lambda_1}\right)^k v_2 + c_3 \left(\frac{\lambda_3}{\lambda_1}\right)^k v_3 + \cdots + c_n \left(\frac{\lambda_n}{\lambda_1}\right)^k v_n \right) \right\|_2} = \\
&\pm \frac{\left(c_1 v_1 + c_2 \left(\frac{\lambda_2}{\lambda_1}\right)^k v_2 + c_3 \left(\frac{\lambda_3}{\lambda_1}\right)^k v_3 + \cdots + c_n \left(\frac{\lambda_n}{\lambda_1}\right)^k v_n \right).}{\left\| \left(c_1 v_1 + c_2 \left(\frac{\lambda_2}{\lambda_1}\right)^k v_2 + c_3 \left(\frac{\lambda_3}{\lambda_1}\right)^k v_3 + \cdots + c_n \left(\frac{\lambda_n}{\lambda_1}\right)^k v_n \right) \right\|_2}.
\end{aligned} \tag{18.13}
$$

Because $c_1 \neq 0$, the denominator in Equation 18.13 is never 0. Since $|\lambda_1| > |\lambda_i|$, $i \geq 2$,

$$
\lim_{k \to \infty} x_k = \pm \frac{v_1}{\|v_1\|_2},
$$

a normalized eigenvector of A corresponding to the largest eigenvalue λ_1. $\qquad\square$

Remark 18.2.

a. If we assume that $|\lambda_1| > |\lambda_2| > |\lambda_3| > \cdots > |\lambda_n|$, then Theorem 5.2 guarantees that there are n linearly independent eigenvectors. Our assumption that A is diagonalizable allows us to require only condition 18.9.

b. The assumption that $c_1 \neq 0$ is critical to the proof. A randomly chosen x_0 will normally guarantee this with high probability.

c. From the proof of Theorem 18.2, we see that the rate of convergence is determined by the ratio $\left|\frac{\lambda_2}{\lambda_1}\right|$, where λ_2 is the second largest eigenvalue in magnitude. The power method will converge quickly if $\left|\frac{\lambda_2}{\lambda_1}\right|$ is small and slowly if $\left|\frac{\lambda_2}{\lambda_1}\right|$ is close to 1.

Example 18.5.

a. Estimate the largest eigenvalue and the corresponding eigenvector for the Leslie matrix $L = \begin{bmatrix} 0 & 6 & 3 & 2 \\ 0.60 & 0 & 0 & 0 \\ 0 & 0.45 & 0 & 0 \\ 0 & 0 & 0.25 & 0 \end{bmatrix}$.

The MATLAB code runs largeeig twice with an error tolerance of 1.0×10^{-10}. The first call uses maxiter = 100, and the iteration failed to attain the tolerance. With maxiter = 300 for the second call, the tolerance was attained in 200 iterations. The two largest eigenvalues of L in magnitude are 2.0091 and 1.7857, and the ratio $\frac{\lambda_2}{\lambda_1} = 0.88879$, so we expect slow convergence.

```
>> x0 = ones(4,1);
>> [lambda x iter] = largeeig(L,x0,1.0e-10,100)

lambda =
       2.0091
x =
       0.95619
       0.28556
       0.063958
       0.0079585
iter =
     -1
```

```
>> [lambda x iter] = largeeig(L,x0,1.0e-10,300)
lambda =
        2.0091
x =
        0.95619
        0.28555
        0.063958
        0.0079584
iter =
    200
```

b. Let

$$A = \begin{bmatrix} 10.995 & -1.6348 & 1.2323 & 3.055 \\ 2.2256 & 5.4029 & 1.3355 & 5.4342 \\ 1.5794 & 1.1108 & 2.2963 & -0.46759 \\ -3.7347 & -2.447 & -1.3122 & -2.6939 \end{bmatrix}.$$

The eigenvalues of A to eight significant figures are

$$\{10.000373, 3.0000000, 2.0000200, 0.99991723\},$$

and the ratio $\frac{\lambda_2}{\lambda_1} = 0.29998782$, so the power method should converge reasonably quickly. In fact, after 27 iterations, the approximate eigenvalue is 10.000373.

∎

18.2.3   Computing the Smallest Eigenvalue and Corresponding Eigenvector

Assume that A is nonsingular and has eigenvalue λ with corresponding eigenvector v. If $Av = \lambda v$, then $A^{-1}v = \frac{1}{\lambda}v$ and $\frac{1}{\lambda}$ is an eigenvalue of A^{-1} with corresponding eigenvector v. (Recall that an invertible matrix cannot have a zero eigenvalue.) Thus, the eigenvalues of A^{-1} are the reciprocals of those for A, but the eigenvectors are the same. If we assume that

$$|\lambda_1| \geq |\lambda_2| \geq |\lambda_3| > \cdots \geq |\lambda_{n-1}| > |\lambda_n|,$$

the smallest eigenvalue of A is the largest eigenvalue of A^{-1}. The power method applied to A^{-1} will yield the smallest eigenvalue of A and is termed the *inverse power method*.

As we have discussed, computing A^{-1} is fraught with problems, so we should avoid the iteration $x_k = A^{-1}x_{k-1}$; however, this is equivalent to solving $Ax_k = x_{k-1}$. Use Gaussian elimination to compute $PA = LU$ and solve the linear systems $Ax_k = x_{k-1}$ using forward and backward substitution. As the iteration progresses, x_k approaches a normalized eigenvector corresponding to eigenvalue $\frac{1}{\lambda_n}$, so $A^{-1}x_k \approx \left(\frac{1}{\lambda_n}\right)x_k$. Therefore, $Ax_k \approx \lambda_n x_k$ and x_k are approximate eigenvectors for the smallest eigenvalue, λ_n, of A. If we keep track of the Rayleigh quotient $\lambda_k = x_k^T A x_k$, then we can apply the convergence test $\|Ax_k - \lambda_k x_k\|_2 < \text{tol}$, at each step and the algorithm is the same as the power method with $x_k = Ax_{k-1}$ replaced by $x_k = \text{lusolve}(L, U, P, x_{k-1})$. The function `smalleig` in the software distribution implements the inverse power method.

Example 18.6.   This problem constructs a matrix with known eigenvalues by generating a random integer matrix, zeroing out all the elements from the diagonal and below using triu, and then adding a diagonal matrix containing the eigenvalues. There is an eigenvalue, 9, of largest magnitude with multiplicity three, and a smallest eigenvalue in magnitude, −1, well separated from the others. As expected, the power method fails, and the inverse power method is successful.

```
>> A = randi([-100,100],10,10);
>> A = triu(A,1);
>> d = [-1 3 5 9 9 9 6 2 -7 4]';
>> A = A + diag(d);
>> iterations = [50 100 1000 10000];
>> for i = 1:length(iterations)
[lambda,v,iter] = largeeig(A,ones(10,1),1.0e-10,iterations(i));
lambda
```

```
iter
end
lambda =
   9.471799775358845
iter =
     -1

lambda =
   9.204649150708208
iter =
     -1

lambda =
   9.018221954378124
iter =
     -1

lambda =
   9.001802197342364
iter =
     -1
>> [lambda,v,iter] = smalleig(A,rand(10,1),1.0e-14,75);
>> lambda

lambda =
  -1.000000000000044

>> iter

iter =
     52
```

■

18.3 THE BASIC *QR* ITERATION

The *QR* method in its various forms is the most-used algorithm for computing all the eigenvalues of a matrix. We will consider a real matrix A whose eigenvalues satisfy

$$|\lambda_1| > |\lambda_2| > |\lambda_3| > \cdots > |\lambda_n| > 0 \tag{18.14}$$

Such a matrix cannot have imaginary eigenvalues, since each member of a complex conjugate pair has the same modulus. For in-depth coverage of the general eigenvalue problem, including matrices with entries having a nonzero imaginary part the interested reader should see Refs. [2, 9]. A real matrix satisfying Equation 18.14 is invertible, since Theorem 5.2 implies the eigenvectors are linearly independent. In addition, Theorem 5.4 guarantees that A is diagonalizable.

We will first discuss what is termed the *basic QR iteration* for computing all the eigenvalues of a matrix satisfying Equation 18.14. This algorithm should be viewed as a starting point, and we will discuss techniques to enhance its performance in later sections. The basic *QR* iteration is very simple to perform. Let $A_0 = A$, apply the *QR* decomposition to A_0 and obtain $A_0 = Q_1 R_1$, and form $A_1 = R_1 Q_1$. Now compute $A_1 = Q_2 R_2$ and let $A_2 = R_2 Q_2$. Continuing in this fashion, we obtain $A_{k-1} = Q_k R_k$ and $A_k = R_k Q_k$. Under the correct conditions, the sequence approaches an upper triangular matrix. What is on the diagonal of this upper triangular matrix? Since $A = A_0 = Q_1 R_1$ and $A_1 = R_1 Q_1$, it follows that $A_1 = Q_1^T A Q_1$. Now, $A_1 = Q_2 R_2$ and $A_2 = R_2 Q_2$, so $A_2 = Q_2^T A_1 Q_2 = Q_2^T Q_1^T A Q_1 Q_2$. If we continue this process k times, we have

$$\begin{aligned}
A_k &= Q_k^T Q_{k-1}^T \ldots Q_2^T Q_1^T A Q_1 Q_2 \ldots Q_{k-1} Q_k \\
&= (Q_1 Q_2 \ldots Q_{k-1} Q_k)^T A \, (Q_1 Q_2 \ldots Q_{k-1} Q_k) \\
&= \overline{Q}_k^T A \overline{Q}_k,
\end{aligned}$$

where $\overline{Q}_k = Q_1 Q_2 \ldots Q_k$ is an orthogonal matrix. A and A_k are similar matrices, and by Theorem 5.1 have the same eigenvalues. Since A_k is approximately an upper triangular matrix, estimates for its eigenvalues lie on the diagonal. Here is an outline of the basic QR algorithm.

Outline of the Basic QR Iteration

```
A₀ = A
for k = 1, 2, ... do
    A_{k-1} = Q_k R_k
    A_k = R_k Q_k
end for
```

- Compute the QR decomposition, multiply the factors Q and R together in the reverse order RQ, and repeat.
- Under suitable conditions, this simple process converges to an upper triangular matrix, so if k is sufficiently large,

$$A_k \approx \begin{bmatrix} r_{11} & r_{12} & \cdots & r_{1,n-1} & r_{1n} \\ & r_{22} & \cdots & r_{2,n-1} & r_{2n} \\ & & \ddots & \ddots & \vdots \\ & & & r_{n-1,n-1} & \vdots \\ & & & & r_{nn} \end{bmatrix}.$$

- The eigenvalues of A are approximately $r_{11}, r_{22}, \ldots, r_{nn}$.

Remark 18.3.   It can be shown that the eigenvalues occur on the diagonal in decreasing order of magnitude, so in the outline of the basic QR iteration $\lambda_1 = r_{11}, \lambda_2 = r_{22}, \ldots, \lambda_n = r_{nn}$.

Example 18.7.   Let $A = \begin{bmatrix} 5 & 1 & 4 \\ -1 & 3 & 1 \\ 3 & -1 & 2 \end{bmatrix}$. Run the following MATLAB script.

```
>> B = A;
>> for i = 1:15
       [Q R] = qr(A);
       A = R*Q;
   end
>> A
>> eig(B)
```

Note that A is transformed into an upper triangular matrix with the eigenvalues of A on its diagonal in descending order of magnitude. ∎

A proof of the following result can be found in Ref. [27, pp. 209-212]. Recall when reading the theorem that a real matrix satisfying Equation 18.14 can be diagonalized. Also, it is assumed that the LU decomposition can be done without row exchanges.

Theorem 18.3.   *Let A be a real matrix satisfying Equation 18.14. Assume also that P^{-1} has an LU decomposition, where P is the matrix of eigenvectors of A, i.e., $A = P\text{diag}(\lambda_1, \lambda_2, \ldots, \lambda_n) P^{-1}$. Then the sequence $\{A_k\}$, $k \geq 1$, generated by the basic QR iteration, converges to an upper triangular matrix whose diagonal entries are the eigenvalues of A.*

The implementation of the basic QR iteration is in the function `eigqrbasic` of the software distribution.

18.4   TRANSFORMATION TO UPPER HESSENBERG FORM

Prior to presenting a more effective approach to eigenvalue computation than the basic QR iteration, we need two definitions.

Definition 18.1. An *orthogonal similarity transformation* is a decomposition of form $B = P^T A P$, where P is an orthogonal matrix. Since $P^T = P^{-1}$, matrices A and B are similar and thus have the same eigenvalues.

We will have many occasions to use Hessenberg matrices in this and the remaining chapters because they have a simpler form than a general matrix.

Definition 18.2. A square matrix H is *upper Hessenberg* if $h_{ij} = 0$ for all $i > j+1$. The transpose of an upper Hessenberg matrix is a *lower Hessenberg* matrix ($h_{ij} = 0$ for all $j > i+1$). The upper and lower Hessenberg matrices are termed "almost triangular." For instance, the matrix

$$A = \begin{bmatrix} 1 & 2 & 3 & 7 & 3 & -1 \\ 2 & 4 & 8 & 1 & 2 & 3 \\ 0 & 5 & -1 & 6 & 4 & 9 \\ 0 & 0 & 9 & 3 & 1 & -1 \\ 0 & 0 & 0 & 4 & 1 & 3 \\ 0 & 0 & 0 & 0 & 5 & 2 \end{bmatrix}$$

is upper Hessenberg, and

$$B = \begin{bmatrix} 1 & 8 & 0 & 0 \\ 2 & 9 & 1 & 0 \\ 3 & 5 & -1 & 7 \\ 4 & 1 & 3 & 4 \end{bmatrix}$$

is lower Hessenberg.

The basic QR iteration is not practical for large matrices. It requires $O(n^3)$ flops for each QR decomposition, so k iterations cost $kO(n^3)$ flops. If $k = n$, the algorithm becomes $O(n^4)$. Our approach is to split the problem into two parts:

a. Transform A into an upper Hessenberg matrix using orthogonal similarity transformations.
b. Make use of the simpler form of an upper Hessenberg matrix to quickly and accurately compute the eigenvalues.

We will develop an algorithm, hhess, that creates the orthogonal similarity transformation $H = P^T A P$, where H is an upper Hessenberg matrix. Theorem 5.1 guarantees that H will have the same eigenvalues as A. We will then convert the upper Hessenberg matrix into upper triangular form using Givens rotations, and the eigenvalues of A will be on the diagonal.

The transformation to upper Hessenberg form uses Householder reflections. The idea is similar to the use of Householder reflections to perform the QR decomposition, but we must leave a diagonal below the main diagonal (a subdiagonal), and a similarity transformation must be used to maintain the same eigenvalues; in other words, we must use a transformation $A_k = H_{u_k} A_{k-1} H_{u_k}^T$ rather than $A_k = H_{u_k} A_{k-1}$. Assuming a 5×5 matrix, the process proceeds as follows:

$$A = A_0 = \begin{bmatrix} * & * & * & * & * \\ * & * & * & * & * \\ * & * & * & * & * \\ * & * & * & * & * \\ * & * & * & * & * \end{bmatrix}, A_1 = H_{u_1} A H_{u_1}^T = \begin{bmatrix} * & * & * & * & * \\ * & * & * & * & * \\ 0 & * & * & * & * \\ 0 & * & * & * & * \\ 0 & * & * & * & * \end{bmatrix}$$

$$A_2 = H_{u_2} A_1 H_{u_2}^T = \begin{bmatrix} * & * & * & * & * \\ * & * & * & * & * \\ 0 & * & * & * & * \\ 0 & 0 & * & * & * \\ 0 & 0 & * & * & * \end{bmatrix}, A_3 = H_{u_3} A_2 H_{u_3}^T = \begin{bmatrix} * & * & * & * & * \\ * & * & * & * & * \\ 0 & * & * & * & * \\ 0 & 0 & * & * & * \\ 0 & 0 & 0 & * & * \end{bmatrix}$$

In each case, $H_{u_i}A_{i-1}$ zeroes out the elements at indices $(i+2,i)\,,(i+3,i)\,,\ldots,(n,i)$ of A_{i-1}. After $k-1$ steps,

$$
A_{k-1} = \begin{bmatrix}
* & * & * & * & \cdots & * & * & * \\
* & * & * & * & \cdots & * & * & * \\
0 & * & \ddots & * & \cdots & * & * & * \\
0 & 0 & \ddots & \overline{a_{k-1,k-1}} & * & \cdots & * & * \\
\vdots & \vdots & \ddots & \overline{a_{k,k-1}} & \overline{a_{kk}} & \cdots & * & * \\
0 & 0 & \ddots & 0 & \overline{a_{k+1,k}} & \cdots & * & * \\
\vdots & \vdots & \cdots & \vdots & \overline{a_{k+2,k}} & \cdots & * & * \\
0 & 0 & \cdots & 0 & \vdots & \cdots & * & * \\
0 & 0 & \cdots & 0 & \overline{a_{nk}} & \cdots & * & *
\end{bmatrix}.
$$

To find the next Householder matrix, let $y = \begin{bmatrix} \overline{a_{k+1,k}} \\ \overline{a_{k+2,k}} \\ \vdots \\ \overline{a_{nk}} \end{bmatrix}$. Determine the $(n-k)\times(n-k)$ Householder matrix $\widetilde{H}_{u_k}$ so

that $\widetilde{H}_{u_k}y = \begin{bmatrix} \hat{a}_{k+1,k} \\ 0 \\ \vdots \\ 0 \end{bmatrix}$ by using the techniques of Section 17.8.1. We must develop a matrix H_{u_k} containing the submatrix

$\widetilde{H}_{u_k}$ such that when forming $H_{u_k}A_{k-1}$ the elements $\begin{bmatrix} \overline{a_{k+1,k}} \\ \overline{a_{k+2,k}} \\ \vdots \\ \overline{a_{nk}} \end{bmatrix}$ become $\begin{bmatrix} \hat{a}_{k+1,k} \\ 0 \\ \vdots \\ 0 \end{bmatrix}$ without destroying any work we have

already done. We choose Householder matrices H_{u_k} of the form

$$
H_{u_k} = \begin{bmatrix} I_k & 0 \\ 0 & \widetilde{H}_{u_k} \end{bmatrix}, \tag{18.15}
$$

where I_k is the $k \times k$ identity matrix, and $\widetilde{H}_{u_k}$ is the $n-k$ Householder matrix. The submatrix $\begin{bmatrix} I_k \\ 0 \end{bmatrix}$ protects the entries

already having their final value, while $\begin{bmatrix} 0 \\ \widetilde{H}_{k_j} \end{bmatrix}$ makes the required modifications. For instance, suppose one elimination step has executed. The matrix A_1 has the form

$$
\begin{bmatrix}
X & X & X & X & X \\
X & X & X & X & X \\
0 & X & X & X & X \\
0 & \mathbf{X} & X & X & X \\
0 & \mathbf{X} & X & X & X
\end{bmatrix},
$$

and we need to zero out the elements at indices $(4,2)$ and $(5,2)$. Let

$$
H_{u_2} = \begin{bmatrix}
1 & 0 & 0 & 0 & 0 \\
0 & 1 & 0 & 0 & 0 \\
0 & 0 & h_{u_2}^{(11)} & h_{u_2}^{(12)} & h_{u_2}^{(13)} \\
0 & 0 & h_{u_2}^{(21)} & h_{u_2}^{(22)} & h_{u_2}^{(23)} \\
0 & 0 & h_{u_2}^{(31)} & h_{u_2}^{(32)} & h_{u_2}^{(33)}
\end{bmatrix},
$$

and

$$
H_{u_2}A_1 = \begin{bmatrix} 1 & 0 & 0 & 0 & 0 \\ 0 & 1 & 0 & 0 & 0 \\ 0 & 0 & h_{u_2}^{(11)} & h_{u_2}^{(12)} & h_{u_2}^{(13)} \\ 0 & 0 & h_{u_2}^{(21)} & h_{u_2}^{(22)} & h_{u_2}^{(23)} \\ 0 & 0 & h_{u_2}^{(31)} & h_{u_2}^{(32)} & h_{u_2}^{(33)} \end{bmatrix} \begin{bmatrix} X & X & X & X & X \\ X & X & X & X & X \\ 0 & \mathbf{X} & X & X & X \\ 0 & \mathbf{X} & X & X & X \\ 0 & \mathbf{X} & X & X & X \end{bmatrix} = \begin{bmatrix} X & X & X & X & X \\ X & X & X & X & X \\ 0 & Y & Y & Y & Y \\ 0 & 0 & Y & Y & Y \\ 0 & 0 & Y & Y & Y \end{bmatrix},
$$

Verify that forming the product $H_{u_2}A_1H_{u_2}^{\mathrm{T}}$ maintains the zeros created by $H_{u_2}A_1$. Example 18.6 illustrates the process in detail for a 5×5 matrix.

Example 18.8. Let $A = \begin{bmatrix} 9 & 5 & 1 & 2 & 1 \\ 9 & 7 & 10 & 5 & 8 \\ 1 & 7 & 2 & 4 & 3 \\ 4 & 3 & 2 & 10 & 5 \\ 6 & 5 & 4 & 10 & 6 \end{bmatrix}$ and let $A_0 = A$.

Step 1: Compute $\widetilde{H}_{u_1}$ so that $\widetilde{H}_{u_1}\begin{bmatrix} 9 \\ 1 \\ 4 \\ 6 \end{bmatrix} = \begin{bmatrix} -11.5758 \\ 0 \\ 0 \\ 0 \end{bmatrix}$. Using Equation 18.15,

$$
H_{u_1} = \begin{bmatrix} 1 & 0 & 0 & 0 & 0 \\ 0 & -0.7775 & -0.0864 & -0.3455 & -0.5183 \\ 0 & -0.0864 & 0.9958 & -0.0168 & -0.0252 \\ 0 & -0.3455 & -0.0168 & 0.9328 & -0.1008 \\ 0 & -0.5183 & -0.0252 & -0.1008 & 0.8489 \end{bmatrix},
$$

and

$$
H_{u_1}A_0 = \begin{bmatrix} 9.0000 & 5.0000 & 1.0000 & 2.0000 & 1.0000 \\ -11.5758 & -9.6753 & -10.7120 & -12.8716 & -11.3167 \\ 0 & 6.1896 & 0.9934 & 3.1314 & 2.0612 \\ 0 & -0.2417 & -2.0265 & 6.5257 & 1.2448 \\ 0 & 0.1374 & -2.0397 & 4.7886 & 0.3672 \end{bmatrix}
$$

$$
A_1 = H_{u_1}A_0H_{u_1}^{\mathrm{T}} = \begin{bmatrix} 9.0000 & -5.1832 & 0.5051 & 0.0204 & -1.9695 \\ -11.5758 & 18.7612 & -9.3299 & -7.3435 & -3.0245 \\ 0 & -7.0485 & 0.3500 & 0.5579 & -1.7991 \\ 0 & -2.5371 & -2.1380 & 6.0795 & 0.5754 \\ 0 & -1.7756 & -2.1327 & 4.4167 & -0.1907 \end{bmatrix}.
$$

Note that $H_{u_1}A_0$ affected rows 2:5 and columns 1:5, and multiplying on the right by $H_{u_1}^{\mathrm{T}}$ affected rows 1:5, columns 2:5; in other words, the first column of $H_{u_1}A_0$ was not modified, so the work done to zero out the elements at indices $(3, 1)$, $(4, 1)$, and $(5, 1)$ was not modified.

Step 2: Using the vector $\begin{bmatrix} -7.0485 \\ -2.5371 \\ -1.7756 \end{bmatrix}$, form

$$
H_{u_2} = \begin{bmatrix} 1 & 0 & 0 & 0 & 0 \\ 0 & 1 & 0 & 0 & 0 \\ 0 & 0 & -0.9155 & -0.3296 & -0.2306 \\ 0 & 0 & -0.3296 & 0.9433 & -0.0397 \\ 0 & 0 & -0.2306 & -0.0397 & 0.9722 \end{bmatrix}
$$

from which we obtain

$$
H_{u_2}A_1 = \begin{bmatrix} 9.0000 & -5.1832 & 0.5051 & 0.0204 & -1.9695 \\ -11.5758 & 18.7612 & -9.3299 & -7.3435 & -3.0245 \\ 0 & 7.6988 & 0.8760 & -3.5329 & 1.5015 \\ 0 & 0 & -2.0475 & 5.3757 & 1.1433 \\ 0 & 0 & -2.0693 & 3.9241 & 0.2067 \end{bmatrix},
$$

$$
A_2 = H_{u_2}A_1H_{u_2}^T = \begin{bmatrix} 9.0000 & -5.1832 & -0.0149 & -0.0691 & -2.0321 \\ -11.5758 & 18.7612 & 11.6595 & -3.7325 & -0.4973 \\ 0 & 7.6988 & 0.0160 & -3.6809 & 1.3979 \\ 0 & 0 & -0.1607 & 5.7003 & 1.3704 \\ 0 & 0 & 0.5537 & 4.3754 & 0.5225 \end{bmatrix}.
$$

$H_{u_2}A_1$ affected rows 3:5 and columns 2:5, and multiplying on the right by $H_{u_2}^T$ affected rows 1:5, columns 3:5.

Step 3: Compute H_{u_3} using the vector $\begin{bmatrix} -0.1607 \\ 0.5537 \end{bmatrix}$, from which there results

$$
H_{u_3}A_2 = \begin{bmatrix} 9.0000 & -5.1832 & -0.0149 & -0.0691 & -2.0321 \\ -11.5758 & 18.7612 & 11.6595 & -3.7325 & -0.4973 \\ 0 & 7.6988 & 0.0160 & -3.6809 & 1.3979 \\ 0 & 0 & 0.5765 & 2.6136 & 0.1200 \\ 0 & 0 & 0 & 6.6938 & 1.4618 \end{bmatrix},
$$

$$
A_3 = H_{u_3}A_2H_{u_3}^T = \begin{bmatrix} 9.0000 & -5.1832 & -0.0149 & -1.9323 & -0.6326 \\ -11.5758 & 18.7612 & 11.6595 & 0.5625 & -3.7232 \\ 0 & 7.6988 & 0.0160 & 2.3683 & -3.1455 \\ 0 & 0 & 0.5765 & -0.6131 & 2.5435 \\ 0 & 0 & 0 & -0.4614 & 6.8360 \end{bmatrix}.
$$

$H_{u_3}A_2$ affected rows 4:5 and columns 3:5, and multiplying on the right by $H_{u_3}^T$ altered rows 1:5, columns 4:5. The eigenvalues of both A and A_3 are $\{ 25.8275 \quad -4.9555 \quad -0.1586 \quad 6.4304 \quad 6.8562 \}$. ∎

In practice, the Householder matrices H_{u_k} are often not determined explicitly, and Example 18.7 builds them to assist with understanding the process. In practice, use Equations 17.11 and 17.12,

$$
H_u A = A - \beta u u^T A
$$
$$
A H_u = A - \beta A u u^T
$$

to compute the product $H_{u_k}A_{k-1}H_{u_k}^T$ with submatrix operations. From Example 18.6, we see that $H_{u_k}A_{k-1}$ affects rows $k+1:n$, columns $k:n$, and post multiplying by $H_{u_k}^T$ modifies rows $1:n$, columns $k+1:n$, so we can use the following statements:

$$
H(k+1:n, k:n) = H(k+1:n, k:n) - \beta (uu^T) H(k+1:n, k:n)
$$
$$
H(1:n, k+1:n) = H(1:n, k+1:n) - \beta H(1:n, k+1:n) (uu^T).
$$

In order to compute eigenvectors from their corresponding eigenvalues using the Hessenberg inverse iteration in Section 18.8.1, we need to compute the orthogonal matrix P such that $A = PHP^T$. Begin with $P = I$ and continue post multiplying P by the current Householder reflection.

Algorithm 18.2 specifies the transformation to upper Hessenberg form. It uses a function $\begin{bmatrix} u & \beta \end{bmatrix} = $ house (A, i, j) that computes the vector u and the scalar β required to form a Householder reflection that will zero out $A(i+1:n, j)$.

Algorithm 18.2 Transformation to Upper Hessenberg Form

```
function HHESS(A)
  % Given the n × n matrix A,
  % [P H]=hhess(A) returns an upper Hessenberg matrix H and
  % orthogonal matrix P such that A = PHP^T using Householder reflections.
  % H and A have the same eigenvalues.

  H = A
  P = I
  for k = 1:n-2 do
    [ u β ] = house(H, k + 1, k)
    H(k + 1 : n, k : n) = H(k + 1 : n, k : n) − β(uu^T)H(k + 1 : n, k : n)
    H(1 : n, k + 1 : n) = H(1 : n, k + 1 : n) − βH(1 : n, k + 1 : n)(uu^T)
    P(1 : n, k + 1 : n) = P(1 : n, k + 1 : n) − βP(1 : n, k + 1 : n)(uu^T)
    H(k + 2 : n, k) = zeros(n − k − 1, 1)
  end for
  return [H]
end function
```

NLALIB: The function hhess implements Algorithm 18.2.

Example 18.9. Let A be the matrix of Example 18.8. The MATLAB commands use hhess to transform A to upper Hessenberg form and then verify that $PAP^T = H$ within expected roundoff error.

```
>> [P H] = hhess(A);
>> H

H =

        9      -5.1832    -0.014905     -1.9323    -0.63263
  -11.576      18.761       11.659       0.5625     -3.7232
        0       7.6988      0.01596      2.3683     -3.1455
        0            0      0.57652     -0.61311     2.5435
        0            0            0     -0.46141      6.836

>> norm(P*H*P' - A)

ans =
  2.1495e-014
```
■

18.4.1 Efficiency and Stability

Algorithm 18.2 requires $\frac{10}{3}n^3$ flops for the computation of H. To build the orthogonal matrix P requires an additional $4n^3/3$ flops.

The stability of this algorithm is very satisfactory. The computed upper Hessenberg matrix $\hat{H}$ satisfies $\hat{H} = Q^T(A + E)Q$, were Q is orthogonal and $\|E\|_F \leq cn^2$ eps $\|A\|_F$, where c is a small constant [9, pp. 350-351].

18.5 THE UNSHIFTED HESSENBERG *QR* ITERATION

After transforming matrix A into an upper Hessenberg matrix H having the same eigenvalues as A, we can apply the basic *QR* iteration and transform H into an upper triangular matrix with the eigenvalues of A on its diagonal. Before presenting the transformation of an upper Hessenberg matrix to upper triangular form, we need the concept of an unreduced upper Hessenberg matrix.

Definition 18.3. An upper Hessenberg matrix whose subdiagonal entries $h_{i+1,i}$, $1 \leq i \leq n − 1$ are all nonzero is said to be *unreduced* or *proper*.

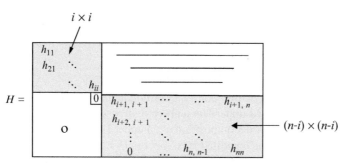

FIGURE 18.7 Reduced Hessenberg matrix.

Figure 18.7 shows a reduced Hessenberg matrix, H.
View the matrix in the form

$$\begin{bmatrix} H_{11} & H_{12} \\ 0 & H_{22} \end{bmatrix},$$

where H_{11} is $i \times i$, H_{22} is $(n - i) \times (n - i)$, and both are upper Hessenberg. The eigenvalues of H are those of H_{11} and H_{22} (Problem 18.8). If either of these submatrices has a zero on its subdiagonal, split it into two submatrices, and so forth. As a result, we only need to consider unreduced matrices.

Remark 18.4.   For the sake of simplicity, we will only deal with unreduced upper Hessenberg matrices in the book.

During the QR iteration, we want the intermediate matrices $A_i = R_{i-1}Q_{i-1}$ to remain upper Hessenberg, and Theorem 18.7 guarantees this.

Theorem 18.4.   *If the $n \times n$ unreduced upper Hessenberg matrix H_k has full column rank and $H_k = Q_kR_k$ is its reduced QR decomposition, then $H_{k+1} = R_kQ_k$ is also an upper Hessenberg matrix.*

Proof. Apply $n - 1$ Givens rotations to transform H_k into upper triangular matrix R_k:

$$J_{n-1}(n - 1, n, c_{n-1}, s_{n-1}) J_{n-2}(n - 2, n - 1, c_{n-2}, s_{n-2}) \ldots J_2(2, 3, c_2, s_2) J_1(1, 2, c_1, s_1) H_k = R_k.$$

Thus, $H_k = Q_kR_k$, where

$$Q_k = J_1(1, 2, c_1, s_1)^{\mathrm{T}} J_2(2, 3, c_2, s_2)^{\mathrm{T}} \ldots J_{n-2}(n - 2, n - 1, c_{n-2}, s_{n-2})^{\mathrm{T}} J_{n-1}(n - 1, n, c_{n-1}, s_{n-1})^{\mathrm{T}}$$

$$= \begin{bmatrix} c_1 & -s_1 & & & \\ s_1 & c_1 & & & \\ & & 1 & & \\ & & & \ddots & \\ & & & & 1 \\ & & & & & 1 \end{bmatrix} \begin{bmatrix} 1 & & & & \\ & c_2 & -s_2 & & \\ & s_2 & c_2 & & \\ & & & 1 & \\ & & & & \ddots \\ & & & & & 1 \end{bmatrix} \cdots \begin{bmatrix} 1 & & & & \\ & 1 & & & \\ & & \ddots & & \\ & & & \ddots & \\ & & & & c_{n-1} & -s_{n-1} \\ & & & & s_{n-1} & c_{n-1} \end{bmatrix},$$

and Q_k is upper Hessenberg. By the results of Problem 17.13, Q_k is unique. Now, since R_k is upper triangular and Q_k is upper Hessenberg, R_kQ_k must be upper Hessenberg.   □

Theorem 18.4 shows us that each new matrix H_{k+1} is upper Hessenberg and that the QR decomposition of an upper Hessenberg matrix H_k is accomplished using $n - 1$ Givens rotations that eliminate the subdiagonal entries. The cost of the decomposition is $O(n^2)$ (Problem 18.12), much better than the $O(n^3)$ flops required for a general square matrix.

In order to implement the algorithm, there must be a criterion for terminating the iteration. It can be shown that as the iteration moves forward, the entry $h_{n, n-1}$ converges to zero rapidly. As suggested in Ref. [2, p. 391], stop the iterations when $|h_{n,n-1}| < \text{tol}(|h_{n-1,n-1}| + |h_{nn}|)$, accept h_{nn} as the approximate eigenvalue, and set $h_{n,n-1}$ to zero. The algorithm then works with the $(n - 1) \times (n - 1)$ submatrix and repeats the process. The final matrix has dimension 2×2, and its diagonal contains the final two eigenvalues. This method is termed *deflation*, and the eigenvalues are on the diagonal of H.

We need to justify deflation by showing that it will yield the same eigenvalues we would obtain by dealing with the whole matrix. Assume we have executed the QR iteration and have reduced the $k \times k$ submatrix, T_k, in the lower right-hand corner to upper triangular form so we now have

$$\overline{H} = \begin{bmatrix} H_{n-k} & X \\ 0 & T_k \end{bmatrix}.$$

H has the same eigenvalues as A, and the QR iteration is an orthogonal similarity transformation, so the eigenvalues of $\overline{H}$ are the same as those of H. $\overline{H}$ is a reduced upper Hessenberg matrix, so its eigenvalues are those of T_k and H_{n-k} (see Problem 18.8). Assuming that the Givens QR decomposition of an upper Hessenberg matrix is implemented in the function givenshessqr, Algorithm 18.3 specifies the unshifted Hessenberg QR iteration.

Algorithm 18.3 Unshifted Hessenberg QR Iteration

```
function EIGQR(A,tol,maxiter)
   % Hessenberg QR iteration for computing all the eigenvalues of
   % a real matrix whose eigenvalues have distinct magnitudes.
   % E=eigqr(A,tol,maxiter), where the tol is error tolerance
   % desired, and maxiter is the maximum number of iterations for
   % computing any single eigenvalue. Vector E contains the eigenvalues.
   % If the desired tolerance is not obtained for any particular eigenvalue,
   % a warning message is printed and computation continues
   % for the remaining eigenvalues.

   H=hhess(A)
   for k=n:-1:2 do
      iter=0
      while |h_{k,k-1}| ≥ tol (|h_{k-1,k-1}| + |h_{kk}|) do
         iter=iter+1
         if iter>maxiter then
            print 'Failure of convergence.'
            print 'Current eigenvalue approximation <value of h_{kk}>'
            break out of inner loop.
         end if
         [ Q_k R_k ] = givenshessqr(H(1 : k, 1 : k))
         H(1 : k, 1 : k) = R_k Q_k
      end while
      H_{k,k-1} = 0
   end for
   E=diag(H)
   return E
end function
```

NLALIB: The function eigqr, supported by givenshessqr, implements Algorithm 18.3. If $|h_{k-1,k-1}| + |h_{kk}| = 0$, the algorithm fails. In the MATLAB implementation, additional code handles this case by changing the convergence criterion to $|h_{k,k-1}| < \text{tol} \|H\|_F$.

Example 18.10. This example demonstrates the behavior of eigqr for the matrix

$$A = \begin{bmatrix} 8 & 6 & 10 & 10 \\ 9 & 1 & 10 & 5 \\ 1 & 3 & 1 & 8 \\ 10 & 6 & 10 & 1 \end{bmatrix}.$$

After reduction to upper Hessenberg form,

$$H = \begin{bmatrix} 8 & -12.156 & -9.0829 & 2.3919 \\ -13.491 & 8.0714 & 13.61 & 0.60082 \\ 0.0000 & 7.4288 & -0.64678 & -3.2612 \\ 0.0000 & 0.0000 & 0.66746 & -4.4246 \end{bmatrix}.$$

The first QR iteration ends with the matrix

$$\begin{bmatrix} 24.348 & 9.2602 & -0.59021 & -5.7314 \\ 1.0076e-006 & -7.5299 & -4.1391 & -2.705 \\ 0 & 0.0010417 & -4.9789 & -0.77024 \\ 0 & 0 & 0 & -0.83907 \end{bmatrix}.$$

Now shift to the upper 3×3 submatrix

$$\begin{bmatrix} 24.348 & 9.2602 & -0.59021 \\ 1.0076e-006 & -7.5299 & -4.1391 \\ 0 & 0.0010417 & -4.9789 \end{bmatrix}.$$

The second QR iteration ends with

$$\begin{bmatrix} 24.348 & 9.2604 & -0.58643 \\ 0 & -7.5282 & -4.1402 \\ 0 & 0 & -4.9806 \end{bmatrix}.$$

The last QR iteration begins with the 2×2 submatrix

$$\begin{bmatrix} 24.348 & 9.2604 \\ 0 & -7.5282 \end{bmatrix},$$

and finishes with

$$\begin{bmatrix} 24.348 & 9.2604 \\ 0 & -7.5282 \end{bmatrix}.$$

The algorithm finds the eigenvalues in the order $\{-0.83907, -4.9806, -7.5282, 24.348\}$. ∎

18.5.1 Efficiency

Reduction to upper Hessenberg form requires $O\left(n^3\right)$ flops. It can be shown that the QR iteration applied to an upper Hessenberg matrix requires $O\left(n^2\right)$ flops [5, p. 92]. First reducing A to upper Hessenberg form and applying the QR iteration to the Hessenberg matrix costs $O\left(n^3\right) + O\left(n^2\right)$ flops, clearly superior to the basic QR iteration.

18.6 THE SHIFTED HESSENBERG QR ITERATION

The rate at which the subdiagonal entries of the Hessenberg matrix converge to zero depends on the ratio $\left|\frac{\lambda_1}{\lambda_2}\right|^p$, where $\lambda_1, \lambda_2, |\lambda_1| < |\lambda_2|$ are the two smallest eigenvalues in the current $k \times k$ submatrix H_k. If λ_2 is close in magnitude to λ_1, convergence will be slow. As a result, we employ a strategy that produces more accurate results. We transform the problem of finding the eigenvalue to finding an eigenvalue of another matrix in which that eigenvalue is more isolated. A lemma that will form the basis for our improved approach.

Lemma 18.1.

1. *If σ is a real number, then the eigenvalues of $A - \sigma I$ are $(\lambda_i - \sigma)$, $1 \le i \le n$, and the eigenvector, v_i, corresponding to eigenvalue λ_i of A is also an eigenvector corresponding to the eigenvalue $(\lambda_i - \sigma)$ of $A - \sigma I$.*
2. *If $\overline{\lambda_i}$ is an eigenvalue of $A - \sigma I$, then $\overline{\lambda_i} + \sigma$ is an eigenvalue of A.*

Proof. For (1), pick any λ_i with eigenvector v_i. Then

$$(A - \sigma I)\, v_i = A v_i - \sigma v_i = \lambda_i v_i - \sigma v_i = (\lambda_i - \sigma)\, v_i,$$

and $\lambda_i - \sigma$ is an eigenvalue of $A - \sigma I$ with corresponding eigenvector v_i.

For (b), if $\overline{\lambda_i}$ is an eigenvalue of $A - \sigma I$ with corresponding eigenvector w_i, then

$$(A - \sigma I)\, w_i = \overline{\lambda} w_i$$
$$A w_i = \left(\overline{\lambda} + \sigma\right) w_i,$$

and $\overline{\lambda_i} + \sigma$ is an eigenvalue of A. □

A solution to the problem of nearly equal eigenvalues and to improve convergence in general is to perform what is called a *shift*. There are two types of shifts, single and double. A single shift is used when computing a real eigenvalue, and a double shift is used when computing a pair of complex conjugate eigenvalues or a pair of real eigenvalues.

18.6.1   A Single Shift

Use a *single shift* to create a new matrix $\overline{H_k} = H_k - \sigma_1 I$, where σ_1 is close to λ_1. By Lemma 18.1, $\lambda_1 - \sigma_1$ and $\lambda_2 - \sigma_1$ are eigenvalues of $H_k - \sigma_1 I$. The rate of convergence to the smallest eigenvalue of $H_k - \sigma_1 I$ then depends on $\left|\frac{\lambda_1 - \sigma_1}{\lambda_2 - \sigma_1}\right|^p$. The numerator $\lambda_1 - \sigma_1$ is small relative to $\lambda_2 - \sigma_2$, and the rate of convergence is improved. Also by Lemma 18.1, if $\overline{\lambda}$ is an eigenvalue of $H_k - \sigma_1 I$ then $\overline{\lambda} + \sigma_1$ is an eigenvalue of H_k. Our strategy is to compute the QR decomposition of the matrix $H_k - \sigma_1 I = Q_k R_k$ and let $H_{k+1} = R_k Q_k + \sigma_1 I$. The reasoning for this is as follows:

$$
\begin{aligned}
H_{k+1} &= R_k Q_k + \sigma_1 I \\
&= Q_k^T (Q_k R_k) Q_k + \sigma_i I \\
&= Q_k^T (H_k - \sigma_1 I) Q_k + \sigma_1 I \\
&= Q_k^T H_k Q_k,
\end{aligned}
$$

and H_{k+1} is similar to H_k and to A. The problem is to choose an optimal σ_1. It practice, the shift is often chosen as $\sigma_1 = H_k(k, k)$, since H_{kk} converges to λ_1. This is known as the *Rayleigh quotient shift.*

Example 18.11.   Let $A = \begin{bmatrix} 8 & 1 \\ -1 & 5 \end{bmatrix}$, which is already in upper Hessenberg form. The example shows the detailed results of three iterations of the single-shifted Hessenberg QR method.

$$
\sigma_1 = 5.0000, \overline{H_1} = H_1 - \begin{bmatrix} 8 & 1 \\ -1 & 5 \end{bmatrix} - \sigma_1 I = \begin{bmatrix} 3 & 1 \\ -1 & 0 \end{bmatrix}.
$$

$$
[\, Q_1, \ R_1 \,] = \left[\begin{bmatrix} 0.9487 & 0.3162 \\ -0.3162 & 0.9487 \end{bmatrix} \begin{bmatrix} 3.1623 & 0.9487 \\ 0 & 0.3162 \end{bmatrix}\right] = qr\left(\overline{H_1}\right),
$$

$$
H_1 = R_1 Q_1 + \sigma_1 I = \begin{bmatrix} 7.7000 & 1.9000 \\ -0.1000 & 5.3000 \end{bmatrix}
$$

$$
\sigma_1 = 5.3000, \overline{H_1} = H_1 - \sigma_1 I = \begin{bmatrix} 2.4000 & 1.9000 \\ -0.1000 & 0 \end{bmatrix}
$$

$$
[\, Q_1, \ R_1 \,] = \left[\begin{bmatrix} 0.9991 & 0.0416 \\ -0.0416 & 0.9991 \end{bmatrix} \begin{bmatrix} 2.4021 & 1.8984 \\ 0 & 0.0791 \end{bmatrix}\right] = qr\left(\overline{H_1}\right)
$$

$$
H_1 = R_1 Q_1 + \sigma_1 I = \begin{bmatrix} 7.6210 & 1.9967 \\ -0.0033 & 5.3790 \end{bmatrix}
$$

One more iteration using $\sigma_1 = 5.3790$ gives $H_1 = \begin{bmatrix} 7.6180 & 2.0000 \\ 0.0000 & 5.3820 \end{bmatrix}$. The eigenvalues accurate to four decimal places are $\{\, 7.6180, \ 5.3820 \,\}$.   ∎

Example 18.12.   Let

$$
A = \begin{bmatrix} -7 & 2 & -1 & 7 & -8 \\ 6 & -5 & -9 & 1 & 10 \\ -4 & 3 & -6 & 10 & -10 \\ 1 & 4 & 9 & -9 & 6 \\ -7 & 5 & -7 & -1 & 7 \end{bmatrix}.
$$

Use the algorithm hhess and transform A to the upper Hessenberg matrix

$$
H = \begin{bmatrix} -7 & -7.8222 & -0.085538 & -5.4477 & 5.2085 \\ -10.1 & -9.6569 & 1.0505 & 0.77123 & 3.0882 \\ 0 & 11.451 & -1.3692 & 13.96 & -5.7396 \\ 0 & 0 & 16.694 & -2.5483 & -4.2948 \\ 0 & 0 & 0 & 4.5531 & 0.57437 \end{bmatrix}.
$$

Using the shift $\sigma = 0.57437$, initially apply the QR iteration to the matrix

$$H - \sigma I = \begin{bmatrix} -7.5744 & -7.8222 & -0.085538 & -5.4477 & 5.2085 \\ -10.1 & -10.231 & 1.0505 & 0.77123 & 3.0882 \\ 0 & 11.451 & -1.9435 & 13.96 & -5.7396 \\ 0 & 0 & 16.694 & -3.1227 & -4.2948 \\ 0 & 0 & 0 & 4.5531 & 0 \end{bmatrix}.$$

After a total of five iterations that involve the additional shifts $\sigma = \{\ 1.6065\ \ 2.3777\ \ 2.3664\ \ 2.3663\ \}$, we obtain the matrix

$$\begin{bmatrix} -21.978 & -3.1496 & -2.8609 & 0.45197 & 5.5643 \\ 1.2298 & -7.0975 & 6.9562 & 8.87 & -0.39277 \\ 0 & 7.9804 & 10.4 & -7.8376 & -9.725 \\ 0 & 0 & 0.2591 & -3.6907 & 0.67703 \\ 0 & 0 & 0 & 0 & 2.3663 \end{bmatrix}.$$

Now choose the shift $\sigma = -3.6907$, and continue with the submatrix

$$R_2 = \begin{bmatrix} -21.978 & -3.1496 & -2.8609 & 0.45197 \\ 1.2298 & -7.0975 & 6.9562 & 8.87 \\ 1.2842e - 016 & 7.9804 & 10.4 & -7.8376 \\ -4.6588e - 016 & -1.5213e - 016 & 0.2591 & -3.6907 \end{bmatrix}.$$

After a total of 14 iterations, we obtain the eigenvalue approximations $\lambda_5 = -21.746$, $\lambda_4 = 13.035$, $\lambda_3 = -9.856$, $\lambda_2 = -3.7993$, and $\lambda_1 = 2.3663$.   ∎

The single-shift Hessenberg QR iteration is implemented by the function `eigqrshift` in the software distribution. Each iteration of a $k \times k$ submatrix during deflation terminates when

$$\left| h_{i,i-1} \right| \leq \mathrm{tol} \left(\left| h_{ii} \right| + \left| h_{i-1,i-1} \right| \right),$$

where tol is larger than the unit roundoff. The only difference between eigqr and eigqrshift is that the statements

```
[Q1, R1] = givenshessqr(H(1:k,1:k));
 H(1:k,1:k) = R1*Q1;
```

are replaced by

```
sigma = H(k,k);
[Q1, R1] = givenshessqr(H(1:k,1:k) - sigma*I);
H(1:k,1:k) = R1*Q1 + sigma*I;
```

Remark 18.5.   The function eigqrshift only applies to a real matrix with eigenvalues having distinct magnitudes. Also, it cannot be used if the eigenvalues consist of complex conjugate pairs. We remedy these problems by developing the implicit double-shift Francis algorithm in Section 18.8.

18.7   SCHUR'S TRIANGULARIZATION

We develop another matrix decomposition called *Schur's triangularization* that involves orthogonal matrices. The decomposition is very useful theoretically because any square matrix can be factored, including singular ones. It is also a good lead-in to the Francis method in Section 18.8. We will restrict ourselves to real matrices with real eigenvalues. The proof involves the use of mathematical induction, and the reader unfamiliar with this proof technique should consult Appendix B.

Theorem 18.5 (Schur's triangularization).   *Every $n \times n$ real matrix A with real eigenvalues can be factored into $A = PTP^{\mathrm{T}}$, where P is an orthogonal matrix, and T is an upper triangular matrix.*

Summary:

If the result is true, then $AP = PT$. Proceeding like we did with the Cholesky decomposition, see what relationships must hold if $AP = PT$ for $n \times n$ orthogonal matrix P. When a pattern evolves, we use mathematical induction to verify the theorem. If the reader chooses to skip the details of the proof, be sure to study Section 19.1, where Schur's triangularization is used to very easily prove the spectral theorem.

Proof. The proof uses construction and induction. If $A = PTP^T$, then $AP = PT$. Let us investigate what we can conclude from this. Equation 18.16 depicts the equation.

$$\begin{bmatrix} a_{11} & a_{12} & \cdots & a_{1n} \\ a_{21} & a_{22} & \cdots & a_{2n} \\ \vdots & \vdots & \ddots & \vdots \\ a_{n1} & a_{n2} & \cdots & a_{nn} \end{bmatrix} \begin{bmatrix} p_{11} & p_{12} & \cdots & p_{1n} \\ p_{21} & p_{22} & \cdots & p_{2n} \\ \vdots & \vdots & \ddots & \vdots \\ p_{n1} & p_{n2} & \cdots & p_{nn} \end{bmatrix} = \begin{bmatrix} p_{11} & p_{12} & \cdots & p_{1n} \\ p_{21} & p_{22} & \cdots & p_{2n} \\ \vdots & \vdots & \ddots & \vdots \\ p_{n1} & p_{n2} & \cdots & p_{nn} \end{bmatrix} \begin{bmatrix} t_{11} & t_{12} & \cdots & t_{1n} \\ 0 & t_{22} & \cdots & t_{2n} \\ \vdots & \vdots & \ddots & \vdots \\ 0 & 0 & \cdots & t_{nn} \end{bmatrix} \tag{18.16}$$

From Equation 18.16, the first column of AP is

$$A \begin{bmatrix} p_{11} \\ p_{21} \\ \vdots \\ p_{n1} \end{bmatrix}, \tag{18.17}$$

and the first column of PT is

$$\begin{bmatrix} p_{11}t_{11} \\ p_{21}t_{11} \\ \vdots \\ p_{n1}t_{11} \end{bmatrix} = t_{11} \begin{bmatrix} p_{11} \\ p_{21} \\ \vdots \\ p_{n1} \end{bmatrix}. \tag{18.18}$$

If we let $\alpha = \begin{bmatrix} p_{11} \\ p_{21} \\ \vdots \\ p_{n1} \end{bmatrix}$, then Equations 18.17 and 18.18 give

$$A\alpha = t_{11}\alpha, \tag{18.19}$$

and t_{11} is an eigenvalue of A with associated normalized eigenvector α (P is orthogonal). The remaining $(n-1)$ columns of P cannot be eigenvectors of A. Pick any other $(n-1)$ linearly independent vectors $w_2, \ldots, w_n$ so that $\{\alpha, w_2, \ldots, w_n\}$ are a basis for $\mathbb{R}^n$. Apply the Gram-Schmidt process to $\{\alpha, w_2, \ldots, w_n\}$ to obtain an orthonormal basis $\{v_1, v_2, \ldots, v_n\}$. The Gram-Schmidt process will not change the first vector if it is already a unit vector, so $v_1 = \alpha$. Let P be the $n \times n$ matrix $P = [v_1 v_2 \ldots v_n]$, where the v_i are the columns of P. P^T can be written as $\begin{bmatrix} v_1^T \\ v_2^T \\ \vdots \\ v_n^T \end{bmatrix}$, where the v_i^T are the rows of P^T. Now form

$$P^T A P = \begin{bmatrix} v_1^T \\ v_2^T \\ \vdots \\ v_n^T \end{bmatrix} A \begin{bmatrix} v_1 & v_2 & \cdots & v_n \end{bmatrix} \tag{18.20}$$

$$= \begin{bmatrix} v_1^T \\ v_2^T \\ \vdots \\ v_n^T \end{bmatrix} \begin{bmatrix} Av_1 & Av_2 & \cdots & Av_n \end{bmatrix} \tag{18.21}$$

$$
= \begin{bmatrix}
v_1^T A v_1 & v_1^T A v_2 & \cdots & v_1^T A v_n \\
v_2^T A v_1 & v_2^T A v_2 & \cdots & v_2^T A v_n \\
\vdots & \vdots & \ddots & \vdots \\
v_n^T A v_1 & v_n^T A v_2 & \cdots & v_n^T A v_n
\end{bmatrix}
\tag{18.22}
$$

$$
= \begin{bmatrix}
t_{11} v_1^T v_1 & v_1^T A v_2 & \cdots & v_1^T A v_n \\
t_{11} v_2^T v_1 & v_2^T A v_2 & \cdots & v_2^T A v_n \\
\vdots & \vdots & \ddots & \vdots \\
t_{11} v_n^T v_1 & v_n^T A v_2 & \cdots & v_n^T A v_n
\end{bmatrix}
\tag{18.23}
$$

$$
= \begin{bmatrix}
v_1^T t_{11} v_1 & \cdots & \cdots \\
0 & \cdots & \cdots \\
\vdots & \cdots & \cdots \\
0 & \cdots & \cdots
\end{bmatrix}.
\tag{18.24}
$$

The first column of Equation 18.24 depends on the inner products $\langle v_1, v_1 \rangle_1, \langle v_1, v_2 \rangle, \ldots, \langle v_1, v_n \rangle$. Since $v_1, v_2, \ldots, v_n$ are orthonormal, the only nonzero inner product is $\langle v_1, v_1 \rangle = 1$. If we exclude row 1 and column 1, the remaining submatrix has size $(n-1) \times (n-1)$, and we will designate it by A_2. $P^T A P$ has the structure shown in Figure 18.8.

Now apply induction on n. If $n = 1$, the theorem is obviously true, since $A = (1) A (1)$. Now assume that Schur's triangularization applies to any $(n-1) \times (n-1)$ matrix. It then applies to A_2, so $A_2 = P_2 T_2 P_2^T$, where P_2 is orthogonal and T_2 is upper triangular. We will show that this assumption implies that the theorem is true for an $n \times n$ matrix. In Figure 18.9, A_2 has been replaced by its decomposition.

Now, we must take this representation of $P^T A P$ and produce the triangularization of A. Using block matrix notation, we can write the matrix in Figure 18.9 as follows:

$$
\begin{bmatrix}
t_{11} & \cdots & \cdots \\
0 & & \\
0 & P_2 T_2 P_2^T & \\
\vdots & & \\
0 & &
\end{bmatrix}
= \begin{bmatrix} 1 & 0 \\ 0 & P_2 \end{bmatrix}
\begin{bmatrix} t_{11} & \cdots \\ 0 & T_2 \end{bmatrix}
\begin{bmatrix} 1 & 0 \\ 0 & P_2 \end{bmatrix}^T.
$$

$$
P^T A P = \begin{bmatrix}
t_{11} & \cdots & \cdots & \cdots \\
0 & & & \\
\vdots & & A_2 & \\
0 & & &
\end{bmatrix}
$$

FIGURE 18.8 Inductive step in Schur's triangularization.

$$
P^T A P = \begin{bmatrix}
t_{11} & \cdots & \cdots & \cdots \\
0 & & & \\
\vdots & & P_2 T_2 P_2^T & \\
0 & & &
\end{bmatrix}
$$

FIGURE 18.9 Schur's triangularization.

Let $\overline{Q} = \begin{bmatrix} 1 & 0 \\ 0 & P_2 \end{bmatrix}$ and $T = \begin{bmatrix} t_{11} & \cdots \\ 0 & T_2 \end{bmatrix}$. This gives $P^{\mathrm{T}}AP = \overline{Q}T\overline{Q}^{\mathrm{T}}$, so $A = P\overline{Q}T\overline{Q}^{\mathrm{T}}P^{\mathrm{T}} = \left(P\overline{Q}\right)T\left(P\overline{Q}\right)^{\mathrm{T}}$. If we show that $P\overline{Q}$ is an orthogonal matrix, the proof is complete. Now,

$$\overline{Q}^{\mathrm{T}}\overline{Q} = \begin{bmatrix} 1 & 0 \\ 0 & P_2 \end{bmatrix}^{\mathrm{T}} \begin{bmatrix} 1 & 0 \\ 0 & P_2 \end{bmatrix} = \begin{bmatrix} 1 & 0 \\ 0 & P_2^{\mathrm{T}} \end{bmatrix} \begin{bmatrix} 1 & 0 \\ 0 & P_2 \end{bmatrix} = \begin{bmatrix} 1 & 0 \\ 0 & P_2^{\mathrm{T}}P_2 \end{bmatrix} = \begin{bmatrix} 1 & 0 \\ 0 & I \end{bmatrix} = I,$$

so $\overline{Q}$ is an orthogonal matrix. Then, $\left(P\overline{Q}\right)^{\mathrm{T}}\left(P\overline{Q}\right) = \overline{Q}^{\mathrm{T}}P^{\mathrm{T}}P\overline{Q} = \overline{Q}^{\mathrm{T}}I\overline{Q} = \overline{Q}^{\mathrm{T}}\overline{Q} = I$, so $P\overline{Q}$ is an orthogonal matrix. If we let $Q = P\overline{Q}$, we have

$$A = QTQ^{\mathrm{T}},$$

and the proof is complete. $\square$

Example 18.13.   Let $A = \begin{bmatrix} 1 & 3 \\ -1 & 5 \end{bmatrix}$. Build the Schur form by starting with Equation 18.19. Find an eigenvalue t_{11} and normalized eigenvector v_1 such that

$$Av_1 = t_{11}v_1$$

and obtain $t_{11} = 2$, $v_1 = \begin{bmatrix} -0.94868 \\ -0.31623 \end{bmatrix}$. Now extend v_1 to an orthonormal basis v_1, v_2 for $\mathbb{R}^2$. Let $v_2 = \begin{bmatrix} 1 & 1 \end{bmatrix}^{\mathrm{T}}$, which is linearly independent of v_1, and apply Gram-Schmidt to obtain the orthonormal basis

$$\begin{bmatrix} -0.94868 \\ -0.31623 \end{bmatrix}, \quad \begin{bmatrix} -0.31623 \\ 0.94868 \end{bmatrix}$$

and matrix

$$P = \begin{bmatrix} -0.94868 & -0.31623 \\ -0.31623 & 0.94868 \end{bmatrix}.$$

Form

$$P^{\mathrm{T}}AP = \begin{bmatrix} 2 & -4 \\ 0 & 4 \end{bmatrix},$$

as indicated in Figure 18.8. In Figure 18.8, $P_2 = 1$, $T_2 = 4$, and $P_2^{\mathrm{T}} = 1$. Follow the remainder of the proof, and you will see that $\overline{Q} = I$, $T = \begin{bmatrix} 2 & -4 \\ 0 & 4 \end{bmatrix}$, and the final value of the orthogonal matrix is $P = \begin{bmatrix} -0.94868 & -0.31623 \\ -0.31623 & 0.94868 \end{bmatrix}$. A Schur's triangularization for A is

$$A = \begin{bmatrix} -0.94868 & -0.31623 \\ -0.31623 & 0.94868 \end{bmatrix} \begin{bmatrix} 2 & -4 \\ 0 & 4 \end{bmatrix} \begin{bmatrix} -0.94868 & -0.31623 \\ -0.31623 & 0.94868 \end{bmatrix}. \qquad \blacksquare$$

Remark 18.6.

- Since t_{11} is any eigenvalue of A, and many choices are possible for performing the extension to an orthonormal basis for $\mathbb{R}^n$, there is no unique Schur's triangularization (Problem 18.21).
- The MATLAB function [U, T] = schur(A) computes the Schur's triangularization of the square matrix A.
- Our proof of Schur's triangularization theorem involved knowing the eigenvalues of the matrix, so it will not help us to compute eigenvalues. However, it gives us a reason to suspect that it is possible to reduce any matrix to upper triangular form using orthogonality similarity transformations. We will see in Section 18.9 that the Francis iteration of degree one or two does exactly that.

The theorem is actually true for any matrix $A \in \mathbb{C}^{n \times n}$. If A is complex, then P and T are complex. If A is real and has complex eigenvalues, it is possible to find a real orthogonal matrix P if T is replaced by a real *quasi-triangular matrix* and obtain what is termed the *real Schur form*. T is a block upper triangular matrix of the form

$$T = \begin{bmatrix} T_{11} & T_{12} & T_{13} & \cdots & T_{1k} \\ 0 & T_{22} & T_{23} & \cdots & T_{2k} \\ 0 & 0 & T_{33} & \cdots & T_{3k} \\ \vdots & \vdots & \vdots & & \vdots \\ 0 & 0 & 0 & \cdots & T_{kk} \end{bmatrix},$$

where the diagonal blocks are of size 1×1 or 2×2. The 1×1 blocks contain the real eigenvalues of A, and the eigenvalues of the 2×2 diagonal blocks contain complex conjugate eigenvalues or two real eigenvalues. For a proof, see Ref. [2, pp. 376-377].

Example 18.14. Compute the real Schur form of a 7×7 matrix and note the 2×2 and 1×1 diagonal blocks that contain the eigenvalues.

```
[U,T] = schur(A);
>> T

T =
     141.6    -86.367    -85.826    -91.634    -14.548    -57.656    -31.229
        0    -167.75     12.805     12.646     -32.73    -14.815     74.447
        0          0     54.834    -93.417     71.143     27.484     21.421
        0          0     80.735     54.834     2.7022     144.71    -2.5564
        0          0          0          0    -9.4734    -117.66     28.839
        0          0          0          0          0    -67.023     130.67
        0          0          0          0          0    -42.112    -67.023

>> eig(A)

ans =
      141.6  +        0i
     54.834  +   86.845i
     54.834  -   86.845i
    -167.75  +        0i
    -9.4734  +        0i
    -67.023  +   74.181i
    -67.023  -   74.181i
```

The diagonal blocks are

$$\begin{bmatrix} -67.023 & 130.67 \\ -42.112 & -67.023 \end{bmatrix}, \quad \begin{bmatrix} -9.4734 \end{bmatrix}, \quad \begin{bmatrix} 54.834 & -93.417 \\ 80.735 & 54.834 \end{bmatrix}, \quad \begin{bmatrix} 141.6 & -86.367 \\ 0 & -167.75 \end{bmatrix}.$$

Block $\begin{bmatrix} -67.023 & 130.67 \\ -42.112 & -67.023 \end{bmatrix}$ gives eigenvalues $-67.023 \pm 74.181i$, block $\begin{bmatrix} 54.834 & -93.417 \\ 80.735 & 54.834 \end{bmatrix}$ gives eigenvalues $54.834 \pm 86.845i$, and block $\begin{bmatrix} 141.6 & -86.367 \\ 0 & -167.75 \end{bmatrix}$ returns the real eigenvalues 141.6 and -167.75. The 1×1 block $\begin{bmatrix} -9.4734 \end{bmatrix}$ produces eigenvalue -9.4734. ∎

18.8 THE FRANCIS ALGORITHM

The *Francis algorithm*, also known as the *implicit QR algorithm*, has been a staple in computing the eigenvalues and eigenvectors of a small- to medium-size general matrix for many years [54, 55]. It begins by reducing A to upper Hessenberg form so that $Q^T A Q = H$, where Q is an orthogonal matrix (Algorithm 18.2). Then the algorithm transforms H to upper triangular form by using a succession of orthogonal similarity transformations rather than directly using shifts and the QR decomposition, and this is the origin of the term implicit. There are two versions of the Francis algorithm, the *single shift* and *double shift*. The single-shift version can be used to determine the eigenvalues of a real matrix whose eigenvalues are real, and this includes symmetric matrices. Chapter 19 discusses eigenvalue computation of a symmetric matrix and uses the single-shift algorithm. The double-shift version computes all eigenvalues, real or complex conjugate pairs, of a nonsymmetric matrix without using complex arithmetic. The development of the double-shift version is complicated, and can be omitted if desired, but the reader should study Section 18.8.1.

18.8.1 Francis Iteration of Degree One

The first step is to apply orthogonal similarity transformations that reduce the matrix A to upper Hessenberg form, $H = Q^T A Q$. With the explicit shifted QR algorithm, we perform a series of QR factorizations on $H - \sigma I$, each of which requires

$O\left(n^2\right)$ flops. Instead, the Francis algorithm executes a sequence of orthogonal similarity transformations to form the QR factorizations.

Recall that an explicit single shift requires executing the following two statements in succession:

$$H - \sigma I = Q_1 R_1, \quad H_1 = R_1 Q_1 + \sigma I,$$

so

$$Q_1^T H - \sigma_1 Q_1^T = R_1$$
$$H_1 Q_1^T - \sigma_1 Q_1^T = R_1,$$

and

$$H_1 = Q_1^T H Q_1. \tag{18.25}$$

See Section 18.6.1 for an alternative derivation of Equation 18.25. The trick is to form H_1 without having to directly compute Q_1 from the QR decomposition. The approach is based upon the *implicit Q theorem*. For a proof see Ref. [2, p. 381].

Theorem 18.6 (The implicit Q theorem). *Let* $Q = \begin{bmatrix} q_1 & q_3 & \cdots & q_{n-1} & q_n \end{bmatrix}$ *and* $V = \begin{bmatrix} v_1 & v_3 & \cdots & v_{n-1} & v_n \end{bmatrix}$ *be orthogonal matrices with the property that both* $Q^T A Q = H$ *and* $V^T A V = K$ *are unreduced upper Hessenberg, where* $A \in \mathbb{R}^{n \times n}$. *If* $q_1 = v_1$, *then* $q_i = \pm v_i$ *and* $\left|h_{i,i-1}\right| = \left|k_{i,i-1}\right|$ *for* $2 \le i \le n$. *In other words, H and K are essentially the same matrix.*

The theorem applies because we are going to execute a series of steps

$$H_{i+1} = Q_i^T H_i Q_i.$$

in lieu of performing a QR decomposition, where Q_i is a orthogonal matrix with an appropriately chosen first column.

Preparation for Understanding the Iteration

Suppose that

$$H = \begin{bmatrix} h_{11} & h_{12} & h_{13} & h_{14} \\ h_{21} & h_{22} & h_{23} & h_{24} \\ \mathbf{h_{31}} & h_{32} & h_{33} & h_{34} \\ 0 & 0 & h_{43} & h_{44} \end{bmatrix}$$

is upper Hessenberg, except for a nonzero entry at position $(3, 1)$. It is important to understand that the element to be zeroed out by $J(i, j, c, s)$ does not have to be a_{ji}. Let c and s be determined by

$$\begin{bmatrix} c, & s \end{bmatrix} = \text{givensparms}\,(h_{21}, h_{31}),$$

and consider the following:

$$
J(2, 3, c, s)\, H = \begin{bmatrix} 1 & 0 & 0 & 0 \\ 0 & c & s & 0 \\ 0 & -s & c & 0 \\ 0 & 0 & 0 & 1 \end{bmatrix} \begin{bmatrix} h_{11} & h_{12} & h_{13} & h_{14} \\ h_{21} & h_{22} & h_{23} & h_{24} \\ \mathbf{h_{31}} & h_{32} & h_{33} & h_{34} \\ 0 & 0 & h_{43} & h_{44} \end{bmatrix}
$$

$$
= \begin{bmatrix} h_{11} & h_{12} & h_{13} & h_{14} \\ ch_{21} + sX & ch_{22} + sh_{32} & ch_{23} + sh_{33} & ch_{24} + sh_{34} \\ -\mathbf{sh_{21}} + \mathbf{ch_{31}} & -sh_{22} + ch_{32} & -sh_{23} + ch_{33} & -sh_{24} + ch_{34} \\ 0 & 0 & h_{43} & h_{44} \end{bmatrix}
$$

$$
= \begin{bmatrix} h_{11} & h_{12} & h_{13} & h_{14} \\ ch_{21} + sX & ch_{22} + sh_{32} & ch_{23} + sh_{33} & ch_{24} + sh_{34} \\ \mathbf{0} & -sh_{22} + ch_{32} & -sh_{23} + ch_{33} & -sh_{24} + ch_{34} \\ 0 & 0 & h_{43} & h_{44} \end{bmatrix}
$$

The action zeroed out position $(3, 1)$. If we multiply by $J(2, 3, c, s)^T$ on the right, here is the form of the product:

$$J(2, 3, c, s) \, HJ(2, 3, c, s)^T = \begin{bmatrix} * & * & * & * \\ * & * & * & * \\ 0 & * & * & * \\ 0 & \mathbf{X} & * & * \end{bmatrix}.$$

By computing $J(2, 3, c, s)$, we remove a nonzero item at position $(3, 1)$, but introduce another at position $(4, 2)$, and the matrix is not upper Hessenberg. We have "chased" a nonzero element from $(3, 1)$ to $(4, 2)$ with an orthogonal similarity transformation.

Demonstration of the Francis Iteration of Degree One

We use a 5×5 matrix to illustrate a step of the Francis algorithm:

$$H = \begin{bmatrix} * & * & * & * & * \\ * & * & * & * & * \\ 0 & * & * & * & * \\ 0 & 0 & * & * & * \\ 0 & 0 & 0 & * & * \end{bmatrix}.$$

Step 1: Assume a real shift, σ, and form a Givens rotation

$$J_1 = J(1, 2, c_1, s_1) = \begin{bmatrix} c_1 & s_1 & & & \\ -s_1 & c_1 & & & \\ & & 1 & & \\ & & & 1 & \\ & & & & 1 \end{bmatrix},$$

where c and s are determined from the vector

$$\begin{bmatrix} h_{11} - \sigma \\ h_{21} \\ 0 \\ \vdots \\ 0 \end{bmatrix}.$$

Note that the computation of c and s using a vector with first element $h_{11} - \sigma$ is critical. It will be a important link to Theorem 18.5 that justifies the implicit approach to the single-shift algorithm. The product $J_1 H J_1^T$ and the resulting matrix have the form

$$H_1 = J_1 H J_1^T = \begin{bmatrix} * & * & * & * & * \\ * & * & * & * & * \\ + & * & * & * & * \\ 0 & 0 & * & * & * \\ 0 & 0 & 0 & * & * \end{bmatrix}.$$

The product disturbs the upper Hessenberg form, leaving a nonzero element at $(3, 1)$. The element at $(3, 1)$ is called a bulge. Our job is to chase the bulge down to the right and off the matrix, leaving the resulting matrix in upper Hessenberg form.

Step 2: Let c_2 and s_2 be formed from the elements $H_1(2, 1)$ and $H_1(3, 1)$. The rotation $J_2 = \begin{bmatrix} 1 & & & & \\ & c_2 & s_2 & & \\ & -s_2 & c_2 & & \\ & & & 1 & \\ & & & & 1 \end{bmatrix}$ applied to H_1

eliminates the bulge at $(3, 1)$, but after postmultiplication by J_2^T, another bulge appears at $(4, 2)$:

$$H_2 = J_2 J_1 H J_1^T J_2^T = \begin{bmatrix} * & * & * & * & * \\ * & * & * & * & * \\ 0 & * & * & * & * \\ 0 & + & * & * & * \\ 0 & 0 & 0 & * & * \end{bmatrix}.$$

Step 3: Let c_3 and s_3 be formed from the elements $H_2 (3, 2)$ and $H_2 (4, 2)$. The rotation $J_3 = \begin{bmatrix} 1 & & & & \\ & 1 & & & \\ & & c_3 & s_3 & \\ & & -s_3 & c_3 & \\ & & & & 1 \end{bmatrix}$ applied to H_2

eliminates the bulge at $(4, 2)$, but after postmultiplication by J_3^T another bulge appears at $(5, 3)$.

$$H_3 = J_3 J_2 J_1 H_2 J_1^T J_2^T J_3^T = \begin{bmatrix} * & * & * & * & * \\ * & * & * & * & * \\ 0 & * & * & * & * \\ 0 & 0 & * & * & * \\ 0 & 0 & + & * & * \end{bmatrix},$$

Step 4: Let c_4 and s_4 be formed from the elements $H_3 (4, 3)$ and $H_3 (5, 3)$. The rotation $J_4 = \begin{bmatrix} 1 & & & & \\ & 1 & & & \\ & & 1 & & \\ & & & c_4 & s_4 \\ & & & -s_4 & c_4 \end{bmatrix}$ applied to H_3

eliminates the bulge at $(5, 3)$, and after postmultiplication by J_4^T no more bulges remain, and the matrix has upper Hessenberg form

$$H_4 = J_4 J_3 J_2 J_1 (H - \sigma I) J_1^T J_2^T J_3^T J_4^T = \begin{bmatrix} * & * & * & * & * \\ * & * & * & * & * \\ 0 & * & * & * & * \\ 0 & 0 & * & * & * \\ 0 & 0 & 0 & * & * \end{bmatrix}$$

In general, each iteration chases the bulge through indices $(3, 1)$, $(4, 2)$, $(5, 3)$, $(6, 4)$, ..., $(n, n - 2)$. If we let K be the orthogonal matrix

$$K = \left(J_1^T J_2^T J_3^T J_4^T \right)^T,$$

then

$$K^T H K = \begin{bmatrix} * & * & * & * & * \\ * & * & * & * & * \\ 0 & * & * & * & * \\ 0 & 0 & * & * & * \\ 0 & 0 & 0 & * & * \end{bmatrix}$$

is an unreduced upper Hessenberg matrix, and a check will show that K has the form

$$K = \begin{bmatrix} c_1 & * & * & * & * \\ s_1 & * & * & * & * \\ 0 & * & * & * & * \\ 0 & 0 & * & * & * \\ 0 & 0 & 0 & * & * \end{bmatrix}.$$

If Givens rotations are used to compute the QR decomposition of $H - \sigma I$, the first column of Q will be the same as that of K. By the implicit Q theorem, the matrices K and the matrix Q are essentially the same matrix, and K can be used to perform the single shift given by Equation 18.16. Except for signs, the matrix after the bulge chase is the matrix we would obtain by doing an explicit shift. Algorithm 18.4 implements the bulge chase. It uses a function givensmult(H,i,j,c,s) that computes $HJ (i, j, c, s)^T$. An implementation of the function is in the book software distribution.

Algorithm 18.4 Single Shift Using the Francis Iteration of Degree One

```
function CHASE(H)
   % Bulge chase in the Francis algorithm.
   % [Q, H1]=chase(H) returns an orthogonal matrix Q such that Q^T HQ = H1,
   % where H and H1 are upper Hessenberg matrices, and Q is an orthogonal matrix.
   Q=I
   % use h_nn as the shift.
   σ = h_nn
   % case (1,2), (2,1) using shift σ.
   [c, s] = givensparms(h_11 − σ, h_21)
   H=givensmul(H,1,2,c,s)
   H=givensmult(H,1,2,c,s)
   Q=givensmul(Q,1,2,c,s)

   % chase the bulge.
   for i=1:n-2 do
      [c, s] = givensparms(h_{i+1,i}, h_{i+2,i})
      H=givensmul(H,i+1,i+2,c,s)
      H=givensmult(H,i+1,i+2,c,s)
      h_{i+2,i} = 0
      Q=givensmul(Q,i+1,i+2,c,s)
   end for
   H1=H
   return [Q,H1]
end function
```

NLALIB: The function `chase` implements Algorithm 18.4.

The single-shift strategy is only practical when we know that all eigenvalues are real, and this is true for a symmetric matrix. Section 19.5 uses the implicit single shift to build a general symmetric matrix eigenvalue/eigenvector solver.

18.8.2   Francis Iteration of Degree Two

When computing a complex conjugate pair of eigenvalues, it is desirable to avoid complex arithmetic, since complex arithmetic requires about four times as many flops. By performing a double shift, we can compute both eigenvalues using only real arithmetic, and the double shift can also be used to approximate two real eigenvalues. The idea is to use the eigenvalues of the 2×2 lower right-hand corner submatrix as the shifts (Figure 18.10).

Let H be the upper Hessenberg matrix whose 2×2 submatrix has eigenvalues σ_1 and σ_2. Shift first by σ_1 to obtain

$$H - \sigma_1 I = Q_1 R_1, \quad H_1 = R_1 Q_1 + \sigma_1 I, \tag{18.26}$$

and then shift H_1 by σ_2 and we have

$$H_1 - \sigma_2 I = Q_2 R_2, \quad H_2 = R_2 Q_2 + \sigma_2 I. \tag{18.27}$$

The initial matrix H is real, but if σ_1 and $\sigma_2 = \overline{\sigma_1}$ are complex conjugates, the double shift as given requires complex arithmetic. By some clever operations, we can perform the shifts entirely with real arithmetic. When dealing with a complex matrix, we use the *Hermitian transpose*, or the *conjugate transpose*, that is, the complex conjugate of the elements in the transpose, indicated by A^*. For instance

FIGURE 18.10 Eigenvalues of a 2×2 matrix as shifts.

$$\begin{bmatrix} 1-i & 2+3i \\ 5-2i & 6+9i \end{bmatrix}^{*} = \begin{bmatrix} 1+i & 5+2i \\ 2-3i & 6-9i \end{bmatrix}.$$

From Equation 18.26, we have

$$Q_1^{*}H - \sigma_1 Q_1^{*} = R_1$$
$$H_1 Q_1^{*} - \sigma_1 Q_1^{*} = R_1,$$

so

$$Q_1^{*}HQ_1 = H_1. \tag{18.28}$$

Similarly, Equation 18.27 gives

$$Q_2^{*}H_1 Q_2 = H_2, \tag{18.29}$$

and by substituting Equation 18.28 into Equation 18.29, we have

$$H_2 = (Q_1 Q_2)^{*} H (Q_1 Q_2) \tag{18.30}$$

Thus, H and H_2 are orthogonally similar and have the same eigenvalues. These relationships allow us to prove the following lemma that leads to an algorithm for performing the double shift using real arithmetic.

Lemma 18.2. *There exist orthogonal matrices Q_1 and Q_2 such that*

a. $Q_1 Q_2$ *is real,*
b. H_2 *is real*

Proof. From Equations 18.26 and 18.27,

$$Q_2 R_2 = H_1 - \sigma_2 I = R_1 Q_1 + (\sigma_1 - \sigma_2) I,$$

and

$$Q_1 Q_2 R_2 R_1 = Q_1 (R_1 Q_1 + (\sigma_1 - \sigma_2) I) R_1 \tag{18.31}$$
$$= Q_1 R_1 Q_1 R_1 + (\sigma_1 - \sigma_2) Q_1 R_1 \tag{18.32}$$
$$= (H - \sigma_1 I)(H - \sigma_1 I) + (\sigma_1 - \sigma_2)(H - \sigma_1 I) \tag{18.33}$$
$$= H^2 - 2\sigma_1 H + \sigma_1^2 I + \sigma_1 H - \sigma_1^2 I - \sigma_2 H + (\sigma_1 \sigma_2) I \tag{18.34}$$
$$= H^2 - (\sigma_1 + \sigma_2) H + \sigma_1 \sigma_2 I \tag{18.35}$$
$$= (H - \sigma_1 I)(H - \sigma_2 I) = S \tag{18.36}$$

If $\sigma_1 = a + ib$ and $\sigma_2 = a - ib$ are complex conjugates, Equation 18.36 takes the form

$$H^2 - 2\mathrm{Real}(\sigma_1) H + |\sigma_1|^2 I.$$

The matrix S is real, and $S = (Q_1 Q_2)(R_2 R_1)$ is a QR decomposition of S, so both $Q_1 Q_2$ and $R_1 R_2$ can be chosen to be real. By Equation 18.30, H_2 is real. $\square$

Now let's see how all this fits together to perform a double shift. The lower right-hand 2×2 matrix is

$$M = \begin{bmatrix} h_{n-1,n-1} & h_{n-1,n} \\ h_{n,n-1} & h_{nn} \end{bmatrix},$$

whose eigenvalues are

$$\lambda = \frac{\left(h_{n-1,n-1} + h_{nn}\right) \pm \sqrt{\left(h_{n-1,n-1} + h_{nn}\right)^2 - 4\left(h_{n-1,n-1} h_{nn} - h_{n,n-1} h_{n-1,n}\right)}}{2}.$$

The matrix S of Equation 18.36 has the value (Problem 18.18)

$$S = H^2 - \left(h_{n-1,n-1} + h_{nn}\right) H + \left(h_{n-1,n-1} h_{nn} - h_{n,n-1} h_{n-1,n}\right) I \tag{18.37}$$
$$= H^2 - \mathrm{trace}(M) + \det(M). \tag{18.38}$$

Find the QR decomposition of $S = QR = (Q_1Q_2)(R_2R_1)$ and compute

$$H_2 = Q^{\mathrm{T}}HQ.$$

In summary,

Computation of a Double Shift

1. Form the real matrix $S = H^2 - (h_{n-1,n-1} + h_{nn})H + (h_{n-1,n-1}h_{nn} - h_{n,n-1}h_{n-1,n})I$.
2. Find the QR decomposition $S = QR$.
3. Compute $H_2 = Q^{\mathrm{T}}HQ$.

It would appear we can simply use the double shift as we have developed it to find complex conjugate eigenvalues or two real eigenvalues. We call this an *explicit double shift*. There is a serious problem! Each execution of step 2 costs $O(n^3)$ flops, so n applications of the double shift will cost $O(n^4)$ flops, and this is not at all satisfactory. We want to compute the upper Hessenberg matrix $H_2 = (Q)^{\mathrm{T}}H(Q)$ without first performing a QR decomposition of S. By using an implicit process, we can compute a double shift using only $O(n^2)$ flops. We will proceed like we did for the implicit single shift, except the rotations used will be a little more complex.

We will find Q and thus H_2 by using the implicit Q theorem. We know that $Q^{\mathrm{T}}HQ = H_2$, where $QR = H^2 - (\sigma_1 + \sigma_2)H + \sigma_1\sigma_2 I$. The implicit Q theorem now tells us that we essentially get H_2 using any orthogonal similarity transformation $Z^{\mathrm{T}}HZ$ provided that $Z^{\mathrm{T}}HZ$ is upper Hessenberg, and Q and Z have the same first column or $Qe_1 = Ze_1$.

A calculation shows that the first column of S is (Problem 18.19)

$$C = \begin{bmatrix} (h_{11} - \sigma_1)(h_{11} - \sigma_2) + h_{12}h_{21} \\ h_{21}((h_{11} + h_{22}) - (\sigma_1 + \sigma_2)) \\ h_{21}h_{32} \\ 0 \\ \vdots \\ 0 \end{bmatrix}. \tag{18.39}$$

We will use a 7×7 matrix to illustrate the sequence of orthogonal transformations. Form Householder reflection H_{u_1} by letting $\overline{H}_{u_1}$ be the 3×3 reflection that zeros out $h_{21}((h_{11} + h_{22}) - (\sigma_1 + \sigma_2))$ and $h_{21}h_{32}$ and adding the 4×4 identity matrix to form

$$H_{u_1} = \begin{bmatrix} \overline{H}_{u_1} & \\ & I \end{bmatrix} = \begin{bmatrix} * & * & * & & & & \\ * & * & * & & & & \\ * & * & * & & & & \\ & & & 1 & & & \\ & & & & 1 & & \\ & & & & & 1 & \\ & & & & & & 1 \end{bmatrix}.$$

Recalling that a Householder reflection is symmetric, now compute

$$H_{u_1}HH_{u_1} = \begin{bmatrix} * & * & * & * & * & * & * \\ * & * & * & * & * & * & * \\ + & * & * & * & * & * & * \\ + & + & * & * & * & * & * \\ & & & * & * & * & * \\ & & & & * & * & * \\ & & & & & * & * \end{bmatrix},$$

and obtain a upper Hessenberg matrix with the exception of a 2×2 bulge. We must chase this bulge down to the right and out. Create a 3×3 Householder reflection $\overline{H}_{u_2}$ that is designed to zero out indices $(3, 1)$ and $(4, 1)$ and form the reflection

$$H_{u_2} = \begin{bmatrix} 1 & \\ & \overline{H}_{u_2} \\ & & I \end{bmatrix} = \begin{bmatrix} 1 & & & & & \\ & * & * & * & & \\ & * & * & * & & \\ & * & * & * & & \\ & & & & 1 & \\ & & & & & 1 \\ & & & & & & 1 \end{bmatrix}.$$

A premultiplication by H_{u_2} zeroes out indices $(3, 1)$ and $(4, 1)$. Compute

$$H_{u_2}H_{u_1}HH_{u_1}H_{u_2} = \left[\begin{array}{c|ccc|ccc} * & * & * & * & * & * & * \\ \hline * & * & * & * & * & * & * \\ & * & * & * & * & * & * \\ + & * & * & * & * & * & * \\ \hline + & + & * & * & * & * & * \\ & & & & * & * & * \\ & & & & & * & * \end{array}\right],$$

and we have chased the bulge down one row and one column to the right. Create a 3×3 reflection $\overline{H}_{u_3}$ that zeros out indices $(4, 2)$ and $(5, 2)$, and form

$$H_{u_3} = \begin{bmatrix} 1 & & & \\ & 1 & & \\ & & \overline{H}_{u_3} & \\ & & & I \end{bmatrix} = \begin{bmatrix} 1 & & & & & \\ & 1 & & & & \\ & & * & * & * & \\ & & * & * & * & \\ & & * & * & * & \\ & & & & & 1 \\ & & & & & & 1 \end{bmatrix}.$$

The computation

$$H_{u_3}H_{u_2}H_{u_1}HH_{u_1}H_{u_2}H_{u_3} = \left[\begin{array}{cc|cccc|cc} * & * & * & * & * & * & * & * \\ * & * & * & * & * & * & * & * \\ \hline & * & * & * & * & * & * & * \\ & & * & * & * & * & * & * \\ & & + & * & * & * & * & * \\ \hline & & + & + & * & * & * & * \\ & & & & & & * & * \end{array}\right],$$

again moves the bulge down. Build a 3×3 reflection $\overline{H}_{u_4}$ that zeros out indices $(5, 3)$ and $(6, 3)$, and form

$$H_{u_4} = \begin{bmatrix} 1 & & & & \\ & 1 & & & \\ & & 1 & & \\ & & & \overline{H}_{u_4} & \\ & & & & I \end{bmatrix} = \begin{bmatrix} 1 & & & & & \\ & 1 & & & & \\ & & 1 & & & \\ & & & * & * & * \\ & & & * & * & * \\ & & & * & * & * \\ & & & & & & 1 \end{bmatrix},$$

and compute

$$H_{u_4}H_{u_3}H_{u_2}H_{u_1}HH_{u_1}H_{u_2}H_{u_3}H_{u_4} = \left[\begin{array}{ccc|cccc|c} * & * & * & * & * & * & * & * \\ * & * & * & * & * & * & * & * \\ & * & * & * & * & * & * & * \\ \hline & & * & * & * & * & * & * \\ & & & * & * & * & * & * \\ & & + & * & * & * & * & * \\ \hline & & + & + & * & * & * & * \end{array}\right],$$

For the next step, build a 3×3 reflection $\overline{H}_{u_5}$ that zeros out indices (6, 4) and (7, 4), and form

$$
H_{u_5} = \begin{bmatrix} 1 & & & & & \\ & 1 & & & & \\ & & 1 & & & \\ & & & 1 & & \\ & & & & \overline{H}_{u_5} & \end{bmatrix} = \begin{bmatrix} 1 & & & & & & \\ & 1 & & & & & \\ & & 1 & & & & \\ & & & 1 & & & \\ & & & & * & * & * \\ & & & & * & * & * \\ & & & & * & * & * \end{bmatrix},
$$

and then

$$
H_{u_5}H_{u_4}H_{u_3}H_{u_2}H_{u_1}HH_{u_1}H_{u_2}H_{u_3}H_{u_4}H_{u_5} = \left[\begin{array}{cccc|ccc} * & * & * & * & * & * & * \\ * & * & * & * & * & * & * \\ & * & * & * & * & * & * \\ & & * & * & * & * & * \\ \hline & & & * & * & * & * \\ & & & & * & * & * \\ & & & & + & * & * \end{array}\right].
$$

To complete the chase, construct a 2×2 Householder reflection, $\overline{H}_{u_6}$ that zeros out index (7, 5), form

$$
H_{u_6} = \begin{bmatrix} 1 & & & & & \\ & 1 & & & & \\ & & 1 & & & \\ & & & 1 & & \\ & & & & 1 & \\ & & & & & \overline{H}_{u_6} \end{bmatrix} = \begin{bmatrix} 1 & & & & & & \\ & 1 & & & & & \\ & & 1 & & & & \\ & & & 1 & & & \\ & & & & 1 & & \\ & & & & & * & * \\ & & & & & * & * \end{bmatrix},
$$

and the final result is an upper Hessenberg matrix

$$
H_{u_6}H_{u_5}H_{u_4}H_{u_3}H_{u_2}H_{u_1}HH_{u_1}H_{u_2}H_{u_3}H_{u_4}H_{u_5}H_{u_6} = QHQ^{\mathrm{T}}
$$

$$
= \left[\begin{array}{ccccc|cc} * & * & * & * & * & * & * \\ * & * & * & * & * & * & * \\ & * & * & * & * & * & * \\ & & * & * & * & * & * \\ & & & * & * & * & * \\ \hline & & & & * & * & * \\ & & & & & * & * \end{array}\right],
$$

where $Q = H_{u_6}H_{u_5}H_{u_4}H_{u_3}H_{u_2}H_{u_1}$.

The 7×7 example motivates the general sequence of $n - 1$ Householder reflections:

$$
\begin{bmatrix} I^{(i-1)\times(i-1)} & & \\ & \overline{H}_{u_i}^{3\times3} & \\ & & I^{(n-i-2)\times(n-i-2)} \end{bmatrix}, \quad 1 \le i \le n - 2
$$

$$
\begin{bmatrix} I^{(n-2)\times(n-2)} & \\ & \overline{H}_{u_{n-1}}^{2\times2} \end{bmatrix}, \quad i = n - 1,
$$

which give rise to

$$
QHQ^{\mathrm{T}} = \overline{H},
$$

where $Q = H_{u_{n-1}}H_{u_{n-2}} \cdots H_{u_2}H_{u_1}$ and $\overline{H}$ is an upper Hessenberg matrix. Now,

$$
Qe_1 = \left(H_{u_{n-1}} \cdots H_{u_3}H_{u_2} \right) H_{u_1}e_1.
$$

The product $\left(H_{u_{n-1}} \ldots H_{u_3} H_{u_2}\right)$ does not affect column 1 of $H_{u_1} e_1$, so

$$Q e_1 = H_{u_1} e_1.$$

In forming the QR decomposition, $S = ZR$, using Householder reflections, the first column of Z is $H_{u_1} e_1$. By the implicit Q theorem, Q^T is a matrix such that

$$H_2 = \left(Q^T\right)^T H \left(Q\right)^T,$$

as desired.

The algorithm requires $O(n)$ flops to apply each reflector, and $n-1$ reflectors are required, so the cost of an iteration is $O\left(n^2\right)$ flops.

Example 18.15. Let

$$H = \begin{bmatrix} 5 & -1 & 1 \\ 6 & 1 & 2 \\ 0 & 3 & 4 \end{bmatrix}.$$

Form the 2×2 matrix $\begin{bmatrix} 1 & 2 \\ 3 & 4 \end{bmatrix}$, $h_{n-1,n-1} + h_{nn} = 5$, $h_{n-1,n-1} h_{nn} - h_{n,n-1} h_{n-1,n} = -2$, so

$$S = H^2 - 5H - 2I = \begin{bmatrix} -8 & 2 & 2 \\ 6 & -6 & 6 \\ 18 & 0 & 0 \end{bmatrix}.$$

The QR decomposition of S gives

$$Q = \begin{bmatrix} -0.3885 & 0.1757 & 0.9045 \\ 0.2914 & -0.9078 & 0.3015 \\ 0.8742 & 0.3807 & 0.3015 \end{bmatrix},$$

and

$$H_2 = Q^T H Q = \begin{bmatrix} 4.2642 & -1.6270 & 1.9328 \\ 1.6896 & -0.9005 & -4.1500 \\ 0 & 1.8719 & 6.6364 \end{bmatrix}.$$

Now we will perform the double shift using the implicit algorithm.

$$v_1 = \begin{bmatrix} -8 \\ 6 \\ 18 \end{bmatrix} \qquad H_{u_1} = \begin{bmatrix} -0.3885 & 0.2914 & 0.8742 \\ .2914 & 0.9389 & -0.1834 \\ 0.8742 & -0.1834 & 0.4497 \end{bmatrix}$$

$$H_{u_1} H = \begin{bmatrix} -0.1943 & 3.3024 & 3.6909 \\ 7.0900 & 0.0971 & 1.4353 \\ 3.2701 & 0.2914 & 2.3059 \end{bmatrix} \qquad H_{u_1} H H_{u_1} = \begin{bmatrix} 4.2642 & 2.3668 & 0.8840 \\ -1.4716 & 1.8938 & 6.8254 \\ 0.8302 & 0.8034 & 3.8420 \end{bmatrix}$$

$$v_2 = \begin{bmatrix} -1.4716 \\ 0.8302 \end{bmatrix} \qquad H_{u_2} = \begin{bmatrix} 1 & 0 & 0 \\ 0 & -0.8710 & 0.4913 \\ 0 & 0.4913 & 0.8710 \end{bmatrix}$$

$$H_{u_2} H_{u_1} H H_{u_1} = \begin{bmatrix} 4.2642 & 2.3668 & 0.8840 \\ 1.6896 & -1.2547 & -4.0570 \\ 0 & 1.6303 & 6.6998 \end{bmatrix} \qquad H_{u_2} H_{u_1} H H_{u_1} H_{u_2} = \begin{bmatrix} 4.2642 & -1.6270 & 1.9328 \\ 1.6896 & -0.9005 & -4.1500 \\ 0 & 1.8719 & 6.6364 \end{bmatrix} = H_2$$

Observe that the product $H_{u_2} H_{u_1} H H_{u_1} H_{u_2}$ removed the bulge at $(3, 1)$, and that the implicit computation produced the same value for H_2. Also, compute the orthogonal matrix

$$Q^T = H_{u_1} H_{u_2} = \begin{bmatrix} -0.3885 & 0.1757 & 0.9045 \\ 0.2914 & -0.9078 & 0.3015 \\ 0.8742 & 0.3807 & 0.3015 \end{bmatrix},$$

which is the same matrix obtained by the QR decomposition of S.   ∎

Building an $n \times n$ Householder reflection like we did in Example 18.15 is not efficient. We should take advantage of the identity and zero block matrices in H_u and do each computation $H_u H H_u$ using submatrix operations.

$$
H_{u_i} H = \begin{bmatrix} I^{(i-1)\times(i-1)} & 0^{(i-1)\times 3} & 0^{(i-1)\times(n-i-2)} \\ 0^{3\times(i-1)} & \overline{H}_{u_i}^{3\times 3} & 0^{3\times(n-i-2)} \\ 0^{(n-i-2)\times(i-1)} & 0^{(n-i-2)\times 3} & I^{(n-i-2)\times(n-i-2)} \end{bmatrix} \begin{bmatrix} H_1^{(i-1)\times(i-1)} & H_2^{(i-1)\times 3} & H_3^{(i-1)\times(n-i-2)} \\ H_4^{3\times(i-1)} & H_5^{3\times 3} & H_6^{3\times(n-i-2)} \\ H_7^{(n-i-2)\times(i-1)} & H_8^{(n-i-2)\times 3} & H_9^{(n-i-2)\times(n-i-2)} \end{bmatrix}
$$

$$
= \begin{bmatrix} H_1^{(i-1)\times(i-1)} & H_2^{(i-1)\times 3} & H_3^{(i-1)\times(n-i-2)} \\ \overline{H}_{u_i}^{3\times 3} H_4^{3\times(i-1)} & \overline{H}_{u_i}^{3\times 3} H_5^{3\times 3} & \overline{H}_{u_i}^{3\times 3} H_6^{3\times(n-i-2)} \\ H_7^{(n-i-2)\times(i-1)} & H_8^{(n-i-2)\times 3} & H_9^{(n-i-2)\times(n-i-2)} \end{bmatrix}
$$

The only computation that need be done involves rows $i : i+2$, and columns $1 : n$. The same type of analysis shows that the product $(H_{u_i} H) H_{u_i}$ is done using rows $1 : n$ and columns $i : i+2$. The final preproduct product using a 2×2 Householder reflection involves rows $n-1 : n$ and columns $1 : n$, followed by the postproduct using rows $1 : n$ and columns $n-1 : n$.

The following algorithm implements the implicit double-shift QR algorithm.

Algorithm 18.5 Implicit Double-Shift QR

```
function IMPDSQR(H)
   % H is an n × n upper Hessenberg matrix.
   % Apply one iteration of the double shift implicit QR algorithm,
   % and return upper Hessenberg matrix H₂ and orthogonal
   % matrix Q such that H₂ = QHQᵀ.
   Q = I

   % trace of lower right-hand 2 × 2 matrix.
   trce = hₙ₋₁,ₙ₋₁ + hₙₙ
   % determinant of lower right-hand 2 × 2 matrix.
   determ = hₙ₋₁,ₙ₋₁ hₙₙ − hₙ,ₙ₋₁ hₙ₋₁,ₙ
   % first nonzero entries of column 1 of
   x = h²₁₁ + h₁₂h₂₁ − h₁₁ × trce + determ
   y = h₁₁h₂₁ + h₂₁(h₂₂ − trce)
   z = h₂₁h₃₂

   for i = 0:n-3 do
      [u β] = houseparms([x y z]ᵀ)
      lengu = length(u)
      H̄ᵤ = I^(lengu×lengu) − βuuᵀ
      H(i+1 : i+3, 1 : n) = H̄ᵤ H(i+1 : i+3, 1 : n)
      H(1 : n, i+1 : i+3) = H(1 : n, i+1 : i+3) H̄ᵤ
      Q(i+1 : i+3, 1 : n) = H̄ᵤ Q(i+1 : i+3, 1 : n)
      x = hᵢ₊₂,ᵢ₊₁
      y = hᵢ₊₃,ᵢ₊₁
      if i < n-3 then
         z = hᵢ₊₄,ᵢ₊₁
      end if
   end for

   [u β] = houseparms([x y]ᵀ)
   lengu = length(u)
   H̄ᵤ = I^(lengu×lengu) − βuuᵀ
   H(n−1 : n, 1 : n) = H̄ᵤ H(n−1 : n, 1 : n)
   H(1 : n, n−1 : n) = H(1 : n, n−1 : n) H̄ᵤ
   H₂ = H
   return [Q, H₂]
end function
```

Now, how do we use *impdsqr* to compute all the eigenvalues, real and complex conjugates, of a nonsymmetric matrix? The answer is to use deflation and apply the function *impdsqr* to the current $k \times k$ deflated matrix. The matrix we build is precisely the quasitriangular matrix of the real Schur form presented in Section 18.7.

We need to know when to terminate the iteration and deflate again. To do this we must have a means of detecting what type of blocks are on the diagonal. If the convergence is indicated by a value tol there are two criteria we must consider. See Problem 18.41 for experiments that exhibit the criteria. If $h_{k-1,k-2} < $ tol, there are two real eigenvalues or a complex conjugate pair in the 2×2 submatrix $T(k-1:k, k-1:k)$.

If $h_{k,k-1} < $ tol, there is a real eigenvalue at $T(k, k)$. This is similar to the criteria we used to determine convergence of the shifted Hessenberg iteration in Section 18.6.1.

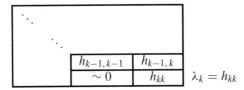

Remark 18.7.   In Ref. [1], it is stated that on the average only two implicit QR iterations per eigenvalue are needed for convergence for most matrices. However, it is possible for convergence to fail (Problem 18.27), and when this happens, a special shift every 10 shifts is applied in production versions of the algorithm, such as the eig function in MATLAB. The iterations may converge slowly (see Problem 18.27), but this is rare. If the iteration appears to be converging slowly, an iteration with random shifts can be used to keep convergence on track.

A study of the mechanisms involved in the Francis algorithm can be found in Ref. [23, Chapter 6]. The analysis involves integrating a study of the power method, similarity transformations, upper Hessenberg form, and Krylov subspaces. Krylov subspaces are discussed in Chapter 21 when we develop iterative methods for large, sparse matrix problems.

Remark 18.8.   The algorithm requires $10n^3$ flops if eigenvalues only are required. It is possible to calculate the corresponding eigenvectors by accumulating transformations (see Ref. [23, pp. 387-389]). In this case, $27n^3$ flops are required.

18.9   COMPUTING EIGENVECTORS

We have discussed ways to compute the eigenvalues of a real matrix, but have not discussed a general method of computing the corresponding eigenvectors. To find an eigenvector corresponding to a given eigenvalue, we use the *shifted inverse iteration*, a variation on the inverse power method for computing the smallest eigenvalue of a matrix. First, we need a lemma that provides a tool needed to develop the inverse iteration algorithm.

Lemma 18.3.   *If* (λ_i, v_i) *are the eigenvalue/eigenvector pairs of A,* $1 \leq i \leq n$, *then* $(A - \sigma I)^{-1}$ *has eigenvalue/eigenvector pairs* $\left(\frac{1}{\lambda_i - \sigma}, v_i \right)$.

Proof.   If A is an $n \times n$ nonsingular matrix, the eigenvalues of A^{-1} are the reciprocals of those for A, and the eigenvectors remain the same. Thus, the result follows immediately from Lemma 18.1.   □

If (λ, v) is an eigenvalue/eigenvector pair of A, and σ is an approximate eigenvalue, then $(\lambda - \sigma)^{-1}$ is likely to be much larger than $\left(\hat{\lambda} - \sigma \right)^{-1}$ for any eigenvalue $\hat{\lambda} \neq \lambda$. Thus, $(\lambda - \sigma)^{-1}$ is the largest eigenvalue of $(A - \sigma I)^{-1}$. Apply the power method to $(A - \sigma I)^{-1}$ to approximate eigenvector v. This is termed the inverse iteration for computing an eigenvector corresponding to an approximate eigenvalue σ.

Just as in Algorithm 18.2, we compute x_{i+1} by solving the system $(A - \sigma I) x_{i+1} = x_i$; however, there appears to be a problem. If σ is very close to an eigenvalue λ, and $v \neq 0$ is close to an eigenvector corresponding to λ, then $(A - \sigma I) v$ is close to zero, and $(A - \sigma I)$ is close to being singular. As a result, there may be serious error in the computation of x_{i+1}. As it turns out, the near singularity of $A - \sigma I$ is a good thing. The error at each iteration causes the approximate eigenvector to move closer and closer to the direction of the actual eigenvector. See Ref. [23, pp. 324-325].

Example 18.16. Let A be the matrix of Example 18.7 that has an eigenvalue 2.2518. Choose $\sigma = 2.2000$, $x_0 = \begin{bmatrix} 1 & 1 & 1 \end{bmatrix}^{\mathrm{T}}$, and execute three shifted inverse iterations. The norm of the difference between the computed and actual result is approximately 0.0018431.

```
>> sigma = 2.2000;
>> x0 = ones(3,1);
>> [L U P] = lu(A - sigma*eye(3));
>> v = x0;
>> for i = 1:3
       v = lusolve(L,U,P,v);
       v = v/norm(v);
   end
>> v'

ans =
      0.25979       0.87965      -0.39841
```

Now that we have intuitively explained the algorithm and given an example, a proof that it works is in order.

Theorem 18.7. *If the eigenvalues of A satisfy Equation 18.14, the sequence $\{x_k\}$ in the shifted inverse iteration converges to an approximate eigenvector corresponding to the approximate eigenvalue σ.*

Proof. The eigenvalues of $(A - \sigma I)^{-1}$ are $(\lambda_1 - \sigma)^{-1}$, $(\lambda_2 - \sigma)^{-1}$, ..., $(\lambda_n - \sigma)^{-1}$, and the eigenvectors are the same as A. Assume our approximate eigenvalue σ approximates λ_1 and that λ_1 corresponds to eigenvector v_1, Just as in the proof of Theorem 18.2, we have

$$\left((A - \sigma I)^{-1} \right)^k x_0 = \frac{c_1}{(\lambda_1 - \sigma)^k} v_1 + \frac{c_2}{(\lambda_2 - \sigma)^k} v_2 + \ldots + \frac{c_n}{(\lambda_n - \sigma)^k} v_n$$

$$= \frac{1}{(\lambda_1 - \sigma)^k} \left(c_1 v_1 + c_2 \left(\frac{\lambda_1 - \sigma}{\lambda_2 - \sigma} \right)^k v_2 + c_3 \left(\frac{\lambda_1 - \sigma}{\lambda_3 - \sigma} \right)^k v_3 + \ldots + c_n \left(\frac{\lambda_1 - \sigma}{\lambda_n - \sigma} \right)^k v_n \right).$$

Since σ approximates λ_1, $\frac{1}{|\lambda_1 - \sigma|} > \frac{1}{|\lambda_i - \sigma|}$, $2 \le i \le n$, it follows that $\left| \frac{\lambda_1 - \sigma}{\lambda_i - \sigma} \right| < 1$. The terms $\left(\frac{\lambda_1 - \sigma}{\lambda_i - \sigma} \right)^k$, $2 \le i \le n$ become small as k increases. If we normalize the sequence as in Equation 18.14, $\left((A - \sigma I)^{-1} \right)^k x_0$ converges to a multiple of v_1. $\square$

A function `inverseiter` that implements inverse iteration is an easy modification of `smalleig` and is left to the exercises.

Remark 18.9.   Given an accurate approximation to an eigenvalue, the inverse iteration converges very quickly.

18.9.1   Hessenberg Inverse Iteration

If we have an isolated approximation to an eigenvalue σ, the *shifted inverse iteration* can be used to compute an approximate eigenvector. However, if we use the Francis iteration to compute all the eigenvalues of an upper Hessenberg matrix H, we should take advantage of the upper Hessenberg structure of the matrix to find the corresponding eigenvectors. H has the same eigenvalues as A but not the same eigenvectors. However, we can use the orthogonal matrix P in the transformation to upper Hessenberg form to compute an eigenvector of A.

Let u be an eigenvector of $H = P^{\mathrm{T}}AP$ corresponding to eigenvalue λ of A. Then $Hu = \lambda u$, so $P^{\mathrm{T}}APu = \lambda u$ and $A(Pu) = \lambda(Pu)$. Thus, Pu is an eigenvector of A corresponding to eigenvalue λ.

Use shifted inverse iteration with matrix H to obtain eigenvector u, and then $v = Pu$ is an eigenvector of A. Since the inverse iteration requires repeatedly solving a linear system, we use the LU decomposition first. The normal LU decomposition with partial pivoting requires $O(n^3)$ flops, but we can take advantage of the upper Hessenberg form of H to perform the decomposition more efficiently. Begin by comparing $|h_{11}|$ and $|h_{21}|$ and exchange rows 1 and 2, if necessary, to place the largest element in magnitude at h_{11}. In general, compare $|h_{ii}|$ and $|h_{i+1,i}|$ and swap rows if necessary. During the process, maintain the lower triangular matrix

$$L = \begin{bmatrix} 1 & & & & \\ l_{21} & 1 & & & \\ & l_{32} & \ddots & & \\ & & \ddots & 1 & \\ & & & l_{n,n-1} & 1 \end{bmatrix}$$

and the permutation matrix P. The algorithm requires $(n-1)$ divisions $\left(\frac{h_{i+1,i}}{h_{ii}}\right)$ and $2\left[(n-1) + (n-2) + \cdots + 1\right] = n(n-1)$ multiplications and subtractions, for a total of $n^2 - 1$ flops. Since the algorithm is very similar to ludecomp (Algorithm 11.2), we will not provide a formal specification. The MATLAB function luhess in the software distribution implements the algorithm.

Example 18.17. The matrix EX18_17 is a 500×500 upper Hessenberg matrix. Time its LU decomposition using ludecomp developed in Chapter 11, and then time its decomposition using luhess. The function ludecomp performs general LU decomposition with pivoting, so it does not take advantage of the upper Hessenberg structure. The execution time of luhess is approximately 13 times faster than that of ludecomp.

```
>> tic;[L1, U1, P1] = ludecomp(EX18_17);toc
Elapsed time is 0.421030 seconds.
>> tic;[L2, U2, P2] = luhess(EX18_17);toc;
Elapsed time is 0.032848 seconds.
```
∎

The algorithm eigvechess uses luhess with inverse iteration to compute an eigenvector of an upper Hessenberg matrix with known eigenvalue σ.

Algorithm 18.6 Inverse Iteration to Find Eigenvector of an Upper Hessenberg Matrix

```
function EIGVECHESS(H,σ,x₀,tol,maxiter)
  % Computes an eigenvector corresponding to the approximate
  % eigenvalue sigma of the upper Hessenberg matrix H
  % using the inverse iteration.
  % [x iter]=eigvechess(H,sigma,x0,tol,maxiter)
  % sigma is the approximate eigenvalue,
  % x₀ is the initial approximation to the eigenvector,
  % tol is the desired error tolerance, and maxiter is
  % the maximum number of iterations allowed.
  % iter=-1 if the method did not converge.

  [L U P]=luhess(A - σ I)
  for i=1:maxiter do
    x̄ᵢ =lusolve(L,U,P,xᵢ₋₁)
    xᵢ = x̄ᵢ/‖x̄ᵢ‖₂
    if ‖(A - σ I) xᵢ‖₂ < tol then
      iter=i
      v = xᵢ
      return [v,i]
    end if
  end for
  iter=-1
  v = xᵢ
  return [v,-1]
end function
```

NLALIB: The function eigvechess implements Algorithm 18.6. If *A* has a multiple eigenvalue σ, Hessenberg inverse iteration can result in vector entries NaN or Inf. The next section discusses a method that attempts to solve this problem.

Remark 18.10.   Although it involves complex arithmetic, eigvechess will compute a complex eigenvector when given a complex eigenvalue σ. There is a way to perform inverse iteration with complex σ using real arithmetic (see Ref. [9, p. 630]).

18.10 COMPUTING BOTH EIGENVALUES AND THEIR CORRESPONDING EIGENVECTORS

We have developed the implicit double-shift *QR* iteration for computing eigenvalues and the Hessenberg inverse iteration for computing eigenvectors. It is now time to develop a function, eigb, that computes both. The function applies the Francis iteration of degree two to compute the eigenvalues, followed by the use of the shifted inverse Hessenberg iteration to determine an eigenvector corresponding to each eigenvalue. Inverse iteration is economical because we do not have to accumulate transformations during the Francis iteration. Inverse iteration deals with $H - \sigma I$ using $O\left(n^2\right)$ flops, and normally one or two iterations will produce a suitable eigenvector [2, pp. 394-395]. If *A* has a multiple eigenvalue, σ, Hessenberg inverse iteration can result in vector entries NaN or Inf. The MATLAB implementation checks for this, perturbs σ slightly, and executes eigvechess again with the perturbed eigenvalue. It is hoped this will produce another eigenvector corresponding to σ. All the pieces are in place, so we will not state the formal algorithm that is implemented in the software distribution. Note that it uses the MATLAB features of variable input arguments and variable output to make its use more flexible.

The function can be called in the following ways:

- [V D] = eigb(A,tol);
  - Returns a diagonal matrix D of eigenvalues and a matrix V whose columns are the corresponding normalized eigenvectors so that $AV = VD$. tol is the error tolerance for the computations.
    * If tol is not present, the default is 1.0×10^{-8}.
- E = eigb(A,tol);
  - Assigns E a column vector containing the eigenvalues. The default for tol is the same as for the previous calling sequence.
- eigb(A,tol)
  - Returns a vector of eigenvalues. The default value of tol is as before.

Example 18.18.   Generate a 75×75 random real matrix that most certainly will have imaginary eigenvalues and use eigb to compute its eigenvalues and eigenvectors. A function, checkeigb, in the software distribution computes the minimum and maximum values of $\|Av_i - \lambda_i v_i\|_2$, $1 \le i \le n$.

```
>> A = randn(75,75);
[V D] = eigb(A,1.0e-12);
[min max] = checkeigb(A,V,D)

min =
    7.4402e-15

max =
    4.3307e-12

>> E = diag(D);
E(1:6)

ans =
      0.25436 +              0i
      0.62739 +              0i
      1.2352 +        1.8955i
      1.2352 -        1.8955i
      2.3356 +        0.90619i
      2.3356 -        0.90619i

>> E(70:75)

ans =
      6.8625 -        1.1636i
```

$$
\begin{array}{rr}
2.7378 + & 7.5951i \\
2.7378 - & 7.5951i \\
7.2777 + & 5.6828i \\
7.2777 - & 5.6828i \\
9.4579 + & 0i
\end{array}
$$

∎

18.11 SENSITIVITY OF EIGENVALUES TO PERTURBATIONS

The eigenvalues of matrix A are the roots of the characteristic equation $p\left(\lambda\right) = \det\left(A - \lambda I\right)$. If A is the 2×2 matrix

$$
A = \begin{bmatrix} a_{11} & a_{12} \\ a_{21} & a_{22} \end{bmatrix},
$$

the eigenvalues are roots of the polynomial

$$
p\left(\lambda\right) = \lambda^2 - \left(a_{11} + a_{22}\right)\lambda + \left(a_{11}a_{22} - a_{12}a_{21}\right).
$$

We can view p as a continuous function of the four variables a_{11}, a_{12}, a_{21}, a_{22}, so the roots of the characteristic equation depend continuously on the matrix coefficients. See Refs. [1, 9] for further discussion and proofs. Under the right conditions this continuity means that if the perturbations of A and δA are small, we can control the perturbations in the eigenvalues. The *Bauer-Fike theorem* [91] is a well-known result that deals with eigenvalue perturbations for diagonalizable matrices. For a proof, see Ref. [23, pp. 472-473].

Theorem 18.8.   *Assume that $A \in \mathbb{R}^n$ is diagonalizable, so that there exists a nonsingular matrix X such that $X^{-1}AX = D$, where $D = \mathrm{diag}\left(\lambda_1, \lambda_2, \ldots, \lambda_n\right)$. If δA is a matrix of perturbations, then for any eigenvalue $\hat{\lambda}_i$ of $A + \delta A$, there is an eigenvalue λ_i of A such that*

$$
\left|\hat{\lambda}_i - \lambda_i\right| \le \kappa\left(X\right)\left\|\delta A\right\|_2,
$$

where X is a subordinate matrix norm.

Theorem 18.8 has consequences for the conditioning of the eigenproblem for A. If $\left\|X\right\|_2 \left\|X^{-1}\right\|_2 = \kappa\left(X\right)$ is large, then a computed eigenvalue $\hat{\lambda}_i$ (an eigenvalue for $A + \delta A$) can be very different from the actual eigenvalue λ_i. The more ill-conditioned X is, the more ill-conditioned the eigenproblem for A will be.

Example 18.19.   Let $A = \begin{bmatrix} 5.0000 & 0 & 0 \\ 2.0000 & 1.0000 & -7.0000 \\ 3.0000 & 0 & 0.9900 \end{bmatrix}$. The eigenvalues of A are

$$
\lambda_1 = 5.000, \quad \lambda_2 = 1.0000, \quad \lambda_3 = 0.9900
$$

so A is diagonalizable. The matrix $X = \begin{bmatrix} 0 & 0 & 0.67198 \\ 1 & 1 & -0.54379 \\ 0 & 0.0014286 & 0.50273 \end{bmatrix}$. The condition number of X is 1922.6, so there could

be a conditioning problem with the eigenvalues of A. If we let $\delta A = \begin{bmatrix} 0 & 0 & 0 \\ 0 & 0 & 0 \\ 0 & 1.0 \times 10^{-5} & 0 \end{bmatrix}$, the eigenvalues of $A + \delta A$ are

$$
\lambda_1 = 5.0000, \quad \lambda_2 = 0.9950 + 0.0067082i, \quad \lambda_3 = 0.9950 - 0.0067082i,
$$

so the high condition number produces instability. But why were only the eigenvalues λ_2 and λ_3 affected? That question is answered by considering the conditioning of individual eigenvalues. ∎

Theorem 18.8 gives one bound for all the eigenvalues, and Example 18.19 demonstrates that some eigenvalues are well conditioned and some are not, so it makes sense to develop a criterion for the conditioning of an individual eigenvalue. To do so requires the concept of a left eigenvector.

Definition 18.4.   The vector y is a *left eigenvector* of A if

$$
y^{\mathrm{T}}A = \lambda y^{\mathrm{T}}.
$$

Remark 18.11.   Recall that the determinant of a matrix and its transpose are equal. As a result, the characteristic equations $\det(A - \lambda I) = 0$ and $\det(A^T - \lambda I) = 0$ have the same roots, so A and A^T have the same eigenvalues. Since $y^T A = \lambda y^T$, it follows that $A^T y = \lambda y$, and a left eigenvector of A is an eigenvector of A^T corresponding to the eigenvalue λ of A.

Assume that all the eigenvalues are distinct, and that during computation an error is made in A causing perturbation δA. The eigenvalue we actually compute, $\hat{\lambda} = \lambda + \delta\lambda$, is an eigenvalue of $A + \delta A$. Let f be the function mapping δA to $\hat{\lambda}$; in other words, as δA varies, $f(\delta A)$ is the resulting eigenvalue $\hat{\lambda}$. Since the eigenvalues are simple, it can be shown that they depend continuously on the entries of the matrix. As a result, the function f is continuous and

$$\lim_{\delta A \to 0} f(\delta A) = \lambda.$$

This says that as a small error δA occurs, the eigenvalue of $A + \delta A$ will be close to λ. It also follows that if $x + \delta x$ is an eigenvector corresponding to $\lambda + \delta\lambda$, as δA becomes small, δx will have a small norm.

This reasoning leads to the following theorem.

Theorem 18.9.   *Assume λ is a simple eigenvalue of the $n \times n$ matrix A having associated right and left eigenvectors x and y, respectively, with $\|x\|_2 = 1$ and $\|y\|_2 = 1$. Let $\lambda + \delta\lambda$ be the corresponding eigenvalue of $A + \delta A$. Then,*

$$|\delta\lambda| \leq \frac{\|\delta A\|_2}{|y^T x|} + O\left(\|\delta A\|_2^2\right).$$

Proof. We have the two equations

$$Ax = \lambda x \tag{18.40}$$

$$(A + \delta A)(x + \delta x) = (\lambda + \delta\lambda)(x + \delta x) \tag{18.41}$$

Subtract Equation 18.40 from Equation 18.41 and obtain

$$A(\delta x) + (\delta A)x + (\delta A)(\delta x) = \lambda(\delta x) + (\delta\lambda)x + (\delta\lambda)(\delta x).$$

Since we know that if $\|\delta A\|_2$ is small $|\delta\lambda|$ and $\|\delta x\|_2$ will be small, we will ignore the terms $(\delta A)(\delta x)$ and $(\delta\lambda)(\delta x)$ involving products of small factors to arrive at

$$A(\delta x) + (\delta A)x \cong \lambda(\delta x) + (\delta\lambda)x \tag{18.42}$$

Multiply both sides of Equation 18.42 by y^T to obtain

$$y^T A(\delta x) + y^T(\delta A)x \cong \lambda y^T(\delta x) + y^T(\delta\lambda)x. \tag{18.43}$$

Since y is a left eigenvector of A, $y^T A = \lambda y^T$, and we have $y^T A(\delta x) = \lambda y^T(\delta x)$. The terms $\lambda y^T(\delta x)$ cancel out in Equation 18.43, leaving

$$y^T(\delta A)x \simeq (\delta\lambda)y^T x,$$

and so

$$\delta\lambda \simeq \frac{y^T(\delta A)x}{y^T x}.$$

It follows that:

$$|\delta\lambda| \leq \|y^T\|_2 \frac{\|\delta A\|_2}{|y^T x|}\|x\|_2 = (1)\frac{\|\delta A\|_2}{|y^T x|}(1) = \frac{\|\delta A\|_2}{|y^T x|}.$$

Throughout the proof, we have used the notation "$\simeq$" to account for ignoring the terms $(\delta A)(\delta x)$ and $(\delta\lambda)(\delta x)$. δy and δx depend on δA so

$$|\delta\lambda| \leq \frac{\|\delta A\|_2}{|y^T x|} + O\left(\|\delta A\|_2^2\right). \qquad \square$$

Theorem 18.7 indicates that the quantity $\frac{1}{|y^T x|}$ affects the accuracy of the computation of a particular eigenvalue λ, leading us to the definition of the condition number for λ.

Definition 18.5. If λ is a simple eigenvalue of matrix A, and x, y are normalized right and left eigenvectors of λ, respectively, then

$$\kappa\left(\lambda\right) = \frac{1}{\left|y^{T}x\right|}$$

is called the *condition number* of the eigenvalue λ.

Now let us determine how to compute $\kappa\left(\lambda\right)$. If A has n distinct eigenvectors, it can be diagonalized, so there is a matrix X such that $D = X^{-1}AX$, where $X = \begin{bmatrix} v_1 & v_2 & \ldots & v_n \end{bmatrix}$ and the v_i are eigenvectors of A corresponding to eigenvalues λ_i. $Av_i = \lambda_i v_i$, and so $x_i = \frac{v_i}{\|v_i\|_2}$ is a normalized right eigenvector of A. By Remark 18.11, if y_i is a normalized right eigenvector of A^{T} corresponding to eigenvalue λ_i, y_i^{T} is a left eigenvector of A. Now, $A = XDX^{-1}$, so $A^{T} = \left(X^{T}\right)^{-1}DX^{T}$ and $\left(\left(X^{T}\right)^{-1}\right)^{-1}A^{T}\left(X^{T}\right)^{-1} = D$. The proof of Theorem 18.1 shows that column i of $\left(X^{-1}\right)^{T}$ is an eigenvector of A^{T} corresponding to λ_i.

Algorithm 18.7 shows how to compute the condition number of every eigenvalue of matrix A, assuming the eigenvalues are distinct.

Algorithm 18.7 Compute the Condition Number of the Eigenvalues of a Matrix

---

```
function EIGCOND(A)
    % Compute the condition number of the distinct eigenvalues
    % of the n x n matrix A.
    % [c λ]=eigcond(A) returns a vectors c and λ such that cᵢ
    % is the condition number of eigenvalue λᵢ.
```

$\begin{bmatrix} X & D \end{bmatrix} = eig(A)$
$\lambda = diag(D)$
$invXT = X^{-1}$
for $i = 1:n$ do
$\quad x = \frac{X(:,i)}{\|X(:,i)\|_2}$
$\quad y = \frac{invXT(:,i)}{\|invXT(:,i)\|_2}$
$\quad c_i = \frac{1}{|y^{Y}x|}$
end for
return $\begin{bmatrix} c & lambda \end{bmatrix}$
end function

NLALIB: The function `eigcond` implements Algorithm 18.7. The MATLAB function `condeig` also computes eigenvalue condition numbers.

Example 18.20. In Example 18.19, we found that the eigenvalues $\lambda_2 = 1.0000$ and $\lambda_3 = 0.9900$ were sensitive to perturbations. The use of `eigcond` shows why. The condition number of the eigenvalue 5.0000 is 1.4881, and it will not be sensitive to perturbations.

```
>> [c lambda] = eigcond(A)

c =
   874.7007
   874.2160
     1.4881

lambda =
     1.0000
     0.9900
     5.0000
```

∎

Example 18.21. The 25×25 matrix EIGBTEST in the book software distribution has distinct integer eigenvalues and a condition number of 4.7250×10^8. The eigenvalues are sensitive to perturbations, as the output shows.

```
>> [V D] = eigb(EIGBTEST);
norm(EIGBTEST*V - V*D)

ans =
  4.0684e-09

>> eigcond(EIGBTEST)

ans =
        412.74
        452.16
          2120
        2389.6
         22744
         33208
         10269
        4115.1
        6047.9
        283.35
    4.2367e+05
     8.313e+06
    1.4719e+07
    1.0591e+07
    3.6274e+06
    2.2256e+05
        3557.7
    2.0129e+06
    2.0363e+06
     1.623e+05
    7.4567e+05
    2.8465e+06
    4.1147e+06
    6.4555e+06
    4.5237e+06

>> [min, max] = checkeigb(EIGBTEST,V,D)

min =
   9.244e-14

max =
   4.066e-09
```

■

Remark 18.12. If λ is a simple eigenvalue of A, then a large condition number implies that A is near a matrix with multiple eigenvalues. If λ is a repeated eigenvalue of a nonsymmetric matrix, the conditioning question is more complicated. The interested reader should consult Ref. [9].

18.11.1 Sensitivity of Eigenvectors

Perturbation analysis for eigenvectors is significantly more complex than that for eigenvalues. The perturbation of a particular eigenvector, v_i, corresponding to eigenvalue, λ_i, is determined by the condition number of the eigenvectors λ_k, $k \neq i$ and $|\lambda_i - \lambda_k|$, $k \neq i$. See Refs. [19, pp. 319-320] and [23, pp. 477-481]. To put it more simply, if the eigenvalues are well separated and well conditioned, then the eigenvectors will be well conditioned. However, if there is a multiple eigenvalue or there is an eigenvalue close to another eigenvalue, then there will be some ill-conditioned eigenvectors. Problem 18.22 deals with eigenvector sensitivity theoretically, and Problems 18.37 and 18.38 are numerical experiments dealing with the problem.

18.12   CHAPTER SUMMARY

Applications of the Eigenvalue Problem

Resonance is the tendency of a system to oscillate at a greater amplitude at some frequencies than at others. With little or no damping, if a periodic driving force matches a natural frequency of vibration, oscillations of large amplitude can occur. The are many types of resonance, including mechanical resonance, acoustic resonance, electromagnetic resonance, and nuclear magnetic resonance. The phenomenon is illustrated by a mass-spring system that is modeled by a system of second-order ordinary differential equations. The solution to the system with no driving force (homogeneous system) depends on eigenvalue and eigenvector computations, that give rise to the natural frequencies of vibration. After adding a driving force and solving the nonhomogeneous system, resonance is illustrated by choosing the frequency of the driving force to be close to a natural frequency.

The model of Leslie is a heavily used tool in population ecology. It models an age-structured population which predicts how distinct populations change over time. The heart of the model is the Leslie matrix, which is irreducible and has an eigenvector with strictly positive entries. Its characteristic function has exactly one real positive root, λ_1, the largest eigenvalue of the Leslie matrix in magnitude. All the other eigenvalues are negative or imaginary. By studying powers of the Leslie matrix, we find that at equilibrium the proportion of individuals belonging to each age class will remain constant, and the number of individuals will increase by λ_1 times each period. The eigenvector with all positive entries can be used to determine the percentage of females in each age class after an extended period of time.

The buckling of a elastic column is determined by solving a boundary value problem. The column will not buckle unless subjected to a critical load, in which case the deflection curve is of the form $k \sin\left(\frac{n\pi x}{L}\right)$, where L is the length of the column and k is a constant. The functions are termed eigenfunctions, and the corresponding eigenvalues are $\lambda_n = \frac{n^2\pi^2}{L^2}$. Associated with the eigenvalues are the critical loads $P_n = \frac{EI\pi^2 n^2}{L^2}$, the only forces that will cause buckling.

Computation of Selected Eigenvalues and Eigenvectors

There are problems for which only selected eigenvalues and associated eigenvectors are needed. If a real matrix has a simple eigenvalue of largest magnitude, the sequence $x_k = Ax_{k-1}$ converges to the eigenvector corresponding to the largest eigenvalue, where x_0 is a normalized initial approximation, and all subsequent x_k are normalized. This is known as the power method. After k iterations, the corresponding eigenvalue is approximately $\lambda_1 = v_k^T (Av_k)$.

Assume A has a simple eigenvalue of smallest magnitude. Since the eigenvalues of A^{-1} are the reciprocals of the eigenvalues of A, the smallest eigenvalue of A is the largest eigenvalue of A^{-1}, and we can compute it by applying the power method to A^{-1}. This is not done by using the iteration $x_k = A^{-1}x_{k-1}$ but by solving the system $Ax_k = x_{k-1}$ for each k.

The Basic *QR* Iteration

The discovery of the *QR* iteration is one of the great accomplishments in numerical linear algebra. Under the right conditions, the following sequence converges to an upper triangular matrix whose diagonal consists of the eigenvalues in decreasing order of magnitude:

$$A_{k-1} = Q_k R_k$$
$$A_k = R_k Q_k.$$

The execution of one *QR* iteration requires $O\left(n^3\right)$ flops, so k iterations requires $O\left(kn^3\right)$ flops. If $k = n$, this is an $O\left(n^4\right)$ algorithm, and this is not satisfactory. However, there are ways to greatly speed it up.

Transformation to Upper Hessenberg Form

Transformation to upper Hessenberg form is the initial step in most algorithms for the computation of eigenvalues. It is accomplished by using orthogonal similarity transformations of the form $A_k = H_{u_k} A_{k-1} H_{u_k}^T$, where H_{u_k} is a Householder reflection. The end result is an upper Hessenberg matrix $H = P^T AP$, where P is an orthogonal matrix comprised of products of Householder matrices. The eigenvalues of H are the same as those of A.

The Unshifted Hessenberg *QR* Iteration

The reduction of a matrix A to upper Hessenberg form requires approximately $\frac{10}{3}n^3$ flops for the computation of H. To build the orthogonal matrix P requires an additional $4n^3/3$ flops. The reduction of H to upper triangular form using the *QR* iteration with deflation requires only $O\left(n^2\right)$ flops. By an initial reduction to upper Hessenberg form followed by the Hessenberg *QR* iteration, we can compute eigenvalues with $O\left(n^3\right) + O\left(n^2\right)$ flops.

The Shifted Hessenberg *QR* Iteration

The *QR* iteration is more effective when it is applied to compute an eigenvalue isolated from other eigenvalues. Let H be an upper Hessenberg matrix. Choose a shift σ close to λ_k, and form $H_k - \sigma I$ that has eigenvalues $\lambda_i - \sigma$, $1 \leq i \leq n$. The eigenvalue $\lambda_k - \sigma$ is smaller than all the other eigenvalues, and the *QR* iteration applied to $H_k - \sigma I$ is very accurate. Compute the *QR* decomposition of the matrix $H_k - \sigma_1 I = Q_k R_k$ and let $H_{k+1} = R_k Q_k + \sigma_1 I$. Repeat this process until $\left|h_{k,k-1}\right|$ is sufficiently small. This technique significantly speeds up the computation of eigenvalues.

The Francis Algorithm

The Francis algorithm has for many years been the staple for eigenvalue computation. By using a double shift, it enables the computation of complex conjugate pairs of eigenvalues without using complex arithmetic. The algorithm is also known as the implicit *QR* iteration because it indirectly computes a single and a double shift in an upper Hessenberg matrix without actually computing the *QR* decomposition. This is done using orthogonal similarity transformations that introduce bulges, disturbing the upper Hessenberg form. By moving down the matrix, the transformations chase the bulge down and off the Hessenberg matrix. The algorithm makes the computation of all the eigenvalues of a matrix run in $O\left(n^3\right)$ flops.

Computing Eigenvectors

If λ_k is an accurate eigenvalue of matrix A, apply the inverse power iteration to the matrix $H_k - \lambda_k I$ to find a normalized eigenvector v_k, which is also an eigenvector of H_k. This means solving linear systems of the form $\overline{H} x_k = x_{k-1}$. Since $\overline{H}$ is upper Hessenberg, a simple modification of the *LU* decomposition enables very rapid computation of $P\overline{H} = LU$.

Computing Both Eigenvalues and Their Corresponding Eigenvectors

An eigenproblem solver, eigb, is easy to build using the algorithms developed in the book. First, reduce matrix A to upper Hessenberg form $H = P^{\mathrm{T}} A P$ and compute the eigenvalues by applying the Francis double-shift *QR* iteration to H. Now find corresponding eigenvectors of H using inverse iteration. For each eigenvector v of H, Pv is an eigenvector of A.

Sensitivity of Eigenvalues and Eigenvectors to Perturbations

Assume that A is diagonalizable so that $X^{-1}AX = D$. The Bauer-Fike theorem says that if δA is a matrix of perturbations, then for any eigenvalue $\hat{\lambda}_i$ of $A + \delta A$, there is an eigenvalue λ_i of A such that

$$\left|\hat{\lambda}_i - \lambda_i\right| \leq \kappa\left(X\right) \|\delta A\|_2$$

This is a global estimate, meaning it applies to all eigenvalues. If $\kappa\left(X\right)$ is large, it is possible that one or more eigenvalues are ill-conditioned.

The condition number of an individual eigenvalue λ is defined by $\kappa\left(\lambda\right) = \frac{1}{|y^{\mathrm{T}}x|}$, where x is a right and y is a left normalized eigenvector, respectively. If λ is a simple eigenvalue of A, then a large condition number implies that A is near a matrix with multiple eigenvalues.

If the eigenvalues are well separated and well conditioned, then the eigenvectors will be well conditioned. However, if there is a multiple eigenvalue or there is an eigenvalue close to another eigenvalue, then there will be some ill-conditioned eigenvectors.

18.13 PROBLEMS

18.1 Consider the coupled mass vibration problem consisting of springs having the same mass attached to each other and fixed outer walls by springs of spring constants k_1 and k_2 (Figure 18.11).

The displacements x_1 and x_2 satisfy the system of second-order differential equations

$$m\frac{d^2 x_1}{dt^2} + k_1 x_1 - k_2 (x_2 - x_1) = F_0 \sin \omega_0 t$$

$$m\frac{d^2 x_2}{dt^2} + k_1 x_2 + k_2 (x_2 - x_1) = 0$$

Leaving ω_0 variable, solve the system given $m = 1\,\text{kg}$, $k_1 = 1\,\text{N/m}$, $k_2 = 2\,\text{N/m}$, $F_0 = 2\,\text{N}$, and initial conditions $x_1(0) = 3$, $x_2(0) = 1$, $x_1'(0) = x_2'(0) = 0$. Demonstrate resonance by making ω_0 near one of the natural frequencies of the system and graphing $x_1(t)$, $x_2(t)$ for $0 \le x \le 5$.

18.2 Using paper and pencil, execute the first three iterations of the power method for computing the largest eigenvalue of $A = \begin{bmatrix} 1 & 3 \\ 1 & 1 \end{bmatrix}$.

18.3 Using paper and pencil, execute the first three iterations of the inverse power method for computing the smallest eigenvalue of the matrix in Problem 18.2.

18.4 Explain the slow rate of convergence of the power method for the matrices

a. $A = \begin{bmatrix} 1 & 0 & 0 \\ 1 & 10 & 0 \\ 1 & 1 & 9.8 \end{bmatrix}$.

b. $\begin{bmatrix} 2.9910 & 1.2104 & 0.7912 \\ -1.4082 & 0.5913 & -2.8296 \\ -0.1996 & -0.3475 & 2.0678 \end{bmatrix}$.

18.5 Explain what happens when the inverse power method is applied to the matrix

$$A = \begin{bmatrix} 0.7674 & 0.2136 & 3.3288 \\ -0.7804 & -0.9519 & -0.4240 \\ 0.5086 & -0.1812 & 2.1899 \end{bmatrix}.$$

18.6 a. Prove that the left eigenvectors of a symmetric matrix are eigenvectors.

b. Let λ and μ be two distinct eigenvalues of A. Show that all left eigenvectors associated with λ are orthogonal to all right eigenvectors associated with μ.

18.7 Let $A = \begin{bmatrix} 3.8865 & 0.29072 & 1.6121 \\ -1.6988 & 2.6922 & -4.3539 \\ -0.50715 & -0.04681 & 1.4712 \end{bmatrix}$. Use MATLAB to compute its eigenvalues. Do you think using the shift strategy will enable more rapid computation of the eigenvalues? Explain.

18.8 If an upper Hessenberg matrix has a zero on the subdiagonal, the problem must be split into two eigenvalue problems. Assume a split has occurred and the matrix has the form

$$\begin{bmatrix} A_{11} & A_{12} \\ 0 & A_{22} \end{bmatrix},$$

where A_{11} is $i \times i$, A_{22} is $(n-i) \times (n-i)$, and both are upper Hessenberg. Show that the eigenvalues of H are those of A_{11} and A_{22}. For the sake of simplicity, assume that A_{11} and A_{22} have no common eigenvalues and that if λ is an eigenvalue of A_{22} then the matrix $A_{11} - I$ is nonsingular.

18.9 Let A be an $n \times n$ matrix.

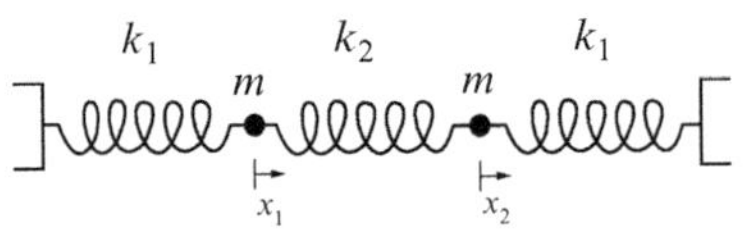

FIGURE 18.11 Springs problem.

a. If λ is an eigenvalue of A and $v = \begin{bmatrix} v_1 & v_2 & \cdots & v_n \end{bmatrix}^T$ is an associated eigenvector, use $Av = \lambda v$ to show that

$$(\lambda - a_{ii}) v_i = \sum_{\substack{j=1 \\ j \neq i}}^{n} a_{ij} v_j, \quad i = 1, 2, \ldots, n.$$

b. Let v_k be the largest component of v in absolute value. Using part (a), show that

$$|\lambda - a_{kk}| \leq \sum_{\substack{j=1 \\ j \neq k}}^{n} |a_{kj}|.$$

c. Using part (b), show that each eigenvalue of A satisfies at least one of the inequalities

$$|\lambda - a_{ii}| \leq r_i, \quad 1 \leq i \leq n,$$

where

$$r_i = \sum_{\substack{j=1 \\ j \neq i}}^{n} |a_{ij}|.$$

d. Argue this means that all the eigenvalues of A lie in the union of the disks $\{z \mid |z - a_{ii}| \leq r_i\}, i = 1, 2, \ldots, n$ in the complex plane. This result is one of *Gergorin's disk theorems* that are used in perturbation theory for eigenvalues.

e. Let

$$A = \begin{bmatrix} 9 & 5 & 4 \\ 10 & 3 & 1 \\ 5 & 9 & 8 \end{bmatrix}.$$

Draw the three Gergorin disks and verify that the three eigenvalues lie in their union.

18.10 A square matrix is strictly row diagonally dominant if $|a_{ii}| > \sum_{\substack{j=1 \\ j \neq i}}^{n} a_{ij}$. Use Gergorin's disk theorem developed in Problem 18.9 to prove that a strictly row diagonally dominant matrix is nonsingular.

18.11 In this problem, you are to show that if Equation 18.14 holds, as the QR iterations progress, the elements of $A_k = \overline{Q}_{k-1}^T A \overline{Q}_{k-1}$ at index $(1, 1)$ converge to the largest eigenvalue λ_1 by the power method.

a. Let $Q^{(k)} = Q_1 Q_2 \ldots Q_{k-1} Q_k$, and show that $AQ^{(k-1)} = Q^{(k)} R_k$. Hint:

$$(Q_1 Q_2 \ldots Q_{k-1} Q_k) R_k = Q_1 Q_2 \ldots Q_{k-1} (Q_k R_k)$$
$$= Q_1 Q_2 \ldots Q_{k-1} A_{k-1}$$
$$= Q_1 Q_2 \ldots Q_{k-1} (R_{k-1} Q_{k-1}) = \cdots$$

b. $Q^{(k-1)}$ has n orthonormal columns, so $Q^{(k-1)} = \begin{bmatrix} q_1^{(k-1)} & q_2^{(k-1)} & \cdots & q_{n-1}^{(k-1)} & q_n^{(k-1)} \end{bmatrix}$. Let $r_{11}^{(k)}$ be entry $R_k(1, 1)$ and equate the first columns on both sides of the result from part (a) to develop the relation

$$Aq_1^{(k-1)} = r_{11}^{(k)} q_1^{(k)}.$$

c. Argue that the result of part (b) shows that $q_1^{(k)}$ converges by the power method to an eigenvector of A with associated eigenvalue $r_{11}^{(k)}$.

18.12 Show that the QR decomposition of an upper Hessenberg matrix using Givens rotations costs $O\left(n^2\right)$ flops.

18.13 Given $A = \begin{bmatrix} 1 & 1 \\ 0 & 1+\epsilon \end{bmatrix}$, find the eigenvector matrix S such that $S^{-1} A S$ is diagonal. Use the Bauer-Fike theorem to show that the eigenvalues are ill-conditioned.

18.14 Show that an orthogonal similarity transformation maintains the condition number of an eigenvalue.

A real matrix is said to be *normal* if $A^T A = A A^T$. Problems 18.15-18.17 deal with properties of a normal matrix.

18.15 There are matrices that can be diagonalized, but are not normal. Let $A = \begin{bmatrix} -1 & 3 \\ 0 & 2 \end{bmatrix}$ and $P = \begin{bmatrix} 1 & \frac{1}{\sqrt{2}} \\ 0 & \frac{1}{\sqrt{2}} \end{bmatrix}$.

    a. P has normalized columns. Is it orthogonal?
    b. Show that A can be diagonalized using the matrix P.
    c. Show that A is not normal.

18.16 Prove that
    a. a symmetric matrix is normal.
    b. an orthogonal matrix is normal.
    c. if A is *skew symmetric* $\left(A^T = -A\right)$, then A is normal.
    d. If there is a polynomial p such that $A^T = p(A)$, then A is normal.

18.17 Using the Bauer-Fike theorem, show that if A is normal, and λ is an eigenvalue of $A + \delta A$, then

$$\min_{1 \le i \le n} |\lambda_i - \lambda| \le \|\delta A\|_2 .$$

    What does this say about the conditioning of the eigenvalues?

18.18 Show that Equation 18.37 is correct.

18.19 Show that Equation 18.39 is correct.

18.20 What are the condition numbers for the eigenvalues of a symmetric matrix?

18.21 Using the matrix of Example 18.13, show that the Schur's triangularization is not unique.

18.22 In Watkins [23, p. 480], it is shown that if λ_k is a simple eigenvalue of A, then

$$\max_{\substack{1 \le i \le n \\ i \ne k}} |\lambda_i - \lambda_k|^{-1} = 1 / \min_{\substack{1 \le i \le n \\ i \ne k}} |\lambda_i - \lambda_k|$$

    is a lower bound for the condition number of an eigenvector corresponding to λ_k. Under what conditions will this formula indicate that an eigenvector is ill-conditioned?

18.23 Given two matrices A and B, the generalized eigenvalue problem is to find nonzero vectors v and a number λ such that $Av = \lambda Bv$. The matrix $A - \lambda B$ is called a *matrix pencil*, usually designated by (A, B). The characteristic equation for the pencil (A, B) is $\det(A - \lambda B) = 0$, and the generalized eigenvalues are the roots of the characteristic polynomial. See Ref. [19, Chapter 11] or [23, pp. 526-541] for a discussion of the problem.

    a. How does this problem relate to the standard eigenvalue problem?
    b. Show that the degree of the characteristic equation is less than or equal to n. Give an example where the degree is less than n. In general, when is the degree guaranteed to be less than n?
    c. Assume that B is nonsingular. Show that λ is an eigenvalue of (A, B) with associated eigenvector v if and only if λ is an eigenvalue of $B^{-1}A$ with associated eigenvector v.
    d. Show that the nonzero eigenvalues of (B, A) are the reciprocals of the nonzero eigenvalues of (A, B).
    e. Find the generalized eigenvalues for (A, B) by hand. Using `eig`, find corresponding eigenvectors.

$$A = \begin{bmatrix} 1 & 2 \\ -1 & 5 \end{bmatrix}, \quad B = \begin{bmatrix} 1 & 5 \\ 1 & 2 \end{bmatrix}$$

18.13.1 MATLAB Problems

18.24 Suppose a population has five age classes, and that the following table specifies the data for the population.

Age Class	$n_i(0)$	f_i	p_i
1	8	0	0.60
2	10	6	0.45
3	12	4	0.25
4	7	3	0.15
5	5	2	

    a. Find the Leslie matrix, L, for this population.
    b. Find the eigenvalues and eigenvectors of L.
    c. Find the rate of increase of the number of individuals at equilibrium for each time period.
    d. Graph the age population for each age group over a 10-year period using a logarithmic scale on the vertical axis.

18.25 Compute the largest eigenvalue of the matrix

$$
A = \begin{bmatrix}
-68 & 20 & -10 & 65 & -79 \\
59 & -48 & -84 & 8 & 93 \\
-38 & 31 & -54 & 100 & -100 \\
6 & 38 & 83 & -85 & 55 \\
-67 & 50 & -70 & -12 & 64
\end{bmatrix}
$$

using the power method.

18.26 In this problem, you will compare the execution time and accuracy of `eigqrbasic`, `eigqr`, `eigqrshift`, and `eigb`.

a. Execute the following:

```
A = gallery('gearmat',35);
tic;E1 = eigqrbasic(A,5000);toc;
tic;E2 = eigqr(A,1.0e-10,1000);toc
tic;E3 = eigqrshift(A,1.0e-10,50);toc;
tic;E4 = eigb(A,1.0e-10,50);toc;
```

Comment on the results.

b. Using the MATLAB `sort` command, sort E1, E2, E3, E4, and compute E = `sort(eig(A))`. Compute the accuracy of each of the four functions against that of `eig` by computing $\|E - E_i\|_2$, $i = 1, 2, 3, 4$. Explain the results.

18.27 **a.** Except for signs, no elements of the $n \times n$ matrix

$$
A = \begin{bmatrix}
0 & & & & 1 \\
1 & 0 & & & \\
 & 1 & \ddots & & \\
 & & \ddots & 0 & \\
 & & & 1 & 0
\end{bmatrix}
$$

change for $m < n$ executions of the Francis iteration of degree two [9]. If n is large, this creates a very slow algorithm. When lack of convergence is detected, the situation can be resolved by replacing the Francis double shift by an exceptional shift after 10-20 iterations [56]. Build a 25×25 version of this matrix, apply `impdsqr` 24 times, and then execute `diag(A,-1)` and `A(1,25)`. Repeat the experiment by applying `impdsqr` 35 times and 100 times. Comment on the results.

b. The function `eigb` terminates if any 2×2 or 1×1 eigenvalue block fails to converge in a default of 1000 iterations. In Ref. [57], it is noted that there are small sets of matrices where the Francis algorithm fails to converge in a reasonable number of iterations. Let

$$
A = \begin{bmatrix}
0 & 1 & 0 & 0 \\
1 & 0 & h & 0 \\
0 & -h & 0 & 1 \\
0 & 0 & 1 & 0
\end{bmatrix},
$$

and apply `eigb` to A for $h = 1.0 \times 10^{-6}$. Comment on the result. Does `eig` give results? If so, why?

18.28 The function `trid` in the book software distribution takes three vectors $a^{(n-1)\times1}$, $b^{n\times1}$, and $c^{(n-1)\times1}$ and builds a tridiagonal matrix with subdiagonal a, diagonal b, and superdiagonal c. Use `trid` to construct the 200×200 matrix

$$
A = \begin{bmatrix}
2 & -1 & 0 & \cdots & 0 \\
-1 & 2 & -1 & \cdots & 0 \\
\vdots & \ddots & \ddots & \ddots & \vdots \\
\vdots & & \ddots & \ddots & -1 \\
0 & 0 & \cdots & -1 & 2
\end{bmatrix}.
$$

a. Is A symmetric?

b. Is A positive definite?

c. Does A have distinct eigenvalues?

d. Apply the power method and the inverse power method to estimate the largest and smallest eigenvalues of A.

e. Is A ill-conditioned?

f. Find the condition number for each eigenvalue of A. Explain the result.

18.29 For this problem, we will use the MATLAB command `eig`. The 20×20 Wilkinson upper bidiagonal matrix has diagonal entries 20, 19, ..., 1. The superdiagonal entries are all equal to 20. The aim of this problem is to compute the condition number of the Wilkinson matrix and study the sensitivity of its eigenvalues by executing the series of steps:

a. Build W.

b. Compute $\kappa(W)$.

c. Find the eigenvalues, $\lambda_1, \lambda_2, \ldots, \lambda_{20}$, and the matrix of eigenvectors X.

d. Compute $\kappa(X)$.

e. Find c, the vector of eigenvalue condition numbers.

f. Let $\hat{W}$ be the matrix obtained from W by assigning $w_{20,1} = 1.0 \times 10^{-10}$ and recompute the eigenvalues $\hat{\lambda}_1, \hat{\lambda}_2, \ldots, \hat{\lambda}_{20}$. Sort the eigenvalues in descending order.

g. For $i = 1, 2, \ldots, 20$ print $\quad \lambda_i, \quad \hat{\lambda}_i, \quad \left| \lambda_i - \hat{\lambda}_i \right|, \quad c_i$

h. Explain the results.

18.30

a. Write a MATLAB function `wilkcondeig(n)` that builds the upper Hessenberg matrix

$$
\begin{bmatrix}
n & n-1 & \cdots & 2 & 1 \\
n-1 & n-1 & n-2 & \ddots & 1 \\
& n-2 & \ddots & \ddots & \\
& & \ddots & 2 & 1 \\
& & & 1 & 1
\end{bmatrix}.
$$

b. Compute the eigenvalue condition numbers of $W = $ `wilkcondeig(20)`.

c. Compute the eigenvalues of W.

d. Let $\hat{W}$ be the matrix W with the value at index $(20,1)$ set to 1.0×10^{-10}. Compute the eigenvalues of $\hat{W}$. Do the perturbations correspond to what you would expect from the condition numbers?

Problem 18.40 studies this matrix in more detail.

18.31 Let A be the matrix of Problem 18.7. Perform the following tests.

a. Execute the statements and explain the results.

```
>> E1 = eigqr(A,1.0e-10,10)
>> E2 = eigqrshift(A,1.0e-10,10)
>> eig(A)
```

b. Beginning with maxiter = 500, determine the smallest value of maxiter so that

```
>> E = eigqr(A,1.0e-10,maxiter)
```

terminates successfully.

c. Beginning with 1, find the smallest value of maxiter so that

```
>> E = eigqrshift(A,1.0e-10,maxiter)
```

terminates successfully.

d. Now execute

```
>> E = eigqrshift(A,1.0e-14,10);
>> EM = eig(A)
>> norm(E-EM)
```

e. Make a general statement about what you have learned from this problem.

18.32 Consider the matrix $A = \begin{bmatrix} 8 & 3 & 0 \\ 0 & 7.9999 & 0 \\ 2 & 5 & 10 \end{bmatrix}$.

a. Diagonalize A to obtain the eigenvector matrix X and the diagonal matrix of eigenvalues, D. Now let $E = 10^{-6} \begin{bmatrix} 0 & 0 & 0 \\ 0 & 0 & 1 \\ 0 & 0 & 0 \end{bmatrix}$, and find the eigenvalues of $A + E$. What happens?

b. Compute $\kappa(X)$ and $\kappa(X)\|E\|_2$. Why is the behavior in part (a) predicted by the Bauer-Fike theorem? Which eigenvalues are actually ill-conditioned, and why?

18.33 Let

$$A = \begin{bmatrix} -3 & 1/2 & 1/3 \\ -36 & 8 & 6 \\ 30 & -15/2 & -6 \end{bmatrix}, \quad \delta A = \begin{bmatrix} 0.005 & 0 & 0 \\ 0 & 0.005 & 0.005 \\ 0 & 0.005 & 0 \end{bmatrix}, \quad \text{and} \quad b = \begin{bmatrix} 1 \\ 1 \\ 1 \end{bmatrix}.$$

a. Compute the eigenvalues of A and the solution to $Ax = b$.
b. Let $\hat{A} = A + \delta A$. Compute the eigenvalues of $\hat{A}$ and the solution of $\hat{A}x = b$.
c. Explain the results.

18.34 Use MATLAB to compute the left eigenvectors of the matrix

$$A = \begin{bmatrix} 1 & -2 & -2 & -2 \\ -4 & 0 & -2 & -4 \\ 1 & 2 & 4 & 2 \\ 3 & 1 & 1 & 5 \end{bmatrix}.$$

18.35 `A = gallery('clement',50)` returns a nonsymmetric 50×50 tridiagonal matrix with zeros on its main diagonal. The eigenvalues are $\pm49, \pm47, \pm45, \ldots, \pm3, \pm1$. Since $|\lambda_1| = |\lambda_2| > |\lambda_3| = |\lambda_4| > \cdots > |\lambda_{49}| = |\lambda_{50}|$, the basic and the unshifted Hessenberg QR iterations may not converge. Fill-in the table and, if a method does not converge, indicate this in the last column. Sort the eigenvalues using `sort` prior to computing $\|E - E_i\|_2$, $i = 1, 2, 3, 4$. Discuss the results.

Number	Function	Time	$\|E - E_i\|_2$
	`E = eig(A);`		
1	`E1 = eigqrbasic(A,1000);`		
2	`E2 = eigqr(A,1.0e-12,500);`		
3	`E3 = eigqrshift(A,1.0e-12,500);`		
4	`E4 = eigb(A,1.0e-12,500);`		

18.36 Implement the shifted inverse iteration with a function

```
[x, iter] = inverseiter(A, sigma, x0, tol, maxiter)
```

where A is any real matrix with distinct eigenvalues. Use it to find eigenvectors corresponding to the given eigenvalues.

a. $A = \begin{bmatrix} 1 & 7 & 8 \\ 3 & 1 & 5 \\ 0 & -1 & -8 \end{bmatrix}$, $\lambda = 5.1942, \lambda_2 = -3.5521$.

b. $A = \begin{bmatrix} -1 & 3 & 5 & 3 \\ 22 & -1 & 8 & 2 \\ 9 & 12 & 3 & 7 \\ 25 & 33 & 55 & 35 \end{bmatrix}$,

$\lambda = 50.7622, \lambda_2 = 0.8774, \lambda_3 = -6.7865, \lambda_4 = -8.8531$

c. $A = \begin{bmatrix} 2 & 7 & 5 & 9 & 2 \\ 8 & 3 & 1 & 6 & 10 \\ 4 & 7 & 3 & 10 & 1 \\ 6 & 7 & 10 & 1 & 8 \\ 2 & 8 & 2 & 5 & 9 \end{bmatrix}$, $\lambda_1 = -9.4497, \lambda_2 = -1.8123$.

18.37 Input the 8×8 matrix `EIGVECTST.mat` in the software distribution. Perform these actions in the order given:

a. Compute the eigenvector matrix `V1` and the eigenvalue matrix `D1` for `EIGVECTST`.

b. Output the eigenvalues of `EIGVECTST`.

c. Create a matrix perturbation as follows:

```
deltaEIGVECTST = ones(8,8);
for i = 1:8
    for j = 1:8
        deltaEIGVECTST(i,j) = rand*1.e-6*deltaEIGVECTST(i,j);
    end
end
```

Perturb `EIGVECTST` by computing `EIGVECTST_hat = EIGVECTST + deltaEIGVECTST;`
Compute $\|EIGVECTST - EIGVECTST\_hat\|_2$.
Find the eigenvector matrix `V2` and the eigenvalue matrix `D2` for `EIGVECTST_hat`.
Output the eigenvalues of `EIGVECTST_hat`.
Output $\|V1 - V2\|_2$.
Explain the results.

18.38 Let $A = \begin{bmatrix} 1 & 0 \\ 0 & 1+\epsilon \end{bmatrix}$. A is symmetric and has two nearly equal eigenvalues if ϵ is small.

a. For $\epsilon = 10^{-3}$, 10^{-5}, 10^{-8}, do the following:

  i. Compute the eigenvalues and eigenvectors of A.

  ii. Let $\delta A = 10^{-8}$ randn (2), where randn(2) is a 2×2 matrix of random entries drawn from the standard normal distribution. Compute the eigenvalues and eigenvectors of $A + \delta A$.

b. Perform the same numerical experiment as in part (a) using the matrix $A = \begin{bmatrix} 1+\epsilon & 0 & 0 \\ 0 & 1-\epsilon & 0 \\ 0 & 0 & 2 \end{bmatrix}$.

c. Make a general statement about the conditioning of eigenvalues and eigenvectors of a symmetric matrix.

18.39 This problem assumes access to the MATLAB Symbolic Math Toolbox. Let

$$A = \begin{bmatrix} 3+\epsilon & 5 \\ 0 & 3 \end{bmatrix}.$$

a. Show that the condition number of both eigenvalues of A is

$$c_1 = c_2 = \sqrt{\frac{25}{\epsilon^2} + 1}$$

by executing the following statements

```
>> syms A epsilon X invX x y c
>> A = sym([3+epsilon 5;0 3])
>> [X D] = eig(A)
>> invX = inv(X)
>> x = X(:,1)/norm(X(:,1))
>> y = invX(:,1)/norm(invX(:,1))
>> c1 = 1/abs(y'*x)
>> x = X(:,2)/norm(X(:,2))
>> y = invX(:,2)/norm(invX(:,2))
>> c2 = 1/abs(y'*x)
```

b. Using the formula, compute the condition number of the eigenvalues for $\epsilon = 1.0 \times 10^{-6}$.

c. By entering

$$B = \begin{bmatrix} 3+\epsilon & 5 \\ 0 & 3 \end{bmatrix},$$

compute the condition number of the eigenvalues using floating point arithmetic.

d. Explain why the matrix A has ill-conditioned eigenvalues.

18.40 This problem is motivated by a discussion in the classic book by Wilkinson [9, pp. 92-93]. Consider the class of matrices

$$
B_n = \begin{bmatrix}
n & (n-1) & (n-2) & \ldots & 3 & 2 & 1 \\
(n-1) & (n-1) & (n-2) & \ldots & 3 & 2 & 1 \\
 & (n-2) & (n-2) & \ldots & 3 & 2 & 1 \\
 & & & \ldots & & & \\
 & & & & 2 & 2 & 1 \\
 & & & & & 1 & 1
\end{bmatrix}.
$$

a. If you have not done Problem 18.30, write a MATLAB function, wilkcondeig, that builds B_n.

b. Using MATLAB, let $n = 20$ and show that the largest eigenvalues of B_{20} are well-conditioned, while the smallest ones are quite ill-conditioned.

c. Show that the determinant of B_n is one for all n.

d. If we assign $B_n(1, n) = 1 + \epsilon$, the determinate becomes

$$1 \pm (n-1)!\epsilon.$$

Let $n = 20$, $B(1, 20) = 1 + 1.0 \times 10^{-10}$, and compute the determinant.

e. Prove that the determinant of a matrix is equal to the product of its eigenvalues.

f. Explain why there must be large eigenvalues in the perturbed B_n?

18.41 This problem illustrates convergence of the double shift implicit QR iteration to a single real or a complex conjugate pair of eigenvalues.

a. Create the upper Hessenberg Toeplitz matrix

$$
\begin{bmatrix}
1 & 2 & 3 & 4 & 5 \\
1 & 1 & 2 & 3 & 4 \\
0 & 1 & 1 & 2 & 3 \\
0 & 0 & 1 & 1 & 2 \\
0 & 0 & 0 & 1 & 1
\end{bmatrix}
$$

as follows:

```
>> H =  toeplitz([1 1 0 0 0], [1 2 3 4 5]);
```

b. Compute its eigenvalues using eig.

c. Execute

```
>> [~, H] = impdsqr(H)
```

six times and observe what happens in the lower right-hand corner of H. Comment.

d. Execute

```
>> [~, H(1:4,1:4)] = impdsqr(H(1:4,1:4))
```

six times and then apply the statement

```
eig(H(3:4,3:4))
```

Comment.

18.42 The execution of the MATLAB command help gallery contains the following entry

"gallery(5) is an interesting eigenvalue problem. Try to find its EXACT eigenvalues and eigenvectors."

Let

$$
A = \text{gallery}(5) = \begin{bmatrix}
-9 & 11 & -21 & 63 & -252 \\
70 & -69 & 141 & -421 & 1684 \\
-575 & 575 & -1149 & 3451 & -13801 \\
3891 & -3891 & 7782 & -23345 & 93365 \\
1024 & -1024 & 2048 & -6144 & 24572
\end{bmatrix},
$$

and request double precision output using format long.

a. Execute eig(A), and compute $\|Av_i - \lambda_i v_i\|_2$, $1 \le i \le 5$.

b. Compute `det(A)` and `rank(A)`. Why does the output of both commands conflict with the results of (a)?

c. Approximate the characteristic polynomial of A using `p = poly(A)`. What do you believe is the actual characteristic polynomial, p?

d. The Cayley-Hamilton theorem states that every matrix satisfies its characteristic equation. Compute $p(A)$, where p is the assumed characteristic polynomial from part (c). Is the result close to zero?

e. Explain why this matrix causes so much trouble for eigenvalue solvers.

18.43 This problem takes another approach to the column buckling application in Section 18.1.3.

a. If the beam is hinged at both ends, we obtain the boundary value problem

$$EI\frac{d^2y}{dx^2} + Py = 0, \quad y(0) = 0, \, y(L) = 0,$$

whose terms are defined in Section 18.1.3. Approach the problem using finite differences, as we did for the heat equation in Section 12.2. Divide the interval $0 \le x \le L$ into n subintervals of length $h = \frac{L}{n}$ with points of division $x_1 = 0, x_2 = h, x_3 = 2h, \ldots, x_n = L - h$, and $x_{n+1} = L$. Use the central finite difference approximation

$$\frac{d^2y}{dx^2}(x_i) \approx \frac{y_{i+1} - 2y_i + y_{i-1}}{h^2}, \quad 2 \le i \le n$$

along with the boundary conditions, to show that the result is a matrix eigenvalue problem of the form

$$Ay = \lambda y,$$

where A is a symmetric tridiagonal matrix. What is the relationship between λ and the critical loads?

b. For copper, $E = 117 \times 10^9 \, \text{N/m}^2$. Assume $I = .0052 \, \text{m}^4$ and that $L = 1.52 \, \text{m}$. Using $h = 1.52/25$ find the smallest three values of λ, compute the critical loads, and graph (x_i, y_i) for each value of λ on the same axes. Relate the results to the discussion in Section 18.1.3.

Note: Suppose that EV is a 26×3 matrix containing the eigenvectors corresponding to $\lambda_1, \lambda_2,$ and λ_3 as well as the zero values at the endpoints. Each row has the format $\begin{bmatrix} 0 & y_2 & y_3 & \ldots & y_{24} & y_{25} & 0 \end{bmatrix}^T$. To create a good looking graph place these statements in your program.

```
minval = min(min(EV));
maxval = max(max(EV));
axis equal;
axis([0 L minval maxval]);
hold on;
graph each displacement using different line styles
add a legend
hold off;
```

18.44 Problem 18.23 introduced the generalized eigenvalue problem. MATLAB can solve these problems using `eig(A,B)` that finds matrices V and D such that

$$AV = BVD.$$

The diagonal of D contains the generalized eigenvectors and the columns of V are the associated eigenvectors.

a. Solve the following generalized eigenvalue problem and verify that $AV = BVD$.

$$\begin{bmatrix} 1 & -7 & 3 \\ 0 & 1 & 5 \\ 3 & 7 & 8 \end{bmatrix} v = \lambda \begin{bmatrix} 1 & 3 & 7 \\ -1 & -8 & 1 \\ 4 & 3 & 1 \end{bmatrix}.$$

b. The basis for the computing the generalized eigenvalues and eigenvectors is the QZ iteration that is analogous to the QR iteration for eigenvalues. Look up the function `qz` in the MATLAB documentation and use it to solve the problem in part (a).

Chapter 19

The Symmetric Eigenvalue Problem

You should be familiar with

- Properties of a symmetric matrix
- *QR* decomposition
- Givens rotations
- Orthogonal similarity transformations
- Householder reflections
- Deflation during transformations
- Using a shift during eigenvalue computation

Symmetric matrices appear naturally in many applications that include the numerical solution to ordinary and partial differential equations, the theory of quadratic forms, rotation of axes, matrix representation of undirected graphs, and principal component analysis in statistics. Symmetry allows the development of algorithms for solving certain problems more efficiently than the same problem can be solved for a general matrix. For instance, the computation of eigenvalues and their associated eigenvectors can be done accurately and with a lower flop count than their determination for a general matrix. We have invoked the spectral theorem many times in the book. It guarantees that every $n \times n$ symmetric matrix has n linearly independent eigenvectors, even in the presence of eigenvalues of multiplicity greater than 1. We have not proved the theorem, and we do so in this chapter by using Schur's triangularization, developed in Section 18.7.

We will discuss the Jacobi, the symmetric QR iteration, the Francis algorithm, and the bisection algorithm for computing the eigenvectors and eigenvalues of a real symmetric matrix. The Jacobi algorithm uses a modification of Givens rotations to create orthogonal similarity transformations that reduce the symmetric matrix into a diagonal matrix containing the eigenvalues, all the while computing the corresponding eigenvectors. The algorithm does not require that the matrix first be brought into upper Hessenberg form. The symmetric QR algorithm is an adaptation of the implicit single shift QR iteration for a general matrix, except that the shift is chosen to take advantage of the matrix symmetry. Note that a symmetric upper Hessenberg matrix is tridiagonal, and that a reduction to upper triangular form creates a diagonal matrix of eigenvalues. As a result, the symmetric QR iteration is faster than the iteration for a general matrix. The Francis algorithm is usually the method of choice for the computation of eigenvalues. First, the matrix is reduced to a tridiagonal matrix. Rather than using the QR decomposition, the Francis algorithm performs orthogonal similarity transformations to reduce the tridiagonal matrix to a diagonal matrix of eigenvalues. As the algorithm progresses, an eigenvector can be computed right along with its eigenvalue. The bisection algorithm for computing eigenvalues of a symmetric tridiagonal matrix is very different from the previous methods. By applying the bisection method for computing the roots of a nonlinear function along with some amazing facts about the eigenvalues and characteristic polynomials of a symmetric matrix, the algorithm can accurately compute one particular eigenvalue, all the eigenvalues in an interval, and so forth.

The chapter concludes with a summary of Cuppen's divide-and-conquer algorithm [58]. It is more than twice as fast as the QR algorithm if both eigenvalues and eigenvectors of a symmetric tridiagonal matrix are required. However, the algorithm is difficult to implement so that it is stable. In fact, it was 11 years before a proper implementation was discovered [59, 60].

19.1 THE SPECTRAL THEOREM AND PROPERTIES OF A SYMMETRIC MATRIX

We have used the spectral theorem a number of times in the book to develop important results, but have never presented a proof. Schur's triangularization allows to easily prove the spectral theorem.

Theorem 19.1 (Spectral theorem). *If A is a real $n \times n$ symmetric matrix, then it can be factored in the form $A = PDP^T$, where P is an orthogonal matrix containing n orthonormal eigenvectors of A, and D is a diagonal matrix containing the corresponding eigenvalues.*

Numerical Linear Algebra with Applications. http://dx.doi.org/10.1016/B978-0-12-394435-1.00019-3

Proof. Since A is real and symmetric, it has real eigenvalues (Lemma 7.3), and applying Schur's triangularization $A = PTP^T$, where T is an upper triangular matrix. We have $T = P^T AP$, so $T^T = P^T A^T P = P^T AP$ and T is symmetric. Since T is upper triangular, $t_{ij} = 0$, $i > j$. The symmetry of T means that the elements $t_{ji} = 0$, $j < i$, and $D = T$ is a diagonal matrix. Since $A = PDP^T$, A and D are similar matrices, and by Theorem 5.1 they have the same eigenvalues. The eigenvalues of D are the elements on its diagonal and thus are eigenvalues of A. Now, if $P = \begin{bmatrix} v_1 & v_2 & \dots & v_{n-1} & v_n \end{bmatrix}$, then

$$AP = \begin{bmatrix} Av_1 & Av_2 & \dots & Av_{n-1} & Av_n \end{bmatrix} = \begin{bmatrix} \lambda_1 v_1 & \dots & \lambda_{n-1} v_{n-1} & \lambda_n v_n \end{bmatrix} = \begin{bmatrix} v_1 & \dots & v_{n-1} & v_n \end{bmatrix} \operatorname{diag}(\lambda_1, \lambda_2, \dots, \lambda_{n-1}, \lambda_n)$$

so $Av_i = \lambda_i v_i$, $1 \leq i \leq n$, and the columns of P are orthonormal eigenvectors of A. $\square$

Remark 19.1.   Since a real symmetric matrix has n orthonormal eigenvectors, we see from the proof of Theorem 19.1 that we can arrange P so its columns contain eigenvectors that correspond to the eigenvalues in decreasing order of magnitude;

in other words, D will have the form $\begin{bmatrix} \lambda_1 & & & 0 \\ & \lambda_2 & & \\ & & \ddots & \\ 0 & & & \lambda_n \end{bmatrix}$, where $|\lambda_1| \geq |\lambda_2| \geq \dots \geq |\lambda_n|$.

19.1.1   Properties of a Symmetric Matrix

We begin by listing some properties of a symmetric matrix we already know and then developing a new result.

- The eigenvalues of a symmetric matrix are real, and the eigenvectors can be assumed to be real (Lemma 7.3).
- If A is symmetric and x, y are $n \times 1$ vectors, then $\langle Ax, y \rangle = \langle x, Ay \rangle$.
- If A is a symmetric matrix, $\|A\|_2 = \rho(A)$, where $\rho(A)$ is the spectral radius of A.
- A symmetric $n \times n$ matrix A is positive definite if and only if all its eigenvalues are positive.
- If A is a real symmetric matrix, then any two eigenvectors corresponding to distinct eigenvalues are orthogonal (Theorem 6.4).
- A symmetric matrix can be diagonalized with an orthogonal matrix (Theorem 19.1—the spectral theorem). Thus, even if there are multiple eigenvalues, there is always an orthonormal basis of n eigenvectors.

Another special property of a real symmetric matrix is that the eigenvalues are well conditioned. Thus, by using a good algorithm we can compute eigenvalues with assurance that the errors will be small.

Theorem 19.2.   *The eigenvalues of a real symmetric matrix are well conditioned.*

Proof. We know by the spectral theorem that any real symmetric matrix can be diagonalized. If x is a normalized right eigenvector of A corresponding to eigenvalue λ, then $Ax = \lambda x$. We have $x^T A^T = x^T A = \lambda x^T$, so x is also a left eigenvector of A. The condition number, κ, of λ is

$$\kappa(\lambda) = \frac{1}{x^T x} = 1,$$

and λ is perfectly conditioned. $\square$

Since this chapter concerns the accurate computation of eigenvalues for a symmetric matrix, this is good news. Unfortunately, the same is not true for the eigenvectors of a symmetric matrix. They can be ill-conditioned (see Example 19.2 and Problem 19.21). The sensitivity of the eigenvalues of a symmetric matrix depends on the separation of the eigenvalues. If a matrix has an eigenvalue of multiplicity greater than 1, or if there is a cluster of closely spaced eigenvalues, the eigenvectors will be ill-conditioned.

19.2   THE JACOBI METHOD

As we have indicated, special algorithms that exploit symmetry have been developed for finding the eigenvalues and eigenvectors of a symmetric matrix. We will confine ourselves to real symmetric matrices and begin with the Jacobi method. The Jacobi method does not first transform the matrix A to upper Hessenberg form, but transforms A directly to a diagonal matrix using orthogonal similarity transformations. The strategy used in the Jacobi algorithm is to develop an iteration that will make the sum of the squares of the entries off the diagonal converge to 0, thus obtaining a diagonal matrix of eigenvalues.

In practical terms, the algorithm continues until $\sqrt{\sum_{i=1}^{n}\sum_{j=1,j\neq i}^{n}a_{ij}^2}$ is sufficiently small. Recall that $\|A\|_F^2 = \sum_{i=1}^{n}\sum_{j=1}^{n}a_{ij}^2$ is the square of the Frobenius norm, and so we can define the function off(A) as follows.

Definition 19.1. $\text{off}(A) = \sqrt{\sum_{i=1}^{n}\sum_{j=1,j\neq i}^{n}a_{ij}^2} = \sqrt{\|A\|_F^2 - \sum_{i=1}^{n}a_{ii}^2}$

The algorithm creates orthogonal matrices $J_0, J_1, \ldots J_{k-1}$ such that

$$\lim_{k\longrightarrow\infty}\text{off}(A_k) = 0,$$

where

$$\begin{cases} A_0 = A \\ A_k = J_{k-1}^{\mathrm{T}}A_{k-1}J_{k-1} \end{cases}$$

Since $A = A_0$ is symmetric and $\left(J_{k-1}^{\mathrm{T}}A_{k-1}J_{k-1}\right)^{\mathrm{T}} = J_{k-1}^{\mathrm{T}}A_{k-1}^{\mathrm{T}}J_{k-1} = J_{k-1}^{\mathrm{T}}A_{k-1}J_{k-1}$, A_k is symmetric. Each orthogonal matrix J_k is a Givens rotation, slightly different from the rotations used in Section 17.2. Those Givens rotations were of the form

$J(i, j, c, s)$

where the numbers c and s were chosen such that the product $J(i,j,c,s)A$ caused entry a_{ji} to become 0. For the Jacobi method, we choose c and s so that $J(i,j,c,s)^{\mathrm{T}}AJ(i,j,c,s)$ zeros out the pair of nonzero elements a_{ij} and a_{ji}.

Suppose we are at step k of the iteration, $A_k = J_{k-1}^{\mathrm{T}}A_{k-1}J_{k-1}$, and want to zero out nonzero entries $a_{ij}^{(k-1)}$ and $a_{ji}^{(k-1)}$ of A_{k-1}. To determine how we should choose c and s, look at the product $\overline{A} = J^{\mathrm{T}}(i,j,c,s)AJ(i,j,c,s)$.

In forming the product $\overline{A} = J(i,j,c,s)^T AJ(i,j,c,s)$, patterns emerge, and are displayed in the following table.

TABLE 19.1 Jacobi Iteration Formulas

$\overline{a_{kl}} = a_{kl},\ k,l \neq i,j$
$\overline{a_{ik}} = \overline{a_{ki}} = ca_{ik} - sa_{jk},\ k \neq i,j$
$\overline{a_{jk}} = \overline{a_{kj}} = sa_{ik} + ca_{jk},\ k \neq i,j$
$\overline{a_{ij}} = \overline{a_{ji}} = \left(c^2 - s^2\right)a_{ij} + cs\left(a_{ii} - a_{jj}\right)$
$\overline{a_{ii}} = c^2 a_{ii} - 2csa_{ij} + s^2 a_{jj}$
$\overline{a_{jj}} = s^2 a_{ii} + 2csa_{ij} + c^2 a_{jj}$

We want $\overline{a_{ij}} = \overline{a_{ji}} = 0$, so let

$$\left(c^2 - s^2\right) a_{ij} + cs \left(a_{ii} - a_{jj}\right) = 0. \tag{19.1}$$

If $a_{ij} = 0$, let $c = 1$, and $s = 0$. This satisfies Equation 19.1. Otherwise, we must find c and s. Rewrite Equation 19.1 as

$$\frac{c^2 - s^2}{cs} = \frac{a_{jj} - a_{ii}}{a_{ij}} \tag{19.2}$$

Note that we have assumed $a_{ij} \neq 0$. Since $J(i, j, c, s)$ is orthogonal, we must have $c^2 + s^2 = 1$, and $c = \cos\theta$, $s = \sin\theta$ for some θ. Substitute these relations into Equation 19.2 to obtain

$$\frac{\cos^2\theta - \sin^2\theta}{\sin\theta\cos\theta} = \frac{a_{jj} - a_{ii}}{a_{ij}} \tag{19.3}$$

Noting that $\cos 2\theta = \cos^2\theta - \sin^2\theta$, and $\sin 2\theta = 2\sin\theta\cos\theta$, Equation 19.3 can be written as

$$\cot 2\theta = \frac{1}{2}\left(\frac{a_{jj} - a_{ii}}{a_{ij}}\right). \tag{19.4}$$

We need the following trigonometric identity:

$$\tan^2\theta + 2\tan\theta\cot 2\theta - 1 = 0. \tag{19.5}$$

Equation 19.5 is verified as follows:

$$\begin{aligned} \tan^2\theta + 2\tan\theta\cot 2\theta - 1 &= \frac{\sin^2\theta}{\cos^2\theta} + 2\frac{\sin\theta}{\cos\theta}\frac{\cos 2\theta}{\sin 2\theta} - 1 \\ &= \frac{\sin^2\theta}{\cos^2\theta} + 2\frac{\sin\theta}{\cos\theta}\frac{\left(\cos^2\theta - \sin^2\theta\right)}{2\sin\theta\cos\theta} - 1 \\ &= \frac{\sin^2\theta}{\cos^2\theta} + \frac{\cos^2\theta - \sin^2\theta}{\cos^2\theta} - 1 = 0. \end{aligned}$$

Noting that $\tan\theta = \frac{s}{c}$, we now use the identity 19.5 with Equations 19.2 and 19.4 to obtain

$$\frac{s^2}{c^2} + 2\frac{s}{c}\frac{1}{2}\left(\frac{a_{jj} - a_{ii}}{a_{ij}}\right) - 1 = \frac{s^2}{c^2} + \frac{s}{c}\left(\frac{c^2 - s^2}{cs}\right) - 1 = 1 - 1 = 0. \tag{19.6}$$

Let $t = \frac{s}{c}$, and we can write Equation 19.6 as

$$t^2 + 2\tau t - 1 = 0, \tag{19.7}$$

where $\tau = \frac{1}{2}\left(\frac{a_{jj} - a_{ii}}{a_{ij}}\right)$. An application of the quadratic equation to Equation 19.7 produces two roots:

$$t = -\tau \pm \sqrt{\tau^2 + 1}.$$

The solutions for t can be rewritten as follows:

$$t_1 = \left(-\tau + \sqrt{\tau^2 + 1}\right)\frac{\left(-\tau - \sqrt{\tau^2 + 1}\right)}{\left(-\tau - \sqrt{\tau^2 + 1}\right)} = \frac{-1}{-\tau - \sqrt{\tau^2 + 1}} = \frac{1}{\tau + \sqrt{\tau^2 + 1}},$$

$$t_2 = \left(-\tau - \sqrt{\tau^2 + 1}\right)\frac{\left(-\tau + \sqrt{\tau^2 + 1}\right)}{\left(-\tau + \sqrt{\tau^2 + 1}\right)} = \frac{-1}{-\tau + \sqrt{\tau^2 + 1}}.$$

When $\tau < 0$, choose t_2, and when $\tau > 0$, choose t_1. This choice means we are adding two positive numbers in the denominator, and so there is no cancelation error. Using t, we now must find values for c and s. When we find an appropriate value for c,

$$s = ct.$$

Now, $1 + \tan^2 \theta = 1 + t^2 = \sec^2 \theta = \frac{1}{\cos^2 \theta} = \frac{1}{c^2}$, so

$$c = \frac{1}{\sqrt{1 + t^2}}.$$

The following is a summary of our results to this point.

TABLE 19.2 Computation of c and s for the Jacobi Method

$$\tau = \frac{a_{jj} - a_{ii}}{2a_{ij}}$$

$$t = \left\{ \begin{array}{ll} \dfrac{1}{\tau + \sqrt{\tau^2 + 1}} & \tau \geq 0 \\ \dfrac{-1}{-\tau + \sqrt{\tau^2 + 1}} & \tau < 0 \end{array} \right.$$

$$c = \frac{1}{\sqrt{1 + t^2}}$$

$$s = ct$$

The question now is the issue of convergence. We will investigate this question by looking at the off-diagonal entries as the iteration progresses. Before proceeding, we need to note the following:

- trace $(AB) =$ trace (BA) for any $n \times n$ matrices A and B (Theorem 1.2).
- If A is symmetric, trace $(A^2) = \sum_{i=1}^{n} \sum_{j=1}^{n} a_{ij}^2$ (Problem 19.2).
- If P is an orthogonal matrix, then $\left\| P^{\mathrm{T}} A P \right\|_F = \|A\|_F$, which follows from Lemma 15.1.

A technical lemma is necessary before we can prove convergence of the Jacobi method. It says that $\mathrm{off}(\overline{A}_k) \leq \mathrm{off}(\overline{A}_{k-1})$ ($\mathrm{off}(\overline{A}_i)$ is monotonically decreasing).

Lemma 19.1.   *If* $\overline{A} = J(i, j, c, s)^{\mathrm{T}} A J(i, j, c, s)$, *then*

$$\mathrm{off}(\overline{A})^2 = \mathrm{off}(A)^2 - 2a_{ij}^2.$$

Proof. By looking at Table 19.1, it is evident that all the entries on the diagonal of $\overline{A}$ except $\overline{a}_{ii}$ and $\overline{a}_{jj}$ are the same as those of A. The elements at indices (i, i) and (j, j) on the diagonal must satisfy the following relation:

$$\begin{bmatrix} \cos \theta & -\sin \theta \\ \sin \theta & \cos \theta \end{bmatrix} \begin{bmatrix} a_{ii} & a_{ij} \\ a_{ij} & a_{jj} \end{bmatrix} \begin{bmatrix} \cos \theta & \sin \theta \\ -\sin \theta & \cos \theta \end{bmatrix} = \begin{bmatrix} \overline{a}_{ii} & 0 \\ 0 & \overline{a}_{jj} \end{bmatrix},$$

and so

$$\left\| \begin{bmatrix} \overline{a}_{ii} & 0 \\ 0 & \overline{a}_{jj} \end{bmatrix} \right\|_F^2 = \left\| \begin{bmatrix} \cos \theta & -\sin \theta \\ \sin \theta & \cos \theta \end{bmatrix} \begin{bmatrix} a_{ii} & a_{ij} \\ a_{ij} & a_{jj} \end{bmatrix} \begin{bmatrix} \cos \theta & \sin \theta \\ -\sin \theta & \cos \theta \end{bmatrix} \right\|_F^2,$$

By Lemma 15.1,

$$\overline{a}_{ii}^2 + \overline{a}_{jj}^2 = a_{ii}^2 + a_{jj}^2 + 2a_{ij}^2.$$

Since all the other diagonal entries of $\overline{A}$ are identical to those of A, we have

$$\sum_{i=1}^{n} \overline{a}_{ii}^2 = \sum_{i=1}^{n} a_{ii}^2 + 2a_{ij}^2. \tag{19.8}$$

Since A and $\overline{A}$ are orthogonally similar, $\|\overline{A}\|_F^2 = \|A\|_F^2$ by Lemma 15.1. This and Equation 19.8 imply

$$\mathrm{off}(\overline{A})^2 = \|\overline{A}\|_F^2 - \sum_{k=1}^{n} \overline{a}_{ii}^2 = \|A\|_F^2 - \sum_{i=1}^{n} a_{ii}^2 - 2a_{ij}^2 = \mathrm{off}(A)^2 - 2a_{ij}^2. \qquad \square$$

Theorem 19.3.   *At each iteration, choose for a_{ij} the off-diagonal element largest in magnitude. With this strategy, the Jacobi method converges.*

Proof. Consider $A_k = J_{k-1}^T A_{k-1} J_{k-1}$. There are $n^2 - n$ entries off the diagonal, and since a_{ij} is largest in magnitude,

$$(\text{off}\,(A_{k-1}))^2 \le n\,(n-1)\,a_{ij}^2.$$

Write the equation in the form

$$a_{ij}^2 \ge \frac{(\text{off}\,(A_{k-1}))^2}{n\,(n-1)}.$$

By Lemma 19.1, $\text{off}\,(A_k)^2 = \text{off}\,(A_{k-1})^2 - 2a_{ij}^2$, and so

$$\text{off}(A_k)^2 \le \text{off}\,(A_{k-1})^2 - 2\frac{(\text{off}\,(A_{k-1}))^2}{n\,(n-1)} = \left(1 - \frac{1}{N}\right)\text{off}\,(A_{k-1})^2 \qquad (19.9)$$

where $N = \frac{n(n-1)}{2}$. Equation 19.9 shows that after k Jacobi iteration steps,

$$\text{off}(A_k) \le \left(\sqrt{1 - \frac{1}{N}}\right)^k \text{off}\,(A),$$

which shows that the Jacobi iteration converges. $\qquad\qquad\square$

The term $\text{off}\,(A_k)$ decreases at a rate of $\sqrt{\left(1 - \frac{1}{N}\right)}$. This rate of convergence is considered linear. However, it can be shown that the average rate of convergence (asymptotic rate) is quadratic [2, pp. 479-480].

Remark 19.2.   Each iteration of the Jacobi algorithm makes a_{ij} and a_{ji} zero, and the computation to do this can destroy pairs of zeros already created. However, as the iteration progresses, $\text{off}\,(A_k)$ decreases, leaving approximations to the eigenvalues on the diagonal. There is a function in the software distribution, `eigsymjdemo`, that allows you to see this behavior happening.

19.2.1   Computing Eigenvectors Using the Jacobi Iteration

Our preceding discussion did not include the computation of eigenvectors, and it is easy to do. The iteration computes

$$J_k^T J_{k-1}^T \ldots J_2^T J_1^T A J_1 J_2 \ldots J_{k-1} J_k = (J_1 J_2 \ldots J_{k-1} J_k)^T A\,(J_1 J_2 \ldots J_{k-1} J_k) \approx D,$$

where D is a diagonal matrix of eigenvalues, and each J_i is a Givens rotation. Thus, the matrix

$$J = J_1 J_2 \ldots J_{k-1} J_k$$

is an orthogonal matrix of eigenvectors. Starting with $J_0 = I$, maintain this product.

19.2.2   The Cyclic-by-Row Jacobi Algorithm

The algorithm we have presented is called the *classical Jacobi method*. The product $A_1 = J\,(i,j,c,s)^T A_{k-1}$ affects only rows i and j of A_{k-1}, and $A_2 = A_1 J\,(i,j,c,s)$ affects only columns i and j of A_1. Thus, each Jacobi iteration costs $O\,(n)$ flops. To find the largest entry in magnitude requires searching $\frac{n(n-1)}{2}$ entries, so it is necessary to execute $O\,(n^2)$ comparisons for one Jacobi iteration. This is simply too expensive, so it is almost never used in practice.

There is a modification of the Jacobi algorithm that is designed to speed it up. It is known as the *cyclic-by-row Jacobi algorithm*. Compute $\text{off}(\overline{A})$ by cycling through the $N = \frac{n(n-1)}{2}$ entries above the diagonal by rows left-to-right as follows:

$$(1,2),(1,3),\ldots,(1,n),(2,3),(2,4),\ldots,(2,n),\ldots,(n-1,n).$$

One cycle of N Jacobi rotations is termed a *sweep*, and sweeps are performed until $\text{off}(A_k)$ is sufficiently small. This algorithm has the same convergence properties as the classical Jacobi algorithm [9, pp. 270-271].

We are now ready to specify Algorithm 19.1, eigsymj, that computes the eigenvalues and corresponding eigenvectors of a real symmetric matrix. We assume the following functions are available:

- jacobics: computes c and s for the Jacobi rotation defined in Table 19.2.
- jacobimul: computes the Jacobi rotation $J\,(i,j,c,s)^T AJ\,(i,j,c,s)$ for an iteration of the Jacobi algorithm as defined in Table 19.1.

- givensmulp: computes AJ (i, j, c, s).
- off: computes off $(A) = \sqrt{\sum_{k=1}^{n} \sum_{p=1, k \neq p}^{n} a_{kp}^2}$.

NLALIB: The function `eigsymj` implements Algorithm 19.1. The supporting functions `jacobics`, `jacobimul`, `givensmulp`, and `off` are in the book software distribution.

Remark 19.3.   The MATLAB implementation uses variable input and output arguments to make `eigsymj` as flexible as possible. Here are the possible calling formats:

- `eigsymj(A)`: returns a column vector of eigenvalues.
- `E = eigsymj(A)`: assigns E a column vector containing the eigenvalues.
- `[V, D] = eigsymj(A)`: assigns the columns of V the eigenvectors corresponding to the eigenvalues on the diagonal matrix D.
- `[V, D, numsweeps] = eigsymj(A)`: adds the number of sweeps required to the output.

Algorithm 19.1 Jacobi Method for Computing All Eigenvalues of a Real Symmetric Matrix

---

```
function EIGSYMJ(A,tol,maxsweeps)
  % executes the cyclic-by-row Jacobi method to approximate the eigenvalues
  % and eigenvectors of a real symmetric matrix A.
  % [V D numsweeps]=eigsymj(A,tol,maxsweeps) returns an orthogonal
  % matrix V and diagonal matrix D of eigenvalues such that
  % Vᵀ AV = D. The algorithm returns when tol < off(Aₖ).
  % If the desired tolerance is not obtained within maxsweeps sweeps,
  % a value of -1 is returned for numsweeps.

  Print error message and return if A is not symmetric.

  desiredAccuracy=false
  numsweeps=1
  V=I
  while (numsweeps ≤ maxsweeps) and (not desiredAccuracy) do
    % execute a cycle of n(n-1)/2 Jacobi rotations.
    for i=1:n-1 do
      for j=i+1:n do
        % compute c and s so that aᵢⱼ = aⱼᵢ = 0.
        [ c  s ] = jacobics(A, i, j)
        % compute A = J(i, j, c, s)ᵀ AJ(i, j, c, s)
        A=jacobimul(A,i,j,c,s)
        % multiply V on the right by the Givens rotation J(i,j,c,s).
        V=givensmulp(V,i,j,c,s);
      end for
    end for
    if off(A)<tol then
      desiredAccuracy=true
    end if
    numsweeps=numsweeps+1
  end while

  if desiredAccuracy=false then
    numsweeps=-1
  end if
  D=diag(diag(A))
  return [ V  D  numsweeps ]
end function
```

Example 19.1. Let $A = \begin{bmatrix} 1 & 5 & 2 \\ 5 & -1 & 3 \\ 2 & 3 & 4 \end{bmatrix}$. Two sweeps of the Jacobi algorithm produces the following sequence of matrices, with the data rounded to four decimal places.

$$\begin{bmatrix} 5.0990 & 0.0000 & 3.4487 \\ 0.0000 & -5.0990 & 1.0520 \\ 3.4487 & 1.0520 & 4.0000 \end{bmatrix}, \begin{bmatrix} 8.0417 & 0.6829 & 0.0000 \\ 0.6829 & -5.0990 & 0.8003 \\ 0.0000 & 0.8003 & 1.0574 \end{bmatrix}, \begin{bmatrix} 8.0417 & 0.6774 & 0.0866 \\ 0.6774 & -5.2014 & 0.0000 \\ 0.0866 & 0.0000 & 1.1597 \end{bmatrix},$$

$$\begin{bmatrix} 8.0762 & 0.0000 & 0.0865 \\ 0.0000 & -5.2359 & -0.0044 \\ 0.0865 & -0.0044 & 1.1597 \end{bmatrix}, \begin{bmatrix} 8.0773 & -0.0001 & 0.0000 \\ -0.0001 & -5.2359 & -0.0044 \\ 0.0000 & -0.0044 & 1.1586 \end{bmatrix}, \begin{bmatrix} 8.0773 & -0.0001 & 0.0000 \\ -0.0001 & -5.2359 & 0.0000 \\ 0.0000 & 0.0000 & 1.1586 \end{bmatrix}.$$

Notice that the first rotation made $a_{12} = a_{21} = 0$, but the second rotation made $a_{13} = a_{31} = 0$, while a_{12} and a_{21} became 0.6829. On the third rotation the entries a_{31} and a_{13} become 0.0866 when a_{32} and a_{23} become 0. Despite this behavior, off(A) for the final matrix is 7.80332×10^{-5}. The values on the diagonal of the final matrix are eigenvalues correct to four decimal places. ∎

Example 19.2. This example demonstrates that a symmetric matrix can have ill-conditioned eigenvectors. The symmetric matrix EIGVECSYMCOND from the software distribution has a condition number of 70.867. The following MATLAB statements clearly demonstrate that the eigenvectors are ill-conditioned.

```
>> [V1,D1] = eigsymj(EIGVECSYMCOND,1.0e-14,20);
>> E = 1.0e-10*rand(25,1);
>> E = diag(E);
>> EIGVECSYMCONDP = EIGVECSYMCOND;
>> EIGVECSYMCONDP = EIGVECSYMCONDP + E;
>> [V2,D2] = eigsymj(EIGVECSYMCONDP,1.0e-14,20);
>> norm(D1-D2)

ans =
   1.091393642127514e-10

>> norm(V1-V2)

ans =
   0.962315272737213
```
∎

19.3 THE SYMMETRIC *QR* ITERATION METHOD

In this section, we will develop the symmetric *QR* iteration method for computing the eigenvalues and eigenvectors of a real symmetric matrix. The method takes advantage of concepts we have already developed, namely, the orthogonal reduction of a matrix to upper Hessenberg form and the shifted Hessenberg *QR* iteration. However, adjustments are made to take advantage of matrix symmetry. A symmetric upper Hessenberg matrix is tridiagonal, so our first task is to develop an efficient way to take advantage of symmetry when reducing A to a tridiagonal matrix. We will use Householder matrices for the orthogonal similarity transformations.

We motivate the process using a general symmetric 3×3 matrix.

$$A = \begin{bmatrix} a & b & c \\ b & d & e \\ c & e & f \end{bmatrix}.$$

Recall that a Householder matrix is orthogonal and symmetric. Using the vector $x_1 = \begin{bmatrix} b \\ c \end{bmatrix}$, create a 2×2 Householder matrix $H_{u_1} = \begin{bmatrix} h_{11} & h_{12} \\ h_{12} & h_{22} \end{bmatrix}$ that zeros out c. Then,

$$H_{u_1} \begin{bmatrix} b \\ c \end{bmatrix} = \begin{bmatrix} h_{11}b + h_{12}c \\ h_{12}b + h_{22}c \end{bmatrix} = \begin{bmatrix} h_{11}b + h_{12}c \\ 0 \end{bmatrix}.$$

Embed H_{u_1} in the 3×3 identity matrix as the lower 2×2 submatrix to form

$$H_1 = \begin{bmatrix} 1 & 0 & 0 \\ 0 & h_{11} & h_{12} \\ 0 & h_{12} & h_{22} \end{bmatrix}.$$

Form the product

$$H_1 A = \begin{bmatrix} 1 & 0 & 0 \\ 0 & h_{11} & h_{12} \\ 0 & h_{12} & h_{22} \end{bmatrix} \begin{bmatrix} a & b & c \\ b & d & e \\ c & e & f \end{bmatrix} = \begin{bmatrix} a & b & c \\ h_{11}b + h_{12}c & h_{11}d + h_{12}e & h_{11}e + h_{12}f \\ h_{12}b + h_{22}c & h_{12}d + h_{22}e & h_{12}e + h_{22}f \end{bmatrix}$$

$$= \begin{bmatrix} a & b & c \\ h_{11}b + h_{12}c & h_{11}d + h_{12}e & h_{11}e + h_{12}f \\ 0 & h_{12}d + h_{22}e & h_{12}e + h_{22}f \end{bmatrix}.$$

Now multiply on the right by H_1 to obtain

$$H_1 A H_1 = \begin{bmatrix} a & bh_{11} + ch_{12} & 0 \\ h_{11}b + ch_{12} & h_{11}(h_{11}d + h_{12}c) + h_{12}(h_{11}e + h_{12}f) & h_{12}(h_{11}d + h_{12}e) + h_{22}(h_{11}e + h_{12}f) \\ 0 & h_{12}(h_{11}d + h_{12}e) + h_{22}(h_{11}e + h_{12}f) & h_{12}(h_{12}e + h_{22}f) \end{bmatrix}.$$

The product is symmetric and tridiagonal.

The algorithm for an $n \times n$ symmetric matrix is a generalization of this 3×3 example, and results in the Algorithm 19.2, trireduce. If the details are not required, the reader can skip to Example 19.4 that demonstrates the use of trireduce.

For an $n \times n$ symmetric matrix, construct a Householder matrix that will zero out all the elements below a_{21}. For this,

choose the vector $x_1 = \begin{bmatrix} a_{21} \\ a_{31} \\ \vdots \\ a_{n1} \end{bmatrix}$ and form the $(n-1) \times (n-1)$ Householder matrix $H_{u_1}^{(n-1)}$. Insert it into the identity matrix

to create matrix H_1 so that

$$H_1 A = \begin{bmatrix} 1 & 0 & 0 & \ldots & 0 \\ 0 & & & & \\ 0 & & H_{u_1}^{(n-1)} & & \\ \vdots & & & & \\ 0 & & & & \end{bmatrix} \begin{bmatrix} a_{11} & a_{12} & a_{13} & \ldots & a_{1n} \\ a_{21} & & & & \\ a_{31} & & & & \\ \vdots & & & & \\ a_{n1} & & & & \end{bmatrix}$$

$$= \begin{bmatrix} a_{11} & a_{12} & a_{13} & \ldots & a_{1n} \\ v_1 & & & & \\ 0 & & X & & \\ \vdots & & & & \\ 0 & & & & \end{bmatrix}.$$

The value $v_1 = \pm \|x_1\|_2$ (Section 17.8.1). The product zeros out all the entries in column 1 in the index range $(3,1) - (n,1)$, alters a_{21}, but leaves the first row of A unchanged. Noting that $H_1^T = H_1$, form the orthogonal similarity transformation

$$A^{(1)} = H_1 A H_1 = \begin{bmatrix} a_{11} & v_1 & 0 & \ldots & 0 \\ v_1 & a_{22}^{(1)} & & & \\ 0 & & X & & \\ \vdots & & & & \\ 0 & & & & \end{bmatrix}$$

that maintains symmetry and zeros out the elements at indices $(1,3)-(1,n)$.

The next step is to zero out the elements in $A^{(1)}$ at indices $(4, 2)$-$(n, 2)$ using the vector $x_2 = \begin{bmatrix} a_{32}^{(1)} \\ a_{42}^{(1)} \\ a_{52}^{(1)} \\ \vdots \\ a_{n2}^{(1)} \end{bmatrix}$ to construct the

Householder $(n - 2) \times (n - 2)$ Householder matrix

$$H_2 = \begin{bmatrix} 1 & 0 & 0 & \ldots & 0 \\ 0 & 1 & 0 & \ldots & 0 \\ 0 & 0 & & & \\ \vdots & \vdots & & H_{u2}^{(n-2)} & \\ 0 & 0 & & & \end{bmatrix}.$$

The identity matrix in the upper left corner maintains the tridiagonal structure already built. Now form $A_2 = H_2 A_1 H_2$ to create the matrix

$$\begin{bmatrix} a_{11} & v_1 & 0 & \ldots & 0 \\ v_1 & a_{22}^{(1)} & v_2 & \ldots & 0 \\ 0 & v_2 & & & \\ \vdots & \vdots & & X & \\ 0 & 0 & & & \end{bmatrix},$$

where $v_2 = \pm \|x_2\|_2$. Continue by forming $H_3 A_2 H_3, \ldots, H_{n-2} A_{n-3} H_{n-2}$ to arrive at a symmetric tridiagonal matrix T. The product $P = H_{n-2} H_{n-3} \ldots H_2 H_1$ is an orthogonal matrix such that $T = PAP$.

An example using a 5×5 matrix will help in understanding the process. The example uses the MATLAB function

```
H = hsym(A,i)
```

in the software distribution. It builds the $n \times n$ Householder matrix with the embedded $(n - i) \times (n - i)$ Householder submatrix used to zero out elements $(i + 2, i), \ldots, (n, i)$.

Example 19.3.   Let

$$A = \begin{bmatrix} 1 & 2 & -1 & 3 & 5 \\ 2 & -1 & -2 & 1 & 0 \\ -1 & -2 & 1 & -7 & 2 \\ 3 & 1 & -7 & 2 & -1 \\ 5 & 0 & 2 & -1 & 1 \end{bmatrix}.$$

```
>> H1 = hsym(A,1)

H1 =
         1         0         0         0         0
         0  -0.32026   0.16013  -0.48038  -0.80064
         0   0.16013   0.98058  0.058264  0.097106
         0  -0.48038  0.058264   0.82521  -0.29132
         0  -0.80064  0.097106  -0.29132   0.51447

>> H1*A

ans =
         1         2        -1         3         5
    -6.245  -0.48038    2.5621   -1.6013         0
         0    -2.063   0.44669   -6.6845         2
         0    1.1891   -5.3401    1.0535        -1
         0   0.31511    4.7666   -2.5775         1

>> A1 = H1*A*H1
```

```
A1 =
            1        -6.245           0            0            0
       -6.245        1.3333       2.3421     -0.94135       1.0999
            0        2.3421    -0.087586      -5.0817       4.6714
            0      -0.94135      -5.0817      0.27834      -2.2919
            0        1.0999       4.6714      -2.2919       1.4759
>> H2 = hsym(A1,2)

H2 =
            1             0            0            0            0
            0             1            0            0            0
            0             0     -0.85061      0.34189     -0.39947
            0             0      0.34189      0.93684     0.073798
            0             0     -0.39947     0.073798      0.91377
>> A2 = H2*A1*H2
    ...
A3 =
            1        -6.245           0            0            0
       -6.245        1.3333      -2.7534            0            0
            0       -2.7534       6.9609      -4.5858            0
            0             0      -4.5858      -3.9358      0.32451
            0             0            0      0.32451      -1.3584
>> P = H3*H2*H1;
>> P*A*P'

ans =
            1        -6.245           0            0            0
       -6.245        1.3333      -2.7534            0            0
            0       -2.7534       6.9609      -4.5858            0
            0             0      -4.5858      -3.9358      0.32451
            0             0            0      0.32451      -1.3584
```

∎

19.3.1 Tridiagonal Reduction of a Symmetric Matrix

While the explanation provided shows how the algorithm works, the computation is not efficient. In particular, it should not be necessary to construct the entire matrix, H_i having the $i \times i$ identity matrix in the upper left-hand corner. The product $H_i A_{i-1} H_i$ should be done implicitly and take advantage of symmetry. Recall that a Householder matrix is of the form $H_u = I - \beta uu^T$, where $\beta = \frac{2}{u^T u}$. Since H_u is symmetric, $H_u A H_u^T = H_u A H_u$, and

$$
\begin{aligned}
H_u A H_u &= \left(I - \beta uu^T\right) A \left(I - \beta uu^T\right) \\
&= A - \beta A uu^T - u\beta u^T A + \beta uu^T \beta A uu^T
\end{aligned}
$$

Define $p = \beta A u$, so

$$
H_u A H_u = A - pu^T - up^T + \beta uu^T pu^T.
$$

Noting that $u^T p$ is a real number, define $K = \frac{\beta u^T p}{2}$ so that

$$
H_u A H_u = A - pu^T - up^T + 2Kuu^T = A - (p - Ku)u^T - u\left(p^T - Ku^T\right).
$$

Define $q = p - Ku$, and we have the final result

$$p = \beta A u \tag{19.10}$$

$$K = \frac{\beta u^T p}{2}, \tag{19.11}$$

$$q = p - Ku \tag{19.12}$$

$$H_u A H_u = A - qu^T - uq^T, \tag{19.13}$$

Equation 19.13 enables a faster computation for the transformation.

Our algorithm for reduction to a tridiagonal matrix replaces A by the tridiagonal matrix, so the remaining discussion will assume we are dealing with matrix A as it changes. As i varies from 1 to $n - 2$, we know that after the product $H_i A H_i$

- the entries at indices $(i + 2, i) \ldots (n, i)$ and $(i, i + 2) \ldots (i, n)$ have value 0.
- $a_{i+1,i} = a_{i,i+1} = \pm \left\| \begin{bmatrix} a_{i+1,i} & a_{i+2,i} & \ldots & a_{n-1,i} & a_{ni} \end{bmatrix}^{\mathrm{T}} \right\|_2$.

Looking at Example 19.3, we see that each iteration $H_i A H_i$ affects only the submatrix $A\,(i + 1 : n, i + 1 : n)$, so there is no need to deal with any other portion of the matrix. Our strategy is to assign the values $a_{i+1,i}$, $a_{i,i+1}$ and then perform the product

$$A\,(i + 1 : n, i + 1 : n) = A\,(i + 1 : n, i + 1 : n) - uq^{\mathrm{T}} - qu^{\mathrm{T}}$$

using Equation 19.13. After completion of the $n - 2$ iterations, the algorithm places zeros above and below the three diagonals to eliminate small entries remaining due to roundoff error. If the orthogonal transforming matrix P is required, compute $P = H_{n-2} H_{n-3} \ldots H_2 H_1 I$ using the formula

$$P\,(2 : n, i + 1 : n) = P\,(2 : n, i + 1 : n) - \beta P\,(2 : n, i + 1 : n)\,uu^{\mathrm{T}}$$

that affects only the portion of P that changes with each iteration. We put all this together in Algorithm 19.2.

Algorithm 19.2 Orthogonal Reduction of a Symmetric Matrix to Tridiagonal Form

---

```
function TRIREDUCE(A)
    % Compute a tridiagonal matrix orthogonally similar to the
    % symmetric matrix A.
    % T=trireduce(A) - assigns to T a symmetric tridiagonal matrix
    % orthogonally similar to A.
    % [P T]=trireduce(A) - assigns to T a symmetric tridiagonal
    % matrix orthogonally similar to A and an orthogonal matrix
    % P such that Pᵀ AP = T.

    if P required then
        P = I
    end if
    for i = 1:n-2 do
        [ u  β ] = houseparms(A(i + 1 : n, i))
        p = βA(i + 1 : n, i + 1 : n) u
        K = βuᵀp / 2
        q = p - Ku
        a_{i+1, i} = ± ‖A(i + 1 : n, i)‖₂
        a_{i, i+1} = a_{i+1, i}
        A(i + 1 : n, i + 1 : n) = A(i + 1 : n, i + 1 : n) - uqᵀ - quᵀ
        if P required then
            P(2 : n, i + 1 : n) = P(2 : n, i + 1 : n) - βP(2 : n, i + 1 : n) uuᵀ
        end if
    end for
    Clear all elements above and below the diagonals.
    if P required then
        return [ P  A ]
    else
        return A
    end if
end function
```

NLALIB: The function `trireduce` implements Algorithm 19.2.

Example 19.4. Example 19.2 dealt with the matrix EIGVECSYMCOND. In this example, we apply `trireduce` to that matrix and compute $\left\| PTP^{\mathrm{T}} - EIGVECSYMCOND \right\|_2$. In addition, MATLAB code compares the eigenvalues obtained from A and T.

```
>> [P T] = trireduce(EIGVECSYMCOND);
>> norm(P*T*P'-EIGVECSYMCOND)

ans =

   1.955492905735752e-10

>> EA = sort(eigsymj(EIGVECSYMCOND,1.0e-10,20));
>> ET = sort(eigsymj(T,1.0e-10,20));
>> norm(EA - ET)

ans =
   5.799990492988926e-10
```

■

Efficiency

The reduction to tridiagonal form requires $\frac{4}{3}n^3$ flops, and the algorithm is stable [2, pp. 458-459].

19.3.2   Orthogonal Transformation to a Diagonal Matrix

The final step of the symmetric shifted QR iteration is the orthogonal reduction of the tridiagonal matrix to a diagonal matrix of eigenvalues. In Section 18.6.1, we discussed a shift strategy that involves computing the QR decomposition of the matrix $T_k - \sigma I = Q_k R_k$ and then forming $T_{k+1} = R_k Q_k + \sigma I$. We need to show that $R_k Q_k$ is tridiagonal and symmetric.

$T = T_0$ is initially symmetric and tridiagonal. Now,

$$T_k - \sigma I = Q_k R_k,$$

so

$$Q_k^{\mathrm{T}} \left(T_k - \sigma I \right) Q_k = R_k Q_k.$$

$R_k Q_k$ is upper Hessenberg by Theorem 18.4, and thus is tridiagonal. We have

$$
\begin{aligned}
(R_k Q_k)^{\mathrm{T}} &= \left(Q_k^{\mathrm{T}} \left(T_k - \sigma I \right) Q_k \right)^{\mathrm{T}} \\
&= Q_k^{\mathrm{T}} \left(T_k - \sigma I \right)^{\mathrm{T}} Q_k \\
&= Q_k^{\mathrm{T}} \left(T_k - \sigma I \right) Q_k \\
&= R_k Q_k,
\end{aligned}
$$

and therefore T_{k+1} is symmetric.

As discussed in Section 18.6.1, $\sigma = h_{kk}$, the Rayleigh quotient shift, is often used with a nonsymmetric matrix. For a symmetric matrix, σ is usually chosen using the *Wilkinson shift*, which is a properly chosen eigenvalue of the 2×2 lower right submatrix,

$$W_k = \begin{bmatrix} t_{k-1,k-1} & t_{k,k-1} \\ t_{k,k-1} & t_{kk} \end{bmatrix}.$$

The eigenvalues of W_k are the roots of the characteristic polynomial

$$\left(t_{k-1,k-1} - \lambda \right) \left(t_{kk} - \lambda \right) - t_{k,k-1}^2.$$

Using the quadratic formula, we have

$$
\begin{aligned}
\lambda &= \frac{\left(t_{k-1,k-1} + t_{kk} \right) \pm \sqrt{\left(t_{k-1,k-1} + t_{kk} \right)^2 - 4 \left(t_{k-1,k-1} t_{kk} - t_{k,k-1}^2 \right)}}{2} \\
&= \frac{\left(t_{k-1,k-1} + t_{kk} \right) \pm \sqrt{\left(t_{k-1,k-1} - t_{kk} \right)^2 + 4 t_{k,k-1}^2}}{2}
\end{aligned}
$$

$$= \frac{2t_{kk} + \left(t_{k-1,k-1} - t_{kk}\right) \pm 2\sqrt{\left(\frac{t_{k-1,k-1}-t_{kk}}{2}\right)^2 + t_{k,k-1}^2}}{2}$$

$$= t_{kk} + \left(\frac{t_{k-1,k-1} - t_{kk}}{2}\right) \pm \sqrt{\left(\frac{t_{k-1,k-1} - t_{kk}}{2}\right)^2 + t_{k,k-1}^2}$$

$$= t_{kk} + r \pm \sqrt{r^2 + t_{k,k-1}^2},$$

where $r = \frac{t_{k-1,k-1} - t_{kk}}{2}$. We want to choose the eigenvalue closest to t_{kk} as the shift, so if $r < 0$, choose "+", and if $r > 0$, choose "−". If $r = 0$, choose "−". Using the function

$$\text{sign}(x) = \begin{cases} 1, & x > 0 \\ 0, & x = 0 \\ -1, & x < 0 \end{cases},$$

the shift is

$$\begin{cases} \sigma = t_{kk} + r - \text{sign}(r)\sqrt{r^2 + t_{k,k-1}^2}, & r \neq 0 \\ \sigma = t_{kk} - \sqrt{t_{k,k-1}^2}, & r = 0 \end{cases} \tag{19.14}$$

The reason for using the Wilkinson shift has to do with the guarantee of convergence. The choice of $\sigma = t_{kk}$ can lead to divergence of the iteration, but convergence when the using the Wilkinson shift is guaranteed to be at least linear, but in most cases is cubic [61].

The MATLAB function eigsymqr computes the eigenvalues and, optionally, the eigenvectors of a real symmetric matrix. After applying trireduce, the eigenvalue computation in eigsymqr method is identical to that for a nonsymmetric matrix, except that the shift is given by Equation 19.14. The eigenvector computation is done by maintaining the orthogonal matrices involved in the transformations.

Example 19.5.   The matrix SYMEIGTST in the book software distribution is a 21×21 symmetric matrix. The matrix is orthogonally similar to the famous symmetric tridiagonal 21×21 Wilkinson matrix used for testing eigenvalue computation. The MATLAB documentation for wilkinson(21) specifies that this matrix is a symmetric with pairs of nearly, but not exactly, equal eigenvalues. Its two largest eigenvalues are both about 10.746; they agree to 14, but not to 15, decimal places. The example uses eigsymqr to compute the eigenvalues.

```
>> load SYMEIGTST
eigsymqr(SYMEIGTST,1.0e-14,50)

ans =
  -1.125441522119974
   0.253805817096685
   0.947534367529301
      . . .
   5.000244425001915
   6.000234031584169
   6.000217522257100
      . . .
  10.746194182903356
  10.746194182903350
```

19.4   THE SYMMETRIC FRANCIS ALGORITHM

As noted in Section 18.8.1, the Francis algorithm has been a staple in eigenvalue computation for many years. The double-shift version will compute all the eigenvalues and eigenvectors of a general real matrix, and will find all the complex eigenvalues without using complex arithmetic. Since the eigenvalues of a symmetric matrix are real, the double-shift version is not necessary. We will present the single-shift version, called the *Francis iteration of degree one*.

The first step is to apply orthogonal similarity transformations that reduce the matrix A to a symmetric tridiagonal matrix, $T = Q^T A Q$. The algorithm now executes the single-shift bulge chase discussed in Section 18.8.1. After k iterations, the sequence approximates a diagonal matrix.

$$(J_k J_{k-1} \ldots J_2 J_1)\, T \left(J_1^T J_2^T \ldots J_k^T \right) = D. \tag{19.15}$$

If we let $P = J_k J_{k-1} \ldots J_2 J_1$ in Equation 19.15 and use the fact that $T = Q^T A Q$ there results

$$P Q^T A Q P^T = D.$$

The orthogonal matrix $P Q^T$ approximates the matrix of eigenvectors, and D approximates the corresponding eigenvalues.

The MATLAB function `eigsymb` finds the eigenvalues and, optionally, the eigenvectors of a symmetric matrix. The function `chase(T)` discussed in Section 18.8.1 performs the bulge chase. Like `eigsymqr`, `eigsymb` uses deflation so it only works on submatrices. The function just replaces the explicit single shift by the implicit single shift, so we will not present the algorithm.

Example 19.6. The Poisson matrix is a symmetric block tridiagonal sparse matrix of order n^2 resulting from discretizing Poisson's equation with the 5-point central difference approximation on an n-by-n mesh. We will discuss the equation in Section 20.5. This examples computes the 100×100 Poisson matrix, computes its eigenvalues and corresponding eigenvectors using eigsymb, and checks the result.

```
>> P = gallery('poisson',10);
>> P = full(P);  % convert from sparse to full matrix format
>> [V,D] = eigsymb(P,1.0e-14,100);
>> norm(V'*P*V-D)

ans =

    8.127291292857505e-14
```
∎

19.4.1 Theoretical Overview and Efficiency

The flop count for the Francis algorithm is the sum of the counts for reduction to tridiagonal form and reduction to a diagonal matrix. These counts are as follows [2, pp. 458-464]:

- The cost of computing just the eigenvalues of A is approximately $\frac{4}{3}n^3$ flops.
- Finding all the eigenvalues and eigenvectors costs approximately $9n^3$ flops.

This is the same order of magnitude as the simpler-shifted Hessenberg QR iteration in presented in Section 18.7. However, eigenvectors are readily found, and the number of multiplications needed is smaller.

19.5 THE BISECTION METHOD

After a symmetric matrix has been reduced to tridiagonal form, T, the bisection method can be used to compute a subset of the eigenvalues; for instance, eigenvalue i of n, the largest 15% of the eigenvalues or the smallest 5. If important eigenvalues lie in an interval $a \le \lambda \le b$, the algorithm can find all of them. If needed, inverse iteration will compute the corresponding eigenvector(s).

The bisection algorithm for computing roots of a nonlinear function is a standard topic in any numerical analysis or numerical methods course (see Ref. [33, pp. 61-64]). If f is a continuous real-valued function on an interval left $\le x \le$ right with f (left) and f (right) having opposite signs (f (left)f (right) < 0), then there must be a value r in the interval such that $f(r) = 0$. r is a *root* of f. Let mid $= \frac{\text{left}+\text{right}}{2}$ be the middle point of the interval. After evaluating $v = f$ (left)f (mid), either you have found a root or know in which of the intervals (left, mid) or (mid, right) of length $\frac{\text{right}-\text{left}}{2}$ the root lies

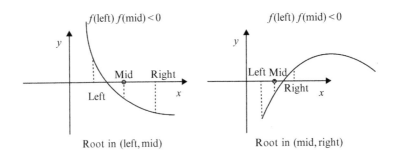

FIGURE 19.1 Bisection.

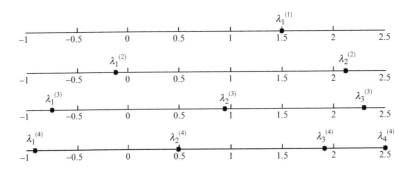

FIGURE 19.2 Interlacing.

(Figure 19.1). Move to the new interval and repeat the process until finding a root or isolating the root in a very small interval.

Outline of the Bisection Method

a. $v = 0$: mid is a root.
b. $v > 0$: root is in the interval mid $< r <$ right.
c. $v < 0$: root is in the interval left $< r <$ mid.

For any $n \times n$ matrix with real distinct eigenvalues, let $f(\mu) = \det(A - \mu I)$, and find two values $\mu_l =$ left and $\mu_r =$ right for which

$$f(\text{left})f(\text{right}) < 0,$$

and apply the bisection algorithm until approximating a root. Of course, the root is an eigenvalue of A. This does not violate the fact that polynomial root finding is unstable, since the algorithm deals only with the polynomial value and never computes any coefficients. There are no problems like evaluation of the quadratic formula (see Section 8.4.2) or perturbing a particular coefficient of a polynomial (see Section 10.3.1). A determinant of an arbitrary matrix can be computed stably using Gaussian elimination with partial pivoting ($PA = LU \Rightarrow A = P^{\mathrm{T}}LU \Rightarrow \det(A) = (-1)^r u_{11}u_{22}\ldots u_{nn}$, where r is the number of row exchanges). What makes the use of bisection extremely effective when applied to a symmetric tridiagonal matrix is some extraordinary properties of its eigenvalues and $f(\mu)$.

Let T be an unreduced symmetric tridiagonal matrix (no lower diagonal element is zero). At the end of the section, we will discuss the method for a matrix with a zero on its lower diagonal. Assume

$$T = \begin{bmatrix} a_1 & b_1 & & & \\ b_1 & a_2 & b_2 & & \\ & b_2 & a_3 & \ddots & \\ & & \ddots & \ddots & b_{n-1} \\ & & & b_{n-1} & a_n \end{bmatrix}, \quad b_i \neq 0.$$

Let $T^{(1)}$ be the 1×1 matrix $[a_1]$, $T^{(2)}$ be the 2×2 matrix $\begin{bmatrix} a_1 & b_1 \\ b_1 & a_2 \end{bmatrix}$ and, in general, $T^{(k)} = T(1:k, 1:k)$ be the upper-left $k \times k$ submatrix of T, $1 \leq k \leq n$. The eigenvalues of $T^{(k)}$ are distinct (Problem 19.1), and assume they are $\lambda_1^{(k)} < \lambda_2^{(k)} < \cdots < \lambda_k^{(k)}$. The essence of the bisection algorithm is that the eigenvalues of two successive matrices $T^{(k)}$ and $T^{(k+1)}$ *strictly interlace* [9, pp. 103-104]. This means that

$$\lambda_i^{(k+1)} < \lambda_i^{(k)} < \lambda_{i+1}^{(k+1)}$$

for $k = 1, 2, \ldots, n-1$ and $i = 1, 2, \ldots, k-1$. This remarkable property enables us to know the precise number of eigenvalues in any interval on the real line.

Example 19.7.   Let $T = \begin{bmatrix} \frac{3}{2} & 1 & 0 & 0 \\ 1 & \frac{1}{2} & 1 & 0 \\ 0 & 1 & \frac{1}{2} & 1 \\ 0 & 0 & 1 & \frac{3}{2} \end{bmatrix}$. Figure 19.2 shows the position of eigenvalues for $T^{(1)}$, $T^{(2)}$, $T^{(3)}$, and $T^{(4)} = T$.

Note the strict interlacing.   ∎

In addition to depending on the interlacing property of $T^{(k)}$, the bisection algorithm requires the computation of the characteristic polynomials, $p_k(\mu)$, of $T^{(k)}$ for a specified μ. The characteristic polynomial of $T^{(1)}$ is $p_1(\mu) = a_1 - \mu$, and define $p_0(\mu) = 1$. Assume we know $p_i(\mu)$, $1 \leq i \leq k-1$ and want to compute

$$p_k(\mu) = \det \left(\begin{bmatrix} a_1 - \mu & b_1 & & & & \\ b_1 & a_2 - \mu & b_2 & & & \\ & b_2 & a_3 - \mu & \ddots & & \\ & & \ddots & \ddots & \ddots & \\ & & & \ddots & a_{k-1} - \mu & b_{k-1} \\ & & & & b_{k-1} & a_k - \mu \end{bmatrix} \right).$$

Expand by minors across row k to obtain

$$\begin{aligned} p_k(\mu) &= (-1)^{k+(k-1)} b_{k-1} \det\left(T^{(k-2)} - \mu I\right) + (-1)^{2k}(a_k - \mu) \det\left(T^{(k-1)} - \mu I\right) \\ &= (a_k - \mu) p_{k-1}(\mu) - b_{k-1}^2 p_{k-2}(\mu). \end{aligned}$$

If you are unsure of the result, draw a 5×5 matrix and verify the equation. Putting things together, we have the three term recurrence relation

$$p_k(\mu) = \begin{cases} 1 & k = 0 \\ a_1 - \mu & k = 1 \\ (a_k - \mu) p_{k-1}(\mu) - b_{k-1}^2 p_{k-2}(\mu) & k \geq 2 \end{cases} \tag{19.16}$$

This recurrence relation is known as a *Sturm sequence* and enables the computation of $p_k(\mu)$ without using Gaussian elimination. The interlacing of eigenvalues gives rise to the following extraordinary result [27, pp. 203-208].

Theorem 19.4.   *The number, $d(\mu)$, of disagreements in sign between consecutive numbers of the sequence $\{p_0(\mu), p_1(\mu), p_2(\mu), \ldots, p_n(\mu)\}$ is equal to the number of eigenvalues smaller than μ.*

Remark 19.4.   If $p_k(\mu) = 0$, define the sign of $p_k(\mu)$ to be the opposite of that for $p_{k-1}(\mu)$. There cannot be two consecutive zero values in the sequence (Problem 19.13).

Example 19.8.   Let T be the matrix of Example 19.7. Its characteristic polynomials are

$p_0(\mu)$	$p_1(\mu)$	$p_2(\mu)$	$p_3(\mu)$	$p_4(\mu)$
1	$\frac{3}{2} - \mu$	$\mu^2 - 2\mu - \frac{1}{4}$	$-\mu^3 + \frac{5}{2}\mu^2 + \frac{1}{4}\mu - \frac{13}{8}$	$\mu^4 - 4\mu^3 + \frac{5}{2}\mu^2 + 4\mu - \frac{35}{16}$

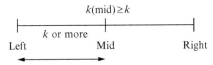

FIGURE 19.3 Bisection method: λ_k located to the left.

1. Let $\mu = 0$, and the sequence is $\left\{\ 1\ \ \frac{3}{2}\ \ -\frac{1}{4}\ \ -\frac{13}{8}\ \ -\frac{35}{16}\ \right\}$, with 1 sign change.
2. Let $\mu = 1$, and the sequence is $\left\{\ 1\ \ \frac{1}{2}\ \ -\frac{5}{4}\ \ \frac{1}{8}\ \ \frac{21}{16}\ \right\}$ with 2 sign changes.
3. Let $\mu = 3$. The sequence is $\left\{\ 1\ \ -\frac{3}{2}\ \ \frac{11}{4}\ \ -\frac{43}{8}\ \ \frac{85}{16}\ \right\}$ with 4 sign changes.

From (1), we see there is one negative eigenvalue, (2) tells us there is an eigenvalue in the range $0 \leq \mu < 1$, and from (3) we conclude that there must be 2 eigenvalues in the interval $1 \leq \mu < 3$.

In fact the eigenvalues are $\left\{\ \frac{1}{2} - \sqrt{2}\ \ \frac{1}{2}\ \ \frac{1}{2} + \sqrt{2}\ \ \frac{5}{2}\ \right\}$.   ■

Assume that the symmetric tridiagonal matrix T has eigenvalues $\lambda_1 \leq \lambda_2 \leq \cdots \lambda_n$ and that we wish to compute eigenvalue λ_i. The eigenvalue must be isolated in an interval (a, b) before generating a series of smaller and smaller intervals enclosing λ_i. The following lemma gives us starting values for a and b.

Lemma 19.2. *If $\|\cdot\|$ is a subordinate norm and $A \neq 0$, $\rho(A) \leq \|A\|$.*

Proof. Let λ_i be an eigenvalue of A with associated eigenvector v_i. Since $Av_i = \lambda_i v_i$,

$$\|Av_i\| = \|\lambda_i v_i\| = |\lambda_i|\,\|v_i\|,$$

and

$$\|Av_i\| \leq \|A\|\,\|v_i\|,$$

so

$$|\lambda_i|\,\|v_i\| \leq \|A\|\,\|v_i\|.$$

Since $\|v_i\| \neq 0$, we have $|\lambda_i| \leq \|A\|$ for all λ_i, and

$$\max_{1 \leq i \leq n} |\lambda_i| = \rho(A) \leq \|A\|.$$ □

Since the infinity norm can be computed quickly, Lemma 19.2 gives us starting values $a = -\|A\|_\infty$, $b = \|A\|_\infty$. We can now outline the bisection method for computing a particular λ_k, $1 \leq k \leq n$.

> **a.** Let left $= -\|A\|_\infty$, right $= \|A\|_\infty$.
> **b.** Compute mid $= \frac{\text{left} + \text{right}}{2}$.
> **c.** Compute $d(\text{mid})$, the number of disagreements in sign between consecutive numbers in the sequence
>
> $$p_0(\text{mid}), p_1(\text{mid}), p_2(\text{mid}), \ldots, p_n(\text{mid}),$$
>
> properly handling a case where $p_k(\text{mid}) = 0$.
> **d.** If $d(\text{mid}) \geq k$, then λ_k is in the interval [left, mid]. Let right $=$ mid (Figure 19.3).
> else
>
> λ_k is in the interval $\left[\text{mid, right}\right]$. Let left $=$ mid (Figure 19.4).
> **e.** Repeat steps 2-4 until (right $-$ left) $<$ tol, where tol is an acceptable length for an interval enclosing the root.

We will not give the algorithm using pseudocode. The MATLAB function `bisection` with calling format `bisection(T,k,tol)` in the book software distribution implements the method. If `tol` is not given, it defaults to 1.0×10^{-12}.

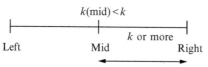

FIGURE 19.4 Bisection method: λ_k located to the right.

Example 19.9. Let A be the matrix of Example 19.7 and perform seven iterations of the bisection method to approximate $\lambda_2 = \frac{1}{2}$.

Iteration	Left	Right	Mid	d (mid)	Action
1	−2.5	2.5	0	1	left = 0
2	0	2.5	1.25	2	right = 1.725
3	0	1.25	0.625	2	right = 0.625
4	0	0.625	0.3125	1	left = 0.3125
5	0.3125	0.625	0.46875	1	left = 0.46875
6	0.46875	0.625	0.546875	2	right = 0.546875
7	0.46875	0.546875	0.507813	2	right = 0.507813

After seven iterations, the eigenvalue is isolated in the interval $0.46875 < \lambda_2 < 0.507813$. If the process completes a total of 16 iterations, $0.499954 < \lambda_2 < 0.500031$, and the approximation for λ_2 is

$$\frac{0.499954 + 0.500031}{2} = 0.49999.$$ ∎

Example 19.10. This example uses the bisection method to compute the two largest eigenvalues of the Wilkinson symmetric 21×21 tridiagonal matrix.

```
>> W = wilkinson(21);
>> lambda20 = bisection(W,20,1.0e-14)

lambda20 =
   10.746194182903317

>> lambda21 = bisection(W,21,1.0e-14)

lambda21 =
   10.746194182903395
```
   ∎

19.5.1   Efficiency

Each evaluation of $\{p_0 \,(\text{mid}), p_1 \,(\text{mid}), p_2 \,(\lambda\text{mid}), \ldots, p_n \,(\text{mid})\}$ costs $O \,(n)$ flops. If tol is the desired size of the subinterval containing the eigenvalue, since each iteration halves the search interval, the number of iterations, k, required is determined by

$$\frac{|\text{right} - \text{left}|}{2^k} < \text{tol}.$$

Thus,

$$k \approx \log_2 |\text{right} - \text{left}| - \log_2 \text{tol}$$

and the algorithm requires $O \,(kn)$ flops. If only a few eigenvalues are required, the bisection method is faster than finding all the eigenvalues using orthogonal similarity reduction to a diagonal matrix.

19.5.2   Matrix *A* Is Not Unreduced

If λ is an eigenvalue of multiplicity $m > 1$, the bisection algorithm for computing a root will find one occurrence of λ if m is odd (point of inflection) and will fail to find λ if m is even (tangent to horizontal axis) (Figure 19.5). This is not a problem, since the bisection method requires that A be unreduced, and a symmetric unreduced tridiagonal matrix has distinct eigenvalues (Problem 19.1). What happens if there are one or more zeros on the subdiagonal of A? We split the

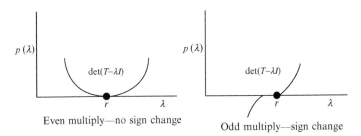

FIGURE 19.5 Bisection and multiple eigenvalues.

problem into finding eigenvalues of the unreduced matrices between the pairs of zeros. Lemma 19.3 shows how to handle the case when the subdiagonal contains one zero.

Lemma 19.3. *Suppose a tridiagonal matrix has the form*

$$T = \begin{bmatrix} a_1 & b_1 & & & & & & & \\ b_1 & a_2 & b_2 & & & & & & \\ & b_2 & \ddots & \ddots & & & & & \\ & & \ddots & a_{i-1} & b_{i-1} & & & & \\ & & & b_{i-1} & a_i & \mathbf{0} & & & \\ & & & & \mathbf{0} & a_{i+1} & b_{i+1} & & \\ & & & & & b_{i+1} & \ddots & & \\ & & & & & & \ddots & b_{n-1} \\ & & & & & & & b_{n-1} & a_n \end{bmatrix}.$$

Then,

$$\det\left(T - \lambda I^{n \times n}\right) = \det\left(T\left(1:i,\,1:i\right) - \lambda I^{i \times i}\right) \det\left(T\left(i+1:n,\,i+1:n\right) - \lambda I^{(n-i) \times (n-i)}\right).$$

The proof is left to the problems.

With one zero on the subdiagonal, Lemma 19.3 says to apply bisection to the unreduced $i \times i$ and $(n-i) \times (n-i)$ submatrices and form the union of the eigenvalues.

19.6 THE DIVIDE-AND-CONQUER METHOD

This presentation is a summary of the divide-and-conquer method. For more in-depth coverage of this algorithm, see Refs. [1, pp. 216-228], [19, pp. 359-363], and [26, pp. 229-232].

The recursive algorithm divides a symmetric tridiagonal matrix into submatrices and then applies the same algorithm to the submatrices. We will illustrate the method of splitting the problem into smaller submatrix problems using a 5×5 matrix

$$T = \begin{bmatrix} a_1 & b_1 & 0 & 0 & 0 \\ b_1 & a_2 & b_2 & 0 & 0 \\ 0 & b_2 & a_3 & b_3 & 0 \\ 0 & 0 & b_3 & a_4 & b_4 \\ 0 & 0 & 0 & b_4 & a_5 \end{bmatrix}.$$

Write A as a sum of two matrices as follows:

$$T = \begin{bmatrix} a_1 & b_1 & 0 & 0 & 0 \\ b_1 & a_2 - b_2 & 0 & 0 & 0 \\ 0 & 0 & a_3 - b_2 & b_3 & 0 \\ 0 & 0 & b_3 & a_4 & b_4 \\ 0 & 0 & 0 & b_4 & a_5 \end{bmatrix} + \begin{bmatrix} 0 & 0 & 0 & 0 & 0 \\ 0 & b_2 & b_2 & 0 & 0 \\ 0 & b_2 & b_2 & 0 & 0 \\ 0 & 0 & 0 & 0 & 0 \\ 0 & 0 & 0 & 0 & 0 \end{bmatrix}.$$

If we let

$$T_1 = \begin{bmatrix} a_1 & b_1 \\ b_1 & a_2 - b_2 \end{bmatrix}, \quad T_2 = \begin{bmatrix} a_3 - b_2 & b_3 & 0 \\ b_3 & a_4 & b_4 \\ 0 & b_4 & a_5 \end{bmatrix},$$

and

$$H = \begin{bmatrix} 0 & 0 & 0 & 0 & 0 \\ 0 & b_2 & b_2 & 0 & 0 \\ 0 & b_2 & b_2 & 0 & 0 \\ 0 & 0 & 0 & 0 & 0 \\ 0 & 0 & 0 & 0 & 0 \end{bmatrix},$$

then

$$T = \begin{bmatrix} T_1 & \\ & T_2 \end{bmatrix} + H.$$

$T_1^{2 \times 2}$ and $T_2^{3 \times 3}$ are symmetric tridiagonal matrices, and $H^{5 \times 5}$ has rank one. The rank-one matrix can be written more simply as $H = b_2 v v^T$, where

$$v = \begin{bmatrix} 0 \\ 1 \\ 1 \\ 0 \\ 0 \end{bmatrix}.$$

In the $n \times n$ case, write T as the sum of a 2×2 block symmetric tridiagonal matrix and a rank one matrix, known as a *rank-one correction*.

Note that $T_1(k, k) = a_k - b_k$ and $T_2(1, 1) = a_{k+1} - b_k$, and the rank-one correction matrix can be written as $t_k v v^T$, where

$$v = \begin{bmatrix} 0 & 0 & \dots & 1 & 1 & 0 & \dots & 0 \end{bmatrix}^T.$$

The two entries of 1 are at indices k and $k + 1$.

Suppose the divide-and-conquer algorithm is named dconquer and returns the eigenvalues and eigenvectors of a symmetric tridiagonal matrix using the format

[V, D] = dconquer(T).

The algorithm dconquer must be able to take any symmetric tridiagonal matrix T, split it into the sum of a 2×2 block symmetric tridiagonal matrix and a rank-one correction matrix and return the eigenvalues and eigenvectors of T. Here is how dconquer must function. Choose $k = \lfloor \frac{n}{2} \rfloor$ and form

$$T = \begin{bmatrix} T_1 & \\ & T_2 \end{bmatrix} + t_k v v^T.$$

Now compute

$$\begin{bmatrix} V_1, & D_1 \end{bmatrix} = \text{dconquer}(T_1), \quad \begin{bmatrix} V_2, & D_2 \end{bmatrix} = \text{dconquer}(T_2),$$

and put $\begin{bmatrix} V_1, & D_1 \end{bmatrix}, \begin{bmatrix} V_2, & D_2 \end{bmatrix}$, and $t_k v v^T$ together to obtain the matrices V and D. Each of the recursive calls dconquer (T_1) and dconquer (T_2) must divide their respective matrix as described and compute eigenvalues and eigenvectors for them. Continue this process until arriving at a set of 1×1 eigenvalue problems, each having a rank-one correction. This is called the *stopping condition*. These are easily solved, and a series of function returns solves all the problems encountered on the way to the stopping condition. Returning from the first recursive call gives the eigenvalues and eigenvectors of the initial matrix.

How do we find the eigenvalues of T from T_1, T_2 and the rank-one correction? By the spectral theorem,

$$T_1 = P_1 D_1 P_1^T, \quad T_2 = P_2 D_2 P_2^T,$$

where P_1, P_2 are orthogonal matrices of eigenvectors, and D_1, D_2 are diagonal matrices of corresponding eigenvalues. Then,

$$T = \begin{bmatrix} T_1 & 0 \\ 0 & T_2 \end{bmatrix} + t_k v v^T = \begin{bmatrix} P_1 D_1 P_1^T & 0 \\ 0 & P_2 D_2 P_2^T \end{bmatrix} + t_k v v^T.$$

Let $D = \begin{bmatrix} D_1 & 0 \\ 0 & D_2 \end{bmatrix}, u = \begin{bmatrix} P_1^T & 0 \\ 0 & P_2^T \end{bmatrix} v$ and form

$$\begin{bmatrix} P_1 & 0 \\ 0 & P_2 \end{bmatrix} \left(\begin{bmatrix} D_1 & 0 \\ 0 & D_2 \end{bmatrix} + t_k u u^T \right) \begin{bmatrix} P_1^T & 0 \\ 0 & P_2^T \end{bmatrix} = \begin{bmatrix} P_1 & 0 \\ 0 & P_2 \end{bmatrix} \begin{bmatrix} D_1 & 0 \\ 0 & D_2 \end{bmatrix} \begin{bmatrix} P_1^T & 0 \\ 0 & P_2^T \end{bmatrix}$$

$$+ t_k \begin{bmatrix} P_1 & 0 \\ 0 & P_2 \end{bmatrix} \begin{bmatrix} P_1^T & 0 \\ 0 & P_2^T \end{bmatrix} v v^T \begin{bmatrix} P_1 & 0 \\ 0 & P_2 \end{bmatrix} \begin{bmatrix} P_1^T & 0 \\ 0 & P_2^T \end{bmatrix}$$

$$= \begin{bmatrix} P_1 D_1 P_1^T & 0 \\ 0 & P_2 D_2 P_2^T \end{bmatrix} + t_k v v^T = T$$

We have shown that T is similar to the matrix

$$D + t_k u u^T,$$

and so it has the same eigenvalues as T. It can be shown [1, p. 218] that its eigenvalues are roots of the function

$$f(\lambda) = 1 + t_k \sum_{i=1}^{n} \frac{u_i^2}{d_i - \lambda}. \tag{19.17}$$

The equation $f(\lambda) = 0$ is known as the *secular equation*, and finding the roots of f accurately is not an easy problem. Figure 19.6 is a graph of f for particular values of t_k, u, and d, where $d = \text{diag}(D)$. It would seem reasonable to use Newton's method [33, pp. 66-71] to compute the roots, which lie between the singularities $\lambda = d_i$, called the *poles*. It is possible that the first iteration of Newton's method will take an initial approximation $\bar{\lambda}_0$ and produce a very large value $\bar{\lambda}_1$. Using the classical Newton's method can cause the algorithm to become unstable (Problem 19.28). The solution is to approximate

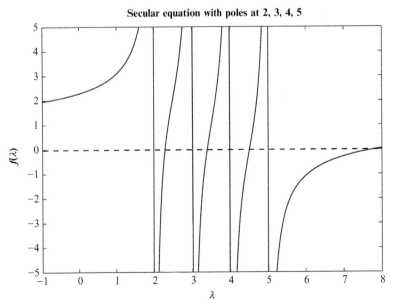

Secular equation with poles at 2, 3, 4, 5

FIGURE 19.6 Secular equation.

$f(\lambda)$, $d_i < \lambda < d_{i+1}$ by another function $h(\lambda)$ that makes the root computation stable. The interested reader should see Ref. [1, pp. 221-223] for the details.

The algorithm can compute an eigenvector from its associated eigenvalue using only $O(n)$ flops, The computation of eigenvectors for the divide-and-conquer algorithm will not be discussed. See Ref. [1, pp. 224-226] for the technique used.

The cost of finding the eigenvalues is $O(n^2)$, the same as for the Francis algorithm. The computation of an eigenvector using the Francis method costs $O(n^2)$ flops, as opposed to $O(n)$ flops for divide-and-conquer. We stated in the introduction to this chapter that this algorithm is more than twice as fast as the QR algorithm if both eigenvalues and eigenvectors of a symmetric tridiagonal matrix are required.

19.6.1   Using dconquer

Implementing dconquer using a MATLAB function is difficult. It is possible to write a function, say *myMEX*, in the programming language C in such a way that the function can be called from MATLAB. The function *myMEX* must written so it conforms with what is termed the MEX interface and must be compiled using the MATLAB command mex (see Ref. [20]). This interface provides access to the input and output parameters when the function is called from MATLAB. Normally, the function serves an interface to a Fortran or C machine code library. For instance, LAPACK [62] is a set of functions written in Fortran that provide methods for solving systems of linear equations, least-squares solutions of linear systems of equations, eigenvalue problems, and executing various factorizations. One of the LAPACK functions, dsyevd, transforms A to a symmetric tridiagonal matrix, and then applies the divide-and-conquer algorithm to compute the eigenvalues and corresponding eigenvectors. It will compute the eigenvalues only, but if we want only eigenvalues there is no advantage to the divide-and-conquer algorithm.

Change into the directory "divide-and-conquer" and then into a subdirectory for your computer architecture, "windows" or "OSX_linux." There you will find a function, dconquer.c, that uses the MEX interface to call dsyevd. Open the file reame.pdf for instructions concerning compiling dconquer.c. Applying mex produces a file of the form "dconquer.mexw64" if the operating system is 64-bit Windows, or "dconquer.mexmaci64" for 64-bit OS X. Call dconquer using the format

```
[V, D] = dconquer(A);
```

Example 19.11.   The following MATLAB sequence builds a 1000×1000 symmetric matrix, computes its eigenvalues and eigenvectors using dconqer, and times its execution. As evidence of accuracy, the code computes $\left\| VDV^T - A \right\|_2$ and $\left\| V^T V - I \right\|_2$.

```
>> A = randi([-1000000 1000000],1000,1000);
>> A = A + A';

>> tic;[Vdconquer Ddconquer] = dconquer(A);toc;
Elapsed time is 0.272290 seconds.
>> norm(Vdconquer*Ddconquer*Vdconquer'-A)

ans =
   3.0434e-07

>> norm(Vdconquer'*Vdconquer - eye(1000))

ans =
   8.7754e-15                                                                    ■
```

Remark 19.5.   In general, MEX is complicated and should be used sparingly. We applied it here to call a specific method whose complexity made it difficult to implement in an m-file.

19.7   CHAPTER SUMMARY

The Spectral Theorem and Properties of a Symmetric Matrix

At this point in the book, we have a great deal of machinery in place and have discussed the accurate computation of eigenvalues. We also are familiar with the Schur's triangularization theorem from Section 18.7, which says that any $n \times n$ matrix A, even a singular one, can be factored as $A = PTP^T$, where P is an orthogonal matrix, and T is upper triangular. We use it to prove the spectral theorem for symmetric matrices, a result we have used without proof until this chapter.

This section also reviews some fundamental facts about real symmetric matrices and shows that the condition number of each eigenvalue is 1 (perfect). Unfortunately, this is not true for eigenvectors. It is possible for a symmetric matrix to have ill-conditioned eigenvectors.

The Jacobi Method

The Jacobi method directly reduces a symmetric matrix to diagonal form without an initial conversion to tridiagonal form. It works by zeroing out pairs of equal entries, a_{ij}, a_{ji} off the diagonal until the off $(A) = \sqrt{\|A\|_F^2 - \sum_{i=1}^{n} a_{ii}^2}$ is sufficiently small. We prove that the rate of convergence is linear. However, it can be shown that the average rate of convergence (asymptotic rate) is quadratic.

The Symmetric QR Iteration Method

The first step of this algorithm is the orthogonal similarity transformation of A to tridiagonal matrix T using Householder reflections. The algorithm presented takes advantage of matrix symmetry. The final step is the application of the QR iteration with the Wilkinson shift that reduces T to a diagonal matrix. As with a nonsymmetric matrix, Givens rotations are used to find the QR decomposition of each shifted matrix. This is an $O\left(n^3\right)$ algorithm.

The Symmetric Francis Algorithm

The single-shift Francis algorithm is the method of choice for computing the eigenvalues and associated eigenvectors of a symmetric matrix. The first phase is reduction to tridiagonal form, as with the symmetric QR iteration. The reduction to diagonal form uses orthogonal similarity transformations produced by Givens rotations applied to a Wilkinson-shifted matrix, and the QR algorithm is not used directly. Using deflation, iterations chase a bulge from the top off the bottom of the matrix until the current diagonal entry is sufficiently close to an eigenvalue. The method is still $O\left(n^3\right)$ but generally performs better than the symmetric QR iteration.

The Bisection Method

This algorithm is very different from the other algorithms we have discussed. Finding the roots of a nonlinear function is covered in a numerical analysis or numerical methods course. One of the methods, bisection, can be applied to compute the eigenvalues of a matrix, which are roots of the nonlinear function $p(\lambda) = \det(A - \lambda I)$, the characteristic polynomial of A. Using Gaussian elimination with partial pivoting, $p(\lambda)$ can be computed in a stable fashion. For a general matrix $A \in \mathbb{R}^{n \times n}$, this method of computing eigenvalues cannot compete with the methods we discussed in Chapter 18. For a symmetric matrix, upper Hessenberg form is a symmetric tridiagonal matrix. After transforming A to such a matrix T, let $T^{(k)}$ be the submatrix $T(1:k, 1:k)$ of T, and $\{p_0, p_1, p_2, \ldots, p_n\}$ be the characteristic polynomials of $T^{(k)}$, $p_0(\lambda) = 1$. The characteristic polynomials are evaluated by a simple recurrence relation. The eigenvalues of $T^{(k)}$ and $T^{(k+1)}$ strictly interlace, which means that $\lambda_i^{(k+1)} < \lambda_i^{(k)} < \lambda_{i+1}^{(k+1)}$. In turn, this property can be used to prove that the number of sign changes between consecutive numbers of the sequence $\{p_0(\lambda), p_1(\lambda), p_2(\lambda), \ldots, p_n(\lambda)\}$ is equal to the number of eigenvalues smaller than λ. This remarkable property enables accurate evaluation of the eigenvalues using the bisection technique.

The Divide-and-Conquer Method

The recursive divide-and-conquer is the fastest algorithm for computing both eigenvalues and eigenvectors of a symmetric matrix A. Like the QR algorithm, it first transforms the A into a tridiagonal matrix T. The computation then proceeds by writing T in the form

$$T = \begin{bmatrix} T_1 & \\ & T_2 \end{bmatrix} + t_k v v^{\mathrm{T}}.$$

T is the sum of a 2×2 block symmetric tridiagonal matrix and a rank-one correction. Using recursion, the algorithm finds the eigenvalues and eigenvectors of T. During the computation, the secular equation $f(\lambda) = 1 + t_k \sum_{i=1}^{n} \frac{u_i^2}{d_i - \lambda}$ must be solved. This secular function has singularities (poles) at the diagonal entries, d_i, of T. Rather than using Newton's method, the best approach to finding the eigenvalues located between poles is to find the roots of a function that approximates f.

19.8 PROBLEMS

In these problems, the term "by hand" means that you must show your work step by step. You can use MATLAB. For instance, if you are required to form $J(1, 2, c, s) AJ(1, 2, c, s)^T$ with a 4×4 matrix do this:

```
[c, s] = givensparms(A(1,1),A(2,1));
J = eye(4);
J(1,1) = c, J(2,2) = c;
J(2,1) = -s, J(1,2) = s;
A1 = J*A*J';
```

19.1 Let A be an $n \times n$ symmetric tridiagonal matrix with its sub- and superdiagonals nonzero. Prove that the eigenvalues of A are distinct by answering parts (a)-(e).

  a. If λ is an eigenvalue, show that the rank of $E = A - \lambda I$ is at most $n - 1$.
  b. Consider the upper triangular $(n - 1) \times n$ submatrix $E(2:n, 1:n)$. Show that E has rank $n - 1$.
  c. Show that rank $(E) = $ rank $(A - \lambda I) = n - 1$.
  d. Show that the null space of E is spanned by an eigenvector corresponding to λ.
  e. Prove that the symmetry of A implies that λ must be distinct.

19.2 If A is symmetric, show that trace $\left(A^2\right) = \sum_{i=1}^{n} \sum_{j=1}^{n} a_{ij}^2$.

19.3 The QR iteration with the Wilkinson shift and the symmetric Francis algorithm both begin by zeroing out $T(2, 1)$ using an orthogonal similarity transformation. Show that the first column is the same for both algorithms.

19.4 For the matrix $T = \begin{bmatrix} 1 & 3 & 0 \\ 3 & 1 & 4 \\ 0 & 4 & 2 \end{bmatrix}$, execute the QR iteration three times using the Wilkinson shift to estimate the eigenvalue 6.3548. Just show the values of σ and $T = RQ + \sigma I$ for each iteration.

19.5 For the matrix $T = \begin{bmatrix} -2 & 4 & 0 & 0 \\ 4 & -6 & 4 & 0 \\ 0 & 4 & 4 & 1 \\ 0 & 0 & 1 & -6 \end{bmatrix}$, execute one bulge chase by hand assuming $\sigma = 0$.

19.6 Let $A = \begin{bmatrix} 2 & 1 & 0 \\ 1 & 2 & 1 \\ 0 & 1 & 6 \end{bmatrix}$. Perform four iterations of the bisection method by hand to estimate the eigenvalue between 2 and 3.

For Problems 19.7–19.9, you will find it useful to write a MATLAB function

```
sign_changes = d(T,lambda)
```

that computes d(λ).

19.7 Let $A = \begin{bmatrix} 1 & -1 & 0 & 0 \\ -1 & 1 & 2 & 0 \\ 0 & 2 & 2 & -1 \\ 0 & 0 & -1 & 1 \end{bmatrix}$.

  a. Show that there must be an eigenvalue greater than or equal to 2 and an eigenvalue less than zero.
  b. Show there are two eigenvalues between 0.12 and 2.

19.8 Let $A = \begin{bmatrix} 3 & -1 & 0 \\ -1 & 1 & -1 \\ 0 & -1 & 1 \end{bmatrix}$.

  a. Show there must be one negative eigenvalue.
  b. Show there must be one eigenvalue in the range $0 \le \lambda < 2$.
  c. Show there is one eigenvalue greater than or equal to 2.

19.9 How many eigenvalues of $A = \begin{bmatrix} 1 & -1 & 0 & 0 \\ -1 & 1 & -1 & 0 \\ 0 & -1 & 1 & -1 \\ 0 & 0 & -1 & 1 \end{bmatrix}$ lie in the interval $0 < \lambda < 2$?

19.10 Let

$$T = \begin{bmatrix} a_1 & b_1 & & & & \\ b_1 & a_2 & \ddots & & & \\ & \ddots & \ddots & b_{n-2} & & \\ & & b_{n-2} & \ddots & 0 \\ & & & 0 & a_n \end{bmatrix},$$

where $a_i \neq 0$, $1 \leq i \leq n$, $b_i \neq 0$, $1 \leq i \leq n-2$.

a. Show that a_n is an eigenvalue of T with associated eigenvector $e_n = \begin{bmatrix} 0 & 0 & \ldots & 0 & 1 \end{bmatrix}^{\mathrm{T}}$.

b. Explain how this result relates to our choice of convergence criteria for the QR and Francis methods.

19.11 Outline an algorithm for finding the eigenvalues of a reduced symmetric tridiagonal matrix (the subdiagonal contains one or more zeros).

19.12 Let $A = \begin{bmatrix} 0 & 1 \\ 1 & 0 \end{bmatrix}$. Show that the QR algorithm with shift $\sigma = t_{k,k}$ fails, but the Wilkinson shift succeeds.

19.13 Let $p_0(\mu), p_1(\mu), \ldots, p_{i-1}(\mu), p_i(\mu), \ldots, p_n(\mu)$ be the Sturm sequence used by the bisection method, with $b_i \neq 0$, $1 \leq i \leq n$. Show it is not possible that $p_{i-1}(\mu) = 0$ and $p_i(\mu) = 0$; in other words, there cannot be two consecutive zero values in the sequence. Hint: What does Equation 19.16 say about $p_{i-2}(\mu)$?

19.14 Prove Lemma 19.3. Hint: Look at it as a problem involving block matrices.

19.15 Let A be a positive definite matrix and $A = R^{\mathrm{T}}R$ be the Cholesky decomposition. Using the singular value decomposition of R, show how to compute the eigenvalues and associated eigenvectors of A.

19.16 If A is positive definite, the Cholesky decomposition can be used to determine the eigenvalues of A. Here is an outline of the algorithm:

```
A = R^T R
A_1 = RR^T
for i = 1:maxiter do
    A_i = R_i^T R_i
    A_{i+1} = R_i R_i^T
end for
return A_maxiter
```

It can be shown that the sequence converges to a diagonal matrix of eigenvalues.

a. Prove that if A is positive definite, then any matrix similar to A is also positive definite. Hint: What can you say about the eigenvalues of a positive definite matrix?

b. Show that each matrix, A_i, is similar to A so that upon termination of the algorithm, A and A_{maxiter} have the same eigenvalues.

19.17 We know that if v is an eigenvector of A, $\frac{v^{\mathrm{T}}Av}{v^{\mathrm{T}}v}$ is called the Rayleigh quotient and that the corresponding eigenvalue $\lambda = \frac{v^{\mathrm{T}}Av}{v^{\mathrm{T}}v}$. Now, suppose v is an approximation to an eigenvector. How well does the Rayleigh quotient approximate λ? We can answer that question when A is symmetric.

a. Without loss of generality, assume that v is a good approximation to eigenvector v_1 of the symmetric matrix A. Argue that $v = c_1 v_1 + c_2 v_2 + \cdots + c_n v_n$, where the v_i are an orthonormal set of eigenvectors of A corresponding to eigenvalues λ_i.

b. Show that

$$\sigma = \frac{v^{\mathrm{T}}Av}{v^{\mathrm{T}}v} = \frac{\lambda_1 c_1^2 + \lambda_2 c_2^2 + \cdots \lambda_n c_n^2}{c_1^2 + c_2^2 + \cdots + c_n^2}.$$

c. Using the result of part (b), show that

$$\sigma = \lambda_1 \left[\frac{1 + \left(\frac{\lambda_2}{\lambda_1}\right)\left(\frac{c_2}{c_1}\right)^2 + \cdots + \left(\frac{\lambda_n}{\lambda_1}\right)\left(\frac{c_n}{c_1}\right)^2}{1 + \left(\frac{c_2}{c_1}\right)^2 + \cdots + \left(\frac{c_n}{c_1}\right)^2} \right],$$

and argue that σ is a close approximation to λ_1.

19.18 We have not discussed another approach to the symmetric eigenvalue problem, the *Rayleigh quotient iteration*. Basically, it is inverse iteration using the Rayleigh quotient as the shift. From Problem 19.17, we know that if v is a good approximation to an eigenvector of the symmetric matrix A, then the Rayleigh quotient $\frac{v^T A v}{v^T v}$ is a good approximation to the corresponding eigenvalue. Start with an initial approximation, v_0, to an eigenvector, compute the Rayleigh quotient, and begin inverse iteration. Unlike the inverse iteration algorithm discussed in Section 18.9, the shift changes at each iteration to the Rayleigh quotient determined by the next approximate eigenvector. A linear system must be solved for each iteration. The Rayleigh quotient iteration computes both an eigenvector and an eigenvalue, and convergence is almost always cubic [1, pp. 214-216]. This convergence rate is very unusual and means that the number of correct digits triples for every iteration.

```
function ROITER(A,v0,tol,maxiter)
    v0 = v0/‖v0‖2
    σ0 = v0ᵀAv0/v0ᵀv0
    iter=0
    i=0
    while (‖Avᵢ − σᵢvᵢ‖2 ≥ tol) and (iter ≤ maxiter) do
        i=i+1
        vᵢ = (A − σᵢ₋₁I)⁻¹ vᵢ₋₁
        vᵢ = vᵢ/‖vᵢ‖2
        σᵢ = vᵢᵀAvᵢ/vᵢᵀvᵢ
        iter=iter+1
    end while

    if iter>maxiter then
        iter=-1
    end if
    return[ vᵢ, σᵢ, iter ]
end function
```

Let

$$A = \begin{bmatrix} 4 & 6 & 7 \\ 6 & 12 & 11 \\ 7 & 11 & 4 \end{bmatrix}.$$

a. Let $v_0 = \begin{bmatrix} 0.9 & -0.8 & 0.4 \end{bmatrix}^T$, and perform two iterations of rqiter by hand.

b. To obtain the starting approximation, perform two iterations of the power method starting with $v_0 = \begin{bmatrix} 1 & 1 & 1 \end{bmatrix}^T$. Using the new v_0, perform one Rayleigh quotient iteration by hand. Start a second and comment on what happens. What eigenvalue did you approximate, and why?

19.8.1 MATLAB Problems

19.19

a. Implement the classical Jacobi method by modifying eigsymj to create a function cjacobi(A,tol,maxiter). You will need to implement a function jacobifind that finds the largest off-diagonal entry in magnitude in the symmetric matrix. The calling format should be the same as eigsymj, except return the number of iterations required to attain the tolerance rather than the number of sweeps. The function eigsymj executes $\frac{n(n-1)}{2}$ Jacobi rotations per sweep, so for the classical Jacobi method the maxiter will be much larger than maxsweeps.

b. Create a random 100×100 symmetric matrix and time the execution of `cjacobi` and `eigsymj`. Compute $\|VDV^\mathrm{T} - A\|_2$ for each method. Compare the results.

19.20 Find all the eigenvalues of the following symmetric matrices using both the Jacobi method and the Francis algorithm. For each method, compute $\|\mathrm{Emat} - E\|_2$, where Emat is the MATLAB result, and E is the result from the method. Sort the eigenvalues before computing the norms.

a. $A = \begin{bmatrix} -12 & 6 & 1 \\ 6 & 3 & 2 \\ 1 & 2 & 15 \end{bmatrix}$

b. $A = \begin{bmatrix} -8 & 16 & 23 & -13 \\ 16 & 9 & 2 & 3 \\ 23 & 2 & 1 & -23 \\ -13 & 3 & -23 & -7 \end{bmatrix}$

19.21 The MATLAB command

```
A = gallery('pei',n,alpha),
```

where `alpha` is a scalar, returns the symmetric matrix `alpha eye(n) + ones(n)`. The default for `alpha` is 1, and the matrix is singular for `alpha` equal to 0. The eigenvalues are

$$\lambda_1 = \lambda_2 = \cdots = \lambda_{n-1} = \mathrm{alpha}, \quad \lambda_n = n + \mathrm{alpha}.$$

For parts (a) and (b), use (1) `eigsymj`, (2) `eigsymqr`, (3) `eigsymb`, and (4) `dconquer`, to compute the eigenvalues. In each case, compute $\|E - E_i\|_2$, where E are the exact eigenvalues. Do these methods handle the eigenvalues of multiplicity $n - 1$ properly?

a. $n = 25$, alpha = 5

b. $n = 50$, alpha = 0

c. Let `A = gallery('pei',5,3)`, and compute the eigenvector matrix `V1` and the corresponding diagonal matrix `D1` of eigenvalues. Perturb `A(5,1)` by 1.0×10^{-8} and find matrices `V2` and `D2`. Compute $\|D1 - D2\|_2$, and list the eigenvectors `V1` and `V2`. Explain the results.

19.22 Use the function `eigsymj` to compute all the eigenvalues of the 50×50 Hilbert matrix. Use tol $= 1.0 \times 10^{-10}$. Compute the norm of difference between that solution and the one obtained using the MATLAB function `eig`. Given that the Hilbert matrices are very ill-conditioned, explain your result.

19.23 Problem 19.16 presents the basics for a method to compute the eigenvalues of a positive definite matrix.

a. Implement a function `choleig` that takes a positive definite matrix A and executes `maxiter` iterations. After each iteration, use the MATLAB function `spy` to show the location of the nonzero entries as follows:

```
A(abs(A)<1.e-7)=0;
spy(A);
pause;
```

This will allow you to "watch" convergence to a diagonal matrix of eigenvalues. The function should verify that A is positive definite.

b. Test `choleig` with the matrix

```
A = gallery('gcdmat',4);
```

c. Create a 3×3 positive definite matrix using the code provided, and then run `choleig` with `maxiter = 75` and watch convergence. Depending on the matrix, it may or may not graphically show as a diagonal matrix. In any case, check the results against `eig (A)`.

```
A = diag(randi([1 10],3,1));
B = randn(3,3);
[Q,R] = qr(B);
A = Q'*A*Q;
```

19.24 The term *eigenvector localization* applies to an eigenvector where the majority of its length is contributed by a small number of entries. This means that majority of its entries are zero or close to zero. This phenomenon is well known in several scientific applications such as quantum mechanics, DNA data, and astronomy. If the eigenvector is normalized, one measure of this property is the *inverse participation ratio* (IPR) that is defined as

$$\text{IPR}(v) = \sum_{i=1}^{n} v_i^4.$$

The larger the value of IPR, the more localized the eigenvector.

a. If the eigenvector values are equally distributed throughout its indices, show that the IPR is $\frac{1}{n}$.

b. Show that if the eigenvector has only one nonzero entry, its IPR is 1.

c. If n is an integer, the following statements generate a random symmetric tridiagonal matrix. Create a 200×200 random symmetric tridiagonal matrix and plot the eigenvector number against IPR (v_i). What do you observe?

```
d = randn(n,1);
sd = randn(n-1,1);
A = trid(sd,d,sd);
```

d. Do part (c) with the 200×200 matrix $B = \begin{bmatrix} 1+2r & -r & 0 & \cdots & & 0 \\ -r & 1+2r & -r & \cdots & & 0 \\ 0 & \ddots & \ddots & \ddots & & \vdots \\ \vdots & \vdots & & -r & 1+2r & -r \\ 0 & 0 & 0 & & -r & 1+2r \end{bmatrix}$ used in the solution to the

heat equation in Section 12.2. Let $r = 0.5$.

19.25 Modify the function bisection so it computes all the eigenvalues in an interval $a < \lambda < b$. Name function bisectinterval having the declaration

```
lambda = bisectinterval(T,a,b,tol)
```

Test it using a random symmetric tridiagonal matrix and wilkinson(21).

19.26 This problem shows how shifts help when computing eigenvalues. Create functions that are simple modifications of eigsymqr and eigsymb as follows:

- eigsymqr0: Remove the shift from the code.
- eigsymqrr: Replace the Wilkinson shift by $\sigma = t_{k,k}$, the value we used for a general matrix.
- eigsymb0, chase0: Remove the shift from the chase code.
- eigsymbr, chaser: Replace the Wilkinson shift by $\sigma = t_{k,k}$.

In the functions eigsymqr0 and eigsymb0, perform the following code replacement so termination occurs the first time the iteration does not converge to an eigenvalue within the allotted iterations.

Replace

```
if iter > maxiter
    fprintf('Failure of convergence. ');
    fprintf('Current eigenvalue approximation %g\n',T(k,k));
    break;
end
```

by

```
if iter > maxiter
    fprintf('Failure of convergence. ');
    varargout{1} = {};
    varargout{2} = {};
    return;
end
```

Generate a random 400×400 symmetric matrix (A = randn(400,400), A = A + A'). Time the use of the original functions eigsymqr and eigsymb to compute the eigenvalues and eigenvectors of A. Now do the same for each of the modified functions, and discuss the results. For the methods that converge, compute $\left\| VDV^T - A \right\|_2$.

19.27

    a. Implement the Rayleigh quotient iteration described in Problem 19.18 in the MATLAB function `rqiter`.

    b. Using a random v_0, test `rqiter` with random matrices of dimensions 5×5 and 25×25. Use tol $= 1.0 \times 10^{-12}$, maxiter $= 25$.

19.28

    a. Implement Newton's method with a MATLAB function having calling syntax

    `root = newton(f,df,x0,tol,maxiter);`

    The argument `f` is the function, `df` is $f'(x)$, `x0` is the initial approximation, `tol` is the required relative error bound $\left(\frac{|x_{i+1} - x_i|}{|x_i|} < \text{tol} \right)$, and `maxiter` is the maximum number of iterations to perform.

    b. Graph $f(\lambda) = 1 + \frac{1}{1-\lambda} + \frac{1}{2-\lambda} + \frac{1}{3-\lambda}$ and use `newton` to estimate all roots of the f. How well did Newton's method do?

    c. Now graph $g(\lambda) = 1 + \frac{0.005}{1-\lambda} + \frac{0.005}{2-\lambda} + \frac{0.005}{3-\lambda}$ over the interval $0.95 < \lambda < 1.05$. Describe the shape of the graph for all points except those very close to the pole at $\lambda = 1$.

    d. Plot $g(\lambda)$ of the interval $1.004 < \lambda < 1.006$. Is there a root in the interval?

    e. Try to compute the root using Newton's method with $x_0 = 1.01$. Explain the results.

Chapter 20

Basic Iterative Methods

You should be familiar with

- Eigenvalues and the spectral radius
- Matrix norms
- Finite difference methods for approximating derivatives
- Block matrices

When using finite difference methods for the solution of partial differential equations, the matrix may be extremely large, possibly 1000×1000 or larger. The main source of large matrix problems is the discretization of partial differential equations, but large linear systems also arise in other applications such as the design and computer analysis of circuits, and chemical engineering processes. Often these matrices are *sparse*, which means that most of the matrix entries are 0. Standard Gaussian elimination turns zeros into nonzeros, reducing the sparsity. For large, sparse matrices, the preferred method of solution is to use an *iterative method*. An iterative method generates a sequence that converges to the solution, and the iteration is continued until a desired error tolerance is satisfied. Unlike Gaussian elimination, iterative methods do not alter the matrix, but use only a small set of vectors obtained from the matrix, so they use far less storage than working directly with the matrix. This chapter presents three classical iterative methods, the Jacobi, Gauss-Seidel, and the successive overrelaxation (SOR) iterations. These methods are relatively slow, but they provide an introduction to the idea of using iteration to solve systems and form a basis for more sophisticated methods such as multigrid. The main idea of multigrid is to accelerate the convergence of a basic iterative method by solving a coarse problem. It takes an $n \times n$ grid and uses every other point on an $n/2 \times n/2$ grid to estimate values on the larger grid using interpolation. Points on an $n/4 \times n/4$ grid are used to approximate values on the $n/2$ grid, and so forth, forming a recursive algorithm. These methods are beyond the scope of this book. The interested reader can consult Refs. [1, 63] for detailed information about these iterative methods.

20.1 JACOBI METHOD

For the Jacobi method, write each unknown in terms of the other unknowns. We illustrate this process using a 4×4 system.

$$a_{11}x_1 + a_{12}x_2 + a_{13}x_3 + a_{14}x_4 = b_1 \implies x_1 = [b_1 - (a_{12}x_2 + a_{13}x_3 + a_{14}x_4)]/a_{11}$$
$$a_{21}x_1 + a_{22}x_2 + a_{23}x_3 + a_{24}x_4 = b_2 \implies x_2 = [b_2 - (a_{21}x_1 + a_{23}x_3 + a_{24}x_4)]/a_{22}$$
$$a_{31}x_1 + a_{32}x_2 + a_{33}x_3 + a_{34}x_4 = b_3 \implies x_3 = [b_3 - (a_{31}x_1 + a_{32}x_2 + a_{34}x_4)]/a_{33}$$
$$a_{41}x_1 + a_{42}x_2 + a_{43}x_3 + a_{44}x_4 = b_4 \implies x_4 = [b_4 - (a_{41}x_1 + a_{42}x_2 + a_{43}x_3)]/a_{44}$$

Assume initial values for x_1, x_2, x_3, and x_4, insert them in the right-hand side of the equations and compute a second set of approximate values for the unknowns. These new values are then substituted into the right-hand side of the equations to obtain a third set of approximate solutions, and so forth, until the iterations produce a relative error estimate that is small enough. For the general case of n unknowns, the iteration is defined by

$$x_i = \frac{1}{a_{ii}} \left[b_i - \left(\sum_{j=1, j \neq i}^{n} a_{ij}x_j \right) \right], \quad 1 \leq i \leq n \tag{20.1}$$

Definition 20.1. Let x be a vector obtained during an iteration. In order to clearly distinguish the number of the iteration and a component of the vector, we use the notation $x_i^{(k)}$, where k refers to the iteration number and i refers to the ith component of $x^{(k)}$.

Numerical Linear Algebra with Applications. http://dx.doi.org/10.1016/B978-0-12-394435-1.00020-X

If a good initial approximation for the solution $x^{(0)}$ is known, use $x_1^{(0)}, x_2^{(0)}, \ldots, x_n^{(0)}$ to begin the iteration. If there is no estimate for the initial value $x^{(0)}$, use $x^{(0)} = 0$. The second approximation for the solution, $x_1^{(1)}, x_2^{(1)}, \ldots, x_n^{(1)}$ is calculated by substituting the first estimate into the right-hand side of Equation 20.1:

$$x_i^{(1)} = \frac{1}{a_{ii}} \left[b_i - \left(\sum_{j=1, j \neq i}^{n} a_{ij} x_j^{(0)} \right) \right], \quad 1 \leq i \leq n$$

In general, the estimate $x^{(k)}$ for the solution is calculated from the estimate x^{k-1} by

$$x_i^{(k)} = \frac{1}{a_{ii}} \left[b_i - \left(\sum_{j=1, j \neq i}^{n} a_{ij} x_j^{(k-1)} \right) \right], \quad 1 \leq i \leq n \tag{20.2}$$

The iterations continue until the error tolerance tol is satisfied. What criterion do we use to determine that the algorithm has met the required tolerance? An obvious choice is $\|Ax_i - b\|_2 < \text{tol}$. However, note that

$$A(x - x_i) = Ax - Ax_i = b - Ax_i,$$

so

$$\|x - x_i\|_2 = \left\| A^{-1} (b - Ax_i) \right\|_2 \leq \left\| A^{-1} \right\|_2 \|b - Ax_i\|_2 \leq \left\| A^{-1} \right\|_2 \text{tol}.$$

If $\left\| A^{-1} \right\|_2$ is large, $\|x - x_i\|$ can be large. The best alternative is to apply a relative error estimate and terminate when

$$\frac{\|b - Ax_i\|_2}{\|b\|} \leq \text{tol}.$$

Thus, terminate the iteration when the current residual relative to b is sufficiently small [16, pp. 335-336]. We will call $\frac{\|b - Ax_i\|_2}{\|b\|}$, the *relative residual*.

Example 20.1. Consider the system $Ax = b$, where $A = \begin{bmatrix} 5 & -1 & 2 \\ -1 & 4 & 1 \\ 1 & 6 & -7 \end{bmatrix}$ and $b = \begin{bmatrix} 1 \\ -2 \\ 5 \end{bmatrix}$. We start with $x^{(0)} = 0$,

execute the first two iterations in detail, continue for a total of 12 iterations, and compute the relative residual.

$$x_1^{(1)} = \frac{1}{5}(1) = 0.2000, \quad x_2^{(1)} = \frac{1}{4}(-2) = -0.5000, \quad x_3^{(1)} = -0.7143$$

$$x_1^{(2)} = \frac{1}{5}\left(1 - \frac{1}{2} + \frac{10}{7}\right) = 0.3857, \quad x_2^{(2)} = \frac{1}{4}\left(-2 + \frac{1}{5} + \frac{5}{7}\right) = -0.2714, x_3^{(2)} = -\frac{1}{7}\left(5 - \frac{1}{5} - 6\left(-\frac{1}{2}\right)\right) = -1.1143$$

$$\vdots$$

$$x_1^{(12)} = 0.4838, \quad x_2^{(12)} = -0.1795, \quad x_3^{(12)} = -0.7998$$

$$\frac{\left\| b - Ax^{(12)} \right\|_2}{\|b\|_2} = 1.1116 \times 10^{-3} \qquad \blacksquare$$

20.2 THE GAUSS-SEIDEL ITERATIVE METHOD

In the *Gauss-Seidel method*, start with approximate values $x_2^{(0)}, \ldots, x_n^{(0)}$ if known; otherwise choose $x^{(0)} = 0$. Use these values to calculate $x_1^{(1)}$. Use $x_1^{(1)}$ and $x_3^{(0)}, \ldots, x_n^{(0)}$ to calculate $x_2^{(1)}$, and so forth. At each step, we are applying new vector component values as soon as we compute them. The hope is that this strategy will improve the convergence rate. Applying this method with Equation 20.1, we have the iteration formula:

$$x_1^{(k)} = \frac{1}{a_{11}} \left[b_1 - \left(\sum_{j=2}^{n} a_{1j} x_j^{(k-1)} \right) \right] \tag{20.3}$$

$$x_i^{(k)} = \frac{1}{a_{ii}}\left[b_i - \left(\sum_{j=1}^{i-1} a_{ij}x_j^{(k)} + \sum_{j=i+1}^{n} a_{ij}x_j^{(k-1)}\right)\right], \quad i = 2,3,\ldots,n-1 \tag{20.4}$$

$$x_n^{(k)} = \frac{1}{a_{nn}}\left[b_n - \sum_{j=1}^{n-1} a_{nj}x_j^{(k)}\right] \tag{20.5}$$

Example 20.2. Use the matrix of Example 20.1 and apply the Gauss-Seidel method, with the iteration defined by Equations 20.3–20.5. Begin with $x^{(0)} = 0$, execute the first two iterations in detail, continue for a total of 12 iterations, and compute the relative residual.

$$x_1^{(1)} = \frac{1}{5}(1) = 0.2000, \quad x_2^{(1)} = \frac{1}{4}\left[-2 - \left(-\frac{1}{5} + (1)0\right)\right] = -0.4500, x_3^{(1)} = -\frac{1}{7}\left[5 - \left((1)\frac{1}{5} + 6\left(-\frac{9}{20}\right)\right)\right] = -1.0714$$

$$x_1^{(2)} = \frac{1}{5}\left[1 - \left((-1)\left(-\frac{9}{20}\right) + 2\left(-\frac{15}{14}\right)\right)\right] = 0.5386, x_2^{(2)} = \frac{1}{4}\left[-2 - \left((-1)\frac{377}{700} + (1)\left(-\frac{15}{14}\right)\right)\right] = -0.0975,$$

$$x_3^{(3)} = -\frac{1}{7}\left[5 - \left((1)\left(\frac{377}{700}\right) + 6\left(-\frac{39}{400}\right)\right)\right] = -0.7209$$

$$\vdots$$

$$x_1^{(12)} = 0.4837, \quad x_2^{(12)} = -0.1794, \quad x_3^{(12)} = -0.7989$$

$$\frac{\|b - Ax^{(12)}\|_2}{\|b\|_2} = 2.8183 \times 10^{-7}$$

If you compare this result with that of Example 20.1, it is clear that the Gauss-Seidel iteration obtained higher accuracy in the same number of iterations. ∎

20.3 THE SOR ITERATION

The SOR method was developed to accelerate the convergence of the Gauss-Seidel method. The idea is to successively form a weighted average between the previously computed value $x_i^{(k-1)}$ and the new value

$$x_i^{(k)} = \frac{1}{a_{ii}}\left[b_i - \left(\sum_{j=1}^{i-1} a_{ij}x_j^{(k)} + \sum_{j=i+1}^{n} a_{ij}x_j^{(k-1)}\right)\right], \quad i = 1,2,\ldots,n.$$

Weight the newly computed value by ω and the previous value by $(1 - \omega)$. By assuming that $\sum_{j=1}^{i-1} a_{ij}x_j^{(k)}$ and $\sum_{j=i+1}^{n} a_{ij}x_j^{(k-1)}$ are ignored when $i = 1$ and n, respectively, we have

$$x_i^{(k)} = \frac{\omega}{a_{ii}}\left[b_i - \left(\sum_{j=1}^{i-1} a_{ij}x_j^{(k)} + \sum_{j=i+1}^{n} a_{ij}x_j^{(k-1)}\right)\right] + (1 - \omega) x_i^{(k-1)}, \quad i = 1,2,\ldots,n \tag{20.6}$$

For this method to provide an improvement over the Gauss-Seidel method, ω must be chosen carefully. It is called the *relaxation parameter*. If $\omega = 1$, the SOR method and the Gauss-Seidel method are identical. If $\omega > 1$, we are said to be using *overrelaxation* and if $\omega < 1$ we are using *underrelaxation*.

Example 20.3. Consider the system with $A = \begin{bmatrix} 1 & 1 & 1 \\ 1 & 2 & 1 \\ 1 & 1 & 3 \end{bmatrix}$ and $b = \begin{bmatrix} -1 & 5 & 7 \end{bmatrix}^T$, whose exact solution is $\begin{bmatrix} -11 & 6 & 4 \end{bmatrix}^T$. Apply SOR with $\omega = 1.1$ in detail for two iterations, continue for a total of 15 iterations, and compute the relative residual. Also show the result of applying the Jacobi and Gauss-Seidel methods for 15 iterations.

$$x_1^{(1)} = 1.1\,(-1) = -1.1, \quad x_2^{(1)} = \frac{1.1}{2}\,[5 + 1.1] = 3.3550, \quad x_3^{(1)} = \frac{1.1}{3}\,[7 + 1.1 - 3.3550] = 1.7398$$

$$x_1^{(2)} = 1.1\,[-1 - 3.3550 - 1.7398] + (-0.1)\,(-1.1) = -6.5943$$

$$x_2^{(2)} = \frac{1.1}{2}\,[5 - (-6.5943) - 1.7398] + (-0.1)\,(3.3550) = 5.0845$$

$$x_3^{(2)} = \frac{1.1}{3}\,[7 - (-6.5943) - 5.0845] + (-0.10)\,(1.7398) = 2.9463$$

$$\vdots$$

$$x_1^{(15)} = -11.0000, x_2^{(15)} = 6.0000, x_3^{(15)} = 4.0000$$

$$\frac{\|b - Ax^{(15)}\|_2}{\|b\|_2} = 8.18045 \times 10^{-7}$$

After 15 iterations of Gauss-Seidel, the relative residual is 4.72×10^{-5}. It is interesting to note that the Jacobi iteration yields a relative residual of 3.8521; in fact, it diverges. In Section 20.4, we will see why. ∎

The choice of $\omega = 1.1$ in Example 20.3 gives a relative residual of 8.18045×10^{-7}, but with $\omega = 1.2$ the residual is 1.4144×10^{-6}. Clearly the accuracy of SOR depends on ω. Later on in this chapter, we will discuss when it is possible to make an optimal choice for ω.

Algorithm 20.1 presents the SOR iteration algorithm. Implementation of the Jacobi and Gauss-Seidel iterations are included in the book software in the functions jacobi and gausseidel.

Algorithm 20.1 SOR Iteration

```
function SOR(A,b,x₀,ω,tol,maxiter)
    % [ x, iter, relresid ] = sor(A,b,x0,omega,tol,maxiter) computes
    % the solution of Ax=b using the SOR iteration.
    % x₀ is the initial approximation, ω is the relaxation parameter,
    % tol is the error tolerance, and maxiter is the maximum number of iterations.
    % x is the approximate solution, and iter is the number of iterations required.
    % iter=-1 if the tolerance was not achieved.
    % relresid is the relative residual obtained by the iteration.

    k=1
    x = x₀
    while k ≤ maxiter do
        x₁ = (ω/a₁₁) (b₁ − A(1, 2 : n)) + (1 − ω) x₁
        for i=2:n-1 do
            xᵢ = (ω/aᵢᵢ) (bᵢ − A(i, 1 : i − 1) x(1 : i − 1) − ⋯
            −A(i, i + 1 : n) x(i + 1 : n)) + (1 − ω) xᵢ
        end for
        xₙ = (ω/aₙₙ) (bₙ − A(n, 1 : n − 1) x(1 : n − 1)) + (1 − ω) xₙ
        relesid = ‖b − Ax‖₂ / ‖b‖₂
        if relresid < tol then
            iter=k
            return[x, iter, relresid]
        end if
        k=k+1
    end while
    iter=-1
    return[x, iter, relresid]
end function
```

NLALIB: The function sor implements Algorithm 20.1.

20.4   CONVERGENCE OF THE BASIC ITERATIVE METHODS

The examples may provide the impression that the Jacobi, Gauss-Seidel, and SOR iterations always converge and that Gauss-Seidel always converges faster than Jacobi. Unfortunately, general statements like this are not true, and we must investigate conditions that will guarantee convergence and enable us to compare convergence rates. For this purpose, we express the iterations in the matrix form

$$x^{(k)} = Bx^{(k-1)} + c,$$

where B is called the *iteration matrix*. In this way, we will have the tools of matrix theory available to us.

20.4.1   Matrix Form of the Jacobi Iteration

In the case of the Jacobi iteration, write the coefficient matrix A as a sum of three matrices, a diagonal matrix, and an upper and lower triangular matrix

$$D = \begin{bmatrix} a_{11} & & & 0 \\ & a_{22} & & \\ & & \ddots & \\ 0 & & & a_{nn} \end{bmatrix}, \quad L = \begin{bmatrix} 0 & 0 & \cdots & 0 \\ a_{21} & 0 & & \\ \vdots & & \ddots & \\ a_{n1} & \cdots & a_{n,n-1} & 0 \end{bmatrix}, \quad U = \begin{bmatrix} 0 & a_{12} & \cdots & a_{1n} \\ 0 & 0 & \cdots & a_{2n} \\ \vdots & & \ddots & \\ 0 & 0 & \cdots & 0 \end{bmatrix}, \quad (20.7)$$

$A = D + L + U$ allows us to write the problem $Ax = b$ in the form

$$Dx + Lx + Ux = b.$$

If we assume that A has no zero entries on its diagonal, then

$$D^{-1} = \begin{bmatrix} \frac{1}{a_{11}} & & & 0 \\ & \frac{1}{a_{22}} & & \\ & & \ddots & \\ 0 & & & \frac{1}{a_{nn}} \end{bmatrix}$$

and

$$x = D^{-1}b - D^{-1}(L+U)x.$$

The Jacobi iterations 20.2 can be written in matrix form as

$$x^{(k)} = B_J x^{(k-1)} + c_J,$$

where

$$B_J = -D^{-1}(L+U), \quad c_J = D^{-1}b$$
$$(20.8)$$

20.4.2   Matrix Form of the Gauss-Seidel Iteration

Formulating the Gauss-Seidel iteration as a matrix problem is more challenging. By assuming that $\sum_{j=1}^{i-1} a_{ij}x_j^{(k)}$ and $\sum_{j=i+1}^{n} a_{ij}x_j^{(k-1)}$ are ignored when $i = 1$ and n, respectively, we can write the Gauss-Seidel iteration as

$$x_i^{(k)} = \frac{1}{a_{ii}}\left[b_i - \left(\sum_{j=1}^{i-1} a_{ij}x_j^{(k)} + \sum_{j=i+1}^{n} a_{ij}x_j^{(k-1)}\right)\right], \quad i = 1, 2, 3, \ldots, n.$$

Rearrange the equation by multiplying both sides by a_{ii} and placing all the (k) terms are on the left-hand side to obtain

$$a_{ii}x_i^{(k)} + \sum_{j=1}^{i-1} a_{ij}x_j^{(k)} = b_i - \sum_{j=i+1}^{n} a_{ij}x_j^{(k-1)}, \quad i = 1, 2, \ldots, n. \qquad (20.9)$$

The left-hand side of Equation 20.9 can be expressed by the matrix $Dx^{(k)} + Lx^{(k)}$, and the right-hand side by $b - Ux^{(k-1)}$ using the matrix definitions in Equation 20.7. We now have the matrix equation

$$(L + D) x^{(k)} = -Ux^{(k-1)} + b. \tag{20.10}$$

If we solve the left-hand side of Equation 20.10 for x, we obtain $x^{(k)} = -(L + D)^{-1} Ux^{(k-1)} + (L + D)^{-1} b$, giving us the matrix form

$$x^{(k)} = B_{\text{GS}}x^{(k-1)} + c_{\text{GS}}$$

where

$$B_{\text{GS}} = -(L + D)^{-1} U, \quad c_{\text{GS}} = (L + D)^{-1} b \tag{20.11}$$

20.4.3 Matrix Form for SOR

The SOR iteration 20.6 is

$$x_i^{(k)} = \frac{\omega}{a_{ii}} \left[b_i - \left(\sum_{j=1}^{i-1} a_{ij} x_j^{(k)} + \sum_{j=i+1}^{n} a_{ij} x_j^{(k-1)} \right) \right] + (1 - \omega) x_i^{(k-1)}, \quad i = 1, 2, 3, \ldots, n. \tag{20.12}$$

Rearrange Equation 20.12 by multiplying both sides by a_{ii} and placing all the k terms on the left-hand side to obtain

$$a_{ii} x_i^{(k)} + \omega \sum_{j=1}^{i-1} a_{ij} x_j^{(k)} = \omega b_i - \omega \sum_{j=i+1}^{n} a_{ij} x_j^{(k-1)} + (1 - \omega) a_{ii} x_i^{(k-1)}, \quad i = 1, 2, \ldots, n. \tag{20.13}$$

Using the matrix definitions in Equation 20.7, we can now express Equation 20.13 in the matrix form

$$(D + \omega L) x^{(k)} = ((1 - \omega) D - \omega U) x^{(k-1)} + \omega b, \tag{20.14}$$

leading us to the specification of the SOR matrix.

$$x^{(k)} = B_{\text{SOR}}x^{(k-1)} + c_{\text{SOR}}$$

where

$$B_{\text{SOR}} = (D + \omega L)^{-1} ((1 - \omega) D - \omega U), \quad c_{\text{SOR}} = \omega (D + \omega L)^{-1} b \tag{20.15}$$

20.4.4 Conditions Guaranteeing Convergence

You may try an initial value, say $x^{(0)} = 0$ and find that your chosen iterative method diverges. Thus, it is helpful to be aware of conditions you know will guarantee convergence for any initial approximation. These conditions are dependent on properties of B in the matrix formulation of the iteration, particularly, the norms and the spectral radius of B. The following theorem provides a simple test for convergence of any iteration with matrix form $x^{(k)} = Bx^{(k-1)} + c$.

Theorem 20.1. *If the matrix B in the iteration $x^{(k)} = Bx^{(k-1)} + c$ has the property that $\|B\| < 1$ for some subordinate norm, then the iteration converges for any choice of $x^{(0)}$.*

Proof. Assume that $\|B\| < 1$ for some subordinate norm. Then $\|B^k\| \leq \|B\|^k \to 0$ as $k \to \infty$, and so $B^k \to 0$ as $k \to \infty$. Now consider the iteration

$$x^{(k)} = Bx^{(k-1)} + c. \tag{20.16}$$

If x is the true solution,

$$x = Bx + c, \tag{20.17}$$

by subtracting Equation 20.16 from Equation 20.17, we get $x - x^{(k)} = (Bx + c) - (Bx^{(k-1)} + c)$, and

$$x - x^{(k)} = B(x - x^{(k-1)}). \tag{20.18}$$

If we let $e^{(k)} = x - x^{(k)}$ be the error at the kth step of the iteration, then Equation 20.18 gives the relation

$$e^{(k)} = Be^{(k-1)}. \tag{20.19}$$

Applying Equation 20.18 repeatedly, $e^{(1)} = Be^{(0)}$, $e^{(2)} = Be^{(1)} = B^2 e^{(0)}$, $e^{(3)} = Be^{(2)} = B^3 e^{(0)}$, and by induction $e^{(k)} = B^k e^{(0)}$. Since $B^k \to 0$ as $k \to \infty$, it follows that $\lim_{k \to \infty} e^{(k)} = 0$, and the iteration converges. □

Example 20.4. For the matrix in Examples 20.1 and 20.2

$$D = \begin{bmatrix} 5 & 0 & 0 \\ 0 & 4 & 0 \\ 0 & 0 & -7 \end{bmatrix}, \quad L = \begin{bmatrix} 0 & 0 & 0 \\ -1 & 0 & 0 \\ 1 & 6 & 0 \end{bmatrix}, \quad U = \begin{bmatrix} 0 & -1 & 2 \\ 0 & 0 & 1 \\ 0 & 0 & 0 \end{bmatrix}.$$

Using $\omega = 1.2$ for SOR, Table 20.1 lists matrix B for each method along with its subordinate norms. In each case, one of the norms is less than 1, so convergence is guaranteed. Note that for the SOR iteration, $\|B\|_\infty = 0.9255$. You might suspect that the iteration converges slowly.

TABLE 20.1 Convergence of Iterative Methods

Method	B	$\|\|B\|\|_1$	$\|\|B\|\|_\infty$	$\|\|B\|\|_2$
Jacobi	$\begin{bmatrix} 0 & 0.2000 & -0.4000 \\ 0.2500 & 0 & -0.2500 \\ 0.1429 & 0.8571 & 0 \end{bmatrix}$	1.0571	1.0000	0.8997
Gauss-Seidel	$\begin{bmatrix} 0 & 0.2000 & -0.4000 \\ 0 & 0.0500 & -0.3500 \\ 0 & 0.0714 & -0.3571 \end{bmatrix}$	1.1071	0.6000	0.6692
SOR	$\begin{bmatrix} -0.2000 & 0.2400 & -0.4800 \\ -0.0600 & -0.1280 & -0.4440 \\ -0.0690 & -0.0905 & -0.7390 \end{bmatrix}$	1.6630	0.9255	1.0063

■

The condition $\|B\| < 1$ for some subordinate matrix norm guarantees convergence. However, it is possible for the iteration to converge even if $\|B\| \geq 1$ for the Jacobi and Gauss-Seidel iterations (Problem 20.6). As a first check to determine if an iteration converges, compute the subordinate norms of B and determine if one has value less than 1. If so, you can proceed; however, if the subordinate norms are all greater than or equal to 1, the iteration may still converge. Recall that the spectral radius of an $n \times n$ square matrix B, $\rho(B)$, is defined by $\rho(B) = \max_{1 \leq i \leq n} |\lambda_i|$, where λ_i are the eigenvalues of B. Find the spectral radius, $\rho(B)$, and verify that $\rho(B) < 1$; in other words, $\rho(B) < 1$ guarantees convergence. It is also the case that if an iteration converges, then $\rho(B) < 1$, so if $\rho(B) \geq 1$, the iterative method will not converge.

Theorem 20.2. *The iteration $x^{(k+1)} = Bx^{(k)} + c$ converges if and only if $\rho(B) < 1$.*

Proof. Assume that the iteration converges. If $e^{(k)} = x^{(k)} - x$, as in the proof of Theorem 20.1, $e^{(k)} = B^k e_0$. Since $\lim_{k \to \infty} e^{(k)} = 0$, it follows that $\lim_{k \to \infty} B^k \to 0$. As a result, $\lim_{k \to \infty} \|B^k x\|_2 \leq \lim_{k \to \infty} \|B^k\|_2 \|x\|_2$, and $\lim_{k \to \infty} B^k x = 0$ for all vectors $x \in \mathbb{R}^n$. If we assume that $\rho(B) \geq 1$, there must be an eigenvector u corresponding to eigenvalue λ with

$|\lambda| \geq 1$. Since $Bu = \lambda u$, $B^2 u = \lambda^2 u$, $B^3 u = \lambda^3 u$, ..., $B^k u = \lambda^k u$, and it is not true that $\lim_{k \to \infty} B^k u = 0$. By contradiction, $\rho(B) < 1$.

Proving that if $\rho(B) < 1$, then $x^{(k+1)} = Bx^{(k)} + c$ converges will be omitted. For a proof, see Refs. [1, p. 280], [2, p. 614], and [27, pp. 143-145]. $\qquad \square$

Example 20.5. In Example 20.3, the Jacobi iteration failed, but the Gauss-Seidel and SOR methods succeeded. The spectral radius of each iteration matrix is

$$\rho(B_{GS}) = 0.5$$
$$\rho(B_{SOR}) = 0.3687$$
$$\rho(B_J) = 1.1372,$$

so these are the results expected. $\qquad\blacksquare$

Example 20.6. For the matrix $A = \begin{bmatrix} 1 & 4 & -1 \\ 2 & -1 & 5 \\ 1 & 0 & 3 \end{bmatrix}$, $B_{GS} = \begin{bmatrix} 0.0000 & -4.0000 & 1.0000 \\ 0.0000 & -8.0000 & 7.0000 \\ 0.0000 & 1.3333 & -0.3333 \end{bmatrix}$, and $\rho(B_{GS}) = 9.0685$. The

Gauss-Seidel iteration will not converge. The same is true for the Jacobi iteration. Verify that $\rho(B_J) = 2.9825$. $\qquad\blacksquare$

20.4.5 The Spectral Radius and Rate of Convergence

Intuitively, there should be a link between the spectral radius of the iteration matrix B and the rate of convergence. Suppose that B has n linearly independent eigenvectors, $v_1, v_2, \ldots, v_n$ and associated eigenvalues $\lambda_1, \lambda_2, \ldots, \lambda_n$. Use the notation of Theorems 20.1 and 20.2 for the error $e^{(k)}$. Since the eigenvectors are a basis,

$$e^{(0)} = \sum_{i=1}^{n} c_i v_i.$$

It follows that:

$$e^{(1)} = Be^{(0)} = \sum_{i=1}^{n} c_i B v_i = \sum_{i=1}^{n} c_i \lambda_i v_i$$

$$e^{(2)} = Be^{(1)} = \sum_{i=1}^{n} c_i \lambda_i B v_i = \sum_{i=1}^{n} c_i \lambda_i^2 v_i.$$

By continuing in this fashion, there results

$$e^{(k)} = \sum_{i=1}^{n} c_i \lambda_i^k v_i.$$

Let $\rho(B) = \lambda_1$ and suppose that $|\lambda_1| > |\lambda_2| \geq |\lambda_3| \geq \cdots \geq \lambda_n$ so that

$$e^{(k)} = c_1 \lambda_1^k v_1 + \sum_{i=2}^{n} c_i \lambda_i^k v_i$$

$$= \lambda_1^k \left(c_1 v_1 + \sum_{i=2}^{n} c_i \left(\frac{\lambda_i}{\lambda_1} \right)^k v_i \right).$$

As k becomes large, $\left(\frac{\lambda_i}{\lambda_1} \right)^k$, $2 \leq i \leq n$ becomes small and we have

$$e^{(k)} \approx \lambda_1^k c_1 v_1.$$

This says that the error varies with the kth power of the spectral radius and that the spectral radius is a good indicator for the rate of convergence.

20.4.6 Convergence of the Jacobi and Gauss-Seidel Methods for Diagonally Dominant Matrices

A matrix is *strictly row diagonally dominant* if the absolute value of the diagonal element is greater than the sum of the absolute values of the off-diagonal elements in its row.

$$|a_{ii}| > \sum_{j=1, j \neq i}^{n} |a_{ij}|, \quad i = 1, 2, \ldots, n \tag{20.20}$$

We will prove that when A is strictly row diagonally dominant, the Jacobi iteration will converge. The reverse is not true. There are matrices that are not strictly row diagonally dominant for which the iteration converges. The matrix of Examples 21.1 and 21.2 is an example.

Theorem 20.3.   *If A is strictly row diagonally dominant, then the Jacobi iteration converges for any choice of the initial approximation $x^{(0)}$.*

Proof.   Recall that the matrix for the Jacobi iterative method is $B_J = -D^{-1}(L+U)$. Then

$$B_J = -D^{-1}(L+U) = \begin{bmatrix} -\dfrac{1}{a_{11}} & & & 0 \\ & -\dfrac{1}{a_{22}} & & \\ & & \ddots & \\ 0 & & & -\dfrac{1}{a_{nn}} \end{bmatrix} \begin{bmatrix} 0 & a_{12} & \ldots & a_{1n} \\ a_{21} & 0 & \ldots & a_{2n} \\ \vdots & \vdots & \ddots & \vdots \\ a_{n1} & a_{n2} & \cdots & 0 \end{bmatrix}$$

$$= \begin{bmatrix} 0 & -\dfrac{a_{12}}{a_{11}} & \ldots & \ldots & -\dfrac{a_{1n}}{a_{11}} \\ -\dfrac{a_{21}}{a_{22}} & 0 & -\dfrac{a_{23}}{a_{22}} & \ldots & -\dfrac{a_{2n}}{a_{22}} \\ \vdots & \ddots & \ddots & \ddots & \vdots \\ \vdots & & \ddots & \ddots & -\dfrac{a_{n-1,n}}{a_{n-1,n-1}} \\ -\dfrac{a_{n1}}{a_{nn}} & \ldots & \ldots & -\dfrac{a_{n,n-1}}{a_{nn}} & 0 \end{bmatrix}$$

Recall that for any square matrix G,

$$\|G\|_{\infty} = \max_{1 \leq i \leq n} \sum_{j=1}^{n} |g_{ij}|.$$

For any row i of B_J,

$$\sum_{j=1, j \neq i}^{n} \left| -\frac{a_{ij}}{a_{ii}} \right| = \frac{1}{|a_{ii}|} \sum_{j=1, j \neq i}^{n} |a_{ij}| < 1$$

by Equation 20.20. Thus, $\|B_J\|_{\infty} < 1$, and by Theorem 20.1, the Jacobi method converges.   □

If A is strictly row diagonally dominant, the Gauss-Seidel iteration converges for any choice of the initial approximation $x^{(0)}$. For a proof, see Ref. [1, pp. 286-288]. In Ref. [2, pp. 615-616], you will find the proof of another result concerning the Gauss-Seidel method. It is particularly useful, since many engineering problems involve symmetric positive definite matrices.

Theorem 20.4.   *Let A be a symmetric positive definite matrix. For any arbitrary choice of initial approximation $x^{(0)}$, the Gauss-Seidel method converges.*

It should be noted that for matrices that are not row diagonally dominant, there are examples where the Jacobi iteration converges and the Gauss-Seidel iteration diverges. Similarly, there are examples of matrices for which the Gauss-Seidel method converges and the Jacobi method diverges (see Ref. [11]).

20.4.7 Choosing ω for SOR

We have seen that poor choice of the relaxation parameter ω can lead to poor convergence rates. On the other hand, a good choice of ω can lead to very fast convergence compared to the Jacobi or Gauss-Seidel methods. The spectral radius of B_{SOR} is the eigenvalue of B_{SOR} with maximum magnitude, and we want to choose ω so that $|\rho(B_{SOR}(\omega))|$ is a minimum. Finding the optimal value of ω is very difficult in general, and the optimal value is known only for special types of matrices. It is known, however, that ω must satisfy $0 < \omega < 2$ [1, p. 290].

Theorem 20.5. *If the SOR iteration converges for every initial approximation $x^{(0)}$, then $0 < \omega < 2$.*

This result says that you never choose a relaxation parameter ω outside the range $(0, 2)$. Normally, the choice is overrelaxation ($\omega > 1$) to put the greatest weight on the newly computed values. If $\|B_{SOR}\| < 1$ for some subordinate norm or $\rho(B_{SOR}) < 1$, the SOR iteration converges. The following theorem [1, p. 290], which we state without proof, provides another criterion that guarantees convergence.

Theorem 20.6. *If A is a symmetric positive definite matrix and $0 < \omega < 2$, the SOR and Gauss-Seidel iterations converge for any choice of $x^{(0)}$.*

Example 20.7. Let A be the 10×10 pentadiagonal matrix

$$
A = \begin{bmatrix}
6 & -2 & -1 & & & & \\
-2 & 6 & -2 & -1 & & & \\
-1 & -2 & 6 & \ddots & \ddots & & \\
& -1 & \ddots & \ddots & \ddots & -1 & \\
& & \ddots & & -2 & 6 & -2 \\
& & & & -1 & -2 & 6
\end{bmatrix}
$$

that can be built using the function `pentd` in the software distribution. The function `optomega` in the book software distribution approximates the optimal ω for a matrix and graphs $\rho(B_{SOR}(\omega))$ as a function of ω, $0 < \omega < 2$. It should be noted that the function `optomega` is provided for demonstration purposes only, and is not intended for use with a large matrix. The following code estimates the optimal ω for A and then applies the SOR iteration to solve the system $Ax = b$, where b is $\begin{bmatrix} 1 & 1 & \dots & 1 & 1 \end{bmatrix}^T$. It prints the number of iterations required to attain a error tolerance of 1.0×10^{-14} and the relative residual. Figure 20.1 is the graph produced by `optomega`.

```
>> woptimal = optomega(A)
woptimal =
    1.4600
>> [x,iter,relresid] = sor(A,rhs,x0,woptimal,1.0e-14,100);
>> iter

iter =
    54

>> relresid

ans =
  9.2334e-015
```

Try another value of ω such as 1.3 and observe a slower convergence rate. ∎

20.5 APPLICATION: POISSON'S EQUATION

Poisson's equation is one of the most important equations in applied mathematics and has applications in such fields as astronomy, heat flow, fluid dynamics, and electromagnetism. Let R be a bounded region in the plane with boundary ∂R (Figure 20.2), $g(x, y)$ be defined on ∂R, and $f(x, y)$ be a function defined in R. Find a function $u(x, y)$ such that

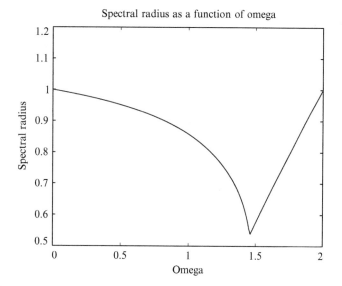

FIGURE 20.1 SOR spectral radius.

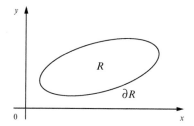

FIGURE 20.2 Region in the plane.

$$-\frac{\partial^2 u}{\partial x^2} - \frac{\partial^2 u}{\partial y^2} = f(x, y),$$

$$u(x, y) = g(x, y) \quad \text{on } \partial R$$

Rarely is an analytical solution known, so approximation techniques must be used. We will use the finite difference method to obtain a numerical solution and assume for simplicity that the region R is the unit square $0 \le x \le 1$, $0 \le y \le 1$. Divide the square into a grid of small squares having sides of length $h = \frac{1}{n}$ (Figure 20.3). In Section 12.2, we studied the heat equation in one spacial variable and approximated the second partial derivative by a difference equation

$$\frac{\partial^2 u}{\partial x^2}(x_i, t_j) \approx \frac{1}{h^2}\left(u_{i-1,j} - 2u_{i,j} + u_{i+1,j}\right),$$

where h is the equal spacing between points. We will do the same thing here for the two partial derivatives to obtain the difference approximation

$$\frac{1}{h^2}\left(-u_{i-1,j} + 2u_{i,j} - u_{i+1,j}\right) + \frac{1}{h^2}\left(-u_{i,j-1} + 2u_{i,j} - u_{i,j+1}\right) = f(x_i, y_j)$$

If we multiply by h^2 and collect terms, the result is the set of equations

$$-u_{i-1,j} - u_{i+1,j} + 4u_{ij} - u_{i,j-1} - u_{i,j+1} = h^2 f(x_i, y_j), \quad 1 \le i, j \le n - 1 \tag{20.21}$$

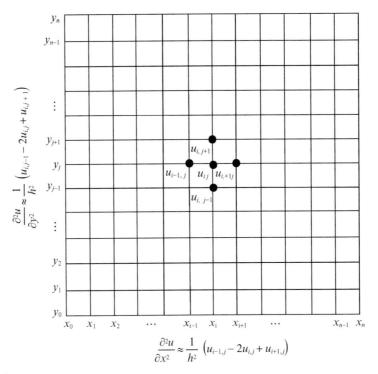

$$\frac{\partial^2 u}{\partial x^2} \approx \frac{1}{h^2}\left(u_{i-1,j} - 2u_{i,j} + u_{i+1,j}\right)$$

FIGURE 20.3 Five-point stencil.

Each equation contains five values at grid points and forms what is called a *five-point central difference approximation*, also called a *five-point stencil* (Figure 20.3).

When one or two of the four points around the center touch the boundary, the boundary condition must be used. For instance, if $i = 1$, then the equation centered at (1, 1) is

$$-u_{0,1} - u_{2,1} + 4u_{11} - u_{1,0} - u_{1,2} = h^2 f(x_1, y_1)$$

or

$$-u_{2,1} + 4u_{11} - u_{1,2} = h^2 f(x_1, y_1) + g(0, y_1) + g(x_1, 0) \tag{20.22}$$

Execute the SOR iteration by solving for u_{ij} using Equation 20.21 to obtain

$$u_{ij} = \omega \left(\frac{u_{i-1,j} + u_{i+1,j} + u_{i,j-1} + u_{i,j+1} + h^2 f(x_i, y_i)}{4} \right) + (1 - \omega) u_{ij}. \tag{20.23}$$

In Ref. [23, p. 564], it is stated that the optimal value of ω is

$$\omega_{\text{opt}} = \frac{2}{1 + \sin \pi h}. \tag{20.24}$$

We have not specified the matrix form of the finite difference approximation, but will do so in Section 21.11. The matrix is symmetric and positive definite, so we know the SOR iteration will converge by Theorem 20.6. The function sorpoisson in the software distribution assigns the boundary values, assigns values of $f(x, y)$ in the interior, and executes the SOR iteration.

```
%SORPOISSON Numerically approximates the solution of the Poisson
%equation on the square 0 <= x,y <= 1
%
%    [x y u] = sorpoisson(n,f,g,omega,numiter) computes the solution.
%    n is the number of subintervals, f is the right-hand side, g is
%    the boundary condition on the square, omega is the relaxation parameter,
%    and numiter is the number of SOR iterations to execute. x and y are
%    the grid of points on the x and y axes, and u is the matrix containing
```

```
%    the numerical solution. Upon building x, y, and u, sorpoisson draws a
%    surface plot of the solution.
%
```

Example 20.8. Consider the Poisson equation

$$-\frac{\partial^2 u}{\partial x^2} - \frac{\partial^2 u}{\partial y^2} = 20\pi^2 \sin(2\pi x) \sin(4\pi y)$$

$$u(x, y) = \sin(2\pi x) \sin(4\pi y) \text{ on } \partial R$$

whose exact solution is $u(x, y) = \sin(2\pi x) \sin(4\pi y)$. Apply the SOR iteration 20.23 with $n = 100$ and numiter $= 125$. Using Equation 20.24, assign $\omega = 1.9196$. Figure 20.4 is a graph of the solution obtained from the SOR iteration, and Figure 20.5 shows the actual solution. ∎

Problems 20.9–20.14 deal with the Jacobi and Gauss-Seidel iterations for the one-dimensional Poisson equation. These problems deal with the coefficient matrix and its relation to the convergence of the iterations.

20.6 CHAPTER SUMMARY

The Jacobi Iteration

The Jacobi iteration is the simplest of the classical iterative methods and, generally, the slowest. However, it forms a basis for the understanding of other methods, such as Gauss-Seidel and SOR. Starting with an initial approximation, $x^{(0)}$, to the solution of $Ax = b$, use the first equation to compute $x_1^{(1)}$ in terms of $x_2^{(0)}, x_3^{(0)}, \ldots, x_n^{(0)}$. Now use the second equation to compute $x_2^{(1)}$ in terms of $x_1^{(0)}, x_3^{(0)}, \ldots, x_n^{(0)}$. In general, the equation for the computation is

$$x_i^{(k)} = \frac{1}{a_{ii}} \left[b_i - \left(\sum_{j=1, j\neq i}^{n} a_{ij} x_j^{(k-1)} \right) \right], \quad 1 \leq i \leq n.$$

$$-\frac{\partial^2 u}{\partial x^2} - \frac{\partial^2 u}{\partial y^2} = 5\pi^2 \cos(\pi x) \sin(2\pi y), \, u(x, y) = \cos(\pi x) \sin(2\pi y) \text{ on } \partial R$$

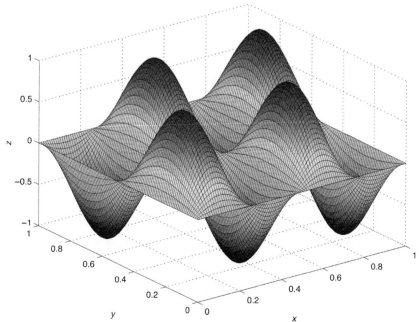

FIGURE 20.4 Poisson's equation. (a) Approximate solution and (b) analytical solution.

Continued

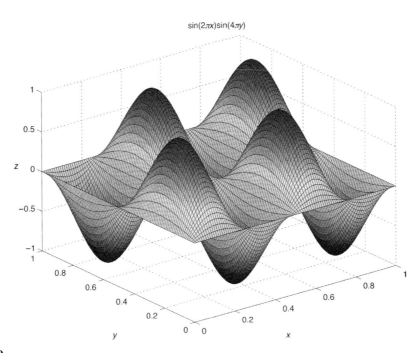

$$\sin(2\pi x)\sin(4\pi y)$$

FIGURE 20.4 , CONT'D

The Jacobi method does not make use of new components of the approximate solution as they are computed. This requires storing both the previous and the current approximations. If A is strictly row diagonally dominant, then the Jacobi iteration converges for any choice of the initial approximation $x^{(0)}$. However, the Jacobi iteration may converge for a matrix that is not strictly row diagonally dominant.

The Gauss-Seidel Iteration

In general, the Gauss-Seidel iteration is an improvement over the Jacobi iteration because it uses the approximation to a component of the solution as soon as it is available. For instance, after computing $x_1^{(1)}$, it is used in the computation of $x_2^{(1)}$. The general formula for the iteration is

$$x_1^{(k)} = \frac{1}{a_{11}}\left[b_1 - \left(\sum_{j=2}^{n} a_{1j}x_j^{(k-1)}\right)\right]$$

$$x_i^{(k)} = \frac{1}{a_{ii}}\left[b_i - \left(\sum_{j=1}^{i-1} a_{ij}x_j^{(k)} + \sum_{j=i+1}^{n} a_{ij}x_j^{(k-1)}\right)\right], \quad i = 2, 3, \ldots, n-1$$

$$x_n^{(k)} = \frac{1}{a_{nn}}\left[b_n - \sum_{j=1}^{n-1} a_{nj}x_j^{(k)}\right]$$

Unlike the Jacobi method, this method requires storing only one vector rather than two. The Gauss-Seidel iteration is guaranteed to converge for any initial approximation if A is strictly diagonally dominant and when A is symmetric and positive definite.

The SOR Iteration

This iteration computes $x_i^{(k)}$ by forming a weighted average between the previous value $x_i^{(k-1)}$ and the value computed by the Gauss-Seidel iteration, and the formula is

$$x_i^{(k)} = \frac{\omega}{a_{ii}} \left[b_i - \left(\sum_{j=1}^{i-1} a_{ij} x_j^{(k)} + \sum_{j=i+1}^{n} a_{ij} x_j^{(k-1)} \right) \right] + (1 - \omega) x_i^{(k-1)}, \quad i = 1, 2, \ldots, n.$$

The idea is to choose a value of ω that will accelerate the rate of convergence of the iteration. It is known that unless $0 < \omega < 2$, the method will not converge. There is no general formula for an optimal ω. Usually $\omega > 1$, but $\omega < 1$ (underrelaxation) sometimes gives better results. SOR converges for any ω if A is positive definite, and positive definite matrices appear in many engineering problems.

Convergence of the Basic Iterative Methods

Analysis of convergence criteria for the basic iterations is based upon writing the equations defining the iteration in the matrix form

$$x^{(k)} = Bx^{(k-1)} + c. \tag{20.25}$$

If $\|B\| < 1$ for any subordinate norm, then any iteration 20.25 converges. The matrices B_J, B_{GS}, and B_{SOR} for the Jacobi, Gauss-Seidel, and SOR methods, respectively, are determined in Sections 20.4.1–20.4.3. Note that this is a sufficient condition only. It is possible for the iteration 20.25 to converge if $\|B\| \geq 1$. A necessary and sufficient condition for an iteration to converge is that the spectral radius of B ($\rho(B)$) is less than 1. While this result is of significant theoretical importance, computing the spectral radius is computationally costly. It is appropriate to first check to see if $\|B\| < 1$. Also, the Jacobi and Gauss-Seidel methods converge if A is strictly diagonally dominant, and the Gauss-Seidel iteration converges if A is positive definite. Convergence of the SOR iteration is guaranteed if $0 < \omega < 2$ and A is positive definite.

The Poisson Equation

Poisson's equation is very important in many areas of application. Section 20.5 develops a five-point central difference approximation for the two-dimensional Poisson equation. The coefficient matrix is positive definite, so the SOR iteration applies. This is one of the rare cases where the optimal choice for ω is known, and a function `sorpoisson` in the book software distribution uses the SOR iteration to numerically approximate and make a surface plot of the solution to the Poisson equation on the unit square $0 \leq x, y \leq 1$.

20.7 PROBLEMS

20.1 Perform three iterations of the Jacobi method using pencil and paper. Use $x^{(0)} = 0$.

$$\begin{bmatrix} 3 & 1 \\ 2 & 5 \end{bmatrix} \begin{bmatrix} x_1 \\ x_2 \end{bmatrix} = \begin{bmatrix} -1 \\ 1 \end{bmatrix}$$

20.2 Do Problem 21.2 using the Gauss-Seidel iteration.

20.3 Do Problem 21.1 using SOR with $\omega = 1.2$.

20.4 Let $A = \begin{bmatrix} \frac{1}{6} & 0 & \frac{1}{8} \\ 0 & \frac{1}{3} & \frac{1}{7} \\ \frac{1}{5} & 0 & \frac{1}{4} \end{bmatrix}$. What can you say about the convergence of the Jacobi and Gauss-Seidel iterations?

20.5 Let A be the general 2×2 matrix $A = \begin{bmatrix} a_{11} & a_{12} \\ a_{21} & a_{22} \end{bmatrix}$.

a. Compute exact formulas for B_J defined by Equation 20.8 and B_{GS} defined by Equation 20.11. Consider using the MATLAB Symbolic Math Toolbox.

b. Determine analytic formulas for the eigenvalues of B_J and B_{GS}.

c. Using the result of (b), determine if either or both of the Jacobi and Gauss-Seidel iterations converge for $A = \begin{bmatrix} 3 & -4 \\ 4 & 5 \end{bmatrix}$.

d. Answer part (c) for the matrix $A = \begin{bmatrix} 2 & 2 \\ 1 & 3 \end{bmatrix}$.

e. Is it possible for one of Jacobi and Gauss-Seidel to converge and the other diverge for a 2×2 matrix?

20.6 Let $A = \begin{bmatrix} 2 & 2 \\ 1 & 3 \end{bmatrix}$.

a. Show that the Jacobi and Gauss-Seidel iterations converge, but that the iteration matrices B_J and B_{GS} have one-, two-, and infinity norms greater than or equal to 1.

b. The matrix

$$B_{SOR}(\omega) = \begin{bmatrix} 1 - \omega & -\omega \\ \frac{\omega(2\omega - 2)}{6} & \frac{\omega^2}{3} - \omega + 1 \end{bmatrix}$$

Let ω range from 0.01 in steps of 0.01-1.99, and graph ω vs. $\rho(B_{SOR}(\omega))$.

c. Do part (b), except graph ω vs. $\|B_{SOR}(\omega)\|_2$.

d. Comment on the results of parts (a)-(c).

20.7 Show that

a. The Jacobi iteration converges for $A_1 = \begin{bmatrix} 2 & 1 & -2 \\ 1 & 1 & 1 \\ 3 & 2 & 1 \end{bmatrix}$ but the Gauss-Seidel iteration does not converge.

b. The Gauss-Seidel iterations converges for $A_2 = \begin{bmatrix} 2 & 1 & 3 \\ 1 & 2 & 1 \\ 1 & 1 & 3 \end{bmatrix}$ but the Jacobi iteration does not.

20.8 If $\|\cdot\|$ is a subordinate matrix norm, prove that

$$\rho(A) < \|A\|.$$

Assume the result of Theorem 20.2, and provide an alternate proof of Theorem 20.1.

Problems 20.9–20.14 deal with the one-dimensional Poisson equation,

$$-\frac{d^2 u}{dx^2} = f(x), \quad 0 \le x \le 1, \ u(0) = u(1) = 0 \tag{20.26}$$

In most cases, it is not possible to find an analytical solution, so an approximation technique must be used. Divide the interval into n subintervals of width $h = \frac{1}{n}$, giving $n + 1$ points

$$\{x_1 = 0, x_2 = h, \ldots, x_i = (i - 1)h, \ldots, x_n = (n - 1)h, x_{n+1} = 1\}.$$

20.9 Using Taylor series, it can be shown that $\frac{d^2 u}{dx^2}(x_i) \approx \frac{u(x_{i+1}) - 2u(x_i) + u(x_{i-1})}{h^2}$. Use the notation $u_i = u(x_i)$ and show that after using the approximation for the second derivative in the differential equation, the result is the system of equations (Figure 20.5)

$$-u_{i+1} + 2u_i - u_{i-1} = h^2 f(x_i), \quad 2 \le i \le n.$$

20.10

a. Noting that $u_1 = 0, u_{n+1} = 0$, show that the system of equations in Problem 21.11 can be written in matrix form $Ax = b$, where

$$A = \frac{1}{h^2} \begin{bmatrix} 2 & -1 & 0 & \cdots & 0 \\ -1 & 2 & -1 & & \vdots \\ 0 & \ddots & \ddots & \ddots & 0 \\ \vdots & & -1 & 2 & -1 \\ 0 & & & -1 & 2 \end{bmatrix}, \quad b = \begin{bmatrix} f(x_2) \\ f(x_3) \\ \vdots \\ f(x_{n-1}) \\ f(x_n) \end{bmatrix}.$$

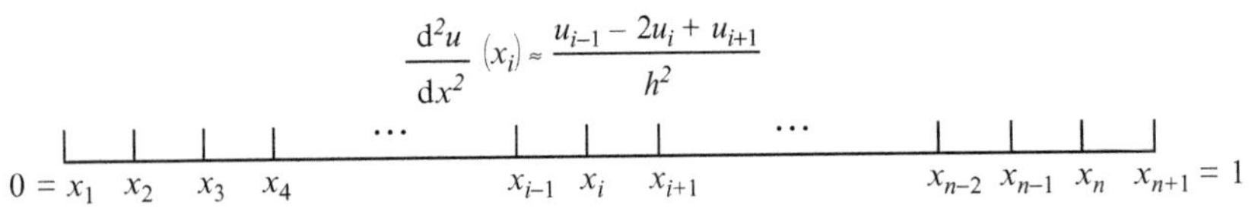

$$\frac{d^2 u}{dx^2}(x_i) \approx \frac{u_{i-1} - 2u_i + u_{i+1}}{h^2}$$

FIGURE 20.5 One-dimensional Poisson equation grid.

b. Show that A is positive definite by proving that $x^T A x > 0$ for all $x \neq 0$. Hint: Show that $x^T A x = \langle Ax, x \rangle$. Multiply this out and look for terms of the form $(x_{i+1} - x_i)^2$.

20.11 In this problem, you will find the eigenvalues of matrix A in Problem 20.10.

a. Matrix A has dimension $(n-1) \times (n-1)$. Let $v = \begin{bmatrix} v_1 \\ v_2 \\ \vdots \\ v_{n-2} \\ v_{n-1} \end{bmatrix}$ be an eigenvector of A. Letting $h = \frac{1}{n}$, show that

$$-v_{i-1} + \left(2 - \lambda h^2\right) v_i - v_{i+1} = 0, \quad 1 \le i \le n-1,$$

where $v_0 = v_n = 0$.

b. Try solutions of the form $v_j = \sin(\omega j)$, $0 \le j \le n$ and show that

$$\lambda_j = \frac{4}{h^2} \sin^2\left(\frac{\pi j}{2n}\right), \quad 1 \le j \le n-1$$

are distinct eigenvalues of A. Hint: Use trigonometric identities including $1 - \cos(\omega) = 2\sin^2\left(\frac{\omega}{2}\right)$.

c. We know that A is positive definite. Are its eigenvalues consistent with that?

d. Find the condition number of A, and show that the matrix is ill-conditioned for large n.

Remark 20.1.   With homogeneous boundary conditions $u(0) = u(1) = 0$, it can be shown due to the form of the right-hand side b that

$$\frac{\left\| \delta u^{(n)} \right\|_2}{\left\| u^{(n)} \right\|_2} \le C \frac{\left\| \delta b^{(n)} \right\|_2}{\left\| b^{(n)} \right\|_2},$$

where C is a constant independent of n. In short, the system $Ax = b$ can be solved reliably (see Ref. [27, pp. 87-88]).

20.12 To solve the problem $Ax = b$ using the Jacobi iteration, the spectral radius of the matrix B_J must be less than one.

a. Using the result of Problem 20.11(b), show that the eigenvalues of B_J are

$$\mu_j = 1 - \frac{\lambda_j}{2n^2},$$

where λ_j are the eigenvalues of A. Hint: Show that $B_J = I - D^{-1}A$, and let v be an eigenvector of B_J with corresponding eigenvalue μ.

b. Show that the spectral radius of B_J is

$$\rho(B_J) = \cos\left(\frac{\pi}{n}\right),$$

and that $\rho(B_J) < 1$. Thus, the Jacobi method converges. Hint: Use the half-angle formula $\sin^2\left(\frac{\theta}{2}\right) = \frac{1}{2}(1 - \cos\theta)$.

c. Using the McLaurin series for $\cos\frac{\pi}{n}$, show that

$$\rho(B_J) = 1 - \frac{\pi^2}{2n^2} + O\left(h^4\right).$$

20.13

a. It can be shown that the spectral radius of the iteration matrix B_{GS} for the Gauss-Seidel iteration is $\rho(B_{GS}) = \cos^2\left(\frac{\pi}{n}\right) < 1$ [27, p. 155]. Show that

$$\rho(B_{GS}) = 1 - \frac{\pi^2}{n^2} + O\left(h^4\right).$$

b. Explain why for large values of n, the Gauss-Seidel method converges faster.

20.14 There is a means of handling the Gauss-Seidel iteration for the one-dimensional Poisson equation that involves coloring half the unknown values red and half black. This method allows parallelism, whereas the iteration 20.4

FIGURE 20.6 One-dimensional red-black GS.

does not, and the modification leads to best case convergence results that depend on it [1, pp. 282-283, 285-294]. Color x_2 light gray (represents red), x_3 black, and so forth, as shown in Figure 20.6.

a. If a grid point is red, what are the colors of its neighbors? If a grid point is black, what is the color of its neighbors?

b. To apply Gauss-Seidel, begin with x_2 and apply the difference formula to all red points. Now start at x_3 and apply the difference formula to all the black points. Sketch an algorithm that uses this coloring to implement the Gauss-Seidel iteration.

c. Explain why this ordering allows the iteration to be parallelized.

20.15 In finding a numerical solution to the two-dimensional Poisson equation, we used row ordering to define the finite difference equations. There is a better way, called *red-black ordering*. Think of the grid as a chessboard. Begin in the lower left corner of the grid of unknowns and color $(1, 1)$ red. Now alter red and black until you have colored $(n - 1, n - 1)$.

a. Draw the board corresponding to $n = 7$, and color the interior grid points.

b. In the five-point central difference scheme, if (i, j) is red, what color are its four neighbors? What color are the neighbors of a black grid point?

c. In terms of indices i, j, when is u_{ij} red, and when is it black?

d. To apply Gauss-Seidel, begin with red grid point $(1, 1)$ and apply the difference formula to all the red points. Now start with the black grid point $(2, 1)$ and apply the difference formula to the black points. Sketch an algorithm that implements this version of Gauss-Seidel.

e. Explain why this ordering allows the iteration to be parallelized.

20.7.1 MATLAB Problems

20.16 Using the Jacobi and Gauss-Seidel methods, solve the system with tol $= 0.5 \times 10^{-14}$ and numiter $= 50$. In each case, print the number of iterations required and the relative residual $\|b - Ax\|_2 / b$.

$$\begin{bmatrix} 6 & -1 & 2 & 1 \\ 1 & 6 & 1 & -1 \\ 0 & 1 & 3 & 1 \\ 1 & -2 & 1 & 5 \end{bmatrix} \begin{bmatrix} 3 \\ 2 \\ -6 \\ 1 \end{bmatrix}.$$

20.17 This problem investigates convergence properties. The matrix $A = \begin{bmatrix} 2 & -1 & 0 \\ -1 & 2 & -1 \\ 0 & -1 & 2 \end{bmatrix}$ is positive definite (Problem 20.10(b)).

a. Is matrix A strictly row diagonally dominant?

b. Is the Jacobi method guaranteed to converge? What about the Gauss-Seidel and SOR iterations?

c. If the Gauss-Seidel method converges, solve $Ax = \begin{bmatrix} 1 \\ 3 \\ 7 \end{bmatrix}$.

d. If the SOR iteration converges, use `optomega` to estimate an optimal ω, and demonstrate that `sor` improves the convergence rate relative to Gauss-Seidel using tol $= 1.0 \times 10^{-10}$ and maxiter $= 100$.

20.18 Let $A = \begin{bmatrix} 0.1000 & 0.5000 & -0.1000 \\ 0.4000 & 0.2000 & 0.6000 \\ 0.2000 & -0.3000 & 0.4000 \end{bmatrix}$. The matrix A is not diagonally dominant.

a. Compute $\|B_J\|_1, \|B_J\|_\infty, \|B_J\|_2$ and $\|B_{GS}\|_1, \|B_{GS}\|_\infty, \|B_{GS}\|_2$.

b. Compute the spectral radius of B_J and B_{GS}.

c. Does either method converge? Demonstrate your answer by using a random b and $x_0 = 0$.

20.19 In Chapter 12, we used finite difference methods to approximate the solution of the heat equation in one spacial variable. We needed to solve a tridiagonal linear system with coefficient matrix

$$
B = \begin{bmatrix}
1+2r & -r & 0 & \cdots & & 0 \\
-r & 1+2r & -r & \cdots & & 0 \\
0 & \ddots & \ddots & \ddots & & \vdots \\
\vdots & & \vdots & -r & 1+2r & -r \\
0 & & 0 & 0 & -r & 1+2r
\end{bmatrix}.
$$

This matrix is strictly row diagonally dominant, and so both the Jacobi and Gauss-Seidel iterations will converge.

a. Let $r = 0.25$ and solve the 100×100 linear system $Bu = c$, with $c = \begin{bmatrix} 1 & 1 & \cdots & 1 & 1 \end{bmatrix}^T$ using the Jacobi iteration with tol $= 0.5 \times 10^{-14}$, numiter $= 100$, and $x_0 = 0$.

b. Do part (a) using the Gauss-Seidel iteration.

c. Do part (a) using SOR with $\omega = \{1.1, 1.2, 1.3, 1.5, 1.9\}$. Which value of ω works best?

d. Solve the system using the MATLAB function thomas introduced in Section 9.4.

e. Solve the system using the MATLAB command "B\c."

f. Compare the results of parts (a)-(e).

20.20

a. Show that the matrix

$$
A = \begin{bmatrix}
4 & -2 & 0 & 0 & 0 & 0 & 0 \\
-2 & 4 & -2 & 0 & 0 & 0 & 0 \\
0 & -2 & 4 & -2 & 0 & 0 & 0 \\
0 & 0 & -2 & 4 & -2 & 0 & 0 \\
0 & 0 & 0 & -2 & 4 & -2 & 0 \\
0 & 0 & 0 & 0 & -2 & 4 & -2 \\
0 & 0 & 0 & 0 & 0 & -2 & 4
\end{bmatrix}
$$

is positive definite.

b. Compute $\rho(B_J)$, $\rho(B_{GS})$, and $\rho(B_{SOR}(1.2))$. Project a ranking for the number of iterations required by the methods.

c. Generate a random 7×1 vector *rhs* and solve $Ax = rhs$ using all three methods. Are your projections in part (b) correct?

d. For the matrix $A = \begin{bmatrix} -1 & 3 & 4 & 0 \\ 5 & 7 & 12 & -1 \\ 1 & 1 & 5 & 2 \\ 5 & -1 & -1 & 2 \end{bmatrix}$, compute the spectral radius of B_J, B_{GS}, and B_{SOR}. Will any of the iterative methods converge. Try one and see what happens.

20.21

a. Write a function x = poisson1dj(f, n, numiter) that uses the Jacobi iterations to numerically approximate the solution to the one-dimensional Poisson equation 20.26, where f is the right-hand side, n is the number of subintervals of $0 \le x \le 1$, and numiter is the number of iterations.

b. Do part (a) using the Gauss-Seidel iteration by writing a function poisson1dgs. If you have done Problem 20.14, use red-black ordering.

c. The exact solution to the Poisson equation

$$
\frac{d^2u}{dx^2} = -x(x-1), \quad u(0) = u(1) = 0
$$

is

$$
u(x) = \frac{x^4}{12} - \frac{x^3}{6} + \frac{x}{12}.
$$

Using $n = 25$ and numiter $= 700$ (Jacobi), numiter $= 350$ (GS) compute the solution using poisson1dj and poisson1dgs. Compare the results to the true solution.

20.22 Let R be the unit square, $0 \le x, y \le 1$. Consider the two-dimensional Laplace equation

$$\frac{\partial^2 u}{\partial x^2} + \frac{\partial^2 u}{\partial y^2} = 0, \quad 0 < x, \, y < 1, \tag{20.27}$$

$$u = g(x, y) \quad \text{on } \partial R. \tag{20.28}$$

a. Write a MATLAB `function [x y u] = laplacej(g,n,maxiter)` that uses the Jacobi iteration to approximate the solution to Equation 20.27 and creates a surface plot.

b. The exact solution with

$$g(x, y) = \begin{cases} 0 & y = 0 \\ \sin(\pi x) & y = 1 \\ 0 & x = 0 \\ 0 & x = 1 \end{cases}$$

is

$$ut(x, y) = \frac{1}{\sinh(\pi)} \sin(\pi x) \sinh(\pi y).$$

Plot the solution using the MATLAB function `ezsurf`.

c. Plot the solution by applying `laplacej` with $n = 25$ and numiter $= 100$.

20.23

a. Write function `jacobiConverge(A)` that returns true if the Jacobi iteration converges for a system with coefficient matrix A and false if it does not converge

b. Do part (a) for the Gauss-Seidel method by writing a function `gsConverge(A)`.

c. For each matrix, use your functions to determine if the Jacobi or Gauss-Seidel iterations converge.

$$A = \begin{bmatrix} 3 & 1 & 1 \\ 2 & 5 & 2 \\ 1 & 5 & 7 \end{bmatrix}, \quad B = \begin{bmatrix} 5 & -1 & 4 \\ 1 & 3 & 1 \\ 2 & 3 & 4 \end{bmatrix}, \quad C = \begin{bmatrix} 6 & 1 & 2 \\ 1 & 1 & 8 \\ 1 & 2 & 5 \end{bmatrix}, \quad D = \begin{bmatrix} 2 & 1 & 1 \\ 3 & -4 & 1 \\ -1 & 3 & 4 \end{bmatrix}$$

$$E = \begin{bmatrix} 1 & 1 & 1 & 1 & 1 & 1 & 1 & 1 \\ 1 & 2 & 2 & 2 & 2 & 2 & 2 & 2 \\ 1 & 2 & 3 & 3 & 3 & 3 & 3 & 3 \\ 1 & 2 & 3 & 4 & 4 & 4 & 4 & 4 \\ 1 & 2 & 3 & 4 & 5 & 5 & 5 & 5 \\ 1 & 2 & 3 & 4 & 5 & 6 & 6 & 6 \\ 1 & 2 & 3 & 4 & 5 & 6 & 7 & 7 \\ 1 & 2 & 3 & 4 & 5 & 6 & 7 & 8 \end{bmatrix}$$

20.24

a. Write a function `diagDom(A)` that determines if A is strictly row diagonally dominant by returning true or false.

b. Apply your function to the matrices

i. $A = \begin{bmatrix} 3 & -1 & 1 \\ 2 & 5 & 2 \\ -8 & 3 & 12 \end{bmatrix}$

ii. $A = \begin{bmatrix} 1.0 & 0.3 & 0.4 & 0.1 & 0.1 \\ 0.00 & 2.8 & 2.00 & 0.10 & 0.40 \\ .30 & -0.10 & 0.70 & 0.20 & 0.05 \\ 0.00 & 0.60 & 1.20 & 4.25 & 2.40 \\ 0.50 & 6.80 & 1.40 & 2.30 & -10.6 \end{bmatrix}$

iii. $A = \begin{bmatrix} 3.0 & 1.0 & 1.0 & 0.90 \\ 2.0 & 8.0 & 1.0 & -4.0 \\ 5.0 & 1.0 & 7.0 & 0.80 \\ 1.0 & 1.0 & -3.0 & 5.0 \end{bmatrix}$

c. The MATLAB function call `gallery('neumann',25)` returns a sparse 25×25 matrix that results from discretizing a partial differential using a five-point finite difference operator on a regular mesh. Execute `diagDom` using the matrix

```
A = full(gallery('neumann', 25)) + diag(0.1*ones(25,1));
```

20.25

a. If A is a sparse matrix of the form

$$
A = \begin{bmatrix}
c_1 & d_1 & e_1 & & & & & \\
b_1 & c_2 & d_2 & e_2 & & & & \\
a_1 & b_2 & c_3 & d_3 & \ddots & & & \\
& a_2 & b_3 & \ddots & \ddots & e_{n-3} & & \\
& & \ddots & \ddots & \ddots & d_{n-2} & e_{n-2} & \\
& & & a_{n-3} & b_{n-2} & c_{n-1} & d_{n-1} & \\
& & & & a_{n-2} & b_{n-1} & c_n &
\end{bmatrix},
$$

what is the maximum number of nonzero elements it contains? What is the density, where density $=\frac{\text{number of nonzero elements}}{n}$. Such a matrix is termed *pentadiagonal*.

b. Code a function

```
[x,iter] = pentsolve(a,b,c,d,e,x0,rhs,tol,maxiter)
```

that applies the Gauss-Seidel iteration to solve a system $Ax = rhs$. Your iteration must use only the elements on the five diagonals. Apply the termination criteria

$$
\frac{\|\text{xnew} - \text{xprev}\|_2}{\|\text{xprev}\|_2} < \text{tol}.
$$

c. Test your function by solving the 1000×1000 problem

$$
\begin{bmatrix}
11 & -4 & -1 & 0 & 0 & \cdots & 0 \\
-4 & 11 & -4 & -1 & 0 & \cdots & 0 \\
-1 & -4 & 11 & -4 & -1 & \cdots & 0 \\
0 & -1 & -4 & 11 & -4 & \ddots & 0 \\
0 & 0 & -1 & -4 & 11 & \ddots & \vdots \\
& & & \ddots & \ddots & \ddots & -1 \\
\vdots & \vdots & \vdots & \ddots & \ddots & \ddots & -4 \\
0 & 0 & 0 & \cdots & -1 & -4 & 11
\end{bmatrix} x = rhs,
$$

where *rhs* is a random matrix generated with the MATLAB command `rhs = 100*rand(1000,1)`. Time the computation. Create the matrix in part (b) using MATLAB, and solve the system using `xmat = A\rhs` and compare the two solutions and the time required.

20.26

a. Develop a MATLAB function `findomega` that plots the number of iterations of the SOR method required as a function of the relaxation parameter ω for the system $Ax = b$, $b = \text{rand}(n, 1)$. For the purpose of graphing, let ω range from 0.01 to 1.99 in steps of 0.005. Use tol $= 1.0 \times 10^{-12}$ and maxiter $= 1000$. Estimate the optimal value of ω by determining the ω corresponding to the minimum number of iterations required to satisfy the error tolerance. Be sure to exclude values of -1 (iteration failure) in the search for the minimum.

b. Apply `findomega` to the matrix E of Problem 20.23(c).

20.27

a. Develop a numerical solution to the two-dimensional Poisson problem using the red-black scheme of Problem 20.15. Name the function `poisson2dgs`.

b. Apply `poisson2dgs` to the problem of Example 20.8.

c. A thin membrane is stretched over a wire bent in the shape of a triangle. The resulting structure satisfies Laplace's equation. Consider the specific problem

$$
\frac{\partial^2 u}{\partial x^2} + \frac{\partial^2 u}{\partial y^2} = 0,
$$

$$u(x, y) = \begin{cases} x, & 0 \le x \le \frac{1}{2}, \ y = 0 \\ 1 - x, & \frac{1}{2} \le x \le 1, \ y = 0 \\ 0, & 0 \le x \le 1, \ y = 1 \\ 0, & x = 0, \quad 0 \le y \le 1 \\ 0, & x = 1, \quad 0 \le y \le 1 \end{cases}$$

whose exact solution is

$$u(x, y) = \frac{4}{\pi^2} \sum_{n=0}^{\infty} (-1)^n \frac{\sin((2n+1)\pi x) \sinh((2n+1)\pi(1-y))}{(2n+1)^2 \sinh((2n+1)\pi)}.$$

i. Graph $u(x, y)$, $1 \le x, y \le 1$, by summing the series until the relative error of the partial sums is less than 10^{-8}.

ii. Apply poisson2dgs and graph the approximate solution.

Chapter 21

Krylov Subspace Methods

You should be familiar with

- Symmetric matrices and their properties
- Inner product
- Vector and matrix norms
- Matrix condition number
- Gradient of a function of n variables
- Cholesky decomposition
- Gram-Schmidt process
- Overdetermined least squares
- *LU* decomposition
- Poisson's equation and five-point finite difference approximation

Chapter 20 discussed the classical iterative methods, Jacobi, Gauss-Seidel, and successive overrelaxation (SOR) for solving linear systems. This chapter introduces more sophisticated methods for solving systems with a sparse coefficient matrix. After discussing the storage of sparse matrices, we introduce one of the most amazing algorithms ever developed in the field of numerical linear algebra, the conjugate gradient (CG) method. Many matrices used in the approximation of the solution to partial differential equations are symmetric positive definite. We saw an example of this situation when we discussed the heat equation in Chapter 12 and the Poisson equation in Chapter 20. The CG method is extremely effective for positive definite matrices. Rather than applying the same iteration scheme over and over, CG makes a decision at each step in order to obtain an optimal result. It uses matrix multiplication, which can be done very efficiently by taking advantage of the sparse structure of the matrices involved.

The CG method is an example of a Krylov subspace method, which is a class of algorithms that project the problem into a lower-dimensional subspace formed by powers of a matrix. We will discuss two additional Krylov subspace methods, minimum residual (MINRES) method that solves symmetric indefinite systems, and the general minimal residual (GMRES) method that applies to general sparse matrices.

In applications, the linear system is frequently ill-conditioned. It is possible using multiplication by an appropriate matrix to transform a system into one with a smaller condition number. This technique is termed preconditioning, and can yield very accurate results for a system with a poorly conditioned coefficient matrix. We will develop preconditioning techniques for CG and GMRES.

The chapter concludes with a discussion of the two-dimensional biharmonic equation, a fourth-order partial differential equation whose positive definite finite-difference matrix is ill-conditioned. Preconditioned CG gives good results for this somewhat difficult problem.

21.1 LARGE, SPARSE MATRICES

The primary use of iterative methods is for computing the solution to large, sparse systems and for finding a few eigenvalues of a large sparse matrix. Along with other problems, such systems occur in the numerical solution of partial differential equations. A good example is our discussion of a finite difference approach to the heat equation in Chapter 12. In that problem, it is necessary to solve a large tridiagonal system of equations, and we were able to use a modification of Gaussian elimination known as the Thomas algorithm. In many cases, matrices have four or more diagonals, are block structured, where nonzero elements exists in blocks throughout the matrix, or have little organized structure. The three types of large, sparse matrices are positive definite, symmetric indefinite, and nonsymmetric. The *sparsity pattern* of a matrix is a plot that shows the presence of nonzeros. Figure 21.1 shows the sparsity pattern of an actual problem for each type. An annotation specifies the computational area for which the matrix was used.

Numerical Linear Algebra with Applications. http://dx.doi.org/10.1016/B978-0-12-394435-1.00021-1

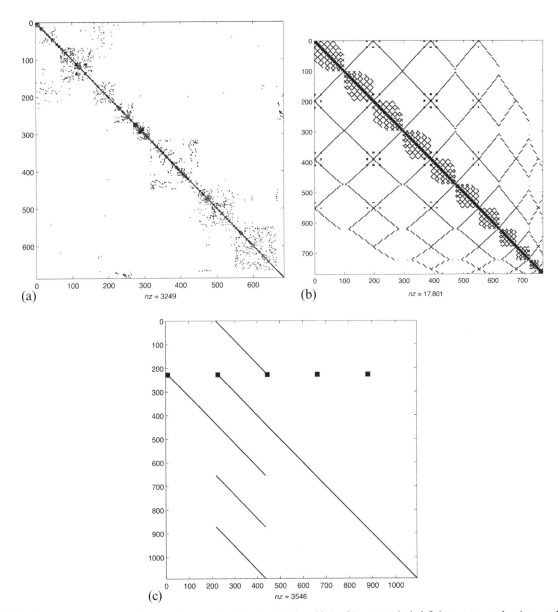

FIGURE 21.1 Examples of sparse matrices. (a) Positive definite: structural problem, (b) symmetric indefinite: quantum chemistry problem, and (c) nonsymmetric: computational fluid dynamics problem.

Algorithms for dealing with large, sparse matrices are an active area of research. The choice of a suitable method for handling a given problem depends on many factors, including the matrix size, structure, and what type of computer is used. Algorithms are different for vector computers and parallel computers. These problems can be very difficult, often requiring much experimentation before finding a suitable method of solution. The text by Saad [64] is an excellent reference for advanced material in this area.

Remark 21.1. The matrices in Figure 21.1 were obtained from the *University of Florida Sparse Matrix Collection*, maintained by Tim Davis, University of Florida, and Yifan Hu of AT&T Research [90]. It is located at http://www.cise. ufl.edu/research/sparse/matrices/. Matrices can be downloaded from the Web site or accessed by obtaining the UFgui Java interface and running the application on your system.

21.1.1 Storage of Sparse Matrices

There are a number of formats used to store sparse matrices. In each case, only the nonzero elements are recorded. We will look at the *compressed row storage (CRS) format* that stores the nonzero elements by rows. Assume we have a nonsymmetric sparse matrix. Create three vectors: one for floating point numbers (AV) and the other two for integers (AI, AI). The AV vector stores the values of the nonzero elements of the matrix as they are traversed in a row-wise manner. The AJ vector stores the column indices of the elements in the AV vector; for instance, if $AV(j) = v$, then v is in column $AJ(j)$. The AI vector stores the locations in the AV vector that begin a row; that is, if $AI(i) = k$, then $AV(k)$ is the first nonzero element in row i. The remaining elements in row i are the elements in AV up to but not including $AI(i + 1)$, so the number of nonzero elements in the ith row is equal to $AI(i + 1) - AI(i)$. In order that this relationship will hold for the last row of the matrix, an additional entry, $nnz + 1$ is added to the end of A, where nnz is the number of nonzero entries in the matrix. The vector AI must have an entry for every row. If all the elements in row k are zeros, then $AI(k)$ must have a value such as -1 to indicate row k has no nonzero elements.

The CRS format saves a significant amount of storage. Instead of storing n^2 elements, we need only storage locations for n elements in AI, nnz elements in both AV and AJ, and one additional entry $nnz+1$ in AI. If we assume that a machine has 8 byte floating point numbers and 4 byte integers, then the requirement for CRS storage is $8nnz + 4nnz + 4(n + 1)$. For instance, without using sparse matrix storage a 1000×1000 matrix of double values requires $8 \times 10^6 = 8,000,000$ bytes of storage. If the matrix has only $nnz = 12,300$ nonzero elements, we need storage for only $8(12,300) + 4(12,300) + 4(1001) = 151,604$ bytes, which is $151,604/8,000,000$, or approximately 1.90% of the space required to store the entire matrix. If the matrix is symmetric, we only need to store the nonzero values on and above the diagonal, saving even more space.

The CRS storage format is very general. It makes no assumptions about the sparsity structure of the matrix. On the other hand, this scheme is not efficient for accessing matrices one element at a time. Inserting or removing a nonzero entry may require extensive data reorganization. However, element-by-element manipulation is rare when dealing with sparse matrices. As we will see, iterative algorithms for the solution of sparse linear systems avoid direct indexing and use matrix operations such as multiplication and addition for which there are algorithms that efficiently use the sparse matrix storage structure. There has been a significant amount of research in the area of sparse matrices, and the interested reader can consult Refs. [64–66].

Example 21.1.   Consider the nonsymmetric matrix

$$A = \begin{bmatrix} 3 & 0 & 0 & 0 & 2 & 0 & 0 \\ 0 & 0 & -1 & 0 & 0 & 1 & 0 \\ 0 & 5 & 0 & 2 & 6 & 0 & 0 \\ 0 & 0 & 0 & 0 & 0 & 0 & 0 \\ 7 & 0 & 0 & 0 & 12 & 0 & 0 \\ 0 & 0 & 0 & 3 & 0 & 0 & 0 \\ 0 & 0 & 1 & 0 & 0 & 8 & 3 \end{bmatrix}.$$

The CRS format for this matrix is then specified by the arrays $\{AV, AJ, AI\}$ as follows:

AV	3	2	-1	1	5	2	6	7	12	3	1	8	3
AJ	1	5	3	6	2	4	5	1	5	4	3	6	7

AI	1	3	5	-1	8	10	11	14

Consider row 5. $AI(5) = 8$, $AJ(8) = 1$, and $AV(8) = 7$. The first element in row 5 is in column 1 and has value 7. There are $AI(7) - AI(6) = 2$ elements in row 5. The remaining element in row 5 is $AV(9) = 12$ in column $AJ(9) = 5$.   ∎

21.2   THE CG METHOD

The *CG* iteration is one of the most important methods in scientific computation, and is the algorithm of choice for solving positive definite systems. We used the SOR iteration for solving a Poisson equation in Section 20.5. The SOR iteration depends on a good choice of the relaxation parameter ω. A poor choice leads to slow convergence or divergence. The CG method does not have this problem. There are a number of approaches to developing CG. We take the approach of minimizing a quadratic function formed from the $n \times n$ matrix A. A *quadratic function* mapping $\mathbb{R}^n$ to $\mathbb{R}$ is a function of n variables $x_1, x_2, \ldots, x_n$ where each term involves the product of at most two variables. This approach allows us to incorporate illustrations that help with understanding this important algorithm.

21.2.1   The Method of Steepest Descent

To develop CG we start with the *method of steepest descent* for solving a positive definite system and then make modifications that improve the convergence rate, leading to CG. To solve the system $Ax = b$, where A is positive definite, we find the minimum of a quadratic function whose minimum is the solution to the system $Ax = b$. Let

$$\phi(x) = \frac{1}{2}x^T A x - x^T b, \tag{21.1}$$

where $A \in \mathbb{R}^{n \times n}$ and $x \in \mathbb{R}^n$. To minimize $\phi(x)$, set each partial derivative $\frac{\partial \phi}{\partial x_i} = 0$, $1 \le i \le n$, and solve for x. The *gradient* of ϕ is the vector $\nabla \phi = \begin{bmatrix} \frac{\partial \phi}{\partial x_1} & \frac{\partial \phi}{\partial x_2} & \cdots & \frac{\partial \phi}{\partial x_{n-1}} & \frac{\partial \phi}{\partial x_n} \end{bmatrix}^T$, so x is a minimum if $\nabla \phi(x) = 0$. Some manipulations (Problem 21.2) will show that

$$\nabla \phi(x) = \frac{1}{2}A^T x + \frac{1}{2}Ax - b.$$

Since A is symmetric,

$$\nabla \phi(x) = Ax - b, \tag{21.2}$$

and the minimum value of $\phi(x)$ occurs when $Ax = b$. Thus, finding the minimum value of $\phi(x)$ is equivalent to solving $Ax = b$. Since A is positive definite, it is nonsingular, and $x_{\min} = A^{-1}b$. Thus,

$$\nabla \phi(x_{\min}) = \frac{1}{2}x_{\min}^T A A^{-1} b - x_{\min}^T b = -\frac{1}{2}x_{\min}^T b.$$

Equation 21.1 also shows us the type of surface generated by $\phi(x)$. By the spectral theorem, there is an orthogonal matrix P such that $A = PDP^T$, where D is a diagonal matrix containing the eigenvalues of A. We have seen that an orthogonal matrix can be used to effect a change of coordinates (Section 7.1.2). Let $u = P^T(x - x_{\min})$, so $x = Pu + x_{\min}$. Substituting this into Equation 21.1, after some matrix algebra and noting that $D = P^T A P$ we obtain

$$\begin{aligned}
\widetilde{\phi}(u) &= \frac{1}{2}(Pu)^T A (Pu) - \frac{1}{2}x_{\min}^T A \bar{x} = \frac{1}{2}u^T D u - \frac{1}{2}x_{\min}^T A x_{\min} \\
&= \sum_{i=1}^{n} \lambda_i u_i^2 - \frac{1}{2}x_{\min}^T b
\end{aligned}$$

Since A is positive definite, $\lambda_i > 0$, $1 \le i \le n$. For $n = 2$, $\widetilde{\phi}(u) = \lambda_1 u_1^2 + \lambda_2 u_2^2 - \frac{1}{2}x_{\min}^T b$ and is a paraboloid. In general, the orthogonal transformation $u = P^T(x - \bar{x})$ maintains lengths and angles, so $\phi(x)$ is a paraboloid in $\mathbb{R}^n$.

Example 21.2. If $Ax = b$, where $A = \begin{bmatrix} 5 & 1 \\ 1 & 4 \end{bmatrix}$ and $b = \begin{bmatrix} 1 \\ 1 \end{bmatrix}$, the eigenvalues of the coefficient matrix are 3.3820 and 5.6180, so A is positive definite. Figure 21.2(a) is a graph of the paraboloid $\phi(x, y) = \frac{5}{2}x^2 + 2y^2 + xy - x - y$. The solution to the system is $\begin{bmatrix} 0.1579 & 0.2105 \end{bmatrix}^T$, and we know the global minimum value is

$$-\frac{1}{2}\begin{bmatrix} 0.1579 & 0.2105 \end{bmatrix}\begin{bmatrix} 1 & 1 \end{bmatrix}^T = -0.1842. \qquad \blacksquare$$

Before continuing to develop the method of steepest descent, we need to define some concepts. A *level set* of a real-valued function of n variables is a curve along which the function has a constant value. For a function of two variables, level sets are often called *contour lines*. They are formed when a plane parallel to the xy-plane passes through the surface and the curve formed is projected onto the xy-plane. Since $\phi(x)$ is a paraboloid, its contour lines are ellipses. Figure 21.2(b) is a graph that shows contour lines for our example function $\phi(x, y) = \frac{5}{2}x^2 + 2y^2 + xy - x - y$. These contour lines picture movement downward to the minimum of $\phi(x)$ located near the middle of the figure. Contour lines and the gradient are related. For a given point x, the gradient always points in the direction of greatest increase of $\phi(x)$, so $-\nabla \phi(x)$ points in the direction of greatest decrease. In addition, the gradient is always orthogonal to the contour line at the point x (Figure 21.2(b)).

In steepest descent, the minimum value $\phi(x_{\min})$ is found by starting at an initial point and descending toward $x_{\min}$ (the solution of $Ax = b$). Assume the current point is x_i, and we want to move from it to the point $x_{\min}$. Since the quadratic function ϕ decreases most rapidly in the direction of the negative gradient, Equation 21.2 specifies that $-\nabla \phi(x_i) = b - Ax_i$. Let

$$r_i = b - Ax_i$$

be the residual at point x_i. The residual itself is pointing in the direction of most rapid decrease from x_i, and is called a *direction vector*. If the residual is zero, we are done. If the residual is nonzero, move along a line specified by the direction vector, so let $x_{i+1} = x_i + \alpha_i r_i$, where α_i is a scalar. We must find α_i so that we minimize ϕ for each step. Let

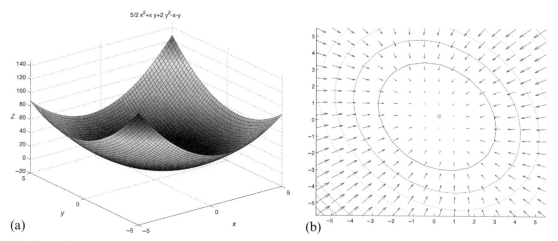

FIGURE 21.2 Steepest descent. (a) Quadratic function in steepest descent and (b) gradient and contour lines.

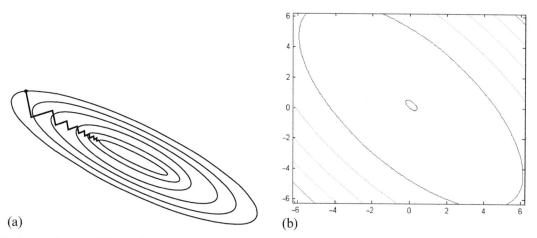

FIGURE 21.3 Steepest descent. (a) Deepest descent zigzag and (b) gradient contour lines.

$$f(\alpha_i) = \phi(x_i + \alpha_i r_i).$$

Using the symmetry of A and the commutativity of the inner product we have,

$$
\begin{aligned}
f(\alpha_i) &= \frac{1}{2}(x_i + \alpha_i r_i)^{\mathrm{T}} A (x_i + \alpha_i r_i) - (x_i + \alpha_i r_i)^{\mathrm{T}} b \\
&= \frac{1}{2} x_i^{\mathrm{T}} A x_i + \frac{\alpha_i^2}{2} r_i^{\mathrm{T}} A r_i + \alpha_i x_i^{\mathrm{T}} A r_i - x_i^{\mathrm{T}} b - \alpha_i r_i^{\mathrm{T}} b.
\end{aligned}
$$

The minimum occurs when $f'(\alpha_i) = 0$, and

$$
\begin{aligned}
f'(\alpha_i) &= \alpha_i r_i^{\mathrm{T}} A r_i + x_i^{\mathrm{T}} A r_i - r_i^{\mathrm{T}} b \\
&= \alpha_i r_i^{\mathrm{T}} A r_i + r_i^{\mathrm{T}} A x_i - r_i^{\mathrm{T}} b \\
&= \alpha_i r_i^{\mathrm{T}} A r_i + r_i^{\mathrm{T}} (A x_i - b) \\
&= \alpha_i r_i^{\mathrm{T}} A r_i - r_i^{\mathrm{T}} r_i.
\end{aligned}
$$

Thus, the minimum occurs when

$$\alpha_i = \frac{r_i^{\mathrm{T}} r_i}{r_i^{\mathrm{T}} A r_i}. \tag{21.3}$$

Step from x_i to $x_{i+1} = x_i + \alpha_i r_i$, and then use x_{i+1} and r_{i+1} to compute $x_{i+2} = x_{i+1} + \alpha_{i+1} r_{i+1}$, the next point downward toward the minimum.

Our discussion gives rise to the steepest descent algorithm.

NLALIB: The function steepestDescent implements Algorithm 21.1.

It can be shown that [2, pp. 625-627]

$$\left(\phi\left(x_i\right) + \frac{1}{2}b^{\mathsf{T}}A^{-1}b \right) \le \left(1 - \frac{1}{\kappa\left(A\right)} \right) \left(\phi\left(x_{i-1}\right) + \frac{1}{2}b^{\mathsf{T}}A^{-1}b \right),$$

Algorithm 21.1 Steepest Descent

```
function STEEPESTDESCENT(A,b,x₀,tol,maxiter)
    % Solve Ax=b using the steepest descent method.
    % Input: Matrix A, right-hand side b, initial approximation x₀,
    % error tolerance tol, and maximum number of iterations maxiter.
    % Output: Approximate solution x, norm of the residual ‖b − Ax‖₂,
    % and the number of iterations required.
    % If the method does not converge, iter=-1.
    r₀=b - Ax₀
    iter=1
    i=0
    while  do(‖rᵢ‖₂ ≥ tol) and (iter ≤ maxiter)
        αⱼ = rᵢᵀrᵢ / rᵢᵀArᵢ
        xᵢ₊₁ = xᵢ + αⱼrᵢ
        rᵢ₊₁ = b − Axᵢ₊₁
        i=i+1
        iter=iter+1
    end while
    iter=iter-1
    if iter ≥ numiter then iter=-1
    end if
end function
```

---

which says that the algorithm converges no matter what initial value we choose. If $\kappa\left(A\right)$ is large, then $\left(1 - \frac{1}{\kappa\left(A\right)} \right)$ is close to 1, and convergence will be slow. We can see this geometrically. For any $n \times n$ matrix, successive search directions, r_i, are orthogonal, as we can verify by a calculation. Noting that $Ax_i = b - r_i$,

$$\begin{aligned}
r_i^T r_{i+1} = \; & r_i^T\left(b - Ax_{i+1}\right) = r_i^T\left[b - A\left(x_i + \alpha_i r_i\right)\right] = \\
& r_i^T b - r_i^T\left(Ax_i + \alpha_i Ar_i\right) = \\
& r_i^T b - r_i^T\left[\left(b - r_i\right) + \alpha_i Ar_i\right] = \\
& r_i^T b - r_i^T b + r_i^T r_i - \alpha_i r_i^T Ar_i = \\
& r_i^T r_i - \alpha_i r_i^T Ar_i.
\end{aligned}$$

From Equation 21.3

$$r_i^{\mathsf{T}} r_i - \alpha_i r_i^{\mathsf{T}} Ar_i = 0,$$

so $r_i^{\mathsf{T}} r_{i+1} = 0$. If we are at x_i in the descent to the minimum, the negative of the gradient vector is orthogonal to the contour line $\phi\left(x_i\right) = k_i$. The next search direction r_{i+1} is orthogonal to r_i and orthogonal to the contour line $\phi\left(x_{i+1}\right) = k_{i+1}$. As illustrated in Figure 21.3(a), for $n = 2$ the approach to the minimum follows a zigzag pattern. Steepest descent always makes a turn of $90°$ as it moves toward the minimum, and it could very well be that a different turn is optimal. In $\mathbb{R}^2$, if the eigenvalues of A are $\lambda_1 = \lambda_2$, the contour lines are circles; otherwise, they are ellipses. The eigenvalues of a positive definite matrix are the same as its singular values (Problem 21.8), so the shape of the ellipses depends on the ratio of the largest to the smallest eigenvalue, which is the condition number. If the condition number of A is small, the ellipses are close to circles, and the steepest descent algorithm steps from contour to contour, quickly moving toward the center. However, as the condition number gets larger, the ellipses become long and narrow. The paraboloid $\phi(x)$ has a steep, narrow canyon near the minimum, and the values of x_i move back and forth across the walls of the canyon, moving down very slowly, and requiring a great many iterations to converge to the minimum. This is illustrated in Figure 21.3(b) for the matrix $A = \begin{bmatrix} 5 & 3 \\ 3 & 4 \end{bmatrix}$. Note that the contour lines are more eccentric ellipses than those for the matrix $A = \begin{bmatrix} 5 & 1 \\ 1 & 4 \end{bmatrix}$ of Example 21.2.

Example 21.3. The MATLAB function `steepestDescent` in the software distribution returns the approximate solution and the number of iterations required. Apply the method to the matrices $A = \begin{bmatrix} 5 & 1 \\ 1 & 4 \end{bmatrix}$, $\kappa(A) = 1.6612$ and $B = \begin{bmatrix} 7 & 6.99 \\ 6.99 & 7 \end{bmatrix}$, $\kappa(B) = 1399$. In both cases, $b = \begin{bmatrix} 46.463 & 17.499 \end{bmatrix}^{\mathrm{T}}$ and $x_0 = \begin{bmatrix} 5 & 5 \end{bmatrix}^{\mathrm{T}}$. For matrix A, 21 iterations were required to attain an error tolerance of 1.0×10^{-12}; however, matrix B required 8462 iterations for the same error tolerance. ∎

21.2.2 From Steepest Descent to CG

The method of steepest descent does a line search based on the gradient by computing $x_{i+1} = x_i + \alpha_i r_i$, where the r_i are the residual vectors. The CG method computes

$$x_{i+1} = x_i + \alpha_i p_i,$$

where the direction vectors $\{p_i\}$ are chosen so $\{x_i\}$ much more accurately and rapidly descends toward the minimum of $\phi(x) = \frac{1}{2}x^{\mathrm{T}}Ax - x^{\mathrm{T}}b$. Repeat the calculations that lead to Equation 21.3, replacing r_i by p_i, to obtain

$$\alpha_i = \frac{p_i^{\mathrm{T}} r_i}{p_i^{\mathrm{T}} A p_i}. \tag{21.4}$$

Before showing how to choose $\{p_i\}$, we must introduce the A-norm.

Definition 21.1. If A is an $n \times n$ positive definite matrix and $x, y \in \mathbb{R}^n$, then $\langle x, y \rangle_A = x^{\mathrm{T}}Ay$ is an inner product, and if $\langle x, y \rangle_A = 0$, x and y are said to be A-*conjugate*. The corresponding norm, $\|x\|_A = \sqrt{x^{\mathrm{T}}Ax}$ is called the A-*norm* or the *energy norm*.

Remark 21.2.   We leave the fact that $\langle \cdot, \cdot \rangle_A$ is an inner product to Problem 21.3. We use the term energy norm because the term $\frac{1}{2}x^{\mathrm{T}}Ax = \frac{1}{2}\|x\|_A^2$ represents physical energy in many problems.

The function $\phi(x) = \frac{1}{2}x^{\mathrm{T}}Ax - x^{\mathrm{T}}b$ can be written using the A-norm as follows:

$$\phi(x) = \frac{1}{2}\|x\|_A^2 - \langle x, b \rangle,$$

where $\langle x, b \rangle$ is the Euclidean inner product. If $\bar{x}$ is the approximation to the minimum of $\phi(x)$ and $e = x - \bar{x}$ is the error, then completing the square gives (Problem 21.4)

$$\phi(x) = \frac{1}{2}e^{\mathrm{T}}Ae - \frac{1}{2}\|x\|_A^2 = \frac{1}{2}\|e\|_A^2 - \frac{1}{2}\|\bar{x}\|_A^2.$$

Thus, to minimize ϕ, we must minimize the A-norm of the error. The steepest descent algorithm minimizes the 2-norm of the error along a gradient line at each step. This is one-dimensional minimization. The CG method uses information from past steps so that it can minimize over higher-dimensional subspaces. As we move from x_{i-1} to x_i, the algorithm minimizes over an i-dimensional subspace. Another way of putting it is that instead of minimizing over a line we minimize over a plane ($n = 2$) or a hyperplane ($n > 2$). The approximations $x_{i+1} = x_i + \alpha_i p_i$ do not zigzag toward the minimum of $\phi(x)$, but follow a much better path. In the following figure, from point A the thin line represents the zigzag of steepest descent and requires four steps. The CG iteration requires only one step (Figure 21.4).

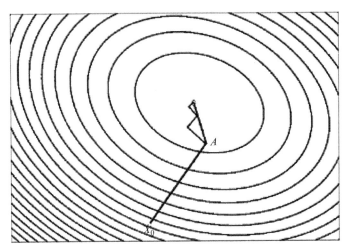

FIGURE 21.4 2-Norm and A-norm convergence.

For the steepest descent method, successive search directions r_i are orthogonal. Recall that the Gram-Schmidt process takes a set of linearly independent vectors and produces a set of mutually orthogonal unit vectors relative to the 2-norm that span the same subspace as the original set. To find $\{p_i\}$, we apply the Gram-Schmidt process to the residual vectors $\{r_i\}$ but use the energy inner product and norm. Start with $r_0 = b - Ax_0$ just like we did with steepest descent. The set of vectors $\{p_i\}$ must be A-conjugate ($p_i^T A p_j = 0$, $i \neq j$), but it is not necessary that they be unit vectors. If we skip the normalization step of Gram-Schmidt at step i we have

$$p_i = r_i - \sum_{j=1}^{i-1} c_{j,i} p_j, \tag{21.5}$$

where

$$c_{ji} = \frac{r_i^T A p_j}{p_j^T A p_j}.$$

The purpose of Equation 21.5 is to determine the next search direction from the current residual and the previous search directions, thus doing far better than steepest descent that uses only the last search direction.

The computation of α_i defined in Equation 21.4 can be simplified by building some additional machinery. The following technical lemmas contains some formulas we will need.

Lemma 21.1. *Let x be the true solution to $Ax = b$, and the error at step i defined by $e_i = x - x_i$. Then,*

1. $r_{i+1} = r_i - \alpha_i A p_i$
2. $A e_{i+1} = r_{i+1}$
3. $e_{i+1} = e_i - \alpha_i p_i$.

Proof. Note that $r_{i+1} = b - Ax_{i+1} = b - A(x_i + \alpha_i p_i) = r_i - \alpha_i A p_i$, and we have property 1.

$Ae_{i+1} = A(x - x_{i+1}) = A(x) - Ax_{i+1} = b - Ax_{i+1} = r_{i+1}$, which proves property 2.

For the proof of property 3, note that $e_{i+1} = x - x_{i+1} = x - (x_i + \alpha_i p_i) = e_i - \alpha_i p_i$. $\qquad\square$

Using Lemma 21.1, we can develop a result that will further aid in simplifying the computation of α_i. It says that the error at step i+1 is A-conjuate to the previous direction vector p_i.

Lemma 21.2. $\langle e_{i+1}, p_i \rangle_A = 0.$

Proof. From property 1 of Lemma 21.1, $r_{i+1}^T p_i = r_i^T p_i - \alpha_i (Ap_i)^T p_i = r_i^T p_i - \alpha_i (p_i^T A p_i) = 0$ by substituting the value of α_i. Now, multiply the equation in property 2 by p_i^T on the left to obtain

$$p_i^T A e_{i+1} = \langle e_{i+1}, p_i \rangle_A = p_i^T r_{i+1} = r_{i+1}^T p_i = 0. \qquad\square$$

We are now in a position to simplify the values of α_i. The algorithm begins with the assignment $p_0 = r_0 = b - Ax_0$ and the computation of x_1. Lemma 21.2 says that $\langle e_1, p_0 \rangle_A = 0$. In the next step, the Gram-Schmidt process creates p_1 so that $\langle p_1, p_0 \rangle_A = 0$, determining x_2. Again, from Lemma 21.2, $\langle e_2, p_1 \rangle_A = 0$. At this point we have $\langle e_1, p_0 \rangle_A = 0$ and $\langle e_2, p_1 \rangle_A = 0$. Noting that $\langle p_1, p_0 \rangle = 0$, apply property 3 of Lemma 21.1 to obtain

$$\langle e_2, p_0 \rangle_A = \langle e_1 - \alpha_1 p_1, p_0 \rangle_A = \langle e_1, p_0 \rangle_A - \alpha_1 \langle p_1, p_0 \rangle_A = 0.$$

Thus, e_2 is A-orthogonal to both p_0 and p_1. Continuing in this fashion, we see that e_i is A-orthogonal to $\{p_0, p_1, \ldots, p_{i-1}\}$, or

$$\langle e_i, p_j \rangle_A = 0, \quad j < i. \tag{21.6}$$

Now,

$$\langle e_i, p_j \rangle_A = p_j^{\mathrm{T}} A e_i = p_j^{\mathrm{T}} (Ax - Ax_i) = p_j^{\mathrm{T}} (b - Ax_i) = \langle r_i, p_j \rangle$$

and from Equation 21.6 it follows that:

$$\langle r_i, p_j \rangle = 0, \quad j < i. \tag{21.7}$$

We can obtain two useful properties from Equation 21.7. Using Equation 21.5, take the inner product of p_j and r_i, $j < i$, and

$$r_i^{\mathrm{T}} p_j = r_i^{\mathrm{T}} r_j - \sum_{k=1}^{j-1} c_{kj} r_i^{\mathrm{T}} p_k = r_i^{\mathrm{T}} r_j = 0.$$

Thus,

$$r_i^{\mathrm{T}} r_j = 0, \quad j < i. \tag{21.8}$$

and all residuals are orthogonal to the previous ones. Similarly

$$\langle p_i, r_i \rangle = \|r_i\|_2^2 - \sum_{j=1}^{i-1} c_{ji} \langle r_i, p_j \rangle = \|r_i\|_2^2 \tag{21.9}$$

from Equation 21.7. Using Equation 21.9 in Equation 21.4 gives us the final value for α_i.

$$\alpha_i = \frac{\|r_i\|_2^2}{\langle Ap_i, p_i \rangle} \tag{21.10}$$

The evaluation of formula 21.5 seems expensive, and it appears that we must retain all the previous search directions. Here is where the CG algorithm is remarkable. We get the benefit of a combination of search directions in Equation 21.5 by computing only $c_{i-1,i}$, the coefficient of the last direction p_{i-1}! Take the inner product of r_i and the relationship in property 1 of Lemma 21.1 to obtain $r_i^{\mathrm{T}} r_{j+1} = r_i^{\mathrm{T}} r_j - \alpha_j r_i^{\mathrm{T}} Ap_j$, and so

$$\alpha_j r_i^{\mathrm{T}} Ap_j = r_i^{\mathrm{T}} r_j - r_i^{\mathrm{T}} r_{j+1}. \tag{21.11}$$

Equations 21.8 and 21.11 give us the results

$$r_i^{\mathrm{T}} Ap_j = \begin{cases} \dfrac{1}{\alpha_i} r_i^{\mathrm{T}} r_i & j = i \\[2mm] -\dfrac{1}{\alpha_{i-1}} r_i^{\mathrm{T}} r_i & j = i - 1 \\[2mm] 0 & j < i - 1 \end{cases} \tag{21.12}$$

Recall that the Gram-Schmidt constants are $c_{ji} = \dfrac{r_i^{\mathrm{T}} Ap_j}{p_j^{\mathrm{T}} Ap_j}$. Using Equations 21.10 and 21.12,

$$c_{i-1,i} = \frac{r_i^{\mathrm{T}} Ap_{i-1}}{p_{i-1}^{\mathrm{T}} Ap_{i-1}} = -\frac{1}{\alpha_{i-1}} \frac{r_i^{\mathrm{T}} r_i}{p_{i-1}^{\mathrm{T}} Ap_{i-1}} = -\frac{\langle Ap_{i-1}, p_{i-1} \rangle}{\|r_{i-1}\|_2^2} \frac{r_i^{\mathrm{T}} r_i}{\langle Ap_{i-1}, p_{i-1} \rangle} = -\frac{\|r_i\|_2^2}{\|r_{i-1}\|_2^2}.$$

From Equation 21.12, $c_{j,i} = 0$, $j < i - 1$. Thus, all the coefficients $\{c_{ji}\}$ in Equation 21.5 are zero except $c_{i-1,i}$. By defining

$$\beta_i = \frac{\|r_i\|_2^2}{\|r_{i-1}\|_2^2},$$

we see that

$$p_i = r_i + \beta_i p_{i-1}. \tag{21.13}$$

We can now give the CG algorithm.

Algorithm 21.2 Conjugate Gradient

```
function cg(A,b,x1,tol,numiter)
    r0 = b - Ax1
    p0 = r0
    for i=1:numiter do
        αi-1 = (rᵀi-1 ri-1) / (pᵀi-1 Api-1)
        xi = xi-1 + αi-1 pi-1
        ri = ri-1 - αi-1 Api-1
        if ‖ri‖2 < tol then
            iter=i
            return[xi, iter]
        end if
        βi = (rᵀi ri) / (rᵀi-1 ri-1)
        pi = ri + βi pi-1
    end for
    iter=-1 return[xnumiter, iter]
end function
```

NLALIB: The function cg implements Algorithm 21.2.

Example 21.4. Let $A = \begin{bmatrix} 5 & 1 & 1 \\ 1 & 4 & 1 \\ 1 & 1 & 6 \end{bmatrix}$, $b = \begin{bmatrix} 1 & 2 & 3 \end{bmatrix}^T$, $x_0 = \begin{bmatrix} 0 & 0 & 0 \end{bmatrix}^T$, trace two iterations of Algorithm 21.2, then apply the function cg to compute the solution. Note that only one more iteration is required to obtain a very accurate result:

$$r_0 = p_0 = b - Ax_0 = \begin{bmatrix} 1 & 2 & 3 \end{bmatrix}^T$$

$i = 1$:

$$\alpha_0 = (r_0^T r_0) / (p_0^T A p_0) = 0.1443 \quad x_1 = x_0 + \alpha_0 p_0 = \begin{bmatrix} 0.1443 & 0.2887 & 0.4330 \end{bmatrix}^T$$

$$r_1 = r_0 - \alpha_0 A p_0 = \begin{bmatrix} -0.4433 & 0.2680 & -0.0309 \end{bmatrix} \quad \beta_1 = (r_1^T r_1) / (r_0^T r_0) = 0.0192$$

$$p_1 = r_1 + \beta_1 p_0 = \begin{bmatrix} -0.4241 & 0.3065 & 0.0268 \end{bmatrix}^T$$

$i = 2$:

$$\alpha_1 = (r_1^T r_1) / (p_1^T A p_1) = 0.2659 \quad x_2 = x_1 + \alpha_1 p_1 = \begin{bmatrix} 0.0316 & 0.3702 & 0.4401 \end{bmatrix}^T$$

```
>> [x r iter] = cg(A,b,x0,1.0e-15,10)
x =
    0.0374
    0.3832
    0.4299
```

```
r =
  1.0839e-016
iter =
  3
```
■

21.2.3 Convergence

We will not prove convergence of the CG method, but will state convergence theorems. For proofs, see Refs. [23, Chapter 8] and [27, Chapter 9].

Theorem 21.1.   *If exact arithmetic is performed, the CG algorithm applied to an $n \times n$ positive definite system $Ax = b$ converges in n steps or less.*

CG converges after n iterations, so why should we care about convergence analysis? CG is commonly used for problems so large it is not feasible to run even n iterations. In practice, floating point errors accumulate and cause the residual to gradually lose accuracy and the search vectors to lose A-orthogonality, so it is not realistic to think about an exact algorithm.

Theorem 21.2 indicates how fast the CG method converges. For a proof, see Ref. [1, pp. 312-314]. The proof uses Chebyshev polynomials, which are defined and discussed in Appendix C.

Theorem 21.2.   *Let the CG method be applied to a symmetric positive definite system $Ax = b$, where the condition number of A is κ. Then the A-norms of the errors satisfy*

$$\frac{\|e_k\|_A}{\|e_0\|_A} \le 2 \left(\frac{\sqrt{\kappa} - 1}{\sqrt{\kappa} + 1} \right)^k .$$

The term $\frac{\sqrt{\kappa}-1}{\sqrt{\kappa}+1}$ can be written as $1 - \frac{2}{\sqrt{\kappa}+1}$. If $\sqrt{\kappa}$ is reasonably small, the algorithm will converge quickly, but if $\sqrt{\kappa}$ is large, convergence slows down. Since κ depends on the largest and smallest eigenvalue of A, if the eigenvalues are clustered closely together, convergence will be good. If the eigenvalues of A are widely separated, CG convergence will be slower. Each iteration of the CG method requires $O\left(n^2\right)$ flops, so n iterations will cost $O\left(n^3\right)$ flops, which is the same as the Cholesky decomposition. However, practice has shown that convergence using floating point arithmetic often occurs in less than n iterations.

We conclude this section with a comparison of the convergence rates of steepest descent and CG. Of course, we expect CG to outperform steepest descent.

Example 21.5.   The software distribution contains a sparse 300×300 symmetric positive definite matrix CGDES in the file CGDES.mat. It has a condition number of 30.0. Figure 21.5(a) is a density plot showing the location of its 858 nonzeros. If b is a random column vector, and $x_0 = \begin{bmatrix} 1 & 1 & \ldots & 1 & 1 \end{bmatrix}^T$, Figure 21.5(b) is a graph of the log of the residual norms for the steepest descent and CG algorithms after $n = 5, 6, \ldots, 50$ iterations. Note the superiority of CG. After 50 iterations, CG has attained a residual of approximately 10^{-7}, but steepest descent is still moving very slowly downward toward a more accurate solution.
■

21.3   PRECONDITIONING

Since the convergence of an iterative method depends on the condition number of the coefficient matrix, it is often advantageous to use a *preconditioner matrix M that* transforms the system $Ax = b$ to one having the same solution but that can solved more rapidly and accurately. The idea is to choose M so that

- M is not too expensive to construct.
- M is easy to invert.
- M approximates A so that the product of A and M^{-1} "is close" to I ($\eta(I) = 1$).
- the preconditioned system is more easily solved with higher accuracy.

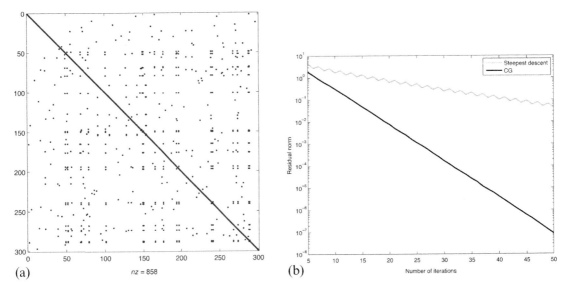

FIGURE 21.5 CG vs. steepest descent. (a) Density plot for symmetric positive definite sparse matrix CGDES and (b) residuals of CG and steepest descent.

If M is a nonsingular matrix, then the system

$$M^{-1}Ax = M^{-1}b \tag{21.14}$$

has the same solution as $Ax = b$. The solution to the system 21.14 depends on the coefficient matrix $M^{-1}A$ instead of A, and we hope that $\kappa\left(M^{-1}A\right)$ is much smaller than $\kappa(A)$.

The computation of $M^{-1}A$ is rarely done. The product may be a dense matrix, and we lose all the advantages of sparsity. Also, there is the expense and potential inaccuracy of computing $M^{-1}A$. We can form the product $M^{-1}Ax$ indirectly as follows:

a. Let $z = M^{-1}Ax$, and compute $w = Ax$.
b. To find z, solve $Mz = w$.

Compute the right-hand side of Equation 21.14 by letting $y = M^{-1}b$ and solving $My = b$. Such a matrix M is called a *left preconditioner*.

A matrix M can also be a *right preconditioner* by computing AM^{-1} in hope that its condition number is much smaller than that of A. In this case, form the equivalent system

$$\left(AM^{-1}\right)Mx = b,$$

and let $u = Mx$, so we need to solve

$$AM^{-1}u = b.$$

If M is of the factored form $M = M_L M_R$, then we can use a *split preconditioner*. First write the system as

$$M_L^{-1}A\left(M_R^{-1}M_R\right)x = M_L^{-1}b,$$

and then solve the system

$$M_L^{-1}AM_R^{-1}u = M_L^{-1}b,$$

where $u = M_R x$.

Remark 21.3.   Note that

$$M^{-1}A = M^{-1}A\left(M^{-1}M\right) = M^{-1}\left(AM^{-1}\right)M$$
$$M_R^{-1}M_L^{-1}AM_R^{-1}M_R = (M_L M_R)^{-1}A = M^{-1}A,$$

and all the matrices for preconditioning are similar, so they have the same eigenvalues. If the eigenvalues of one of these matrices, say Mat, are close to 1 and $\|\text{Mat} - I\|_2$ is small, then the preconditioned system will converge quickly. If a preconditioner does not satisfy this criterion, its distribution of eigenvalues may be favorable to fast convergence.

21.4 PRECONDITIONING FOR CG

We will discuss two methods of finding a preconditioner for CG. In each case, we apply the CG method to the preconditioned system in the hope that the CG iteration will converge faster. Of course, in addition to wanting a system with a smaller condition number, M must be chosen so that the preconditioned system is positive definite. The two methods we will present are the incomplete Cholesky decomposition and the construction of an SSOR preconditioner.

21.4.1 Incomplete Cholesky Decomposition

For the moment, assume we have a positive definite left preconditioner matrix M. M has a Cholesky decomposition $M = R^{\mathrm{T}}R$. We will use this decomposition to obtain an equivalent system $\overline{A}\overline{x} = \overline{b}$, where $\overline{A}$ is positive definite so the CG method applies.

$$M^{-1}Ax = M^{-1}b$$

$$\left(R^{\mathrm{T}}R\right)^{-1} Ax = \left(R^{\mathrm{T}}R\right)^{-1} b$$

$$R^{-1}\left(R^{\mathrm{T}}\right)^{-1} Ax = R^{-1}\left(R^{\mathrm{T}}\right)^{-1} b$$

$$\left(R^{\mathrm{T}}\right)^{-1} A \left(R^{-1}R\right) x = \left(R^{\mathrm{T}}\right)^{-1} b$$

$$\left(\left(R^{\mathrm{T}}\right)^{-1} AR^{-1}\right) Rx = \left(R^{\mathrm{T}}\right)^{-1} b$$

System 21.15 is equivalent to the original system $Ax = b$.

$$\overline{A}\overline{x} = \overline{b}, \quad \text{where } \overline{A} = \left(R^{-1}\right)^{\mathrm{T}} AR^{-1}, \ \overline{x} = Rx, \ \overline{b} = \left(R^{-1}\right)^{\mathrm{T}} b \tag{21.15}$$

The fact that $\overline{A}$ is symmetric is left to the exercises. Note that $R^{-1}x = 0$ only if $x = 0$. Thus, $\overline{A}$ is positive definite because

$$x^{\mathrm{T}} \left(R^{-1}\right)^{\mathrm{T}} AR^{-1}x = \left(R^{-1}x\right)^{\mathrm{T}} A \left(R^{-1}x\right) > 0, \quad x \neq 0.$$

Use the CG method to solve $\overline{A}\overline{x} = \overline{b}$. After obtaining $\overline{x}$, find x by solving $Rx = \overline{x}$.

The *incomplete Cholesky decomposition* is frequently used for preconditioning the CG method. As stated in the introduction, iterative methods are applied primarily to large, sparse systems. However, the Cholesky factor R used in system 21.15 is usually less sparse than M. Figure 21.6(a) shows the distribution of nonzero entries in a 685×685 positive definite sparse matrix, POWERMAT.mat, used in a power network problem, and Figure 21.6(b) shows the distribution in its Cholesky factor. Note the significant loss of zeros in the Cholesky factor.

The incomplete Cholesky decomposition is a modification of the original Cholesky algorithm. If an element a_{ij} off the diagonal of A is zero, the corresponding element r_{ij} is set to zero. The factor returned, R, has the same distribution of nonzeros as A above the diagonal. Form $M = R^{\mathrm{T}}R$ from a modified Cholesky factor of A with the hope that the condition number of $M^{-1}A$ is considerably smaller than that of A. The function icholesky in the software distribution implements it with the calling sequence R = icholesky(A). Implementation involves replacing the body of the inner for loop with

```
if A(i,j) == 0
   R(i,j) = 0;
else
    R(i,j) = (A(i,j) - sum(R(1:i-1,i).*R(1:i-1,j)))/R(i,i);
end
```

We can directly apply CG to system 21.15 by implementing the following statements.
Initialize

$$\overline{r_0} = \overline{b} - \overline{A}\overline{x_1}$$

$$\overline{p_0} = \overline{r_0}$$

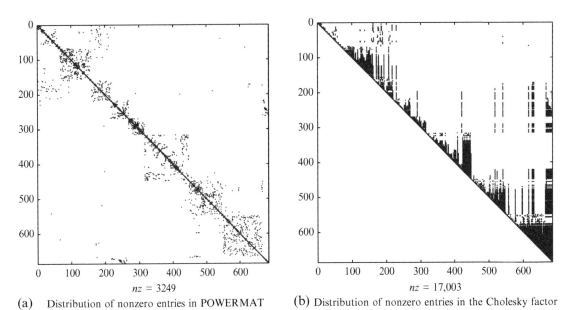

(a)   Distribution of nonzero entries in POWERMAT   (b) Distribution of nonzero entries in the Cholesky factor

FIGURE 21.6 Cholesky decomposition of a sparse symmetric positive definite matrix.

Loop

$$
\begin{aligned}
\overline{\alpha_{i-1}} &= (\overline{r_{i-1}}^{\mathrm{T}}\overline{r_{i-1}})/(\overline{p_{i-1}}^{\mathrm{T}}\overline{A}\overline{p_{i-1}}) \\
\overline{x_i} &= \overline{x_{i-1}} + \overline{\alpha_{i-1}}\overline{p_{i-1}} \\
\overline{r_i} &= \overline{r_{i-1}} - \overline{\alpha_{i-1}}A\overline{p_{i-1}} \\
\overline{\beta_i} &= (\overline{r_i}^{\mathrm{T}}\overline{r_i})/(\overline{r_{i-1}}^{\mathrm{T}}\overline{r_{i-1}}) \\
\overline{p_i} &= \overline{r_i} + \overline{\beta_i}\overline{p_{i-1}}
\end{aligned}
$$

Upon completion, solve the upper triangular system $Rx = \overline{x}$. However, this is not an efficient implementation. Improvement can be made by determining relationships between the transformed variables and the original ones.

We can express the residual, $\overline{r_i}$, in terms of the original residual, r_i, by

$$
\overline{r_i} = \left(R^{-1}\right)^{\mathrm{T}} r_i \tag{21.16}
$$

using the steps

$$
\begin{aligned}
\overline{r_i} &= \overline{b} - \overline{A}\overline{x_i} \\
&= \left(R^{-1}\right)^{\mathrm{T}} b - \left[\left(R^{-1}\right)^{\mathrm{T}} AR^{-1}\right] Rx_i \\
&= \left(R^{-1}\right)^{\mathrm{T}} b - \left(R^{-1}\right)^{\mathrm{T}} Ax_i \\
&= \left(R^{-1}\right)^{\mathrm{T}} (b - Ax_i) \\
&= \left(R^{-1}\right)^{\mathrm{T}} r_i.
\end{aligned}
$$

Make a change of variable by letting $\overline{p_i} = Rp_i$. This leads to

$$
\langle \overline{p_i}, \overline{p_j} \rangle_A = \langle p_i, p_j \rangle_A \tag{21.17}
$$

as follows:

$$
\begin{aligned}
\langle \overline{p_i}, \overline{p_j} \rangle_A &= \langle \overline{A}\overline{p_i}, \overline{p_j} \rangle \\
&= \left\langle \left(R^{-1}\right)^{\mathrm{T}} AR^{-1}Rp_i, Rp_j \right\rangle
\end{aligned}
$$

$$= \left\langle \left(R^{-1}\right)^{\mathrm{T}} Ap_i, Rp_j \right\rangle$$
$$= p_i^{\mathrm{T}} AR^{-1} Rp_j$$
$$= p_i^{\mathrm{T}} Ap_j$$
$$= \langle p_i, p_j \rangle_A .$$

By applying Equations 21.16 and 21.17, it follows that (Problem 21.11):

$$\overline{\alpha_{i-1}} = \frac{r_{i-1}^{\mathrm{T}} M^{-1} r_{i-1}}{p_{i-1}^{\mathrm{T}} Ap_{i-1}} \tag{21.18}$$

$$\overline{\beta_i} = \frac{r_i^{\mathrm{T}} M^{-1} r_i}{r_{i-1}^{\mathrm{T}} M^{-1} r_{i-1}} \tag{21.19}$$

Applying the identities to the CG algorithm for the computation of $\overline{x}$, gives the preconditioned CG algorithm 21.3. Note that the algorithm requires the computation of the incomplete Cholesky factor R and then the solution of systems $Mz_i = r_i$, which can be done using the incomplete Cholesky factor (Section 13.3.3). It is never necessary to compute M or M^{-1}.

Algorithm 21.3 Preconditioned Conjugate Gradient

```
function PRECG(A,b,x1,tol,numiter)

  R = icholesky(A)
  r₁ = b − Ax₁
  z₁ = cholsolve(R, r₁)
  p₁ = z₁
  for i = 1:numiter do
    αᵢ = zᵢᵀrᵢ / pᵢᵀApᵢ
    xᵢ₊₁ = xᵢ + αᵢpᵢ
    rᵢ₊₁ = rᵢ − αᵢApᵢ
    if ‖rᵢ₊₁‖₂ < tol then
      iter = i
      return[xᵢ₊₁, iter]
    end if
    zᵢ₊₁ = cholsolve(R, rᵢ₊₁)
    βᵢ = rᵢ₊₁ᵀzᵢ₊₁ / rᵢᵀzᵢ
    pᵢ₊₁ = zᵢ₊₁ + βᵢpᵢ
  end for
  iter = -1 return[xₙᵤₘᵢₜₑᵣ₊₁, iter]
end function
```

For a particularly difficult matrix, it is possible that the incomplete Cholesky decomposition will break down due to cancelation error. One of the more sophisticated techniques used to improve the algorithm is the *drop tolerance*-based incomplete Cholesky decomposition. This method keeps the off-diagonal element r_{ij} computed by the Cholesky algorithm if a condition applies and retains the original value a_{ij} otherwise. For instance, in Ref. [67], the following criterion is suggested:

$$r_{ij} = \begin{cases} \frac{a_{ij} - \sum_{k=1}^{i-1} r_{ki} r_{kj}}{r_{ii}} & a_{ij}^2 > \mathrm{tol}^2 a_{ii}\, b_{jj} \\ a_{ij} & \text{otherwise} \end{cases}$$

As the drop tolerance decreases, the incomplete Cholesky factor becomes more dense (Problem 21.25). Even with more advanced techniques, it still can be difficult to find an incomplete Cholesky preconditioner that works. The sophisticated MATLAB function `ichol` computes the incomplete Cholesky decomposition (see the documentation for `ichol`). The function `precg` in the software distribution uses `ichol` with selective drop tolerances.

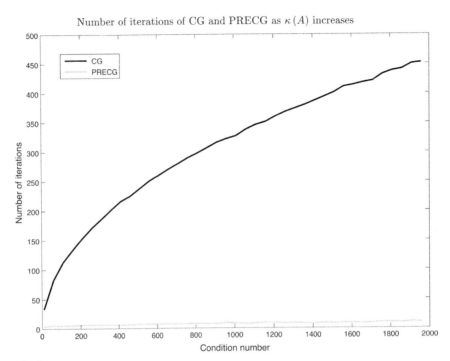

FIGURE 21.7 CG vs. PRECG.

Example 21.6.   The MATLAB function call

`R = sprandsym(n,n,density,rc,1)`

generates a sparse random symmetric positive definite matrix of size $n \times n$, where `density` is the fraction of nonzeros, and the reciprocal of `rc` is the exact condition number. The number 1 as the last argument instructs the function to make the matrix positive definite. In the software distribution demos directory is a program `cg_vs_precg` that uses `sprandsym` to generate symmetric positive definite matrix of size 800×800 with density 0.05 and condition numbers varying from 10 to 200. The program applies CG and preconditioned CG to each problem and graphs (Figure 21.7) the number of iterations required to attain a minimum error tolerance of 1.0×10^{-6}. ∎

21.4.2 SSOR Preconditioner

The SOR method can be used as a basis for building a CG preconditioner that normally one uses if incomplete Cholesky preconditioning is not effective or fails. Assume the diagonal elements of A are nonzero. The formula

$$(D + \omega L)\, x^{(k+1)} = ((1 - \omega)\, D - \omega U)\, x^{(k)} + \omega b$$

specifies the SOR iteration in matrix form (see Equation 20.15). Using the equation $U = A - L - D$ along with some matrix algebra, it follows that (Problem 21.12):

$$
\begin{aligned}
x^{(k+1)} &= x^{(k)} + \left(\frac{D}{\omega} + L\right)^{-1} \left(b - Ax^{(k)}\right) \\
&= x^{(k)} + \left(\frac{D}{\omega} + L\right)^{-1} r^{(k)}
\end{aligned}
\tag{21.20}
$$

The matrix $M = \left(\frac{D}{\omega} + L\right)$ can serve as a preconditioner. Assume that $M \approx A$. Then,

$$x^{(k+1)} \approx x^{(k)} + M^{-1}b - x^{(k)} = M^{-1}b.$$

Unfortunately, an SOR preconditioner is not symmetric and cannot be used for a symmetric positive definite matrix. As a result, we will use Equation 21.20 to construct a symmetric positive definite matrix termed the *SSOR preconditioning matrix.* Starting with the approximation $x^{(k)}$, designate $x^{\left(k+\frac{1}{2}\right)}$ as the result of a forward SOR sweep. Now perform a backward sweep by reversing the iteration direction to obtain $x^{(k+1)}$. The reader can verify that in the backward sweep, the roles of U and L are swapped, so we have the two equations

$$x^{\left(k+\frac{1}{2}\right)} = x^{(k)} + \left(\frac{D}{\omega} + L\right)^{-1} r^{(k)}$$

$$x^{(k+1)} = x^{\left(k+\frac{1}{2}\right)} + \left(\frac{D}{\omega} + U\right)^{-1} r^{\left(k+\frac{1}{2}\right)}, \tag{21.21}$$

where $r^{(k)}$, $r^{(k+1)}$ are residuals. Eliminating $x^{\left(k+\frac{1}{2}\right)}$ from Equation 21.21 gives the SSOR preconditioning matrix. The somewhat involved computations follow:

$$x^{\left(k+\frac{1}{2}\right)} - x^{(k)} = \left(\frac{D}{\omega} + L\right)^{-1} r^{(k)} = \left(\frac{D}{\omega} + L\right)^{-1} \left(b - Ax^{(k)}\right)$$

$$r^{\left(k+\frac{1}{2}\right)} = b - Ax^{\left(k+\frac{1}{2}\right)} = b - Ax^{(k)} + Ax^{(k)} - Ax^{\left(k+\frac{1}{2}\right)} = r^{(k)} - A\left(x^{\left(k+\frac{1}{2}\right)} - x^{(k)}\right) =$$

$$\left(L + \frac{1}{\omega}D - A\right)\left(x^{\left(k+\frac{1}{2}\right)} - x^{(k)}\right) = \left(\left(\frac{1}{\omega} - 1\right)D - U\right)\left(\frac{D}{\omega} + L\right)^{-1} r^{(k)}$$

$$x^{(k+1)} - x^{\left(k+\frac{1}{2}\right)} = \left(\frac{D}{\omega} + U\right)^{-1} r^{\left(k+\frac{1}{2}\right)} = \left(\frac{D}{\omega} + U\right)^{-1}\left(\left(\frac{1}{\omega} - 1\right)D - U\right)\left(\frac{D}{\omega} + L\right)^{-1} r^{(k)}$$

$$x^{(k+1)} - x^{(k)} = \left(\frac{D}{\omega} + L\right)^{-1} r^{(k)} + \left(\frac{D}{\omega} + U\right)^{-1}\left(\left(\frac{1}{\omega} - 1\right)D - U\right)\left(\frac{D}{\omega} + L\right)^{-1} r^{(k)} =$$

$$\left(I + \left(\frac{D}{\omega} + U\right)^{-1}\left(\left(\frac{1}{\omega} - 1\right)D - U\right)\right)\left(\frac{D}{\omega} + L\right)^{-1} r^{(k)} =$$

$$\left(\left(\frac{D}{\omega} + U\right)^{-1}\left(\frac{D}{\omega} + U\right) + \left(\frac{D}{\omega} + U\right)^{-1}\left(\left(\frac{1}{\omega} - 1\right)D - U\right)\right)\left(\frac{D}{\omega} + L\right)^{-1} r^{(k)} =$$

$$\left(\frac{D}{\omega} + U\right)^{-1}\left(\left(\frac{D}{\omega} + U\right) + \left(\left(\frac{1}{\omega} - 1\right)D - U\right)\right)\left(\frac{D}{\omega} + L\right)^{-1} r^{(k)} = \left(\frac{D}{\omega} + U\right)^{-1}\left(\frac{2}{\omega} - 1\right)D\left(\frac{D}{\omega} + L\right)^{-1} r^{(k)}$$

Thus,

$$M_{\text{SSOR}}^{-1} = \left(U + \frac{1}{\omega}D\right)^{-1}\left(\frac{2}{\omega} - 1\right)D\left(L + \frac{1}{\omega}D\right)^{-1}$$

and

$$M_{\text{SSOR}} = \frac{\omega}{2 - \omega}\left(L + \frac{1}{\omega}D\right)D^{-1}\left(U + \frac{1}{\omega}D\right). \tag{21.22}$$

The reader should verify that M_{SSOR} is symmetric positive definite (Problem 21.9). Using $\omega = 1$ in Equation 21.22 gives

$$M_{\text{PGS}} = (D + L)D^{-1}(D + U).$$

The matrix M_{PGS} corresponds to the Gauss-Seidel method. The value of ω is not as critical as the choice of ω for the SOR iteration, and $\omega = 1$ can be quite effective in many cases. Solve $M_{\text{PGS}}\, z = r$, or $(D + L)D^{-1}(D + U)\, z = r$ in two stages.

a. Solve $(D + L)\, y_1 = r$ for y_1.
b. Solve $(D + U)\, z = Dy_1$ for z.

Remark 21.4.   In MATLAB, compute z as follows:

```
z = (U+D)\(D*((L+D)\r));
```

The function `precg` in the book software distribution implements Algorithm 21.3 and adds the option of using the SSOR preconditioner M_{PGS}. Its calling format is

```
function [x residual iter] =
precg(A,b,x0,tol,maxiter,method,droptol)
```

where method = 'incomplete Cholesky' or 'SSOR' and droptol is the drop tolerance. The drop tolerance is only applicable when method is 'incomplete Cholesky'. If method is omitted, incomplete Cholesky is assumed with zero-fill. If method is specified and droptol is not, droptol defaults to 1.0e−4.

There is an extensive literature concerning preconditioning that includes the CG method as well as other iterations. We will present a preconditioner for the GMRES method in Section 21.7.1. For a general discussion of preconditioning, see Refs. [2, pp. 650-669], [64, pp. 283-351], and the book by Ref. [68].

Example 21.7. The matrix PRECGTEST in the software distribution is a 10000×10000 block pentadiagonal matrix having an approximate condition number of 8.4311×10^6. The following table shows four attempts to solve the system with $b = \begin{bmatrix} 1 & 1 & \dots & 1 & 1 \end{bmatrix}^{\mathrm{T}}$, $x_0 = 0$, tol $= 1.0 \times 10^{-6}$, and maxiter $= 2000$. The incomplete Cholesky decomposition failed with zero-fill and also with a drop tolerance of 1.0×10^{-2} and 1.0×10^{-3}. Using a drop tolerance of 1.0×10^{-4} it succeeded very quickly. The SSOR preconditioned system required 1159 iterations and gave a slightly better result.

Method	droptol	Iterations	Residual	Time
Preconditioned Cholesky	Nofill	–	Fail	–
Preconditioned Cholesky	1.0×10^{-2}	–	Fail	–
Preconditioned Cholesky	1.0×10^{-3}	–	Fail	–
Preconditioned Cholesky	1.0×10^{-4}	59	9.7353×10^{-7}	0.324 s
SSOR	–	1159	9.4987×10^{-7}	2.594 s

■

21.5 KRYLOV SUBSPACES

If the coefficient matrix is large, sparse, and positive definite, the method of choice is normally CG. However, in applications it is often the case that the coefficient matrix is symmetric but not positive definite or may not even be symmetric. There are sophisticated iterations for these problems, most of which are based on the concept of a Krylov subspace. A Krylov subspace-based method does not access the elements of the matrix directly, but rather performs matrix-vector multiplication to obtain vectors that are projections into a lower-dimensional Krylov subspace, where a corresponding problem is solved. The solution is then converted into a solution of the original problem. These methods can give a good result after a relatively small number of iterations.

Definition 21.2. Assume $A \in \mathbb{R}^{n \times n}$, $u \in \mathbb{R}^n$, and k is an integer, the *Krylov sequence* is the set of vectors

$$u, Au, A^2 u, \dots, A^{k-1} u$$

The *Krylov subspace* $\mathcal{K}_n(A, u)$ generated by A and u is

$$\text{span} \left\{ u, Au, A^2 u, \dots, A^{k-1} u \right\}.$$

It is of dimension k if the vectors are linearly independent.

Although we approached the CG method using an optimization approach, CG is also a Krylov subspace method. Theorem 21.3 shows the connection.

Theorem 21.3. *Assume A is nonsingular. With an initial guess $x_0 = 0$, after i iterations of the CG method,*

$$\begin{aligned} \text{span} \{x_1, x_2, \dots, x_i\} &= \text{span} \{p_0, p_1, \dots, p_i\} = \text{span} \{r_0, r_2, \dots, r_i\} \\ &= \mathcal{K}_i(A, p_0) = \text{span} \left\{p_0, Ap_0, \dots, A^{i-1} p_0\right\} = \text{span} \left\{r_0, Ar_0, \dots, A^{i-1} r_0\right\} \end{aligned}$$

Proof. Let

$$S_i = \text{span} \{p_0, p_1, \dots, p_{i-1}\}.$$

Now, $x_i = x_{i-1} + \alpha_{i-1} p_{i-1}$, and so $x_1 = 0 + \alpha_0 p_0$, $x_2 = x_1 + \alpha_1 p_1 = \alpha_0 p_0 + \alpha_1 p_1$. In general, $x_i = \sum_{k=0}^{i-1} \alpha_k p_k$, and

$$\text{span} \{x_1, x_2, \dots, x_i\} = S_i.$$

We have $r_0 = p_0$ and from Equation 21.13 $r_i = p_i - \beta_i p_{i-1}$, so

$$\text{span} \{r_0, r_1, \dots, r_{i-1}\} = S_i.$$

From property 1 of Lemma 21.1

$$r_i = r_{i-1} - \alpha_{i-1} A p_{i-1}.$$

Since r_i and r_{i-1} are in S_i,

$$A p_{i-1} \in S_i.$$

As a result,

$$S_i = \mathcal{K}_i(A, p_0) = \mathrm{span}\left\{p_0, A p_0, \ldots, A^{i-1} p_0\right\} = \mathcal{K}_i(A, r_0) = \mathrm{span}\left\{r_0, A r_0, \ldots, A^{i-1} r_0\right\}.$$

The p_i are A-orthogonal, the r_i are orthogonal, and so both sequences are linearly independent. Since A is nonsingular, the vectors in $\mathcal{K}_i(A, p_0)$ and $\mathcal{K}_i(A, \dot{r}_0)$ are linearly independent. $\quad\square$

There are a number of highly successful iterative methods based on Krylov subspaces that work with the full range of matrix types. We will discuss the GMRES algorithm that applies to general matrices. We will also develop the MINRES method for symmetric nonpositive definite problems, since it can be effective and its derivation is very similar to the GMRES method. Other methods include the biconjugate gradient (Bi-CG) method, and the quasi-minimal residual (QMR) method for nonsymmetric matrices. The reader will find a presentation of these Krylov subspace methods in Refs. [2, pp. 639-647], [26, pp. 303-312], and [69, Chapters 7 and 9].

21.6   THE ARNOLDI METHOD

Our aim is to develop a method of solving a large, sparse system $Ax = b$, where $A^{n \times n}$ is a general nonsymmetric matrix. We approximate the solution by projecting the problem into a Krylov subspace $\mathcal{K}_m(A, r_0) = \left\{r_0, A r_0, \ldots, A^{m-1} r_0\right\}$, where m is much smaller than n ($m \ll n$). Obtaining the solution to a related problem will be practical in this much smaller subspace $\mathbb{R}^m \subset \mathbb{R}^n$. We then convert the solution to the problem in $\mathbb{R}^m$ to the solution we want in $\mathbb{R}^n$ (Figure 21.8).

Our approach is to develop the *Arnoldi decomposition* of A and use it to solve a problem in $\mathbb{R}^m$ that, in turn, will be used to approximate the solution x to $Ax = b$ in $\mathbb{R}^n$. The decomposition creates an $n \times (m+1)$ matrix Q_{m+1} with orthonormal columns and an $(m+1) \times m$ upper Hessenberg matrix $\overline{H}_m$ such that

$$A Q_m = Q_{m+1} \overline{H}_m,$$

where $Q_m = Q_{m+1}(:, 1:m)$. We build the decomposition by using a Krylov subspace

$$K_{m+1}(A, x_1) = \left\{x_1, A x_1, \ldots, A^m x_1\right\},$$

where x_1 is an initial vector. Generally, the set of vectors $\left\{x_1, A x_1, \ldots, A^{i-1} x_1\right\}$ is not a well-conditioned basis for $\mathcal{K}_i(A, x_1)$ since, as we showed in Chapter 18 when presenting the power method, the sequence approaches the dominant eigenvector of A. Thus the last few vectors in the sequence may be very close to pointing in the same direction. In order to fix this problem, we compute an orthonormal basis for $\mathcal{K}_i(A, x_1)$ using the modified Gram-Schmidt process. After i steps of the Gram-Schmidt process we have vectors $q_1, q_2, \ldots, q_i$ and do not retain the original set $\left\{x_1, A x_1, \ldots, A^{i-1} x_1\right\}$. If we had $A^i x_1$, we could apply Gram-Schmidt to extend the orthonormal sequence to $q_1, q_2, \ldots, q_i, q_{i+1}$. We must find an alternative for the determination of q_{i+1}. We do not have A^i, but we do have A, so we compute $A q_i$ and apply Gram-Schmidt to extend the orthonormal sequence by 1. It can be shown [23, pp. 439-441] that even though we used $A q_i$ as the next vector rather

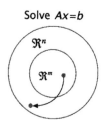

Solve $Ax=b$

Solve for $y \in \mathfrak{R}^m$
Transform y to a solution $x \in \mathfrak{R}^n$

FIGURE 21.8 Arnoldi projection from $\mathbb{R}^n$ into $\mathbb{R}^m$, $m \ll n$.

than $A^i x_1$,

$$\text{span}\left\{x_1, Ax_1, \ldots, A^i x_1\right\} = \text{span}\left\{q_1, q_2, \ldots, q_{i+1}\right\}.$$

The first step is to normalize the initial vector x_1 by computing

$$q_1 = \frac{x_1}{\|x_1\|_2}.$$

For $1 \leq i \leq m$,

$$\overline{q_{i+1}} = Aq_i - \sum_{j=1}^{i} q_j h_{ji},$$

where h_{ji} is the Gram-Schmidt coefficient

$$h_{ji} = \langle q_j, Aq_i \rangle = q_j^T Aq_i, \, j \leq i$$

Assign $h_{i+1,i} = \|\overline{q_{i+1}}\|_2$, and normalize $\overline{q_{i+1}}$ to obtain

$$q_{i+1} = \frac{\overline{q_{i+1}}}{h_{i+1,i}}.$$

The Gram-Schmidt coefficients are computed as follows:

$$h_{11}, h_{21}$$
$$h_{12}, h_{22}, h_{32}$$
$$h_{13}, h_{23}, h_{33}, h_{43}$$
$$\vdots$$
$$h_{1m}, h_{2m}, \ldots h_{m+1,m}$$

and form the $(m+1) \times m$ upper Hessenberg matrix

$$\overline{H_m} = \begin{bmatrix} h_{11} & h_{12} & \cdots & & \cdots & h_{1,m} \\ h_{21} & h_{22} & \cdots & & \cdots & h_{2,m} \\ 0 & h_{32} & \ddots & & \cdots & \vdots \\ \vdots & \vdots & \ddots & h_{m-1,m-1} & & \vdots \\ \vdots & \vdots & \ddots & h_{m,m-1} & & h_{m,m} \\ 0 & 0 & \cdots & & 0 & h_{m+1,m} \end{bmatrix}.$$

From the two equations

$$\overline{q_{i+1}} = Aq_i - \sum_{j=1}^{i} q_j h_{ji}$$

$$q_{i+1} = \frac{\overline{q_{i+1}}}{h_{i+1,i}},$$

we have

$$Aq_i = \sum_{j=1}^{i} q_j h_{ji} + \overline{q_{i+1}} = \sum_{j=1}^{i} q_j h_{ji} + h_{i+1,i}q_{i+1} = \sum_{j=1}^{i+1} q_j h_{ji}. \tag{21.23}$$

Form the $n \times m$ matrix $Q_m = \begin{bmatrix} q_1 & q_2 & \cdots & q_m \end{bmatrix}$ and the $n \times (m+1)$ matrix $Q_{m+1} = \begin{bmatrix} q_1 & q_2 & \cdots & q_m & q_{m+1} \end{bmatrix}$ from the column vectors $\{q_i\}$. Matrix equation 21.24 follows directly from Equation 21.23:

$$AQ_m = Q_{m+1}\overline{H_m} \tag{21.24}$$

Figure 21.9 depicts the decomposition. Algorithm 21.4 specifies the Arnoldi process that generates an orthonormal basis $\{q_1, q_2, \ldots, q_{m+1}\}$ for the subspace spanned by $K_{m+1}(A, x_1)$ and builds the matrix $\overline{H_m}$.

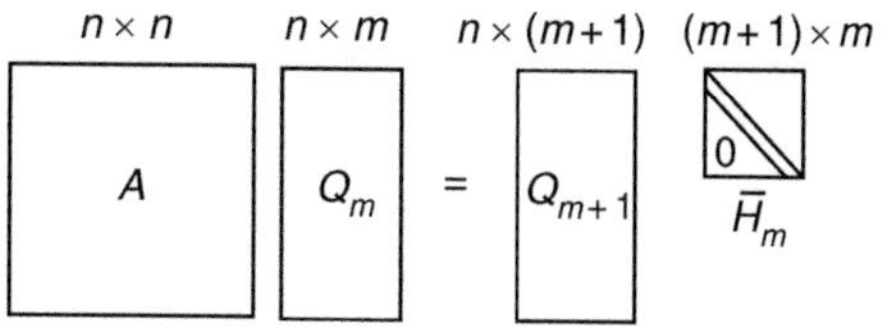

FIGURE 21.9 Arnoldi decomposition form 1.

Algorithm 21.4 Arnoldi Process

```
function ARNOLDI(A, x₁, m)
    % Input: n × n matrix A, n × 1 column vector x₁, and integer m
    % Output: m+1 orthogonal vectors q₁, q₂, ... , qₘ₊₁
    % and an (m + 1) × m matrix H̄
    q₁ = x₁/‖x₁‖₂
    for k = 1:m do
        w = Aqₖ
        for j = 1:k do
            hⱼₖ = qⱼᵀw
            w = w − hⱼₖqⱼ
        end for
        hₖ₊₁,ₖ = ‖w‖₂
        if hₖ₊₁,ₖ = 0 then
            return [Q = [qᵢ], H̄ = [hᵢⱼ]]
        end if
        qₖ₊₁ = w/hₖ₊₁,ₖ
    end for
    return [Q = [qᵢ], H̄ = [hᵢⱼ]]
end function
```

NLALIB: The function `arnoldi` implements Algorithm 21.4.

Efficiency

Since the Arnoldi method uses Gram-Schmidt, the approximate number of flops for the algorithm is $2m^2n$.

Notice that when $h_{k+1,k} = 0$, the Arnoldi algorithm stops. Clearly we cannot compute q_{k+1} due to division by zero, but there is a more to it than that, as indicated in Definition 21.3.

Definition 21.3. Let T be a linear transformation mapping a vector space V to V. A subspace $W \subseteq V$ is said to be an invariant subspace of T if for every $x \in W$, $Tx \in W$.

If $h_{i+1,i} = \|\overline{q_{i+1}}\|_2 = 0$, then the subspace span $\{q_1, q_2, \ldots, q_i\}$ is an invariant subspace of A, and further iterations produce nothing, and so the algorithm terminates. This means that a method for solving $Ax = b$ based on projecting onto the subspace $\mathcal{K}(A, j)$ will be exact. This is what is termed a "lucky breakdown" [64, pp. 154-156].

21.6.1 An Alternative Formulation of the Arnoldi Decomposition

Equation 21.24 can be written as (Problem 21.6)

$$AQ_m - Q_mH_m = h_{m+1,m} \begin{bmatrix} 0 & 0 & \ldots & 0 & q_{1m} \\ 0 & 0 & \ldots & 0 & q_{2m} \\ \vdots & \vdots & \vdots & \vdots & \vdots \\ 0 & 0 & \ldots & 0 & q_{nm} \end{bmatrix},$$

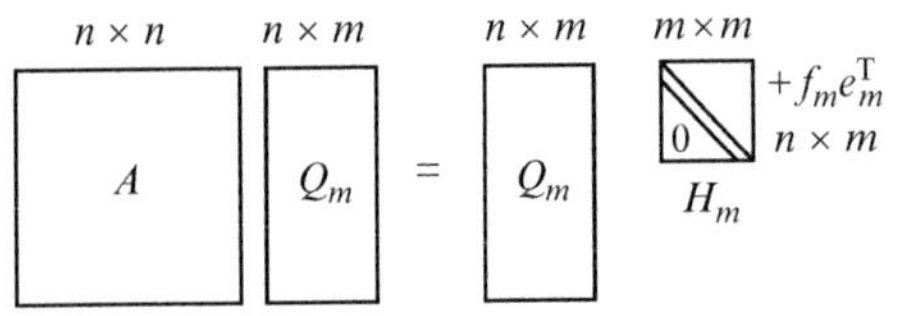

FIGURE 21.10 Arnoldi decomposition form 2.

or

$$AQ_m = Q_m H_m + h_{m+1,m} q_{m+1} e_m^{\mathrm{T}},\tag{21.25}$$

where H_m is $\overline{H}_m$ with the last row removed, $f_m = h_{m+1,m} q_{m+1}$, and $e_m^{\mathrm{T}} = \begin{bmatrix} 0 & 0 & \dots & 0 & 1 \end{bmatrix}$. Figure 21.10 depicts the decomposition. This form of the Arnoldi decomposition will be very useful in Chapter 22 when we discuss the computation of eigenvalues of sparse matrices.

21.7 GMRES

Assume A is a real $n \times n$ matrix, b is an $n \times 1$ vector, and we want to solve the system $Ax = b$. Assume that x_0 is an initial guess for the solution, and $r_0 = b - Ax_0$ is the corresponding residual. The *GMRES method* looks for a solution of the form $x_m = x_0 + Q_m y_m$, $y_m \in \mathbb{R}^m$ where the columns of Q_m are an n-dimensional orthogonal basis for the Krylov subspace $\mathcal{K}_m(A, r_0) = \{r_0, Ar_0, \dots, A^{m-1} r_0\}$. The vector y_m is chosen so the residual

$$\|r_m\|_2 = \|b - A(x_0 + Q_m y_m)\|_2 = \|r_0 - AQ_m y_m\|_2$$

has minimal norm over $\mathcal{K}_m(A, r_0)$. This is a least-squares problem. We must find a vector y_m that minimizes the residual specified in Equation 21.26,

$$\|r_m\|_2 = \|r_0 - AQ_m y_m\|_2 .\tag{21.26}$$

Let $\beta = \|r_0\|_2$. The first Arnoldi vector is $q_1 = \frac{r_0}{\|r_0\|}$, so $r_0 = \beta q_1$. By using Equations 21.24 and 21.26

$$\|r_m\|_2 = \left\|\beta q_1 - Q_{m+1} \overline{H}_m y_m\right\|_2 .\tag{21.27}$$

The Arnoldi vector q_1 is the first column of Q_{m+1}, so $q_1 = Q_{m+1} \begin{bmatrix} 1 & 0 & \dots & 0 \end{bmatrix}^{\mathrm{T}} = Q_{m+1} e_1$, and from Equation 21.27

$$\|r_m\|_2 = \left\|Q_{m+1} \left(\beta e_1 - \overline{H}_m y_m\right)\right\|_2 .$$

Since the columns of Q_{m+1} are orthonormal, we must minimize

$$\left\|\beta e_1 - \overline{H}_m y_m\right\|_2 .$$

This means that after solving the $(m + 1) \times m$ least-squares problem

$$\overline{H}_m y_m = \beta e_1,\tag{21.28}$$

the approximate solution to $Ax = b$ is

$$x_m = x_0 + Q_m y_m.$$

Use the QR decomposition approach to solving overdetermined least-squares problems (Section 16.2.2). Since $\overline{H}_m$ in Equation 21.28 is an upper Hessenberg matrix, the system can be solved in $O\left(m^2\right)$ flops. In the practical implementation of GMRES, one estimate x_m is often not sufficient to obtain the error tolerance desired. Use x_m as an improved initial vector and repeat the process until satisfying the error tolerance. Algorithm 21.5 presents the GMRES method.

Algorithm 21.5 GMRES

```
function GMRESB(A,b,x₀,m,tol,maxiter)
  % Solve Ax = b using the GMRES method
  % Input: n × n matrix A, n × 1 vector b,
  % initial approximation x₀, integer m < n,
  % error tolerance tol, and the maximum number of iterations, maxiter.
  % Output: Approximate solution xₘ, associated residual r,
  % and iter, the number of iterations required.
  % iter = −1 if the tolerance was not satisfied.

  iter = 1
  while iter ≤ maxiter do
      r = b − Ax₀
      [ Q_{m+1}  H̄ₘ ] = arnoldi(A, r, m)
      β = ‖r‖₂
      Solve the (m + 1) × m least-squares problem H̄ₘyₘ = βe₁
      using Givens rotations that take advantage of the upper
      Hessenberg structure of H̄ₘ
      xₘ = x₀ + Qₘyₘ
      r = ‖b − Axₘ‖₂
      if r < tol then
          return [xₘ,r,iter]
      end if
      x₀ = xₘ
      iter = iter + 1
  end while
  iter = −1
  return [xₘ, r, iter]
end function
```

NLALIB: The function gmresb implements Algorithm 21.5. The function hesslqsolve in the software distribution implements the *QR* decomposition of an upper Hessenberg matrix. The implementation simply calls givenshess to perform the *QR* decomposition rather than qr. The name contains the trailing "*b*" because MATLAB supplies the function gmres that implements the GMRES method.

The choice of m is experimental. Try a small value and see if the desired residual norm can be attained. If not, increase m until obtaining convergence or finding that GMRES simply does not work. Of course, as you increase m, memory and computational effort increase. It may happen that the problem is not tractable using GMRES. In that case, there are other methods such as Bi-CG and QMR that may work [64, pp. 217-244].

Example 21.8. A *Toeplitz matrix* is a matrix in which each diagonal from left to right is constant. For instance,

$$\begin{bmatrix} 1 & 2 & 8 & -1 \\ 3 & 1 & 2 & 8 \\ 8 & 3 & 1 & 2 \\ 7 & 8 & 3 & 1 \end{bmatrix}$$

is a Toeplitz matrix. The following MATLAB statements create a 1000×1000 pentadiagonal sparse Toeplitz matrix with a small condition number. As you can see, gmresb works quite well with $m = 50$, tol $= 1.0 \times 10^{-14}$ and a maximum of 25 iterations. The method actually required 11 iterations and obtained a residual of 9.6893×10^{-15}.

```
>> P = gallery('toeppen',1000);
>> condest(P)

ans =
    23.0495

>> b = rand(1000,1);
```

```
>> x0 = ones(1000,1);
>> [x r iter] = gmresb(P,b,x0,50,1.0e-14,25);
>> r

r =
  9.6893e-015

>> iter

iter =
    11
```

■

Convergence

If A is positive definite, then GMRES converges for any $m \geq 1$ [64, p. 205]. There is no other simple result for convergence. A theorem that specifies some conditions under which convergence will occur can be found in Ref. [64, p. 206].

21.7.1 Preconditioned GMRES

In the same way that we used incomplete Cholesky decomposition to precondition A when A is positive definite, we can use the *incomplete LU decomposition* to precondition a general matrix. Compute factors L and U so that if element $a_{ij} \neq 0$ then the element at index (i, j) of $A - LU$ is zero. To do this, compute the entries of L and U at location (i, j) only if $a_{ij} \neq 0$. It is hoped that if $M = LU$, then $M^{-1}A$ will have a smaller condition number than A. Algorithm 21.6 describes the incomplete LU decomposition. Rather than using vectorization, it is convenient for the algorithm to use a triply nested loop. For more details see Ref. [64, pp. 287-296].

Algorithm 21.6 Incomplete *LU* Decomposition

```
function ILUB(A)
  % Compute an incomplete LU decomposition
  % Input: n × n matrix A
  % Output: lower triangular matrix L and upper triangular matrix U.
  for i=2:n do
    for j=1:i-1 do
      if aij ≠ 0 then
        aij = aij/ajj
        for k = j+1:n do
          if aik ≠ 0 then
            aik = aik − aijajk
          end if
        end for
      end if
    end for
  end for
  U=upper triangular portion of A
  L=portion of A below the main diagonal

  for i=1:n do
    lii = 1
  end for
end function
```

NLALIB: The function ilub implements Algorithm 21.6.

Proceeding as we did with incomplete Cholesky, there results

$$(LU)^{-1} Ax = (LU)^{-1} b$$

$$\left(L^{-1}AU^{-1} \right) (Ux) = L^{-1}b$$

and

$$\overline{A}\overline{x} = \overline{b},$$

where

$$\overline{A} = L^{-1}AU^{-1}$$
$$\overline{b} = L^{-1}b \qquad\qquad (21.29)$$
$$\overline{x} = Ux$$

The function `pregmres` in the software distribution approximates the solution to $Ax = b$ using Equation 21.29.

Remark 21.5.   Algorithm 21.6 will fail if there is a zero on the diagonal of U. In this case, it is necessary to use Gaussian elimination with partial pivoting. We will not discuss this, but the interested reader will find a presentation in Ref. [64, pp. 287-320]. The software distribution contains a function `mpregmres` that computes the incomplete LU decomposition with partial pivoting by using the MATLAB function `ilu`. It returns a decomposition such that $P\overline{A} = LU$, so $\overline{A} = P^{T}LU$. It is recommended that, in practice, `mpregmres` be used rather than pregmres.

Example 21.9.   The 903×903 nonsymmetric matrix, `DK01R`, in Figure 21.11 was used to solve a computational fluid dynamics problem. `DK01R` was obtained from the University of Florida Sparse Matrix Collection. A right-hand side, `b_DK01R`, and an approximate solution, `x_DK01R`, were supplied with the matrix. The approximate condition number of the matrix is 2.78×10^8, so it is ill-conditioned. Using $x_0 = \begin{bmatrix} 0 & \dots & 0 \end{bmatrix}^{T}$, $m = 300$, and tol $= 1.0 \times 10^{-15}$, niter $= 20$, the solution was obtained using `gmresb` and `mpregmres`. Table 21.1 gives the results of comparing the solutions from `mpregmres` and `gmresb` to `x_DK01R`.

TABLE 21.1 Comparing gmresb and mpregmres

	iter	r	Time	$\|x\_DK01R - x\|_2$
Solution supplied	–	6.29×10^{-16}	–	–
gmresb	-1(failure)	5.39×10^{-10}	6.63	9.93×10^{-11}
mpregmres	1	1.04×10^{-15}	0.91	5.20×10^{-17}

In a second experiment, the function `gmresb` required 13.56 s and 41 iterations to attain a residual of 8.10×10^{-16}. Clearly, preconditioning GMRES is superior to normal GMRES for this problem.   ■

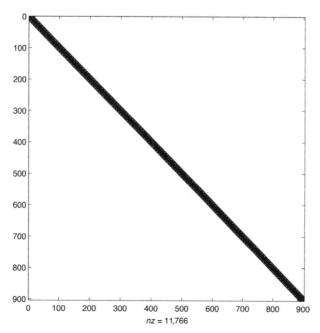

FIGURE 21.11 Large nonsymmetric matrix.

21.8 THE SYMMETRIC LANCZOS METHOD

The *Lanczos method* is the Arnoldi method applied to a symmetric matrix. For reasons that will become evident, replace the matrix name H by T. Like the Arnoldi decomposition, the Lanczos decomposition can be written in two ways

$$AT_m = Q_{m+1}\overline{T_m}$$
$$AQ_m = Q_mT_m + t_{m+1,m}q_{m+1}e_m^T$$

Figure 21.12 depicts the two formulations. Now, noting that the columns of Q_m are orthogonal to q_{m+1},

$$Q_m^TAQ_m = Q_m^TQ_mT_m + t_{m+1,m}\left(Q_m^Tq_{m+1}e_m^T\right) = T_m + 0 = T_m,$$

and

$$T_m^T = \left(Q_m^TAQ_m\right)^T = Q_m^TA^TQ_m = Q_m^TAQ_m = T_m.$$

Thus, T_m is symmetric, and a symmetric upper Hessenberg matrix must be tridiagonal, so T_m and $\overline{T_m}$ have the form

$$T_m = \begin{bmatrix} \alpha_1 & \beta_1 & & & \\ \beta_1 & \alpha_2 & \beta_2 & & \\ & \beta_2 & \ddots & \ddots & \\ & & \ddots & \alpha_{m-1} & \beta_{m-1} \\ & & & \beta_{m-1} & \alpha_m \end{bmatrix}, \quad \overline{T_m} = \begin{bmatrix} \alpha_1 & \beta_1 & & & \\ \beta_1 & \alpha_2 & \beta_2 & & \\ & \beta_2 & \ddots & \ddots & \\ & & \ddots & \alpha_{m-1} & \beta_{m-1} \\ & & & \beta_{m-1} & \alpha_m \\ 0 & 0 & \cdots & 0 & t_{m+1,} \end{bmatrix}$$

The interior loop of the Arnoldi iteration uses the array entries h_{jm}, $1 \le j \le m$. From the tridiagonal structure of T_m, the only entries in column k from rows 1 through k are β_{k-1} in row $k-1$ and α_k in row k. Trace the inner Arnoldi loop for $j = k-1$ and $j = k$. You will determine that the Arnoldi inner loop can be replaced by two statements:

$$\alpha_k = q_k^TAq_k$$
$$w = w - \beta_{k-1}q_{k-1} - \alpha_kq_k,$$

so the computation of $\overline{T_m}$ is much faster than for the Arnoldi process. Algorithm 21.7 defines the Lanczos process.

NLALIB: The function `lanczos` implements Algorithm 21.7.

21.8.1 Loss of Orthogonality with the Lanczos Process

Using the Lanczos process is not as simple as it may seem. Poor convergence can result from a bad choice of the starting vector, so a random vector is a good choice. Using exact arithmetic, the vectors $q_1, q_2, \ldots q_m, q_{m+1}$ are mutually

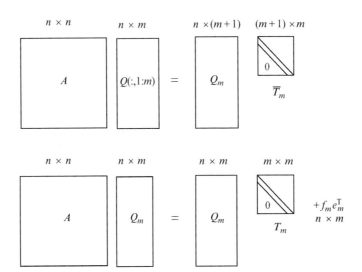

FIGURE 21.12 Lanczos decomposition.

orthogonal. When implemented in finite precision arithmetic, the Lanczos algorithm does not behave as expected. Roundoff error can cause lack or orthogonality among the Lanczos vectors with serious consequences. We will outline why this happens in Chapter 22 when we discuss computing a few eigenvalues of a large sparse symmetric matrix. One approach to fix this problem is to reorthogonalize as each new Lanczos vector is computed; in other words, we orthogonalize twice.

Algorithm 21.7 Lanczos Method

```
function LANCZOS(A, x₀, m)
    q₀ = 0
    β₀ = 0
    q₁ = x₀/‖x₀‖₂

    for k=1:m do
        w = Aqₖ
        αₖ = qₖᵀw
        w = w − βₖ₋₁qₖ₋₁ − αₖqₖ
        βₖ = ‖w‖₂
        qₖ₊₁ = w/βₖ
    end for
    T(1:m,1:m)=diag(β(1:m-1),-1)+diag(α)+diag(β(1:m-1),1);
    T(m+1,m)=β(m);
    return[ Q, T ]
end function
```

This is termed *full reorthogonalization* and is time-consuming but, as we will see in Example 21.10, can be of great benefit. Do the reorthogonalization by following

$$\alpha_k = q_k^T A q_k$$
$$w = w - \beta_{k-1}q_{k-1} - \alpha_k q_k$$

with

```
for i=1:k-1 do
    h = qᵢᵀw
    w = w − qᵢh
end for
```

In the software distribution, the function lanczos will perform full reorthogonalization by adding a fourth parameter of 1. If speed is critical, try the algorithm without full reorthogonalization and add it if the results are not satisfactory. It is possible to perform selective reorthogonalization. For a discussion of these issues, see Refs. [1, pp. 366-383] and [2, pp. 565-566].

Example 21.10. Ref. [70, pp. 498-499], a numerical example using a matrix introduced by Strakoŝ [71] demonstrates lack of orthogonality of the Lanczos vectors. Recall that the eigenvalues of a diagonal matrix are its diagonal elements. Define a diagonal matrix by the eigenvalues

$$\lambda_i = \lambda_1 + \left(\frac{i-1}{n-1}\right)(\lambda_n - \lambda_1)\rho^{n-i}, 1 \leq i \leq n.$$

The parameter ρ controls the distribution of the eigenvalues within the interval $[\lambda_1 \ \lambda_n]$. Create the matrix with $n = 30$, $\lambda_1 = 0.1$, $\lambda_n = 100$, and $\rho = 0.8$, that has well-separated large eigenvalues. Execute $m = n$ steps of the Lanczos process and compute $q_i^T q_j$, $1 \leq i, j \leq n$. Using exact arithmetic, the Lanczos vectors comprising Q should form an orthonormal set, so $\|q_i^T q_j\|_2 = 0$, $i \neq j$ and $\|q_i^T q_i\|_2 = 1$. Figure 21.13(a) is a plot of $\|q_i^T q_j\|_2$ over the grid $1 \leq i, j \leq n$ without reorthogonalization, and Figure 21.13(b) plots the same data with reorthogonalization. Note the large difference between the two plots. Clearly, Figure 21.3(a) shows that the Lanczos vectors lost orthogonality. As further evidence, if we name the variable Q in the first plot $Q1$ and name Q in the second plot $Q2$, we have

```
norm(Q1)=1.41843
norm(Q2)=1.00000
```

Problem 21.36 asks you to reproduce these results.   ∎

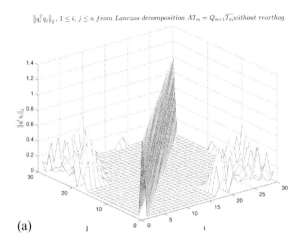

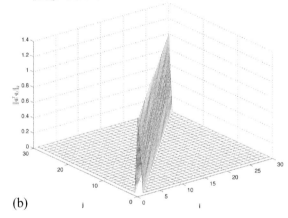

FIGURE 21.13 Lanczos process with and without reorthogonalization. (a) Lanczos without reorthogonalization and (b) Lanczos with reorthogonalization.

21.9   THE MINRES METHOD

If a symmetric matrix is indefinite, the CG method does not apply. The minimum residual method (MINRES) is designed to apply in this case. In the same fashion as we developed the GMRES algorithm using the Arnoldi iteration, Algorithm 21.8 implements the MINRES method using the Lanczos iteration. In the resulting least-squares problem, the coefficient matrix is tridiagonal, and we compute the QR decomposition using Givens rotations.

Algorithm 21.8 MINRES

```
function MINRESB(A,b,x₀,m,tol,maxiter)
   % Solve Ax = b using the MINRES method
   % Input: n × n matrix A, n × 1 vector b,
   % initial approximation x₀, integer m < n,
   % error tolerance tol, and the maximum number of iterations, maxiter.
   % Output: Approximate solution xₘ, associated residual r,
   % and iter, the number of iterations required.
   % iter = −1 if the tolerance was not satisfied.

   iter = 1
   while iter ≤ maxiter do
      r = b − Ax₀
      [ Qₘ₊₁  T̄ₘ ] = lanczos(A, r, m)
      β = ‖r‖₂
      Solve the (m + 1) × m least-squares problem T̄ₘyₘ = βe₁
      using Givens rotations that take advantage of the tridiagonal
      structure of T̄ₘ
      xₘ = x₀ + Qₘyₘ
      r = ‖b − Axₘ‖₂
      if r < tol then
         return [xₘ,r,iter]
      end if
      x₀ = xₘ
      iter = iter + 1
   end while
   iter = −1
   return [xₘ, r, iter]
end function
```

NLALIB: The function `minresb` implements Algorithm 21.8. The Lanczos process uses full reorthogonalization.

MINRES does well when a symmetric matrix is well conditioned. The tridiagonal structure of T_k makes MINRES vulnerable to rounding errors [69, pp. 84-86], [72]. It has been shown that the rounding errors propagate to the approximate solution as the square of $\kappa(A)$. For GMRES, the errors propagate as a function of the $\kappa(A)$. Thus, if A is badly conditioned, try `mpregmres`.

Example 21.11.   The MINRES method was applied to three systems whose matrices are shown in Figure 21.14. In each case, $x_0 = 0$, and b was a matrix with random integer values. Matrix (a) has a small condition number. Using $m = 50$ and tol $= 1.0 \times 10^{-6}$, one iteration gave a residual of 3.5×10^{-10}. Matrix (b) has a condition number of approximately 772, but with the same parameters, MINRES yielded a residual of 2.5×10^{-8} in three iterations. Matrix (c) is another story. It has an approximate condition number of 2.3×10^4 and so is ill-conditioned. Using the parameters $m = 1000$ and tol $= 1.0 \times 10^{-6}$, MINRES gave a residual of 8.79×10^{-7} using 42 iterations. On the author's system, this required approximately 4 min, 13 s of computation. In this situation, it is appropriate to try preconditioned GMRES. Using $m = 50$ and tol $= 1.0 \times 10^{-6}$, `mpregmres` produced a residual of 2.75×10^{-12} in one iteration requiring approximately 1.8 s of computation.   ∎

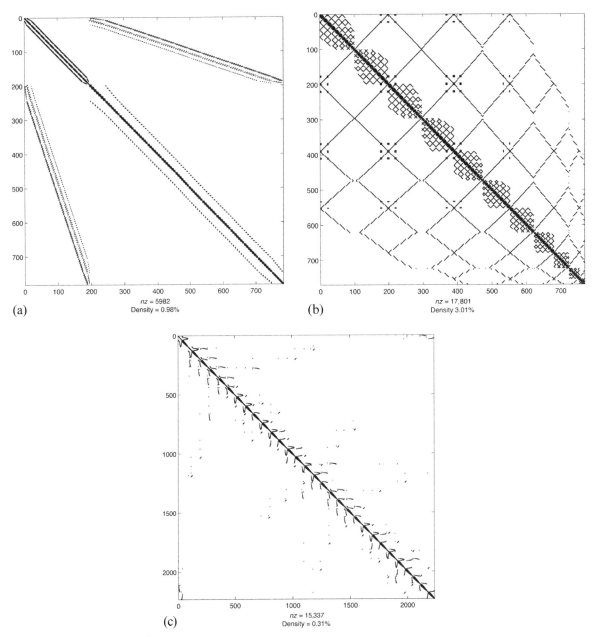

FIGURE 21.14 Large sparse symmetric matrices.

Convergence

Like GMRES, there is no simple set of properties that guarantee convergence. A theorem that specifies some conditions under which convergence will occur can be found in Ref. [73, pp. 50-51].

Remark 21.6.   If A is positive definite, one normally uses CG or preconditioned CG. If A is symmetric indefinite and ill-conditioned, it is not safe to use a symmetric preconditioner K with MINRES if $K^{-1}A$ is not symmetric. Finding a preconditioner for a symmetric indefinite matrix is difficult, and in this case the use of GMRES is recommended.

21.10   COMPARISON OF ITERATIVE METHODS

We have developed three iterative methods for large, sparse matrices, CG, GMRES, and MINRES. In the case of CG and GMRES, algorithms using preconditioning were presented. Given the tools we have developed, the following decision tree suggests (Figure 21.15) an approach for choosing an iterative algorithm. Note that sometimes sparse matrix problems will

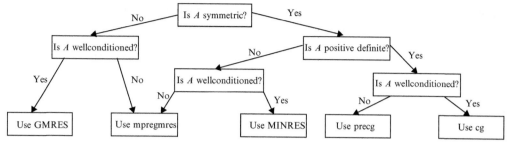

FIGURE 21.15 Iterative method decision tree.

defy proper solution using these techniques, and either other methods should be tried or the user must develop a custom preconditioner.

21.11 POISSON'S EQUATION REVISITED

In Section 20.5, we discussed the numerical solution of the two-dimensional Poisson equation

$$-\frac{\partial^2 u}{\partial x^2} - \frac{\partial^2 u}{\partial y^2} = f(x, y)$$

$$u(x, y) = g(x, y) \quad \text{on } \partial R, \quad \text{where } R = [0, 1] \times [0, 1]$$

using finite difference equations

$$-u_{i-1,j} - u_{i+1,j} + 4u_{ij} - u_{i,j-1} - u_{i,j+1} = h^2 f(x_i, y_j), \quad 1 \le i, j \le n - 1. \tag{21.30}$$

The resulting system was solved using the SOR iteration. In this section, we will solve the system of equations using preconditioned CG. There are $(n - 1)^2$ unknown values inside the grid, so we must solve a system $Au = b$ of dimension $(n - 1)^2 \times (n - 1)^2$. We need to determine the structure of A and b. The matrix A is sparse, since each unknown is connected only to its four closest neighbors. We let $n = 4$, explicitly draw the grid (Figure 21.16) and, from that, determine the form of A and b. The form for larger values of n follows the same pattern.

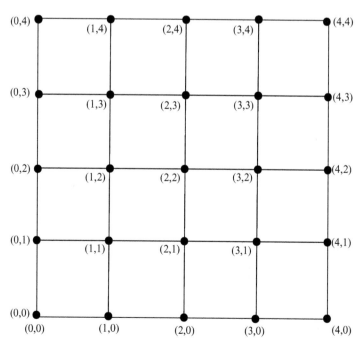

FIGURE 21.16 Poisson's equation grid for $n = 4$.

We adopt the notation $f(x_i, y_j) = f_{ij}$. Cycle through the grid by rows, applying Equation 21.30, and obtain nine equations in the nine unknowns $\begin{bmatrix} u_{11} & u_{21} & u_{31} & \dots & u_{13} & u_{23} & u_{33} \end{bmatrix}^T$:

$$4u_{11} - u_{21} - u_{12} = h^2 f_{11} + g(0, y_1) + g(x_1, 0)$$

$$-u_{11} + 4u_{2,1} - u_{31} - u_{22} = h^2 f_{21} + g(x_2, 0)$$

$$-u_{2,1} + 4u_{3,1} - u_{3,2} = h^2 f_{31} + g(x_3, 0) + g(x_4, y_1)$$

$$-u_{11} + 4u_{12} - u_{22} - u_{1,3} = h^2 f_{12} + g(0, y_2)$$

$$-u_{21} - u_{12} + 4u_{22} - u_{32} - u_{2,3} = h^2 f_{22}$$

$$-u_{31} - u_{22} + 4u_{32} - u_{33} = h^2 f_{32} + g(x_4, y_2)$$

$$-u_{12} + 4u_{13} - u_{23} = h^2 f_{13} + g(0, y_3) + g(x_1, y_4)$$

$$-u_{22} - u_{13} + 4u_{23} - u_{33} = h^2 f_{23} + g(x_2, y_4)$$

$$-u_{32} - u_{23} + 4u_{33} = h^2 f_{33} + g(x_4, y_3) + g(x_3, y_4)$$

In matrix form, the set of equations is

$$\begin{bmatrix} 4 & -1 & 0 & -1 & 0 & 0 & 0 & 0 & 0 \\ -1 & 4 & -1 & 0 & -1 & 0 & 0 & 0 & 0 \\ 0 & -1 & -4 & 0 & 0 & -1 & 0 & 0 & 0 \\ -1 & 0 & 0 & 4 & -1 & 0 & -1 & 0 & 0 \\ 0 & -1 & 0 & -1 & 4 & -1 & 0 & -1 & 0 \\ 0 & 0 & -1 & 0 & -1 & 4 & 0 & 0 & -1 \\ 0 & 0 & 0 & -1 & 0 & 0 & 4 & -1 & 0 \\ 0 & 0 & 0 & 0 & -1 & 0 & -1 & 4 & -1 \\ 0 & 0 & 0 & 0 & 0 & -1 & 0 & -1 & 4 \end{bmatrix} \begin{bmatrix} u_{11} \\ u_{21} \\ u_{31} \\ u_{12} \\ u_{22} \\ u_{32} \\ u_{13} \\ u_{23} \\ u_{33} \end{bmatrix} = \begin{bmatrix} h^2 f_{11} + g(0, y_1) + g(x_1, 0) \\ h^2 f_{21} + g(x_2, 0) \\ h^2 f_{31} + g(x_3, 0) + g(x_4, y_1) \\ h^2 f_{12} + g(0, y_2) \\ h^2 f_{22} \\ h^2 f_{32} + g(x_4, y_2) \\ h^2 f_{13} + g(0, y_3) + g(x_1, y_4) \\ h^2 f_{23} + g(x_2, y_4) \\ h^2 f_{33} + g(x_4, y_3) + g(x_3, y_4) \end{bmatrix}$$

The coefficient matrix has a definite pattern. It is called a *block tridiagonal matrix*. Let

$$T = \begin{bmatrix} 4 & -1 & 0 \\ -1 & 4 & -1 \\ 0 & -1 & 4 \end{bmatrix}.$$

If I is the 3×3 identity matrix, we can write the coefficient matrix using block matrix notation as

$$\begin{bmatrix} T & -I & 0 \\ -I & T & -I \\ 0 & -I & T \end{bmatrix}.$$

In general, if $h = \frac{1}{n}$, the $(n-1)^2 \times (n-1)^2$ matrix for the numerical solution of Poisson's equation has the form

$$\begin{bmatrix} T & -I & & & & \\ -I & T & -I & & & \\ & -I & T & -I & & \\ & & & \ddots & & \\ & & & -I & T & -I \\ & & & & -I & T \end{bmatrix}, \tag{21.31}$$

where I is the $(n-1) \times (n-1)$ identity matrix and T is the $(n-1) \times (n-1)$ tridiagonal matrix

$$T = \begin{bmatrix} 4 & -1 & & & & \\ -1 & 4 & -1 & & & \\ & -1 & 4 & -1 & & \\ & & & \ddots & & \\ & & & -1 & 4 & -1 \\ & & & & -1 & 4 \end{bmatrix}.$$

This matrix can grow very large. For instance, if $n = 50$, the problem requires solving $49^2 = 2401$ equations. If we double n, there are approximately four times as many equations to solve. Standard Gaussian elimination will destroy the structure of the matrix, so iterative methods are normally used for this positive definite, ill-conditioned matrix. Problem 21.31 presents an interesting situation involving the Poisson equation.

21.12   THE BIHARMONIC EQUATION

The *biharmonic equation* is a fourth-order partial differential equation that is important in applied mechanics. It has applications in the theory of elasticity, mechanics of elastic plates, and the slow flow of viscous fluids [74]. The two-dimensional equation takes the form

$$\frac{\partial^4 u}{\partial x^4} + 2\frac{\partial^4 u}{\partial x^2 \partial y^2} + \frac{\partial^4 u}{\partial y^4} = f(x, y), \tag{21.32}$$

with specified boundary conditions on a bounded domain. For our purposes, we assume the domain is the square $R = [0, 1] \times [0, 1]$ and that

$$u(x, y) = 0 \quad \text{and} \quad \frac{\partial u}{\partial n} = 0 \quad \text{on } \partial R.$$

The notation $\frac{\partial u}{\partial n}$ refers to the *normal derivative*, or $\langle \nabla u, n \rangle$, where n is a unit vector orthogonal to the surface. In our case, this becomes

$$\frac{\partial u}{\partial n} = \begin{cases} \frac{\partial u}{\partial x} = 0, & x = 0, \ x = 1 \\ \frac{\partial u}{\partial y} = 0, & y = 0, \ y = 1 \end{cases}.$$

The square can be partitioned into an $(n + 1) \times (n + 1)$ grid with equal steps, $h = \frac{1}{n}$, in the x and y directions. Use a 13-point central difference formula

$$\begin{aligned} [\ 20u_{ij} &- 8\left(u_{i-1,j} + u_{i+1,j} + u_{i,j-1} + u_{i,j+1}\right) \\ &+ 2\left(u_{i-1,j} + u_{i-1,j+1} + u_{i+1,j-1} + u_{i+1,j+1}\right) \\ &+ \left(u_{i-2,j} + u_{i+2,j} + u_{i,j-2} + u_{i,j+2}\right)]\ /h^4 \end{aligned}$$

to approximate $\frac{\partial^4 u}{\partial x^4} + 2\frac{\partial^4 u}{\partial x^2 \partial y^2} + \frac{\partial^4 u}{\partial y^4}$, and approximate the normal derivative on each side of the square by reflecting the first interior set of grid points across the boundary (Figure 21.17).

This results in an ill-conditioned positive definite $(n - 1)^2 \times (n - 1)^2$ *block pentadiagonal* coefficient matrix of the form

$$A = \frac{1}{h^4} \begin{bmatrix} \ddots & \ddots & & \ddots & \\ & \ddots & \ddots & & \ddots \\ \ddots & & \ddots & \ddots & \ddots \\ & D & C & B & C & D \\ & \ddots & \ddots & \ddots & & \ddots \\ & & & \ddots & \ddots \end{bmatrix}.$$

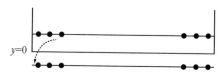

FIGURE 21.17 Estimating the normal derivative.

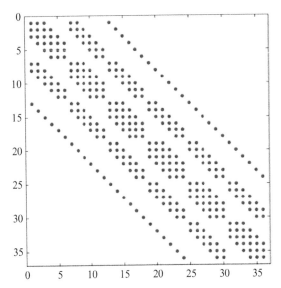

FIGURE 21.18 36×36 biharmonic matrix density plot.

Each block is of size $(n-1) \times (n-1)$. Block D is the identity matrix, C has the pattern $\begin{bmatrix} 2 & -8 & 2 \end{bmatrix}$, and B has the pattern $\begin{bmatrix} 1 & -8 & 21 & -8 & 1 \end{bmatrix}$. For $n = 7$, the blocks are

$$D = \begin{bmatrix} 1 & 0 & 0 & 0 & 0 & 0 \\ 0 & 1 & 0 & 0 & 0 & 0 \\ 0 & 0 & 1 & 0 & 0 & 0 \\ 0 & 0 & 0 & 1 & 0 & 0 \\ 0 & 0 & 0 & 0 & 1 & 0 \\ 0 & 0 & 0 & 0 & 0 & 1 \end{bmatrix}, \quad C = \begin{bmatrix} -8 & 2 & 0 & 0 & 0 & 0 \\ 2 & -8 & 2 & 0 & 0 & 0 \\ 0 & 2 & -8 & 2 & 0 & 0 \\ 0 & 0 & 2 & -8 & 2 & 0 \\ 0 & 0 & 0 & 2 & -8 & 2 \\ 0 & 0 & 0 & 0 & 2 & -8 \end{bmatrix},$$

$$B = \begin{bmatrix} 21 & -8 & 1 & 0 & 0 & 0 \\ -8 & 20 & -8 & 1 & 0 & 0 \\ 1 & -8 & 20 & -8 & 1 & 0 \\ 0 & 1 & -8 & 20 & -8 & 1 \\ 0 & 0 & 1 & -8 & 20 & -8 \\ 0 & 0 & 0 & 1 & -8 & 21 \end{bmatrix}$$

In the first and last occurrence of B, each diagonal element is incremented by 1. You can see the sparsity pattern in Figure 21.18, a density plot of the 36×36 biharmonic matrix.

In the software distribution, the functions `biharmonic_op` and `biharmonic` build the biharmonic coefficient matrix and use the preconditioned CG method to approximate the solution, respectively. Ref. [75, p. 385] provides an exact solution, $u(x, y) = 2350x^4 (x-1)^2 y^4 (y-1)^2$ to Equation 21.32 when

$$f(x, y) = 56,400 \left(1 - 10x + 15x^2\right)(1-y)^2 y^4 + 18,800x^2 \left(6 - 20x + 15x^2\right) y^2 \left(6 - 20y + 15y^2\right)$$
$$+56,400 (1-x)^2 x^4 \left(1 - 10y + 15y^2\right)$$

Figure 21.19(a) and (b) shows a surface plot of the approximate and exact solutions, respectively. MATLAB code for this problem is in the directory "Text Examples/Chapter 21/biharmonicprob.m".

For more information about the biharmonic equation, see Refs. [29, 74].

21.13 CHAPTER SUMMARY

Large, Sparse Matrices

The primary use of iterative methods is for computing the solution to large, sparse systems. Along with other problems, such systems occur in the numerical solution of partial differential equations. There are a number of formats used to store

$$\frac{\partial^4 u}{\partial x^4} + 2\frac{\partial^4 u}{\partial x^2 \partial y^2} + \frac{\partial^4 u}{\partial y^4} = f(x, y)$$
$$u = 0, \frac{\partial u}{\partial n} = 0 \text{ on } \partial R = \{0 \le x, y \le 1\}$$

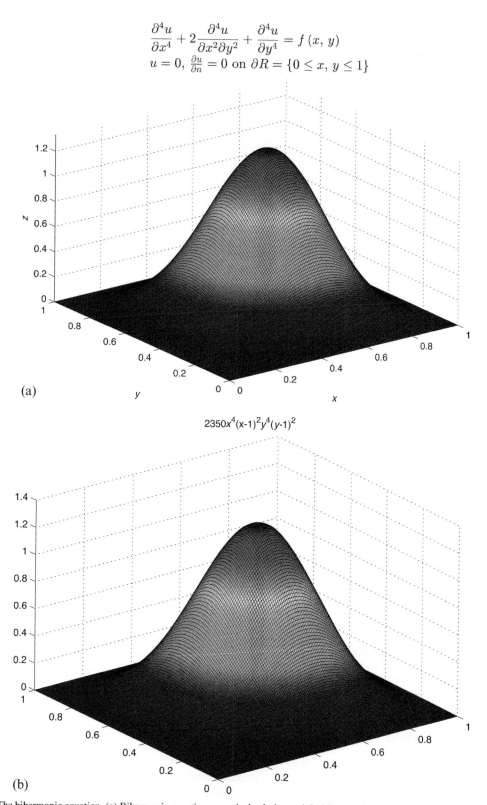

FIGURE 21.19 The biharmonic equation. (a) Biharmonic equation numerical solution and (b) biharmonic equation true solution.

sparse matrices. In each case, only the nonzero elements are recorded. We look at the *CRS format* that stores the nonzero elements by rows. The algorithm uses three vectors that specify the row, column, and value of each nonzero matrix entry.

The CG Method

The CG method approximates the solution to $Ax = b$, where A is symmetric positive definite. The CG method is an extremely clever improvement of the method of steepest descent. The minimum of the quadratic function $\phi(x) = \frac{1}{2}x^T A x - x^T b$ is the solution to $Ax = b$. The negative of the gradient, $-\nabla\phi$, points in the direction of greatest decrease of ϕ. In addition, the gradient is always orthogonal to the contour line at each point. The method of steepest descent takes a step toward the solution by moving in the direction of the gradient line using $x_{i+1} = x_i + \alpha_i r_i$, where α_i is a scalar chosen to minimize ϕ for each step, and r_i is the residual for the current step. Successive residuals are orthogonal, which creates a "zigzag" toward the minimum of $\phi(x)$. The CG method uses an iteration of the same form, $x_{i+1} = x_i + p_i r_i$, but determines the next search direction from the current residual and the previous search directions. Successive approximations are orthogonal relative to the A norm, $\|x\|_A = \sqrt{x^T A x}$, do not "zigzag," and move much faster to the minimum of $\phi(x)$ than steepest descent.

Preconditioning

Iterative methods can converge very slowly if the coefficient matrix is ill-conditioned. Preconditioning involves finding a matrix, M, that approximates A. In this case, $M^{-1}A$ will have a lower condition number than A, and we can solve the equivalent system $M^{-1}Ax = M^{-1}b$, which is termed left-preconditioning. Other approaches are right- and split-preconditioning. Determination of a good matrix M can be difficult.

Preconditioning for CG

Any preconditioner for CG must be symmetric positive definite. We discuss two methods for preconditioning, the incomplete Cholesky decomposition and the SSOR preconditioner. The Cholesky approach can suffer from roundoff error problems, requiring the use of a drop tolerance, which decreases the sparsity of M. The SSOR preconditioner is inexpensive to compute and does not suffer from roundoff error to the same extent. However, incomplete Cholesky normally results in faster convergence.

Krylov Subspace Methods

The Krylov subspace $\mathcal{K}_m$ generated by A and u is span $\{ u \ Au \ A^2u \ \ldots \ A^{m-1}u \}$. It is of dimension m if the vectors are linearly independent. The Krylov subspace methods project the solution to the $n \times n$ problem, $Ax = b$, into a Krylov subspace $\mathcal{K}_m = $ span $\{ r \ Ar \ A^2r \ \ldots \ A^{m-1}r \}$, where r is the residual and $m < n$.

Two Krylov subspace methods are discussed, the GMRES for the solution of any system and the MINRES for the solution of a symmetric indefinite system. The Arnoldi and Lanczos factorizations are required for the development of GMRES and MINRES, respectively. The solution to a least-squares problem in the Krylov subspace yields the solution to the $n \times n$ problem.; If the symmetric matrix is positive definite, CG or preconditioned CG is normally used. In the case of GMRES, the incomplete LU decomposition preconditioner is developed.

Comparison of Iterative Methods

Solving a large, sparse system can be very difficult, and there are many methods available. In the book, we have presented some of the most frequently used methods. Figure 21.15 is a diagram of possible paths that can be used for solving a particular problem.

The Biharmonic Equation

The biharmonic equation is a fourth-order partial differential equation that is important in applied mechanics. A 13-point central finite difference approximation results in a block pentadiagonal matrix that is symmetric positive definite and ill-conditioned. Preconditioned CG is effective in computing a solution to the biharmonic equation. A problem with a known solution is solved for the boundary conditions $u(x, y) = 0$ and $\frac{\partial u}{\partial n} = 0$ on ∂R.

21.14 PROBLEMS

21.1 Fill-in the table by describing the characteristics of each iterative method. Briefly note when convergence is guaranteed and its speed relative to the other methods.

Method	Properties
Jacobi	
Gauss-Seidel	
SOR	
CG	
GMRES	
MINRES	

21.2 If $A \in \mathbb{R}^{n \times n}$, $x \in \mathbb{R}^n$, and $\phi(x) = \frac{1}{2}x^{\mathrm{T}}Ax - x^{\mathrm{T}}b$, show that the gradient of ϕ,

$$\nabla \phi = \left[\begin{array}{ccccc} \frac{\partial \phi}{\partial x_1} & \frac{\partial \phi}{\partial x_2} & \cdots & \frac{\partial \phi}{\partial x_{n-1}} & \frac{\partial \phi}{\partial x_n} \end{array} \right]^{\mathrm{T}},$$

is

$$\nabla \phi(x) = \frac{1}{2}A^{\mathrm{T}}x + \frac{1}{2}Ax - b.$$

21.3

a. Show that $\langle x, y \rangle_A = x^{\mathrm{T}}Ay$ is an inner product, where A is a positive definite matrix.

b. If A is symmetric, there is an orthonormal basis, v_i, $1 \le i \le n$, of eigenvectors. Show that any two distinct basis vectors are A-orthogonal.

21.4 The technique of completing the square is another way to show that the solution, $\bar{x}$, to $Ax = b$ minimizes ϕ (Equation 21.1).

a. Show that for arbitrary vectors x, y, and symmetric matrix A, $x^{\mathrm{T}}Ay = (Ay)^{\mathrm{T}}x = y^{\mathrm{T}}Ax$.

b. From part (a), we know that $\bar{x}^{\mathrm{T}}Ax = x^{\mathrm{T}}A\bar{x}$, where $\bar{x}$ is the solution to $Ax = b$. By completing the square show that

$$\phi(x) = \frac{1}{2}(x - \bar{x})^{\mathrm{T}}A(x - \bar{x}) - \frac{1}{2}\bar{x}^{\mathrm{T}}A\bar{x}.$$

c. Assume A is positive definite. Using the result of part (b), argue that the minimum for $\phi(x)$ occurs when $x = \bar{x}$.

21.5 Show that if $A = R^{\mathrm{T}}R$ is the Cholesky decomposition of the positive definite matrix A, then $\bar{A} = \left(R^{-1}\right)^{\mathrm{T}}AR^{-1}$ is symmetric.

21.6 Verify Equation 21.25, $AQ_m = Q_m H_m + h_{m+1,m}q_{m+1}e_m^{\mathrm{T}}$, for the Arnoldi iteration.

21.7 The diagonal preconditioner, also called the *Jacobi preconditioner*, is one of the simplest means for reducing the condition number of the coefficient matrix. The method is particularly effective for a diagonally dominant matrix or a matrix with widely different diagonal elements but, of course, the method does not always help.

a. Let M be the diagonal of A and use M^{-1} as the preconditioner. Develop a formula for the method that does not involve computing M^{-1}.

b. Let $A = \begin{bmatrix} 20 & 20.01 \\ 9.99 & 10 \end{bmatrix}$. Compute the condition number of the coefficient matrix before and after using diagonal preconditioning.

c. What happens if all elements of the diagonal are constant?

21.8 Show that the eigenvalues of a positive definite matrix are the same as its singular values.

21.9 Show that the SSOR preconditioning matrix $M_{\mathrm{SSOR}}(\omega)$ (Equation 21.22) is symmetric positive definite.

21.10 SSOR can be used as a preconditioner for GMRES. If we choose $\omega = 1$, the matrix $M_{\mathrm{PGS}} = (D + L)D^{-1}(D + U)$ serves as a preconditioner. Using the development of pregmres as a guide, develop equations for the preconditioned matrix, $\bar{A}$, the right-hand side, $\bar{b}$, and the solution, x. HINT: The matrix expression

$$D(D + L)^{-1}A(D + U)^{-1}$$

can be evaluated without computing an inverse using the matrix operators \ and / as follows:

$$(I + L/D) \backslash (A/(D + U)).$$

The definition of the operators is:

$$x = A \backslash b$$

solves

$$Ax = b,$$

and

$$x = b/A$$

solves

$$xA = b.$$

In each case, x and b can be matrices. The book has not discussed the solution to a system of the form $xA = b$, since this type of system is rarely seen. A look at the 2×2 or 3×3 case for A and b will reveal a method of solution.

21.11 Verify Equations 21.18 and 21.19.

21.12 Develop Equation 21.20.

21.13 Two bases $V = \{ v_1 \ \ldots \ v_m \}$ and $W = \{ w_1 \ \ldots \ w_m \}$ are *biorthogonal* if

$$\langle v_i, w_j \rangle = \begin{cases} 0 & i \neq j \\ 1 & i = j \end{cases}.$$

Show that if the $m \times m$ matrices P and Q have columns formed from the vectors in V and W, respectively, then $P^{\mathrm{T}}Q = I$. Biorthogonality is a primary component in the Bi-CG iteration for general sparse matrices. See Problem 21.34.

21.14.1 MATLAB Problems

21.14 Load the positive definite 100×100 sparse matrix ACG from the software distribution. Let b be a random 100×1 vector, $x_0 = 0$, and use `cg` to approximate the solution to $ACGx = b$. Using the same tolerance and maximum number of iterations, approximate the solution using `precg`. Compute the residual for each iteration.

21.15 When a nonsymmetric matrix is well conditioned and it is feasible to compute A^{T}, then CG can be applied to the normal equations.

  a. Write a MATLAB `function [x r iter] = normalsolve(A,b,x0,tol,numiter)` that uses the normal equation approach to solving $Ax = b$.

  b. The MATLAB command `R = sprandn(m,n,density,rc)` generates a random sparse matrix of size $m \times n$, where `density` is the fraction of nonzeros and the reciprocal of `rc` is the approximate condition number. Generate b randomly, and set `x0 = ones(size,1)`. Using an error tolerance of 1.0×10^{-10} and maximum number of iterations set to 500, compute the residual and number of iterations required by `normalsolve` for each sparse system.

  i. R1 = sprandn(1000,1000,0.03,0.10)

  ii. R2 = sprandn(1500,1500,0.05,0.01)

  iii. R3 = sprandn(2000,2000,0.2,0.001)

  c. Add the normal equation approach to the appropriate locations in Figure 21.15.

21.16 The MATLAB function `sprandsym` generates a random symmetric matrix and will additionally make the matrix positive definite. The statement

```
R = sprandsym(n,density,rc,1);
```

generates a sparse random symmetric positive definite matrix of size $n \times n$, where `density` is the fraction of nonzeros, and the reciprocal of `rc` is the exact condition number. Using `spransym`, generate a sparse symmetric positive definite matrix of dimension 500×500, with 3% zeros, and condition number of 50. Execute CG and preconditioned CG using random b, random x_0, tol $= 1.0 \times 10^{-12}$, and numiter $= 100$. For preconditioned CG, use incomplete Cholesky with droptol $= 1.0 \times 10^{-4}$. Compare the time required and the resulting residual.

For the remaining problems, unless stated otherwise, use $x_0 = 0$ and let b be a random $n \times 1$ vector.

21.17 Load the 1138×1138 matrix HB1138 used in the solution of a power network problem.

  a. Verify it is positive definite, and approximate its condition number.

  b. Apply CG with tol $= 1.0 \times 10^{-10}$ and maxiter $= \{500, 1000, 3000, 5000, 10000\}$.

  c. Apply preconditioned CG with 500 iterations.

  d. Apply SSOR preconditioning. You will need to experiment with the maximum number of iterations required.

21.18 Using $m = 3$, compute the Arnoldi decomposition for the matrix

$$A = \begin{bmatrix} 1 & 3 & -1 & 7 & 2 \\ 5 & -8 & 25 & 3 & 12 \\ 0 & -1 & 0 & 3 & 7 \\ 8 & -3 & 23 & 6 & 9 \\ 56 & 13 & 8 & -9 & 1 \end{bmatrix}.$$

21.19 Using $m = 3$, compute the Lanczos decomposition for the symmetric matrix

$$A = \begin{bmatrix} 1 & 5 & 3 & -1 & 6 \\ 5 & 1 & 7 & -8 & 2 \\ 3 & 7 & 12 & -1 & 3 \\ -1 & -8 & -1 & 9 & 4 \\ 6 & 2 & 3 & 4 & 1 \end{bmatrix}.$$

21.20 Load the 2500×2500 matrix SYM2500.
   a. Verify it is symmetric but not positive definite, and approximate its condition number.
   b. Apply `minresb` with tol $= 1.0 \times 10^{-6}$, niter = 10, and $m = \{50, 100, 150, 200, 250\}$. Time each execution, and comment on the result

21.21 Load the 1024×1024 matrix DIMACS10 used in an undirected graph problem.
   a. Verify it is symmetric but not positive definite, and approximate its condition number.
   b. Apply `minresb` with tol $= 1.0 \times 10^{-6}$, maxiter = 50, and $m = \{50, 100, 250, 350\}$. Using $m = 50$ with all the other parameters the same, estimate the solution using mpregmres. Time each execution, and comment on the results.

21.22 Load the 1080×1080 matrix SHERMAN2 used in a computational fluid dynamics problem.
   a. Verify it is not symmetric, and approximate its condition number.
   b. Apply `gmresb` with tol $= 1.0 \times 10^{-6}$, maxiter = 10, and $m = \{500, 800\}$. Time each execution.
   c. Apply `mpregmres` with tol $= 1.0 \times 10^{-6}$, niter = 10, and $m = 500$. Time the execution, and comment on the result and those of part (b).

21.23 The software distribution contains a MATLAB sparse matrix file for each of three matrices:
   a. ACG.mat   **b.** Si.mat   **c.** west0479.mat
   Solve each system as best you can without preconditioning. Then solve each system using preconditioning.

21.24
   a. Using the statement `A = sprandsym(2000,0.05,.01)`, create a 2000×2000 symmetric indefinite matrix with a condition number of 100. Solve the system using `minresb`.
   b. Using `A = sprandsym(2000,0.05,.001)`, create a symmetric indefinite matrix with a condition number of 1000. Solve the system using both `minresb` and `mpregmres`. Use $m = 500$ and tol $= 1.0 \times 10^{-6}$.

21.25 As the drop tolerance decreases when using Cholesky preconditioning, the matrix R becomes more dense.
   a. Using `spy`, graph the complete Cholesky factor for the matrix ROOF in the software distribution.
   b. Computing an incomplete Cholesky factorization of ROOF is particularly difficult. Try `icholesky` and then `ichol` with no drop tolerance.
   c. Using a drop tolerance will make `ichol` successful. Using `spy`, demonstrate the increase in density using the MATLAB function `ichol` with drop tolerances of 1.0×10^{-4} and 1.0×10^{-5}.
   d. What happens when you try drop tolerances of 1.0×10^{-2} and 1.0×10^{-3}?

21.26 The function `spdiags` extracts and creates sparse band and diagonal matrices and is a generalization of the MATLAB function `diag`. We will only use one of its various forms. To create a pentadiagonal matrix, create vectors that specify the diagonals. We will call these diagonals d_1, d_2, d_3, d_4, d_5, where d_1, d_2 are the subdiagonals and d_4, d_5 are the superdiagonals. For an $n \times n$ matrix, each vector must have length n. To create the matrix, execute

   ```
   >> A = spdiags([d₁ d₂ d₃ d₄ d₅], -2:2, n, n);
   ```

   The parameter `-2:2` specifies that the diagonals are located at offsets -2, -1, 0, 1, and 2 from the main diagonal.
   a. Create a $10,000 \times 10,000$ pentadiagonal matrix with sub- and superdiagonals $\begin{bmatrix} -1 & -1 & \dots & -1 & -1 \end{bmatrix}^T$ and diagonal $\begin{bmatrix} 4 & 4 & \dots & 4 & 4 \end{bmatrix}^T$. Show the matrix is positive definite, with an approximate condition number of 2.0006×10^7.
   b. Using $x_0 = 0$ and a random b, apply `cg` using tol $= 1.0 \times 10^{-14}$ and maxiter = [10 100 1000 10000 20000]. In each case, output the residual.
   c. Use `precg` with the same tolerance and niter = 100. Output the residual.

21.27 The biharmonic matrix is ill-conditioned. Verify this for matrices with $n = [10\ 50\ 100\ 500]$ over the interval $0 \le x, y \le 1$.

21.28 We will investigate the one-dimensional version of the biharmonic equation:

$$\frac{d^4 u}{dx^4} = f(x). \tag{21.33}$$

Assuming the boundary conditions

$$u(0) = u(1) = \frac{d^2u}{dx^2}(0) = \frac{d^2u}{dx^2} = 0, \quad 0 \le x \le 1$$

and using a five-point central finite-difference approximation over $0 \le x \le 1$ with uniform step size $h = 1/n$, gives rise to the following pentadiagonal $(n-1) \times (n-1)$ matrix [29]:

$$A = \frac{1}{h^4} \begin{bmatrix} 5 & -4 & 1 & & & & & 0 \\ -4 & 6 & -4 & 1 & & \ddots & & \\ 1 & -4 & 6 & -4 & 1 & & \ddots & \\ \vdots & \ddots & & & & & \ddots & \\ \cdots & 1 & -4 & 6 & -4 & 1 & & \\ \vdots & & & & & & \ddots & \\ 0 & & \cdots & & 1 & -4 & 5 \end{bmatrix}$$

A is symmetric positive definite and ill-conditioned, so the preconditioned CG method should be used.

a. Write a function `[x, y, u, residual, iter] = biharmonic1D(n, f, tol, maxiter)` that uses preconditioned CG with n subintervals, specified tolerance and maximum number of iterations to approximate a solution to Equation 21.33 and graph it.

b. Consider the specific problem

$$\frac{d^4u}{dx^4} = x, \quad 0 \le x \le 1,$$

$$u(0) = u(1) = \frac{d^2u}{dx^2}(0) = \frac{d^2u}{dx^2}(1) = 0,$$

whose exact solution is $u(x) = \frac{1}{120}x^5 - \frac{1}{36}x^3 + \frac{7}{360}x$. Use $h = \frac{1}{100}$, that requires solving a 99×99 pentadiagonal system having 489 diagonal entries, so the density of nonzero entries is 4.99%.

21.29

a. The Jacobi diagonal preconditioner (Problem 21.7) performs best when the coefficient matrix is strictly diagonally dominant and has widely varying diagonal entries. Write a function `jacobipregmres` that implements Jacobi preconditioning for the GMRES method. Return with a message if the preconditioned matrix does not have a condition number that is at most 80% of the condition number for A.

b. Build the 1000×1000 matrix, T, with seven diagonals:

$$a_i = -1, \ b_i = 2, \ c_i = 4, \ d_i = 20 + 100(i-1), \ e_i = 4, f_i = 2, \ g_i = -1, \ 1 \le i \le 1000.$$

Compute the condition number of T. Using rhs $= \begin{bmatrix} 1 & 1 & \cdots & 1 & 1 \end{bmatrix}^T$, tol $= 1.0 \times 10^{-6}$, and $m = 100$, compute the solution using both `gmresb` and `jacobipregmres`. Display the residuals.

21.30 Example 21.7 applies `precg` to the matrix `PRECGTEST` in the software distribution. Modify `precg` and create the function `precgomega` with the following calling sequence:

```
% If method = 'incomplete Cholesky', parm is the drop tolerance.
% If method = 'SSOR', parm is omega.
[x, residual, iter] = precgomega(A,b,x0,tol,maxiter,method,parm);
```

Rather than assuming $\omega = 1$ in Equation 21.22, maintain the equation as a function of ω. Apply `precgomega` with the SSOR preconditioner to PRECGTEST using different values of ω and see if you can improve upon the result in Example 21.7.

21.31 The ill-conditioned Poisson matrix 21.11 developed for the five-point central finite difference approximation to the solution of Poisson's equation is positive definite, so the preconditioned CG method applies. To use CG, we will need to construct the block tridiagonal matrix. Fortunately, MATLAB builds the sparse $n^2 \times n^2$ matrix with the command

```
P = gallery('poisson', n);
```

a. Modify the Poisson equation solver `sorpoisson` referenced in Section 20.5 to use the preconditioned CG method, where $\partial R = \{0 \le x, y \le 1\}$. The function declaration should be `function [x y u] = cgpoisson(n,f,g,numiter)`.

b. Test cgpoisson using the following problem with $n = 75$, maxiter $= 100$, and tol $= 1.0 \times 10^{-10}$.

$$-\frac{\partial^2 u}{\partial x^2} - \frac{\partial^2 u}{\partial y^2} = xy, u(x, y) = 0 \quad \text{on } \partial R = [0, 1] \times [0, 1]$$

c. The following Poisson problem describes the electrostatic potential field induced by charges in space, where u is a potential field and ρ is a charge density function.

$$-\frac{\partial^2 u}{\partial x^2} - \frac{\partial^2 u}{\partial y^2} = 4\pi\rho$$
$$u(x, y) = g(x, y) \quad \text{on } \partial R.$$

Assume that the electrostatic potential fields are induced by approximately 15 randomly placed point charges with strength 1 ($\rho = 1$). The edges are grounded, so $u(x, y) = 0$ on $\partial R = \{0 \le x, y \le 1\}$. Note that the right-hand side function $4\pi\rho$ affects only the points inside the boundary. Solve the problem using $n = 100$, and draw a surface plot and a contour plot of the result (Figure 21.20 shows sample plots). One way of coding the right-hand side is

```
function r = rho(~,~)
%r = rho(x,y) generates a point charge at random points.
%
%    generate a point charge of strength 1 at approximately
%    3/2000 points (x,y) in the plane. this causes a point charge
%    of approximately 1/15 of the points in the plot of the voltage
%    produced by 'potentialsolve.m'.

p = randi([1 2000],1,1);
if p == 500 || p == 250 || p == 750
    r = 4 * pi;
else
    r = 0;
end
```

21.32

a. Using the results of Problem 21.10, develop function [x, r, iter] = ssorpregmres(A,b,x0,m,tol, maxiter) that uses SSOR as a preconditioner for GMRES.

b. Ref. [64, p. 97] discusses the nonsymmetric matrix ORSIIR_1 with dimension 1030×1030 that arises from a reservoir engineering problem. Let $x_0 = 0$, $b = \begin{bmatrix} 1 & 1 & \dots & 1 & 1 \end{bmatrix}$, $m = 150$, and tol $= 1.0 \times 10^{-6}$. Solve the system ORSII_1$x = b$ using the functions gmresb, mpregmres, and ssorpregmres.

c. The positive definite matrix ROOF presented in Problem 21.25 required experimenting with a drop tolerance when preconditioning using the incomplete Cholesky decomposition. Using $b = \begin{bmatrix} 1 & 1 & \dots & 1 & 1 \end{bmatrix}^T$, $x_0 = 0$, solve ROOF$x = b$ using ssorpregmres with tol $= 1.0 \times 10^{-6}$, $m = 100$, and maxiter $= 15$.

21.33 MATLAB provides various solvers for large, sparse problems.

a. The MATLAB function gmres implements the GMRES algorithm. Look it up using the help system and apply it to the nonsymmetric matrix DK01R from the software distribution, using the right-hand side b_DK01R. You will need to precondition the matrix using a drop tolerance with the statement

```
[L,U] = ilu(DK01R,struct('type','ilutp','droptol',droptol));
```

b. The MATLAB function minres implements the MINRES algorithm. Use it with the matrix bcsstm10 from the software distribution with $b = \begin{bmatrix} 1 & 1 & \dots & 1 & 1 \end{bmatrix}^T$, $x_0 = 0$. Use sufficiently many iterations so it succeeds with an error tolerance of 1.0×10^{-6}.

c. The MATLAB function symmlq is a general sparse symmetric matrix solver based upon solving a symmetric tridiagonal system. Use it with the matrix bcsstm10 from the software distribution with $b = \begin{bmatrix} 1 & 1 & \dots & 1 & 1 \end{bmatrix}^T$, $x_0 = 0$. Use sufficiently many iterations so it succeeds with an error tolerance of 1.0×10^{-6}.

21.34 MATLAB provides functions we have not discussed in the book that implement sparse system solvers for general matrices. In this problem, you will investigate two such algorithms.

a. The bi-CG method is a CG-type method. The bi-CG method generates two sets of residual sequences, $\{r_i\}$ and $\{\bar{r}_i\}$, which are biorthogonal (Problem 2.13). Two sets of direction vectors, $\{p_i\}$ and $\{\bar{p}_i\}$ are computed from the

$$-\frac{\partial^2 V}{\partial x^2} - \frac{\partial^2 V}{\partial y^2} = 4\pi\rho$$
$$V(x, y) = 0 \text{ on } \partial R = \{0 \le x, y \le 1\}$$

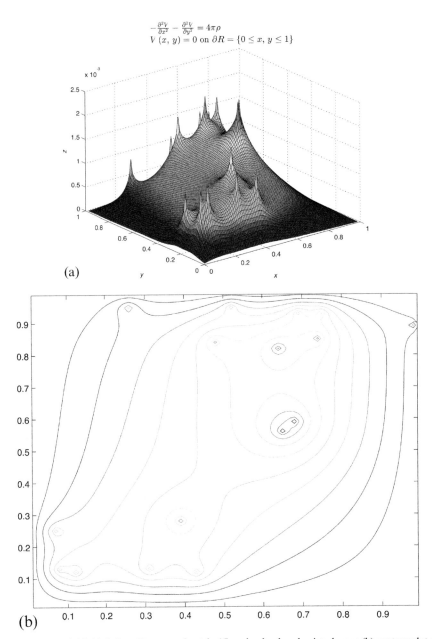

(a)

(b)

FIGURE 21.20 (a) Electrostatic potential fields induced by approximately 15 randomly placed point charges (b) contour plot of randomly placed point charges.

residuals. The function `bicg` implements the algorithm. Try it with the matrix TOLS1090 from the software distribution. Preconditioning is necessary.

b. The QMR method applies to general matrices and functions by working with a nonsymmetric tridiagonal matrix. Repeat part (a) using `qmr`.

21.35 Let $R = [0, 1] \times [0, 1]$, and graph the approximate solution to the biharmonic equation

$$\frac{\partial^4 u}{\partial x^4} + 2\frac{\partial^4 u}{\partial x^2 \partial y^2} + \frac{\partial^4 u}{\partial y^4} = \sin(\pi x)\sin(\pi y),$$
$$u(x, y) = 0 \quad \text{on } \partial R$$
$$\frac{\partial u}{\partial n} = 0 \quad \text{on } \partial R$$

21.36 Reproduce Figures 21.13(a), (b), and console output for Example 21.10.

Chapter 22

Large Sparse Eigenvalue Problems

You should be familiar with

- Methods for computing eigenvalues of nonsparse matrices, both symmetric and nonsymmetric
- Arnoldi method
- Lanczos method

Chapters 18 and 19 present direct methods for computing eigenvalues of real matrices. These methods require $O\left(n^3\right)$ flops, and for large n it is unrealistic to use them. For instance, suppose a sparse matrix A is of dimension 500×500. It has 500 eigenvalues, accounting for multiplicities. Often in applications only certain eigenvalues and associated eigenvectors are required, so what we really need is a means of computing just those eigenvalues and eigenvectors. After studying this chapter, the interested reader will be prepared for more extensive discussions in such works as [3–5].

We begin our discussion with the power (Algorithm 18.1) method to compute the largest eigenvalue in magnitude. This method generates a Krylov sequence $x_0, Ax_0, A^2x_0, \ldots, A^{k-1}x_0$, and Krylov subspace methods form the basis for our iterative approach. To find a small collection of eigenvalues, we will make use of two methods presented in Chapter 21, the Arnoldi and Lanczos methods, both of which find an orthogonal basis for a Krylov subspace. The Arnoldi method provides a means for estimating a few eigenvalues and eigenvectors for a large sparse nonsymmetric matrix. There are two approaches, explicit and implicit Arnoldi. In practice, the implicit method is used, but it is necessary to first understand explicit methods. For symmetric matrices, the Lanczos decomposition is the tool of choice. The algorithms are similar to those for nonsymmetric matrices, but symmetry provides a much clearer picture of convergence properties.

22.1 THE POWER METHOD

The power method computes the largest eigenvalue in magnitude and an associated eigenvector. Recall that the statements

$$x_k = Ax_{k-1},$$
$$x_k = x_k / \|x_k\|_2$$

are computed in a loop, so all that is required is matrix-vector multiplication and normalization. The initial guess should be randomly generated, and the convergence rate depends on the ratio $\left|\frac{\lambda_2}{\lambda_1}\right|$, where $|\lambda_1| > |\lambda_2|$ are two largest eigenvalues in magnitude.

Example 22.1. The matrix `rdb200` in the book software is a 200×200 nonsymmetric sparse matrix obtained from Matrix Market (Figure 22.1). The "rdb" refers to the reaction-diffusion Brusselator model used in chemical engineering. The MATLAB function `eigs` computes a few eigenvalues and associated eigenvectors of a large, sparse, matrix; in particular, E = eigs(A) returns a vector containing the six largest eigenvalues of A in magnitude. Apply `eigs` to `rdb200` and find the two largest eigenvalues. After computing the ratio $\left|\frac{\lambda_2}{\lambda_1}\right|$, use the function `largeeig` from the book software distribution to estimate the largest eigenvalue and compare the result with that of `eigs`. The ratio $\left|\frac{\lambda_2}{\lambda_1}\right|$ is 0.9720 and is not very favorable, but an error tolerance of 1.0×10^{-6} was obtained in 520 iterations, requiring 0.029465 s on the author's computer. ■

```
>> v = eigs(rdb200);
>> v(1:2)
ans =
   -33.3867   -32.4503
```

Numerical Linear Algebra with Applications. http://dx.doi.org/10.1016/B978-0-12-394435-1.00022-3

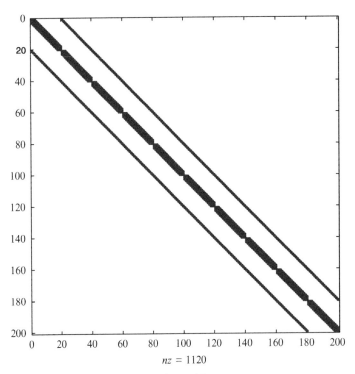

$nz = 1120$

FIGURE 22.1 Nonsymmetric sparse matrix used in a chemical engineering model

```
>> abs(v(2)/v(1))
ans =
    0.9720
>> [lambda x iter] = largeeig (rdb200,rand(200,1),1.0 e-6,600);
>> lambda
lambda =
  -33.3867
>> abs(lambda - v(1))
ans =
    1.1795e-12
```

Starting with x_0, the power method generates the Krylov sequence $x_0 \ Ax_0 \ A^2 x_0 \ \dots \ A^{k-2} x_0 \ A^{k-1} x_0$. Of these vectors, we only use $A^{k-1} x_0$ as our approximate eigenvector. It seems reasonable that we can do better by choosing our approximate eigenvector as a linear combination $y = \sum_{i=0}^{k-1} c_i \left(A^i x_0 \right)$ of vectors from the Krylov sequence. This is indeed the case as we will see in subsequent sections when discussing the use of the Arnoldi and Lanczos methods.

22.2 EIGENVALUE COMPUTATION USING THE ARNOLDI PROCESS

Recall from Chapter 21 that the Arnoldi process applied to an $n \times n$ matrix A produces a decomposition of the form (Equation 21.25)

$$AQ_m = Q_m H_m + h_{m+1,m} \, q_{m+1} e_m^{\mathrm{T}}, \tag{22.1}$$

where H is $m \times m$ upper Hessenberg matrix. The $n \times m$ matrix $Q_m = \begin{bmatrix} q_1 & q_2 & \dots & q_{m-1} & q_m \end{bmatrix}$ is formed from column vectors $\{q_i\}$, and $\{q_1, q_2, \dots, q_m\}$ form an orthogonal basis for the subspace spanned by $K_{m+1}(A, x_1) = \{x_1, Ax_1, \dots, A^m x_1\}$. The additional vector, q_{m+1}, is orthogonal to $\{q_i\}$, $1 \le i \le m$. We used the Arnoldi process to develop the general minimal residual (GMRES) method (Algorithm 21.5) for approximating the solution to a large sparse system $Ax = b$. To approximate the solution we chose m to be smaller than n and project the problem into a Krylov subspace of dimension m. If $m = n$, $h_{m+1,m} = 0$, and the Arnoldi process gives a Schur decomposition (Theorem 18.5)

$$Q^{\mathrm{T}} AQ = H,$$

where Q is an $n \times n$ orthogonal matrix of Schur vectors, and H is $n \times n$ quasi-upper-triangular matrix with the eigenvalues of A on its diagonal. For large n, finding the Schur decomposition is impractical. Might it make sense to try the same type of approach as that of GMRES and approximate the eigenvalues of H_m in a Krylov subspace of dimension $m < n$? The question then becomes can we use the eigenvalues of H_m in Equation 22.1 to approximate the eigenvalues of A? If we can, then we will be able to compute a maximum of m eigenvalues and their associated eigenvectors. In reality, only a few of those m eigenvalues will lead to good approximations to the eigenvalues of A, and those eigenvalues tend to be the largest and the smallest eigenvalues of H_m.

Adopt the notation that $\lambda_i^{(m)}$ is eigenvalue i of H_m, and $u_i^{(m)}$ is the corresponding eigenvector. We call $\lambda_i^{(m)}$ a *Ritz value*, the vector $v_i^{(m)} = Q_m u_i^{(m)}$ a *Ritz eigenvector*, and the pair (λ_i, v_i) a *Ritz pair*. Now, from Equation 22.1

$$
\begin{aligned}
AQ_m u_i^{(m)} &= Q_m H_m u_i^{(m)} + h_{m+1,m} q_{m+1} e_m^{\mathrm{T}} u_i^{(m)} \\
AQ_m u_i^{(m)} &= Q_m \lambda_i^{(m)} u_i^{(m)} + h_{m+1,m} q_{m+1} e_m^{\mathrm{T}} u_i^{(m)} \\
A v_i^{(m)} &= \lambda_i^{(m)} v_i^{(m)} + h_{m+1,m} q_{m+1} e_m^{\mathrm{T}} u_i^{(m)}.
\end{aligned}
\tag{22.2}
$$

Define the residual

$$
r_i^{(m)} = \left(A - \lambda_i^{(m)} I \right) v_i^{(m)}.
$$

Noting that $\|q_{m+1}\|_2 = 1$, $h_{m+1,n} \geq 0$, and $e_m^{\mathrm{T}} u_i^{(m)}$ is a real number, Equation 22.2 gives

$$
\left\| r_i^{(m)} \right\|_2 = h_{m+1,m} \left| \left(u_i^{(m)} \right)_m \right|,
\tag{22.3}
$$

where $\left| \left(u_i^{(m)} \right)_m \right|$ is component m of $u_i^{(m)}$. If $h_{m+1,m}$ or $\left| \left(u_i^{(m)} \right)_m \right|$ is small, Equation 22.3 gives us a means of estimating the error of using the Ritz pair $\left(\lambda_i^{(m)}, v_i^{(m)} \right)$ as an eigenpair of A.

We can now begin to develop an algorithm for computing a few eigenvalues of A. If m is small, We had not such guarantee for a nonsymmetric matrix. then we can quickly compute the eigenvalues and associated cigenvectors of H_m and compute the value $\left\| r_i^{(m)} \right\|_2 = h_{m+1,m} \left| \left(u_i^{(m)} \right)_m \right|$ for all eigenvectors $u_i^{(m)}$. If any of the residuals is small, then we hope $\lambda_i^{(m)}/v_i^{(m)}$ is a good approximation to an eigenpair of A. Computational experience has shown that some of the eigenvalues $\lambda_i^{(m)}$ will be good approximations to eigenvalues of A well before m gets close to n [23, p. 444]. In practice, we want to use the smallest m possible.

Example 22.2. Recall that the function i, not 1 builds the biharmonic matrix discussed in Section 21.12. The reader should consult the MATLAB documentation to determine the action of `sprandn`. Generate a 144×144 random sparse matrix using the MATLAB statements

```
B = biharmonic_op(12,1,12,1);
A = sprandn(B);
```

The eigenvalues of A are complex. Apply the Arnoldi process to generate a matrix H_{35}, and compute its eigenvalues, the Ritz values. Then compute the eigenvalues of A using `eig` and plot them marked with a circle, followed by a plot on the same axes of the Ritz values marked by "x". Figure 22.2 displays the results. Note that the largest and smallest Ritz values in magnitude are close to eigenvalues of A. This is typical behavior. ∎

22.2.1 Estimating Eigenvalues Without Restart or Deflation

The following algorithm outline chooses a random vector v_0, and performs an Arnoldi decomposition. It then computes eigenvalues (Ritz values) of the factor H_m, creating Ritz vectors $v_i^{(m)} = Q_m u_i^{(m)}$, sorts the Ritz pairs $\left(\lambda_i^{(m)}, v_i^{(m)} \right)$ in descending order of eigenvalue magnitude, and cycles through the Ritz pairs retaining up to *nev* of those for which the residual given by Equation 22.3 is sufficiently small. It is possible that none will be found.

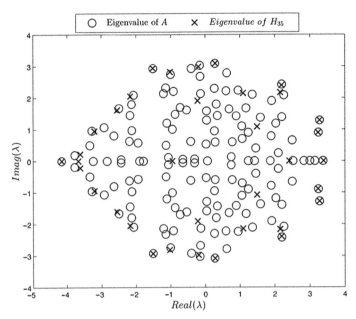

FIGURE 22.2 Eigenvalues and Ritz values of a random sparse matrix.

Simple Use of Arnoldi to Approximate a Few Eigenpairs of a Large, Sparse Matrix

1. Compute the Arnoldi decomposition $AQ_m = Q_m H_m + h_{m+1,m} q_{m+1} e_m^T$ using initial vector v_0.
2. Compute the eigenvectors $\left\{ u_1^{(m)}, u_2^{(m)}, \ldots, u_m^{(m)} \right\}$ and eigenvalues $\left\{ \lambda_1^{(m)}, \lambda_2^{(m)}, \ldots, \lambda_m^{(m)} \right\}$ of H_m.
3. Sort the Ritz pairs in descending order of eigenvalue magnitude.
4. Loop through the sorted Ritz pairs $\left(\lambda_i^{(m)}, v_i^{(m)} = Q_m u_i^{(m)} \right)$ finding up to *nev* of those such that $h_{m+1,m} \left| \left(u_i^{(m)} \right)_m \right| < tol$.

With the GMRES and MINRES methods for solving sparse systems of equations, we found it necessary to restart the Arnoldi or Lanczos process using the most recently estimated solution as the initial vector. Recall that our approach to the eigenvalue problem for a dense matrix was to compute one eigenvalue at a time using deflation. After computing eigenvalue λ_k, we "deflated" the problem to computing an eigenvalue in the $(k-1) \times (k-1)$ submatrix. Both restart and deflation are components of production quality algorithms for computing a few eigenpairs of a sparse matrix.

22.2.2 Estimating Eigenvalues Using Restart

Assume *nev* eigenvalues are required. If $k < nev$ of the current Ritz pairs are sufficiently accurate, we can retain those and then restart Arnoldi with an improved initial vector in hopes of estimating the remaining eigenvalues. Such an algorithm is said to use *explicit restart*. In the next section, we will discuss more sophisticated restart strategies. For now, we use the simple strategy of restarting with the current Ritz vector. The following outline summarizes this algorithm.

Outline of Explicit Arnoldi with Restart

1. Compute the initial starting vector v_0, and let $k = 1$.
2. Using initial vector v_0, compute the Arnoldi decomposition $AQ_m = Q_m H_m + h_{m+1,m} q_{m+1} e_m^T$.
3. Compute the eigenvectors $\left\{ u_1^{(m)}, u_2^{(m)}, \ldots, u_m^{(m)} \right\}$ and eigenvalues $\left\{ \lambda_1^{(m)}, \lambda_2^{(m)}, \ldots, \lambda_m^{(m)} \right\}$ of H_m.
4. Sort the eigenvalues and associated eigenvectors in descending order of eigenvalue magnitude.
5. See if Ritz pair k satisfies the error tolerance $h_{m+1,m} \left| \left(u_k^{(m)} \right)_m \right| < tol$. If so, increase k and return if $k \geq nev$.
6. Let v_0 be the normalized current Ritz vector, $\frac{Q_m u_k}{\|Q_m u_k\|_2}$, and go to step 2, unless some maximum number of iterations are exceeded.

The success of this algorithm depends on the matrix and the initial approximation v_0. The implementation of this method is left to the problems.

22.2.3   A Restart Method Using Deflation

We assume that we want a few of the largest eigenvalues in magnitude and that those eigenvalues are real. We will deal with complex eigenvalues in Section 22.3. Suppose we have found $k - 1$ approximate unit eigenvectors of A, $\left\{ v_1 \; v_2 \; \ldots \; v_{k-2} \; v_{k-1} \right\}$ and an additional unit eigenvector, v_k, orthogonal to the $k - 1$ vectors. Since $h_{m+1, m} \left| \left(u_k^{(m)} \right)_m \right|$ is small, the vectors $\left\{ v_1 \; v_2 \; \ldots \; v_{k-2} \; v_{k-1} \; v_k \right\}$ satisfy

$$H_m \left(1 : k, \; 1 : k \right) \simeq Q_m^T \left(:, \; 1 : k \right) A Q_m \left(:, \; 1 : k \right). \tag{22.4}$$

In other words, span $\left\{ v_1 \; v_2 \; \ldots \; v_{k-2} \; v_{k-1} \; v_k \right\}$ is approximately an invariant subspace of A. We need this result in the following algorithm development.

Deflation was a primary tool in computing eigenvalues for dense matrices (see Sections 18.5 and 18.6). It allowed us to gradually reduce the size of the problem, saving computing time, and yielding higher accuracy. Assume that Arnoldi is started using the initial vector, v_0, to obtain the decomposition $A Q_m = Q_m H_m + h_{m+1, m} q_{m+1} e_m^T$. Compute the eigenvalues and eigenvectors of H_m, and choose the pair $\left(\lambda_1^{(m)}, u_1^{(m)} \right)$ with the largest eigenvalue in magnitude. Then compute the corresponding Ritz pair $\left(\lambda_1^{(m)}, v_1^{(m)} = Q_m u_1^{(m)} \right)$, normalize $v_1^{(m)}$ and compute the residual

$$\left\| r_1^{(m)} \right\|_2 = h_{m+1, m} \left| \left(u_1^{(m)} \right)_m \right|.$$

If $\left\| r_1^{(m)} \right\|_2 \geq$ tol, restart the Arnoldi process, and continue until $\left\| r_1^{(m)} \right\|_2 <$ tol. At this point, we have an approximate eigenpair $\left(\lambda_1^{(m)}, v_1^{(m)} \right)$, and are ready to compute the next pair $\left(\lambda_2^{(m)}, v_2^{(m)} \right)$. Apply Equation 22.4 and obtain

$$H_m \left(1 : 1, \; 1 : 1 \right) = Q_m \left(:, \; 1 : 1 \right)^T A Q_m \left(:, \; 1 : 1 \right),$$

so that $H_m \left(1 : 1, 1 : 1 \right)$ is a 1×1 matrix containing the estimated eigenvalue $\lambda_1^{(m)}$. H has the approximate form

$$\begin{bmatrix} \boxed{X} & X & X & \ldots & & X & X \\ 0 & X & X & \ldots & & X & X \\ 0 & X & X & X & & \ldots & X \\ 0 & 0 & X & X & & \ddots & \\ \vdots & \vdots & \vdots & \ddots & & \ddots & X \\ 0 & 0 & 0 & & \ldots & X & X \end{bmatrix}$$
$$k = 1$$

where h_{11} is the estimated eigenvalue. The Ritz pair $\left(\lambda_1^{(m)}, v_1^{(m)} \right)$ is said to be *locked*, since no more operations will affect column 1 of H_m. Execute the Arnoldi iteration beginning with $k = 2$, compute the eigenvalues and eigenvectors of H_m, and sort them. Select the pair $\left(\lambda_2^{(m)}, u_2^{(m)} \right)$ and compute $v_2^{(m)} = Q_m u_2^{(m)}$. It is essential that each new vector be orthogonal to all the previous ones, so make the vector $v_2^{(m)}$ be orthogonal to $v_1^{(m)}$ by using the Gram-Schmidt process. Determine if $\left(\lambda_2^{(m)}, v_2^{(m)} \right)$ satisfies

$$h_{m+1, m} \left| \left(u_2^{(m)} \right)_m \right| < \text{tol}.$$

If not, restart and repeat the process until it does. At that point, apply Equation 22.4,

$$H_m \left(1 : 2, \; 1 : 2 \right) = Q_m \left(:, \; 1 : 2 \right)^T A Q_m \left(:, \; 1 : 2 \right),$$

and H_m has the approximate form

$$\begin{bmatrix} X & X & X & \ldots & & X & X \\ 0 & X & X & \ldots & & X & X \\ 0 & 0 & X & X & & \ldots & X \\ 0 & 0 & X & X & & \ddots & \\ \vdots & \vdots & \vdots & \ddots & & \ddots & X \\ 0 & 0 & 0 & & \ldots & X & X \end{bmatrix}$$
$$k = 2$$

Now lock $\left(\lambda_1^{(m)}, v_1^{(m)}\right)$ and $\left(\lambda_2^{(m)}, v_2^{(m)}\right)$ by leaving columns 1 and 2 of H_m alone, increase k to 3, and continue until the error tolerance is satisfied. After executing

$$H_m\left(1:3,\, 1:3\right) = Q_m\left(:,\, 1:3\right)^{\mathrm{T}} A Q_m\left(:,\, 1:3\right),$$

H_m approximately has the form (Table 22.1)

TABLE 22.1 Restart Method Using Deflation

X	X	X	...	X	X
0	X	X	...	X	X
0	0	X	X	...	X
0	0	0	X	⋱	
⋮	⋮	⋮	⋱	⋱	X
0	0	0	...	X	X

$$k = 3$$

Our ultimate aim is to reduce the upper $nev \times nev$ submatrix of H_m to an upper-triangular matrix with the estimated eigenvalues of A on its diagonal.

$$
nev \quad
\begin{bmatrix}
X & X & X & \cdots & X & X & \cdots & X \\
  & X & \cdots & X & X & X & \cdots & X \\
  &   & \ddots & \vdots & \cdots & X & \cdots & X \\
  &   &   & X & \cdots & X & X & X \\
\hline
  &   & nev &   & X & X & \cdots & X \\
  &   &   &   & X & X & X & X \\
  &   &   &   &   & X & \cdots & X \\
  &   &   &   &   &   & X & X \\
  &   &   &   &   &   &   & X
\end{bmatrix}
$$

Using the preceding discussion, we are now in a position to outline an algorithm for estimating nev eigenpairs of matrix A.

Outline of Arnoldi with Deflation and Explicit Restart

1. Select a starting vector v_0 with $\|v_0\|_2 = 1$. Use a random vector if a good starting value is not known. Let $k = 1$.
2. Loop
   a. Execute the Arnoldi iteration

   ```
   % k—1 columns are good. Start with column k.

   for j = k:m do
       ...
   end for
   ```

   b. Compute the eigenvalues and eigenvectors of H_m and sort them. Pick the eigenpair $\left(\lambda_k^{(m)}, u_k^{(m)}\right)$, and let $\tilde{v}_k^{(m)} = V_m u_k^{(m)}$.

   c. Othogonalize $\tilde{v}_k^{(m)}$ against all the previous Ritz vectors $\left\{v_1^{(m)}, v_2^{(m)}, \ldots, v_{k-1}^{(m)}\right\}$ and let $v_k^{(m)} = \tilde{v}_k^{(m)} / \left\|\tilde{v}_k^{(m)}\right\|_2$.

   d. Compute $e_k = h_{m+1,m}\left|\left(u_k^{(m)}\right)_m\right|$.

   e. If $e_k < tol$,
      i   Compute $H_m\left(1:k,\, 1:k\right) = Q_m\left(:,\, 1:k\right)^T A Q\left(:,\, 1:k\right)$.
      ii Let $k = k + 1$ (deflate)
      iii If $k \geq nev$, then return else go to 2 (restart).

   f. Go to 2 (restart)

The function `eigsbexplicit` in the book software distribution implements Algorithm 22.3. Type `help eigsbexplicit` to determine the calling sequence.

Remark 22.1.   Recall from Section 21.8.1 that the Lanczos vectors can lose orthogonality due to roundoff error. To a lesser extent, this is also true of the Arnoldi process. For that reason, the NLALIB function `arnoldi` has a fourth parameter, `reorthog`. When `reorthog = 1`, reorthogonalization is performed.

Example 22.3.   We test eigsbexplicit first with the 2233×2233 symmetric matrix `lshp2233.mat`, followed by using eigsbexplicit with the 1030×1030 nonsymmetric matrix ORSIIR_1.mat. This matrix presents a much greater challenge, and we are able to adequately compute only three eigenvalues. The function `residchk` in the software distribution takes arguments A, V, D and for each eigenvalue $D(i,i)$ with associated eigenvector $V(:,i)$, and prints $\|AV(:, i) - D(i, i) V(:, i)\|_2$, $1 \leq i \leq \text{size}(D, 1)$. In addition, it returns the average of the residuals. This examples suppresses the output of the average.   ∎

```
>> x0 = rand(2233,1);
>> nev = 6;
>> m = 50;
>> tol = 1.0e-6;
>> maxiter = 100;
>> reorthog = 1;
[V,D] = eigsbexplicit(lshp2233, x0, nev, m, tol, maxiter, reorthog);
diag(D)

ans =
    6.9860
    6.9717
    6.9547
    6.9533
    6.9331
    6.9238

>> residchk(lshp2233,V,D)
Eigenpair 1 residual = 9.51284e-08
Eigenpair 2 residual = 6.16388e-08
Eigenpair 3 residual = 8.59643e-07
Eigenpair 4 residual = 5.34262e-07
Eigenpair 5 residual = 1.53693e-07
Eigenpair 6 residual = 3.43595e-08
>> norm(V'*lshp2233*V-D)
ans =
    2.6078e-07
>> x0 = rand(1030,1);
>> nev = 6;
>> m = 20;
>> [V,D] = eigsbexplicit(ORSIIR_1, x0, nev, m, tol, maxiter, reorthog);
>> residchk(ORSIIR_1,V,D)
Eigenpair 1 residual = 1.62177e-10
Eigenpair 2 residual = 6.45912e-09
Eigenpair 3 residual = 1.96075e-05
Eigenpair 4 residual = 9597.58
Eigenpair 5 residual = 9586.87
Eigenpair 6 residual = 9588.03
>> norm(V'*ORSIIR_1*V - D)
ans =
    9.597579997195124e+03
```

22.2.4   Restart Strategies

In our discussion so far, we have restarted with the current Ritz eigenvector $v_k^{(m)}$. We now discuss better strategies for the Arnoldi restart. Clearly, our best strategy is to select a vector v such that $H_m(i+1, i) = 0$. In this case, $Q(:, 1:i)$ is in an invariant subspace of dimension i. The Arnoldi iteration stops at iteration i, and the i eigenvalues are exact [3,

p. 126, 99]. Therefore, a reasonable strategy is to select a vector close to being in an invariant subspace of dimension less than or equal to m. One approach is to choose to restart with a vector that is a linear combination of the current Ritz vectors

$$\widetilde{v}_k^{(m)} = \sum_{i=1}^{m} c_i v_k^{(m)}.$$

This is a special case of the approach we will take that involves constructing a polynomial which, when applied to A, can improve the current Ritz vector prior to restarting Arnoldi. To understand how we choose such a polynomial, suppose that matrix A has n linearly independent eigenvectors $v_1, v_2, \ldots, v_n$ that form a basis for $\mathbb{R}^n$. If we choose any vector $w \in R^n$, then there are constants c_i such that

$$w = \sum_{i=1}^{n} c_i v_i.$$

We are interested in approximating k of these eigenvectors, so write the sum as

$$w = \sum_{i=1}^{k} c_i v_i + \sum_{i=k+1}^{n} c_i v_i.$$

Let p_r be a polynomial, compute $p_r(A)$, apply it to w and obtain

$$p_r(A)w = \sum_{i=1}^{k} c_i p_r(A) v_i + \sum_{i=k+1}^{n} c_i p_r(A) v_i.$$

This can be written as (Problem 22.2)

$$p_r(A)w = \sum_{i=1}^{k} c_i p_r(\lambda_i) v_i + \sum_{i=k+1}^{n} c_i p_r(\lambda_i) v_i. \tag{22.5}$$

If we build p_r so that $p_r(\lambda_i), k+1 \leq i \leq n$, are small relative to $p_r(\lambda_i), 1 \leq i \leq k$, then $p_r(A) w$ will be dominated by the eigenvalues in which we are interested. This is termed a *filtering polynomial*. There are a number of choices for the filtering polynomial. Saad [3, pp. 165-169] and Stewart [5, pp. 321-325] discuss some choices. Consider the polynomial

$$p_r(A) = \left(A - \lambda_{nev+1}^{(m)} I\right) \left(A - \lambda_{nev+2}^{(m)} I\right) \ldots \left(A - \lambda_m^{(m)} I\right), \tag{22.6}$$

where the $\{\lambda_i^{(m)}\}$, $nev + 1 \leq i \leq m$ are the Ritz values we do not want. This has the effect of making $p_r(A) w$ dominated by a linear combination of the eigenvectors we care about (Equation 22.5). However, its computation is too costly because we include all the unwanted eigenvalues $\{\lambda_i^{(m)}\}$, $nev + 1 \leq i \leq m$. Consider the factors $(A - \lambda_i^{(m)} I), nev + 1 \leq i \leq m$ as shifts away from the eigenvalues we do not want, as opposed to shifts closer to eigenvalues we do want. It turns out that Equation 22.6 can be evaluated quickly by using the implicit QR iteration described in Section 18.8, and this is the basis for the implicitly restarted Arnoldi method discussed in Section 22.3. Another approach is to use Chebyshev polynomial filters (see Ref. [3, pp. 169-178]).

22.3 THE IMPLICITLY RESTARTED ARNOLDI METHOD

In practice, the *implicitly restarted Arnoldi method* is used to find some eigenvalues/eigenvectors of a large sparse nonsymmetric matrix. The MATLAB function `eigs` uses a version of this algorithm. We will discuss the method, provide a simplified algorithm, and a MATLAB implementation. This implementation gives good results in many cases, but in no way competes with `eigs`.

In Section 22.2, we saw that an effective approach to estimating a few eigenpairs involves both restarting and deflation. A filter polynomial enhances convergence, and by using the Francis algorithm (implicitly shifted QR), the filtering polynomial 22.6 can be evaluated in only $O\left(m^2\right)$ flops. The following presentation derives from Saad [3, pp. 166-169]. We begin with an outline of the algorithm, and follow it with some of the details.

Outline of the Implicitly Restarted Arnoldi Method

Assume we are interested in *nev* eigenvalues, leaving $p = m - nev$ eigenvalues in which we are not interested. Begin with a unit vector v.

a. Compute the m step Arnoldi decomposition, $AQ_m = Q_m H_m + h_{m+1,m} q_{m+1} e_m^{\mathrm{T}}$.

b. Loop until computing *nev* eigenvalues or exceeding a maximum number of iterations

> *Find the m eigenpairs* $\left(\lambda_i, v_i^{(m)}\right)$ *of* H_m.
>
> *Compute the implicit QR decomposition using the $m - nev$ unwanted eigenvalues as the shifts. This evaluates the filter polynomial* $p\left(A\right) v_k^{(m)} = \left(A - \lambda_{nev+1} I\right)\left(A - \lambda_{nev+2} I\right) \ldots \left(A - \lambda_m I\right) v_k^{(m)} = \tilde{v}_k^{(m)}$ *that enhances the approximate eigenvector* $v_k^{(m)}$, *and at the same time builds an nev step Arnoldi decomposition. This decomposition is said to be compressed.*

c. If the eigenvalue λ_k corresponding to the eigenvector $\tilde{v}_k^{(m)}$ has converged, lock it. If convergence has not occurred or if eigenvalues remain to be found, extend the factorization to an m step factorization by applying $m - nev$ additional Arnoldi steps.

Assume that we have computed the Arnoldi decomposition

$$AV_m = V_m H_m + h_{m+1,m} v_{m+1} e_m^{\mathrm{T}}. \tag{22.7}$$

Using Equation 22.7, we have

$$\begin{aligned}\left(A - \lambda_{u_1}^{(m)} I\right) V_m &= AV_m - \lambda_{u_1}^{(m)} V_m \\ &= V_m H_m + h_{m+1,m} v_{m+1} e_m^{\mathrm{T}} - \lambda_{u_1}^{(m)} V_m \\ &= V_m \left(H_m - \lambda_{u_1}^{(m)} I\right) + h_{m+1,m} v_{m+1} e_m^{\mathrm{T}}. \end{aligned} \tag{22.8}$$

Compute the implicit QR decomposition of H_m with shift $\lambda_{u_1}^{(m)}$

$$H_m - \lambda_{u_1}^{(m)} I = Q_1 R_1.$$

Using it in Equation 22.8 results in

$$\left(A - \lambda_{u_1}^{(m)} I\right) V_m = V_m Q_1 R_1 + h_{m+1,m} v_{m+1} e_m^{\mathrm{T}} \tag{22.9}$$

$$\left(A - \lambda_{u_1}^{(m)} I\right)\left(V_m Q_1\right) = \left(V_m Q_1\right) R_1 Q_1 + h_{m+1,m} v_{m+1} e_m^{\mathrm{T}} Q_1 \tag{22.10}$$

$$A\left(V_m Q_1\right) = \left(V_m Q_1\right)\left(R_1 Q_1 + \lambda_{u_1}^{(m)} I\right) + h_{m+1,m} v_{m+1} e_m^{\mathrm{T}} Q_1 \tag{22.11}$$

Let

$$\begin{aligned} H_m^{(1)} &= R_1 Q_1 + \lambda_{u_1}^{(m)} I, \\ \left(b_{m+1}^{(1)}\right)^{\mathrm{T}} &= e_m^{\mathrm{T}} Q_1, \\ V_m^{(1)} &= V_m Q_1. \end{aligned}$$

Using these definitions in Equation 22.11, we have

$$AV_m^{(1)} = V_m^{(1)} H_m^{(1)} + h_{m+1,m} v_{m+1} \left(b_{m+1}^{(1)}\right)^{\mathrm{T}}. \tag{22.12}$$

We now make some observations about Equation 22.12:

- $H_m^{(1)} = R_1 Q_1 + \lambda_{u_1}^{(m)} I$ is the matrix that results from a step of the standard QR algorithm applied to H_m with shift $\lambda_{u_1}^{(m)}$.

- $H_m^{(1)}$ is an upper Hessenberg matrix.
- Equation 22.12 is essentially an Arnoldi decomposition with e_m^T replaced by $\left(b_{m+1}^{(1)}\right)^T$.
- The first column of $V_m^{(1)}$ is a multiple of $\left(A - \lambda_{u_1}^{(m)}I\right) v_1^{(m)}$, where $v_1^{(m)}$ is the first column of V_m. We determine this as follows:
  - Multiply Equation 22.9 by e_1:

$$\left(A - \lambda_{u_1}^{(m)}I\right) V_m e_1 = (V_m Q_1) R_1 e_1 + h_{m+1,m} v_{m+1} e_m^T e_1.$$

$V_m e_1 = v_1$, and R_1 is upper triangular, so

$$\left(A - \lambda_{u_1}^{(m)}I\right) v_1 = V_m^{(1)} \begin{bmatrix} r_{11} \\ 0 \\ \vdots \\ 0 \end{bmatrix} + 0 = r_{11} v_1^{(m)},$$

and the first column of $V_1^{(m)}$, $v_1^{(m)}$, is a multiple, $\frac{1}{r_{11}}$, of $\left(A - \lambda_{u_1}^{(m)}I\right) v_1$.

- The columns of V_m are orthonormal, and Q_1 is an orthogonal matrix, so the columns of $V_m^{(1)}$ are orthonormal (Problem 22.1).

Now apply the second shift to Equation 22.12 in the same fashion as we did in Equation 22.8:

$$\left(A - \lambda_{u_2}^{(m)}\right) V_m^{(1)} = V_m^{(1)} \left(H_m^{(1)} - \lambda_{u_2}^{(m)}I\right) + h_{m+1,m} v_{m+1} \left(b_{m+1}^{(1)}\right)^T. \tag{22.13}$$

Now, apply the implicit QR decomposition and obtain

$$H_m^{(1)} - \lambda_{u_2}^{(m)}I = Q_2 R_2,$$

and multiply Equation 22.13 by Q_2 on the right. This gives (Problem 22.3)

$$AV_m^{(2)} = V_m^{(2)} H_m^{(2)} + h_{m+1,m} v_{m+1} \left(b_{m+1}^{(2)}\right)^T, \tag{22.14}$$

where $H_m^{(2)} = R_2 Q_2 + \lambda_{u_2}^{(m)}I$ and $V_m^{(2)} = V_m^{(1)} Q_2$. Continuing the argument presented in Ref. [3, pp. 167-168], there results a decomposition of the form

$$A\hat{V}_{m-2} = \hat{V}_{m-2}\hat{H}_{m-2} + \hat{h}_{m+1,m}\hat{v}_{m-1}e_m^T.$$

This decomposition is exactly the one that would be obtained by executing $(m-2)$ steps of the Arnoldi process to the unit vector obtained from $\left(A - \lambda_{u_1}^{(m)}I\right)\left(A - \lambda_{u_2}^{(m)}I\right) v_1$. At this point, we can accept an eigenvalue or perform two more Arnoldi steps to obtain an m step Arnoldi decomposition.

The process we have described applies to a filter of degree 2. If a filter of degree $m - nev > 2$ is required, the algorithm results in an nev step Arnoldi decomposition. If necessary, apply $m - nev$ more Arnoldi steps to obtain an m step decomposition. Algorithm 22.1 specifies the implicit Arnoldi process. It features a version of the Arnoldi decomposition that returns V, H, and f such that $AV_m = V_m H_m + f e_m^T$. The algorithm either begins with a decomposition extending m steps or continues from a partial decomposition.

Remark 22.2. If an eigenvalue is real, perform a single shift during the implicit QR decomposition; otherwise, perform a double shift to obtain a complex conjugate pair of eigenvalues.

Algorithm 22.1 The Implicitly Restarted Arnoldi Process

```
function EIGSB(A,nev,m,tol,maxiter)
   k=1
   Allocate matrices Vⁿˣᵐ and Hᵐˣᵐ
   f=rand(n,1)
   f = f/‖f‖₂
   % perform an m step Arnoldi decomposition of A.
   [V, H, f]=arnoldif(A,V,H,f,k,m)
   iter=0
   while true do
      iter=iter+1
      % find the eigenvalues and eigenvectors of H
      [ UH, DH ] = eig(H)
      sigma=diag(UH)
      sort the eigenvalues from largest to smallest in magnitude
      Q = Iᵐˣᵐ
      % use sigma(nev+1), ..., sigma(m) as the shifts
      j=m
      while j ≥ nev+1 do
         lambda=sigma(j)
         alpha=imag(lambda)
         if |alpha| > 0 then
            % the eigenvalue is complex. use a double shift.
            beta=real(lambda)
            [ Qj, Rj ] = implicit double shift QR (beta,alpha)
            j=j-2
         else
            % the eigenvalue is real. use a single shift.
            [Qj, Rj] = implicit single shift QR (lambda)
            j=j-1
         end if
         H = QjᵀＨ*Qj
         Q = Q*Qj
      end while
      % compute the residual norm for the kth eigenpair.
      u=UH(:,k)
      residnorm = ‖f‖₂ |u(m)|
      % lock vₖ if the tolerance is obtained
      if residnorm < tol then
         if k<nev then
            k=k+1
         else
            return [ Vₘ(:,1 : nev), diag(sigma(1 : nev)) ]
         end if
      end if
      % build an m step decomposition from the nev step one.
      betak=H(nev+1,nev)
      sigmak=Q(m,nev)
      fk=V(:,nev+1)*betak+f*sigmak
      V(:,1:nev)=V(:,1:m)*Q(:,1:nev)
      [V, H, f]=arnoldif(A,V(:,1:nev),H(1:nev,1:nev),fk,nev+1,m)

      if iter ≥ maxiter then
         print(Error: could not compute nev eigenvalues within specified number of iterations.)
         terminate
      end if
   end while
end function
```

NLALIB: The function `eigsb` implements Algorithm 22.1. It is supported by the function `arnoldif` that computes the Arnoldi decomposition in the form given by Equation 22.1. `eigsb` returns a vector of `nev` eigenvalues or the eigenvectors in a matrix `V` and a diagonal matrix `D` with the `nev` corresponding eigenvalues. Recall that the function avgresid = residchk(A,V,D) prints the residuals $\|AV(:,i) - D(i,i)V(:,i)\|_2$, $1 \leq i \leq$ size $(D, 1)$, and returns the average of the residuals.

Example 22.4. The following table lists four nonsymmetric matrices from actual applications. `eigsb` was applied to each matrix using the defaults, tol $= 1.0 \times 10^{-6}$, maxiter $= 100$, *nevs* $= 6$. Experimentation was required to determine a value of m that resulted in the average residual given in the table.

Matrix	Dimensions	Application Area
rotor2	791×791	Large helicopter rotor model
qh882	961×961	Power systems simulations
dw256A	317×317	Square dielectric waveguide
TOLS1090	1090×1090	Aeroelasticity

Matrix	rotor2	qh882	dw256A	TOLS1090
m	35	40	35	16
Average residual	1.1125e−11	1.3725e−08	6.8321e−14	6.8211e−7

∎

22.3.1 Convergence of the Arnoldi Iteration

Convergence analysis of the Arnoldi iteration is complex. See Saad [3, pp. 151-159].

22.4 EIGENVALUE COMPUTATION USING THE LANCZOS PROCESS

As expected, a sparse symmetric matrix A has properties that will enable us to compute eigenvalues and eigenvectors more efficiently than we are able to do with a nonsymmetric sparse matrix. Also, much more is known about convergence properties for the eigenvalue computations. We begin with the following lemma and then use it to investigate approximate eigenpairs of A.

Lemma 22.1. *Let A be an $n \times n$ symmetric matrix. Let θ be a real number and x be an arbitrary vector in $\mathbb{R}^n$ with $x \neq 0$. Let $\mu = \|(A - \theta I)x\|_2 / \|x\|_2$. Then there is an eigenvalue of A in the interval $\theta - \mu \leq \lambda \leq \theta + \mu$.*

Proof. Let

$$A = PDP^{\mathrm{T}} = \sum_{i=1}^{n} \lambda_i p_i p_i^{\mathrm{T}}$$

be the spectral decomposition of A (Theorem 19.1). It follows that:

$$(A - \theta I)x = \left(PDP^{\mathrm{T}} - \theta PP^{\mathrm{T}}\right)x = \sum_{i=1}^{n}\left(\lambda_i p_i p_i^{\mathrm{T}} - \theta p_i p_i^{\mathrm{T}}\right)x$$

$$= \sum_{i=1}^{n}\left(\lambda_i \left(p_i^{\mathrm{T}}x\right)p_i - \theta\left(p_i^{\mathrm{T}}x\right)p_i\right)$$

$$= \sum_{i=1}^{n}(\lambda_i - \theta)\left(p_i^{\mathrm{T}}x\right)p_i$$

Taking the norm of the equality and noting that the $\{p_i\}$ are orthonormal, we obtain

$$\|(A - \theta I)x\|_2^2 = \sum_{i=1}^{n}(\lambda_i - \theta)^2\left(p_i^{\mathrm{T}}x\right)^2.$$

Note that $\sum_{i=1}^{n} p_i p_i^T = I$ (Problem 22.4). Let λ_k be the eigenvalue closest to θ, i.e., $|\lambda_k - \theta| \leq |\lambda_i - \theta|$ for all i, and we have

$$
\begin{aligned}
\|(A - \theta I) x\|_2^2 &\geq (\lambda_k - \theta)^2 \sum_{i=1}^{n} (p_i^T x)^2 \\
&= (\lambda_k - \theta)^2 \sum_{i=1}^{n} (p_i^T x p_i^T x) = (\lambda_k - \theta)^2 \sum_{i=1}^{n} (x^T p_i p_i^T x) \\
&= (\lambda_k - \theta)^2 x^T \left(\sum_{i=1}^{n} p_i p_i^T \right) x = (\lambda_k - \theta)^2 x^T I x \\
&= (\lambda_k - \theta)^2 \|x\|_2^2
\end{aligned}
$$

This implies that

$$
\mu = \|(A - \theta I) x\|_2 / \|x\|_2 \geq |\lambda_k - \theta|,
$$

and so there is an eigenvalue λ_k in the interval $\theta - \mu \leq \lambda \leq \theta + \mu$. $\qquad\square$

Recall that the Lanczos process for a symmetric matrix discussed in Section 21.8 is the Arnoldi process for a symmetric matrix and takes the form

$$
AQ_m = Q_m T_m + h_{m+1,m} q_{m+1} e_m^T,
$$

where

$$
T_m = \begin{bmatrix}
\alpha_1 & \beta_1 & & & \\
\beta_1 & \alpha_2 & \beta_2 & & \\
& \beta_2 & \ddots & \ddots & \\
& & \ddots & \alpha_{m-1} & \beta_{m-1} \\
& & & \beta_{m-1} & \alpha_m
\end{bmatrix},
$$

is symmetric tridiagonal, and Q_m is orthogonal. We will proceed like we did for nonsymmetric matrices and use Ritz pairs of T_m to approximate eigenpairs of A. Let $\mu = \lambda_i^{(m)}$ be a Ritz value and $u_i^{(m)}$ be a corresponding eigenvector obtained from T_m so that $T_m u_i^{(m)} = \lambda_i^{(m)} u_i^{(m)}$, and let $v_i^{(m)} = Q_m u_i^{(m)}$ be the Ritz vector. Applying the same operations we used to derive Equation 22.3, we obtain

$$
\left\| \left(A - \lambda_i^{(m)} I \right) v_i^{(m)} \right\|_2 = h_{m+1,m} \left| \left(u_i^{(m)} \right)_m \right|.
$$

Since $v_i^{(m)} = Q_m u_i^{(m)}$, $\left\| u_i^{(m)} \right\|_2 = 1$, and Q_m is orthogonal, we have $\left\| v_i^{(m)} \right\|_2 = 1$, and so

$$
\mu = \left\| \left(A - \lambda_i^{(m)} I \right) v_i^{(m)} \right\| / \left\| v_i^{(m)} \right\| = h_{m+1,m} \left| \left(u_i^{(m)} \right)_m \right|.
$$

It follows from Lemma 22.1 that there is an eigenvalue λ such that

$$
\lambda_i^{(m)} - \mu \leq \lambda \leq \lambda_i^{(m)} + \mu,
$$

so

$$
-\mu \leq \lambda - \lambda_i^{(m)} \leq \mu,
$$

and

$$
\left| \lambda - \lambda_i^{(m)} \right| \leq h_{m+1,m} \left| \left(u_i^{(m)} \right)_m \right|. \tag{22.15}
$$

Equation 22.15 indicates we will have a good approximation to an eigenpair of A *as* long as $h_{m+1,m}\left|\left(u_i^{(m)}\right)_m\right|$ is small. We had not such a guarantee for a nonsymmetric matrix.

Poor convergence can result from a bad choice of the starting vector, so a random vector is a good choice. Multiple eigenvalues or eigenvalues that are very close to each other particularly cause problems. As discussed in Section 21.8.1, roundoff error can cause lack or orthogonality among the Lanczos vectors, and this happens as soon as Ritz vectors have converged accurately enough to eigenvectors [2, p. 565]. The loss of orthogonality can cause simple eigenvalues to appear as multiple eigenvalues, and these are called *ghost eigenvalues* [2, p. 566] (Problems 22.6 and 22.14). In our implementation of the implicitly restarted Lanczos process, we will perform full reorthogonalization. The function eigssymb in the software distribution implements the implicitly restarted Lanczos process. The only real differences between this function and eigsb is the use of the Lanczos decomposition instead of Arnoldi and the fact that only a single shift is necessary since the matrix has real eigenvalues.

Example 22.5. The very ill-conditioned (approximate condition number 2.5538×10^{17}) symmetric 60000×60000 matrix Andrews obtained from the Florida sparse matrix collection was used in a computer graphics/vision problem. As we know, even though the matrix is ill-conditioned, its eigenvalues are well conditioned (Theorem 19.2). The MATLAB statements time the approximation of the six largest eigenvalues and corresponding eigenvectors using eigsymb. A call to residchk outputs the residuals.

```
>> tic;[V, D] = eigssymb(Andrews, 6, 50);toc;
Elapsed time is 5.494509 seconds.
>> residchk(Andrews,V,D)
Eigenpair 1 residual = 3.59008e-08
Eigenpair 2 residual = 1.86217e-08
Eigenpair 3 residual = 2.31836e-08
Eigenpair 4 residual = 7.68169e-08
Eigenpair 5 residual = 4.10453e-08
Eigenpair 6 residual = 6.04127e-07

ans =
    1.332825500776470e-07
```
∎

22.4.1 Mathematically Provable Properties

This section presents some theoretical properties that shed light on the use and convergence properties of the Lanczos method.

In Ref. [76, p. 245], Scott proves the following theorem:

Theorem 22.1. *Let A be a symmetric matrix with eigenvalues $\lambda_1 \leq \lambda_2 \leq \ldots \leq \lambda_n$, $\delta_A = min_{k\neq i}|\lambda_i - \lambda_k|$. Then there exists a starting vector v_0 such that for the exact Lanczos algorithm applied to A with v_0, at any step $j < n$ the residual norm*

$$\|Av_i - \lambda_i v_i\|_2$$

of any Ritz pair (λ_i, v_i) will be larger than $\delta_A/4$.

This theorem says that if the spectrum of A is such that $\delta_A/4$ is larger than some given convergence tolerance, then there exist poor starting vectors which delay convergence until the nth step. Thus, the starting vector is a critical component in the performance of the algorithm. If a good starting vector is not known, then use a random vector. We have used this approach in our implementation of eigssymb.

The expository paper by Meurant and Stratkos [70, Theorem 4.3, p. 504] supports the conclusion that orthogonality can be lost only in the direction of converged Ritz vectors. This result allows the development of sophisticated methods for maintaining orthogonality such as selective reorthogonalization [2, pp. 565-566], [6, pp. 116-123].

There are significant results concerning the rate of convergence of the Lanczos algorithm. Kaniel [77] began investigating these problems. Subsequently, the finite precision behavior of the Lanczos algorithm was analyzed in great depth by Chris Paige in his Ph.D. thesis [78]; see also Paige [79–81]. He described the effects of rounding errors in the Lanczos algorithm using rigorous and elegant mathematical theory. The results are beyond the scope of this book, so we will assume exact arithmetic in the following result concerning convergence proved in Saad [3, pp. 147-150].

Theorem 22.2. *Let A be an n-by-n symmetric matrix. The difference between the ith exact and approximate eigenvalues* λ_i *and* $\lambda_i^{(m)}$ *satisfies the double inequality*

$$0 \le \lambda_i - \lambda_i^{(m)} \le (\lambda_1 - \lambda_n) \left(\frac{\kappa_i^{(m)} \tan \theta \, \langle v_1, u_i \rangle}{C_{m-i} (1 + 2\gamma_i)} \right)^2,$$

where

$$C_{m-i}(x)$$

is the Chebyshev polynomial of degree $m - i$,

$$\gamma_i = \frac{\lambda_i - \lambda_{i+1}}{\lambda_{i+1} - \lambda_n},$$

$\kappa_i^{(m)}$ *is given by*

$$\kappa_1^{(m)} = 1, \quad \kappa_i^{(m)} = \prod_{j=1}^{i-1} \frac{\lambda_j^{(m)} - \lambda_n}{\lambda_j^{(m)} - \lambda_i}, \ i > 1,$$

and θ *is the angle defined in Ref. [3, p. 147].*

Remark 22.3. Error bounds indicate that for many matrices and for relatively small m, several of the largest or smallest of the eigenvalues of A are well approximated by eigenvalues of the corresponding Lanczos matrices. In practice, it is not always the case that both ends of the spectrum of a symmetric matrix are approximated accurately. However, it is generally true that at least one end of the spectrum is approximated well.

22.5 CHAPTER SUMMARY

The Power Method

The power method generates the Krylov subspace $\mathcal{K}_m (A, x_0) = \text{span} \left\{ x_0 \ Ax_0 \ A^2x_0 \ \dots \ A^{k-2}x_0 \ A^{k-1}x_0 \right\}$, where $A^{k-1}x_0$ is the approximate eigenvector corresponding to the largest eigenvalue. This method can be effective in some cases; however, its main importance is that it leads to ideas that use a combination of vectors from the Krylov subspace.

Eigenvalue Computation Using the Arnoldi Process

Given a nonsymmetric matrix A, the basic idea is simple:

Perform an Arnoldi decomposition, $AV_m = V_m H_m + h_{m+1,m} v_{m+1} e_m^{\mathrm{T}}$, *and use some eigenvalues* $\left\{ \lambda_1^{(m)}, \lambda_2^{(m)}, \dots, \lambda_k^{(m)} \right\}$ *of* H_m *as approximations to the eigenvalues of* A. *The corresponding eigenvectors are* $V_m u_i^{(m)}$, *where* $u_i^{(m)}$ *is the eigenvector of* H_m *corresponding to* $\lambda_i^{(m)}$.

However, the implementation is not simple. The section discusses three approaches:

- Compute a few eigenvalues of H_m, sort them in descending order, and estimate the error of each using Equation 22.3. Accept the eigenvalues and optionally the corresponding eigenvectors that satisfy an error tolerance. Basically, "You get what you get."
- Accept eigenvalues satisfying the error tolerance and otherwise restart Arnoldi with the current Ritz vector or some improvement of it.
- Restart until computing eigenvalue λ_1, deflate the matrix and search for λ_2, and continue until computing the desired eigenvalues.

The Implicitly Restarted Arnoldi Method

To approximate *nev* eigenvalues and corresponding eigenvectors, use deflation and a filter polynomial such as $p_r (A) v_k^{(m)} = \left(A - \lambda_{nev+1}^{(m)}I\right) \left(A - \lambda_{nev+2}^{(m)}I\right) \dots \left(A - \lambda_m^{(m)}I\right) v_k^{(m)}$ to provide a better restart vector. Evaluate this polynomial using the implicitly shifted QR algorithm, which requires only $O\left(m^2\right)$ flops. This is the most commonly used method for computing a few eigenvalues of a large sparse matrix.

Eigenvalue Computation Using the Lanczos Process

This section discusses the implicitly shifted QR algorithm for computing a few eigenvalues of a large sparse symmetric matrix using the Lanczos decomposition. The Lanczos matrix is symmetric tridiagonal, and its columns tend to lose orthogonality during the Lanczos iteration, resulting in inaccurate eigenvalues. Full reorthogonalization of each new vector against the already computed ones is computationally expensive but solves the problem. Other methods such as partial reorthogonalization can be used to speed up the algorithm and retain accuracy.

Matrix symmetry has led to extensive results about the algorithm performance. For instance, the choice of the starting vector is critical, and orthogonality can be lost only in the direction of converged Ritz vectors. Also, there are results that specify convergence properties of the method.

22.6 PROBLEMS

22.1 Assume that the columns of matrix V are orthonormal and Q is an orthogonal matrix. Prove that the columns of VQ are orthonormal.

22.2 Develop Equation 22.5.

22.3 Develop Equation 22.14.

22.4 Show that $\sum_{i=1}^{n} u_i u_i^{\mathrm{T}} = I$ if $\{u_i\}$ is an orthonormal basis for $\mathbb{R}^n$. Proceed in stages.

Let v be an arbitrary vector in $\mathbb{R}^n$ and consider the product $\left(\sum_{j=1}^{n} u_i u_i^{\mathrm{T}}\right) v$. There exist constants $\{c_j\}$ so that $v = \sum_{j=1}^{n} c_j u_j$.

  a. Show that $\left(\sum_{i=1}^{n} u_i u_i^{\mathrm{T}}\right) v = v$.

  b. Argue that (a) implies $\sum u_i u_i^{\mathrm{T}} = I$.

22.5 This is a restatement of Exercise 6.4.36 in Ref. [23].

  a. Assume that λ is an approximate eigenvalue for A with corresponding approximate eigenvector v, $\|v\|_2 = 1$. Form the residual $r = Av - \lambda v$, let $\epsilon = \|r\|_2$ and $E = -r v^{\mathrm{T}}$. Show that (λ, v) is an eigenpair of $A + E$ and $\|E\|_2 = \epsilon$.

  b. Argue that (λ, v) is an exact eigenpair of a matrix that is close to A.

  c. Is (λ, v) a good approximate eigenpair of A in the sense of backward error?

22.6 If a symmetric tridiagonal matrix is unreduced (no zeros on the subdiagonal and thus none on the superdiagonal), it must have distinct real eigenvalues (Problem 19.1). When testing for ghost eigenvalues, why is knowing this result important?

22.6.1 MATLAB Problems

22.7 Using the function `biharmonic_op` in the software distribution, build a block pentadiagonal matrix of size $10,000 \times 10,000$. Use `eigs` to compute the six largest eigenvalues in magnitude, and then apply the power method in an attempt to compute the largest eigenvalue. Explain why you have great difficulty or fail to compute an accurate result.

22.8 Carry out the numerical experiment described in Example 22.2 and construct a plot like that in Figure 22.2.

22.9 Implement a function, `eigsimple`, following the outline in Section 22.2.1. It should return just approximate eigenvalues or eigenvectors and a diagonal matrix of eigenvalues, depending on the number of output arguments. Test it on the nonsymmetric matrices qh882 and TOLS340 by estimating up to six eigenpairs and checking them with `residchk`. Hint: For qh882, use $m = 200$, and for TOLS1090, use $m = 350$.

22.10 Implement a function, `eigsrestart`, following the outline in Section 22.2.2. Test it on the nonsymmetric matrices bfwa398 and west2021 by estimating maximum of six eigenpairs and checking them with `residchk`. Use $m = 100$.

22.11 Test the function eigsbexplicit on the nonsymmetric matrices steam2, ORSIIR_1, and the symmetric matrix DIMACS10. You will need to determine appropriate values of m for each matrix. Try to estimate six eigenpairs and check them with `residchk`. You might not be able to find six.

22.12 Look at the code for eigssymb and lanczosf and determine how to turn off reorthogonalization. Run your modified code with the symmetric matrix lshp2233 and test the results using `residchk`. Turn reorthogonalization back on and execute the same steps. Explain the results.

22.13 This problem investigates ghost eigenvalues.

    a. The matrix `ghosttest` in the book software distribution is a 100×100 diagonal matrix with `ghosttest(1,1)` = 100 and `ghosttest(100,100)` = 10. The remaining diagonal elements are in the range (0, 1). The function `lanczosfplot` produces a plot of the Lanczos iteration number vs. the eigenvalues of T. Run it with `m` = 35 and `reorthog` = 0 and see if you can identify ghost eigenvalues. Run it again with `reorthog` = 1 and compare the results.

    b. Create the diagonal matrix defined in Example 21.10 using $n = 50$, lambda_1 = 100, lambda_n = 0.2, and $\rho = 0.9$. Repeat part (a) for this matrix using `m` = 50.

22.14 Modify `eigssymb` so it computes either the *nev* largest or the *nev* smallest eigenvalues of a large sparse symmetric matrix by adding a parameter, direction, that has values "*L*" or "*S*." Name the function `eigssymb2`. Test your implementation by finding the six largest and the six smallest eigenvalues of the symmetric matrices SHERMAN1.mat and Andrews.mat in the software distribution. You will need to experiment with *m* and maxiter to obtain satisfactory results.

Chapter 23

Computing the Singular Value Decomposition

You should be familiar with

- The SVD theorem
- Jacobi rotations

In Chapter 15, we proved the singular value decomposition (SVD) by construction, discussed some information it provides about a matrix and showed how to use the SVD in image compression. In subsequent chapters, we applied the SVD to least squares and other problems. However, a significant issue remains. How do we efficiently compute the SVD? The chapter develops two algorithms for its computation. We begin with the one-sided Jacobi method, since it is based upon the use of Jacobi rotations, very similar to those we used in Chapter 19 in the computation of the eigenvalues of a symmetric matrix. We will then discuss the Demmel and Kahan Zero-shift QR Downward Sweep algorithm. This method involves using Householder reflections to transform any $m \times n$ matrix to a bidiagonal matrix. The bidiagonal matrix is then reduced to a diagonal matrix containing the singular values using bulge chasing, a technique presented in Section 18.8.

23.1 DEVELOPMENT OF THE ONE-SIDED JACOBI METHOD FOR COMPUTING THE REDUCED SVD

The SVD can be computed in the following way:

Find the singular values of A by computing the eigenvalues and orthonormal eigenvectors for $A^{\mathrm{T}}A$. Place the square roots of the positive eigenvalues on the diagonal of the matrix $\widetilde{\Sigma}$ in order from greatest to least and fill all the other entries with zeros. These normalized eigenvectors form V. Find orthonormal eigenvectors of AA^{T}. These form the columns of U.

This is a slow and potentially inaccurate means of finding the SVD. Roundoff errors can be introduced into the computation of $A^{\mathrm{T}}A$ that alter the correct eigenvalues. Here is an example.

Example 23.1. Let $A = \begin{bmatrix} 3.0556 & 3.0550 \\ 3.0550 & 3.0556 \end{bmatrix}$. The singular values of A are 6.1106 and 0.0006. Now compute $A^{\mathrm{T}}A$ using six-digit arithmetic and obtain

$$A^{\mathrm{T}}A = \begin{bmatrix} 18.6697 & 18.6697 \\ 18.6697 & 18.6697 \end{bmatrix}.$$

The eigenvalues of $A^{\mathrm{T}}A$ are 6.1106 and 0.0000, as opposed to 6.1106 and 0.0006. ∎

In this section, we will develop an algorithm for the reduced SVD based on Jacobi rotations, since we are already familiar with this approach to compute the eigenvalues of a symmetric matrix. The algorithm, known as the *one-sided Jacobi algorithm*, will generally give good results.

We need to avoid having to compute $A^{\mathrm{T}}A$ so we take an approach that will indirectly perform computations on $A^{\mathrm{T}}A$ while actually working with a sequence of seemingly different problems. We use a sequence of Jacobi rotations that will make columns $i, j, i < j$ of $AJ(i, j, c, s)$ orthogonal. The presence of $J(i, j, c, s)$ to the right of A is the reason the algorithm is called one-sided Jacobi. Consider the product

Numerical Linear Algebra with Applications. http://dx.doi.org/10.1016/B978-0-12-394435-1.00023-5

which yields the matrix

Require that the vectors in columns i and j be orthogonal.

$$\left\langle \begin{bmatrix} ca_{1i} - sa_{1j} \\ \vdots \\ ca_{ii} - sa_{ij} \\ \vdots \\ ca_{ji} - sa_{ji} \\ \vdots \\ ca_{ni} - sa_{nj} \end{bmatrix}, \begin{bmatrix} sa_{1i} + ca_{1j} \\ \vdots \\ sa_{ii} + ca_{ij} \\ \vdots \\ sa_{ji} + ca_{jj} \\ \vdots \\ sa_{ni} + ca_{nj} \end{bmatrix} \right\rangle = 0. \tag{23.1}$$

Form the inner product in Equation 23.1 to obtain

$$(ca_{1i} - sa_{1j})(sa_{1i} + ca_{1j}) + \cdots + (ca_{ii} - sa_{ij})(sa_{ii} + ca_{ij}) + \cdots + (ca_{ji} - sa_{ji})(sa_{ji} + ca_{jj}) + \cdots \tag{23.2}$$

$$+ (ca_{ni} - sa_{nj})(sa_{ni} + ca_{nj}) = 0. \tag{23.3}$$

After some algebra, Equation 23.3 transforms to

$$(c^2 - s^2) \sum_{k=1}^{n} a_{ki}a_{kj} + cs \left[\sum_{k=1}^{n} a_{ki}^2 - \sum_{k=1}^{n} a_{kj}^2 \right] = 0,$$

and so

$$\frac{c^2 - s^2}{cs} = \frac{\sum_{k=1}^{n} a_{kj}^2 - \sum_{k=1}^{n} a_{ki}^2}{\sum_{k=1}^{n} a_{ki}a_{kj}}. \tag{23.4}$$

Proceed like we did with Equation 19.2, except that the right-hand side is different. The result is

$$t^2 + 2\tau t - 1 = 0,$$

where

$$\tau = \frac{1}{2} \frac{\sum_{k=1}^{n} a_{kj}^2 - \sum_{k=1}^{n} a_{ki}^2}{\sum_{k=1}^{n} a_{ki}a_{kj}},$$

and $s = ct$. Table 23.1 provides a summary of the required computations.

TABLE 23.1 Computation of c and s for the Jacobi One-Sided Method

$$\tau = \frac{1}{2} \frac{\sum_{k=1}^{n} a_{kj}^2 - \sum_{k=1}^{n} a_{ki}^2}{\sum_{k=1}^{n} a_{ki}a_{kj}}$$

$$t = \begin{cases} \dfrac{1}{\tau+\sqrt{\tau^2+1}}, & \tau \geq 0 \\[2ex] \dfrac{-1}{-\tau+\sqrt{\tau^2+1}}, & \tau < 0 \end{cases}$$

$$c = \frac{1}{\sqrt{1+t^2}}$$

$$s = ct$$

Now, what does this computation have to do with $A^{\mathrm{T}}A$? Require that the rotation $J(i,j,c,s)^{\mathrm{T}} A^{\mathrm{T}}AJ(i,j,c,s)$ zero-out the off diagonal entries at indices (i,j) and (j,i) of the symmetric matrix $A^{\mathrm{T}}A$. The entries of $A^{\mathrm{T}}A$ at indices $(i,i), (i,j), (j,i)$, and (j,j) are shown in the following matrix:

$$A^{\mathrm{T}}A = \begin{array}{c} \\ i \\ \\ j \\ \\ \end{array} \overset{\displaystyle i \qquad\qquad j}{\left[\begin{array}{ccc} \cdots & \cdots & \cdots \\ \sum_{k=1}^{n} a_{ki}^2 & & \sum_{k=1}^{n} a_{ki}a_{kj} \\ \cdots & \cdots & \cdots \\ \sum_{k=1}^{n} a_{ki}a_{kj} & & \sum_{k=1}^{n} a_{kj}^2 \\ & \cdots & \end{array} \right]}$$

To zero-out $\left(A^{\mathrm{T}}A\right)_{ji}$ and $\left(A^{\mathrm{T}}A\right)_{ij}$ by forming $J(i,j,c,s)^{\mathrm{T}} A^{\mathrm{T}}AJ(i,j,c,s)$, proceed just as we did in Section 19.1, substituting $\sum_{k=1}^{n} a_{ki}^2$ for a_{ii}, $\sum_{k=1}^{n} a_{kj}^2$ for a_{jj}, and $\sum_{k=1}^{n} a_{ki}a_{kj}$ for a_{ji} and a_{ij}, and apply the results in Table 19.2. The values obtained are the same as those in Table 23.1. Choosing c and s so that columns i and j of $AJ(i,j,c,s)$ are orthogonal zeros-out the entries at indices (i,j) and (j,i) of $A^{\mathrm{T}}A$.

The algorithm now proceeds as follows. Start with A, and apply a sequence of right Jacobi rotations until the result is a matrix $\overline{U}$ with "nearly orthogonal" columns

$$AJ_1J_2J_3\ldots J_k = \overline{U}. \tag{23.5}$$

Performing the Jacobi rotations given in Equation 23.5 is actually performing orthogonal similarity transformations on $A^{\mathrm{T}}A$, producing a matrix with the eigenvalues of $A^{\mathrm{T}}A$ on its diagonal.

$$J_k^{\mathrm{T}}\ldots J_2^{\mathrm{T}}J_1^{\mathrm{T}}A^{\mathrm{T}}AJ_1J_2\ldots J_k \approx \Sigma^2, \tag{23.6}$$

$$\Sigma = \begin{bmatrix} \sigma_1 & & & 0 \\ & \sigma_2 & & \\ & & \ddots & \\ 0 & & & \sigma_n \end{bmatrix},$$

where the σ_i, $1 \leq i \leq n$, are the singular values of A. Let V be the orthogonal matrix $V = J_1J_2J_3\ldots J_k$, so Equation 23.5 can be written as

$$AV = \overline{U}, \tag{23.7}$$

and

$$A = \overline{U}V^{\mathrm{T}}. \tag{23.8}$$

From Equation 23.8, $A^{\mathrm{T}} = V\overline{U}^{\mathrm{T}}$. Use this result in Equation 23.6 to obtain

$$J_k^{\mathrm{T}}\ldots J_2^{\mathrm{T}}J_1^{\mathrm{T}}V\overline{U}^{\mathrm{T}}AJ_1J_2\ldots J_k \approx \Sigma^2.$$

Now, $AJ_1J_2\ldots J_k = AV = \overline{U}$ from Equation 23.7, so

$$J_k^{\mathrm{T}}\ldots J_2^{\mathrm{T}}J_1^{\mathrm{T}}V\overline{U}^{\mathrm{T}}\overline{U} \approx \Sigma^2$$

Since $V = J_1J_2J_3\ldots J_k$, we have

$$J_k^{\mathrm{T}}\ldots J_2^{\mathrm{T}}J_1^{\mathrm{T}}J_1J_2J_3\ldots J_k\overline{U}^{\mathrm{T}}\overline{U} \approx \Sigma^2$$

and

$$\overline{U}^{\mathrm{T}}\overline{U} = \Sigma^2. \tag{23.9}$$

Assuming that the columns of $\overline{U}$ are orthogonal, write it in the form $(\overline{u}_1\overline{u}_2\ldots\overline{u}_n)$, where the $\overline{u}_i$ are orthogonal, and Equation 23.9 can be written as follows:

$$\begin{bmatrix} \|\overline{u}_1\|_2^2 & & & 0 \\ & \|\overline{u}_2\|_2^2 & & \\ & & \ddots & \\ 0 & & & \|\overline{u}_n\|_2^2 \end{bmatrix} = \begin{bmatrix} \sigma_1^2 & & & 0 \\ & \sigma_2^2 & & \\ & & \ddots & \\ 0 & & & \sigma_n^2 \end{bmatrix}. \tag{23.10}$$

Keep in mind that the columns of $\overline{U}$ are actually nearly orthogonal, so there will most likely be small entries off the diagonal. Equation 23.10 says that the 2-norm of the columns of $\overline{U}$ is approximately the singular values of A. Since $\frac{\overline{u}_i}{\sigma_i}$ is a unit vector, $U = \left(\frac{\overline{u}_1}{\sigma_1}\ \frac{\overline{u}_2}{\sigma_2}\ \cdots\ \frac{\overline{u}_n}{\sigma_n}\right)$ is an orthogonal matrix, and $\overline{U} = U\Sigma$. Note that $\overline{U} = U\Sigma$, and by using this in Equation 23.8, we have

$$A = U\Sigma V^{\mathrm{T}},$$

the SVD of A.

We know that the Jacobi method applied to the symmetric matrix $A^{\mathrm{T}}A$ converges to a diagonal matrix containing its eigenvalues (Theorem 19.3). We stated to continue the Jacobi algorithm for the SVD until $AJ_1J_2J_3\ldots J_k$ is "nearly orthogonal." What test do we use to measure the extent of orthogonality, and will this test guarantee that the eigenvalues of $A^{\mathrm{T}}A$ are computed accurately? Let $\overline{u}_i$ and $\overline{u}_j$ be column vectors of $\overline{U}$. The error tolerance test is that the maximum value of expression 23.11 for all i, j in the current sweep is less than a prescribed tolerance.

$$\left|\left\langle \frac{u_i}{\|u_i\|_2}, \frac{u_j}{\|u_j\|_2} \right\rangle\right|. \tag{23.11}$$

This says that the inner product of the normalized columns of $\overline{U}$ should be small. A proof that this criterion leads to convergence can be found in the paper *Jacobi's method is more accurate than QR*, by Demmel and Veselić [82] and in a 1989 report by the same authors that can be found at http://www.netlib.org/lapack/lawnspdf/lawn15.pdf. This paper shows that Jacobi can compute small singular values with better relative accuracy than other commonly used methods.

23.1.1 Stability of Singular Value Computation

We have seen that the computation of the eigenvalues of a nonsymmetric matrix A can be ill-conditioned. A natural question to ask is whether the same is true for the computation of singular values. $A^{\mathrm{T}}A$ is symmetric, and so we know that the condition numbers of the eigenvalues of $A^{\mathrm{T}}A$ are one. However, theoretically we have to deal with a product of two matrices, and roundoff error will be present. Assuming that U and V have orthonormal columns, suppose we introduce errors δA into A, resulting in errors $\delta\Sigma$ in Σ. Then, $A + \delta A = U(\Sigma + \delta\Sigma)V^{\mathrm{T}}$, and $\Sigma + \delta\Sigma = U^{\mathrm{T}}(A + \delta A)V$. Orthogonal matrices preserve norms, so $\|\Sigma + \delta\Sigma\|_2 = \|A + \delta A\|_2$, and perturbations in A cause perturbations of roughly the same size in its singular values, so the computation of singular values is well conditioned. To this effect, see Ref. [19, pp. 366-367], where a proof of the following theorem is provided.

Theorem 23.1. *Let A and $A + E$ be $m \times n$ matrices, $m \geq n$. Let $\sigma_1 \geq \sigma_2 \geq \cdots \geq \sigma_n$ and $\tilde{\sigma}_1 \geq \tilde{\sigma}_2 \geq \cdots \tilde{\sigma}_n$ be, respectively, the singular values of A and $A + E$. Then $|\sigma_i - \tilde{\sigma}_i| \leq \|E\|_2$ for each i.*

Of course for Theorem 23.1 to be useful, E must be small. In addition, there are issues with small singular values. For a discussion of problems with small singular values and other singular value perturbation results, see Ref. [83].

23.2 THE ONE-SIDED JACOBI ALGORITHM

Algorithm 23.1 implements the one-sided Jacobi method for computing the reduced SVD. There are some notes you will need before reading the algorithm.

- Like the Jacobi algorithm for finding the eigenvalues of a real symmetric matrix, Algorithm 23.1 uses the cyclic-by-row method.
- Before performing an orthogonalization step, the norms of columns i and j of U are compared. If the norm of column i is less than that of column j, the two columns are switched. This necessitates swapping the same columns of V as well. This action assures that the singular values in S appear in decreasing order.
- The norms of the final columns of U are the approximation to the singular values. If a norm is less than the machine precision eps, it is assumed that the singular value is zero.
- If the matrix A has a row or column of zeros, the algorithm produces a decomposition $A = U\Sigma V^{\mathrm{T}}$, but U is not orthogonal, since it will have a row or column of zeros.

Algorithm 23.1 One-Sided Jacobi Algorithm

```
function JACOBISVD(A,tol,maxsweeps)
  % One-sided Jacobi method for computing the reduced SVD of
  % an m × n matrix.
  % Input: Matrix A, error tolerance tol, and the
  % maximum number of sweeps, maxsweeps.
  % Output: m × n orthogonal matrix U, n × n diagonal matrix Σ
  % containing the singular values of A in decreasing order, an
  % n × n orthogonal matrix V, and numsweeps, the number of sweeps required.
  % If the error tolerance is not obtained in maxsweeps, numsweeps = -1.
  U = A
  V = I
  singvals = 0
```
$$tmp = \begin{bmatrix} 1 & 1 & \dots & 1 & 1 \end{bmatrix}^{\mathrm{T}}$$
```
  errormeasure = tol + 1
  numsweeps = 0
  while (errormeasure ≥ tol) and (numsweeps ≤ maxsweeps) do
    numsweeps = numsweeps + 1
    for i = 1:n-1 do
      errormeasure = 0
      for j = i+1:n do
        normcoli = ‖U(:, i)‖₂
        normcolj = ‖U(:, j)‖₂
        if normcoli < normcolj then
          % Assure the singular values will appear in decreasing order in S.
          swap columns i and j of U and V
        end if
```
$$\alpha = \sum_{k=1}^{m} u_{ki}^2$$
$$\beta = \sum_{k=1}^{m} u_{kj}^2$$
$$\gamma = \sum_{k=1}^{m} u_{ki} u_{kj}$$
```
        if αβ ≠ 0 then
```
$$errormeasure = \max\left(errormeasure, \frac{|\gamma|}{\sqrt{\alpha\beta}}\right)$$
```
        end if
        % compute Jacobi rotation that makes columns i and j of U
        % orthogonal and also zeros-out (AᵀA)ᵢⱼ and (AᵀA)ⱼᵢ
```

```
             if γ ≠ 0 then
                 ζ = β-α/2γ
                 if ζ ≥ 0 then
                     t = 1/|ζ|+√(1+ζ²)
                 else
                     t = - 1/|ζ|+√(1+ζ²)
                 end if
                 c = 1/√(1+t²)   s = ct
             else
                 c = 1
                 s = 0
             end if
             % update columns i and j of U.
             t = U(:,i)
             U(:,i) = ct - s*U(:,j)
             U(:,j) = st + c*U(:,j)
             % update matrix V of right singular vectors.
             t = V(:,i)
             V(:,i) = ct - sV(:,j)
             V(:,j) = st + cV(:,j)
         end for
     end for
 end while
 % The singular values are the norms of the columns of U.
 % The left singular vectors are the normalized columns of U.
 for j = 1:n do
     singvalsⱼ = ‖U(:, j)‖₂
     if singvalsⱼ > eps then
         U(:, j) = U(:, j)/singvalsⱼ
     end if
 end for
 Σ = diag(singvals)
 if errormeasure ≥ tol then
     numsweeps = -1
 end if
end function
```

NLALIB: The function jacobisvd implements Algorithm 23.1. Its return values can be one of three forms:

a. [U, S, V, maxsweeps] = jacobisvd(A,tol,maxsweeps)
b. S = jacobisvd(A,tol,maxsweeps)
c. jacobisvd(A,tol,maxsweeps)

The default values of tol and maxsweeps are 1.0×10^{-10} and 10, respectively.

Example 23.2. The first part of the example finds the SVD for the Hanowa matrix of order 500. This matrix is often used as a test matrix for eigenvalue algorithms because all its eigenvalues lie on a line in the complex plane. We will apply jacobisvd to the matrix and compute $\left\| A - USV^{\mathrm{T}} \right\|_2$.

```
>> A = gallery('hanowa', 500);
[U S V] = jacobisvd(A, 1.0e-14);
norm(A - U*S*V')

ans =
      1.110223024625157e-16
```

For the second part, load the 20×20 matrix SMLSINGVAL.mat from the software distribution. It has singular values σ_i, $1 \le i \le 15$ that range from 5.0 down to 1.0. The last five singular values are

$$\sigma_{16} = 1.0 \times 10^{-12}, \quad \sigma_{17} = 1.0 \times 10^{-13}, \quad \sigma_{18} = 1.0 \times 10^{-14}, \quad \sigma_{19} = 1.0 \times 10^{-15}, \quad \sigma_{20} = 0.5 \times 10^{-15}.$$

Compute the singular values of SMLSINGVAL and output the smallest six with 16 significant digits.

```
>> S = jacobisvd(SMLSINGVAL,1.0e-15);
>> for i = 15:20
      fprintf('S(%d) = %.16g\f',i,S(i));
   end
S(15) = 1.1000000000000001
S(16) = 0.0000000000010000
S(17) = 0.0000000000001000
S(18) = 0.0000000000000100
S(19) = 0.0000000000000010
S(20) = 0.0000000000000005
```

■

23.2.1 Faster and More Accurate Jacobi Algorithm

A variant of the one-sided Jacobi algorithm described in Refs. [84, 85] provides higher accuracy and speed than the algorithm we have described. The algorithm uses rank-revealing QR with column pivoting [2, pp. 276-280] that generates a decomposition $AP = QR$, where P is a permutation matrix. The algorithm described in the two papers delivers outstanding performance, and very rapidly computes the SVD of a dense matrix with high relative accuracy. The speed of the algorithm is comparable to the classical methods. The algorithm is said to be a preconditioned Jacobi SVD algorithm. Computation of singular values is well conditioned; however, there are some classes of matrices for which the computation of singular values appears ill-conditioned [84, p. 1323]. This is termed *artificial ill-conditioning*, and the algorithm handles this phenomenon correctly, while bidiagonalization-based methods do not. This algorithm is too complex for presentation in the text, but there are some interesting facets of the algorithm we can present.

After computing the QR decomposition of $m \times n$ matrix A with partial pivoting $m \ge n$, the SVD of A and the upper-triangular matrix R have the same singular values. Let

$$AP = QR, \tag{23.12}$$

and then

$$A^{\mathrm{T}}A = \left(QRP^{\mathrm{T}}\right)^{\mathrm{T}}\left(QRP^{\mathrm{T}}\right) =$$
$$PR^{\mathrm{T}}Q^{\mathrm{T}}QRP^{\mathrm{T}} = P\left(R^{\mathrm{T}}R\right)P^{\mathrm{T}}.$$

P is an orthogonal matrix, so $A^{\mathrm{T}}A$ and $R^{\mathrm{T}}R$ have the same eigenvalues. As we will see, the only SVD computation is for the upper $n \times n$ submatrix of R.

The algorithm deals with two cases, rank $(A) = n$, and rank $(A) = rA < n$. If rank $(A) = n$, we can compute the SVD of A using the following steps:

a. Compute $AP = QR$ using column pivoting.
b. Let $U = I^{m \times m}$ and $V = I^{n \times n}$
c. Compute the SVD of $R^{T}(1:n, 1:n)$ using the enhanced Jacobi method:
$$\left[\hat{V}, \ \hat{\Sigma}, \ U(1:n, 1:n)\right] = \text{enhanced Jacobi}\left(R(1:n, 1:n)^{\mathrm{T}}\right)$$
d. Form $U = QU$ and $V = P\hat{V}$.
e. Let Σ be the $m \times n$ zero matrix with $\hat{\Sigma}$ placed in its upper left-hand corner.

To see that this works, note that

$$R(1:n, 1:n)^{\mathrm{T}} = \hat{V}\hat{\Sigma}U(1:n, 1:n)^{\mathrm{T}},$$
$$R(1:n, 1:n) = U(1:n, 1:n)\hat{\Sigma}\hat{V}^{\mathrm{T}},$$
$$U(1:n, 1:n)^{\mathrm{T}}R(1:n, 1:n) = \hat{\Sigma}\hat{V}^{\mathrm{T}}.$$

Form

$$
U\Sigma V^{\mathrm{T}} = Q
\begin{bmatrix}
U(1:n,1:n) & 0 & \cdots & 0 \\
0 & 1 & & \\
\vdots & & \ddots & \\
0 & & & 1
\end{bmatrix}
\begin{bmatrix}
\tilde{\Sigma} \\
0 \\
\vdots \\
0
\end{bmatrix}
\hat{V}^{\mathrm{T}} P^{\mathrm{T}}
$$

$$
= Q
\begin{bmatrix}
U(1:n,1:n) & 0 & \cdots & 0 \\
0 & 1 & & \\
\vdots & & \ddots & \\
0 & & & 1
\end{bmatrix}
\begin{bmatrix}
\hat{\Sigma}\hat{V}^{\mathrm{T}} \\
0 \\
\vdots \\
0
\end{bmatrix}
P^{\mathrm{T}}
$$

$$
= Q
\begin{bmatrix}
R(1:n,1:n) \\
0 \\
\vdots \\
0
\end{bmatrix}
P^{\mathrm{T}} = A.
$$

The case where rank $(A) = rA < n$ is somewhat more involved. Problem 23.9 asks you to implement a simplified version of this algorithm, using `jacobisvd` to compute the required decomposition for a submatrix of R. The problem provides the code to handle the rank-deficient case.

23.3 TRANSFORMING A MATRIX TO UPPER-BIDIAGONAL FORM

The Demmel and Kahan Zero-shift QR Downward Sweep algorithm for computing the SVD first reduces A to a bidiagonal matrix. The outline of an algorithm for transforming an $m \times n$ matrix to upper-bidiagonal form is easy to understand graphically. Let $k = \min(m-1, n)$. First, use premultiplication by a Householder matrix to zero-out $a_{21}, a_{31}, \ldots, a_{m1}$. Now zero-out elements $a_{13}, a_{14}, \ldots, a_{1n}$ of A using postmultiplication by a Householder matrix.

$$
A =
\begin{bmatrix}
* & * & * & * & * & * \\
* & * & * & * & * & * \\
* & * & * & * & * & * \\
* & * & * & * & * & * \\
* & * & * & * & * & * \\
* & * & * & * & * & *
\end{bmatrix}
\xrightarrow{H_{u_1}A}
\begin{bmatrix}
X & * & * & * & * & * \\
0 & X & * & * & * & * \\
0 & * & X & * & * & * \\
0 & * & * & X & * & * \\
0 & * & * & * & X & * \\
0 & * & * & * & * & X
\end{bmatrix}
\xrightarrow{H_{u_1}AH_{v_1}}
\begin{bmatrix}
X & X & 0 & 0 & 0 & 0 \\
0 & X & * & * & * & * \\
0 & * & X & * & * & * \\
0 & * & * & X & * & * \\
0 & * & * & * & X & * \\
0 & * & * & * & * & X
\end{bmatrix}
= A_1.
$$

Repeat the process by using Householder matrices to zero-out elements $a_{32}, a_{42}, \ldots, a_{m2}$ and $a_{24}, a_{25}, \ldots, a_{2n}$.

$$
A_1 =
\begin{bmatrix}
X & X & 0 & 0 & 0 & 0 \\
0 & X & * & * & * & * \\
0 & * & X & * & * & * \\
0 & * & * & X & * & * \\
0 & * & * & * & X & * \\
0 & * & * & * & * & X
\end{bmatrix}
\xrightarrow{H_{u_2}H_{u_1}AH_{v_1}}
\begin{bmatrix}
X & X & 0 & 0 & 0 & 0 \\
0 & X & * & * & * & * \\
0 & 0 & X & * & * & * \\
0 & 0 & * & X & * & * \\
0 & 0 & * & * & X & * \\
0 & 0 & * & * & * & X
\end{bmatrix}
\xrightarrow{H_{u_2}H_{u_1}AH_{v_1}H_{v_2}}
\begin{bmatrix}
X & X & 0 & 0 & 0 & 0 \\
0 & X & X & 0 & 0 & 0 \\
0 & 0 & X & * & * & * \\
0 & 0 & * & X & * & * \\
0 & 0 & * & * & X & * \\
0 & 0 & * & * & * & X
\end{bmatrix}
= A_2.
$$

Execute the pre- and postmultiplication $k - 1$ times, and finish with one more premultiplication. Through the series of Householder reflections, we have formed the upper-bidiagonal matrix B as follows:

$$
B = H_{u_k}H_{u_{k-1}} \ldots H_{u_1}AH_{v_1}H_{v_2} \ldots H_{v_{k-1}}.
$$

Since the Householder matrices are orthogonal, the singular values of B are the same as those of A.

Generating the postproduct by H_{v_i} requires an explanation. If we compute A^{T}, then $a_{i,i+1}, a_{i,i+2}, a_{i,i+3}, \ldots, a_{i,n}$ are in column i, and we can compute a Householder reflection that zeros them out. By again taking the transpose, the required elements of row i are zero. Taking the transpose is inefficient, so we proceed as follows:

Let $B = A^{\mathrm{T}}$.

Compute Householder reflection H_{v_i} that zeros-out $b_{i+2,i}, b_{i+3,i}, \ldots, b_{n,i}$ and form $H_{v_i}B = H_{v_i}A^{\mathrm{T}}$. Recalling that $H_{v_i}^{\mathrm{T}} = H_{v_i}$, take the transpose of $H_{v_i}A^{\mathrm{T}}$, and we have AH_{v_i}, a matrix in which the elements $a_{i,i+2}, a_{i,i+3}, \ldots, a_{i,k}$ are zero. Compute AH_{v_i} implicitly using Equation 17.12.

Example 23.3. This example illustrates the conversion to bidiagonal form step by step for the matrix $A = \begin{bmatrix} 1 & 5 & 3 \\ 1 & 0 & -7 \\ 3 & 8 & 9 \end{bmatrix}$.
Of course, the Householder matrices are not actually formed.

$$H_{u_1} \qquad\qquad H_{u_1}A \qquad\qquad H_{v_1}$$

$$\begin{bmatrix} -0.3015 & -0.3015 & -0.9045 \\ -0.3015 & 0.9302 & -0.2095 \\ -0.9045 & -0.2095 & 0.3714 \end{bmatrix} \quad \begin{bmatrix} -3.3166 & -8.7438 & -6.9348 \\ 0 & -3.1839 & -9.3015 \\ 0 & -1.5518 & 2.0955 \end{bmatrix} \quad \begin{bmatrix} 1.0000 & 0 & 0 \\ 0 & -0.7835 & -0.6214 \\ 0 & -0.6214 & 0.7835 \end{bmatrix}$$

$$H_{u_1}AH_{v_1} \qquad\qquad H_{u_2} \qquad\qquad B = H_{u_2}H_{u_1}AH_{v_1}$$

$$\begin{bmatrix} -3.3166 & 11.1600 & 0 \\ 0 & 8.2745 & -5.3092 \\ 0 & -0.0863 & 2.6061 \end{bmatrix} \quad \begin{bmatrix} 1 & 0 & 0 \\ 0 & -0.9999 & 0.0104 \\ 0 & 0.0104 & 0.9999 \end{bmatrix} \quad \begin{bmatrix} -3.3166 & 11.1600 & 0 \\ 0 & -8.2750 & 5.3361 \\ 0 & 0 & 2.5506 \end{bmatrix}$$

■

Algorithm 23.2 describes the reduction to upper-bidiagonal form. Note that the function

$$\begin{bmatrix} A, & u \end{bmatrix} = \text{hzero2}\,(A, i, j, \text{row})$$

zeros-out the required column elements if row $= 0$ and the required row elements if row $= 1$. Its implementation is in the book software distribution.

Algorithm 23.2 Reduction of a Matrix to Upper-bidiagonal Form

---

```
function BIDIAG(A)
  % Reduces the m × n matrix A to bidiagonal form.
  % Input: matrix A
  % Output: matrix B in upper-bidiagonal form.
  k = min (m − 1,  n)
  for i = 1:k do
    A = hzero2(A,i,i)
    if i ≤ k then
      A = hzero2 (A, i, i + 1, 1)
    end if
  end for
  return A
end function
```

NLALIB: The function `bidiag` implements Algorithm 23.2.

Remark 23.1.   The book software distribution contains a function `bidiagdemo` that illustrates the algorithm. A press of the space bar graphically shows the location of the nonzero elements. NLALIB contains a 4×4 matrix SVALSDEMO that serves well for this purpose.

23.4   DEMMEL AND KAHAN ZERO-SHIFT *QR* DOWNWARD SWEEP ALGORITHM

We presented the one-sided Jacobi algorithm because it is based on ideas we have discussed previously and because research has proven it is capable of high accuracy. For the one-sided Jacobi method, our MATLAB implementation returned U, S, and V or a vector containing the singular values.

For many years, the Golub-Kahan-Reinsch algorithm has been the standard for SVD computation [25, 86]. It involves working implicitly with $A^{T}A$. We will not discuss this algorithm but, instead, present the Demmel and Kahan zero-shift QR downward sweep algorithm, since it has excellent performance, and it reinforces our understanding of bulge chasing introduced in Section 18.8 [87]. A paper describing the algorithm can be accessed from the Internet at http://www.netlib. org/lapack/lawnspdf/lawn03.pdf. We will develop the algorithm to return only a vector of singular values. The algorithm executes in two stages. The first stage transforms an $m \times n$ matrix, $m \geq n$, into an upper-bidiagonal matrix using Householder reflections, and then this matrix is transformed into a diagonal matrix of singular values, again using products of orthogonal matrices.

$$\text{Stage 1} \quad A \Rightarrow B = \begin{bmatrix} b_{11} & b_{12} & & & & 0 \\ & b_{22} & b_{23} & & & \\ & & \ddots & \ddots & & \\ & & & \ddots & \ddots & \\ & & & & \ddots & b_{n-1,n} \\ 0 & & & & & b_{nn} \end{bmatrix}$$

$$\text{Stage 2} \quad B \Longrightarrow \tilde{\Sigma} = \begin{bmatrix} \sigma_1 & & & & & 0 \\ & \ddots & & & & \\ & & \sigma_r & & & \\ & & & 0 & & \\ & & & & \ddots & \\ 0 & & & & & 0 \end{bmatrix}$$

Phase 2 is similar to the implicit QR algorithm bulge chasing, and its ultimate aim is to eliminate the superdiagonal entries at indices $(1, 2), (2, 3), (3, 4), \ldots, (n - 1, n)$, leaving the singular values on the diagonal. In each pass, a rotation is applied on the right to zero-out an element of the superdiagonal. In the process, a nonzero element is introduced in a location where we don't want it (the bulge), but another element is zeroed-out as a side effect. The algorithm then applies a rotation on the left to remove the nonzero element created from the previous rotation but creates nonzeros in two other locations. After the last pass, the matrix remains in upper-bidiagonal form. By repeating the $k - 1$ passes repeatedly, convergence to a diagonal matrix of singular values occurs. We will not attempt to explain why this algorithm works, but will just demonstrate the process. The interested reader should refer to http://www.netlib.org/lapack/lawnspdf/lawn03.pdf.

Actions in a Pass

Step $i = 1$:

Zero-out the entry at $(1, 2)$ *by multiplying on the right by a rotation matrix. This action introduces a non-zero value at* $(2, 1)$ *immediately below the diagonal.*

Multiply by a rotation on the left to zero-out $(2, 1)$. *This introduces nonzeros at indices* $(1, 2)$ *and* $(1, 3)$.

Steps $i = 2$ through $(k - 2)$:

Multiply by a rotation on the right that zeros-out $(i, i + 1)$ *and, as a side effect,* $(i - 1, i + 1)$. *This leaves a nonzero value at index* $(i + 1, i)$.

Zero-out $(i + 1, i)$. *This leaves nonzeros at indices* $(i, i + 1)$ *and* $(i, i + 2)$.

Step $i = k - 1$:

Multiply by a rotation on the right that zeros-out $(i, i + 1)$ *and, as a side effect,* $(i - 1, i + 1)$. *This leaves a nonzero value at index* $(i + 1, i)$.

Zero-out $(i + 1, i)$. *The matrix remains in upper-bidiagonal form.*

We use a 5×5 matrix to illustrate one pass of the algorithm.

$$A = \begin{bmatrix} * & * & 0 & 0 & 0 \\ 0 & * & * & 0 & 0 \\ 0 & 0 & * & * & 0 \\ 0 & 0 & 0 & * & * \\ 0 & 0 & 0 & 0 & * \end{bmatrix}.$$

Step 1: *Develop a Givens rotation, J_{r_1}, that zeros-out (1, 2) but generates a nonzero value at index (2, 1).*

$$A = \begin{bmatrix} * & 0 & 0 & 0 & 0 \\ X & * & * & 0 & 0 \\ 0 & 0 & * & * & 0 \\ 0 & 0 & 0 & * & * \\ 0 & 0 & 0 & 0 & * \end{bmatrix}.$$

Develop a rotation, J_{l_1}, that zeros-out (2, 1). It introduces nonzeros at (1, 2) and (1, 3).

$$A = \begin{bmatrix} * & X & X & 0 & 0 \\ 0 & * & * & 0 & 0 \\ 0 & 0 & * & * & 0 \\ 0 & 0 & 0 & * & * \\ 0 & 0 & 0 & 0 & * \end{bmatrix}.$$

Step 2: *Create a rotation, J_{r_2}, that zeros-out (2, 3). It leaves a nonzero at (3, 1) but zeros out (1, 3).*

$$A = \begin{bmatrix} * & * & 0 & 0 & 0 \\ 0 & * & 0 & 0 & 0 \\ X & 0 & * & * & 0 \\ 0 & 0 & 0 & * & * \\ 0 & 0 & 0 & 0 & * \end{bmatrix}.$$

Develop a rotation, J_{l_2}, that zeros-out (3, 1) and leaves nonzeros at (2, 3) and (2, 4)

$$A = \begin{bmatrix} * & * & 0 & 0 & 0 \\ 0 & * & X & X & 0 \\ 0 & 0 & * & * & 0 \\ 0 & 0 & 0 & * & * \\ 0 & 0 & 0 & 0 & * \end{bmatrix}.$$

Step 3: *Compute a rotation J_{r_3} to zero-out $(3, 4)$ that leaves a nonzero at $(4, 3)$ but zeros-out $(2, 4)$.*

$$A = \begin{bmatrix} * & * & 0 & 0 & 0 \\ 0 & * & * & 0 & 0 \\ 0 & 0 & * & 0 & 0 \\ 0 & 0 & X & * & * \\ 0 & 0 & 0 & 0 & * \end{bmatrix},$$

Develop a rotation, J_{l_3}, that zeros-out (4, 3), leaving nonzeros at (3, 4) and (3, 5).

$$A = \begin{bmatrix} * & * & 0 & 0 & 0 \\ 0 & * & * & 0 & 0 \\ 0 & 0 & * & X & X \\ 0 & 0 & 0 & * & * \\ 0 & 0 & 0 & 0 & * \end{bmatrix}.$$

Step 4: *Multiply by a rotation J_{r_4} that zeros-out (4, 5), leaves a nonzero at (5, 4), and zeros-out (3, 5).*

$$A = \begin{bmatrix} * & * & 0 & 0 & 0 \\ 0 & * & * & 0 & 0 \\ 0 & 0 & * & * & 0 \\ 0 & 0 & 0 & * & 0 \\ 0 & 0 & 0 & X & * \end{bmatrix}.$$

As the final operation, multiply by a rotation J_{l_4} that zeros-out (5, 4) and places a nonzero value at (4, 5).

$$A = \begin{bmatrix} * & * & 0 & 0 & 0 \\ 0 & * & * & 0 & 0 \\ 0 & 0 & * & * & 0 \\ 0 & 0 & 0 & * & * \\ 0 & 0 & 0 & 0 & * \end{bmatrix}.$$

A pictorial view of a downward sweep is useful, and Figure 23.1 depicts each step graphically using a 4×4 matrix.

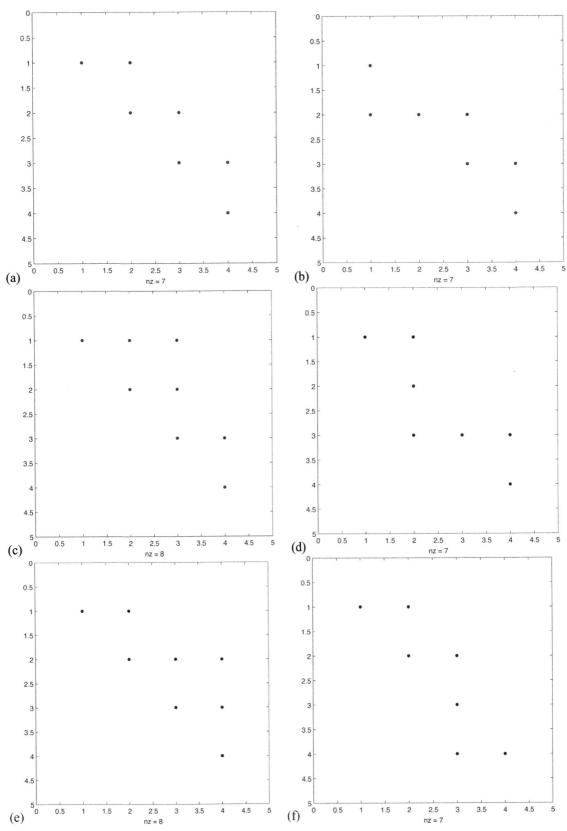

FIGURE 23.1 Demmel and Kahan zero-shift QR downward sweep.

Continued

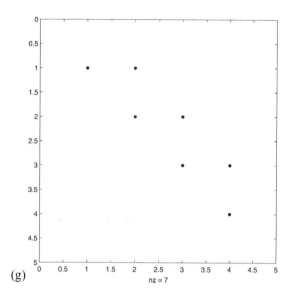

(g)

nz = 7

FIGURE 23.1, CONT'D

The algorithm as described in Ref. [87] is of production quality. It describes very fast rotations to speed up the algorithm, switches between downward and upward sweeps depending on conditions, applies sophisticated convergence criteria, and so forth. A basic version of the algorithm is not difficult to understand, and we present it in Algorithm 23.3. Note that it uses deflation for efficiency and accuracy and features a function givensmulpsvd that performs the right-hand side product with a rotation matrix.

Algorithm 23.3 Demmel and Kahan Zero-Shift QR Downward Sweep.

```
function SINGVALS(A, tol)
   % Computes the singular values of an m × n matrix.
   % Input: real matrix A and error tolerance tol.
   % Output: a vector S of the singular values of A ordered
   % largest to smallest.
   if m < n then A = Aᵀ
      tmp = m
      m = n
      n = tmp
   end if

   A = bidiag(A)
   k = min(m, n)

   while k ≥ 2 do
      % convergence test
      if |aₖ₋₁,ₖ| < tol(|aₖ₋₁,ₖ₋₁| + |aₖₖ|) then
         aₖ₋₁,ₖ = 0
         k = k-1
      else
         for i = 1:k-1 do % Compute the Givens parameters for a rotation
            % that will zero-out A(i,i+1) and A(i-1,i+1),
            % but makes A(i+1,i) non-zero.
            [c, s] = givensparms(aᵢᵢ, aᵢ,ᵢ₊₁)
            % Apply the rotation by performing a postproduct.
            A(1:k,1:k) = givensmulpsvd(A(1:k,1:k),i,i+1,c,s)
            % Compute the Givens parameters for a rotation
```

```
        % that will zero-out a_{i+1,i} to correct the result
        % of the previous rotation. The rotation makes
        % a_{i,i+2} and A(i,i+1) non-zero.
        [ c, s ] = givensparms(a_{ii}, a_{i+1,i})
        % Apply the rotation as a preproduct.
        A(1:k,1:k) = givensmul(A(1:k,1:k),i,i+1,c,s)
      end for
    end if
  end while
  return S = diag(A)
end function
```

NLALIB: The function `singvals` implements Algorithm 23.3.

Remark 23.2.   The book software distribution contains a function `singvalsdemo(A)` that demonstrates convergence to the diagonal matrix of singular values. Initially, a graphic showing the bidiagonal matrix appears. A press of the space bar creates graphics like those in Figure 23.1. Continue pressing the space bar and see convergence taking place. At the conclusion, the function returns the computed singular values. NLALIB contains a 4×4 matrix SVALSDEMO that serves well with `singvalsdemo`.

Example 23.4.   The matrix

$$A = gallery\,(5) = \begin{bmatrix} -9 & 11 & -21 & 63 & -252 \\ 70 & -69 & 141 & -421 & 1684 \\ -575 & 575 & -1149 & 3451 & -13,801 \\ 3891 & -3891 & 7782 & -23,345 & 93,365 \\ 1024 & -1025 & 2048 & -6144 & 24,572 \end{bmatrix}$$

is particularly interesting. Apply the function `eigb` to A:

```
>> A = gallery(5);
>> eigb(A)

ans =
   0.021860170045529 + 0.015660137292070i
   0.021860170045529 - 0.015660137292070i
  -0.008136735236891 + 0.025992813783568i
  -0.008136735236891 - 0.025992813783568i
  -0.027446869619491 + 0.000000000000000i
```

All the eigenvalues but one are complex; however, the characteristic equation of A is $p(\lambda) = \lambda^5$, so in fact all its eigenvalues are 0. To explain the results, compute the condition numbers of the eigenvalues.

```
>> eigcond(A)

ans =
  1.0e+10 *

  2.196851076143216
  2.146816343479836
  2.146816260054680
  2.068763020180955
  2.068762702670772
```

All the eigenvalues of A are ill-conditioned, so the failure of `eigb` is not unexpected. The MATLAB function `eig` fails as well.

MATLAB finds the rank of a matrix by computing the SVD and determining the number of singular values larger than the default tolerance **max(size(A))*eps(norm(A))**. As we have shown, the computation of singular values is well conditioned. The following sequence computes the singular values using `singvals`. Since all the eigenvalues of A are 0, A is not invertible, so it must have at least one zero singular value, and the last computed singular value is approximately 8.713×10^{-14}.

```
>> A = gallery(5);
>> rank(A)

ans =
     4

>> S = singvals(A,1.0e-15);
>> S(5)

ans =
     8.713452466443371e-14

>> max(size(A))*eps(norm(A))

ans =
     7.275957614183426e-11
```

Because S(5) is less than max(size(A))*eps(norm(A)), the rank is determined to be four.   ∎

23.5   CHAPTER SUMMARY

The One-Sided Jacobi Method for Computing the SVD

The use of Jacobi rotations is one of the first methods for computing the SVD but was replaced by the Golub-Kahan-Reinsch and Demmel-Kahan algorithms. Recently, the one-sided Jacobi method, with proper stopping conditions, was shown to compute small singular values with high relative accuracy. The method uses a sequence of postproducts of Jacobi rotation matrices that cause $AJ_1J_2 \ldots J_k$ to have approximately orthogonal columns. The norms of the columns of this matrix are the singular values. It turns out that the sequence of one-sided products implicitly reduces A^TA to a diagonal matrix using orthogonal similarity transformations, where the diagonal entries are the eigenvalues of A^TA. Thus, it is never necessary to compute A^TA and deal with the computational time and rounding errors this will cause.

Transforming a Matrix to Upper-Bidiagonal Form

The first step in the standard algorithms for computing the SVD first reduce the matrix to upper-bidiagonal form using a sequence of Householder matrices. A left multiplication by a Householder matrix zeros-out elements below the diagonal, and a right multiplication zeros-out elements $(i, i + 2)$ through (i, n) of row i.

The Demmel and Kahan Zero-Shift QR Downward Sweep Algorithm

The first step of this algorithm is reducing the matrix to upper-bidiagonal form. The algorithm then continues by bulge chasing that converges to a diagonal matrix of singular values. In the book software, the function singvals estimates singular values by continually chasing the bulge downward. In the production quality algorithm, chasing varies from downward to upward as convergence conditions change.

23.6   PROBLEMS

23.1 Let $A = \begin{bmatrix} 1 & 1 \\ 0.000001 & 0 \\ 0 & 0.000001 \end{bmatrix}$. Find the singular values of A using exact arithmetic and show that A has rank 2 but is close to a matrix of rank 1. Use the Symbolic Toolbox if available.

23.2 Let $A = \begin{bmatrix} 1 & 2 & -1 \\ 3 & 0 & 4 \\ -1 & 5 & 6 \end{bmatrix}$.

   a. Use bidiag and convert A to an upper-bidiagonal matrix.
   b. Carry out one downward sweep of the Demmel-Kahan algorithm.

23.3 The computation of singular values is well conditioned, but the same is not true of singular vectors. Singular vectors corresponding to close singular values are ill-conditioned. This exercise derives from an example in Ref. [83].

Let

$$A = \begin{bmatrix} 1 & 0 \\ 0 & 1+\epsilon \end{bmatrix},$$

and show that the right singular vectors of A are

$$V = \begin{bmatrix} 1 & 0 \\ 0 & 1 \end{bmatrix}.$$

Let

$$\hat{A} = \begin{bmatrix} 1 & \epsilon \\ \epsilon & 1 \end{bmatrix}$$

be a perturbation of A. Show that the right singular vectors of $\hat{A}$ are

$$\hat{V} = \frac{1}{\sqrt{2}} \begin{bmatrix} 1 & 1 \\ 1 & -1 \end{bmatrix}.$$

23.4 Develop an algorithm that takes a tridiagonal matrix A and transforms it to an upper bidiagonal matrix B using orthogonal matrices U and V such that $UAV = B$. HINT: Use Givens rotations with bulge chasing. First, eliminate $(2, 1)$ and locate the bulge. Remove it with a column rotation, and look for the next bulge. Eliminate it with a row rotation, and so forth. To determine the pattern of rotations, experiment with a 4×4 matrix.

23.5

    a. Let A be an upper-bidiagonal matrix having a multiple singular value. Prove that A must have a zero on either its diagonal or superdiagonal.

    b. Is part (a) true for a lower-bidiagonal matrix.

    c. Assume that the diagonal and superdiagonal of a bidiagonal matrix are nonzero. Show that the singular values of the matrix are distinct.

23.6 The NLA implementation of the Demmel and Kahan zero-shift QR downward sweep algorithm does not compute U and V of the SVD.

    a. Upon convergence of the singular values, some will be negative. Recall that each column $V(:, i)$ is an eigenvector of $A^{\mathrm{T}}A$ corresponding to singular value σ_i^2, so that $A^{\mathrm{T}}A(V(:, i)) = \sigma_i^2 V(:, i)$. If $\sigma_i < 0$, show that it is necessary to negate $V(:, i)$.

    b. Show how to modify the algorithm so it computes the full SVD for A, $m \geq n$. You will need to maintain the products of the right and left Householder reflections used to bidiagonalize A. During bulge sweeping, maintain the products of the left and right Givens rotations.

23.6.1 MATLAB Problems

23.7 Randomly generate a matrix A of order 16×4 by using the MATLAB command `rand(16,4)`. Then verify using the MATLAB command rank that rank(A) = 4. Now run the following MATLAB commands:

```
[U S V] = svd(A);
S(4,4) = 0;
B = U*S*V';
```

What is the rank of B? Explain.

23.8 The execution of `A = gallery('kahan',n,theta)` returns an $n \times n$ upper-triangular matrix that has interesting properties regarding estimation of condition number and rank. The default value of theta is 1.2.

    Let $n = 90$, and compute singvals(A). What is the smallest singular value? Verify that this is correct to five significant digits using `svd`. Try to compute the inverse of A. What is the true rank of A? What is the result of computing the rank using MATLAB? If there is a difference, explain.

23.9 This problem asks for a simplified implementation the modified Jacobi SVD algorithm presented in Section 23.2.1. When an SVD is required, use jacobisvd. The QR factorization used is rank revealing, so compute the rank of A as follows:

```
[Q,R,P] = qr(A);
rA = 0 ;
for i = 1 : n
    if abs(R(i,i)) > max(size(A))*eps(norm(A))
        rA = rA + 1 ;
    end
end
```

Section 23.2.1 presented the algorithm for the case of full rank. Use the following code when $rA < n$:

```
[Q1,R1] = qr(R(1:rA,1:n)') ;
[U(1:rA,1:rA),S,V(1:rA,1:rA)] = jacobisvd(R1(1:rA,1:rA)',tol,maxsweeps);
U = Q * U;
V = P*Q1*V;
```

Name the function `svdj`, and test it using the matrices wilkinson(21), gallery(5), a 10×6 matrix with full rank, and a 10×6 rank deficient matrix.

23.10

    a. Implement the algorithm described in Problem 23.4 as the function `tritobidiag`.

    b. In a loop that executes five times, generate a random 100×100 tridiagonal matrix A as indicated, and compute its singular values using S1 = svd(A). Use `tritobidiag` to transform A to a matrix B in upper-bidiagonal form. Compute its singular values using S2 = `svd(B)`. Check the result by computing $\|S1 - S2\|_2$.

        `>> a = randn(99,1);`
        `>> b = randn(100,1);`
        `>> c = randn(99,1);`
        `>> A = trid(a,b,c);`

23.11 Using your results from Problem 23.6, modify `singvals` so it optionally returns the full SVD $A = U\tilde{\Sigma}V^{T}$. Name the function `svd0shift`, and test it with gallery(5), the rosser matrix, and a random 50×30 matrix.

Appendix A

Complex Numbers

Complex numbers are very important in engineering and science. Engineers use complex numbers in analyzing stresses and strains on beams and in studying resonance phenomena in structures as different as tall buildings and suspension bridges. There are many other applications of complex numbers, including control theory, signal analysis, quantum mechanics, relativity, and fluid dynamics.

You have probably dealt with complex numbers before. If so this appendix will serve as a review; otherwise, there is sufficient material here for you to understand complex numbers when they arise in the book. Vectors and matrices of complex numbers are not dealt with in a formal fashion. Occasionally they will arise as eigenvectors or eigenvalues. You will encounter complex roots of polynomials when dealing with eigenvalues and a small number of proofs that involve complex numbers.

A.1 CONSTRUCTING THE COMPLEX NUMBERS

It is clear that the equation $x^2 = -1$ has no real solution, so mathematics defines $i = \sqrt{-1}$, and $i^2 = -1$. The solutions to $x^2 = -1$ are then $x = i$ and $x = -i$. The complex number i forms the basis for the set of complex numbers we call $\mathbb{C}$.

Definition A.1. A complex number has the form $z = x + iy$, where x and y are real numbers and $i = \sqrt{-1}$. We can express a real number x as a complex number $z = x + i0$.

When $z = x + iy$, we call x the *real part* of z and y the *imaginary part.*.

Two complex numbers are equal if they have the same real and imaginary parts:

$$x_1 + iy_1 = x_2 + iy_2 \Rightarrow x_1 = x_2 \quad and \quad y_1 = y_2,$$

where x_1, x_2, y_1, y_2 are real numbers.

The sum of two complex numbers is a complex number:

$$(x_1 + iy_1) + (x_2 + iy_2) = (x_1 + x_2) + i(y_1 + y_2)$$

The product of two complex numbers is a complex number.

$$(x_1 + iy_1)(x_2 + iy_2) = (x_1 x_2 - y_1 y_2) + i(x_1 y_2 + y_1 x_2)$$

The easy way to perform this calculation is to proceed just like you are computing $(a+b)(c+d) = ac + ad + bc + bd$, except that $i^2 = -1$.

$$(x_1 + iy_1)(x_2 + iy_2) = x_1 x_2 + ix_1 y_2 + ix_2 y_1 + i^2 y_1 y_2 = x_1 x_2 + ix_1 y_2 + ix_2 y_1 - y_1 y_2 = (x_1 x_2 - y_1 y_2) + i(x_1 y_2 + y_1 x_2)$$

A useful identity satisfied by complex numbers is

$$(x + iy)(x - iy) = x^2 + y^2.$$

This leads to a method of computing the quotient of two complex numbers.

$$\frac{x_1+iy_1}{x_2+iy_2}=\frac{(x_1+iy_1)(x_2-iy_2)}{(x_2+iy_2)(x_2-iy_2)}$$

$$=\frac{(x_1x_2+y_1y_2)+i(-x_1y_2+y_1x_2)}{x_2^2+y_2^2}.$$

The process is known as *rationalization of the denominator.*

A.2 CALCULATING WITH COMPLEX NUMBERS

We can now do all the standard linear algebra calculations with complex numbers - find the upper triangular form of a matrix whose elements are complex numbers, solve systems of linear equations, find inverses and calculate determinants.

For example, solve the system

$$(1+i)z+(2-i)w= 2 + 7i$$
$$7z+(8 - 2i)w= 4 - 9i.$$

The coefficient determinant is

$$\begin{vmatrix} 1+i & 2-i \\ 7 & 8 - 2i \end{vmatrix} = (1+i)(8 - 2i) - 7(2-i)= (8 - 2i)+i(8 - 2i) - 14 + 7i = = -4 + 13i \neq 0.$$

Hence by Cramer's rule, there is a unique solution:

$$z = \frac{\begin{vmatrix} 2 + 7i & 2-i \\ 4 - 9i & 8 - 2i \end{vmatrix}}{-4 + 13i}=\frac{(2 + 7i)(8 - 2i) - (4 - 9i)(2-i)}{-4 + 13i} =$$

$$= \frac{2(8 - 2i) + (7i)(8 - 2i)-\{(4(2-i) - 9i(2-i)\}}{-4 + 13i}$$

$$= \frac{16 - 4i+56i-14i^2-\{8 - 4i-18i+9i^2\}}{-4 + 13i} =$$

$$\frac{31 + 74i}{-4 + 13i} = \frac{(31+74i)(-4 - 13i)}{(-4)^2+13^2}=\frac{838 - 699i}{(-4)^2+13^2}=\frac{838}{185} - \frac{699}{185}i.$$

Similarly $w=\frac{-698}{185}+\frac{229}{185}i.$

A property enjoyed by complex numbers is that every complex number has a square root.

Theorem A.1. *If w is a non-zero complex number, then the equation $z^2=w$ has a solution $z \in \mathbb{C}$.*

Proof. Let $w=a+ib$, $a, b \in \mathbb{R}$.
Case 1. Suppose $b= 0$. Then if $a>0$, $z=\sqrt{a}$ is a solution, while if $a<0$, $i\sqrt{-a}$ is a solution.
Case 2. Suppose $b \neq 0$. Let $z=x+iy$, $x, y \in \mathbb{R}$. Then the equation $z^2=w$ becomes

$$(x+iy)^2=x^2-y^2+2xyi=a+ib,$$

so equating real and imaginary parts gives

$$x^2-y^2=a \quad and \quad 2xy=b.$$

Hence $x \neq 0$ and $y=b/(2x)$. Consequently

$$x^2-\left(\frac{b}{2x}\right)^2=a,$$

so $4x^4 - 4ax^2 - b^2 = 0$ and $4(x^2)^2 - 4a(x^2) - b^2 = 0$. By the quadratic equation,

$$x^2 = \frac{4a \pm \sqrt{16a^2 + 16b^2}}{8} = \frac{a \pm \sqrt{a^2 + b^2}}{2}.$$

However $x^2 > 0$, so we must take the $+$ sign, since $a - \sqrt{a^2 + b^2} < 0$. Then $x^2 = \frac{a + \sqrt{a^2 + b^2}}{2}$, and the solutions are

$$x = \pm \sqrt{\frac{a + \sqrt{a^2 + b^2}}{2}}, \quad y = b/(2x). \tag{A.1}$$

$\square$

Example A.1.   Find the solutions $z = x + iy$ to the equation $z^2 = 1 + i$ using equation A.1.

For our problem, $w = 1 + i$, so a = 1 and b = 1, and $x = \pm\sqrt{\frac{1 + \sqrt{2}}{2}}$, $y = \pm\frac{1}{\sqrt{2}\sqrt{1 + \sqrt{2}}}$. The solution is

$$z = \pm \left(\sqrt{\frac{1 + \sqrt{2}}{2}} + \frac{i}{\sqrt{2}\sqrt{1 + \sqrt{2}}} \right).$$

$\blacksquare$

Example A.2.   Find the cube roots of 1.

We have to solve the equation $z^3 = 1$, or $z^3 - 1 = 0$. Now
$z^3 - 1 = (z-1)(z^2 + z + 1)$. So $z^3 - 1 = 0 \Rightarrow z - 1 = 0$ or $z^2 + z + 1 = 0$.
But

$$z^2 + z + 1 = 0 \Rightarrow z = \frac{-1 \pm \sqrt{1^2 - 4}}{2} = \frac{-1 \pm \sqrt{3}i}{2}.$$

So there are 3 cube roots of 1, namely 1 and $(-1 \pm \sqrt{3}i)/2$.

$\blacksquare$

A.3   GEOMETRIC REPRESENTATION OF $\mathbb{C}$

Complex numbers can be represented as points in the plane, using the correspondence $x + iy \leftrightarrow (x, y)$. The representation is known as the *Argand diagram* or *complex plane*. The real parts lie on the x-axis, which is then called the *real axis*, while the imaginary parts lie on the y-axis, which is known as the *imaginary axis*. The complex numbers with positive imaginary part lie in the *upper half plane*, while those with negative imaginary part lie in the *lower half plane*.

Because of the equation

$$(x_1 + iy_1) + (x_2 + iy_2) = (x_1 + x_2) + i(y_1 + y_2),$$

complex numbers add vectorially, using the parallelogram law. Similarly, the complex number $z_1 - z_2$ can be represented by the vector from (x_2, y_2) to (x_1, y_1), where $z_1 = x_1 + iy_1$ and $z_2 = x_2 + iy_2$ (Figure A.1).

The geometric representation of complex numbers can be very useful when complex number methods are used to investigate properties of triangles and circles. It is useful in the branch of calculus known as Complex Function theory, where geometric methods play an important role.

A.4   COMPLEX CONJUGATE

Definition A.2.   (**Complex conjugate**) If $z = x + iy$, the *complex conjugate* of z is the complex number defined by $\bar{z} = x - iy$. Geometrically, the complex conjugate of z is obtained by reflecting z across the real axis (Figure A.2).

The following properties of the complex conjugate are easy to verify:

a. $\overline{z_1 + z_2} = \overline{z_1} + \overline{z_2}$;
b. $\overline{-z} = -\bar{z}$;
c. $\overline{z_1 - z_2} = \overline{z_1} - \overline{z_2}$;
d. $\overline{z_1 z_2} = \overline{z_1}\,\overline{z_2}$;
e. $\overline{1/z} = 1/\bar{z}$;

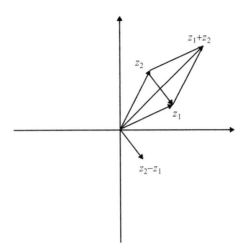

FIGURE A.1 Complex addition and subtraction.

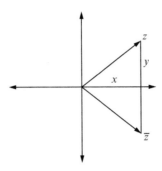

FIGURE A.2 Complex conjugate.

f. $\overline{z_1/z_2} = \overline{z_1}/\overline{z_2}$;

g. z is real if and only if $\bar{z} = z$;

h. With the standard convention that the real and imaginary parts are denoted by Re z and Im z, we have Re $z = \frac{z+\bar{z}}{2}$, Im $z = \frac{z-\bar{z}}{2}$;

i. if $z = x + iy$, then $z\bar{z} = x^2 + y^2$.

The following is an interesting and useful result concerning the roots of polynomials.

Theorem A.2. *Let* $f(z) = a_n z^n + a_{n-1} z^{n-1} + \cdots + a_1 z + a_0 = 0$, *where* $a_n, \ldots, a_0$ *are real. The complex roots occur in complex-conjugate pairs, i.e. if* $f(z) = 0$, *then* $f(\bar{z}) = 0$.

Proof. If $f(z) = 0$, then $0 = \bar{0} = \overline{f(z)} = \overline{a_n z^n} + \overline{a_{n-1} z^{n-1}} + \cdots + \overline{a_1 z} + \overline{a_0}$

$= \overline{a_n} \overline{z^n} + \overline{a_{n-1}} \overline{z^{n-1}} + \cdots + \overline{a_1}\, \bar{z} + \overline{a_0} = a_n \bar{z}^n + a_{n-1} \bar{z}^{n-1} + \cdots + a_1 \bar{z} + a_0 = f(\bar{z})$. $\square$

The computation of the roots of a polynomial play an important role in applications. Some applications include

- Representing Geometric Figures
- Modeling of Steel Corrosion
- Electrical Circuits
- Depth of Flow in Rivers
- Numerical Integration

The computation of polynomial roots is a complex process for polynomials of degree greater than or equal to 3 and is normally done using carefully crafted computer algorithms. Roots of polynomials are also important in theory. Eigenvalues of a square matrix are defined in terms of polynomial roots, but eigenvalues are seldom computed by directly finding the roots.

A real matrix has a transpose, and a real matrix A such that $A^T = A$ is said to be symmetric. There are equivalents of these concepts for a complex matrix.

Definition A.3.   The *conjugate transpose* of a complex matrix A, written A^*, is obtained from A by taking the transpose and then taking the complex conjugate of each entry. The conjugate transpose is the equivalent of the transpose of a real matrix.

Example A.3.   Let

$$A = \begin{bmatrix} 1-i & \frac{1}{2}+2i & 3-5i \\ 6+i & 7+5i & 1+i \\ -1+8i & i & -i \end{bmatrix}.$$

Then,

$$A^* = \begin{bmatrix} 1+i & 6-i & -1-8i \\ \frac{1}{2}-2i & 7-5i & -i \\ 3+5i & 1-i & i \end{bmatrix}. \qquad \blacksquare$$

Recall that if A and B are real matrices, then $(AB)^T = B^T A^T$. If A and B are complex matrices, then $(AB)^* = B^* A^*$.

Definition A.4.   A complex matrix A is said to be *Hermitian* if $A^* = A$, or if $a_{ij} = \overline{a_{ij}}$ for $1 \le i, j \le n$. If $i = j$, then $a_{ii} = \overline{a_{ii}}$, so the diagonal entries of Hermitian matrix are real. A Hermitian matrix is the equivalent of a real symmetric matrix.

Example A.4.   The matrix

$$A = \begin{bmatrix} 1 & i & 6-2i \\ -i & 2 & 4+i \\ 6+2i & 4-i & 3 \end{bmatrix}$$

is Hermitian. $\blacksquare$

A.5   COMPLEX NUMBERS IN MATLAB

MATLAB implements the full range of calculations with complex numbers. For instance, you can assign a complex number to a variable as follows:

```
>> z = 3 + 2i
z =
   3 + 2i
```

Alternatively, you can use the function `complex`.

```
z = complex(5,7)
z =
   5 + 7i
```

You can use any function that accepts a complex variable, and can create and solve complex systems of equations.

```
>> z1 = 4 -i;
>> z2 = 1 +i;
```

```
>> z3 = i;
>> z1^2 + 7*z2 - 8*z3
ans =
   22.0000 - 9.0000i
>> sin(i*pi)
ans =
  0 + 11.549i
>> exp(i*pi)

ans =
-1.0000 + 0.0000i

>> A = [1-i 2+3i -7;-1+i 16+4i i;3+8i -1 7+5i]

A =
   1.0000 - 1.0000i    2.0000 + 3.0000i   -7.0000
  -1.0000 + 1.0000i   16.0000 + 4.0000i         0 + 1.0000i
   3.0000 + 8.0000i   -1.0000              7.0000 + 5.0000i

>> b = [12+2i -1-9i -i]'

b =

  12.0000 - 2.0000i
  -1.0000 + 9.0000i
        0 + 1.0000i

>> z = A\b

z =

   1.6294 - 1.0376i
   0.1008 + 0.4836i
  -1.8082 + 0.0861i

>> A*z

ans =

  12.0000 - 2.0000i
  -1.0000 + 9.0000i
  -0.0000 + 1.0000i

>> E = eig(A)

E =

   0.7504 + 7.7000i
   7.6887 - 3.7542i
  15.5609 + 4.0542i
```

When the matrix is real, complex eigenvalues occur in conjugate pairs. For a complex matrix, this is not true. Recall that a real symmetric matrix has real eigenvalues. The same is true for a Hermitian matrix (Problem A.10).

```
>> A = [1 i 6-2i;-i 2 4+i;6+2i 4-i 3]

A =

            1              0 +       1i       6 -       2i
            0 -       1i       2              4 +       1i
            6 +       2i       4 -       1i       3

>> eig(A)

ans =

      -5.7809
       2.2035
       9.5774
```

A.6   EULER'S FORMULA

If ω is a real constant, what is $e^{i\omega}$? Let's make mathematical sense out of it. It is well known that the McLaurin series for e^x for any real number $-\infty < x < \infty$ is

$$e^x = 1 + \frac{x}{1!} + \frac{x^2}{2!} + \frac{x^3}{3!} + \frac{x^4}{4!} + \frac{x^5}{5!} + \frac{x^6}{6!} \cdots = \sum_{i=0}^{\infty} \frac{x^n}{n!}$$

Now let's apply this same result to a complex number $z = ix$, where x is any real number. Note that $i^{2k} = (-1)^k, k \geq 1$ and $i^{2k+1} = i(-1)^k, k \geq 1$. Now,

$$e^{ix} = 1 + \frac{(ix)}{1!} + \frac{(ix)^2}{2!} + \frac{(ix)^3}{3!} + \frac{(ix)^4}{4!} + \frac{(ix)^5}{5!} + \frac{(ix)^6}{6!} \cdots =$$

$$\left(1 - \frac{x^2}{2!} + \frac{x^4}{4!} - \frac{x^6}{6!} + \cdots\right) + \left(\frac{x}{1!} - \frac{x^3}{3!} + \frac{x^5}{5!} + \cdots\right)i$$

Recall that the McLaurin series for $\cos x$ and $\sin x$, $-\infty < x < \infty$ are

$$\cos x = 1 - \frac{x^2}{2!} + \frac{x^4}{4!} - \frac{x^6}{6!} + \cdots$$

$$\sin x = \frac{x}{1!} - \frac{x^3}{3!} + \frac{x^5}{5!} + \cdots$$

We how have Euler's formula

$$e^{ix} = \cos x + i \sin x \tag{A.2}$$

Remark A.1.   The equation

$$e^{ix} = \cos x + i \sin x$$

is Euler's formula. Now let $x = \pi$ to obtain the very famous *Euler's identity*.

$$e^{i\pi} = -1$$

A poll of readers that was conducted by *Physics World* magazine in 2004 chose Euler's Identity as the "greatest equation ever", in a dead heat with the four Maxwell's equations of electromagnetism.

Richard Feynman called Euler's formula "our jewel" and "one of the most remarkable, almost astounding, formulas in all of mathematics."

After proving Euler's Identity during a lecture, Benjamin Peirce, a noted American 19th-century philosopher/mathematician and a professor at Harvard University, stated that "It is absolutely paradoxical; we cannot understand it, and we don't know what it means, but we have proved it, and therefore we know it must be the truth."

The Stanford University mathematics professor, Dr. Keith Devlin, said, "Like a Shakespearean sonnet that captures the very essence of love, or a painting that brings out the beauty of the human form that is far more than just skin deep, Euler's Equation reaches down into the very depths of existence."

A.7   PROBLEMS

A.1 Evaluate the following expressions.

   a. $(-3+i)(14 - 2i)$

   i. $\dfrac{2 + 3i}{1 - 4i}$

   ii. $\dfrac{(1 + 2i)^2}{1-i}$

A.2 Find the roots of $8z^2 + 2z + 1$.

A.3 x = 2 is a real root of the polynomial $x^3 - x^2 - x - 2$. Find the remaining two roots.

A.4 Verify that $-\frac{\sqrt{2}}{2} + \frac{\sqrt{2}}{2}i$ and $\frac{\sqrt{2}}{2} + \frac{\sqrt{2}}{2}i$ are roots of the polynomial $x^4 + 1$. Find the other two roots.

A.5 Express $1 + (1 + i) + (1 + i)^2 + (1 + i)^3 + \ldots + (1 + i)^{99}$ in the form $x + iy$.

A.6 Solve the system $Ax = b$, where

$$A = \begin{bmatrix} 3+i & -1+2i & 2 \\ 1+i & -1+i & 1 \\ 1+2i & -2+i & 1+i \end{bmatrix}, \ b = \begin{bmatrix} 2+3i \\ 1-i \\ 4+3i \end{bmatrix}.$$

A.7 Find the inverse of the matrix

$$A = \begin{bmatrix} 4-6i & 1+i \\ 12-7i & -i \end{bmatrix}.$$

A.8 Find the conjugate transpose of each matrix.

a. $\begin{bmatrix} 2+i & -1+2i & 2 \\ 1+i & -1+i & 1 \\ 1+2i & -2+i & 1+i \end{bmatrix}$

b. $\begin{bmatrix} -3+i & 2+i & 6-8i \\ 5+0i & 9-i & 16+3i \\ -6+12i & 14-0i & 6+5i \\ 4-i & 8+2i & 1+i \\ 18+3i & 7-i & 1+i \end{bmatrix}$

A.9 If A is a real $m \times n$ matrix, then $A^T A$ is an $n \times n$ symmetric matrix. Prove that if A is an $m \times n$ complex matrix, then A^*A is Hermitian.

A.10 Prove that the eigenvalues of a Hermitian matrix are real. HINT: If $Av = \lambda v$, where λ is an eigenvalue with corresponding eigenvector v, then $v^*Av = \lambda v^*v$. What type of matrices are v^*Av and v^*v?

A.11

a. Prove that $x^*x \geq 0$ for any $n \times 1$ vector x.
  i. Prove that A^*A is positive definite.
  ii. Prove that the eigenvalues of A^*A are nonnegative.

A.7.1 MATLAB Problems

A.12 Using MATLAB, compute

a. $(2 + 3i)(-2 + i)$

b. $\dfrac{6+i}{2+5i}$

c. e^{2+i}

d. $(1 - 2i)^{5i}$

A.13 A *unitary matrix* is the complex equivalent of an orthogonal matrix. A complex matrix is unitary if $A^*A = AA^* = I$. The QR decomposition applies to a complex matrix, and the matrix Q is unitary. Let

$$A = \begin{bmatrix} 1-i & -i & 3+i & 0 \\ 2+3i & -1+2i & i & 3i \\ 5-6i & 1+7i & 3 & 5+i \\ 12 & 4+9i & 1-4i & i \end{bmatrix}$$

Compute the QR decomposition of A and verify that Q is unitary.

A.14 A complex matrix has a singular value decomposition $A = U\widetilde{\Sigma}V^*$, where U and V are unitary.

a. Using the MATLAB command $[U \ S \ V] = \text{svd}$, find the SVD of the matrix in part (b) of Problem A.8. Verify that U and V are unitary.

b. Explain why a complex matrix has real, nonnegative singular values.

A.15 One of the most famous functions in all of mathematics is the Riemann zeta function

$$\zeta(z) = \sum_{n=1}^{\infty} \frac{1}{n^z}.$$

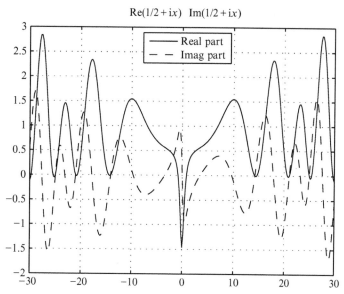

FIGURE A.3 Riemann zeta function.

The Riemann hypothesis states that all non-trivial zeros of the Riemann zeta function have real part 1/2. The trivial zeros are $-2 + 0i$, $-4 + 0i$, ... Proving the Riemann hypothesis has been an open problem for a very long time. The MATLAB function `zeta(s)` computes the Riemann zeta function for a complex variable `s`. Figure A.3 is a MATLAB-generated graph of $Re\left(\frac{1}{2} + ix\right)$ and $Im\left(\frac{1}{2} + ix\right)$ for $\zeta\left(\frac{1}{2} + ix\right)$. Recreate the graph.

Note that the first few zeros with the imaginary part $\frac{1}{2}$ rounded to an integer are:

$$\frac{1}{2} \pm 14i, \ \frac{1}{2} \pm 21i, \frac{1}{2} \pm 25i$$

Appendix B

Mathematical Induction

This appendix is a brief disussion of the topic, and is intended to be sufficient for the times in the book that a proof uses mathematical induction.

Suppose you are given a statement, S, that depends on a variable n; for instance,

$$1 + 2^2 + \ldots + n^2 = \frac{n(n+1)(2n+1)}{6}, n \geq 1$$

Let n_0 be the first value of n for which S applies, and prove the statement true. This is called the *basis step*. For our example, $n_0 = 1$. Now, assume S is true for any $n \geq n_0$ and prove that this implies S is true for $n + 1$. This is called the *inductive step*. Then,

- S is true for n_0, so S is true for $n_1 = n_0 + 1$.
- S is true for n_1, so S is true for $n_2 = n_1 + 1$.
- S is true for n_2, so S is true for $n_2 = n_2 + 1$.
- $\ldots$

This sequence can be continued indefinitely, so S is true for all $n \geq n_0$.

Example B.1. Prove that for $n \geq 1$, $1^2 + 2^2 + \ldots + n^2 = \frac{n(n+1)(2n+1)}{6}$.

Basis step: For $n_0 = 1$, $\frac{1(2)(3)}{6} = 1^2$.

Inductive step: Assume that $1^2 + 2^2 + \ldots + n^2 = \frac{n(n+1)(2n+1)}{6}$. We need to show that

$$1^2 + 2^2 + \ldots + n^2 + (n+1)^2 = \frac{(n+1)((n+1)+1)(2(n+1)+1)}{6} = \frac{(n+1)(n+2)(2n+3)}{6}. \quad (B.1)$$

Now,

$$1^2 + 2^2 + \ldots + n^2 + (n+1)^2 = \left[1^2 + 2^2 + \ldots + n^2\right] + (n+1)^2 = \left[\frac{n(n+1)(2n+1)}{6}\right] + (n+1)^2$$

by the induction assumption. Then,

$$\left[\frac{n(n+1)(2n+1)}{6}\right] + (n+1)^2 = \frac{n+1}{6}\left(2n^2 + n + 6(n+1)\right) = \frac{(n+1)(n+2)(2n+3)}{6},$$

and the proof is complete. ∎

Suppose you have an eigenvalue/eigenvector pair, λ/u, so that $Au = \lambda u$, and you need a way to compute powers $A^n u$. Do some experimenting:

$$A^2 u = A(Au) = A(\lambda u) = \lambda Au = \lambda(\lambda u) = \lambda^2 u,$$
$$A^3 u = A\left(A^2 u\right) = A\left(\lambda^2 u\right) = \lambda^2 Au = \lambda^3 u$$

There is a clear pattern:

$$A^n u = \lambda^n u.$$

When some experimentation yields a pattern, mathematical induction is often the easiest way to prove a result.

Example B.2. Prove that if A is an $n \times n$ matrix, and λ is an eigenvalue with corresponding eigenvector u, then

$$A^n u = \lambda^n u, \ n \geq 1.$$

Basis step: $A^1 u = Au = \lambda u = \lambda^1 u.$
  Inductive step: Assume that $A^n u = \lambda^n u$. Then,

$$A^{n+1} u = A \left(A^n u \right) = A \left(\lambda^n u \right) = \lambda^n Au = \lambda^n \left(\lambda u \right) = \lambda^{n+1} u,$$

and the statement is true for $n + 1$. ■

A *geometric series* is a series with a constant ratio between successive terms. Since geometric series have important applications in science and engineering, the formula for the sum of a geometric series is a very useful result.

Example B.3. If a and r are numbers, $r \neq 1$, then

$$a + ar + ar^2 + ar^{n-1} = \frac{a - ar^n}{1 - r}.$$

Basis step: $\dfrac{a - ar^1}{1 - r} = a$, so the statement if true for $n = 1$.
  Inductive step: Assume that

$$a + ar + ar^2 + ar^{n-1} = \frac{a - ar^n}{1 - r}.$$

Thus,

$$a + ar + ar^2 + ar^{n-1} + ar^n = \left[\frac{a - ar^n}{1 - r} \right] + ar^n =$$
$$\frac{a - ar^n + (1 - r) ar^n}{1 - r} =$$
$$\frac{a - ar^n}{1 - r},$$

and the proof is complete. ■

Strong Induction

It is sometimes necessary to use a variant of mathematical induction called *strong induction*. The basis case is as before

  Let n_0 be the first value of n for which S applies, and prove the statement true.

The inductive step is

  Assume that S is true for all $n_0 \leq k \leq n$. Prove it is true for $n + 1$.

Use this form of induction when the assumed truth for n is not enough. This occurs when several instances of the inductive hypothesis are required to prove the statment true for $n + 1$.

Example B.4. Prove that any positive integer $n \geq 2$ is either prime or a product of primes.
  Basis: $n = 2$ is prime.
  Inductive step: Assume that for all $2 \leq k \leq n$, k is either prime or a product of primes. Consider $n + 1$. If it is prime, we are done; otherwise, it must be a composite number $n + 1 = ab$, where both a and b are in the range $2 \leq k \leq n$. By the inductive hypothesis, a and b are either prime or a product of primes, and the proof is complete. ■

B.1 PROBLEMS

B.1 Prove that

$$1^3 + 2^3 + 3^3 + \ldots + n^3 = \frac{n^2 (n + 1)^2}{4}.$$

B.2 Assume A is an $n \times n$ matrix, X is an invertible matrix, and D is a diagonal matrix such that

$$X^{-1}AX = D.$$

Prove that

$$A^n = XD^nX^{-1}, \, n \geq 1.$$

B.3 Assume that any $n \times n$ matrix M can be factored into the product of an $n \times n$ orthogonal matrix Q, and an $n \times n$ upper triangular matrix R so that $M = QR$. Let A be an $n \times n$ matrix. Prove that there exist orthogonal matrices $Q_i, 1 \leq i \leq k$ and an upper triangular matrix R_k such that

$$(Q_0Q_1 \ldots Q_k)^T A (Q_0Q_1 \ldots Q_k) = R_kQ_k,$$

for any $k \geq 0$.

B.4 Prove that every amount of postage of 12 cents or more can be formed using just 4-cent and 5-cent stamps. HINT: First show that 12, 13, 14, and 15 cents can be formed using 4-cent and 5-cent stamps.

Appendix C

Chebyshev Polynomials

Chebyshev polynomials have important uses in developing convergence results for algorithms; for instance, the analysis for the convergence of the conjugate gradient method in Chapter 21 can be done using these polynomials (see Ref. [1, pp. 312-316]). They also play a large role in convergence results for methods that compute eigenvalues of large sparse matrices (see Ref. [3, pp. 151-159]). There are additional applications to least-squares and interpolation. The book does not actually use Chebyshev polynomials in proofs, but the statement of certain theorems involve Chebyshev polynomials, and so we will give a brief overview of their definition and properties.

C.1 DEFINITION

The definition begins with the form of these polynomials on the interval $-1 \leq t \leq 1$, and then their definition is extended to all real numbers.

Definition C.1. On the interval $-1 \leq x \leq 1$, the Chebyshev polynomial of degree $n \geq 1$ is defined as $T_n(x) = \cos(n\theta)$, where $\theta \in [0, \pi]$ and $\cos(\theta) = x$. More compactly, $T_n(x) = \cos(n\cos^{-1}(x))$.

Looking at the definition, it is not clear that $T_n(x)$ is a polynomial. From the definition, $T_0(x) = 1$ and $T_1(x) = x$, and we can use some trigonometry to find $T_n(x)$, $n \geq 2$.

Recall that

$$\cos(\alpha \pm \beta) = \cos(\alpha)\cos(\beta) \mp \sin(\alpha)\sin(\beta)$$

Now,

$$T_{n+1}(x) = \cos[(n+1)\theta] = \cos(n\theta)\cos(\theta) - \sin(n\theta)\sin(\theta)$$
$$T_{n-1}(x) = \cos[(n-1)\theta] = \cos(n\theta)\cos(\theta) + \sin(n\theta)\sin(\theta)$$

Add the two equations and obtain

$$T_{n+1}(x) + T_{n-1}(x) = 2xT_n(x),$$

and after rearrangement

$$T_{n+1}(x) = 2xT_n(x) - T_{n-1}(x), \, n \geq 1,$$
$$T_0(x) = 1, \, T_1(x) = x \qquad \text{(C.1)}$$

Using Equation C.1, we can extend $T_n(x)$ for all $x \in \mathbb{R}$, and by application of the recurrence relation we can determine a Chebyshev polynomial of any degree. For instance,

$$T_2(x) = 2xT_1(x) - T_0(x) = 2x^2 - 1$$
$$T_3(x) = 2x\left(2x^2 - 1\right) - x = 4x^3 - 3x$$
$$T_4(x) = 2x\left(4x^3 - 3x\right) - \left(2x^2 - 1\right) = 8x^4 - 8x^2 + 1 \qquad \text{(C.2)}$$

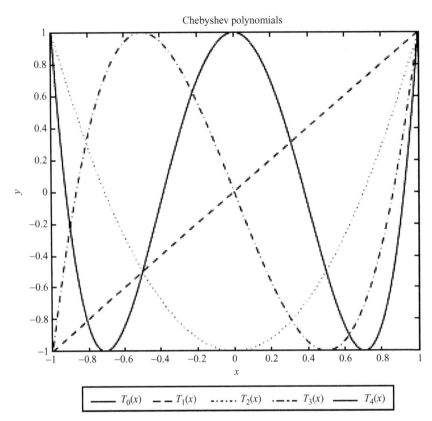

FIGURE C.1 The first five Chebyshev polynomials.

C.2   PROPERTIES

There are many interesting and useful properties of the Chebyshev polynomials, and we mention just a few.

- Since $T_n(x) = \cos(n\theta)$, it follows that

$$|T_n(x)| \leq 1, \ -1 \leq x \leq 1, \ n \geq 0.$$

- We can discern a pattern by looking at Equations C.2. For $n = 2, 3, 4$, the Chebyshev polynomials have the form

$$T_n(x) = 2^{n-1}x^n + O\left(x^{n-1}\right) \tag{C.3}$$

   This is actually true for all $n \geq 1$ (Problem C.2). As a result of Equation C.3, the Chebyshev polynomials grow very quickly with increasing n.
- The roots of $T_n(t)$ are

$$x_i = \cos\left((2i - 1)\,\pi/(2n)\right), \ for \ i = 1, 2, \ldots, n.$$

A polynomial whose highest degree term has a coefficient of 1 is called a *monic polynomial*. If we define $\hat{T}_n(x) = \left(\frac{1}{2^{n-1}}\right) T_n(x)$, then $\hat{T}_n(x)$ is a monic polynomial. By application of these and other properties, we have following theorem (see [88], Chapter 3, Theorem 3.3)

Theorem C.1.   *Consider all possible monic polynomials of degree n. Of all these, the monic polynomial $\hat{T}_n(x)$ has the smallest maximum over $-1 \leq x \leq 1$, and its maximum is $\frac{1}{2^{n-1}}$.*

C.3   PROBLEMS

C.1 Using the recurrence relation C.1, find the Chebyshev polynomials $T_5(x)$ and $T_6(x)$.

C.2 Using the recurrence relation C.1 and mathematical induction, prove relationship C.3.

C.3.1 MATLAB Problems

C.3

    a. Write a function, chebyshev, that takes the degree n as an argument and returns the MATLAB form of the polynomial.

    b. Use your function to graph $T_5(x)$ and $T_6(x)$.

    c. We have said the Chebyshev polynomials grow quickly as n increases. Let $x = 1+\epsilon, \epsilon = \left\{ 10^{-6}, \; 10^{-5}, \; 10^{-4}, \; \right\}$. For each ϵ, evaluate $T_n(x)$, $n = \left\{ 5 \quad 10 \quad 50 \quad 75 \right\}$.

C.4

    a. Let $p(x) = x^3 + 0.00001x^2 - 0.00001x$. Graph $p(x)$ and $T_3(x)/4$ over $\left[-1, \; 1 \right]$.

    b. Compute the maximum of $p(x)$ and $T_3(x)/4$.

Glossary

Absolute error The absolute value of the difference between the true value and the approximate value; for instance, $|x - \mathrm{fl}\,(x)|$.

Adaptive algorithm An algorithm that changes its behavior based on the resources available.

Adjoint The transpose of the matrix of cofactors.

Algorithm A sequence of steps that solve a problem in a finite amount of time.

Argand diagram The plane in which the real part of a complex number lies on the real axis, while the imaginary part lies on the imaginary axis.

Arnoldi method A matrix decomposition of the form $AQ_m = Q_{m+1}\overline{H_m}$, where A is $n \times n$, Q_m is $n \times m$, Q_{m+1} is $n \times (m+1)$, and $\overline{H_m}$ is an $(m+1) \times m$ upper Hessenberg matrix. Q_m is orthogonal, and Q_{m+1} has orthonormal columns. The Arnoldi method is used as a portion of the GMRES algorithm. It is also used in the computation of eigenvalues and their corresponding eigenvectors for a large sparse matrix.

Augmented matrix When solving $Ax = b$ using Gaussian elimination, the matrix formed by attaching the right-hand side vector b as column $n+1$.

Back substitution Solve an upper-triangular system in reverse order from x_n to x_1.

Backward error Roundoff or other errors in the data have produced the result $\hat{y}$. The *backward error* is the smallest Δx for which $\hat{y} = f(x + \Delta x)$; in other words, backward error tells us what problem we actually solved.

Banded matrix A sparse matrix whose nonzero entries appear in a diagonal band, consisting of the main diagonal and zero or more diagonals on either side.

Basic QR iteration A straightforward method of finding all the eigenvalues of a real matrix whose eigenvalues satisfy the relation $|\lambda_1| > |\lambda_2| > \cdots > |\lambda_n|$. There are much better, but more complex, methods of computing the eigenvalues.

Basis A collection of linearly independent vectors. The set of all linear combinations of the basis vectors generates the subspace spanned by the basis. The dimension of the subspace is the number of vectors in the basis.

Bidiagonal matrix A matrix with nonzero entries along the main diagonal and either the diagonal above or the diagonal below.

Big-O A notation that provides an upper bound on the growth rate of a function; for instance, $f(x) = x^3 + x^2 + 5x + 1$ is $O(x^3)$ and also $O(x^4)$, but not $O(x^2)$.

Biharmonic equation The two-dimensional equation takes the form

$$\frac{\partial^4 u}{\partial x^4} + 2\frac{\partial^4 u}{\partial x^2 \partial y^2} + \frac{\partial^4 u}{\partial y^4} = f(x, y),$$

with specified boundary conditions on a bounded domain. The equation has applications in the theory of elasticity, mechanics of elastic plates and the slow flow of viscous fluids.

Boundary value problem An ordinary or partial differential equation or system of equations with prescribed values on a boundary; for instance $(d^2y/dx^2) + 5(dy/dx) + x = 0$, $y(0) = 1$, $y(2\pi) = 3$ is a boundary value problem.

Cancellation error An error in floating point arithmetic that occurs when two unequal numbers are close enough together that their difference is 0.

Cauchy-Schwartz inequality An important result in numerical linear algebra: $|\langle x, y \rangle| \leq \|x\|_2 \|y\|_2$.

Characteristic equation The equation that defines the eigenvalues of a matrix A: $\det(A - \lambda I) = 0$.

Characteristic polynomial The polynomial whose roots are the eigenvalues of the associated matrix A: $p(\lambda) = \det(A - \lambda I)$.

Cholesky decomposition If A is a real positive definite $n \times n$ matrix, there is exactly one upper-triangular matrix R such that $A = R^T R$.

Coefficient matrix The matrix of coefficients, A, for a linear algebraic system $Ax = b$.

Cofactor $C_{ij} = (-1)^{i+j} M_{ij}$, where M_{ij} is the minor for row i, column j of a square matrix. The 2-norm is commonly used.

Column rank The number of linear independent columns in an $m \times n$ matrix.

Column space The subspace generated by the columns of an $m \times n$ matrix.

Column vector An $n \times 1$ matrix.

Complex conjugate If $z = x + iy$, its conjugate is $\bar{z} = x - iy$.

Complex plane The plane in which the real part of a complex number lies on the real axis, while the imaginary part lies on the imaginary axis.

Condition number For an $n \times n$ matrix A, the condition number is $\eta(A) = \|A\| \|A^{-1}\|$ and measures the sensitivity of errors in computing the solutions to problems involving the matrix. The 2-norm is commonly used.

Conjugate gradient method The iterative method of choice for solving a large, sparse, system $Ax = b$, where A is symmetric positive definite.

Cramer's rule A method of solving a square system $Ax = b$. Let B_j be the matrix obtained by replacing column j of A by vector b. Then $x_j = \det(B_j)/\det(A)$. Cramer's rule should not be used, except for very small systems.

Crank-Nicholson method A finite difference scheme for approximating the solution to a partial differential equation over a rectangular grid.

Cross product Given two vectors u, v in $\mathbb{R}^3$, $u \times v = \det \begin{bmatrix} i & j & k \\ u_1 & u_2 & u_3 \\ v_1 & v_2 & v_3 \end{bmatrix}$, where i, j, k are the standard basis vectors. $u \times v$ is orthogonal to both u and v.

Crout's method An method of performing the LU decomposition of an $n \times n$ matrix. The elements of L and U are determined using formulas that are easily programmed.

Cubic spline interpolation An algorithm that fits a cubic polynomial between every pair of adjacent points $\{[a = x_1, x_2], [x_2, x_3], \ldots, [x_n, x_{n+1} = b]\}$. The piecewise polynomial function is twice differentiable. The value of the piecewise polynomial at any $a \le x \le b$ provides accurate interpolation.

Data perturbations Small changes in data that may cause large changes in the solution of a problem using that data.

Demmel and Kahan zero-shift QR downward sweep algorithm Finds the SVD of a matrix. The first step is transformation to upper bidiagonal form, followed by bulge chasing to compute U, V, and the singular values.

Determinant A real number defined recursively using expansion by minors. Row elimination techniques provide a practical means of computing a determinant. Note that

$$\det(AB) = \det(A)\det(B), \quad \det(A^T) = \det(A), \quad \text{and} \quad \det(A^{-1}) = 1/\det(A).$$

Diagonal dominance The absolute value of the diagonal element in a square matrix is greater than the sum of the absolute values of the off-diagonal elements. $|a_{ii}| > \sum_{j=1, j \ne i}^{n} |a_{ij}|$, $1 \le i \le n$.

Diagonal matrix A matrix A whose only nonzero elements are on the diagonal: $a_{ij} = 0$, $i \ne j$.

Diagonalization A matrix A can be diagonalized if there exists an invertible matrix S such that $D = S^{-1}AS$, where D is a diagonal matrix. A and D are similar matrices.

Dimension of a subspace The number of elements in a basis for the subspace.

Digraph A set of vertices and directed edges, with or without weights.

Dominant eigenvalue The eigenvalue of a matrix that is largest in magnitude.

Dominant operations The most expensive operations performed during the execution of an algorithm.

Eigenpair A pair (λ, v), where v is an eigenvector of matrix A associated with eigenvalue λ.

Eigenproblem Finding the eigenvalues and associated eigenvectors of an $n \times n$ matrix.

Eigenvalue A real or complex number such that $Ax = \lambda x$, where A is an $n \times n$ matrix and x is a vector in $\mathbb{R}^n$.

Eigenvector A vector associated with an eigenvalue. If λ is an eigenvalue of A, then v is an eigenvector if $Av = \lambda v$.

Elementary row matrix A matrix E such that if A is a square matrix EA performs an elementary row operation.

Elementary row operations In a matrix, adding a multiple of one row to another, multiplying a row by a scalar, and exchanging two rows.

Encryption The process of transforming information using an algorithm to make it unreadable to anyone except those possessing a key.

Euler's identity $e^{i\pi} = -1$.

Euler's formula $e^{ix} = \cos(x) + i \sin(x)$.

Expansion by minors Computing the value of a determinant by adding multiples of the cofactors in any row or column.

Exponent In a floating point representation $\pm (0.d_1 d_2 \cdots d_p) \times b^n$, n is the exponent, and the d_i are the significant digits.

Extrapolation Taking data in an interval $a \le x \le b$ and using them to approximate values outside that interval. Least-squares can be used for this purpose.

Filtering polynomial A polynomial function designed for restarting the implicit Arnoldi or Lanczos methods for computing eigenvalues and eigenvectors of large sparse matrices.

Finite difference A quotient that approximates a derivative by using a number of nearby points in a grid.

First-row Laplace expansion Evaluation of a determinant using expansion by minors across the first row.

fl The floating-point number associated with the real number x.

Floating point arithmetic Finite-precision arithmetic performed on a computer.

Flop count The number of floating point operations ($\oplus, \ominus, \otimes, \oslash$) required by an algorithm.

Forward error The forward error in computing $f(x)$ is $|\hat{f}(x) - f(x)|$. This measures errors in computation for input x.

Forward substitution Solving a lower-triangular system, $Lx = b$, from x_1 to x_n.

Four fundamental subspaces Let U and V be the orthogonal matrices in the SVD, and r the number of the smallest singular value. The table specifies an orthogonal basis for the range and null space of A and A^T.

	$A = U\widetilde{\Sigma}V^T$	
	Range	Null space
A	$u_i, 1 \le i \le r$	$v_i, r+1 \le i \le n$
A^T	$v_i, 1 \le i \le r$	$u_i, r+1 \le i \le m$

Fourier coefficients The coefficients of the trigonometric functions in a Fourier series expansion of a function.

Fourier series An expansion of a periodic function in terms of an infinite sum of sines and cosines, $f(x) = a_0 \left(1/\sqrt{2\pi}\right) + (1/\sqrt{\pi}) \sum_{i=1}^{\infty} (a_i \cos ix + b_i \sin ix)$

Francis algorithm Often called the implicit QR iteration. Using orthogonal similarity transformations, produce an upper Hessenberg matrix with the same eigenvalues as matrix A. Using bulge chasing, implicitly perform a single or double shift QR step using Givens rotations and Householder reflections, respectively. The end result is an upper triangular matrix whose diagonal contains the eigenvalues of A.

Frobenius norm A matrix norm defined by $\|A\|_F = \sqrt{\text{trace}\left(A^T A\right)} = \left(\sum_{i=1}^{m} \sum_{j=1n} \left|a_{ij}^2\right|\right)^{1/2}$. It is not induced by any vector norm.

Function condition number The limiting behavior of $\dfrac{\frac{|f(x)-f(\bar{x})|}{|f(x)|}}{\frac{|x-\bar{x}|}{|x|}}$ as δx becomes small.

Gauss-Seidel iterative method An iterative method for solving a linear system. Starting with an initial approximation x_0, the iteration produces a new value at each step by using the most recently computed values and the remaining previous values.

Gaussian elimination Use of row elimination operations to solve a linear system or perform some other matrix operation.

Gaussian elimination with partial pivoting During Gaussian elimination, the diagonal element is made largest in magnitude by exchanging rows, if necessary. It is done to help minimize round off error.

Gaussian elimination with complete pivoting In Gaussian elimination, the pivoting strategy exchanges both rows and columns.

Geometric interpretation of the SVD If A is an $m \times n$ matrix, then Ax applied to the unit sphere $\|x\|_2 \leq 1$ in $\mathbb{R}^n$ is a rotated ellipsoid in $\mathbb{R}^m$ with semiaxes σ_i, $1 \leq i \leq r$, where the σ_i are the nonzero singular values of A.

Givens matrix (rotation) An orthogonal matrix, $J(i, j, c, s)$, designed to zero-out a_{ji} when it multiplies another matrix on the left or on the right. c and s must be chosen properly.

GMRES The general minimum residual method for computing the solution to a system $Ax = b$, where A is an $n \times n$ large, sparse, matrix. The method applies to any sparse matrix, but should not be used when A is symmetric positive definite.

Gram-Schmidt algorithm An algorithm that takes a set of n linearly independent column vectors and produces an orthonormal basis for the subspace spanned by the vectors. It also gives rise to a reduced QR decomposition of the matrix formed by the column vectors.

Heat equation A partial differential equation describing heat flow. In one space dimension, the problem is to solve

$$\frac{\partial u}{\partial t} = c \frac{\partial^2 u}{\partial x^2}, \quad 0 \leq x \leq L, \quad u(0, t) = g_1(t), \quad u(L, t) = g_2(t), \quad u(x, 0) = f(x).$$

Hessenberg inverse iteration An algorithm to find an eigenvector of matrix A corresponding to eigenvalue λ. Use an orthogonal similarity transformation to reduce matrix A to upper Hessenberg form, H. Find an eigenvector u of $H = Q^T A Q$ corresponding to eigenvalue λ of A using the inverse iteration. Then Qu is an eigenvector of A corresponding to eigenvalue λ.

Hessenberg matrix A square matrix is upper Hessenberg if $a_{ij} = 0$ for $i > j + 1$. The transpose of an upper Hessenberg matrix is a lower Hessenberg matrix ($a_{ij} = 0$ for $j > i + 1$). A Hessenberg matrix is "almost triangular."

Hilbert matrices Notoriously ill-conditioned matrices defined by $H(i, j) = 1/(i + j - 1)$, $1 \leq i, j, \leq n$.

Homogeneous linear system An $n \times n$ system of the form $Ax = 0$. The system has a unique solution $x = 0$ if an only if A is nonsingular.

Householder matrix (reflection) A symmetric orthogonal matrix, H_u, that takes a vector u and reflects it about a plane in $\mathbb{R}^n$. The transformation has the form

$$H_u = I - \frac{2uu^T}{u^T u}, \quad u \neq 0.$$

Householder matrices are used to compute the QR decomposition, reduction to an upper Hessenberg matrix, and many other things.

Identity matrix An $n \times n$ diagonal matrix whose diagonal consists entirely of ones. If A is an $n \times n$ matrix, $AI = IA = A$.

IEEE arithmetic The standard for 32- and 64-bit hardware-implemented floating point representations.

Ill-conditioned matrix A matrix with a large condition number.

Ill-conditioned problem A problem where small errors in the data may produce large errors in the solution.

Implicit Q theorem If $Q^T A Q = H$ and $Z^T A Z = G$ are both unreduced Hessenberg matrices where Q and Z have the same first column, then Q and Z are essentially the same up to signs.

Implicit QR iteration (see the Francis method).

Implicitly Restarted Arnoldi Method A method for computing eigenvalues and eigenvectors of a large sparse nonsymmetric matrix using the Arnoldi decomposition. Implicit shifts are used to evaluate a filter function that enhances convergence.

Implicitly Restarted Lanczos Method A method for computing eigenvalues and eigenvectors of a large sparse symmetric matrix using the Lanczos decomposition. Implicit shifts are used to evaluate a filter function that enhances convergence.

Inf The MATLAB constant inf returns the IEEE arithmetic representation for positive infinity, and in some situations its use is valid. Infinity is also produced by operations like dividing by zero (1.0/0.0), or from overflow (exp(750)).

Infinity norm A vector norm defined by $\|x\|_\infty = \max_{1 \leq i \leq n} |x_i|$. It induces the matrix infinity norm

$$\|A\|_\infty = \max_{1 \leq k \leq m} \sum_{j=1}^{n} |a_{kj}|.$$

Inner product Given two vectors $x = \begin{bmatrix} x_1 \\ \vdots \\ x_n \end{bmatrix}$ and $\begin{bmatrix} y_1 \\ \vdots \\ y_n \end{bmatrix}$ in $\mathbb{R}^n$, we define the *inner product* of x and y, written $\langle x, y \rangle$, to be the real number

$\langle x, y \rangle = x_1 y_1 + x_2 y_2 + \cdots + x_n y_n = \sum_{i=1}^{n} x_i y_i$. If f and g are functions defined on $a \leq x \leq b$, the L^2 inner product is $\langle f, g \rangle = \int_a^b f(x)g(x)dx$.

Interpolation A method of estimating new data points within the range of a discrete set of known data points.

Inverse Iteration An algorithm to compute an eigenvector from its eigenvalue.

Inverse matrix The unique matrix B such that $BA = AB = I$, where A is a square matrix, and I is the identity matrix. It is normally written as A^{-1}.

Inverse power method A method for computing the smallest eigenvalue in magnitude and an associated eigenvector.

Irreducible matrix Beginning at any vertex of the directed graph formed from the nonzero entries of a matrix, edges can be followed to any other vertex.

Iterative refinement An iteration designed to enhance the values obtained from Gaussian elimination.

Jacobi iterative method An iterative method for solving a linear system. Starting with an initial approximation x_0, the iteration produces a new value at each step by using the previous value.

Jacobi method for computing the eigenvalues of a symmetric matrix Using Jacobi rotations to perform similarity transformations, systematically eliminate a_{ij} and a_{ji} at each step. Even though some zeros may be destroyed, the method converges to a diagonal matrix of eigenvalues.

Jacobi rotation A form of Givens rotation, $J(i, j, c, s)$, such that $J(i,j,c,s)^{\mathrm{T}} AJ(i,j,c,s)$ zeros-out a_{ij} and a_{ji}, $i \neq j$. c and s must be chosen properly.

Kirchhoff's rules Rules governing an electrical circuit which state that

  a. At any junction point in a circuit where the current can divide, the sum of the currents into the junction must equal the sum of the currents out of the junction.

  b. When any closed loop in the circuit is traversed, the sum of the changes in voltage must equal zero.

Krylov subspace methods The Krylov subspace $\mathcal{K}_{\backslash}$ generated by A and u is $span\left\{ u \ Au \ A^2u \ \ldots \ A^{k-1}u \right\}$. It is of dimension k if the vectors are linearly independent. CG is a Krylov subspace method, as are GMRES and MINRES. For GMRES and MINRES, the idea is to find the solution to $Ax = b$ by solving a least-squares problem in a k-dimensional Krylov subspace, where $k < n$. Hopefully k is much smaller than n.

norm (1-norm) A vector norm defined by $\|x\|_1 = \sum_{i=1}^{n} |x_i|$. It induces the matrix 1-norm $\|A\|_1 = \max_{1 \leq k \leq n} \sum_{i=1}^{m} |a_{ik}|$.

norm (2-norm) The vector norm defined by $\|x\|_2 = \sqrt{x_1^2 + x_2^2 + \cdots + x_n^2}$. It induces the matrix 2-norm $\|A\|_2$, that is the square root of the largest eigenvalue of $A^{\mathrm{T}}A$.

Lagrange's identity If u and v are vectors in $\mathbb{R}^3$, then $\|u \times v\|_2^2 = \|u\|_2^2 \|v\|_2^2 - \langle u, v \rangle^2$..

Least-squares Given a real $m \times n$ matrix A of rank $k \leq \min(m, n)$ and a real vector b, find a real vector $x \in \mathbb{R}^n$ such that the residual function $r(x) = \|Ax - b\|_2$ is minimized. Among other applications, the method can be used to fit a polynomial of a specified degree to data.

Lanczos method The Arnoldi method applied to a symmetric matrix. The result is

$$AQ_m = Q_m T_m + t_{m+1, m} q_{m+1} e_m^{\mathrm{T}},$$

where $A^{n \times n}$, $Q_m^{n \times m}$, $T_m^{m \times m}$, and $t_{m+1, m} q_{m+1} e_m^{\mathrm{T}}$ is an $n \times m$ matrix. T_m is symmetric tridiagonal, and Q_m is orthogonal. The Lanczos method is used as a portion of the MINRES algorithm. It is also used to compute some eigenvalues and eigenvectors of a large sparse symmetric matrix.

Left eigenvector If λ is an eigenvalue of matrix A, a left eigenvector associated with λ is a vector x such that $x^{\mathrm{T}}A = \lambda x^{\mathrm{T}}$.

Leslie model A heavily used model in population ecology. It is a model of an age-structured population which predicts how distinct populations change over time.

Linear combination Given a collection of k vectors $v_1, v_2, \ldots, v_k$, a linear combination is the set of all vectors of the form $c_1 v_1 + c_2 v_2 + \cdots + c_k v_k$, where the c_i are scalars.

Linear transformation If A is an $m \times n$ matrix and x is an $n \times 1$ vector, Ax is a linear transformation from $\mathbb{R}^n$ to $\mathbb{R}^m$.

Linearly dependent A set of vectors is linearly dependent if one vector can be written as a linear combination of the others.

Linearly independent A set of vectors is linearly independent if no vector can be written as a linear combination of the others. Equivalently, $v_1, v_2, \ldots, v_k$ are linearly independent when $c_1 v_1 + c_2 v_2 + \cdots + c_k v_k = 0$ if and only if $c_1 = c_2 = \cdots = c_k = 0$.

Lower-triangular matrix An $n \times n$ matrix having zeros above its diagonal; in other words, $a_{ij} = 0$, $j \geq i$.

LU decomposition Using Gaussian elimination to find a lower-triangular matrix, L, an upper-triangular matrix, U, and a permutation matrix, P, such that $PA = LU$.

Machine precision The expression $eps = \frac{1}{2} b^{1-p}$, where b is the base of the number system used, and p is the number of significant digits. It is the distance from 1 to the next largest floating point number.

Mantissa In a floating point representation $\pm (0.d_1 d_2 \ldots d_p) \times b^n$, $m = .d_1 d_2 \ldots d_p$ is the mantissa.

Matrix A rectangular array of rows and columns.

Matrix diagonalization The process of taking a square matrix, A, and finding an invertible matrix, X, such that $D = X^{-1}AX$. Diagonalizing a matrix is also equivalent to finding the eigenvalues and eigenvectors of A. The eigenvalues are on the diagonal of D, and the corresponding eigenvectors are the columns of X.

Matrix inverse The unique matrix B such that $BA = AB = I$, where A is a square matrix. It is normally written as A^{-1}.

Matrix norm A function $\|\cdot\|:\mathbb{R}^{m \times n} \to \mathbb{R}$ is a *matrix norm* provided:

  - $\|A\| \geq 0$ for all $A \in \mathbb{R}^{m \times n}$, and $\|A\| = 0$ if and only if $A = 0$;
  - $\|\alpha A\| = |\alpha| \|A\|$ for all scalars α.
  - $\|A + B\| \leq \|A\| + \|B\|$ for all $A, B \in \mathbb{R}^{m \times n}$.

Matrix product If A is an $m \times k$ matrix, and B is an $k \times n$ matrix, then the product $C = AB$ is the $m \times n$ matrix such that

$$c_{ij} = \sum_{k=1}^{n} a_{ik}b_{kj} = a_{i1}b_{1j} + \cdots + a_{in}b_{nj}.$$

In general, matrix multiplication is not commutative.

Minor The minor $M_{ij}(A)$ of an $n \times n$ matrix A is the determinant of the $(n-1) \times (n-1)$ submatrix of A formed by deleting the ith row and jth column of A.

MINRES The minimum residual method for computing the solution to a system $Ax = b$, where A is an $n \times n$ large, sparse, symmetric, matrix. The method applies to any symmetric matrix, but should not be used when A is symmetric positive definite.

Modified Gram-Schmidt A modification of the Gram-Schmidt method that helps minimize round-off errors.

Modified Gram-Schmidt for QR decomposition A modification of the Gram-Schmidt QR decomposition method that helps minimize round-off errors.

Modulus The absolute value of real number and the magnitude $|z| = |x + iy| = \sqrt{x^2 + y^2}$ of a complex number.

NaN Stands for "not a number." Occurs when an illegal operation such as 0/0 occurs during floating point computation. It is a sure sign that something is wrong with the algorithm.

Nonsingular matrix A matrix having an inverse. An $n \times n$ matrix whose rank is n is nonsingular. A nonsingular matrix cannot have an eigenvalue $\lambda = 0$.

Normal equations If A is an $m \times n$ matrix, the $n \times n$ system $A^{T}Ax = A^{T}x$.

Normal matrix A real matrix A is normal if $A^{T}A = AA^{T}$. All symmetric matrices are normal, and any normal matrix can be diagonalized.

Null space The set of all vectors for which $Ax = 0$. If A is nonsingular, the null space is empty.

One norm A vector norm defined by $\|x\|_1 = \sum_{i=1}^{n} |x_i|$ It induces the matrix 1-norm

$$\|A\|_1 = \max_{1 \le k \le m} \sum_{i=1}^{m} |a_{ik}|.$$

One-sided Jacobi iteration An algorithm involving Jacobi rotations that computes the singular value decomposition of a matrix.

Orthogonal invariance For any orthogonal matrices U and V, $\|UAV\|_2 = \|A\|_2$.

Orthogonal matrix A square matrix P such that $PP^{T} = P^{T}P = I$. The columns of P are an orthonormal basis for $\mathbb{R}^n$.

Orthogonal projection An orthogonal projection of v onto u is defined by $proj_u(v) = \left(\frac{\langle v, u \rangle}{\|u\|_2^2} \right) u$. See Figure 14.1 for a graphical depiction.

Orthogonal vectors Two vectors u and v for which $\langle u, v \rangle = u^{T}v = v^{T}u = 0$.

Orthonormal A set of vectors $v_1, v_2, \ldots, v_k$ are orthonormal if they are orthogonal and each has unit length.

Orthonormal basis A basis for a subspace in which the basis vectors are orthonormal.

Orthonormalization The process of converting a set of linearly independent vectors into an orthonormal basis for the same subspace.

Overdetermined system An $m \times n$ linear system in which $m > n$; in other words, there are more equations than unknowns. Overdetermined systems occur in least-squares problems.

Overflow Occurs when a computer operation generates a number having a magnitude too large to represent; for instance, integer overflow occurs when two positive integers m and n are added and the result is negative. Floating point overflow occurs when an operation produces a result that cannot be represented by the fixed number of bits used to represent a floating point number.

p-norm $\|x\|_p = (|x_1|^p + |x_2|^p + \cdots + |x_n|^p)^{1/p} = \left(\sum_{i=1}^{n} |x_i|^p \right)^{1/p}$. The most commonly used p-norms are $p = 2$ and $p = 1$. The infinity norm $\|x\|_\infty = \max_{1 \le i \le n} |x_i|$ is also considered a p-norm with $p = \infty$.

Pentadiagonal matrix A matrix with five diagonals, all other entries being zero. There are two sub-subdiagonals, the main diagonal, and two super diagonals. These matrices often appear in finite difference methods for the solution of partial differential equations.

Permutation matrix A matrix whose rows are permutations of the identity matrix. If A is an $n \times n$ matrix PA permutes rows of A.

Perturbation analysis A mathematical study of how small changes in the data of a problem affect the solution.

Pivot The element in row i, column i that is used during Gaussian elimination to zero-out all the elements in column i, rows $i + 1$ to n.

Poisson's equation One of the most important equations in applied mathematics with applications in such fields as astronomy, heat flow, fluid dynamics, and electromagnetism. In two dimensions, let R be a bounded region in the plane with boundary ∂R, $f(x, y)$ be a function defined in R, and $g(x, y)$ be defined on ∂R. Find a function $u(x, y)$ such that

$$-\frac{\partial^2 u}{\partial x^2} - \frac{\partial^2 u}{\partial y^2} = f(x, y),$$
$$u(x, y) = g(x, y) \quad \text{on } \partial R.$$

The equation can be defined similarly in n-dimensions.

Positive definite An $n \times n$ symmetric matrix A such that $x^{T}Ax > 0$ for all $x \ne 0$. If A is a positive definite matrix, then there exists an upper-triangular matrix R such that $A = R^{T}R$, $r_{ii} > 0$, (the Cholesky decomposition), and all the eigenvalues of A are real and greater than zero.

Positive semidefinite An $n \times n$ matrix A such that $x^{T}Ax \ge 0$ for all $x \ne 0$.

Power method An algorithm for computing the largest eigenvalue in magnitude and an associated eigenvector by computing successive matrix powers.

Preconditioning   A technique designed to solve a linear system whose matrix is ill-conditioned. Choose a preconditioner, P, whose inverse is close enough to A^{-1} so that the system

$$P^{-1}Ax = P^{-1}b$$

is not as ill-conditioned. Preconditioning can be used effectively with the conjugate gradient and GMRES methods.

Pseudocode   An informal language for describing algorithms.

Pseudoinverse   If A is an $m \times n$ matrix, the pseudoinverse $A^{\ddagger} = \left(A^{\mathrm{T}}A\right)^{-1}A^{\mathrm{T}}$. The pseudoinverse generalizes the concept of an inverse. When $m = n$, $A^{\ddagger} = A^{-1}$.

QR decomposition   A matrix decomposition of an $m \times n$ matrix A such that $A = QR$, where Q is an $m \times m$ orthogonal matrix and R is an $m \times n$ upper-triangular matrix.

QR iteration   An iterative algorithm that computes the eigenvalues of a real matrix A with distinct eigenvalues using the QR decomposition. If A_i is the current matrix in the iteration, compute $A_i = Q_iR_i$, and then set $A_{i+1} = R_iQ_i$. The sequence of matrices converges to an upper-triangular matrix with all the eigenvalues of A on the diagonal. Convergence tends to be slow, so a shift, σ_i, is normally applied as follows: $A_i - \sigma_iI = Q_iR_i$ and $A_{i+1} = R_iQ_i + \sigma_iI$. Choose σ_i to better isolate eigenvalue λ_i.

Quadratic form   An expression in real variables x and y of the form $ax^2 + 2hxy + by^2$. If $X = \begin{bmatrix} x \\ y \end{bmatrix}$, the expression can be written as

$$ax^2 + 2hxy + by^2 = \begin{bmatrix} x & y \end{bmatrix}\begin{bmatrix} a & h \\ h & b \end{bmatrix}\begin{bmatrix} x \\ y \end{bmatrix} = X^{\mathrm{T}}AX.$$

There is a more general definition of a quadratic form, but general quadratic forms are not discussed in the book.

Rank deficient   An $m \times n$ matrix that has a zero singular value. Equivalently, the rank is less than $\min(m, n)$.

Rank 1 matrix   A matrix with only one linearly independent column or row.

Rayleigh quotient   Given an eigenvector v of matrix A, the Rayleigh quotient $(Av)^{\mathrm{T}} v/\|v\|_2^2$ is the eigenvalue corresponding to v.

Reduced QR decomposition   If $m \geq n$, a reduced QR decomposition of matrix A can be performed. In this decomposition $A = QR$, where $A^{m \times n} = Q^{m \times n}R^{n \times n}$. If m is quite a bit larger than n, this computation is considerably faster, and uses less memory.

Reduced SVD   The singular value decomposition, $A = U\tilde{\Sigma}V^{\mathrm{T}}$, where U is $m \times n$, $\tilde{\Sigma}$ is $n \times n$, and V is $n \times n$. If m is quite a bit larger than n, this computation is considerably faster, and uses less memory. See "Singular value decomposition" for more information.

Regression line   A straight line fit to a set of data points using least squares.

Relative error   In floating point conversion, the relative error in converting x is $|\mathrm{fl}(x) - x|/|x|, x \neq 0$. In an iteration, the relative error is $|x_{\mathrm{new}} - x_{\mathrm{prev}}|/|x_{\mathrm{prev}}|$. Relative error is used in many other situations.

Residual $r = b - Ax$, where A is $m \times n$, x is $n \times 1$, and b is $m \times 1$. The residual measures error in an iterative method for solving $Ax = b$, and in least squares the residual is minimized.

Resonance   The tendency of a system to oscillate at a greater amplitude at some frequencies than at others. These are known as the system's resonant frequencies. At these frequencies, even small periodic driving forces can produce oscillations of large amplitude.

Rosser matrix   Symmetric eigenvalue test matrix. It has eigenvalues with particular properties that challenge a symmetric eigenvalue solver.

Rotation matrix   A linear transformation that performs a rotation of an object. $P = \begin{bmatrix} \cos\theta & -\sin\theta \\ \sin\theta & \cos\theta \end{bmatrix}$ is a 2×2 rotation matrix.

Round-off error   The error introduced when a real number, x, is approximated using finite-precision computer arithmetic and when floating point operations are performed.

Row-equivalent matrices   Matrix B is row-equivalent to matrix A if B can be obtained from A using elementary row operations.

Scalar multiple   Multiplying a vector or a row of a matrix by a scalar.

Schur's Triangularization   Every $n \times n$ real matrix A with real eigenvalues can be factored into $A = PTP^{\mathrm{T}}$, where P is an orthogonal matrix, and T is an upper-triangular matrix.

Sensitivity of eigenvalues   A measure of how small changes in matrix entries affect the ability to accurately compute an eigenvalue. If λ is an eigenvalue and x, y are right and left eigenvectors corresponding to λ, then the condition number of λ is $1/y^{\mathrm{T}}x$. The condition numbers for the eigenvalues of a symmetric matrix are one.

Similar matrices   Matrices A and B are similar if there exists a nonsingular matrix X such $B = X^{-1}AX$.

Singular value decomposition   If $A \in \mathbb{R}^{m \times n}$, then there exist orthogonal matrices $U \in \mathbb{R}^{m \times m}$ and $V \in \mathbb{R}^{n \times n}$ such that $A = U\tilde{\Sigma}V^{\mathrm{T}}$, where $\tilde{\Sigma}$ is an $m \times n$ diagonal matrix. The diagonal entries of $\tilde{\Sigma}$ are all nonnegative and are arranged as follows: $\sigma_1 \geq \sigma_2 \geq \cdots \geq \sigma_r > 0$, with $\sigma_{r+1} = \cdots = \sigma_n = 0$.

Singular values   The square root of the eigenvalues of $A^{\mathrm{T}}A$ for any matrix A. The singular values are the entries on the diagonal of the matrix $\tilde{\Sigma}$ in the singular value decomposition.

Sparse matrix   A matrix most of whose entries are 0.

Spectral radius   If A is an $n \times n$ matrix, the spectral radius of A, written $\rho(A)$, is the maximum eigenvalue in magnitude; in other words, $\rho(A) = \max_{1 \leq i \leq n} |\lambda_i|$.

Spectral theorem   If A is a real symmetric matrix, there exists an orthogonal matrix P such that $D = P^{\mathrm{T}}AP$, where D is a diagonal matrix containing the eigenvalues of A, and the columns of P are an orthonormal set of eigenvalues that form a basis for $\mathbb{R}^n$.

Stable algorithm An algorithm is stable if it performs well in general, and an algorithm is unstable if it performs badly in significant cases. In particular, an algorithm should not be unduly sensitive to errors in its input or errors during its execution.

Sub-multiplicative norm A matrix norm is sub-multiplicative if $\|AB\| \leq \|A\| \|B\|$. The induced matrix norms and the Frobenius norm are sub-multiplicative.

Successive overrelaxation (SOR) An iterative method for solving the linear system $Ax = b$. A relaxation parameter, ω, $0 < \omega < 2$, provides a weighted average of the newest value, $x_i^{(k)}$, and the previous one, $x_{i-1}^{(k)}$, $1 \leq i \leq n$, $k = 1, 2, 3, \ldots$, until meeting an error tolerance.

SVD See "Singular value decomposition."

Symmetric matrix A square matrix such that $a_{ij} = a_{ji}$, $i \neq j$. In other words, $A^T = A$.

Symmetric matrix eigenvalue problem The eigenvectors and eigenvalues of a real symmetric matrix are real and can be computed more efficiently than those of a general matrix. The Jacobi iteration, the symmetric QR iteration, the Francis algorithm, bisection, and divide-and-conquer algorithms are discussed in the book.

Symmetric QR iteration Using orthogonal similarity transformation, create a tridiagonal matrix with the same eigenvalues as A. Using the QR iteration with the Wilkinson shift, transform a symmetric matrix to a diagonal matrix of eigenvalues.

Thomas algorithm An algorithm for solving an $n \times n$ tridiagonal system of equations $Ax = b$ with flop count $O(n)$.

Transpose of a matrix If A is an $m \times n$ matrix, then A^T is the $n \times m$ matrix obtained by exchanging the rows and columns of A; in other words, $a_{ij}^T = a_{ji}$, $1 \leq i \leq m$, $1 \leq j \leq n$.

Triangle inequality If $\|\cdot\|$ is a vector or matrix norm, then $\|x + y\| \leq \|x\| + \|y\|$.

Tridiagonal matrix A banded matrix whose only nonzero entries are on the main diagonal, the lower diagonal, and the super diagonal. A tridiagonal matrix is both a lower and an upper Hessenberg matrix, and a tridiagonal matrix can be factored into a product of two bidiagonal matrices.

Truncation Convert to finite precision by dropping all digits past the last valid digit without rounding.

Truncation error The error introduced when an operation, like summing a series, is cut off.

Truss A structure normally containing triangular units constructed of straight members with ends connected at joints referred to as pins. Trusses are the primary structural component of many bridges.

Underdetermined system An $m \times n$ linear system for which $n > m$; in other words, there are more unknowns than equations. These are a type of least-squares problems.

Underflow Occurs when a floating point operation produces a result too small for the precision of the computer.

Upper bidiagonal form A matrix having the main diagonal and the super diagonal, with all other entries equal to zero. An orthogonal transformation to upper bidiagonal form is the first step of the Demmel and Kahan zero-shift QR downward sweep algorithm for computing the SVD.

Upper-triangular matrix A linear system whose coefficient matrix has zeros below the main diagonal; in other words, $a_{ij} = 0$, $j < i$, $1 \leq i, j, \leq n$.

Vandermonde matrix An $m \times n$ matrix of the form:

$$V = \begin{bmatrix} 1 & t_1 & t_1^2 & \cdots & t_1^{n-1} \\ 1 & t_2 & t_2^2 & \cdots & t_2^{n-1} \\ 1 & t_3 & t_3^2 & \cdots & t_3^{n-1} \\ \vdots & \vdots & \vdots & \ddots & \\ 1 & t_m & t_m^2 & \cdots & t_m^{n-1} \end{bmatrix}$$

. The elements of V are represented by the formula $v_{ij} = t_i^{j-1}$. The Vandermonde matrix plays a role in least-squares problems. Note also that if

$$p(x) = a_{n-1}x^{n-1} + a_{n-2}x^{n-2} + \cdots + a_2x^2 + a_1x + a_0, \text{ then } V \begin{bmatrix} a_0 \\ a_1 \\ a_3 \\ \vdots \\ a_{n-1} \end{bmatrix}$$

evaluates $p(x)$ at the points $t_1, t_2, \ldots, t_m$.

Vector norm $\|\cdot\| : \mathbb{R}^n \to \mathbb{R}$ is a vector norm provided:

- $\|x\| \geq 0$ for all $x \in \mathbb{R}^n$. $\|x\| = 0$ if and only if $x = 0$;

- $\|\alpha x\| = |\alpha| \|x\|$ for all $\alpha \in \mathbb{R}$;

- $\|x + y\| \leq \|x\| + \|y\|$ for all $x, y \in \mathbb{R}^n$.

Well-conditioned problem If small perturbations in problem data lead to small relative errors in the solution, a problem is said to be well-conditioned.

Wilkinson bidiagonal matrix The Wilkinson-bidiagonal matrix is $A = \begin{bmatrix} 20 & 20 & 0 & \cdots & 0 \\ 0 & 19 & 20 & \cdots & 0 \\ \vdots & \vdots & 18 & \cdots & 0 \\ \vdots & \vdots & \vdots & \ddots & 20 \\ 0 & 0 & 0 & \cdots & 1 \end{bmatrix}$. This matrix illustrates that even though the eigenvalues of a matrix are not equal or even close to each other, an eigenvalue problem can very ill-conditioned.

Wilkinson shift A shift used in the computation of the eigenvalues of a symmetric matrix. The shift is the eigenvalue closest to h_{kk} of the 2×2 matrix

$$\begin{bmatrix} h_{k-1,k-1} & h_{k,k-1} \\ h_{k,k-1} & h_{kk} \end{bmatrix},$$

where the entries are from the lower right-hand corner of the tridiagonal matrix being reduced to a diagonal matrix.

Wilkinson test matrices These are symmetric and tridiagonal, with pairs of nearly, but not exactly, equal eigenvalues. The most frequently used case is wilkinson(21). Its two largest eigenvalues are both about 10.746; they agree to 14, but not to 15, decimal places.

Zero matrix An $m \times n$ matrix all of whose entries are 0.

Bibliography

1. J.W. Demmel, Applied Numerical Linear Algebra, SIAM, Philadelphia, 1997.
2. G.H. Golub, C.F. Van Loan, Matrix Computations, fourth ed.,The Johns Hopkins University Press, Baltimore, 2013.
3. Y. Saad, Numerical Methods for Large Eigenvalue Problems, Revised ed., SIAM, Philadelphia, 2011.
4. D.S. Watkins, The Matrix Eigenvalue Problem, GR and Krylov Subspace Methods, SIAM, Philadelphia, 2007.
5. G.W. Stewart, Matrix Algorithms, Volume II: Eigensystems, SIAM, Philadelphia, 2001.
6. Z. Bai, J. Demmel, J. Dongarra, A. Ruhe, H. van der Vorst (Eds.), Templates for the Solution of Algebraic Eigenvalue Problems: A Practical Guide, SIAM, Philadelphia, 2000.
7. N. Magnenat-Thalman, D. Thalmann, State of the Art in Computer Animation, Animation 2 (1989) 82–90.
8. G. Strang, Introduction to Linear Algebra, fourth ed., Wellesley-Cambridge Press, Wellesley, MA, 2009.
9. J.H. Wilkinson, The Algebraic Eigenvalue Problem, Oxford University Press, New York, 1965.
10. W.E. Boyce, R.C. DiPrima, Elementary Differential Equations, ninth ed., Wiley, Hoboken, NJ, 2009.
11. R.S. Varga, Matrix Iterative Analysis, Prentice-Hall, Englewood Cliffs, NJ, 1962.
12. J.P. Keener, The Perron-Frobenius theorem and the ranking of football teams, SIAM Rev. 35 (1) (1993) 80–93.
13. S. Brin, L. Page, The anatomy of a large-scale hypertextual web search engine, Comput. Netw. ISDN Syst. 30 (1998) 107–117.
14. T.H. Wei, The Algebraic Foundations of Ranking Theory, Cambridge University Press, University of Cambridge, 1952.
15. R. Horn, C. Johnson, Matrix Analysis, second ed., Cambridge University Press, New York, 2013.
16. N.J. Higham, Accuracy and Stability of Numerical Algorithms, second ed., SIAM, Philadelphia, 2002.
17. G.W. Stewart, Matrix Algorithms, Volume I: Basic Decompositions, SIAM, Philadelphia, 1998.
18. D. Goldberg, What every computer scientist should know about floating-point arithmetic, Comput. Surv. 23(1) (1991) 5–48.
19. B.N. Datta, Numerical Linear Algebra and Applications, second ed., SIAM, Philadelphia, 2010.
20. Mathworks, Create MEX-files, http://www.mathworks.com/help/matlab/create-mex-files.html.
21. G.A. Baker, Jr., P. Graves-Morris, Padé Approximants, Cambridge University Press, New York, 1996.
22. A.J. Laub, Computational Matrix Analysis, SIAM, Philadelphia, 2012.
23. D.S. Watkins, Fundamentals of Matrix Computations, third ed., Wiley, Hoboken, NJ, 2010.
24. A. Levitin, Introduction to the Design and Analysis of Algorithms, third ed., Pearson, Upper Saddle River, NJ, 2012.
25. G.H. Golub, C. Reinsch, Singular value decomposition and least squares solutions, Numer. Math. 14 (1970) 403–420.
26. L. Trefethen, David Bau, III, Numerical Linear Algebra, SIAM, Philadelphia, 1997.
27. G. Allaire, S.M. Kaber, Numerical Linear Algebra (Texts in Applied Mathematics), Springer, New York, 2007.
28. W.W. Hager, Condition estimators, SIAM J. Sci. Stat. Comput. 5 (2) (1984) 311–316.
29. G. Rodrigue, R. Varga, Convergence rate estimates for iterative solutions to the biharmonic equation, J. Comput. Appl. Math. 24 (1988) 129–146.
30. J.R. Winkler, Condition numbers of a nearly singular simple root of a polynomial, Appl. Numer. Math. 38 (3) (2001) 275–285.
31. L.V. Foster, Gaussian elimination with partial pivoting can fail in practice, SIAM J. Matrix Anal. Appl. 15 (1994) 1354–1362.
32. N.J. Higham, Efficient algorithms for computing the condition number of a tridiagonal matrix, SIAM J. Sci. Stat. Comput. 7 (1986) 150–165.
33. A. Gilat, V. Subramaniam, Numerical Methods for Engineers and Scientists: An Introduction with Applications Using MATLAB, second ed., Wiley, Hoboken, NJ, 2011.
34. R.H. Bartels, J.C. Beatty, B.A. Barsky, An Introduction to Splines for Use in Computer Graphics and Geometric Modelling, Morgan Kaufmann, San Francisco, 1995.
35. E.W. Weisstein, Cubic spline, http://mathworld.wolfram.com/CubicSpline.html.
36. T.A. Grandine, The extensive use of splines at Boeing, SIAM News 38 (4) (2005) 1–3.
37. A.J. Jerri, The Gibbs Phenomenon in Fourier Analysis, Splines and Wavelet Approximations, Springer, New York, 1998.
38. G.T. Gilbert, Positive definite matrices and Sylvester's criterion, Am. Math. Mon. 98 (1) (1991) 44–46.
39. R.s. Ran, T.z. Huang, X.p. Liu, T.x. Gu, An inversion algorithm for general tridiagonal matrix, Appl. Math. Mech. Engl. Ed. 30 (2009) 247–253.
40. J.W. Lewis, Inversion of tridiagonal matrices, Numer. Math. 38 (1982) 333–345.
41. Q. Al-Hassan, An algorithm for computing inverses of tridiagonal matrices with applications, Soochow J. Math. 31 (3) (2005) 449–466.
42. E. Kiliç, Explicit formula for the inverse of a tridiagonal matrix by backward continued fractions, Appl. Math. Comput. 197 (2008) 345–357.
43. M. El-Mikkawy, A. Karawia, Inversion of general tridiagonal matrices, Appl. Math. Lett. 19 (8) (2006) 712–720.
44. MIT course 18.335J, Difference in results between the classical and modified Gram-Schmidt methods, http://ocw.mit.edu/courses/mathematics/.

45. L. Giraud, J. Langou, M. Rozloznik, The loss of orthogonality in the Gram-Schmidt orthogonalization process, Comput. Math. Appl. 50 (2005) 1069–1075.

46. C.B. Moler, Numerical Computing with MATLAB, SIAM, Philadelphia, 2004.

47. I. Ipsen, Numerical Matrix Analysis—Linear Systems and Least Squares, SIAM, Philadelphia, 2009.

48. N.J. Higham, Computing the polar decomposition with applications, SIAM J. Sci. Stat. Comput. 7 (1986) 1160–1174.

49. G.H. Golub, Numerical methods for solving linear least squares problems, Numer. Math. 7 (1965) 206–216.

50. K.A. Gallivan, S. Thirumalai, P. Van Dooren, V. Vermaut, High performance algorithms for Toeplitz and block Toeplitz matrices, Linear Algebra Appl. 241 (1996) 343–388.

51. A. Björck, Numerical Methods for Least Squares Problems, SIAM, Philadelphia, 1996.

52. A. Björck, Solving linear least-squares by Gram-Schmidt orthogonalization, BIT 7 (1967) 1–21.

53. D.G. Zill, W.S. Wright, Advanced Engineering Mathematics, fifth ed., Jones & Bartlett Learning, Burlington, MA, 2014.

54. J.G.F. Francis, The QR transformation, part I, Comput. J. 4 (1961) 265–272.

55. J.G.F. Francis, The QR transformation, part II, Comput. J. 4 (1961) 332–345.

56. R.S. Martin, G. Peters, J.H. Wilkinson, The QR algorithm for real Hessenberg matrices, Numer. Math. 14 (1970) 219–231.

57. D. Day, How the QR algorithm fails to converge and how to fix it, Technical report 96–0913J, Sandia National Laboratory, Albuquerque, NM, April 1996.

58. J.J.M. Cuppen, A divide and conquer method for the symmetric tridiagonal eigenproblem, Numer. Math. 36 (1981) 177–195.

59. M. Gu, S.C. Eisenstat, A divide-and-conquer algorithm for the symmetric tridiagonal eigenproblem, SIAM J. Matrix Anal. Appl. 16 (1995) 172–191.

60. M. Gu, S.C. Eisenstat, A stable algorithm for the rank-1 modification of the symmetric eigenproblem, Computer Science Department report YALEU/DCS/RR-916, Yale University, 1992.

61. B.E. Parlett, The Symmetric Eigenvalue Problem, SIAM, Philadelphia, 1997.

62. University of Tennessee, Berkeley University of California, University of Colorado Denver, and NAG Ltd., LAPACK documentation, http://www.netlib.org/lapack/.

63. W.L. Briggs, V.E. Henson, S.F. McCormick, A Multigrid Tutorial, second ed., SIAM, Philadelphia, 2000.

64. Y. Saad, Iterative Methods for Sparse Linear Systems, second ed., SIAM, Philadelphia, 2003.

65. Z. Zlatev, Computational Methods for General Sparse Matrices, Springer, New York, 1991.

66. I.S. Duff, A.M. Erisman, J.K. Reid, Direct Methods for Sparse Matrices, Oxford University Press, New York, 1989.

67. N. Munksgaard, Solving sparse symmetric sets of linear equations by preconditioned conjugate gradients, ACM Trans. Math. Softw. 6 (1980) 206–219.

68. K. Chen, Matrix Preconditioning Techniques and Applications, Cambridge University Press, Cambridge, 2005.

69. H.A. van der Vorst, Iterative Krylov Methods for Large Linear Systems, Cambridge University Press, New York, 2009.

70. G. Meurant, Z. Strakos, The Lanczos and conjugate gradient algorithms in finite precision arithmetic, Acta Numer. 15 (2006) 471–542.

71. Z. Strakos, On the real convergence rate of the conjugate gradient method, Linear Algebra Appl. 154–156 (1991) 535–549.

72. G.L.G. Sleijpen, H.A. van der Votst, J. Modersitzki, Differences in the effects of rounding errors in Krylov solvers for symmetric indefinite linear systems, SIAM J. Matrix Anal. Appl. 22 (3) (2000) 726–751.

73. A. Greenbaum, Iterative Methods for Solving Linear Systems, SIAM, Philadelphia, 1997.

74. A.P.S. Selvadurai, Partial Differential Equations in Mechanics 2: The Biharmonic Equation, Poisson's Equation, Springer, Berlin, 2000.

75. M. Arad, A. Yakhot, G. Ben-Dor, Highly accurate numerical solution of a biharmonic equation, Num. Meth. Partial Diff. Equations 13 (1997) 375–391.

76. D.S. Scott, How to make the Lanczos algorithm converge slowly, Math. Comp. 33 (1979) 239–247.

77. S. Kaniel, Estimates for some computational techniques in linear algebra, Math. Comp. 20 (1966) 369–378.

78. C.C. Paige, The computation of eigenvalues and eigenvectors of very large sparse matrices (Ph.D. thesis), University of London, 1971.

79. C.C. Paige, Computational variants of the Lanczos method for the eigenproblem, J. Inst. Math. Appl. 10 (1972) 373–381.

80. C.C. Paige, Error analysis of the Lanczos algorithm for tridiagonalizing a symmetric matrix, J. Inst. Math. Appl. 18 (1976) 341–349.

81. C.C. Paige, Accuracy and effectiveness of the Lanczos algorithm for the symmetric eigenproblem, Linear Algebra Appl. 34 (1980) 235–258.

82. J. Demmel, K. Veselic, Jacobi's method is more accurate than QR, SIAM J. Matrix Anal. Appl. 13 (1992) 1204–1245.

83. G.W. Stewart, Perturbation theory for the singular value decomposition, in: R.J. Vaccaro (Ed.), SVD and Signal Processing, II: Algorithms, Analysis and Applications, Elsevier, Amsterdam, 1990, pp. 99–109.

84. Z. Drmač, K. Veselic, New fast and accurate Jacobi SVD algorithm: I, SIAM J. Matrix Anal. Appl. 29 (2008) 1322–1342.

85. Z. Drmač, K. Veselic, New fast and accurate Jacobi SVD algorithm: II, SIAM J. Matrix Anal. Appl. 29 (2008) 1343–1362.

86. G.H. Golub, W. Kahan, Calculating the singular values and pseudo-inverse of a matrix, SIAM J. Numer. Anal. 2 (1965) 205–224.

87. J. Demmel, W. Kahan, Accurate singular values of bidiagonal matrices, SIAM J. Sci. Stat. Comput. 11 (1990) 873–912.

88. A. Gil, J. Segura, N.M. Temme, Numerical Methods for Special Functions, SIAM, Philadelphia, 2007.

89. Haag, Michael, Justin Romberg, Stephen Kruzick, Dan Calderon, and Catherine Elder "Cauchy-Schwarz Inequality." OpenStax-CNX. 2013. http://cnx.org/content/m10757/2.8/.

90. T.A. Davis, Y. Hu, ACM transactions on mathematical software, The University of Florida sparse matrix collection, 38 (2011) 1:1–1:25.

91. F.L. Bauer, C.T. Fike, Norms and exclusion theorems, Numer. Math. (1960) 137–141.

Index

1s-complement, 146

A

A-conjugate, 497
A-norm, 275, 497
absolute error, 150
adaptive procedure, 314
adjacency matrix, 13
adjoint, 63, 71
algorithm, 163
  backward stable, 185
  Big-O notation, 167
  Cholesky decomposition, 271
  colon notation, 214
  computing the Givens parameters, 358
  cubic, 167
  forward stable, 186
  Frobenius norm, 164
  Givens QR decomposition, 360
  Gram-Schmidt, 282
  implicit QR, 409
  inner product, 164
  inner product of two vectors, 164
  iterative improvement, 229
  linear, 167
  LU decomposition, 214
  LU decomposition of a tridiagonal matrix, 264
  LU decomposition without a zero on the diagonal, 216
  modified Gram-Schmidt, 285
  product of a Givens matrix with a general matrix, 356
  product of two matrices, 164
  pseudocode, 163
  quadratic, 167
  recursion stopping condition, 459
  solve AX = B using LU factorization, 225
  solve $Ax = b_i, 1 \leq i \leq k$, 225
  solving a lower-triangular system, 170
  solving an upper-triangular system, 169
  solving the least-squares problem using the QR decomposition, 328
  solving the least-squares problem using the SVD, 330
  stability, 186
  steepest descent, 496
  the power method, 391
  unstable, 185
angle between vectors, 107
applications
  biharmonic equation, 523

conductance matrix, 269
counting paths in graphs, 12
cubic splines, 252
electric circuit with inductors, 89
electrical circuit, 39
encryption, 71
finite difference approximations, 244
Fourier series, 241
image compression using the SVD, 310
instability of the Cauchy problem, 188
least-squares fitting, 247
least-squares to fit a power function, 328
Leslie model, 383
Poisson's equation, 478
signal comparison, 112
team ranking, 94
truss, 37
velocity of an enzymatic reaction, 331
vibrating masses on springs, 380
Argand diagram, 571
Arnoldi decomposition, 509
Arnoldi process, 534
augmented matrix, 27

B

back substitution, 29, 168, 207
backward
  error, 184, 587
  stable, 185
band matrix, 238
basic QR iteration, 394
basis, 50
  local, 125
  orthonormal, 122
Bauer-Fike theorem, 424
bi-conjugate gradient method, 531
bidiagonal matrix, 44, 263
Big-O notation, 167
biharmonic equation, 523
biorthogonal bases, 528
bit, 145
block
  pentadiagonal matrix, 523
  tridiagonal matrix, 522
bulge, 411

C

cancellation error, 155
carry, 146
Cauchy problem, 188

Cauchy-Schwarz inequality, 111, 121, 242
  matched filter, 112
characteristic
  equation, 79
  polynomial, 79, 182
Chebyshev
  polynomial, 258, 583
Cholesky
  decomposition, 270
  incomplete decomposition, 503
Chris Paige, 546
clamped cubic spline, 255
classical
  Gram-Schmidt, 284
  Jacobi method, 444, 465
coefficient matrix, 4, 27
cofactor, 62
colon notation, 214
column
  rank, 322
  space, 48
  vector, 7
compatible norms, 141
complex numbers, 569
  Argand diagram, 571
  complex plane, 571
  conjugate, 571
  imaginary part, 569
  rationalization of the denominator, 570
  real part, 569
  square root, 570
complex plane, 571
  imaginary axis, 571
  lower half, 571
  real axis, 571
  upper half, 571
compressed row storage format, 492
condition number, 143, 189, 192
  estimate, 195
conjugate gradient method, 493
conjugate transpose, 117
contour lines, 494
Cramer's rule, 69, 182
critical loads, 388
cross product, 115, 177
Crout's Method, 233
CRS format, 492
cubic algorithm, 167
cubic spline, 252
  clamped, 255

cubic spline *(Continued)*
    natural, 255
    not-a-knot condition, 255
cyclic-by-row Jacobi algorithm, 444

D
deflation, 401
determinant, 15
    and the matrix inverse, 67
    Cramer's rule, 69
    diagonal matrix, 61
    first-row Laplace expansion, 60
    general expansion by minors, 62
    lower triangular matrix, 61
    matrix with row of zeros, 61
    properties for evaluation, 64
    upper triangular matrix, 61
    using QR decomposition, 291
Devlin, Keith, 575
diagonal
    dominance, 172, 477
    matrix, 61
diagonalizable matrix, 84
diagonally dominant matrix, 263
digraph, 92
    strongly connected, 93
dimension of a subspace, 51
directed line segments, 103
direction vector, 494
distance between points, 103
dominant eigenvalue, 389
dot product, 105
drop tolerance, 505

E
eigenfunctions, 388
eigenspace, 83
eigenvalue
    and vibrations, 380
    Bauer-Fike theorem, 424
    characteristic
        equation, 79
        polynomial, 79, 182
    classical Jacobi method, 444, 465
    condition number, 426
    cyclic-by-row Jacobi algorithm, 444
    definition, 79
    deflation, 401
    deflation using the implicit double
        shift, 420
    dominant eigenvalue, 389
    eig function, 95
    eigb function, 423
    eigenspace, 83
    eigqr function, 402
    eigqrshift function, 405
    eigsb function, 542
    eigssymb function, 546
    explicit
        double shift, 415
            restart for large sparse problem, 536
    filtering polynomial, 540
    Francis algorithm, 409
    Francis iteration of degree one, 409

generalized problem, 432
    ghost, 546, 549
    implicit double shift, 415
    implicit Q theorem, 410
    implicitly restarted Arnoldi
        method, 540
    inverse power method, 393
    Jacobi method, 440
    of a symmetric matrix, 134
    power method, 390, 428, 533
        convergence, 391
    QR iteration, 394
    rate of convergence of power
        method, 392
    Rayleigh
        quotient, 389
        quotient iteration, 465
        quotient shift, 404
    sensitivity of perturbations, 424
    shift, 404
    single shift QR algorithm, 404
    strictly interlacing, 455
eigenvector
    computing
        from eigenvalue, 420
        the largest, 391
        the smallest, 393
    definition, 80
    eigenspace, 83
    left, 425
    localization, 466
    matrix, 86, 389
    sensitivity, 427
    shifted inverse iteration, 421
electric circuit
    Kirchhoff's current law, 39
    Kirchhoff's voltage law, 39
elementary
    row matrix, 208
    row operations, 27
ellipsoid, 307
encryption, 71
    example, 72
energy norm, 275, 497
eps, 149
error
    absolute, 150
    backward, 184, 587
    cancellation, 155
    forward, 183
    relative, 150
    round-off, 117, 145
    truncation, 145
Euclidean norm, 120
Euler's
    formula, 575
    identity, 575
Euler-Lotka equation, 385
expansion by minors, 60, 62
explicit
    restart, 536
    shift, 415
extrapolation, 248

F
Feynman, Richard, 575
Fibonacci
    matrix, 12
    sequence, 12
filtering polynomial, 540
finite difference equations, 246
first buckling mode, 388
five-point
    difference approximation, 480
    stencil, 480
fl, 148
floating point, 147
    absolute error, 150
    arithmetic, 150
    avoiding error, 155
    base, 147
    bias, 148
    granularity, 149
    IEEE arithmetic, 151
    mantissa, 147
    normalization, 155
    overflow, 148
    relative error, 150
    round-off error, 151
    rounding, 149, 151
    rounding error in matrix multiplication, 154
    rounding error in scalar multiplication, 154
    truncation, 151
    underflow, 148
flop, 167
    count, 167
force vector, 381
forward
    error, 183
    stable, 186
    substitution, 169, 206
Fourier
    coefficients, 242
Fourier series, 241
    Gibbs phenomenon, 260
    sawtooth wave, 257
    square wave, 243
    triangle wave, 258
Francis algorithm, 409
    degree one, 409, 452
    degree two, 413
    double shift, 409
    single shift, 409
Frobenius norm, 127, 128
full reorthogonalization, 517
fundamental frequency, 380

G
Gauss transformation matrix, 235
Gauss-Seidel
    iteration, 470
    matrix form, 473
Gaussian elimination, 26, 29
    procedure, 29
    stability, 227
    with complete pivoting, 226
    with partial pivoting, 218
GECP, 226

generalized eigenvalue problem, 432
geometric series, 580
GEPP, 218
Gergorin's disk theorem, 431
ghost eigenvalues, 546, 549
Gibbs phenomenon, 260
Givens
   matrix, 352
   method, 351
   parameters, 358
   QR decomposition algorithm, 360
   rotation, 353
GMRES, 512
golden ratio, 88
gradient of a function, 494
Gram-Schmidt
   orthonormalization, 283
   process, 281, 282
   QR decomposition, 288, 351
granularity of floating point numbers, 149
graph, 12
   adjacency matrix, 13
   edges, 12
   number of paths between two vertices, 13
   path, 12
   vertices, 12
growth factor, 227
guard digit, 150

H

Hanowa matrix, 101
heat equation, 245
Hermitian
   matrix, 117, 573
   transpose, 413
Hessenberg
   lower, 396
   upper, 396
Hilbert
   matrices, 22, 77
   space, 242
Hilbert-Schmidt norm, 127
homogeneous system, 15, 36
   nontrivial solution, 36
   trivial solution, 36
Householder
   method, 351
   reflection, 362

I

identity matrix, 11
IEEE
   arithmetic, 151
   floating point standard, 148
ill-conditioned
   matrix, 119
   problem, 187
imaginary part of complex number, 569
implicit
   double shift, 415
   Q theorem, 410
   QR algorithm, 409
   single shift, 409
implicitly restarted Arnoldi method, 540

incomplete
   Cholesky decomposition, 503
   LU decomposition, 514
inconsistent system of linear equations, 25, 31
induced matrix norm, 127
induction
   basis step, 579
   inductive step, 579
initial-boundary value problems, 245
inner product, 104, 589
   geometric interpretation, 105
   properties, 105
integer, 7
   largest positive, 146
   most negative, 147
   overflow, 147
   representation, 145
interpolation, 248
inverse
   computation, 34
   computation using the adjoint, 63
   iteration, 420
   of a matrix product, 14
   participation ratio, 467
   power method, 393
invertible matrix, 13
irrational number, 7
irreducible matrix, 92
iterative methods, 469
   conjugate gradient, 497
   drop tolerance, 505
   Gauss-Seidel, 470
   GMRES, 512
   implicitly restarted Arnoldi, 540
   implicitly restarted Lanczos, 544
   Jacobi, 469
   MINRES, 519
   power method, 390, 533
   SOR, 471
   steepest descent, 493
iterative refinement, 228

J

Jacobi
   iteration, 469
   matrix form of iterative method, 473
   method for computing eigenvalues, 440
   preconditioner, 527
   sweep, 444

K

Kirchhoff's rules, 39
knot, 252
Krylov
   sequence, 508
   subspace, 508

L

L^2
   inner product, 110
   norm, 281
Lagrange interpolation, 252
Lanczos method, 516

least-squares, 248, 321
   geometric interpretation, 321
   linear, 321
   normal equations, 249, 322
   overdetermined system, 249
   rank deficient, 333
   regression line, 250
   residual, 248
   residual equation, 324
   solution using normal equations, 326
   underdetermined system, 249
   use of Cholesky decomposition, 326
   using QR decomposition, 327
   using the SVD, 329
   weighted, 343
left
   eigenvector, 425
   nullspace, 57
   preconditioner, 502
   singular vectors, 301
Leslie
   matrix, 383
   model, 383
level set, 494
linear
   algorithm, 167
   combination, 48
   equation, 25
   interpolation, 252
   splines, 252
   transformation, 7
   translation, 9
linear system
   consistent, 25
   homogeneous, 15
   inconsistent, 25, 31
   trivial solution, 15
linearly
   dependent, 49
   independent, 49
local basis, 125
lower-triangular matrix, 169
LU
   decomposition, 205, 206
   decomposition algorithm, 214

M

machine precision, 149
mantissa, 147
MATLAB function
   chol, 271
   cond, 195, 325
   condest, 196
   det, 62
   eig, 95
   fix, 162
   gallery, 192
   gmres, 513
   help, 162
   ichol, 505
   ilu, 515
   inv, 15
   lu, 225
   minres, 531

MATLAB function (*Continued*)
norm, 121
qr, 289
rand, 57
rank, 53
realmin, 162
schur, 408
svd, 308
symmlq, 531
trace, 6
transpose, 17
vectorization, 6
wilkinson, 452
zeta, 577
matrix, 1
1-norm, 131
2-norm, 132
addition, 2
adjoint, 63
and orthogonal vectors, 108
and system of equations, 4
back substitution, 29, 207
band, 238
bidiagonal, 44, 263
block
pentadiagonal, 523
structured, 101
tridiagonal, 522
Cholesky decomposition, 270
coefficient matrix, 4
cofactor, 62
column rank, 322
computational expense, 3
computing the inverse, 34
condition number, 143, 192
conjugate transpose, 117
Crout 's Method, 233
decomposition, 206
diagonal dominance, 172
diagonalizable, 84
diagonally dominant, 263
elementary row, 208
equality of matrices, 2
factoring a tridiagonal matrix, 264
factorization, 43
forward substitution, 206
Frobenius norm, 128
full rank, 287
Gauss transformation, 235
Gaussian elimination with partial pivoting, 218
Givens, 352
rotation, 353
growth factor, 227
Hanowa, 101
Hermetian transpose, 413
Hermitian, 117, 573
identity, 11
ill-conditioned, 22, 119
inverse, 13
of a product, 14
invertible, 13
irreducible, 92
laws of arithmetic, 3

left
nullspace, 57
preconditioner, 502
lower triangular, 169
LU decomposition, 205, 206
negative of a matrix, 2
nonnegative, 93
nonsingular, 13
normal, 431
null space, 47
null space using SVD, 303
off diagonal entries of a square matrix, 20
orthogonal
and conditioning, 195
invariance, 136
overdetermined system, 321
partial pivoting, 218
pencil, 432
pentadiagonal, 489
permutation, 218
Perron-Frobenius theorem, 93
pivot, 218
positive definite, 44, 267
positive semidefinite, 267
power, 11
preconditioner, 501
product, 2
proper
Hessenberg, 400
orthogonal, 109
properties of the condition number, 193
pseudoinverse, 324
QR decomposition, 351
quasi-triangular, 408
range, 57
range using SVD, 303
rank, 51, 322
deficient, 57
revealing QR decomposition, 296
using SVD, 302
rank 1, 310
real Schur form, 408
reduced QL decomposition, 375
reducible, 92
reversal matrix, 375
reverse identity, 375
right preconditioner, 502
rotation, 7, 8, 10
row
rank, 322
space, 48
row-equivalent matrices, 28
scalar multiple of a matrix, 2
Schur's triangularization, 405
similar, 84
simple eigenvector, 93
singular, 13
singular value decomposition, 53, 299
singular values, 133, 135, 182, 307
sparse, 38, 171, 247, 469
sparsity pattern, 491
spectral theorem, 136
split preconditioner, 502
square matrix diagonal, 20

square root, 317
SSOR preconditioner, 506
stiffness, 381
strictly column diagonally dominant, 234
strictly row diagonally dominant, 477
subordinate norm, 127
subtraction of matrices, 2
superdiagonal entries, 263
SVD, 53, 299
symmetric, 17
the zero matrix, 2
Toeplitz, 203, 513
trace, 5
transpose, 16
tridiagonal, 20, 44, 171, 236, 247, 263
underdetermined system, 321, 338
unitary, 576
unreduced Hessenberg, 400
upper
bidiagonal, 172
triangular, 29, 168, 359
upper triangular, 72
Vandermonde, 67, 203, 248
well-conditioned, 440
McLaurin series, 575
MGS, 288
Millennium Bridge, 79
minimum residual method (MINRES), 519
minor, 60
mode of an image, 311
modified Gram-Schmidt
process, 285
QR decomposition and ill-conditioned matrices, 351
monic polynomial, 584
Moore-Penrose generalized inverse, 324

N
natural
cubic splines, 255
frequency, 381
nonnegative matrix, 93
norm
1-norm, 120
A-norm, 275, 497
energy norm, 275, 497
Frobenius, 127, 128
Hilbert-Schmidt, 127
induced matrix norm, 127
infinity norm, 120
p-norm, 120
Schatten p-norm, 141
sub-multiplicative, 131
normal
derivative, 523
equation of a plane, 115
equations, 249, 322
matrix, 431
not-a-knot condition, 255
null space, 47

O
one's-complement, 159
one-sided Jacobi algorithm, 551

optimal upper bound, 192
orthogonal
  complement, 316
  invariance, 121, 136
  matrix, 195
  projection, 282
  similarity transformation, 396
  vectors, 107, 281
orthonormal
  basis, 122
  vectors, 108
outer product, 116, 184
overdetermined system, 249, 321
overflow, 147, 148

P
p-norms, 120
PageRank process, 94
parallelogram law, 103
partial pivoting, 218
Peirce, Benjamin, 575
permutation matrix, 116, 218
perpendicular vectors, 107
Perron-Frobenius theorem, 93
pessimistic upper bound, 192
pivot, 211
  matrix, 218
Poisson's equation, 478
polar decomposition, 320
poles in divide-and-conquer, 460
position vector, 103
positive definite matrix, 44, 267
positive semidefinite matrix, 267
positivity, 120, 126
power method, 390, 428, 533
  convergence, 391
  inverse, 393
  rate of convergence, 392
power of a matrix, 11
preconditioned
  CG, 503
  GMRES, 514
preconditioner, 501
  CG
    incomplete Cholesky, 503
    SSOR, 506
  GMRES
    incomplete LU decomposition, 514
  Jacobi, 527
preconditioning, 501
projection operator, 281
proper
  Hessenberg, 400
  orthogonal matrix, 109
pseudocode, 163
pseudoinverse, 324
Pythagorean Theorem, 121, 141

Q
QMR method, 532
QR decomposition, 287, 351
  full, 289
  Givens method for computing, 351

Gram-Schmidt method for
    computing, 287
  Householder's method for
    computing, 351
  iteration for computing
    eigenvalues, 394
  reduced, 289
  theorem, 351
quadratic
  algorithm, 167
  form, 22, 267
  formula, 155
  function, 493
quadratic spline, 259
quasi-triangular matrix, 408

R
rank, 51
  1 correction, 459
  1 matrix, 310
  1 update, 374
  deficient matrix, 57
  revealing QR
    decomposition, 296
rational number, 7
Rayleigh
  quotient, 389
  quotient iteration, 465
  quotient shift, 404
real part of complex number, 569
real Schur form, 408
recursive definition, 60
red-black ordering, 486
reduced
  QL decomposition, 375
  QR decomposition, 289
  SVD, 329
reducible matrix, 92
regression line, 250
relative
  error, 150
  residual, 470
relaxation parameter, 471
reorthogonalization, 297
  full, 517
residual, 248
resonance, 382
reversal matrix, 375
reverse identity matrix, 375
Riemann
  hypothesis, 577
  zeta function, 576
right
  preconditioner, 502
  singular vectors, 301
Ritz
  eigenvector, 535
  locked pair, 537
  pair, 535
  value, 535
root, 453
rotation matrix, 7, 8, 10
round-off error, 117, 145, 149
rounding, 148, 149, 151

row
  equivalent matrices, 28
  rank, 322
  space, 48
Runge's phenomenon, 262

S
sawtooth wave, 257
scaling, 120, 126
Schatten p-norm, 141
Schur's triangularization, 405
secular equation, 460
shifted inverse iteration, 420
signal comparison, 112
significant digits, 147
similar matrices, 84
simple eigenvector, 93
single shift, 404
singular matrix, 13
singular value decomposition, 53
  geometric interpretation, 307
  left singular vectors, 301
  right singular vectors, 301
  theorem, 299
singular values, 133, 135, 182, 307
  artificial ill-conditioning, 557
  Demmel and Kahan zero-shift QR downward
    sweep, 558
  one-sided Jacobi algorithm, 551
SOR iteration, 471
sparse matrix, 38, 171, 247
  Arnoldi decomposition, 509
  CRS format, 492
  GMRES, 512
  Lanczos decomposition, 516
  minimum residual method (MINRES), 519
sparsity pattern, 491
spectral
  norm, 132
  radius, 137
  theorem, 136, 439
spline
  cubic, 252
  knot, 252
  quadratic, 259
split preconditioner, 502
square root of a matrix, 317
square wave, 243
SSOR preconditioner, 506
  for GMRES, 527
stability, 186
standard basis vectors, 49
steady state, 91
steepest descent, 493, 496
stiffness matrix, 381
stopping condition, 459
strictly
  column diagonally dominant, 234
  interlacing eigenvalues, 455
strong induction, 580
strongly connected digraph, 93
Sturm sequence, 455
sub-multiplicative norm, 131
subordinate matrix norm, 127

subspace, 47
  basis, 50
  dimension, 51
  Krylov, 508
  orthogonal complement, 316
  spanned by, 48
superdiagonal matrix entries, 263
supremum, 189
SVD, 53, *See* singular value decomposition, 299
Sylvester's criterion, 267
symmetric matrix, 17
  $A^{\mathrm{T}}A$, 133
  biharmonic, 523
  bisection method for computing eigenvalues, 453
  divide-and-conquer method for computing eigenvalues, 458
  eigenvalues, 134
  Francis method for computing eigenvalues, 452
  has n linearly independent eigenvalues, 83
  has real eigenvalues, 134
  Hilbert, 201
  indefinite, 268
  Jacobi method for computing eigenvalues, 440
  Lanczos decomposition of, 516
  listing of important properties, 440
  Poisson, 478
  positive definite, 267
  QR method for computing eigenvalues, 446
  reduction to a tridiagonal matrix, 449
  skew, 432
  spectral theorem for, 439
  Wilkinson 21×21 matrix, 452
system of linear equations, 4, 25, 29
  as a matrix equation, 4
  augmented matrix, 27
  coefficient matrix, 27

  conjugate gradient method, 493
  elementary row operations, 27
  Gauss-Seidel iteration, 470
  Gaussian elimination, 26, 29
  GMRES method, 512
  homogeneous, 36
  Jacobi iteration, 469
  LU decomposition, 205
  MINRES method, 519
  relative residual, 470
  solution using the inverse, 15
  SOR iteration, 471
  vector of constants, 4
  vector of unknowns, 4

T

Tacoma Narrows Bridge, 79
  collapse, 380
tensor product, 116, 176
thermal diffusivity, 245
Toeplitz matrix, 203, 513
trace, 5
transient solution, 91
transpose of a matrix, 16
triangle inequality, 111, 120, 126
triangle wave, 258
tridiagonal matrix, 20, 44, 236, 247, 263
trivial solution, 15
truncation, 148, 151
  error, 145
truss problem, 37
two's-complement, 145

U

underdetermined system, 249, 321, 338
underflow, 148
unit
  sphere, 307

sphere of a norm, 128
  vector, 108
unitary matrix, 576
Univ. of Florida Sparse Matrix Collection, 492
unreduced Hessenberg, 400
unstable algorithm, 185
upper
  bidiagonal matrix, 172
  triangular matrix, 29, 168, 359

V

Vandermonde matrix, 67, 203, 248
vector, 1
  column, 7
  cross product, 115
  norm, 119
  of constants, 4
  of unknowns, 4
  operations, 103
  orthogonal vectors, 107, 281
  orthonormal set of, 108
  outer product, 116
  parallel vectors, 107
  perpendicular vectors, 107
  unit, 108
vectorization, 6
vibration problem, 380

W

weighted least-squares, 343
well-conditioned problem, 187
Wilkinson
  bidiagonal matrix, 201, 434
  polynomial, 187
  shift, 451
  test matrices, 99
Wilkinson, J. H., 99, 237

Printed and bound by CPI Group (UK) Ltd, Croydon, CR0 4YY

08/05/2025

01864928-0004